# Lecture Notes in Nanoscale Science and Technology

Volume 20

*Series Editors*



For further volumes:
http://www.springer.com/series/7544

Zhiming M. Wang
Editor

# FIB Nanostructures

Springer

*Editor*
Zhiming M. Wang
State Key Laboratory of Electronic
Thin Film and Integrated Devices
University of Electronic Science and Technology
Chengdu, People's Republic of China

ISSN 2195-2159 ISSN 2195-2167 (electronic)
ISBN 978-3-319-02873-6 ISBN 978-3-319-02874-3 (eBook)
DOI 10.1007/978-3-319-02874-3
Springer Cham Heidelberg New York Dordrecht London

Library of Congress Control Number: 2013956620



Printed on acid-free paper

Springer is part of Springer Science+Business Media (www.springer.com)

# Preface

The focused ion beam (FIB) has become one of the most popular tools enabling scientific and technological advances in the creation and characterization of objects at the nanometer scale. In many respects, the importance of FIB to nanoscience and nanotechnology is comparable to the scanning tunneling microscope (STM). While STM is considered to be the eyes and hands of needed scientists to manipulate atoms and molecules, the dual-beam or other advanced FIB configurations allow scientists not only to fabricate various nanostructures with a powerful "hand" holding a "drill" but also to "see" nanoscale fabricating processes in real time. These capabilities have allowed FIB nanostructures to find a wide spectrum of applications in areas ranging from the microelectronic industry and materials science to biology and medicine. The fabrication of FIB nanostructures, their applications, and special features are extensively covered in this book with 19 chapters by 60 authors from 12 countries.

The FIB nanofabrication process is reviewed in Chaps. 1–7. Chapter 1 begins with a description of the FIB instrument and then covers FIB nanofabrication by milling and deposition. Chapters 2–4 focus on the milling process. Chapter 2 concludes that FIB milling is a viable approach for fabrication of ferroelectric nanostructures. Chapter 3 demonstrates that low-ion-beam-current FIB milling can create and characterize freestanding nano-objects in a well-controlled manner. FIB milling to reveal the internal structures of carbon nanotube yarns and bucky-papers is described in Chap. 4. A detailed description of the use of FIB-induced deposition to grow nanoscale electrical contacts is presented in Chap. 5. Chapter 6 recapitulates a systematic investigation on FIB-induced deposition of ultrathin amorphous carbon patterns as a route toward the easy integration of carbon nanomaterials. Chapter 7 emphasizes the precision and accuracy of FIB fabrication for both micro- and nanostructures.

Various applications of the FIB technique are described in Chaps. 8–16. Chapter 8 reviews the application of ion beam to scanning probe microscopy. Chapter 9 reports on the FIB fabrication of needle-shaped specimens containing subsurface nanostructures for electron tomography. The FIB technique is used to create gratings and speckle patterns for micro/nanoscale deformation measurements in Chap. 10. Often the application of FIB nanostructures to the fabrication of nanostructures can be enriched by the combined use of another technique. Chapter 11 shows how to combine FIB with self-assembly of epitaxial quantum dots in order to produce regular quantum dot arrays. In Chap. 12, the combination of mold fabrication by FIB and nano-imprinting enables development of functional metallic glassy materials with micro- and nano-sized periodical structures. Chapter 13 describes the feasibility of combined use of dielectrophoresis and FIB for realizing nanodevices based on nanostructured materials. Combining FIB and nano-manipulator, Chap. 14 reports on characterization of thermoelectrical properties of individual one-dimensional nanosystems such as metallic or semiconducting carbon nanotubes and nanowires. In Chap. 15, the FIB technique is introduced as a useful and versatile tool for nanofluidic applications. Chapter 16 discusses the results of FIB use in developing nanostructured stent surfaces for cardiovascular applications.

Potential damage and instability during FIB processing are also considered in this book. Chapter 17 investigates the damage induced by $Ga^+$ FIB in piezoelectric nanostructures. The physical factors that can lead to instabilities are discussed in Chap. 18. In Chap. 19, the mass-separated FIB system is introduced and adapted to several fields of application in nanotechnology beyond classical Ga instruments.

The editor would foremost like to thank all the contributors for their outstanding chapters. I would also like to acknowledge editorial support from Springer and in particular Dr. David Packer and Sara Kate Heukerott. Finally I am grateful to Mr. Xingliang Xu for his contribution in providing helpful editorial assistance.

Beijing, China Zhiming M. Wang, Ph.D.

# Contents

# Contributors

**Fahad Al-Marzouki** Physics Department, King Abdulaziz University, Jeddah, Saudi Arabia

**A.-L. Biance** Institut Lumière Matière, UMR5306 Université Lyon 1-CNRS, Université de Lyon, Villeurbanne, France

**Lothar Bischoff** Helmholtz-Zentrum Dresden-Rossendorf, Institute of Ion-Beam Physics and Materials Research, Dresden, Germany

**N.P. Blanchard** Institut Lumière Matière, UMR5306 Université Lyon 1-CNRS, Université de Lyon, Villeurbanne, France

**L. Bocquet** Department of Civil and Environmental Engineering, Massachusetts Institute of Technology, Cambridge, Massachusetts, USA

**Xavier Borrise** Institut Catala de Nanotecnologia, Bellaterra, Spain

Institut de Microelectronica de Barcelona-CSIC, Bellaterra, Spain

**Roman Bottger** Helmholtz-Zentrum Dresden-Rossendorf, Institute of Ion-Beam Physics and Materials Research, Dresden, Germany

**Tae-Youl Choi** Department of Mechanical and Energy Engineering, University of North Texas, Denton, TX, USA

**R. Cordoba** Laboratorio de Microscopías Avanzadas (LMA), Instituto de Nanociencia de Aragón (INA), Universidad de Zaragoza, Zaragoza, Spain

**Ajuan Cui** Beijing National Lab of Condensed Matter Physics, The Institute of Physics, Chinese Academy of Sciences, Beijing, China

**A. Da Costa** CNRS UMR 8181, Unité de Catalyse et de Chimie du Solide – UCCS, Lens, France

UArtois, UCCS, Lens, France

Univ Lille Nord de France, Lille, France

**D. Deresmes** Univ Lille Nord de France, Lille, France

Institut d'Électronique, de Microélectronique et de Nanotechnologie – IEMN, CNRS UMR 8520, Villeneuve d'Ascq, France

Institut d'Électronique, de Microélectronique et de Nanotechnologie – IEMN, F-59652 Villeneuve d'Ascq, France

**R. Desfeux** CNRS UMR 8181, Unité de Catalyse et de Chimie du Solide – UCCS, Lens, France

UArtois, UCCS, Lens, France

Univ Lille Nord de France, Lille, France

**J.M. De Teresa** Departamento de Física de la Materia Condensada, Instituto de Ciencia de Materiales de Aragón (ICMA), Universidad de Zaragoza-CSIC, Zaragoza, Spain

Laboratorio de Microscopías Avanzadas (LMA), Instituto de Nanociencia de Aragón (INA), Universidad de Zaragoza, Zaragoza, Spain

**Ludovic F. Dumée** Institute for Frontier Materials, Deakin University, Waurn Ponds, VIC, Australia

Institute for Sustainability and Innovation, Victoria University, Hoppers Lane, Werribee, VIC, Australia

**A. Fernandez-Pacheco** Cavendish Laboratory, Department of Physics, University of Cambridge, Cambridge, UK

**A. Ferri** Univ Lille Nord de France, Lille, France

CNRS UMR 8181, Unité de Catalyse et de Chimie du Solide – UCCS, Lens, France

UArtois, UCCS, Lens, France

**Niall Finn** CSIRO Materials Science and Engineering, Waurn Ponds, VIC, Australia

**R. Fulcrand** Institut Lumière Matière, UMR5306 Université Lyon 1-CNRS, Université de Lyon, Villeurbanne, France

**Paul Galvin** Life Sciences Interface (LSI) group, Tyndall National Institute, University College Cork, Cork, Ireland

**Christopher T. Gibson** Flinders Centre for NanoScale Science and Technology, School of Chemical and Physical Sciences, Flinders University of South Australia, Adelaide, SA, Australia

**Changzhi Gu** Beijing National Lab of Condensed Matter Physics, The Institute of Physics, Chinese Academy of Sciences, Beijing, China

**Jesús Hernández-Saz** INNANOMAT Group, Departamento de Ciencia de los Materiales e I.M. y Q.I., Facultad de Ciencias, Universidad de Cádiz, Cádiz, Spain

**Miriam Herrera** INNANOMAT Group, Departamento de Ciencia de los Materiales e I.M. y Q.I., Facultad de Ciencias, Universidad de Cádiz, Cádiz, Spain

**Justin D. Holmes** Materials Chemistry and Analysis Group (MCAG), Electron Microscopy and Analysis Facility (EMAF), Tyndall National Institute, University College Cork, Cork, Ireland

**William Humphries** CSIRO Materials Science and Engineering, Waurn Ponds, VIC, Australia

**M.R. Ibarra** Laboratorio de Microscopías Avanzadas (LMA), Instituto de Nanociencia de Aragón (INA), Universidad de Zaragoza, Zaragoza, Spain

**A. Inoue** Tohoku University, Sendai, Japan

Physics Department, King Abdulaziz University, Jeddah, Saudi Arabia

**Krishna Kant** School of Chemical and physical Sciences, Flinders University of South Australia, Bedford Park, SA, Australia

School of Chemical Engineering, The University of Adelaide, North Engineering Building, Adelaide, SA, Australia

**Vera La Ferrara** Portici Research Center, Portici, NA, Italy

**Kyung-Min Lee** Department of Mechanical and Energy Engineering, University of North Texas, Denton, TX, USA

**Yanjie Li** School of Civil Engineering and Architecture, University of Jinan, Jinan, China

AML, Department of Engineering Mechanics, Tsinghua University, Beijing, China

**Wuxia Li** Beijing National Lab of Condensed Matter Physics, The Institute of Physics, Chinese Academy of Sciences, Beijing, China

**Zhanwei W. Liu** School of Aerospace Engineering, Beijing Institute of Technology, Beijing, China

**Dusan Losic** School of Chemical Engineering, University of Adelaide, North Engineering Building, Adelaide, SA, Australia

**D.V. Louzguine-Luzgin** WPI Advanced Institute for Materials Research, Tohoku University, Sendai, Japan

**Xichun Luo** Department of Design, Manufacture and Management, Faculty of Engineering, Universityof Strathclyde, Glasgow, UK

**Narcis Mestres** Institut de Ciencia dels Materials de Barcelona (ICMAB) Consejo Superior de Investigaciones Científicas, Bellaterra, Spain

**Sergio I. Molina** INNANOMAT Group, Departamento de Ciencia de los Materiales e I.M. y Q.I., Facultad de Ciencias, Universidad de Cádiz, Cádiz, Spain

**Alessio Morelli** Max Planck Institute of Microstructure Physics, Halle, Germany

**Feroze Nazneen** Life Sciences Interface (LSI) group, Tyndall National Institute, University College Cork, Cork, Ireland

**Nikolay Petkov** Materials Chemistry and Analysis Group (MCAG), Electron Microscopy and Analysis Facility (EMAF), Tyndall National Institute, University College Cork, Cork, Ireland

**Peter Philipp** Helmholtz-Zentrum Dresden-Rossendorf, Institute of Ion-Beam Physics and Materials Research, Dresden, Germany

**P. Poncharal** Institut Lumière Matière, UMR5306 Université Lyon 1-CNRS, Université de Lyon, Villeurbanne, France

**Jamie S. Quinton** Flinders Centre for NanoScale Science and Technology, School of Chemical and Physical Sciences, Flinders University of South Australia, Adelaide, SA, Australia

**A.K. Raychaudhuri** Theme Unit of Excellence in Nanodevice Technology, S.N.Bose National centre for Basic Sciences, Salt Lake, Sector-III, Kolkata, India

**D. Rémiens** UVHC, IEMN, Valenciennes, France

Univ Lille Nord de France, Lille, France

Institut d'Électronique, de Microélectronique et de Nanotechnologie – IEMN, CNRS UMR 8520, Villeneuve d'Ascq, France

**Gemma Rius** Nagoya Institute of Technology, Gokiso cho Showa ku, Nagoya, Japan

**S. Sangiao** Laboratorio de Microscopías Avanzadas (LMA), Instituto de Nanociencia de Aragón (INA), Universidad de Zaragoza, Zaragoza, Spain

**Bernd Schmidt** Helmholtz-Zentrum Dresden-Rossendorf, Institute of Ion-Beam Physics and Materials Research, Dresden, Germany

**Michael Schmidt** Materials Chemistry and Analysis Group (MCAG), Electron Microscopy and Analysis Facility (EMAF), Tyndall National Institute, University College Cork, Cork, Ireland

**Kallista Sears** CSIRO Materials Science and Engineering, Bayview Ave., Clayton, VIC, Australia

**A. Siria** Institut Lumière Matière, UMR5306 Université Lyon 1-CNRS, Université de Lyon, Villeurbanne, France

**Ashley D. Slattery** Flinders Centre for NanoScale Science and Technology, School of Chemical and Physical Sciences, Flinders University of South Australia, Adelaide, SA, Australia

**Jining Sun** Institute of Mechanical, Process and Energy Engineering School of Engineering and Physical Sciences Heriot-Watt University, Edinburgh, UK

**D. Troadec** Univ Lille Nord de France, Lille, France

Institut d'Électronique, de Microélectronique et de Nanotechnologie – IEMN, CNRS UMR 8520, Villeneuve d'Ascq, France

Institut d'Électronique, de Microélectronique et de Nanotechnologie – IEMN, F-59652 Villeneuve d'Ascq, France

**Ionela Vrejoiu** Max Planck Institute for solid State Research, Stuttgart, Germany

**Thomas Walther** Department of Electronic & Electrical Engineering, University of Sheffield, Sheffield, UK

**Qinghua H. Wang** National Institute for materials Science, Tsukuba, Ibaraki, Japan

**Huimin M. Xie** AML, Department of Engineering Mechanics, Tsinghua University, Beijing, China

**Xingliang Xu** State Key Laboratory of Electronic Thin Film and Integrated Devices, University of Electronic Science and Technology of China, Chengdu, Sichuan, People's Republic of China

**Haoyu Zhang** Department of Electronic & Electrical Engineering, University of Sheffield, Sheffield, UK

# Chapter 1
# Focused Ion Beam (FIB) Technology for Micro- and Nanoscale Fabrications

**Krishna Kant and Dusan Losic**

**Abstract** The focused ion beam (FIB) technology has become increasingly popular fabrication and characterization tool across many research areas from nanotechnology, material science, microelectronic industry, life science, biology, and medicine. FIB was specially recognized as an attractive tool or the fabrication of micro- and nanostructures with complex geometries and shapes. This chapter presents the basic introduction of FIB dual-beam system and its operation modes, followed by description of instrument in more details. The review has emphasis on FIB fabrication of nanostructures by milling and deposition methods with particular focus on fabrication of nanopatterns and nanopore arrays. Various efforts to fabricate micro- and nanoscale structure and geometrically complex structure are described. Finally, recent developments of applications of FIB in different areas of material science and life science are briefly reviewed.

K. Kant, Ph.D.
School of Chemical Engineering, The University of Adelaide,
North Engineering Building, Adelaide, SA 5005, Australia

School of Chemical and Physical Sciences, Flinders University of South Australia,
Bedford Park, SA 5042, Australia
e-mail: krishna.kant@flinders.edu.au

D. Losic (✉)
School of Chemical Engineering, The University of Adelaide,
North Engineering Building, Adelaide, SA 5005, Australia
e-mail: dusan.losic@adelaide.edu.au

Z.M. Wang (ed.), *FIB Nanostructures*, Lecture Notes in Nanoscale
Science and Technology 20, DOI 10.1007/978-3-319-02874-3_1,

## 1.1 Introduction

Fabricating materials with micro- and nano-dimensions is one of the biggest challenges in the material science and nanotechnology, and focused ion beam (FIB) is an important technology to address these challenges. The FIB technique not only offers the fabrication of nanostructures by the removal of material from the solid surface (milling) or their deposition or surface modifications but also at the same time has the capability to provide their topographical analysis. The dual-beam FIB or more advanced FIB instruments combined with the scanning electron microscope (SEM) and other imaging technologies are recognized currently as the most powerful tools for fabrication and characterization of nanostructured materials.

The FIB technique was introduced in the 1970 and the early 1980s by the invention of liquid metal ion source (LMIS), but the first commercial instruments were introduced more than two decades ago [1]. The FIBs initially have been mostly used in the semiconductor industry for repairing lithographic masks, correction and failure control of electronic circuits, and debugging integrated circuits (ICs) with increasing circuit density and decreasing feature dimension [2]. Since the 1990s an enormous growth in nanoscience and nanotechnology took place and the FIB started to be a very demanding technology for rapid prototyping of micro- and nanoscale structures and development of new areas of applications including microelectromechanical and bio-microelectromechanical systems (MEMS and bio-MEMS) for fabrication of micro-chips, microfluidic devices, miniaturized sensors, biosensors, photonic devices, scanning probe microscope (SPM) tips, magnetic head trimming and other micro-tools [3, 4]. The FIB showed to have many advantages for these applications over electron beam lithography, as it is a mask-free process, has the capability of very fine focusing with choice of a wide variety of ion masses and various ion species, and provides shorter penetration depth in solid. By controlling FIBs, it is possible to achieve an improved reproducibility and precision in fabrication of nanostructures with capability to create very complicated 3-D structures [5, 6]. Ion beam is produced by the high-energy ionized atoms with a relatively high mass of atoms and directed to the sample surface with the help of lenses and apertures provide high resolution in nanofabrication. In general FIB has four types of functionality including milling, deposition imaging, and implantation with high-energy ions [7, 8]. The collision of ion beam on sample surface produces the elastic and inelastic collision, and dual beam uses both of them for milling and imaging. Apart from the fabrication by milling FIB is used for the formation of nanostructures by the metal deposition with the help of gases from metal sources (LMIS) including the ion ($Ga^+$) directed as a beam with high electric field [9–11].

The development of FIB technology is a very active research area over the past years to improve the performance of FIB instruments and extend their applications into new research areas and industry. Even though the FIB is an expensive instrument it is today widely used in laboratories for material research and development, physical and chemical analysis, biological research sectors, and manufacturing companies. While the mainstream of FIB usage still remains within the

**Fig. 1.1** Schematic illustration of a dual-beam FIB–SEM instrument. Expanded view shows the electron and ion beam sample interaction [13]

semiconductor industry and nanofabrication, the FIB usage has significantly expanded for applications in metallurgy, earth science, ceramics, polymers, geology, art, biology, pharmaceuticals, and forensics [12]. The FIB provides huge advantages to the preparation of specimens for microscopy analysis including SEM, transmission electron spectroscopy (TEM), and other characterization techniques.

In this chapter we firstly show the basic introduction of dual-beam system to understand the basic principle of FIB and difference and advantages compared with SEM. Then, basic operation modes of FIB followed with description of instrument with more details are discussed. The review has emphasis on FIB fabrication of nanostructures by milling and deposition methods with particular focus on fabrication of nanopores and nanopore arrays. Finally, recent developments of applications of FIB in different areas of material science and biological applications are briefly reviewed.

## 1.2 Principle of FIB System

The basic components of an FIB system consist of an ion source, ion optics, a substrate stage, and a vacuum chamber with auxiliary equipment (Fig. 1.1) [13]. An FIB instrument looks and operates as an SEM as both instruments are based on a focused electron beam for image and an ion beam for the FIB fabrications.

In the FIB secondary ions can be detected and used to construct image of the sample with magnifications up to 100,000 times. The operating ion beam in FIB is based on LMIS of gallium ($Ga^+$) positioned in contact with a sharp tungsten (W) needle. The $Ga^+$ wets up the needle and flows to the W tip. A high extraction field (>$10^8$ V/cm) is used to pull the liquid $Ga^+$ into a sharp cone of up to 5–10 nm radius as it is shown in Fig. 1.1. The use of $Ga^+$ has its advantages: (1) $Ga^+$ has a low melting point, so it almost exists in the liquid state near room temperature, and (2) $Ga^+$ can be focused to a very fine probe size <10 nm in diameter. The FIB generally operates with an accelerating voltage between 5 and 50 keV. By a fine controlling of the strength of the electrostatic lenses and adjusting of the effective aperture sizes, the ion probe current density may be altered from tens of pA up to several nA. The ions of beam collide with the surface atoms of a solid sample, and they lose energy to the surface electrons. The important physical effects of collision of ions on the substrate are as follows: sputtering of neutral and ionized substrate atoms which lead to the substrate milling, and the electron which is emitted out as secondary electron provides the imaging of surface and also makes sample charged. In the presence of gases chemical interactions between the surface and breaking of their chemical bonds help in the deposition process of ion milling [13].

## 1.2.1 *Topographical Imaging by FIB*

The finely FIB is used to scan over a substrate, and secondary electrons are generated from the surface of the sample and used for imaging purpose (Fig. 1.2a) [14]. The electrons or the ions are collected on a biased detector for secondary electrons. The detector bias is a positive or a negative voltage with respect to the sample for collecting secondary electrons or ions. These secondary ions can be used for secondary ion mass spectroscopy (SIMS) of the target material in a mass spectrometer attached to the system. During the FIB operations, a small amount of $Ga^+$ ions collide on the sample surface and large numbers of secondary electrons leave the sample. The system also prevents the excess of electron collecting when it is used only for imaging due to electrostatic discharge, and it enables the reliable imaging of nonconducting materials such as glass which is largely used in microsystems.

The resolution of FIB images depends on the ion beam spot size, i.e., below 10 nm. In crystalline materials such as aluminum and copper, the ion penetration depth varies due to channeling along open columns in the lattice structure of the metals and its oxides. The rate of secondary electron emission depends on the penetration depth of ions. It is worth noting that imaging with FIB inevitably induces some damage to the sample. Most of the $Ga^+$ ions that arrive at the sample surface enter the sample; thus, the ion implantation takes place; besides implantation some milling also occurs when the ion beam scans over the sample surface. This type of milling effect can be reduced or avoided by using a fine ion beam with a fine spot and low ion current.

**Fig. 1.2** Principle of FIB (**a**) imaging, (**b**) milling, and (**c**) deposition [14]

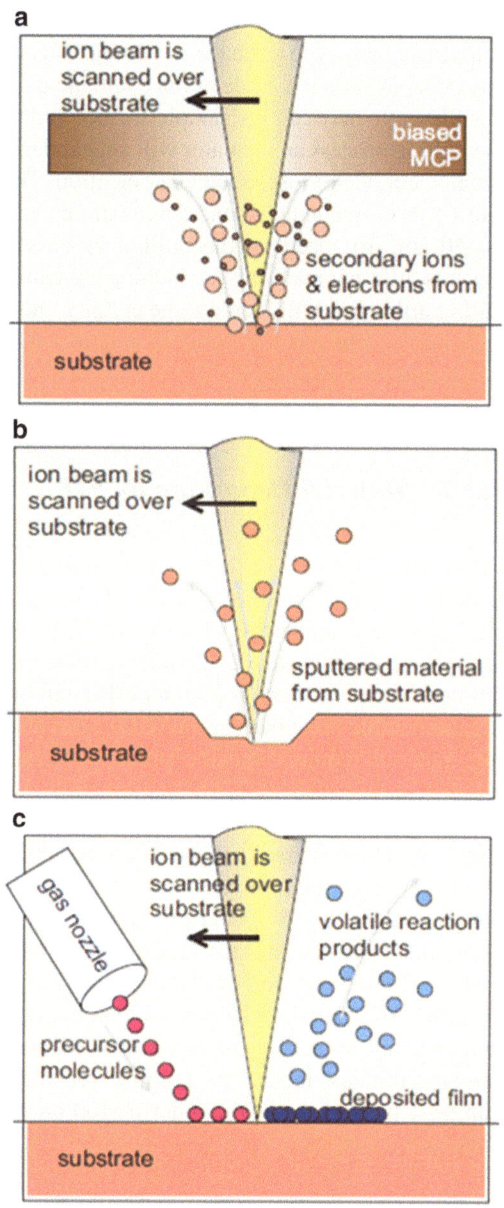


## 1.2.2 *Milling of Solid Structures by FIB*

The removal of sample material and making of patterns and structures can be achieved by using a high-current ion beam. The result of high-current FIB is that the high-mass atom of material comes out and leaves empty space at the site as shown in schematic in Fig. 1.2b. Scanning the beam over the substrate in a particular manner can achieve any type of shape or structure in surface. The sputtering yield is generally dependent on the angle of ion beam incidence; it increases with $1/\cos(\theta)$, where $\theta$ is the angle between the sample surface and the ion beam direction. At the end of the milling process an insulator will appear dark in the SEM image than a conductor because conductor loses secondary electron. The resolution of the milling process is a few tens of nanometers, and the maximum aspect ratio of the milled holes is up to 10–20 nm. To speed up the milling process or increase the selectivity of beam towards different materials, an etching gas can be introduced into the work chamber during milling. It will increase the etching rate and the selectivity towards different materials by chemical removal of reaction products. This technique is called gas-assisted etching (GAE).

### *1.2.3 Material Deposition by FIB*

FIB has the capability of localized mask-free deposition of both metal and insulator materials. It works on the principle of chemical vapor deposition (CVD), and it is similar as laser-induced CVD [15, 16]. The only difference is the better resolution with FIB but with lower deposition rate. The metals, which generally used for deposition on sample surface is platinum (Pt), but in the case of tungsten (W) deposition organometallic precursor gas ($W(CO)_6$) can also be used. The deposition of insulating material requires specific precursors; for example, silica ($SiO_2$) deposition requires gases such as 1,3,5,7-tetramethyl-cyclotetrasiloxane (TMCTS), oxygen ($O_2$) or water vapour are required. The deposition process is shown in schematic in Fig. 1.2c. The precursor gases flow on the surface of sample with the help of gas jet, and it gets adsorbed on surface molecules. When the ion beam hits the sample surface it decomposes the adsorbed precursor gases. The volatile reaction products come out from the surface and are removed through the high-vacuum system, and the desired reaction products (W or $SiO_2$) remain on the sample surface as a thin film. In general the deposited material is not very pure because some organic contaminants as well as $Ga^+$ ions of ion beam are also present. The deposition has a bitmapped shape instead of a simple rectangle shape. The smallest features that can be deposited are of the order of 100 nm (lateral dimension) and up to 10 nm thicknesses [14].

## 1.3 The FIB Instrument Description

A schematic diagram of a FIB/SEM dual-column instrument is shown in Fig. 1.3 [17]. The instrument consists of a high-vacuum system, a chamber with sample stage, LMIS, ion column, gas jet, and a computer to run the complete instrument. The configuration is almost similar to the SEM, and the only difference is the use of a

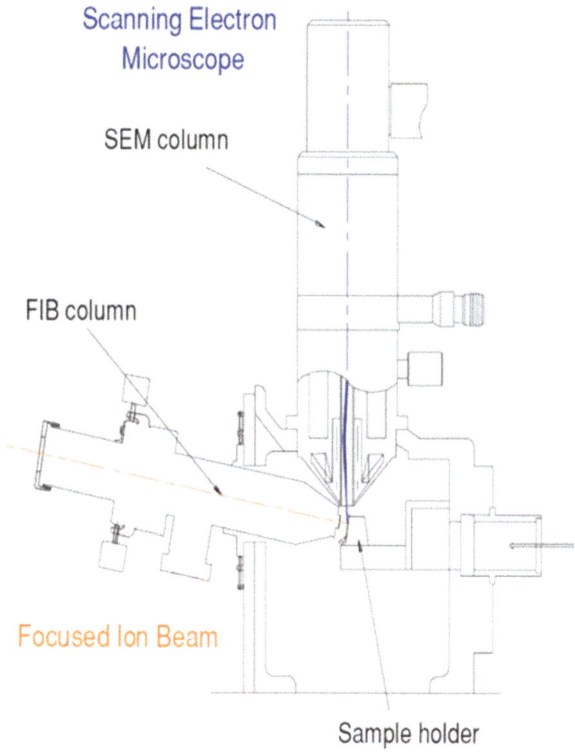


**Fig. 1.3** Schematic diagram of combined FIB/SEM [17]

gallium ion ($Ga^+$) beam instead of electron beam. The ion beam is generated from LMIS by the application of a strong electric field. This electric field causes the emission of positively charged ions from a liquid gallium cone. The operating extraction current is 2 μA, which is formed on the tip of a tungsten needle and has an extraction voltage of 7,000 V. The ion beam energy is typically between 10 and 50 keV, with beam currents varying between 1 pA and 10 nA. Using the variable aperture mechanism, the beam current can be varied, allowing fine beam for high-resolution imaging on sensitive samples and a heavy beam for fast and rough milling.

The sample is mounted on a motorized stage, inside the chamber. The chamber is divided into two parts with the valve: one is beam line chamber, and the other one is working chamber for sample loading and unloading. Under normal operating conditions, inside this chamber a high vacuum up to $10^{-7}$ m bar range is maintained. In order to preserve this vacuum the valve gets closed while loading or unloading the sample. A system of high vacuum is needed to maintain the vacuum inside the column and the work chamber where a rotary pump is used in combination with a turbo pump. The ion column is separated with one pump. Most of the FIB systems are available for delivering a variety of gases in working stage to help milling to the sample surface. To provide gas for working chamber a gas cabinet containing all applicable gases is present outside the vacuum chamber. The gas containers are connected to a gas jet nozzle assembled inside the working chamber through an

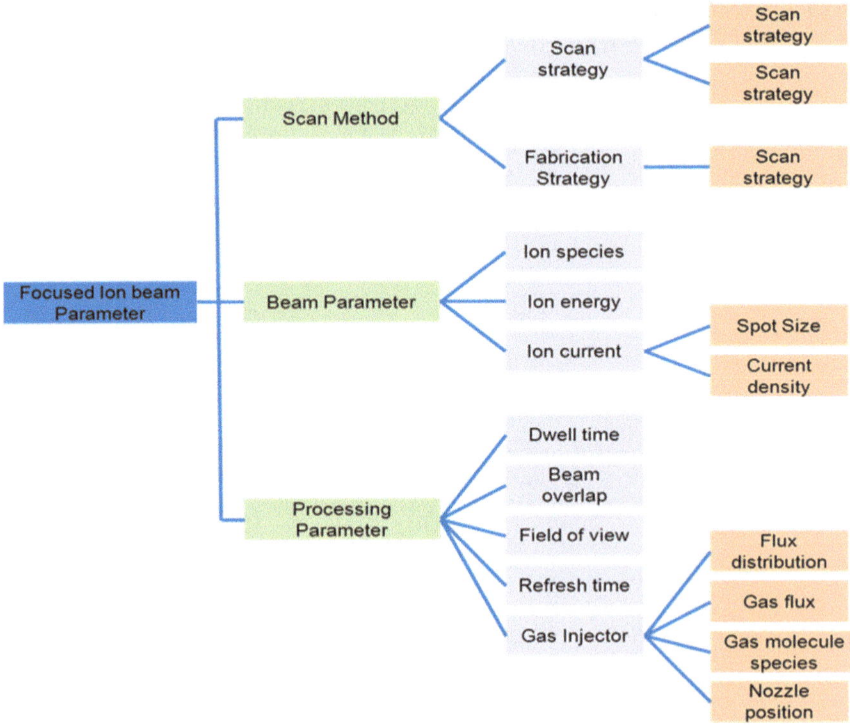


**Scheme 1.1** Classification of FIB parameters important for performances of FIB system

appropriate piping system. The gases are used for faster and more selective etching as well as for the deposition of materials [17]. All operations in FIB instrument such as loading of samples (partly), adjustment of sample stage, manipulation of the ion beam and beam energy, and control of the vacuum system and gas jets for gas delivery are controlled with the computer and software. Indeed, the complete user interface is run by means of a computer workstation. There are a number of FIB parameters which influence performance and efficiency of FIB system which are summarized in Scheme 1.1. The influence of these parameters using different FIB modes and applications was investigated by many authors and presented in a recent review [18].

## 1.4 Fabrication of Nanostructures by FIB and Their Application

### *1.4.1 Nanopatterning and Nanolithography*

The FIB technology is widely used in the nanopatterning or mask-free lithography providing many advantages compared with conventional and electron beam

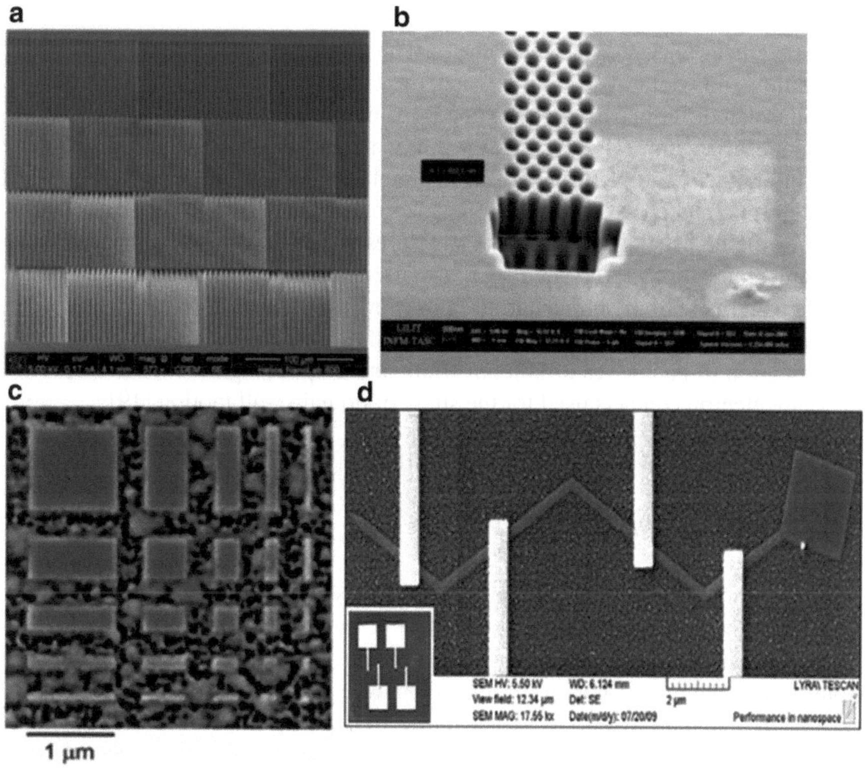


**Fig. 1.4** Examples of patterned nanostructures fabricated by FIB milling and deposition: (**a**) Aligned 2-D structures used as a diffractive optical element of a fast parallel processing spectrometer were milled into Si with a 21 nA FIB beam which was [19]. (**b**) 2-D periodic pattern of closed nanoholes directly written on wave guide slab by the FIB milling. The cross section of the small pattern shows the 600 nm deepness [20]. (**c**) SEM images of square patterns fabricated by milling into a 30 nm permalloy thin film by the 30 keV $Ga^+$ ion beam perpendicularly on substrate. (**b**) Spin-valve nanowire of 200 nm width and 16 μm length with FIB-deposited platinum electric contacts used for the study of domain wall motion [19]

lithography. A mask-less lithography is a one-step process based on direct ion beam writing, i.e., transferring a pattern by direct printing of the FIB onto the sample. The technique can be used for any type of materials including silicon, metal oxides, metals, polymers, and soft and hard biological samples. Most of the FIB applications have aimed to achieve the smallest possible sizes for pattern elemental structures. FIBs with high beam current, small spot size and high accuracy in patterning open the opportunities to expand FIB prototyping to larger area and complex structure. The dual-beam FIB instrument has control not only on the lateral dimensions but also on the depth of each individual pattern element independently. These 2-D patterns on the surface can be fabricated using both milling and deposition mode depending on the required application.

Figure 1.4a shows an example of FIB machining for fabrication of a diffractive optical element for UV–Vis spectrometer. The 2-dimensional trenches with length of 125 mm and width of 2.5 mm are patterned in area of $1 \times 0.5$ $mm^2$ and arranged in groups that differ in pattern depth of 35–3.5 mm in step of 35 nm. These trenches were patterned with a 21 nA current beam, which usually takes about 7 h to write [19]. In another example a pattern of nanohole arrays was made with the 10 pA beam current, and cross section is milled at a particular region to check the correct depth of the holes (Fig. 1.4b). These structures were used on wave guide slab for photonic applications [20]. The FIB is also used as a useful tool for the fabrication of 2-D patterns with the implantation of the ions in silicon nitride for various types of semiconductor device fabrication which is shown in Fig. 1.4c. Another example is fabrication of nanowires of 200 nm width and 16 μm length with FIB-deposited platinum electric contacts used for the study of domain wall motion [19].

### *1.4.2 FIB Nanopore Fabrications*

Nanopores, defined broadly as holes or channels that range from less than 1 to 100 nm in diameter, have attracted considerable interest in recent years for molecular separations, biosensing, molecular electronics, optical devices, and nanofluidics. Solid-state in comparison with soft materials were seen more advantageous for fabrication of nanopores due to its high stability and rigidity, controllable diameter and channel length, and adjustable surface properties [20, 21]. Several approaches for fabrication of solid-state nanopores such as FIB, electron beam drilling, electron sculpting, reactive ion etching, and nuclear track etching have been explored in previous years [22–25]. However, the FIB milling has been particularly accepted as the most popular method for preparing nanopores in thin film because it provides superior resolution and the ability to perform direct patterning of pore arrays [26, 27]. In this case if we look towards the formation of nanopores into silicon wafer it needs to have very high control in accuracy and beam focus.

A typical example of nanopore array electrode fabricated on oxidized silicon wafer substrates with the FIB method is shown in Fig. 1.5 [27]. To fabricate the single nanopore or nanopore array with controlled-pore diameters and inter-pore distance the sequential FIB milling is the best choice. The fabrication of single or array nanopores was particularly attractive for fabrication of DNA sensing devices and biosensors [26]. In Fig. 1.5 shows the disk-shaped pores produced with different size by sequential milling in the silicon nitride layer. The fabrication process requires two steps including making of the pattern by the software which will work as the guide mark for the ion beam to do the milling over the surface. The process provides the flexibility to make the number of patterns and run the instrument in auto mode for long-time milling on sample. The ion beam current is the key factor to know the milling time of the pattern; if the ion beam current is high the milling time will be lower. The lower ion current is useful but has some disadvantages: long time of milling and beam displacement if it is not very stable with sample.

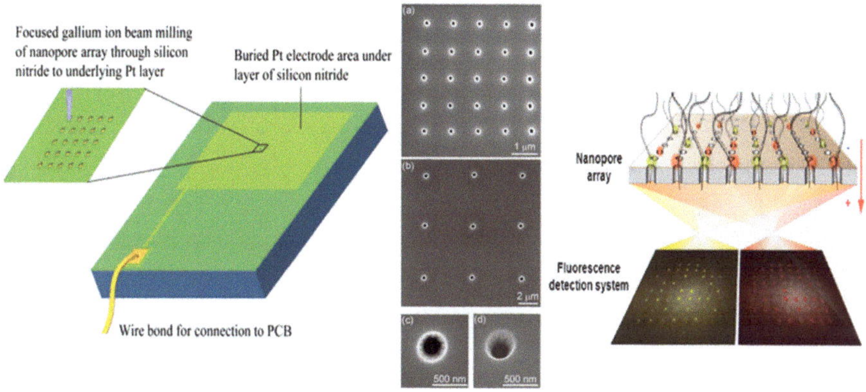


**Fig. 1.5** Schematic of the fabrication strategy (*left*). The Pt-coated surface covered with silicon nitride is milled by focused ion beam to open up nanopores through to the underlying Pt. SEM images from a selection of arrays and single electrodes with varying diameters. Schematic (*right*) illustration of the concept of threading of the DNA/beacon complex through a nanochannel that allows optical detection of the target DNA sequence [27]

The milling rate varies with the ion beam current, and a small beam current is always preferred to use for milling process of structures with nanoscale dimensions. It is reported that the limitation of 150 nm is the lowest pore diameter achieved for longer pores due to the high aspect ratios of the pores and the beam diameter [27]. Close examination of these nanopores fabricated by FIB method shows to have a conical rather than cylindrical shape. The formation of conical pores by FIB milling has been explained to be caused by redeposition of milled material [28, 29]. To address this problem, the use of XeF2-enhanced FIB milling is demonstrated to reduce the effects of material redeposition in the fabrication of inter-digitated nanogap nanoelectrodes [30]. These nanopore arrays produced by FIB were explored by many groups to mimic cell membrane nanopores and for DNA biosensing applications [31].

### *1.4.3 FIB Combined with Anodization Process for Nanopore Array Fabrication*

In recent years, porous anodic alumina (PAA) fabricated by electrochemical anodization process with self-organized nanopores in hexagonal arrangement has created great interest due to its potential to be used for many applications including molecular separations, biosensing, energy storage, photonics, drug delivery, and template synthesis. However this fabrication process has limitations regarding pore diameters, pore inter-distance, and pore patterns and shapes. To address this problem, the FIB and FIB-based lithography is combined with anodization process. The FIB lithography on silicon has been previously demonstrated to produce

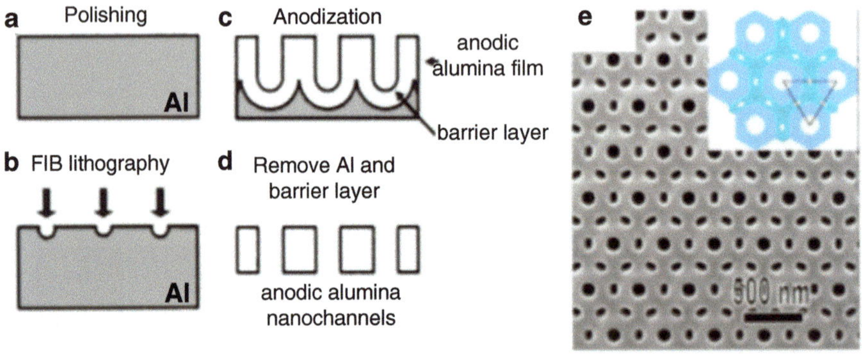


**Fig. 1.6** (**a–d**) Schematic diagram showing the process for fabricating anodic alumina film with ordered nanochannels [35]. (**e**) Alternating FIB pattern after the anodization, and the inset is the schematic of the pore shape and oxide wall shape development [37]

hexagonal patterns to guide the growth of highly ordered PAA with controlled inter-pore distances and patterns. The depths of the FIB patterns range from 5 to 20 nm for effective pore growth during anodization. With the FIB gradation in the size and shape (rectangle, square, and circle) can be easily achieved with the array of the nanopatterns [32–34]. Another approach is used by FIB to create arrays of hexagonally close-packed concaves with variable depths on polished Al surfaces and study their effects on guiding the growth of the nanopores. Figure 1.6 [35] shows the schematic of the fabrication of PAA membranes by FIB patterning process. It was found that concaves as shallow as few nanometers can effectively guide the growth of the PAA nanopores during anodization process. Therefore the fabrication of pores with different pore sizes, inter-pore distance, shapes, and patterns is possible by combining FIB and anodization method [36]. It was found that the $Ga^+$ implantation in Al, during the FIB patterning, has important role in growth of nanopore in the anodization process; the pores start growing from the bottom side of the FIB-patterned concaves [37]. At the same time, the pores have a tendency to expand, and the neighboring walls of these pores approach each other until two alumina layers merge. Since all the nanopores develop at the same rate, the neighboring pores restrain the change of the inter-pore distance. The nanopores developed from the FIB-patterned Al grow faster and have thicker oxide walls. Despite lot of advantages in patterning with FIB for shape and size, the *z*-direction pore growth is not well controlled and it depends on the anodization conditions. If we monitor the effect of FIB patterning on the anodization process we can find that the shallow pattern created by the FIB can effectively guide the growth of the anopores during the anodization. Based on this concept the FIB is used combined with other synthetic approaches including chemical vapor, atomic layer, and electrochemical deposition to grow highly ordered nanorods, nanowires, and nanotubes in a hexagonal pattern on a substrate [38].

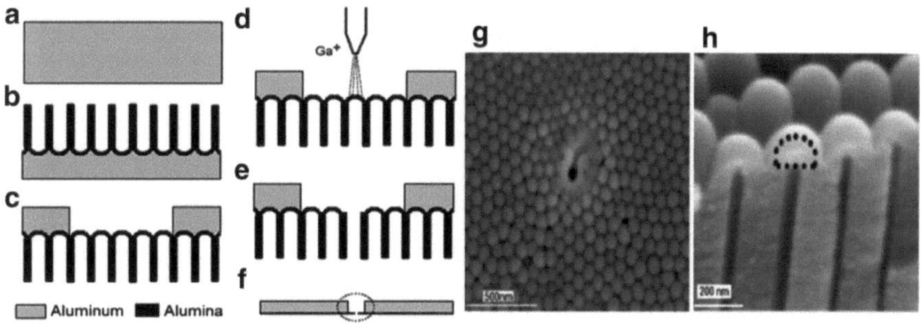


**Fig. 1.7** Schematic diagram of the concept of using focused ion beam (FIB) technique for fabrication of single nanopore and nanopore arrays using porous anodic alumina (PAA) as substrate. (**a, b**) Fabrication of PAA by anodization of aluminum. (**c**) PAA substrate with exposed bottom surface for FIB milling. (**d**) The pore opening of PAA barrier film by FIB milling. (**e**) PAA with opened pore and (**f**) PAA with single pore or nanopore arrays. SEM images of nanopore fabricated on PAA substrate by removing of barrier film using FIB milling. (**g**) Single pore 50 nm fabricated by FIB technique with (**h**) corresponding model of the barrier film removal [38]

In addition to FIB-guided anodization method there is another approach to use FIB for fabrication of PAA with single or nanopore arrays. The method is based on the pore opening from the bottom side of PAA by FIB milling and presented in Fig. 1.7 [38]. To obtain a single nanopore opening on PAA the ion beam milling is done at the central part of a single hexagonal PAA cell using a 30 keV $Ga^+$ focused beam and very low beam current of 1.5 pA for a period of 5 s. A single pore with a diameter that corresponds to the diameter of an internal pore structure is obtained and presented in an SEM image in Fig. 1.7g. The pore was opened by removing the barrier oxide layer from a single PAA cell. The milling of this smallest pore opening is more delicate because very fine adjustments of position and ion beam are required to accomplish milling in exactly the middle of the PAA cell. It was demonstrated that the FIB milling can be successfully used for controlled removal of the oxide barrier film and pore opening of PAA to form a single nanopore or nanopore arrays.

### *1.4.4 Fabrication of Specific Micro- and Nanostructures*

FIB-based fabrication methods were successfully used in recent years for preparation of numerous morphologies and forms from different materials including metal sheets, fibers, powders, and films. The FIB is largely used for sample and tip preparation for TEM and AFM where two types of milling processes were applied: the annular milling process and the site-specific milling. One specific example is the fabrication of atom probe which is a smoothly tapered needle with a circular cross section and an end radius of 50–150 nm, Fig. 1.8 [39]. In this type of milling method, the sample is mounted in a copper tube and oriented towards end on to the ion beam. An annular mask is used to mill away the edge; the outer diameter of the mask is

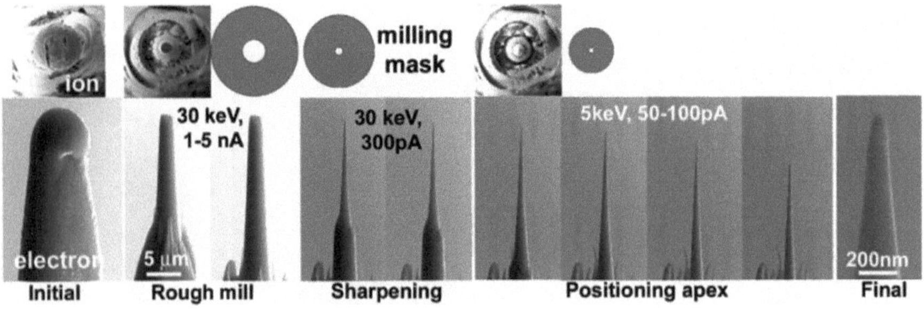


**Fig. 1.8** Annular milling stage of atom probe specimen preparation. The size of the annular mask and the ion current are decreased as shown during the annular milling procedure [39]

chosen to be slightly larger than the maximum diameter of the sample to avoid the generation of multiple sharp edges with the sample. It is a multiple-step process with decreasing inner and outer diameters and ion currents. To maximize the milling rate, milling should be performed concentrically from the outer diameter to the inner diameter of the mask as the milling rate is greater on an exposed edge. This procedure can be completed in about 30 min. Annular milling method is useful in correcting wedge-shaped electropolished samples or simply re-sharpening blunt or previously analyzed samples [40, 41]. At lower ion beam energy 2–5 keV, it may also be used to remove contamination or surface oxide films from sample surface and to prepare tips for scanning tunneling microscope [42].

### *1.4.5 Fabrication of 3-D and Complex Micro- and Nanostructures*

The most attractive application of FIB technique is to fabricate complex 3-D micro- and nanostructures which are not possible to make by any other existing technologies [43]. The high density and uniform supply of reactive gases are commonly used along with FIB milling for these applications. When the sample is exposed to the ion beam and surface of the sample started etching, the deposition of material takes place with higher deposition rate than etching rate. It is also important that ion beam current should be lower for deposition and etching. The ion beam-induced CVD is introduced with the help of excited secondary electrons as well as the primary ions used in reaction. The principle is presented in Fig. 1.9 [43]. The ion beams itself produce secondary electrons at the beam point, and the ion beam CVD reactions take place at lower source-gas pressure levels than the pressure which is required in electron beam CVD processes. It is also observed that $Ga^+$ beam ions are available on the surface of the structures up to 40 nm depth. A lower current beam is caused as low lateral growth rather than a perpendicular growth, so it helps to form an

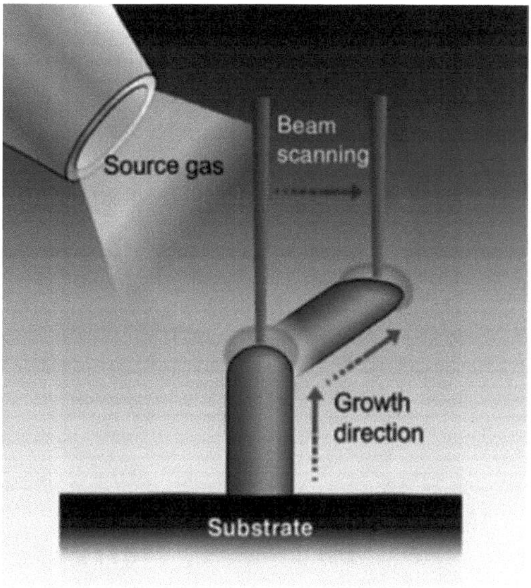


**Fig. 1.9** Schematic illustration of FIB-CVD growth [43]

upstairs. By combining the lateral growth with the rotating beam scanning, complex 3-D structures can be fabricated.

In addition to the low scan, the faster FIB scanning enhances both the lateral and bottom growth to produce downstairs structures. It is reported that with change in gas pressure these two growth modes share their boundary in the same space, where the upstairs growth at the bottom and the downstairs growth at the higher region take place. The secondary electrons which are produced by the ion beam by sample surface decomposition enhance the lateral growth on sample. A slight shift in beam position causes shifts in secondary electrons and its distribution, so the growth moves from its place and starts from the side of the fabricated structure. The combination of this lateral growth mode with shifting beam scanning provides the growth of 3-D structures [45]. At the same time with the continuous rotation in the lateral growth it is possible to achieve a rotational symmetry like a wineglass or a boll. Based on the growth mechanisms very complex 3-D nanostructures such as wineglass and coils were produced (Fig. 1.10) [44].

## 1.5 Emerging Applications of FIB Technology

The main applications of FIB methods based on milling, deposition, patterning, implementation, doping, and imaging are summarized in Scheme 1.2 [18] and Table 1.1. These applications are broad and include a number of areas from

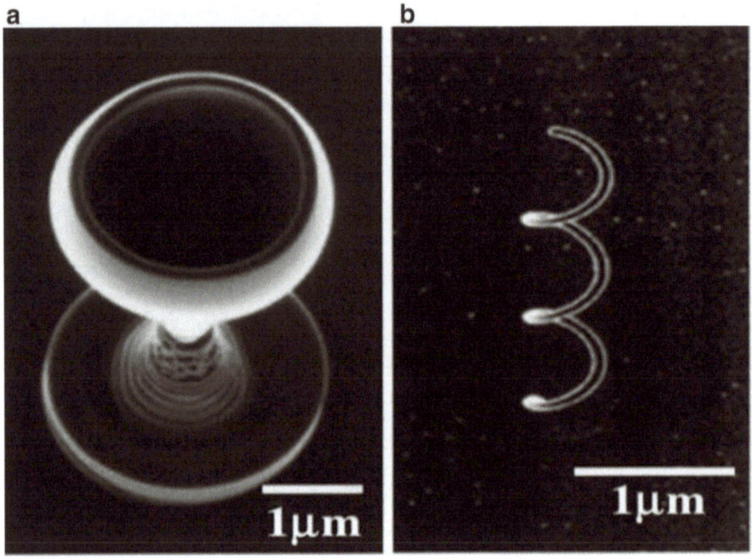


**Fig. 1.10** Examples of FIB-prepared three-dimensional (3-D) structures: (**a**) wineglass and (**b**) coil [44]

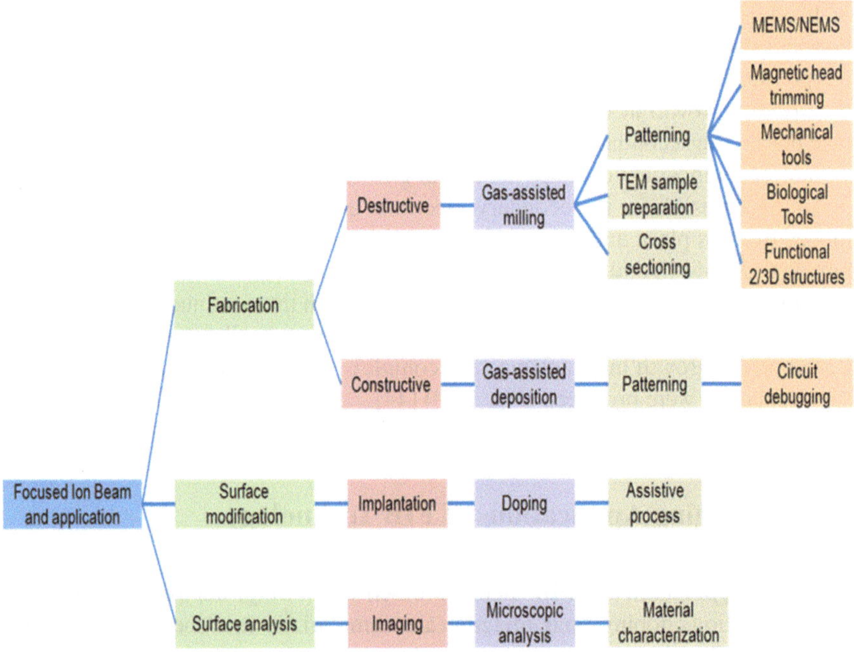


**Scheme 1.2** Classification of FIB methods and their applications

**Table 1.1** FIB technology and its application in various fields

| FIB methods | Applications | Reference |
|---|---|---|
| Nanopatterning | Optoelectronics | [46–48] |
| Preparation of TEM samples | Geosciences/mineralogy | [49–52] |
| Metal grain boundary examination | Metal research | [53–57] |
| Device fabrication | Semiconductors | [58–61] |
| Imaging and sample preparation | Life science | [62–66] |
| Sample analysis | Secondary ion mass spectrometry | [12, 67, 68] |
| Fabrication of nanopores | Nanopore biosensing | [69–71] |
| Sample preparations | AFM, TEM sample | [72–74] |
| Patterning and nanostructure growth | Nanowire fabrication | [75–78] |

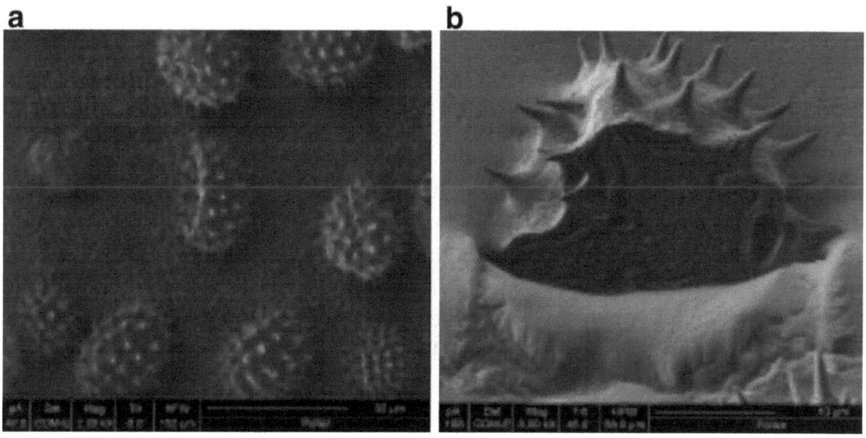


**Fig. 1.11** (**a**) Pollen grains shown by SE imaging with the FIB. (**b**) Cross section by the ion beam (FEI Company)

microelectronic industry, materials, life and earth science, biology, and medicine. A number of researchers working in the field of nanoscience and nanotechnology are using the FIB in their daily work as a basic fabrication or characterization tool. Some of the selected application examples in the life science are presented.

### *1.5.1 Application of FIB in Life Science*

The FIB for imaging, cross-sectioning, and 3-D machining is a rapidly growing application in life science and biological research. Many of the conventional methods of sample preparation used within life sciences are directly compatible with investigation by the FIB. It is difficult to examine biological material in its natural state with scanning or transmission electron microscopy, because biological samples are not compatible with high vacuum. To overcome this problem there are few techniques which can maintain the original state of the sample in such conditions. The chemical fixing, drying, and cryogenic preparation such as high-pressure freezing are generally being used for biological sample preparation. Some of these samples are robust enough to be used directly in the FIB for site-specific investigation. Specimens with hard cell walls such as plants or with an outer skeleton of insects can be applied in the FIB without any sample preparation. The imaging of pollen structure which includes external and cross-sectional imaging after FIB cross-sectioning is shown in Fig. 1.11. A stamen was removed from flower and pollen tapped onto double-sided carbon tape loaded in the chamber, cross-sectioned, and imaged by FIB. The same method is used for many other biological samples (algae, leaves, tissues, etc.) showing the capability of this method for direct and in situ examination of internal structures of biological samples.

## 1.6 Conclusion

The FIB technology has shown tremendous development in last decade with extensive spread of applications in many areas from microelectronic industry, materials, life and earth science, biology, and medicine. This was possible because the FIB technology adopted the developments from all the previous advances in related fields like field emission, charged particle optics, and applied material science. The optimization of specific gallium-based FIB instrument, together with innovative patterning and milling process, was a key progress in these developments. All these advancements make FIB currently the most powerful tool for characterization and fabrication in modern laboratories and for small-scale post-processing or prototype fabrications. The nanofabrication by FIB has been mainly based on the functions of milling, depositing, implanting, and imaging. Specially combination of FIB and CVD has recently become to be attractive for fabrication of complex 3-D micro- and nanoscale structures. The FIB technique is now well established for the diverse nanotechnological applications especially with the aim to explore "bottom-up" approach and use for fabrication of advanced materials and devices. The technique is of major interest for future applications to nano-electronics, nanomagnetism, spin-electronics, nano-optics, and biosensing. However, still there are some obstacles to address the limitations of physical and chemical phenomena in having structural stability, especially in milling mode due to its inherent error and material redeposition. The FIB technique, as one of the most powerful nanofabrication technologies, is expected to lead to numerous new breakthroughs and applications in nanotechnology research in the coming decade.

**Acknowledgments** The authors acknowledge the financial support of the Australian Research Council (FT 110100711, LP 100100272) and the University of Adelaide for this work. The author also thanks Mr. L. Green from Adelaide Microscopy and FEI Company for technical and FIB imaging support.

## References


1. Melngailis, J.: Focused ion beam technology and applications. J. Vac. Sci. Technol. B **5**(2), 469–495 (1987)
2. Stewart, D., Doyle, K., Andrew, F., Casey, Jr. J., David: Focused ion beam deposition of new materials: dielectric films for device modification and mask repair, and tantalum films for X-ray mask repair. In: Proc. SPIE. vol. 276, pp. 276–283. (1995)
3. Smidt, F.A., Hubler, G.K.: Recent advances in ion beam modification of metals. Nucl. Instr. Meth. Phys. Res. B80/81 **1**, 207–216 (1993)
4. Reyntjens, S., De Bruyker, D., Puers, R.: Focused ion beam as an inspection tool for microsystem technology. In: Proc. Microsystem Symp. p. 125. Delft, The Netherlands (1998)
5. Ward, B.W., Economou, N.P., Shaver, D.C., Ivory, J.E., Ward, M.L., Stern, L.A.: Microcircuit modification using focused ion beams. In: Proc. SPIE. p. 923. (1988)
6. Glanville, J.: Focused ion beam technology for integrated circuit modification. Solid State Tech. **32**, 270 (1989)
7. Stewart, D.K., Stern, L.A., Foss, G., Hughes, G., Govil, P.: Focused ion beam induced tungsten deposition for repair of clear defects on X-ray masks. In: Proc. SPIE. p. 1263. (1990)
8. Walker, J.F., Reiner, J.C., Solenthaler, C.: Focused ion beam sample preparation for TEM. Proc. Microscop Semiconductor Material Conf., Oxford (1995)
9. Athas, G.J., Noll, K.E., Mello, R., Hill, R., Yansen, D., Wenners, F.F., Nadeau, J.P., Ngo, T., Siebers, M.: Focused ion beam system for automated MEMS prototyping and processing. In: Proc. SPIE, vol. 3223, p. 198. (1997)
10. Daniel, J.H., Moore, D.F.: A microaccelerometer structure fabricated in silicon-on-insulator using a focused ion beam process. Sens. Actuator A Phys. **73**(3), 201–209 (1999)
11. Brugger, J., Beljakovic, G., Despont, M., de Rooij, N.F., Vettiger, P.: Silicon micro/nanomechanical device fabrication based on focused ion beam surface modification and KOH etching. Microelectron. Eng. **35**(1–4), 401–404 (1997)
12. Giannuzzi, Lucille, A., Stevie, F.A.: Introduction to focused ion beams: Instrumentation, theory, techniques, and practice. New York: Springer (2005)
13. Volkert, C., Minor, A.: Focused ion beam microscopy and micromachining. Mrs Bull **32**(5), 389–395 (2007)
14. Reyntjens, S., Puers, R.: A review of focused ion beam applications in microsystem technology. J. Micromech. Microeng. **11**(287), 300 (2001)
15. Thornell, G., Johansson, S.: Microprocessing at the fingertips. J. Micromech. Microeng. **8**, 251 (1998)
16. Johansson, S., Schweitz, J.A., Westberg, H., Boman, M.: Microfabrication of three-dimensional boron structures by laser chemical processing. J. Appl. Phys. **72**, 5956 (1992)
17. Gierak, J.: Focused ion beam technology and ultimate applications. Semicond. Sci. Tech. **24**, 23 (2009)
18. Kim, C.S., Ahn, S.H., Jang, D.Y.: Review: developments in micro/nanoscale fabrication by focused ion beams. Vacuum **86**(8), 1014–1035 (2012)
19. Urbanek, M., Uhlíř, V., Bábor, P., Kolíbalová, E., Hrničř, T., Spousta, J., Šikola, T.: Focused ion beam fabrication of spintronic nanostructures: an optimization of the milling process. Nanotechnology **21**, 145305 (2010)
20. Storm, A.J., Chen, J.H., Ling, X.S., Zandbergen, H.W., Dekker, C.: Fabrication of solid-state nanopores with single-nanometre precision. Nat. Mater. **2**(8), 537–540 (2003)
21. Chang, H., Iqbal, S.M., Stach, E.A., King, A.H., Zaluzec, N.J., Bashir, R.: Fabrication and characterization of solid-state nanopores using a field emission scanning electron microscope. Appl. Phys. Lett. **88** (2006)
22. Wu, M., Krapf, D., Zandbergen, M., Zandbergen, H.W., Batson, P.E.: Formation of nanopores in a SiN/$SiO_2$ membrane with an electron beam. Appl. Phys. Lett. **87**, 113106-1–113106-3 (2005)

23. Apel, P.Y., Akimenko, A., Blonskaya, I., Cornelius, T., Neumann, R., Schwartz, K., Spohr, R., Trautmann, C.: Etching of nanopores in polycarbonate irradiated with swift heavy ions at 15K. Nucl. Instrum. Methods Phys. Res., Sect. B: Beam Interactions with Materials and Atoms **245**(1), 284–287 (2006)
24. Fu, Y.B., Ngoi, K., Ann, F., Loh, T.: Fabrication and characterization of nanopore array. J. Nanosci. Nanotechnol. **6**(7), 1954–1960 (2006)
25. Li, J., Stein, D., McMullan, C., Branton, D., Aziz, M.J., Golovchenko, J.A.: Ion-beam sculpting at nanometre length scales. Nature **412**(6843), 166–169 (2001)
26. Kim, M.J., Wanunu, M., Bell, D.C., Meller, A.: Rapid fabrication of uniformly sized nanopores and nanopore arrays for parallel DNA analysis. Adv. Mater. **18**(23), 3149–3153 (2006)
27. Lanyon, Y.H., De Marzi, G., Watson, Y.E., Quinn, A.J., Gleeson, J.P., Redmond, G., Arrigan, D.W.M.: Fabrication of nanopore array electrodes by focused ion beam milling. Anal. Chem. **79**(8), 3048–3055 (2007)
28. Langford, R.M.: Focused ion beam nanofabrication: a comparison with conventional processing techniques. J. Nanosci. Nanotechnol. **6**(3), 661–668 (2006)
29. Wei, H.-X., Langford, R.M., Han, X., Coey, J.M.D.: Controlled fabrication of nickel perpendicular nanocontacts using focused ion beam milling. J. Appl. Phys. **99**(8), 8C501–8C503 (2006)
30. Ch Santschi, M.J., Hoffmann, P., Brugger, J.: Interdigitated 50 nm Ti electrode arrays fabricated using XeF2 enhanced focused ion beam etching. Nanotechnology **17**, 2722 (2006)
31. McNally, B., Singer, A., Yu, Z., Sun, Y., Weng, Z., Meller, A.: Optical recognition of converted DNA nucleotides for single-molecule DNA sequencing using nanopore arrays. Nano Lett. **10**(6), 2237–2244 (2010)
32. Kwon, N., Kim, K., Heo, J., Chung, I.: Fabrication of ordered anodic aluminum oxide with matrix arrays of pores using nanoimprint. In: AVS, pp. 803–807. (2009)
33. Shingubara, S., Maruo, S., Yamashita, T., Nakao, M., Shimizu, T.: Reduction of pitch of nanohole array by self-organizing anodic oxidation after nanoimprinting. Microelectron. Eng. **87**(5–8), 1451–1454 (2010)
34. Choi, J., Wehrspohn, R.B., Gösele, U.: Mechanism of guided self-organization producing quasi-monodomain porous alumina. Electrochim. Acta **50**(13), 2591–2595 (2005)
35. Liu, C.Y., Datta, A., Wang, Y.L.: Ordered anodic alumina nanochannels on focused-ion-beam-prepatterned aluminum surfaces. Appl. Phys. Lett. **78**(1), 120–122 (2001)
36. Bo Chen, K.L., Zhipeng, T.: Understanding focused ion beam guided anodic alumina nanopore development. Electrochim. Acta **56**, 9802–9807 (2011)
37. Chen, B., Lu, K., Tian, Z.: Novel patterns by focused ion beam guided anodization. Langmuir **27**(2), 800–808 (2010)
38. Lillo, M., Losic, D.: Ion-beam pore opening of porous anodic alumina: the formation of single nanopore and nanopore arrays. Mater. Lett. **63**(3–4), 457–460 (2009)
39. Miller, M.K., Russell, K.F., Thompson, K., Alvis, R., Larson, D.J.: Review of atom probe FIB-based specimen preparation methods. Microsc. Microanal. **13**(6), 428–436 (2007)
40. Larson, D.J., Foord, D.T., Petford-Long, A.K., Anthony, T.C., Rozdilsky, I.M., Cerezo, A., Smith, G.W.D.: Focused ion-beam milling for field-ion specimen preparation: preliminary investigations. Ultramicroscopy **75**(3), 147–159 (1998)
41. Miller, M.K., Russell, K.F.: FIB-based atom probe specimen preparation of powders. Microsc. Microanal. **12**, 1294–1295 (2006)
42. Vasile, M.J., Grigg, D., Griffith, J.E., Fitzgerald, E., Russell, P.E.: Scanning probe tip geometry optimized for metrology by focused ion beam ion milling. J. Vac. Sci. Tech. B **9**(6), 3569–3572 (1991)
43. Matsui, S., Kaito, T., Fujita, J.-i., Komuro, M., Kanda, K., Haruyama, Y.: Three-dimensional nanostructure fabrication by focused-ion-beam chemical vapor deposition. J. Vac. Sci. Technol. B: Microelectron Nanomet, **18**(6), 3181–3184 (2000)
44. Fujita, J., Ishida, M., Ichihashi, T., Ochiai, Y., Kaito, T., Matsui, S.: Growth of three-dimensional nano-structures using FIB-CVD and its mechanical properties. Nucl. Instrum. Methods Phys. Res., Sect. B **206**, 472–477 (2003)

45. Fujita, J., Ishida, M., Sakamoto, T., Ochiai, Y., Kaito, T., Matsui, S.: Observation and characteristics of mechanical vibration in three-dimensional nanostructures and pillars grown by focused ion beam chemical vapor deposition. J. Vac. Sci. Technol. B, **19**(6), 2834–2837 (2001)
46. Kim, Y.K., Danner, A.J., Raftery, J.J., Choquette, K.D.: Focused ion beam nanopatterning for optoelectronic device fabrication. IEEE J. Quant. Electron. **11**(6), 1292–1298 (2005)
47. Nordseth, O., Kjellman, J.O., You, C.C., Royset, A., Tybell, T., Grepstad, J.K.: The case for electro-optic waveguide devices from ferroelectric (Pb,La) (Zr,Ti)$O_3$ thin film epilayers. In: SPIE China pp. 73810F. (2009)
48. Steckl, A.J.: Exploring the frontiers of optoelectronics with FIB technology. In: Frontiers in Electronics, Advanced Workshop. pp. 47–50. (1997)
49. Wirth, R.: Focused ion beam (FIB): a novel technology for advanced application of micro- and nanoanalysis in geosciences and applied mineralogy. Eur. J. Mineral. **16**(6), 863–876 (2004)
50. Lemmens, H., Butcher, A., Richards, D., Laughrey, C., Dixon, M.: Imaging techniques for 2D and 3D characterization of unconventional reservoirs core and cuttings samples; and how to integrate them. Society of Petroleum Engineers—SPE Americas Unconventional Gas Conference. vol. 1, pp. 555–557. (2011)
51. Lemmens, H.J., Butcher, A.R., Botha, P.W.S.K.: FIB/SEM and automated mineralogy for core and cuttings analysis. Society of Petroleum Engineers—SPE Russian Oil and Gas Technical Conference and Exhibition. vol. 2, pp. 881–884. (2010)
52. Heaney, P.J., Vicenzi, E.P., Giannuzzi, L.A., Livi, K.J.T.: Focused ion beam milling: a method of site-specific sample extraction for microanalysis of earth and planetary materials. Am. Mineral. **86**(9), 1094–1099 (2001)
53. Sivel, V.G.M., Van Den Brand, J., Wang, W.R., Mohdadi, H., Tichelaar, F.D., Alkemade, P.F.A., Zandbergen, H.W.: Application of the dual-beam FIB/SEM to metals research. J. Microsc. **214**(3), 237–245 (2004)
54. Xhoffer, C., Dhont, A., Willems, A.: The use of focused ion beam in metal research. Yejin Fenxi/Metallurg. Anal. **32**(7), 7–11 (2012)
55. Mathew, J.V., Bhattacharjee, S.: Multi-element focused ion beams using compact microwave plasma ion source. Indian J. Phys. **85**(12), 1863–1870 (2011)
56. Schulz, R.: Focused ion beam (FIB)—applications as a research tool in surface finishing. Focused ion beam (FIB). Galvanotechnik **102**(10), 2174–2181 (2011)
57. Schindler, G., Klandzevski, V., Steinlesberger, G., Steinhögl, W., Traving, M., Engelhardt, M.A.: Morphology study of copper and aluminum interconnects. Advanced Metallization Conference. pp. 213–217. (2003)
58. Shinji, M., Yukinori, O.: Focused ion beam applications to solid state devices. Nanotechnology **7**, 247 (1996)
59. Nan, L.K., Lung, L.M.: TEM sample preparation by single-sided low-energy ion beam etching. 19th IEEE International Symposium on the Physical and Failure Analysis of Integrated Circuits, IPFA (2012)
60. Oh, J., Kim, J., Lee, J., Kim, J., Back, T., Kim, W., Kim, H., Kim, C.: Evaluation of electrical properties of cell area on the semiconductor devices by FIB technique. 37th International Symposium for Testing and Failure Analysis, ISTFA. pp. 443–445. (2011)
61. Dhara, S., Lu, C.Y., Wu, C.T., Hsu, C.W., Tu, W.S., Chen, K.H., Wang, Y.L., Chen, L.C., Raj, B.: Focused ion beam induced nanojunction and defect doping as a building block for nanoscale electronics in GaN nanowires. J. Phys. Chem. C **114**(36), 15260–15265 (2010)
62. Kasianowicz, J.J.B.E., Branton, D., Deamer, D.W.: Characterization of individual polynucleotide molecules using a membrane channel. Proc. Natl. Acad. Sci. **93**, 13770–13773 (1996)
63. Mathé, J.A.A., Nelson, D.R., Schulten, K., Meller, A.: Orientation discrimination of single stranded DNA inside the α-Hemolysin membrane channel. Proc. Natl. Acad. Sci. **102**, 12377–12382 (2005)
64. Terada, D., Hattori, S., Honda, T., Iitake, M., Kobayashi, H.: Embossed-carving processing of cytoskeletons of cultured cells by using focused ion beam technology. Microsc. Res. Tech. **76**(3), 290 (2013)

65. Grandfield, K., Engqvist, H.: Focused ion beam in the study of biomaterials and biological matter. Adv. Mater. Sci. Eng. **6**, 841961 (2012)
66. Schertel, A., Kraft, D., Rauscher, M.: Applications of SEM/FIB technology in life science. Microsc. Microanal. **15**(2), 344–345 (2009)
67. Yamazaki, A., Morita, Y., Kim, Y., Owari, M.: Cross-section analysis of a laminated film by dual FIB ToF-SIMS. E. J. Surfac. Sci. Nanotechnol. **9**, 426–429 (2011)
68. Sezen, M., Plank, H., Fisslthaler, E., Chernev, B., Zankel, A., Tchernychova, E., Blümel, A., List, E.J.W., Grogger, W., Pölt, P.: An investigation on focused electron/ion beam induced degradation mechanisms of conjugated polymers. Phys. Chem. Chem. Phys. **13**(45), 20235–20240 (2011)
69. Mariam, A.A.I., Jongin, H., Phillip, K., Emanuele, I., Joshua, B.E., Tim, A.: Precise electrochemical fabrication of sub-20 nm solid-state nanopores for single-molecule biosensing. J. Phys. Condens. Matter **22**, 454128 (2010)
70. Mulero, R., Prabhu, A.S., Freedman, K.J., Kim, M.J.: Nanopore-based devices for bioanalytical applications. JALA **15**(3), 243–252 (2010)
71. Ayub, M., Ivanov, A., Hong, J., Kuhn, P., Instuli, E., Edel, J.B., Albrecht, T.: Precise electrochemical fabrication of sub-20 nm solid-state nanopores for single-molecule biosensing. J. Phys. Condens. Matter **22**(45) (2010)
72. Stevie, F.A., Irwin, R.B., Shofner, T.L., Brown, S.R., Drown, J.L., Giannuzzi, L.A.: Plan view TEM sample preparation using the focused ion beam lift-out technique. In: AIP Conf. Proc. p. 868. (1998)
73. Gnauck, P., Hoffrogge, P.A.: New SEM/FIB crossbeam inspection tool for high resolution materials and device characterization. In: Proc. SPIE. pp. 106–113. (2003)
74. Proff, C., Abolhassani, S., Dadras, M.M., Lemaignan, C.: In situ oxidation of zirconium binary alloys by environmental SEM and analysis by AFM, FIB, and TEM. J. Nucl. Mater. **404**(2), 97–108 (2010)
75. Nam, C.Y., Kim, J.Y., Fischer, J.E.: Focused-ion-beam platinum nanopatterning for GaN nanowires: ohmic contacts and patterned growth. Appl. Phys. Lett. **86**(19), 193112–193113 (2005)
76. Guo, D., Kometani, R., Warisawa, S.i., Ishihara, S.: Three-Dimensional nanostructure fabrication by controlling downward growth on focused-ion-beam chemical vapour deposition. Jpn. J. Appl. Phys. **51**(6), 5001 (2012)
77. Ke, J.J., Tsai, K.T., Dai, Y.A., He, J.H.: Contact transport of focused ion beam-deposited Pt to Si nanowires: from measurement to understanding. Appl. Phys. Lett. **5**, 100 (2012)
78. Naik, J.P., Prewett, P.D., Das, K., Raychaudhuri, A.K.: Instabilities in focused ion beam-patterned Au nanowires. Microelectron. Eng. **88**(9), 2840–2843 (2011)

# Chapter 2
# Epitaxial Ferroelectric Nanostructures Fabricated by FIB Milling

**Alessio Morelli and Ionela Vrejoiu**

**Abstract** In this chapter we explore the possibilities offered by focused ion beam milling to fabricate epitaxial islands out of bismuth ferrite thin films. Three processes are described: first one consisting of high-energy direct milling of a film followed by annealing procedures, a second one of high-energy milling of a film coated with a sacrificial layer which is removed successively, and a third one of low-energy milling of a film previously structured with sacrificial layer islands. Ferroelectric properties of the obtained structures have been investigated at a nanoscale level by piezoresponse force microscopy, in order to evaluate the impact of fabrication-induced damages on the functional properties. Structures produced by direct milling display ferroelectricity only after a post-fabrication annealing procedure, exhibiting polarization pinning only for structures with lateral sizes <500 nm. Detailed investigations have revealed occurrence of pinning and/or high imprint in an area within 100 nm distance from the structure edges, independently of the structure size. On the contrary, 250 nm large islands produced by milling through a sacrificial layer and over the structured sacrificial layer display ferroelectricity, without diminution of functional properties. Hence we conclude that focused ion beam milling is a viable technique for fabrication of ferroelectric structures, in particular bismuth ferrite nanoislands.

A. Morelli (✉)
Max Planck Institute of Microstructure Physics, Weinberg 2, D-06120 Halle, Germany
e-mail: alessio.morelli.pfm@gmail.com

I. Vrejoiu
Max Planck Institute for Solid State Research,
Heisenbergstraße1, D-70569 Stuttgart, Germany
e-mail: i.vrejoiu@fkf.mpg.de

Z.M. Wang (ed.), *FIB Nanostructures*, Lecture Notes in Nanoscale Science and Technology 20, DOI 10.1007/978-3-319-02874-3_2,

## 2.1 Introduction

Ferroelectrics are materials displaying spontaneous dielectric polarization, the direction of which can be reversed upon application of an external electric field [1, 2]. Due to the wide range of properties they display, they are currently used in normal life applications, as constitutive elements for transducers and actuators, sensors, and memory devices [3, 4].

In the past few years $BiFeO_3$ (BFO) has attracted wide attention, being one of the few materials that is multiferroic at room temperature ($T_C \sim 1{,}100$ K and $T_N \sim 650$ K) [5]. Its ferroelectric polarization lies along the diagonal of the unit cell ($[111]_{pseudocubic}$), giving therefore eight possible polarization directions [6]. It is a G-type antiferromagnet [7], and in thin films the magnetic easy plane is orthogonal to the polarization vector [8]. It has been demonstrated that rotating the polarization yields rotation of the magnetic planes [8–10], which gives the possibility for application in magnetoelectric devices. As well, the high remnant polarization value ($P_r \sim 100$ μC/cm$^2$) [5] makes it suitable as an active material in memory applications [4].

The need of miniaturization in current applications requires improvement of fabrication methods and a deeper understanding of the influence of size reduction on the functional properties. Ferroelectrics in the form of thin films have been widely studied, and it has been shown that the limit for preserving functional properties can be pushed down to few nanometers in 2D structures [11]. The most common configuration in which ferroelectrics are used in applications is in the form of capacitors, with the ferroelectric material sandwiched between electrodes. For such reason extensive studies have been performed on top electrodes deposited on ferroelectric thin films, examining in detail nucleation process during switching procedure [12, 13], retention [14], and local switching characteristics [15]. In particular it has been shown that the reduction in size of the electrode leads to different nucleation processes [16]. Additionally it has been shown that in view of memory application, the increase in areal density of top electrodes leads to the detrimental effect of cross talk during writing procedure [17]. Therefore, it is mandatory to use and understand the functional properties of ferroelectric nanoislands.

Focused ion beam (FIB) milling has been employed over the last two decades to fabricate nanostructures in a number of ferroelectric materials, such as $PbNb_{0.04}Zr_{0.28}Ti_{0.68}O_3$ [18–20], $BaTiO_3$ (BTO) [21–26], $PbZr_{0.3}Ti_{0.7}O_3$ [26], and $PbZr_{0.2}Ti_{0.8}O_3$ (PZT) [20, 28]. The investigations on such structures provided experimental evidence of piezo-activity enhancement due to release of clamping from the surrounding material [29], insight into the dynamics of nanoscale polarization domains [26, 28], and understanding of the scaling of domain size with lateral dimension reduction [21, 25].

However, the process of FIB milling itself, involving high-energy ion sputtering material, gives rise to ion implantation and crystal amorphization. Such damages have been thoroughly investigated, and strategies have been devised for recovering the functional properties after the milling process.

Stanishevsky et al. [19, 20] analyzed the impact of FIB fabrication on the properties of PZT islands observing significant modifications due to oxygen and lead loss and gallium implantation. A thermal annealing at 600 °C in $O_2$ atmosphere partially recovered the properties of the fabricated structure. Still a 5–10 nm sidewall damaged layer was present, which affected the functionality of 100 nm islands.

A more elaborated procedure was introduced by Schilling et al. [30] for nanostructuring BTO.

They showed that a simple annealing procedure would lead to the presence of gallium platelets on the surface of the structured material. In order to avoid that, they performed annealing in vacuum in order to avoid oxidation of the desorbed gallium, followed by plasma cleaning to further remove contaminations.

Annealing in oxygen was performed next in order to compensate possible oxygen loss during the previous procedure. This process led to lamellas with healed damages and absence of gallium platelets.

A different approach for PZT was suggested by Hambe et al. [27], namely, using a protective layer of photoresist during milling, to be removed by chemical etching. The resulting structures possessed ferroelectric properties up to the edges, still experiencing a 10–20 nm damaged layer at the sidewalls.

An alternative way to obtain nanostructures was proposed by Remiens et al. [31], by milling amorphous PZT and then crystallizing the obtained islands by annealing. Structures fabricated by milling the crystallized material showed no piezoelectric response, while the ones obtained by Remiens' method display ferroelectricity.

BFO machining by FIB has been reported by Hong et al. [32]. Piezoelectric response of the obtained structures was not free from artifacts. Moreover, switching curves displayed no clear polarization reversal; therefore, no presence of ferroelectricity could be proved. As well, a single annealing procedure improved only slightly the piezoelectric signal from the structures, without any proof of polarization switching.

In this chapter we suggest possible procedures to use FIB to fabricate ferroelectric structures out of epitaxial BFO thin films. Especially we show the impact of fabrication-induced defects on the functional properties of the obtained islands, investigated at the nanoscale level by piezoresponse force microscopy (PFM). In particular we demonstrate, by use of spatially resolved PFM investigations, the detrimental effect of a direct milling on the properties of the islands at their edges, which limits the minimal size for which the structures display switchable polarization. We suggest an alternative strategy involving milling through a sacrificial layer, which reduces greatly the impact of fabrication-induced damages and gives the possibility to reduce the size of islands displaying ferroelectricity.

### 2.1.1 *Piezoresponse Force Microscopy*

PFM has been developed starting in the early 1990s [33] following the need of nondestructive nanoscale characterization of piezoelectric and ferroelectric materials. It emerged soon as the technique of choice for nanoscale investigation of ferroelectrics [34–36]. Later it evolved along with the increasing utilization of ferroelectrics for miniaturized systems and memories [37–45].

PFM is an atomic force microscope (AFM)-based technique, and its operating principle relies on the converse piezoelectric effect, by which a piezoelectric material changes its dimension under an applied external electric field. An alternated electric field is generated through the material by applying a bias (modulation bias) to the AFM tip in contact with it (and acting as a movable top electrode), while the sample is grounded. Such field induces the sample to vibrate, and such vibration (detected by the AFM photodetector) is extracted via lock-in techniques. The amplitude of the vibration (PR amplitude) is proportional to the piezoelectric coefficient normal to the surface ($d_{zz}$) and the amplitude of the modulation bias ($V_{AC}$), while the phase offset with the modulation frequency ($\Phi$, PR phase) gives information on the direction of the polarization's out-of-plane component. The combined signal called piezoresponse (PR) can be expressed as

$$PR = d_{zz} V_{AC} \cos(\Phi)$$

so that PR is proportional to the PR amplitude and positive or negative depending on the polarization direction in the volume underneath the AFM tip. In this way the polarization domain configuration can be investigated at the nanoscale by monitoring the PR phase. Additionally, ramping up the modulation bias and detecting the PR amplitude, the slope of the obtained curve gives a quantitative evaluation of the piezoelectric coefficient [46].

The above-described procedure, which gives information on the out-of-plane component of polarization, is named vertical PFM (VPFM). In the same way, by detecting the lateral vibration of the sample, the in-plane components are explored, and such procedure is commonly known as lateral PFM (LPFM).

A way to investigate the switching behavior of the polarization in the material under investigation is offered by PFM, in the form of acquiring piezo-hysteresis loops [46]. They are obtained by sweeping up and down an applied DC bias superimposed to the modulation voltage, performed in "in-field" mode, i.e., acquiring the piezoresponse while the DC bias is applied, or in "remnant" mode, i.e., acquiring the data while the DC bias is set to null. A change in PR phase of 180° indicates polarization switching, allowing the determination of the polarization switching bias for the material.

A recent development of PFM is the so-called switching spectroscopy piezoresponse force microscopy (SS-PFM) [47], by which the switching properties can be spatially resolved, thus mapping locally the relevant ferroelectric properties. SS-PFM consists in a series of piezo-hysteresis acquired on an array of points. Given the amount of acquired data, this procedure usually employs an automated procedure to fit the curves and extracts the characteristic data, such as switching and nucleation biases, imprint, and work of switching [48].

**Fig. 2.1** Schematic of the "direct milling" fabrication procedure. (**a**) After FIB patterning, (**b**) structures are obtained. (**c**) Thermal treatments aim at the recovery of oxygen content and crystallinity of the structures

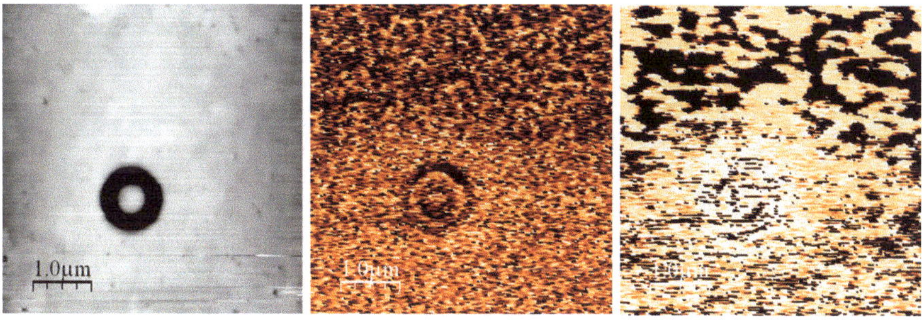


**Fig. 2.2** Topography, VPFM phase, and LPFM phase (*left* to *right*) of an as-fabricated structure (500 nm diameter). Piezoactivity is absent in the structure and the surrounding region

## 2.2 Direct Milling

The fabrication method which will be named for simplicity, "direct milling" (Fig. 2.1), consists of material removal from a bare BFO epitaxial thin film in order to obtain the desired structures [49]. This is done by milling grooves around the structures to isolate them from the rest of the BFO film. We employed ion beam with energy of 30 kV and currents between 1.5 and 28 pA, yielding ion doses of about $1 \times 10^{17}$ ion/cm$^2$. The ion milling process is carried on until a change in contrast in the SEM imaging gives evidence of the passage from an insulating material to a conductive one, meaning that the BFO layer is fully removed exposing the underneath electrode. Round and square structures with size ranging from 250 nm to 5 μm were fabricated. As it will be discussed later, a two-step post-fabrication annealing was performed: first vacuum annealing at 300 °C and $10^{-5}$ mbar, followed by annealing in $O_2$ atmosphere at 500 °C.

PFM imaging performed on as-fabricated structures showed degradation of piezoelectric properties, with reduced or absent piezoresponse (Fig. 2.2). Such effect can be attributed to Ga ion implantation and backscattering together with material amorphization and redeposition. Its influence on the functional properties extends to about 1 μm away from the milled grooves, therefore inhibiting any ferroelectricity in structures with lateral size <2 μm.

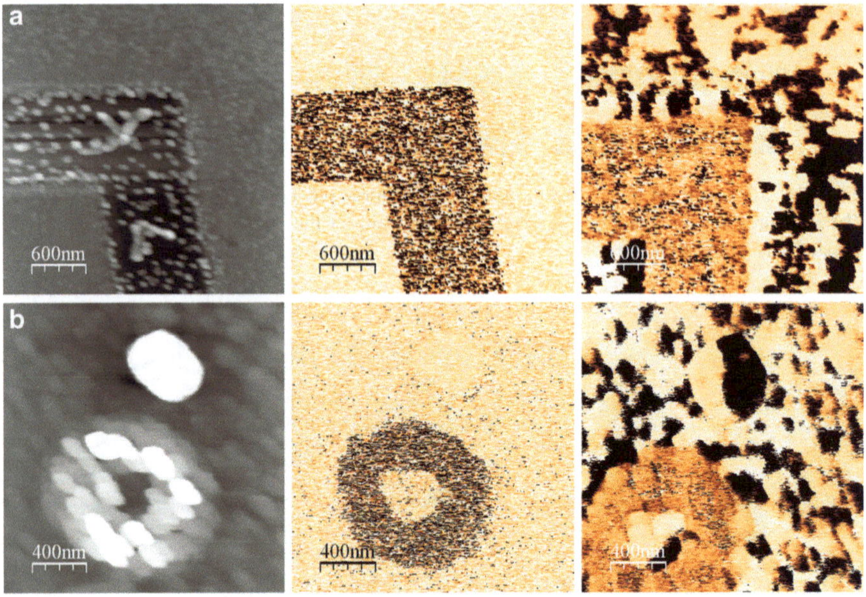


**Fig. 2.3** Topography, VPFM phase, and LPFM phase (*left* to *right*) of FIB-fabricated structures after annealing in oxygen at 600 °C. In (**a**) part of a 5×5 μm$^2$ square structure is shown, while (**b**) exhibits a round structure with diameter 500 nm (adapted from Ref. 49)

For this reason, as performed also by other groups on different ferroelectric materials processed by FIB [20, 27, 32], thermal treatment procedures were considered in order to desorb the implanted ions and recrystallize the amorphized material.

Following the approach suggested by Hong [32], a sample with fabricated structures was annealed at 600 °C in oxygen atmosphere. Successive investigations showed recovery of piezoresponse up to the edges of the structures (Fig. 2.3). However, topographic imaging shows that the surface of the film after annealing is covered by particles of about 100 nm in size. Such particles do not inhibit piezoelectricity but, especially for the smaller structures (~250 nm size), hinder the access to measuring the structures, since they tend to clog in the milled grooves. Moreover for the smaller structures they can represent a problem for the interpretation of the piezoresponse signal, since they can lead to topographic cross talk. It is supposed that these particles may be gallium oxide [30], which is formed by the desorbing gallium and oxidizing it during annealing in oxygen atmosphere.

In order to avoid the presence of particles, a more elaborated approach has been performed on a structured sample. As suggested in [30], an intermediate step was introduced, consisting of annealing in vacuum at a lower temperature (300 °C, $10^{-5}$ mbar). Such step would ideally lead to Ga desorbtion and avoid the Ga to recombine and deposit on the surface. Subsequently, annealing in oxygen (500 °C) was performed in order to replete oxygen after the annealing in vacuum and to further repair amorphized material.

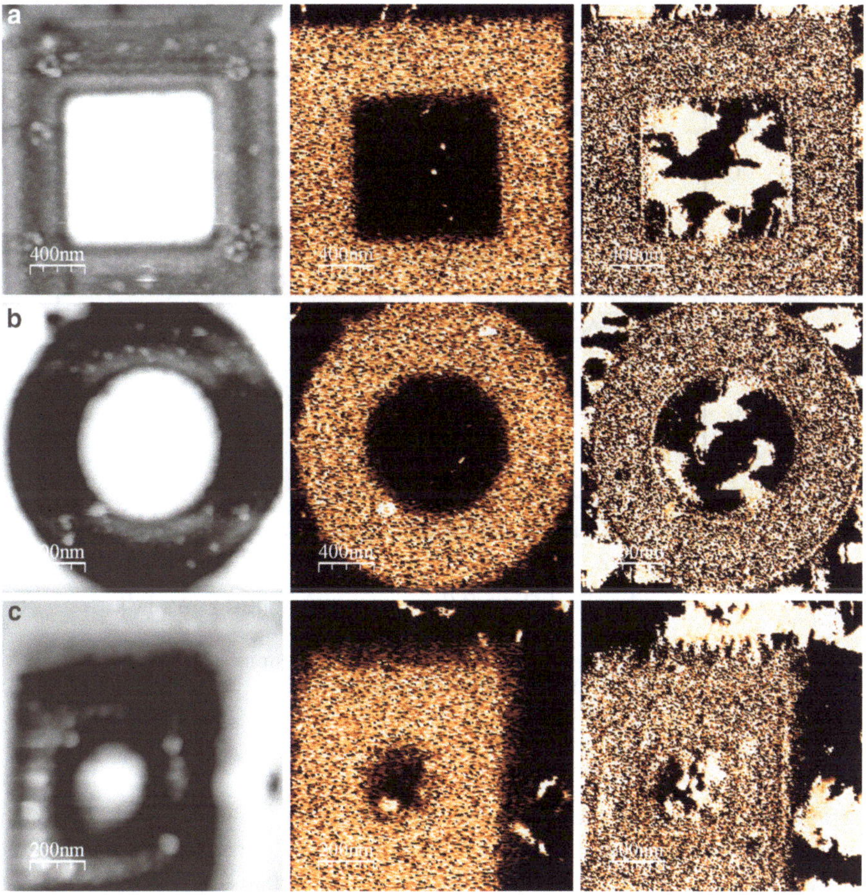


**Fig. 2.4** Topography, VPFM phase, and LPFM phase (*left* to *right*) of FIB-fabricated structures after two-step annealing: (**a**) $1 \times 1$ μm$^2$ square structure, (**b**) round structure 1 μm in diameter, (**c**) $250 \times 250$ nm square structure

PFM imaging demonstrates the effectiveness of this treatment (Fig. 2.4): well-defined structures, no particles, and piezoresponse exhibited up to the edges of the structures, with magnitude of PR amplitude as high as the one of the parent BFO film.

Once proved that the piezoelectricity was recovered in the islands, in order to check the occurrence of ferroelectric switching, we acquired piezo-hysteresis curves, in the middle of the structures of different sizes. As can be observed in Fig. 2.5, the polarization is switchable for structures with lateral size down to 500 nm, therefore proving recovery of ferroelectricity. However, in the smallest structures (~250 nm), although the piezoresponse amplitude was still as high as in the parent BFO film, no switching was observed, independent of the applied bias, indicating polarization pinning. This may be due to the fact that in all structures

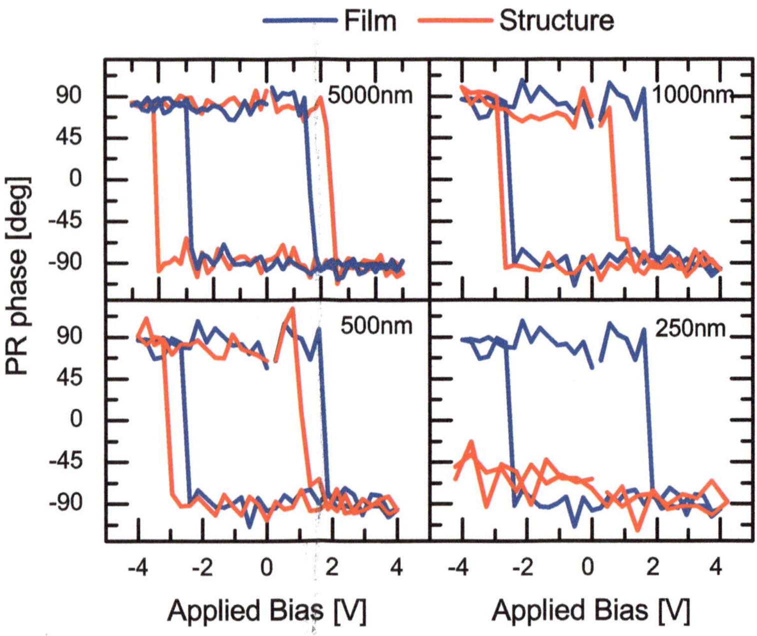


**Fig. 2.5** Local remanent piezo-hysteresis loops as resulting from measurements of the VPFM phase on round structures of different size (reported in figure), after two-step annealing (adapted from Ref. 49)

certain volume close to the edges was expected to be rich in defects [27], which eventually pin the polarization. In the case of the smallest structures the ratio of pinned polarization volume to total volume is big enough to show pinning throughout all the structure [45].

In order to understand the impact of such defects on the functional properties, retention loss experiments were performed. Retention loss is a problem to face and avoid when utilizing ferroelectrics in memory devices. It has been investigated for various materials in the form of thin films [42]. The operation consists of switching uniformly the polarization over an area of the sample, in the way a bit of information would be written, by applying a voltage high enough to overcome the coercive voltage and subsequently monitoring the stability of the switched polarization in time. By PFM such operation can be conducted at a nanoscale level, and in this case it allows the monitoring of the retention loss of a single nanostructure by imaging the VPFM phase.

Figure 2.6 displays the VPFM-phase images taken immediately after switching and then after several hours and more than a day. From the images acquired right after switching we can observe that, irrespective of structure size, some areas at the edges did not switch (dark contrast), therefore supporting the hypothesis of presence of defects pinning the polarization. Observation of polarization domain evolution in time shows that a rapid back-switching phenomenon takes place starting at

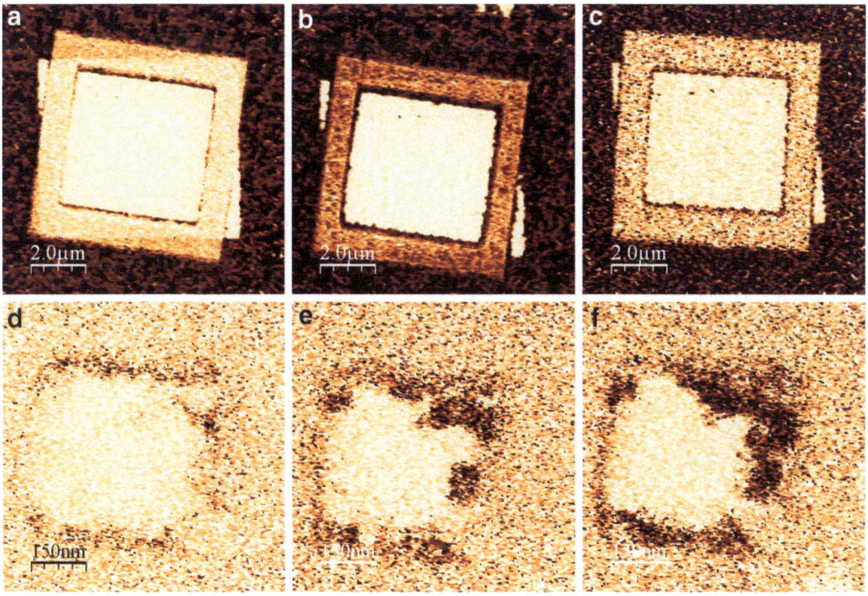


**Fig. 2.6** Evolution of retention loss by screening vertical PFM phase in (**a**, **b**, **c**) 5 μm and (**d**, **e**, **f**) 500 nm lateral size structures fabricated by direct FIB. (**a**) and (**d**) show the domain configuration after switching, (**b**) and (**e**) after few hours, and (**c**) and (**f**) after more than 1 day

the edges and continuing for few hours. Further monitoring shows no change in domain configuration, meaning that the back-switching phenomenon comes to a halt, as observed in the images acquired after several days. Closer image analysis shows that such phenomenon is limited to areas within 100 nm from the edges for structures of all the investigated sizes.

To have a clear view of the retention loss evolution, in Fig. 2.7 the percentage of switched (stable) polarization domain area is plotted versus time, comparing the behavior of structures of different sizes and of the parent BFO film itself. It is clear that a fast back-switching takes place, which halts after few hours, to get stable until the domains are monitored. The percentage of back-switched area decreases with increasing structure size, since the distance from the edges for all the back-switched areas is roughly the same, independent of the structure size.

It can be inferred that the fabrication produces defects, which extend in the structures about 100 nm from the edges and which can lead to charge-trapping. Such phenomenon would be the cause of pinning (non-switchable areas) and imprint (fast back-switching areas). Imprint is the phenomenon by which one direction of polarization is energetically favorable, and therefore the reversal to the inverse direction would yield a metastable configuration. In the case of the structures under investigation, the original vertical polarization state points inward the structure (dark contrast in Fig. 2.4). By effect of imprint, after switching to polarization upward

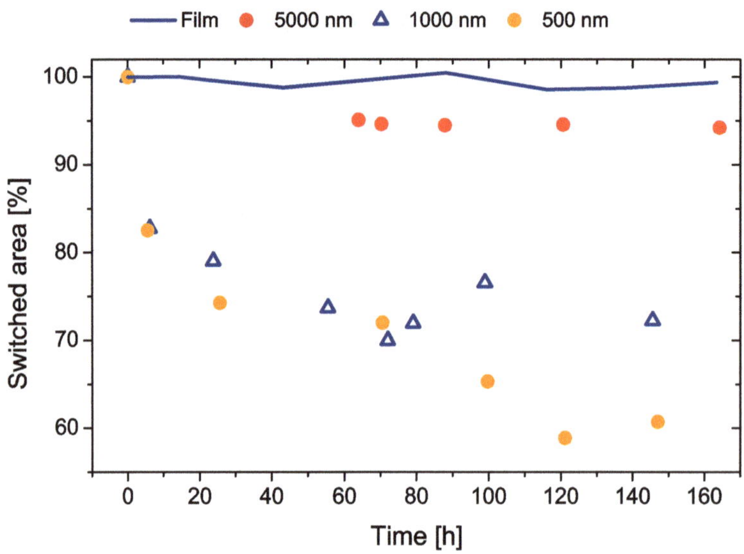


**Fig. 2.7** Switched area versus time for structures with different sizes, normalized to the area with polarization upwards after switching

(bright contrast in Fig. 2.6a, d) the areas at the edges experience a metastable state of polarization, which eventually returns to the original configuration (Fig. 2.6c, dark contrast).

As described in the introduction, PFM offers a way to obtain spatially resolved mapping of the polarization switching characteristics, technique known as SS-PFM. Here we use such a procedure to obtain information about the evolution of imprint in relation to the distance from the edges.

We performed such an experiment through a 1 μm island, with distance between each measured point of 30 nm, which allowed enough lateral resolution to map the decaying of functional properties. This spacing was also larger than the tip radius of curvature, which is the uppermost limit for lateral resolution in PFM. By extracting the positive and negative switching voltages from each curve via a fitting function, it was possible to calculate the imprint values and map them versus distance, along with the topographic height data, as shown in Fig. 2.8a. Given the fact that the scan was performed across a structure and adjacent grooves, the curves acquired on the latter did not exhibit any ferroelectric behavior. This was reflected in the extracted values in the form of unreasonable magnitudes, i.e., values out of the DC voltage range over which the measurement was performed, or negative coercive bias value larger than the positive one. In order to distinguish real imprint values from fictitious ones, the latter were replaced by an assigned negative number, which would be clearly discernible from actual values.

By considering the obtained results (Fig. 2.8a), it is evident how the imprint value is rather constant over the surface of the island and decreases with approaching the edges. The last three hysteresis measured on the structure surface and the

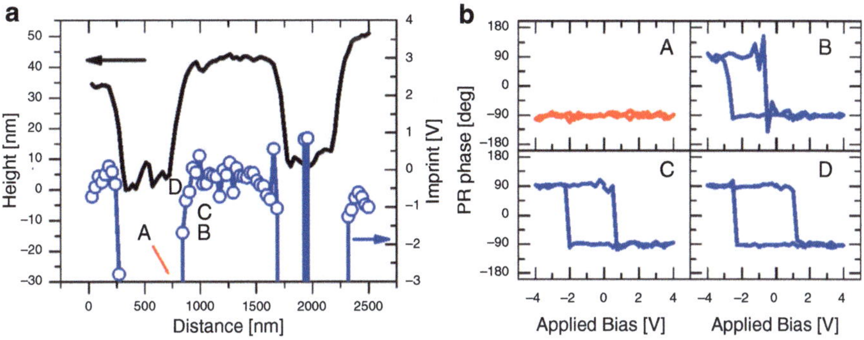


**Fig. 2.8** (**a**) Ferroelectric imprint (*open symbols*) and topography (*line*) scan line through a structure with 1 μm lateral size obtained by direct FIB, and (**b**) in-field hysteresis loops acquired near the edge, from which the imprint values (marked by *letters*) were extracted

first one in the groove are reported in Fig. 2.8b, by which the shifting towards negative values while approaching the edge is more evident.

By considering that the step between points is 30 nm, the area affected by stronger imprint can be evaluated to be at a distance of about 90–120 nm from the edges. Such measurement agrees with the back-switched area distance of 100 nm obtained by retention loss measurements and explains that phenomenon. As well, the exaggerated imprint value at the very edge explains the presence of non-switched areas immediately after writing procedure, which would back-switch right after the applied voltage is removed.

In order to estimate the actual extent of high-energy FIB-induced damage at the sidewalls of the structures, stopping and range of ions in material (SRIM) simulations [50] were performed. The results show that the lateral distance at which Ga penetration and amorphization of the material occur extends up to 80.4 nm from the point of impact. Considering the assumptions on which SRIM is based, such result is in agreement with the experimental data.

In conclusion, the fabrication procedure named "direct milling" proves to be effective in producing ferroelectric structures down to 500 nm in lateral size, while in smaller structures the polarization is pinned.

We showed, by a combination of PFM investigations, that the pinning in the smaller structure is correlated with the fact that an area within 100 nm from the edges displays reduction of functional properties even after annealing treatments.

## 2.3 Sacrificial Layer Milling

Given the drawback of the "direct milling" method, a second method involving a sacrificial layer was designed with the aim to reduce the fabrication-induced defects and thus be able to reduce the size of FIB-fabricated BFO islands retaining ferroelectric properties.

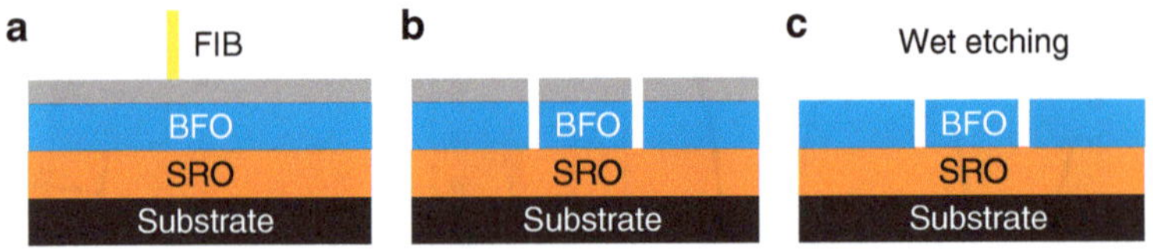


**Fig. 2.9** Schematic of the fabrication procedure through sacrificial layer. (**a**) After FIB patterning, (**b**) structures capped by sacrificial layer are obtained. (**c**) Sacrificial layer is finally removed by chemical etching

A layer of few nanometers of aluminum was evaporated over the bare surface of an epitaxial BFO thin film (Fig. 2.9a). Al was chosen based on SRIM simulations [50], being the material that scatters less the Ga ions and thus reducing the possibility of scattering towards the area which will be the surface of the milled structures. The FIB procedure was carried out in the same way as for the "direct milling," until the bottom electrode (i.e., a $SrRuO_3$ layer) was reached, and ultimately obtaining islands of BFO covered by an Al layer (Fig. 2.9b). The top Al layer was then removed by chemical wet etching in a 10 % solution of KOH, lastly resulting in bare BFO islands (Fig. 2.9c). No annealing process was needed anymore, as it will be discussed in the following.

PFM imaging on the fabricated islands right after sacrificial layer removal is shown in Fig. 2.10. The piezoresponse was measurable up to the edges of the structures and had a magnitude as high as in the parent film, without requiring any post-fabrication annealing procedures. Evidently the debris of sputtered material redeposited, and the backscattered Ga ions during fabrication are collected onto and into the sacrificial layer present over the BFO film. Final removal of the sacrificial layer after FIB, by chemical etching, gets rid of the contaminants.

The functionality of the islands was checked by performing piezo-hysteresis loops on all fabricated islands (Fig. 2.11), with the result that even in 250 nm structures produced by the aid of sacrificial layer the vertical polarization is switchable. Eventually the presence of a mask during fabrication reduces enough the sidewall implantation and amorphization, so that ferroelectricity is preserved and imprint not present.

Analysis of retention loss on structures obtained by direct FIB gave information about the extent of the area affected by degradation of functional properties. For such reason, retention loss was monitored over the sacrificial-layer-method-made structures. In Fig. 2.12 the VPFM-phase images of a 1 μm island are displayed at different times after polarization reversal was performed to switch the polarization upwards. First of all it can be seen that the switching occurs throughout all the structure area (Fig. 2.12a). Secondly it can be seen that the back-switching takes place with a moderate speed, with less than 5 % of polarization upwards area lost after few hours (Fig. 2.12b) and still very little through several days (Fig. 2.12c). In particular, the PFM imaging highlights that the back-switching nucleates at random spots over the structure surface, as it usually happens in BFO thin films.

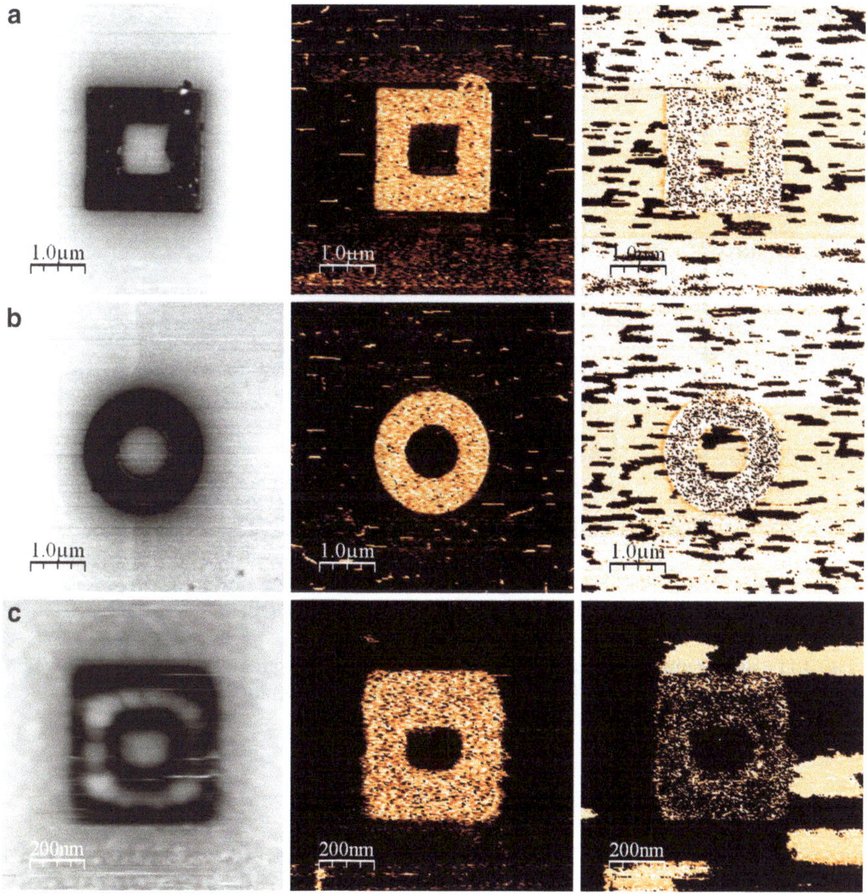


**Fig. 2.10** Topography, VPFM phase, and LPFM phase (*left* to *right*) of structures fabricated with the aid of a sacrificial layer: (**a**) $1 \times 1$ $\mu m^2$ square structure, (**b**) round structure 1 μm in diameter, (**c**) $250 \times 250$ nm square structure

Therefore it can be assumed that fabrication defects have no influence on the functional properties, not even in proximity of the edges.

The above-discussed results led us to suppose that in this case switching characteristics remain unaffected by the fabrication process. To confirm that, SS-PFM measurements with 30 nm distance step were performed on a 1 μm island in order to obtain spatially resolved imprint mapping. As seen in Fig. 2.13a, b the imprint value stays constant until the edges of the structures are reached, after which the curves lose ferroelectric signature (Fig. 2.13b). Therefore it is evident that if defects are present at the sidewalls, they affect minimally the ferroelectric imprint, and as a consequence production of BFO nanostructures retaining ferroelectricity is possible with such procedure.

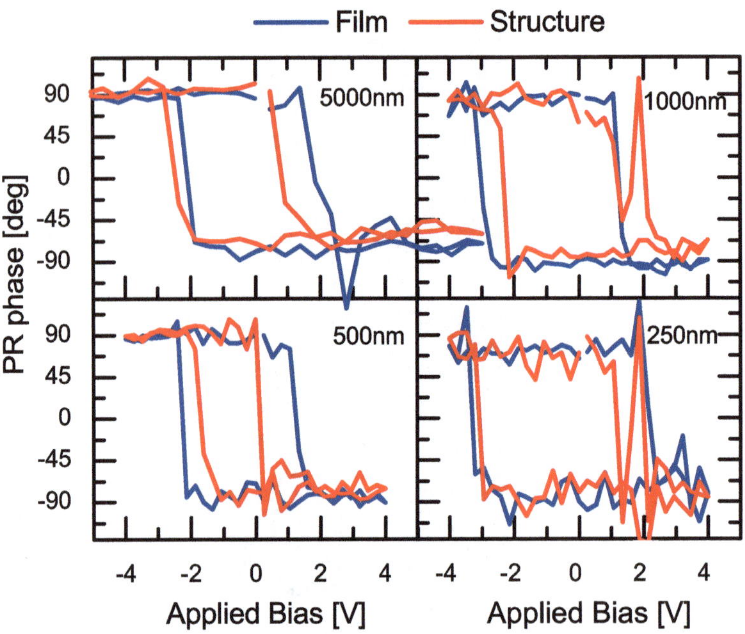


**Fig. 2.11** Local remanent piezo-hysteresis loops as resulting from measurements of the VPFM phase on round structures of different sizes (reported in figure), after removal of the sacrificial layer

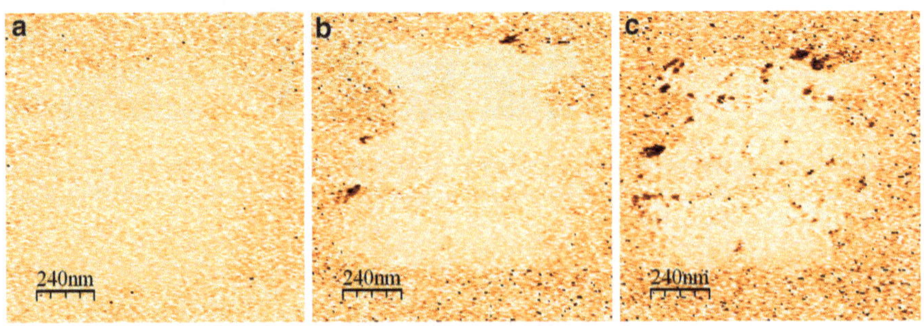


**Fig. 2.12** Evolution of retention loss by screening vertical PFM phase in 1 μm lateral size structure fabricated by FIB milling through a sacrificial layer. (**a**) shows the domain configuration immediately after switching, (**b**) after few hours, and (**c**) after more than 1 day

## 2.4 Mask-Assisted FIB Milling

As demonstrated in the previous section, employment of a protective layer during milling is beneficial for the preservation of functional properties in FIB-fabricated structures. In fact such layer collects the debris which is represented by redeposited material and backscattered ions naturally occurring during milling.

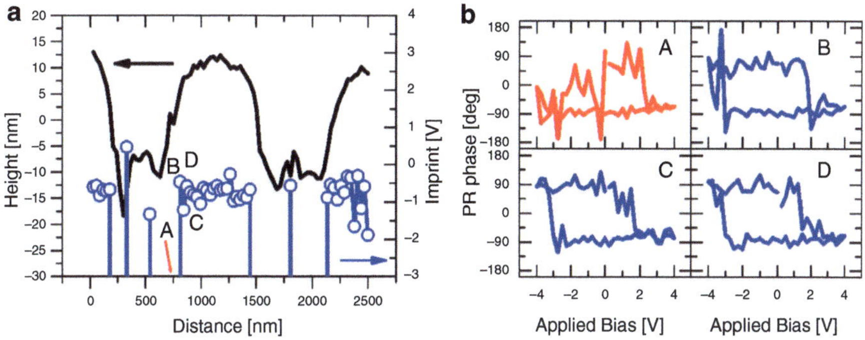


**Fig. 2.13** (**a**) Ferroelectric imprint (*open symbols*) and topography (*line*) scan line through a structure with 1 μm lateral size obtained by milling through an aluminum sacrificial layer, and (**b**) in-field hysteresis loops acquired near the edge, from which the imprint values (marked by *letters*) were extracted

Though it was shown that such expedient resulted in ferroelectric islands down to 250 nm in lateral dimension, it is known [20, 27] that anyway a layer of amorphous material is present at the edges of the fabricated islands, which would be critical with the downscaling of the size.

One way to decrease the extent of damage is to reduce the beam energy, so that implantation and amorphization would be more confined, with decrease in damage extent from 80.4 nm for high-energy FIB to 30.6 nm for 5 keV accelerating voltage predicted by SRIM simulations. However, reduction of FIB energy leads also to a loss of patterning resolution by direct writing, since the beam diameter increases from 7 nm at 30 keV to more than 100 nm at low energies, therefore inducing size limitation for production of well-defined structures.

A solution to such inconvenience is represented by patterning a protective sacrificial layer over the surface to be structured and successively proceeding with low-energy milling. In this way the material is removed where the layer is not present, therefore producing defined islands without utilization of high-energy beam [51].

We used for this purpose aluminum stencil masks, through which Al was evaporated on the sample to be structured (Fig. 2.14a). Stencil masks are currently used for production of ferroelectric islands directly by growth techniques [52], but in the case of $BiFeO_3$ the process would be difficult to tackle, given the complexity of BFO-phase diagram [53]. The evaporation through a mask with aperture diameter of 400 nm resulted in Al islands with a base diameter of 500 nm and a width at half height of 250 nm.

After removal of the mask (Fig. 2.14b), FIB milling (5 keV accelerating voltage, 1.6 pA current) of an area of the sample covered by Al array leads to BFO islands capped with Al (Fig. 2.14c). The removal of the Al capping, as for the previous fabrication method, is obtained by chemical etching, yielding freestanding BFO islands (Fig. 2.14d).

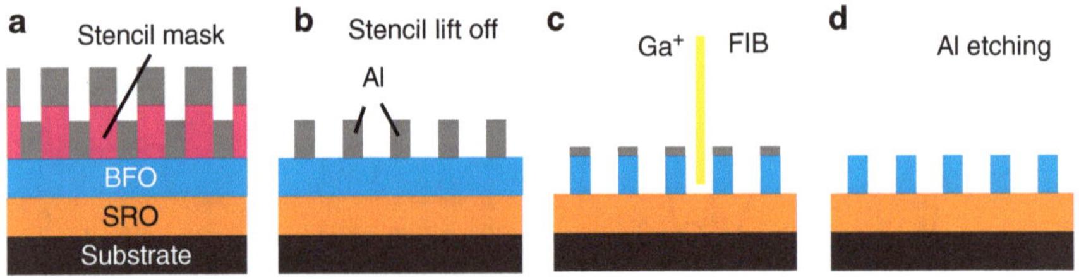


**Fig. 2.14** Schematic diagram of the procedure performed in order to fabricate BFO nanostructures. (**a**) Al evaporation through stencil mask over the BFO; (**b**) mask lift-off resulting in an array of Al dots on the film; (**c**) ion beam milling, yielding BFO islands capped with Al; (**d**) Al chemical etching resulting in BFO nanoislands

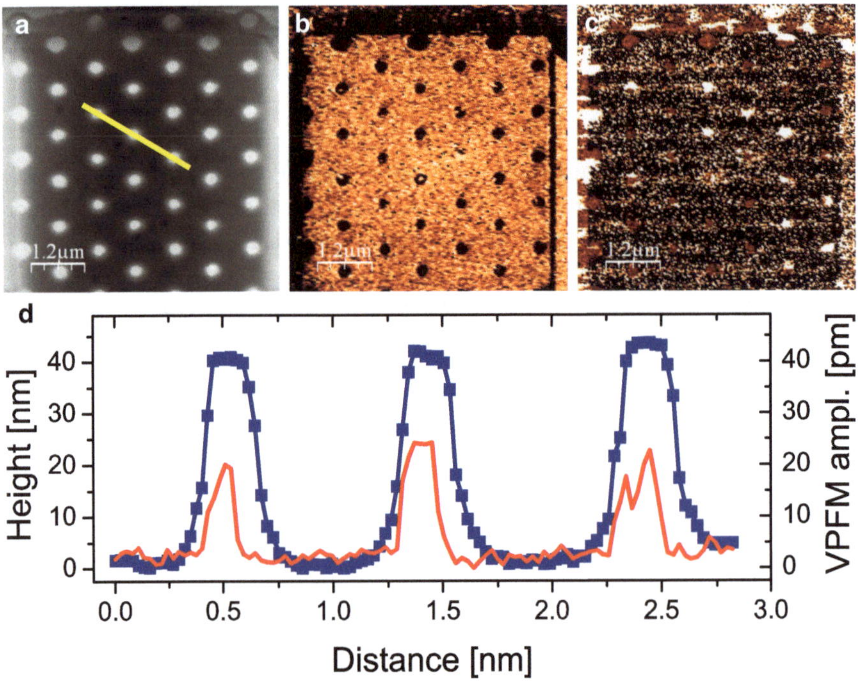


**Fig. 2.15** PFM imaging of the array of BFO islands obtained by mask-assisted method: (**a**) topography, (**b**) VPFM phase, and (**c**) LPFM phase. (**d**) Topographic profile (*line* and *symbol*) as indicated in (**a**) and corresponding VPFM amplitude profile (*line*)

The investigations on the produced islands showed distinct piezoresponse on the top surface of the islands. Figure 2.15a shows the $5\times5\ \mu m^2$ over which the beam was scanned until the bottom electrode was exposed, after Al removal. The islands display a well-defined circular shape, with approximate diameter of 250 nm (Fig. 2.15a, d). The polarization domain configuration resembles the one of the parent film, with predominance of out-of-plane polarization component pointing

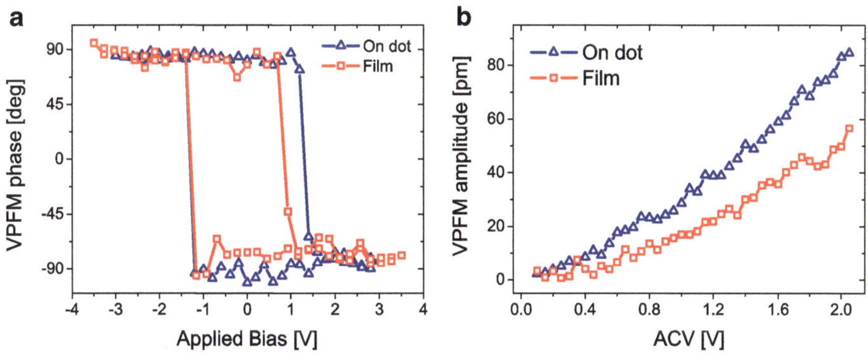


**Fig. 2.16** (**a**) In-field piezo-hysteresis loop of VPFM phase and (**b**) piezoelectric coefficient extrapolation as acquired over the parent film (*squares*) and over an island obtained by mask-assisted method (*triangles*)

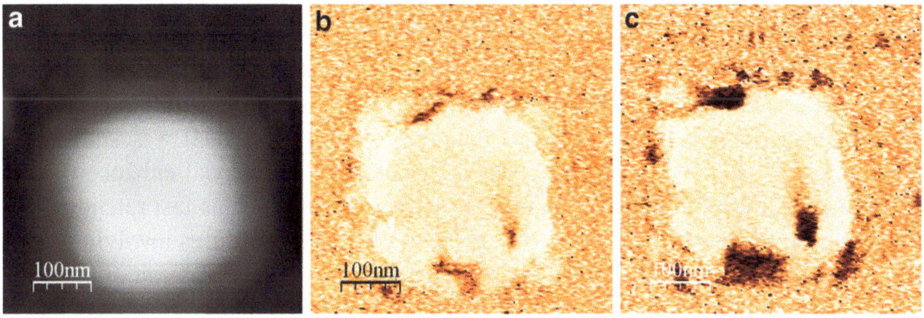


**Fig. 2.17** Retention measurements on an island: (**a**) topography and VPFM phase (**b**) after switching and (**c**) 3 days later. Reprinted with permission from Ref. 51. © 2013, AIP Publishing LLC

downwards (Fig. 2.15b, c). As for islands produced by the methods previously described, the piezoresponse is preserved, but their topographic shape is better defined, displaying a flat top surface.

Preservation of ferroelectric properties is proved by piezo-hysteresis experiments (Fig. 2.16a) showing switchable polarization and therefore preservation of ferroelectric characteristics. Comparison with the behavior of the parent film (Fig. 2.16a) demonstrates that coercive bias and imprint are not strongly affected by the fabrication process. Moreover, piezoelectric coefficient extrapolation (Fig. 2.16b) gives a value ($d_{zz}$ ~ 30 pm/V) higher than the one recorded for the parent film, a fact possibly deriving from the release of constraint from the surrounding material [29].

In order to check indirectly the homogeneity in the imprint behavior over the islands, and to prove the basic idea of possible employment for memory devices, the retention loss of the islands was tested. Switching the polarization upwards did not occur uniformly over the island (Fig. 2.17b), with small areas of the structure displaying polarization downward. Although this occurrence in structures produced by direct FIB is accompanied by fast back-switching from the pinned regions, in this case the propagation of back-switching is very slow, leading to a loss of less than 5 % after 3 days (Fig. 2.17c).

Therefore we can conclude that this latter fabrication method produces defect-free islands, preserving all the ferroelectricity without encountering any of the drawbacks experienced with previous methods. Moreover, the limit for the lateral size reduction is represented exclusively by the Al island size, which can be down-scaled by employing finer masks (such as aluminum anodic oxide masks) [54], and therefore we can assume that production of sub-100 nm islands by FIB will be possible.

## 2.5 Conclusion

In conclusion we presented a detailed study of ferroelectric bismuth ferrite epitaxial structures fabricated by FIB employing different procedures. By means of a combination of PFM investigations, we were able to analyze the impact of fabrication-induced damages on the functional properties. The cause of polarization pinning in smaller structures in the presence of damages was elucidated. It was shown that defects were localized at the structure sidewalls and led to local enhancement of polarization imprint, with all the deriving consequences, such as fast retention loss and pinning in smaller structures. We devised different procedures involving a sacrificial protective layer and mask-assisted fabrication. Characterization of the properties of the produced structures showed that no degradation of ferroelectric properties occurred, implying that the lateral size limit of BFO islands retaining ferroelectricity produced by FIB can be pushed down to tens of nanometers in the near future.

Our investigations pave the way to successful FIB nanostructuring of ferroelectric and multiferroic materials.

**Acknowledgments** We thank Norbert Schammelt for the FIB fabrication of the BFO structures, Florian Johann for the growth of the BFO films and fruitful assistance, and German Science Foundation for financial support in the framework of SFB762.

## References

1. Valasek, J.: Piezo-electric and allied phenomena in Rochelle salt. Phys. Rev. **17**, 475–481 (1921)
2. Lines, M.E., Glass, A.M.: Principles and Applications of Ferroelectrics and Related Materials. Clarendon Press, Oxford (1977)
3. Scott, J.F., Paz de Araujo, C.A.: Ferroelectric memories. Science **246**, 1400–1405 (1989)
4. Maruyama, K., Kondo, M., Singh, S.K., Ishiwara, H.: New ferroelectric material for embedded FRAM LSIs. Fujitsu Sci. Tech. J. **43**, 502–507 (2007)

5. Wang, J., Neaton, J.B., Zheng, H., Nagarajan, V., Ogale, S.B., Liu, B., Viehland, D., Vaithyanathan, V., Schlom, D.G., Waghmare, U.V., Spaldin, N.A., Rabe, K.M., Wuttig, M., Ramesh, R.: Epitaxial $BiFeO_3$ multiferroic thin film heterostructures. Science **299**, 1719–1722 (2003)
6. Kubel, F., Schmid, H.: Structure of a ferroelectric and ferroelastic monodomain crystal of the perovskite $BiFeO_3$. Acta Crystallogr. Sect. B **46**, 698–702 (1990)
7. Sosnowska, I., Peterlin-Neumaier, T., Steichele, E.: Spiral magnetic ordering in bismuth ferrite. J. Phys. C **15**, 4835–4846 (1982)
8. Zhao, T., Scholl, A., Zavaliche, F., Lee, K., Barry, M., Doran, A., Cruz, M.P., Chu, Y.H., Ederer, C., Spaldin, N.A., Das, R.R., Kim, D.M., Baek, S.H., Eom, C.B., Ramesh, R.: Electrical control of antiferromagnetic domains in multiferroic $BiFeO_3$ films at room temperature. Nat. Mater. **5**, 823–829 (2006)
9. Lebeugle, D., Colson, D., Forget, A., Viret, M., Bataille, A.M., Gukasov, A.: Electric-field-induced spin flop in $BiFeO_3$ single crystals at room temperature. Phys. Rev. Lett. **100**, 227602 (2008)
10. Lee, S., Ratcliff, W., Cheong, S.W., Kiryukhin, V.: Electric field control of the magnetic state in $BiFeO_3$ single crystals. Appl. Phys. Lett. **92**, 192906 (2008)
11. Chu, Y.H., Zhao, T., Cruz, M.P., Zhan, Q., Yang, P.L., Martin, L.W., Huijben, M., Yang, C.H., Zavaliche, F., Zheng, H., Ramesh, R.: Ferroelectric size effects in multiferroic $BiFeO_3$ thin films. Appl. Phys. Lett. **90**, 252906 (2007)
12. Rodriguez, B.J., Gruverman, A., Kingon, A.I., Nemanich, R.J., Cross, J.S.: Investigation of the mechanism of polarization switching in ferroelectric capacitors by three-dimensional piezoresponse force microscopy. Appl. Phys. A **80**, 99–103 (2005)
13. Li, W., Alexe, M.: Investigation on switching kinetics in epitaxial $Pb(Zr_{0.2}Ti_{0.8})O_3$ ferroelectric thin films: role of the 90 degrees domain walls. Appl. Phys. Lett. **91**, 262903 (2007)
14. Scott, J.F., Gruverman, A., Wu, D., Vrejoiu, I., Alexe, M.: Nanodomain faceting in ferroelectrics. J. Phys. Condens. Matter **20**, 425222 (2008)
15. Kim, Y.S., Kumar, A., Ovchinnikov, O., Jesse, S., Han, H., Pantel, D., Vrejoiu, I., Lee, W., Hesse, D., Alexe, M.: First-order reversal curve probing of spatially resolved polarization switching dynamics in ferroelectric nanocapacitors. ACS Nano **6**, 491–500 (2011)
16. Kim, Y.S., Han, H., Lee, W., Baik, S., Hesse, D., Alexe, M.: Non-Kolmogorov-Avrami-Ishibashi switching dynamics of nanoscale ferroelectric capacitors. Nano Lett. **10**, 1266–1270 (2010)
17. Kim, Y.S., Han, H., Vrejoiu, I., Lee, W., Hesse, D., Alexe, M.: Cross talk by extensive domain wall motion in arrays of ferroelectric nanocapacitors. Appl. Phys. Lett. **99**, 202901 (2011)
18. Ganpule, C.S., Stanishevsky, A., Su, Q., Aggarwal, S., Melngailis, J., Williams, E., Ramesh, R.: Scaling of ferroelectric properties in thin films. Appl. Phys. Lett. **75**, 409–411 (1999)
19. Stanishevsky, A., Aggarwal, S., Prakash, A.S., Melngailis, J., Ramesh, R.: Focused ion-beam patterning of nanoscale ferroelectric capacitors. J. Vac. Sci. Technol. B **16**, 3899–3902 (1998)
20. Stanishevsky, A., Nagaraj, B., Melngailis, J., Ramesh, R., Khriachtchev, L., McDaniel, E.: Radiation damage and its recovery in focused ion beam fabricated ferroelectric capacitors. J. Appl. Phys. **92**, 3275–3278 (2002)
21. Catalan, G., Schilling, A., Scott, J.F., Gregg, J.M.: Domains in three-dimensional ferroelectric nanostructures: theory and experiment. J. Phys. Condens. Matter **19**, 132201 (2007)
22. Saad, M.M., Bowman, R.M., Gregg, J.M.: Characteristics of single crystal "thin film" capacitor structures made using a focused ion beam microscope. Appl. Phys. Lett. **84**, 1159–1161 (2004)
23. Schilling, A., Adams, T.B., Bowman, R.M., Gregg, J.M., Catalan, G., Scott, J.F.: Scaling of domain periodicity with thickness measured in $BaTiO_3$ single crystal lamellae and comparison with other ferroics. Phys. Rev. B **74**, 024115 (2006)
24. Schilling, A., Byrne, D., Catalan, G., Webber, K.G., Genenko, Y.A., Wu, G.S., Scott, J.F., Gregg, J.M.: Domains in ferroelectric nanodots. Nano Lett. **9**, 3359–3364 (2009)
25. McGilly, L.J., Gregg, J.M.: Scaling of superdomain bands in ferroelectric dots. Appl. Phys. Lett. **98**, 132902 (2011)

26. McQuaid, R.G.P., McMillen, M., Chang, L.W., Gruverman, A., Gregg, J.M.: Domain wall propagation in meso- and nanoscale ferroelectrics. J. Phys. Condens. Matter **24**, 024204 (2012)
27. Hambe, M., Wicks, S., Gregg, J.M., Nagarajan, V.: Creation of damage-free ferroelectric nanostructures via focused ion beam milling. Nanotechnology **19**, 175302 (2008)
28. Nagarajan, V., Roytburd, A., Stanishevsky, A., Prasertchoung, A., Zhao, T., Chen, L., Melngailis, J., Auciello, O., Ramesh, R.: Dynamics of ferroelastic domains in ferroelectric thin films. Nat. Mater. **2**, 43–47 (2003)
29. Nagarajan, V., Stanishevsky, A., Chen, L., Zhao, T., Liu, B.T., Melngailis, J., Roytburd, A.L., Ramesh, R.: Realizing intrinsic piezoresponse in epitaxial submicron lead zirconate titanate capacitors on Si. Appl. Phys. Lett. **81**, 4215–4217 (2002)
30. Schilling, A., Adams, T.B., Bowman, R.M., Gregg, J.M.: Strategies for gallium removal after focused ion beam patterning of ferroelectric oxide nanostructures. Nanotechnology **18**, 035301 (2007)
31. Rémiens, D., Liang, R.H., Soyer, C., Deresmes, D., Troadec, D., Quignon, S., Da Costa, A., Desfeux, R.: Analysis of the degradation induced by focused ion $Ga^{3+}$ beam for the realization of piezoelectric nanostructures. J. Appl. Phys. **108**, 042008 (2010)
32. Hong, S., Klug, J.A., Park, M., Imre, A., Bedzyk, M.J., No, K., Petford-Long, A., Auciello, O.: Nanoscale piezoresponse studies of ferroelectric domains in epitaxial $BiFeO_3$ nanostructures. J. Appl. Phys. **105**, 061619 (2009)
33. Güthner, P., Dransfeld, K.: Local poling of ferroelectric polymers by scanning force microscopy. Appl. Phys. Lett. **61**, 1137–1139 (1992)
34. Franke, K., Besold, J., Haessler, W., Seegebarth, C.: Modification and detection of domains on ferroelectric PZT films by scanning force microscopy. Surf. Sci. Lett. **302**, 283–288 (1994)
35. Gruverman, A., Auciello, O., Tokumoto, H.: Scanning force microscopy for the study of domain structure in ferroelectric thin films. J. Vac. Sci. Technol. B **14**, 602–605 (1996)
36. Gruverman, A., Auciello, O., Tokumoto, H.: Nanoscale investigation of fatigue effects in $Pb(Zr, Ti)O_3$ films. Appl. Phys. Lett. **69**, 3191–3193 (1996)
37. Hidaka, T., Maruyama, T., Saitoh, M., Mikoshiba, N., Shimizu, M., Shiosaki, T., Wills, L.A., Hiskes, R., Dicarolis, S.A., Amano, J.: Formation and observation of 50 nm polarized domains in $PbZr_{1-x}Ti_xO_3$ thin film using scanning probe microscope. Appl. Phys. Lett. **68**, 2358–2359 (1996)
38. Eng, L.M., Güntherodt, H.J., Rosenman, G., Skliar, A., Oron, M., Katz, M., Eger, D.: Nondestructive imaging and characterization of ferroelectric domains in periodically poled crystals. J. Appl. Phys. **83**, 5973–5977 (1998)
39. Eng, L.M., Güntherodt, H.J., Schneider, G.A., Kopke, U., Saldana, J.M.: Nanoscale reconstruction of surface crystallography from three-dimensional polarization distribution in ferroelectric barium-titanate ceramics. Appl. Phys. Lett. **74**, 233–235 (1999)
40. Abplanalp, M., Eng, L., Günter, P.: Mapping the domain distribution at ferroelectric surfaces by scanning force microscopy. Appl. Phys. A **66**, 231–234 (1998)
41. Roelofs, A., Böttger, U., Waser, R., Schlaphof, F., Trogisch, S., Eng, L.M.: Differentiating 180° and 90° switching of ferroelectric domains with three-dimensional piezoresponse force microscopy. Appl. Phys. Lett. **77**, 3444–3446 (2000)
42. Gruverman, A., Tokumoto, H., Prakash, A.S., Aggarwal, S., Yang, B., Wuttig, M., Ramesh, R., Auciello, O., Venkatesan, T.: Nanoscale imaging of domain dynamics and retention in ferroelectric thin films. Appl. Phys. Lett. **71**, 3492–3494 (1997)
43. Tybell, T., Paruch, P., Giamarchi, T., Triscone, J.M.: Domain wall creep in epitaxial ferroelectric $Pb(Zr_{0.2}Ti_{0.8})O_3$ thin films. Phys. Rev. Lett. **89**, 097601 (2002)
44. Alexe, M., Gruverman, A., Harnagea, C., Zakharov, N.D., Pignolet, A., Hesse, D., Scott, J.F.: Switching properties of self-assembled ferroelectric memory cells. Appl. Phys. Lett. **75**, 1158–1160 (1999)
45. Alexe, M., Harnagea, C., Hesse, D., Gösele, U.: Polarization imprint and size effects in mesoscopic ferroelectric structures. Appl. Phys. Lett. **79**, 242–244 (2001)

46. Harnagea, C., Pignolet, A., Alexe, M., Hesse, D., Goesele, U.: Quantitative ferroelectric characterization of single submicron grains in Bi-layered perovskite thin films. Appl. Phys. A **70**, 261–267 (2000)
47. Jesse, S., Baddorf, A.P., Kalinin, S.V.: Switching spectroscopy piezoresponse force microscopy of ferroelectric materials. Appl. Phys. Lett. **88**, 062908 (2006)
48. Jesse, S., Lee, H.N., Kalinin, S.V.: Quantitative mapping of switching behavior in piezoresponse force microscopy. Rev. Sci. Instrum. **77**, 073702 (2006)
49. Morelli, A., Johann, F., Schammelt, N., Vrejoiu, I.: Ferroelectric nanostructures fabricated by focused-ion-beam milling in epitaxial $BiFeO_3$ thin films. Nanotechnology **22**, 265303 (2011)
50. Ziegler, J.F., Ziegler, M.D., Biersack, J.P.: SRIM—The stopping and range of ions in matter (2010). Nucl. Instr. Meth. Phys. Res. B **268**, 1818–1823 (2010)
51. Morelli, A., Johann, F., Schammelt, N., McGrouther, D., Vrejoiu, I.: Mask assisted fabrication of nanoislands of $BiFeO_3$ by ion beam milling. J. Appl. Phys. 113, 154101 (2013)
52. Vrejoiu, I., Morelli, A., Biggemann, D., Pippel, E.: Ordered arrays of multiferroic epitaxial nanostructures. Nano Rev. **2**, 7364 (2011)
53. Palai, R., Katiyar, R.S., Schmid, H., Tissot, P., Clark, S.J., Robertson, J., Redfern, S.A.T., Catalan, G., Scott, J.F.: β phase and γ-β metal-insulator transition in multiferroic $BiFeO_3$. Phys. Rev. B. **77**, 014110 (2008)
54. Lee, W., Han, H., Lotnyk, A., Schubert, M.A., Senz, S., Alexe, M., Hesse, D., Baik, S., Gösele, U.: Individually addressable epitaxial ferroelectric nanocapacitor arrays with near Tb inch$^{-2}$ density. Nat. Nanotechnol. **3**, 402–407 (2008)

# Chapter 3
# Low-Current Focused Ion Beam Milling for Freestanding Nanomaterial Characterization

**Wuxia Li, Ajuan Cui, and Changzhi Gu**

**Abstract** Techniques for the fabrication of one-dimensional or quasi-one-dimensional nanowires as well as to enable studies of the compositional, structural, and electrical properties of freestanding individual nanoobjects are of great importance, for instance, to observe quantum size effects and build quantum information devices. In this chapter, a technique of low-current focused ion beam (FIB) milling for two kinds of applications, namely, the selective felling of freestanding nanoobjects in a well-controllable way and the size reduction of both lateral and freestanding tungsten composite nanostructures, is presented. To fell a freestanding nanoobject, parameters such as the angle of the incident ion beam with respect to the nanoobject long axis, the cutting point, and the ion beam current were carefully chosen for controllable felling. To thin a nanoobject, different exposure times and ion beam currents were used to control the final size, thinning rate, and accuracy of a group of nanowires, or an individual nanowire, or even a portion of a nanowire by site-specific milling. After felling or thinning with low-current ion beam, no obvious structural and electrical property changes were observed for nanoobject grown with FIB-induced deposition. These results suggest that FIB milling is applicable in felling freestanding nanoobjects for the investigation of the properties of selected individuals, and it is also a potential approach for controllable size reduction enabling the observation of size and quantum effects.

W. Li (✉) • A. Cui • C. Gu
Beijing National Lab of Condensed Matter Physics, Institute of Physics,
Chinese Academy of Sciences, Beijing 100190, China
e-mail: liwuxia@aphy.iphy.ac.cn; ajuancui@aphy.iphy.ac.cn; czgu@aphy.iphy.ac.cn

Z.M. Wang (ed.), *FIB Nanostructures*, Lecture Notes in Nanoscale Science and Technology 20, DOI 10.1007/978-3-319-02874-3_3,

## 3.1 Introduction

Recently, with the downscaling of electronics, nanomaterials with various shapes have been synthesized, and their properties have been explored by several different methods [1–7], with a view to building novel nanodevices and new functional logic circuit architectures. In particular, three-dimensional (3D) nanostructures and nanodevices have attracted tremendous interest in the past few years due to their excellent mechanical and physical properties that planar nanodevices cannot achieve. For instance, novel optoelectronic devices [8], nanosensors [9], biological information detectors [10], plasmonics [11], and quantum devices [12] have all been achieved based on 3D nanostructures. However, depending on the shape and the distribution of nanoobjects, the characterization of structural, compositional, and electrical properties may not be possible for some characterization techniques. Up to now, for the characterization of the properties of nanomaterials, the most popular method has been to use a blade to scratch the support substrate and release the nanomaterials, sonicate in ethanol for dispersion, and then transfer to either a transmission electron microscope grid for structural and compositional analysis [13] or a substrate covered with a layer of dielectric material such as $SiO_2$ for characterization of electrical and optical properties [14]. However, the scratch–pick–dispersion–transfer technique has some disadvantages, which include the following: (1) mechanical damage of the original shape of the target, especially for limbed and complex nanostructures; (2) it is time-consuming for contact fabrication; (3) it is extremely difficult, even impossible, to find the target after scratching. For practical applications, of critical importance is the ability to measure the properties of an individual freestanding nanoobject, to explore the size effect and quantum phenomena, and thus to provide feedback for improving their properties to obtain the nanomaterials desired for constructing high-density 3D circuits [15–17].

Meanwhile, size effect and quantum phenomenon observation require nanomaterial with scalable size in nanometer and approaching the physical critical value. For instance, when a superconducting nanowire is reduced smaller than the Ginzburg–Landau-phase coherence length and the magnetic penetration depth, below or near the superconducting transition temperature, the resistance of a 1D wire is no longer zero [18]. In addition, a 3D superconducting quantum interference device (SQUID), formed by integrating a conventional SQUID with freestanding multiple pickup loops, potentially could overcome the present planar SQUID limitation of only being able to detect the field perpendicular to the substrate; however, to realize single-spin detection resolution, the size of the 3D superconducting pickup loop must be submicron, necessitating nanoscale superconducting wires [12].

Thus a precise and controllable technique for high-resolution size reduction and releasing a freestanding nanoobject from the support substrate is clearly demanded. In the following sections, we describe a technique we have developed for atomic resolution thinning and felling of freestanding nanowires by means of low-current focused ion beam (FIB) milling.

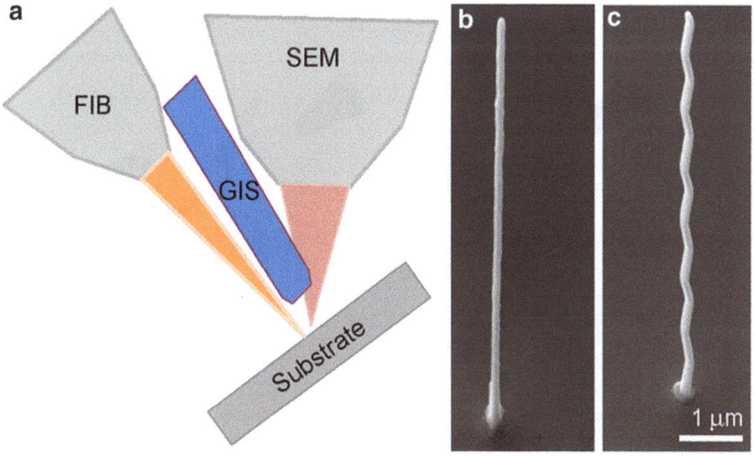


**Fig. 3.1** (**a**) The configuration of FIB/SEM plus GIS used for vertical nanowire growth; (**b**) and (**c**) SEM images of the FIB-grown as-deposited freestanding vertical and zigzag tungsten nanowires, respectively

## 3.2 Felling of Individual Freestanding Nanoobjects by FIB Milling

### *3.2.1 Growth of Vertical Nanowires and Felling Experimental Setup*

A commercially available SEM/FIB system utilizing a beam of 30 keV singly charged $Ga^+$ ions can be used to grow and fell freestanding nanowires. As the felling targets, the tungsten and platinum composite nanowires can be grown by FIB-induced deposition using a 1 pA ion beam current with $W(CO)_6$ and $(CH_3)_3Pt(CpCH_3)$, respectively, as gas precursors. For vertical nanowire growth, the stage was titled so that the ion beam was a normal incident to the substrate surface; the gas was introduced to the substrates through a nozzle, creating a local high pressure in the region scanned by the ion beam without substantial pressure rise in the rest of the work chamber. The base pressure before introducing the precursor gas was $2.4 \times 10^{-6}$ mbar, and during deposition the pressure was in the range of $7.4–8.3 \times 10^{-6}$ mbar. Deposition can be induced by irradiating one point on the substrate using spot mode. In that case, a straight wire is grown. Alternatively, the ion beam may be raster-scanned repeatedly over a rectangular area, first across the area in one direction and then back in the other direction; if the scan speed in one of the dimensions across the wire is slow, then a zigzag profile results. The growth details can be found in [19]. Figure 3.1a shows the schematic configuration of the relative position of the FIB and SEM columns, substrate, and GIS. The typical SEM images of the as-deposited nanowires are shown in Fig. 3.1b, c. From Fig. 3.1b, c, it can be seen that these wires have very smooth surface and are very uniform in size and geometry.

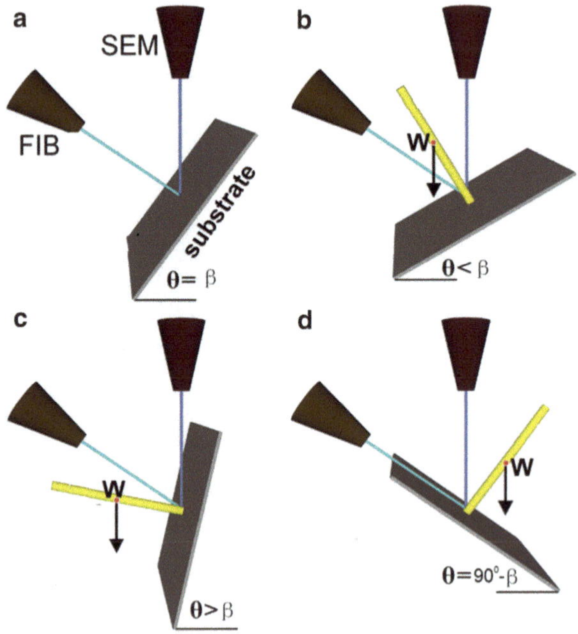


**Fig. 3.2** Schematics of the experimental geometry illustrating the relative position of the incident ion beam and electron beam ($\beta$ is the angle between these two columns), the freestanding nanoobject, and the direction on which the object's weight ($W$) acts from its center of gravity and the substrate: (**a**) ion beam normal to substrate, the tilt angle ($\theta$) of the substrate is $\beta$ from horizontal; (**b**) $\theta<\beta$ and (**c**) $\theta>\beta$ ion beam at a general angle for cutting; (**d**) using a sample holder with a tilt facet of 45° tilt the stage by $-(90°-\beta)$; the ion beam is set to be normal to wire

Although the ion beam direction in FIB systems is generally fixed, in order to fell nanoobjects, the angle of the ion beam with respect to the nanoobject can be modified by tilting the sample stage or using a tilt sample holder or by a combined utilization of these two to offer flexible controlling of both the cutting angle and the cutting position along the wire length. The experimental geometry was designed, and different possible arrangements are schematically shown in Fig. 3.2. For a system in which the angle between the electron beam column and the ion beam is $\beta$ (54° for Carl Ziess system and 52° for FEI systems) and the ion beam is at an angle of $\beta$ from horizontal, when the stage is tilted by $\beta$, then the ion beam is set to be normal to the substrate, a configuration usually used for vertical nanorod growth (as in Fig. 3.2a), or parallel to the substrate surface (as in Fig. 3.2d) by employing a tilted sample holder with a facet tilt angle of 45° and tilting the stage by $(\beta-45°)$. Alternatively, felling may be carried out with the stage tilted at a different angle; for example, as for the Carl Ziess systems, a stage-tilt angle of 32° leads to an angle between the incident beam and the wire of 20° (as in Fig. 3.2b). The maximum relative angle between the incident ion beam and the vertically grown nanoobject is 90° since at larger angles the ion beam is blocked by the upper edge of the substrate.

In fact, the felling direction largely depends on the experimental geometry, which can be controlled by the relative position of the incident ion beam and the freestanding nanoobject, the direction on which the object's weight ($W$) acts from its center of gravity

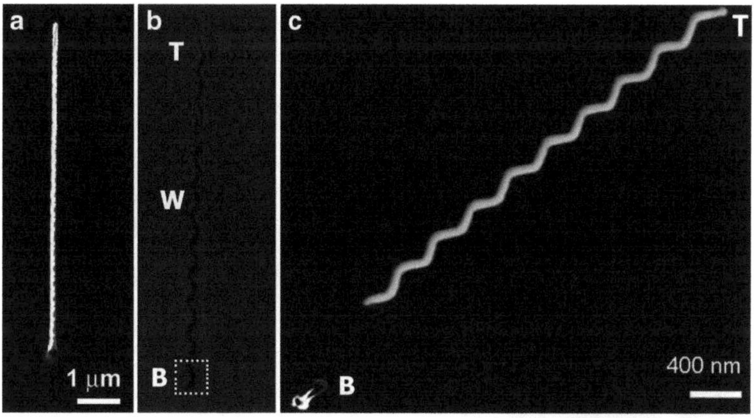


**Fig. 3.3** Images of vertically grown tungsten nanowires illustrating the felling process using a 1 pA ion beam normally incident to the wire length direction: (**a**) is an SEM image of the as-deposited nanowire with a viewing angle of $\beta$ (the zigzags are in a plane parallel to the viewing direction and so are not visible); (**b**) is an FIB image of the freestanding nanowire shown in (**a**) with the ion beam normal to its length and parallel to the support substrate surface; (**c**) the nanowire has been felled away from the incident ion beam and is lying on the substrate surface close to its base, with the top downwards [19]

and substrate. As illustrated in Fig. 3.2, $\theta$ is denoted to the tilt angle of the substrate. Note that the angle between the nanorod and the incident ion beam may have positive or negative sign. When $\theta<\beta$ as shown in Fig. 3.2b, the ion beam is at a general angle for cutting, a case that the nanorod's weight acting from its center of gravity may act towards the incident ion beam; while when $\theta<\beta$ as shown in Fig. 3.2c, then the center of gravity acts to the opposite direction from the incident ion beam. In addition, when a sample holder with a tilt facet of 45° is used as in Fig. 3.2d, the ion beam is set to be normal to wire—oblique incidence for cutting with $\theta=-(90°-\beta)$; by using in such a geometry the cutting area is minimized and milling of the substrate can be virtually avoided.

To fell the wire with the ion beam normal incident to the long axis of the wire, a stage geometry as shown in Fig. 3.2d can be used. To do this, first of all, use SEM to locate the target wire on the sample (Fig. 3.3a); then tilt the stage, find the nanowire with the FIB, and grab one image by FIB as shown in Fig. 3.3b. After that, a 1 pA FIB was first used to scan an area of about $(2–3)d$ nm $\times$ $(2–3)d$ nm near the base of the nanowire for 2 min, as shown by the dotted white rectangle with $d$ denoted to the width or the diameter of the wire. It can be seen from the SEM plan-view image shown in Fig. 3.3c that the wire was felled with the tip away from the incident ion beam (indicated by letter T), that the wire landed less than 2 μm from its base (indicated by letter B), and that the shape and size of the wire are unchanged.

### 3.2.2 *Felling of Freestanding Nanowires for Transport Investigations*

For measurements of transport properties, patterns for the electrical contacts can be fabricated by using electron beam lithography and/or photolithography [14] onto a selected individual nanowire and then metallic layers can be deposited by electron

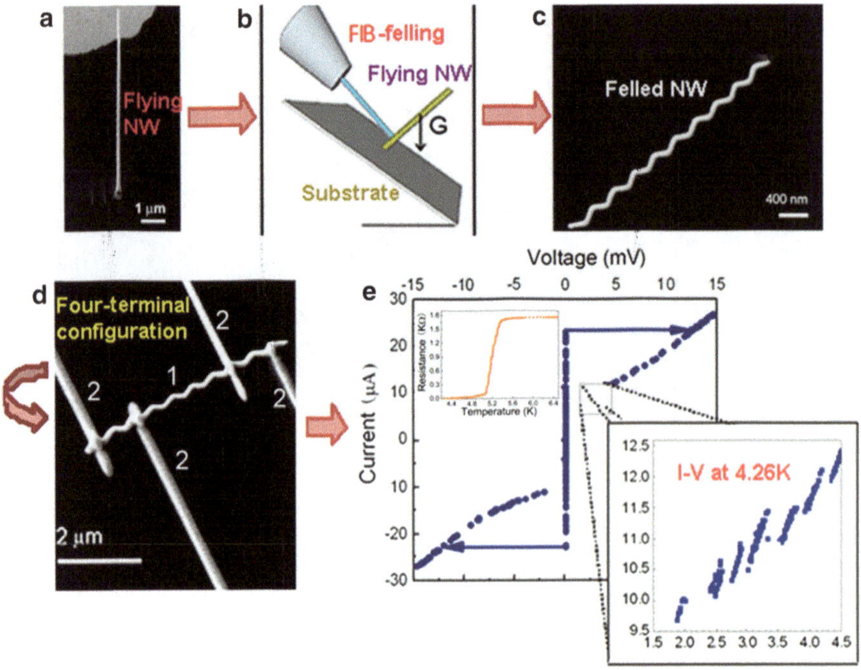


**Fig. 3.4** Felling of selected individual freestanding nanoobjects by focused ion beam milling and ion beam-induced deposition of electrical connections for characterization of transport properties: (**a**) SEM image of the as-deposited vertical tungsten nanowires; (**b**) schematic diagram of normal incident ion milling through the base for nanowire felling; (**c**) SEM image of the felled nanowire; (**d**) four-terminal configuration fabricated by FIB-CVD based on the felled nanowire; (**e**) zero-field *I–V* characteristics at 4.26 K for the felled tungsten nanowire shown in (**d**); in the *inset* is the temperature dependent resistivity of the nanowire as shown in (**d**) [19]

beam evaporation, thermal evaporation, or sputtering, followed by lift-off. FIB-induced deposition of Pt [7, 20] and W [7] composites can be used to direct-write electrical connection and contact pads on individual nanowires. By this technique, patterns with resolution comparable to that made by electron beam lithography can be made in a single step, thus avoiding resist patterning and metal deposition processes.

To demonstrate the felling of freestanding nanoobjects for analysis of their electrical properties, straight and zigzag tungsten nanowires were vertically grown on 200-nm-thick gold test patterns previously deposited on a 200-nm-thick $SiO_2$ layer on a Si substrate using conventional microelectronic techniques. Figure 3.4a is an SEM image of the vertically grown zigzag nanowire with a viewing angle of 54°. Figure 3.4b is the experimental setup with the ion beam normal to the wire length through the base of the nanowire. Then a 1 pA FIB was first used to scan an area of 400 nm × 400 nm near the base of the nanowire for 2 min. It can be seen from the SEM plan-view image shown in Fig. 3.4c that the shape and size of the wire are unchanged. After felling onto the substrate surface, FIB-induced deposition was used to fabricate contact pads and connectors for the measurement of electrical properties. Figure 3.4d shows a typical such structure with a four-terminal configuration. The numbers show the felled wire (1) and the tungsten connector (2) deposited in situ by using an ion beam current of 30 pA. Low-temperature transport measurements were performed with a helium dip-probe and home-made current source and voltage amplifiers. A constant current of 1.0 μA was applied to the two outer terminals, and the voltage was measured between the other two terminals. A temperature-dependent resistance measurement on a felled nanowire is shown in Fig. 3.4e. In the inset is the *R–T* curve for the low-temperature region, from which the superconducting transition temperature ($T_c$) was derived to be of 5.2 K. Here $T_c$ is defined as the temperature at which the resistance falls to 50 % of its value at the onset of the transition. The room-temperature resistivity of the zigzag nanowire was calculated to be about 520 μΩ cm, a value similar to that of an air-bridge structure also grown by 1 pA ion beam current [19]. The results reveal that felling freestanding nanowires using 1 pA ion beam current does not affect their intrinsic electrical properties. For comparison, bulk tungsten has a resistivity of 5 μΩ cm at room temperature. Resistivity values from about 100 to 300 μΩ cm have been reported for tungsten composites grown by FIB-induced deposition [21–23]. The variation in the resistivity of FIB-deposited materials is a consequence of the microstructure and composition difference due to the operating parameters, such as the ion beam current, dose, substrate material, substrate temperature, scanning speed, and gas flux.

To further understand the electrical properties in FIB-deposited nanowires, current–voltage measurements were performed at 4.26 K. The zero-field *I–V* curve is shown in Fig. 3.4e. This shows a critical current of 23 μA, and thus these results demonstrate the use of FIB-CVD to direct-write superconducting nanostructures. Distinct linear branches can also be seen when the current is reduced. These features of the *I–V* curve suggest the occurrence of thermally activated phase slip processes as they are strongly reminiscent of the *I–V* characteristics of phase-slip centers [24, 25]. A phase slip results from the fluctuation of the superconducting order parameter at some point along the wire and leads to a voltage pulse; this means that a thin superconducting wire below $T_c$ can show finite electrical resistance.

### 3.2.3 Wire Felling Direction

It is well known that most freestanding objects undergo plastic deformation when irradiated with ion beams under proper conditions [26–30]. Thus to control the felling direction, conditions that could cause bending of the target must be avoided. Thus unless beam-induced bending of the wire is significant (discussed below, in Sect. 3.2.4), and the felling direction (towards or away from the incident ion beam) is mainly governed by the relative position of the incident ion beam and the center of mass of the nanowire.

For a particular freestanding nanoobject, the felling direction is closely related to the geometry used for felling. The geometries used to grow and fell the nanowires are illustrated in Fig. 3.2. Figure 3.2a shows the geometry used for vertical nanowire growth with the beam normal incident to the substrate. It was found that under conditions with the beam incident obliquely on the nanowire ($-36° < \theta < 62°$), as shown in Fig. 3.2b–d, using raster scans over an area of order 2 μm × 0.5 μm, with 1 pA beam current, the wires were most often felled with the tip pointing to the lower part of the substrate, i.e., in the same direction as the moment due to the nanoobject's weight—"downhill"; this could be either towards (Fig. 3.2b) or away from (Fig. 3.2c, d) the incident ion beam because varying the tilt angle of the substrate as shown in Fig. 3.2b–d changes both the angle at which the ion beam is incident on the freestanding nanoobject and the moment due to the object's weight (*W*) about its base.

One advantage of felling at normal incidence (Fig. 3.2d) is that a minimum cutting cross section is obtained when the ion beam used for felling is normal to the wire length. This geometry minimizes the milling time and also minimizes ion irradiation of the substrate, avoiding possible damage to the top dielectric layer of the substrate. For almost-normal incidence, if felling was finished by one run of cutting with 1 pA and using a rectangle area smaller than 2 μm × 0.5 μm, the nanowire was always felled downhill. For nanowires not cut through by one run of felling, imaging (especially using large (e.g., 50 pA) ion beam current) can also fell the wire towards the incident ion beam direction. We discovered that there is also some dependence of the felling direction on the details of the felling. When felling was carried out at normal incidence (Fig. 3.2d) with a larger 50 pA beam current using a single-scan over a smaller area, the nanowires were always observed to fall towards the beam.

In general, the landing position of the felled wire on the substrate is cutting position and tilt angle dependent. The further the beam scanning area from the base of the wire, the more distant the landing position; however, on some occasions when wires were cut higher up from the base, they would tend to fall over but remain attached at the cutting position. This leads to the practical conclusion that the selected nanowire should be felled near its base. Although the felling might be towards or away from the direction of the incident ion beam, the landing position of the felled object in most cases is then close to its base. This is then easy to be relocated for electrode fabrication.

### 3.2.4 *Investigation of Ion-Induced Wire Bending*

Irradiation of nanomaterials and surfaces with energetic ions can result in either unwanted or desired changes of the morphology due to several different atomistic effects [29, 30]. The most well-known effect is that of sputtering—the removal of the irradiated surface atoms when the impact energy of the incident ions is transferred to them. The felling of freestanding nanoobjects such as described above is based on the sputtering effect. Another very recently reported effect caused by the beam–matter interaction is the bending and alignment of suspended nanoobjects with ion irradiation [29, 30].

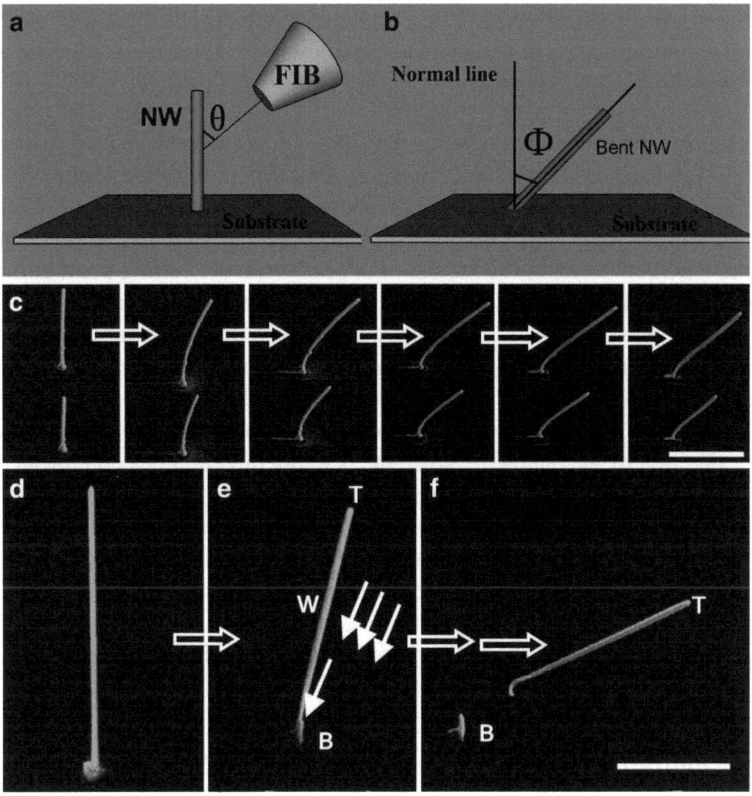


**Fig. 3.5** (**a**, **b**) The schematic diagram for the geometry of ion beam-induced bending; (**c**) SEM images showing the bending process of vertically grown tungsten composite nanowires; (**d**–**f**) felling of tungsten nanowire irradiated with a 30 pA ion beam oblique incidence to the wire length direction ($\theta = 60°$): with (**d**) showing an SEM image of the as-deposited nanowire, (**e**) the nanowire after being cut at the base for 30 s, and (**f**) nanowire felled towards the incident ion beam with the top upwards lying on the substrate surface close to its base. *W*, *T*, and *B* denote, respectively, the center of mass, tip, and base of nanowire; the *three arrows* show the beam direction; and the *one arrow* shows the FIB cutting position. The scale bar is 2 μm [19]

Figure 3.5a, b shows the experimental setup and the definitions of the ion beam incident angle ($\theta$) and the bending angle ($\Phi$). Figure 3.5c presents the bending process of the vertically grown nanowires when successive ion sweeping cycle was performed. The reasons behind the ion beam-induced plastic deformation are complex and mainly can be governed by factors such as (1) the geometry, size, thermal expansion coefficient, Young's modulus, and crystalline properties of the freestanding nanoobjects; (2) the electrical and thermal conductivity of the support substrate; and (3) the ion irradiation conditions, which include the ion beam energy, incident angle, ion beam current, and ion beam scanning strategy [29, 30]. All these factors compete against or combine with each other to cause the freestanding nanoobject to bend away from or towards the incident ion beam and with component normal to the beam incident direction.

The bending effect could also affect the felling process and the final felling direction of the nanoobject under investigation. To fell the nanowire, an FIB image was first grabbed using a frame scanning time of 60 s, and then an area of 400 nm × 400 nm near the base of the nanowire was scanned with 30 pA for 30 s. An SEM image was taken at this point; this is shown in Fig. 3.5e. After this, another FIB image was taken, and this was used as the reference image for further milling for 1 min. An SEM image after this further milling, Fig. 3.5f, shows that the nanowire has been felled.

It can be seen from Fig. 3.5e that after FIB scanning with 30 pA ion beam current, the wire bent towards the incident ion beam with the cutting point at the base being thinned. Further imaging and cutting caused the tip of the nanowire finally to point to the higher part of the substrate, which shows that the up-dragging moment induced by the ion beam-induced bending of the whole nanowire has overtaken that imposed by the down-pulling effect of the weight of the nanowire.

In general, it is the combined effect of bending and gravity that determines the felling direction of the nanowire, since the bending direction is a complicated phenomenon that is controlled by many factors. The center of mass of the irradiated nanowire dynamically changes with the evolution of the shape of the nanoobject during bending, so the nanoobject could be felled either towards or away from the incident ion beam with its tip pointed upwards or downwards, depending on the relative strength of the moment due to the object's weight and that resulting from the bending effect caused by ion beam irradiation during imaging and cutting.

### 3.2.5 *Felling of Freestanding Nanowires for Structural and Compositional Studies*

Nowadays, focused ion/electron beam-induced deposition has been a widely used method to grow freestanding nanostructures of various materials, such as tungsten, platinum, gold, cobalt, copper, and iron composites, with a broad range of applications. The compositional and structural details of these as-grown and thermal-treated structures are of great importance in the determination of the related electrical, optical, and mechanical properties; thus, it is very necessary to develop a reliable technology for such investigations.

To demonstrate the suitability of the FIB felling technique for examining the structural and compositional properties of (previously) freestanding nanomaterials, firstly, tungsten-containing vertical zigzag nanowires were grown on a holey carbon transmission electron microscope (TEM) grid by FIB-induced chemical vapor deposition. The rods were grown with the beam normal to the substrate surface by scanning the beam in an area 100 nm × 100 nm, using 1 pA ion beam current. Then the TEM grid was attached to an SEM sample holder by conductive carbon tape stuck from the topside of the grid to the sample holder. To fell the nanowires, interlinks of the supporting carbon film in the vicinity of a freestanding nanowire were cut through with a 1 pA ion beam normal to the membrane surface. Figure 3.6a shows a plan-view SEM image of a holey carbon TEM grid attached onto an SEM

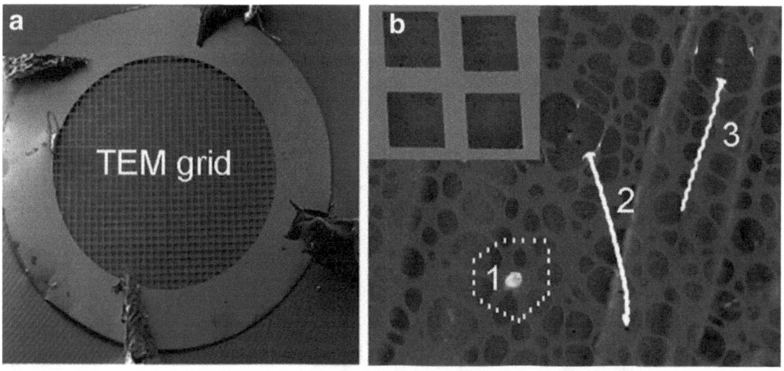


**Fig. 3.6** (**a**) Transmission electron microscope (TEM) grid fixed on a scanning electron microscope (SEM) sample holder; (**b**) SEM top-view image of freestanding tungsten composite wires: an as-deposited wire (*1*), wires lying on the TEM grid felled by FIB (*2* and *3*). The *inset* shows the square grid that supports the holey carbon film; the *dotted line* schematically shows the cutting position of carbon network that supports the as-deposited wires for felling [19]

sample holder; Fig. 3.6b shows an SEM image of the felled tungsten nanowires lying on the holey carbon grid; the top view of the as-grown nanowire is also indicated. The dotted white line schematically shows the FIB cutting positions of the supporting carbon networks. A 1 pA ion beam current was used to cut the interlinks that support the nanowire. Upon losing its support, the wire was felled and laid on the holey carbon grid as shown in Fig. 3.6b.

After being felled, the nanowires lie on the carbon grid and are suitable for chemical composition analysis. For the felled wires shown in Fig. 3.7, their composition and structure were examined by electron energy loss and energy-dispersive X-ray spectroscopy (EDS), scanning electron microscope, and high-resolution TEM. An example of a wire resulting from this method is shown by the scanning transmission electron microscope (STEM) images and high-resolution TEM images in Fig. 3.7. Figure 3.7a, b shows the STEM images of the zigzag and vertical straight nanowire, respectively. It can be seen that the wires are surface smooth and cross-section uniform. The HRTEM images in (c) and (d) indicate that FIB-grown tungsten nanowires consist of carbon skin, which are different for these two kinds of nanowires. HRTEM images in high magnification show that W clusters about 1 nm in size can be observed to be embedded in a carbon matrix, and these nanowires do not display any long-range order—rather there are nanocrystallites with grain size on the order of 1 nm. Electron energy loss and EDS show that the composition is 48 at.% tungsten, 30 at.% carbon, 16 at.% gallium, and 6 at.% oxygen. This composition is similar to that reported previously [12, 18, 19], indicating that FIB felling of the as-deposited objects leads to no changes in their structural properties.

In summary, it has been demonstrated that low-current FIB milling may be used to fell freestanding nanoobjects for compositional, structural, and electrical property investigations. The TEM results and electrical property analysis shows that, firstly, the felling process exhibits negligible side effects on the structures of

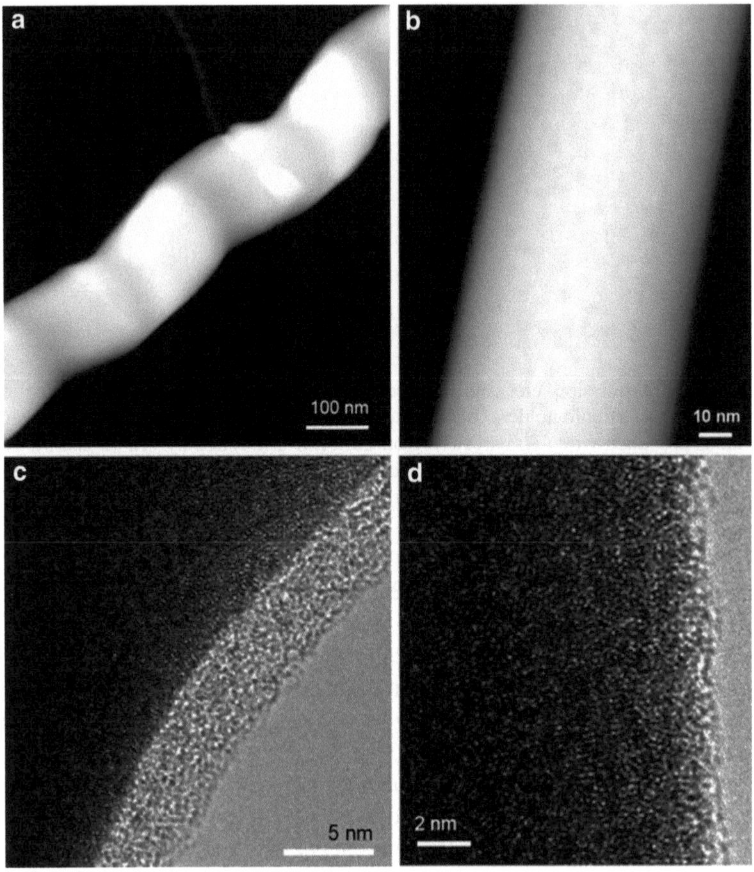


**Fig. 3.7** (**a**, **b**) STEM images of the zigzag and straight nanowires after being felled by FIB milling; (**c**, **d**) HRTEM images on the edges of wires shown in (**a**) and (**b**), respectively

nanowires; secondly, the technique provides a particularly flexible, controllable, and target-specific approach for the exploration of the properties of freestanding nanoobjects; and thirdly, the current–voltage (*I*–*V*) characteristics of the felled nanowire show a series of discrete steps in approaching the normal state, indicating that FIB-induced deposition of vertical nanowires might be a potential approach to probe the quantum phenomena in nanoscale tungsten with significantly enhanced superconductivity—this opens up the possibility of fabricating novel superconducting devices without a mask or other conventional micro-fabrication techniques. By investigating a large number of nanowires, it has been revealed that the phenomenon of bending is a complex process dependent on many factors. For felling nanowires for transport investigations, a 1 pA beam scanned close to the base of the nanowire is recommended.

## 3.3 Atomic Resolution Top-Down Size Reduction with FIB Milling

Various fabrication technologies for nanometer-scale feature growth have been developed [31–38]. Among them, a combination of e-beam lithography, metal evaporation, and lift-off process has been widely used to top-down fabricate under 20 nm metal track (wires) in large scale [35]. However, nanowires based on materials that cannot be grown by deposition and lift-off techniques are unachievable by this method. For example, for the iron-based superconductor, $Cs_{0.8}(FeSe_{0.98})_2$, a kind of material usually grown at temperatures over 1,000 °C, in the shape of well-formed black crystal rods a few mm in diameter [39], an alternative direct and site-specific milling approach is thus required in order to study its superconductivity and the related size effect.

In addition it has been reported that vertical tungsten nanowires grown by FIB-induced deposition are superconducting with much enhanced superconductivity [40] and could be used to form 3D nano-SQUIDs that potentially are able to detect the magnetic field both parallel and perpendicular to the substrate surface [12]. However, the size of the as-deposited 3D structures must be reduced if single-spin resolution and sensitivity are to be achieved. Thus it is of great importance to explore a reliable technique for fabricating nanometer-scale freestanding superconducting nanostructures, for better understanding of superconductivity at nanometer scale, as well as for realization of high-performance 3D superconducting nanodevices.

FIB milling is a technique that can meet this requirement [38]. The advantages of FIB milling include that it is a maskless and resistless process, applicable to most materials in any form, i.e., films, tracks, wires, particles, and flakes. It provides many aspects of flexibility and controllability for processing, e.g., with stage tilting and changing of the ion beam incident angle, as defined in Fig. 3.5a, combined with the easy controlling of different ion beam currents and thinning rate; meanwhile, the field of view can be easily adjusted by magnification, and thus the thinned area and the number of nanowires to be thinned can be designed; by using different scanning strategies, for instance, raster scanning or reduced raster scanning, as shown in Fig. 3.8, the whole wire or a specific portion of the target can be selectively thinned precisely.

To demonstrate the applicability of this technique to thinning of freestanding nanowires, two vertical tungsten nanowires with diameter of about 180 nm were deposited, as shown in Fig. 3.9a. Since FIB irradiation could lead to the bending of freestanding nanowires as discussed above, systematic experiments were performed for optimized condition for FIB milling-resulted size reduction. In particular, the influence of the ion beam current and the ion incident angle during thinning need to be considered to achieve only size reduction of the wire or to thin the wire as well as deform it simultaneously. The ion beam current is a critic parameter for different purposes. To thin a pair of vertical nanowires as well as to obtain a solid contact between them, a reduced raster scan was performed so that the beam scanned in an area that covered the wire to be bent and a portion of the other wire with an ion beam current in the range of 150–300 pA. The stage rotation and tilting angles were also optimized based on the values of the interspacing and heights of wires to be

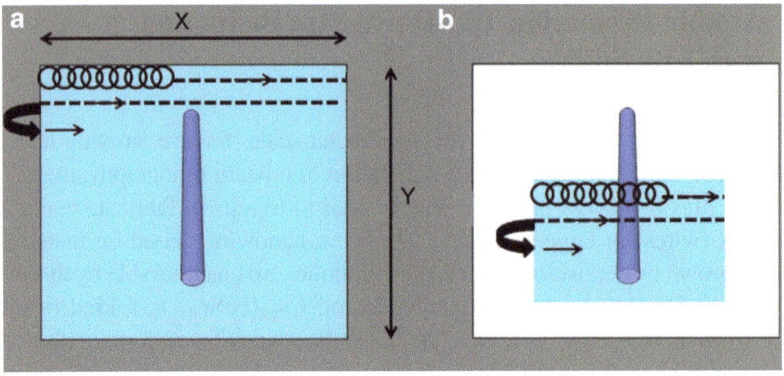


**Fig. 3.8** Ion beam scanning strategy: (**a**) raster scanning; (**b**) reduced raster scanning

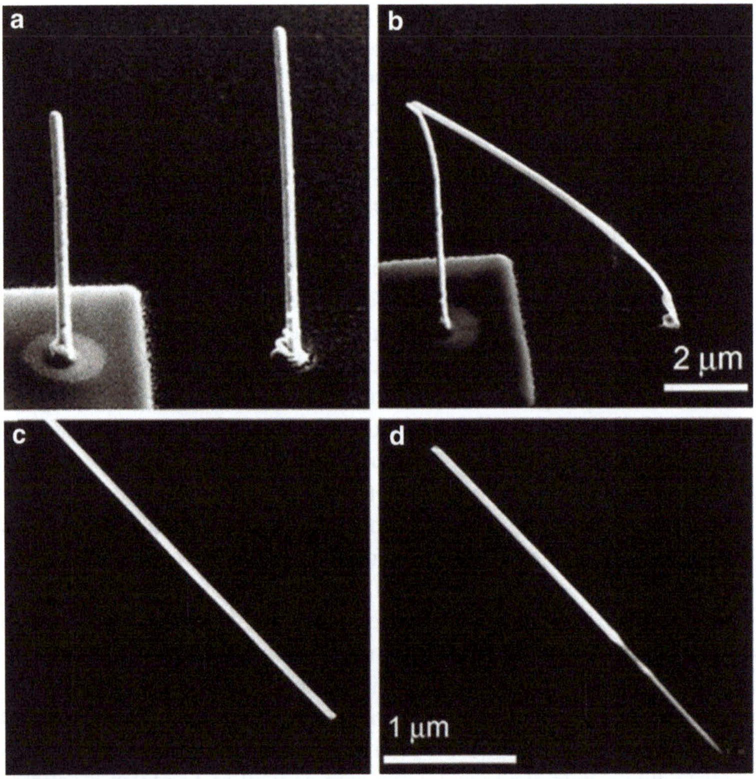


**Fig. 3.9** SEM images of (**a**) two as-deposited freestanding tungsten nanowires (330 nm in width); (**b**) joint contact formed based on the pair of nanowires shown in (**a**) with one bent and both of them with width halved by the irradiation using a 200 pA ion beam during the bending process [41]; (**c**) as-grown ZnO nanowire lying on the substrate; (**d**) tapered ZnO nanowires thinned with ion beam irradiation using reduced rater scanning

bent. Figure 3.9b shows the joint contact formed between these two wires with the width having been halved by using a 200 pA ion beam current for irradiation. To achieve such structures, the stage was rotated by 80° so that the two wires were slightly misaligned, and then reduced raster scanning was performed to irradiate the whole of the longer wire and a small section of the shorter wire with the stage tilted by 45°; by doing this, the longer wire bent towards the incident ion beam and finally made contact with the shorter one. During this process, both wires were thinned. This confirms that FIB milling can be used for controllable size reduction as well as to manipulate the orientation of freestanding nanowires at the same time, which possibly can be used to construct freestanding superconducting pickup loops with desired cross section and neat surface to integrate with a planar SQUID for super-performance 3D nano-SQUID fabrication [41].

Meanwhile, FIB milling can be used to reduce the size of an object that lies on the substrate. Figure 3.9c, d is the plan-view image of an as-deposited ZnO nanowire, being thinned using reduced raster scanning with a 1 pA ion beam. The segment indicated by number 3 is the as-deposited portion. The tapered shape can be achieved by setting different ion beam irradiation doses in different areas, which can be modulated simply by either the irradiation time or the ion beam current or by changing the ion beam incident angle. For accurate control of the size reduction, it is necessary to use a low ion beam current and low magnification. An increased size reduction rate can be obtained with larger ion beam currents and magnifications, though this sacrifices controllability. A high milling rate may also damage the supporting substrate. To obtain a nanowire with evenly reduced size along the length direction, raster scanning with a field of view that barely covers the nanowire was used to avoid unnecessarily harsh thinning.

Furthermore, in situ experiment can be done on the thinned nanowires for electrical property characterization. For instance, to explore the size effect of FIB-grown vertical tungsten nanowires, FIB-induced deposition, milling-resulted thinning, and felling process can be done in one run. Firstly, to grow vertical tungsten nanowires with $W(CO)_6$ as the gas precursor by 1 pA ion beam current, the vertically grown nanowires were felled by lateral FIB milling through the base of the nanowires. Then the size of the nanowire was thinned by raster and/or reduced raster scanning with various ion beam exposure durations to reduce the size of the whole wire evenly or a portion of it, as indicated in Fig. 3.10 [41].

To form a four-terminal device configuration (Fig. 3.10c) for electrical property measurement, a larger ion beam current of 10 pA was then used to deposit strips to connect the thinned wire to large Au contact pads, which had been previously formed on a 200-nm-thick $SiO_2$ layer on a Si substrate by conventional photolithography-based processes. Current-biased transport measurements were performed with a quantum design physics property measurement system (PPMS). Temperature-dependent resistance and *I–V* transport measurements were conducted on the as-deposited nanowire with width thinned to 50 nm. Figure 3.10d shows the *I–V* curves. The transition width of the 50 nm wide nanowire is slightly wider than that of the as-deposited one (80 nm). From the *I–V* curves, it can be derived that the

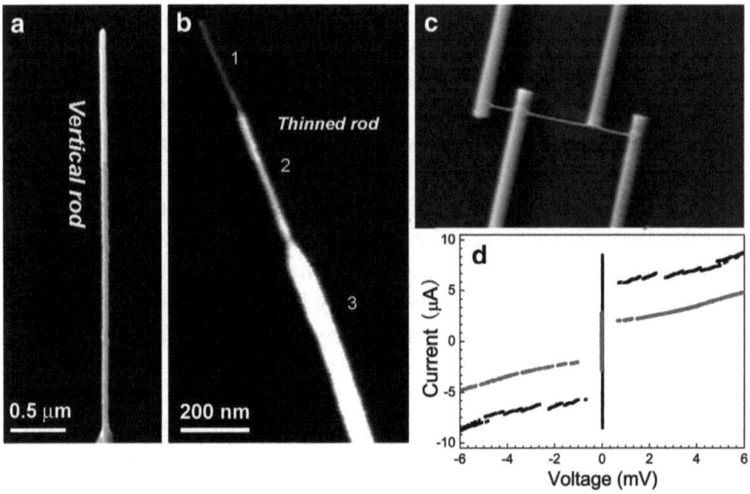


**Fig. 3.10** (**a**) SEM side-view image of a freestanding tungsten nanowire grown by FIB-induced deposition and (**b**) SEM top-view image of a tapered tungsten nanowire thinned by FIB after being felled by FIB at the base; (**c**) SEM image of a typical four-terminal configuration based on the thinned nanowire after being felled; (**d**) the *I*–*V* curve of the thinned nanowire measured at 4.26 K [41]

critical current density upon thinning is insignificant, being $1.7\times10^5$ A/cm$^2$ and $1.4\times10^5$ A/cm$^2$ at 4.26 K for the as-deposited wire (black curve) and wire with width reduced to 50 nm (red curve), respectively. Previously, it has been found that for as-deposited lateral nanowires with width of 19 nm, a residual resistive tail extending down to the low-temperature region exists, possibly as a consequence of a thermally activated phase slip (TAPS) process near $T_c$ [25, 42]. However, the coherence length of FIB-deposited tungsten is about 6 nm. Using most of the currently available commercial FIB systems, it is still a technical challenge to deposit freestanding nanowires with diameter less than 80 nm and to deposit lateral continuous thin nanowires with width and/or thickness less than 10 nm. So the approach of using low-current FIB thinning to reduce the size of the as-deposited nanofeature to the physical critical values might find potential application in the research fields of quantum science as well as nanoscience and nanotechnology.

The width, thickness, and height of nanowires were measured by in situ SEM and by atomic force microscope (AFM). After thinning, nanowires were transferred onto holey carbon TEM grids by an in situ nano-manipulator. The chemical composition was obtained by EDS and electron energy loss spectroscopy attached to the TEM. The TEM image of a thinned tungsten nanowire (not shown here) is promptly inserted into a TEM facility for structural property examination. The bright field image shows that the width was reduced to about 50 nm. An image at a higher magnification confirms that the thinned nanowires do not display any long-range crystalline order—rather there are nanocrystallites with grain size on the order of 1 nm, a result similar to that observed for the as-deposited nanowire [43]. This further indicates that low-current ion beam milling did not bring observable change to the microstructure of the tungsten nanowires. However, the outer amorphous layer is slightly thinner compared to the as-deposited nanowire (not shown here), confirming the removal of the outer amorphous layer during thinning. The electron energy loss and EDS show that the composition of the thinned nanowire is 49 at.% tungsten, 29 at.% carbon, 16 at.% gallium, and 6 at.% oxygen, indicating that FIB thinning of the as-deposited objects leads to no obvious changes in their chemical composition.

TEM images of a thinned tungsten composite nanowire with width reduced from 80 to 20 nm show uniform shrinking along the length of the wire, and high-resolution images show no obvious changes of the morphology after thinning. The critical current densities of the as-deposited wire and one thinned to a width of 50 nm are $1.7 \times 10^5$ A/cm$^2$ and $1.4 \times 10^5$ A/cm$^2$ at 4.26 K, respectively, suggesting insignificant modulation of the electrical properties during thinning. These results suggest that FIB milling is a potential approach for controllable size reduction with high resolution towards the observation of size and quantum effects as well as for construction of 3D superconducting nanodevices.

## 3.4 Conclusions

In summary, by low-ion-beam-current FIB milling, individual freestanding nanoobjects can be felled as desired; also, site-specific size reduction can be achieved for both freestanding and lateral nanowires in a well-controllable way. Parameters such as the angle of the incident ion beam, ion beam scanning area, ion beam current, and scanning strategy and scanning time were carefully chosen for controllable felling and thinning. After felling or thinning, no obvious structural and electrical property changes were observed for nanoobject grown with FIB-induced deposition. These results suggest that FIB milling is an effective method that could be used for the characterization of size and quantum effects in nanomaterials, especially freestanding nanomaterials, as well as for fabrication of novel 3D nanodevices.

## References

1. Katsaros, G., Spathis, P., Stoffel, M., Fournel, F., Mongillo, M., Bouchiat, V., Lefloch, F., Rastelli, A., Schmidt, O.G., De Franceschi, S.: Nat. Nanotechnol. **5**, 458–464 (2010). doi:10.1038/nnano.2010.84
2. Kim, J.R., Kim, B.K., OLee, J., Kim, J., Seo, H.J., Lee, J., Kim, J.J.: Nanotechnology **15**, 1397 (2004)
3. Qu, Y.Q., Liao, L., Li, Y.J., Zhang, H., Huang, Y., Duan, X.F.: Nano Lett. **9**, 4539 (2009)
4. Bruhn, B., Valenta, J., Linnros, J.: Nanotechnology **20**, 505301 (2009)
5. Lai, C.X., Wu, Q.B., Chen, J., Wen, L.S., Ren, S.: Nanotechnology **21**, 215602 (2010)
6. Qu, Y.Q., Liao, L., Cheng, R., Wang, Y., Lin, Y.C., Huang, Y., Duan, X.F.: Nano Lett. **10**, 1941 (2010). doi:10.1021/nl101010m

7. Newton, M.C., Firth, S., Warburton, P.A.: Appl. Phys. Lett. **89**, 072194 (2006)
8. Valentine, J., Zhang, S., Zentgraf, T., Ulin-Avila, E., Genov, D.A., Bartal, G., Zhang, X.: Nature **455**, 376 (2008)
9. Gouma, P., Kalyanasundaram, K., Yun, X., Stanacevic, M., Wang, L.: IEEE Sensor. J. **10**, 49 (2010)
10. Tian, B., Cohen-Karni, T., Qing, Q., Duan, X., Xie, P., Lieber, C.M.: Science **329**, 830 (2010)
11. Noda, S., Tomoda, K., Yamamoto, N., Chutinan, A.: Science **289**, 604 (2000)
12. Romans, E.J., Osley, E.J., Young, L., Warburton, P.A., Li, W.: Appl. Phys. Lett. **97**, 222506 (2010)
13. He, Y., Wang, J.A., Chen, X.B., Zhang, W.F., Zeng, X.Y., Gu, Q.W.: J. Nanopart. Res. **12**, 169 (2010)
14. Long, R.H., Chen, J.J., Lim, J.H., Bwiley, J., Zhou, W.L.: Nanotechnology **20**, 285306 (2009)
15. Magasinski, A., Dixon, P., Hertzberg, B., Kvi, A., Ayala, J., Yushin, G.: Nat. Mater. **9**, 353 (2010)
16. Assefa, S., Xia, F.N., Vlasov, Y.A.: Nature **464**, 80 (2010)
17. Xu, S., Qin, Y., Xu, C., Wei, Y.G., Yang, R.M., Wang, Z.L.: Nat. Nanotechnol. **5**, 366 (2010). doi:10.1038/nnano.2010.46
18. Bezryadin, A.: J. Phys. Condens. Matter **20**, 043202 (2008)
19. Li, W., Fenton, J.C., Cui, A., Wang, H., Wang, Y.Q., Gu, C.Z., McComb, D.W., Warburton, P.A.: Nanotechnology **23**, 105301 (2012)
20. Lu, S.L., Song, Z.T., Liu, Y., Liu, Y., Chia, S.L.: Phys. Lett. **27**, 028401 (2010)
21. Ross, I.M., Luxmoor, I.J., Gullis, A.G., Orr, J., Buckle, P.D., Jefferson, J.H.: J. Phys. Conf. **26**, 363 (2006)
22. Prestigiacomo, M., Bedu, F., Jandard, F., Tonneau, D., Dallaporta, H., Roussel, L., Sudraud, P.: Appl. Phys. Lett. **86**, 192112 (2005)
23. Nakamatsu, K., Igaki, J., Nagase, M., Ichihashi, T., Matsui, S.: Microelectron. Eng. **83**, 808 (2006)
24. Tian, M., Wang, J., Kurtz, J.S., Liu, Y., Chan, M.H.W., Mayer, T.S., Mallouk, T.E.: Phys. Rew. B **71**, 104521 (2005)
25. Lau, N., Markovic, N., Bockrath, M., Bezryadin, A., Tinkham, M.: Phys. Rev. Lett. **87**, 217003 (2001)
26. Romano, L., Rudawski, N.G., Holzworth, M.R., Jones, K.S., Choi, S.G., Picraux, S.T.: J. Appl. Phys. **106**, 114316 (2009)
27. Park, B. C., Jung, K.Y., Song, W. Y., Beom-Hoan, O., Ahn, S. J.: Adv. Mater. **18**, 95 (2006)
28. Xia, L., Wu, W., Xu, J., Hao Y., Wang, Y.: MEMS **22**, 118 (2006)
29. Borschel, C., Niepelt, R., Geburt, S., Gutsche, C., Regolin, I., Prost, W., Tegude, F.-J., Stichtenoth, D., Schwen, D., Ronning, C.: Small **5**, 2576–2580 (2009)
30. Borschel, C., Spindler, S., Lerose, D., Bochmann, A., Christiansen, S.H., Nietzsche, S., Oertel, M., Ronning, C.: Nanotechnology **22**, 185307 (2011)
31. Martin, C.R.: Science **266**, 1961 (1994)
32. Butko, V.Y., Ditusa, J.F., Adams, P.W.: Phys. Rev. Lett. **84**, 1543 (2000)
33. Abadal, G., Perez-Murano, F., Barniol, N., Aymerich, X.: Appl. Phys. A **66**, S791 (1998)
34. Bezryadin, A., Lau, C.N., Tinkham, M.: Nature **404**, 971 (2000)
35. Zgirski, M., Arutyunov, K.Y.: Phys. Rev. B **75**, 172509 (2007)
36. Cheng, H.H., Alkaisi, M.M., Wu, S.E., Liu, C.P.: AIP Conf. Proc. **1151**, 48 (2009)
37. Troeman, A.P.G., Derking, H., Borger, B., Pleikies, J., Veldhius, D., Hilgenkamp, H.: Nano Lett. **7**, 2152 (2007)
38. Tettamanzi, G.C., Pakes, C.I., Lam, S.K.H., Prawer, S.: Supercond. Sci. Tech. **22**, 064006 (2009)
39. Krzton-Maziopa, A., Shermadini, Z., Pomjakushina, E., Pomjakushin, V., Bendele, M., Amato, A., Khasanov, R., Luetkens, H., Conder, K.: J. Phys. Condens. Matter **23**, 052203 (2011)
40. Li, W., Fenton, J.C., Gu, C.Z., Warburton, P.A.: Microelectron. Eng. **88**, 2636 (2011)
41. Li, W., Cui, A., Gu, C.Z., Warburton, P.A.: Microelectron. Eng. **98**, 301 (2012)
42. Rogachev, A., Bezryadin, A.: Appl. Phys. Lett. **83**, 512 (2003)
43. Li, W., Gu, C.Z., Warburton, P.A.: J. Nanosci. Nanotechnol. **10**, 7436 (2010)

# Chapter 4
# Focused Ion Beam Milling of Carbon Nanotube Yarns and Bucky-Papers: Correlating Their Internal Structure with Their Macro-Properties

**Kallista Sears, Ludovic F. Dumée, Niall Finn, and William Humphries**

**Abstract** Focused ion beam (FIB) milling through carbon nanotube (CNT) yarns and bucky-papers followed by scanning electron microscopy has recently emerged as a powerful tool for eliciting details of their internal structure. The internal arrangement of CNTs in bucky-papers and yarns directly affects their performance and characteristics. Consequently this information is critical for further optimisation of these structures and to tailor their properties for specific applications. This chapter describes in detail FIB milling of CNT yarns and bucky-papers and gives a range of examples where FIB milling has enabled a better understanding of how processing conditions and treatments affect the internal structure. Emphasis is placed on how FIB milling elucidates the influence of fabrication conditions on the internal arrangement of CNTs and how this influences the material's macroscopic properties.

K. Sears (✉)
CSIRO Materials Science and Engineering, Bayview Ave.,
Clayton, VIC 3168, Australia
e-mail: kallista.sears@csiro.au

L.F. Dumée
Institute for Frontier Materials, Deakin University,
Pigdons Rd., Waurn Ponds, VIC 3216, Australia

Institute for Sustainability and Innovation, Victoria University,
Hoppers Lane, Werribee, VIC 3030, Australia

N. Finn • W. Humphries
CSIRO Materials Science and Engineering,
Waurn Ponds, VIC 3216, Australia

Z.M. Wang (ed.), *FIB Nanostructures*, Lecture Notes in Nanoscale Science and Technology 20, DOI 10.1007/978-3-319-02874-3_4,

## 4.1 Introduction

This chapter gives an overview of how focused ion beam (FIB) milling has been used to reveal the internal structure of carbon nanotube (CNT) yarns and bucky-papers. CNTs are nanoscale hollow cylinders of rolled up graphene, with inner diameters as small as 0.7 nm and lengths up to several millimetres, and can be capped at one or both ends by half a fullerene. They can consist of a single cylinder (single walled) or a number of concentric cylinders (double walled, few walled, and multi-walled). Most importantly, they exhibit exceptional intrinsic properties such as high tensile strength (20–40 GPa for single-walled CNTs), high electrical and thermal conductivity and excellent chemical and electrical stability, making them of interest for a wide range of applications [1–5].

In order to be useful for practical applications, CNTs need to be manipulated into macroscopic structures, often with a highly ordered arrangement. However this is a major challenge, and the formation of macrostructures that take full advantage of the exceptional, individual CNT properties remains an elusive goal. Two promising CNT macrostructures are dry-spun yarns and bucky-papers, which have been considered for a range of applications including sensors [6–8], membranes [9–11], actuators [12], composites [13, 14], advanced textiles [15] and biological scaffolds [16].

Pure CNT yarns can be formed via two main routes. The first involves dispersing the CNTs into solutions of acids or surfactants followed by wet-spinning into fibres [17–19]. The second approach is a solid-state process where the yarns are spun directly from either a "spinnable" forest of vertically aligned CNTs [20–25] or an aerogel of nanotubes in the reaction zone of a chemical vapour deposition (CVD) furnace [26–28]. CNT nanocomposite fibres can also be spun from mixtures of polymers and CNTs via more conventional wet-spinning or melt-spinning routes. This chapter focuses on FIB milling of dry-spun CNT yarns formed from vertically aligned spinnable CNT forests. However the same FIB milling technique should be equally applicable to CNT yarns formed by the other routes. The characteristic of spinnability refers to the ability to draw a continuous, free-standing web directly from the edge of a CNT forest as illustrated in Fig. 4.1a. This web can be used as is or further twisted into a yarn (Fig. 4.2) in a manner similar to conventional textile yarns. These yarns exhibit high breaking stresses of up to 2 GPa and high thermal and chemical stability [23, 29–33]. Furthermore their excellent mechanical properties enable knottability and weavability as well as the incorporation of various additives for multifunctional yarns and fabrics [20, 23].

The term "bucky-paper" (BP) generally refers to a mat of randomly entangled CNTs prepared by vacuum filtration (Sect. 4.2.3) [17, 18]. CNTs within the bucky-paper are held together through strong van der Waals interactions leading to a cohesive structure. Bucky-paper properties are highly dependent on the type and morphology of CNTs used and on the processing conditions. They have been shown to be mechanically robust and flexible with bulk modulus of up to ~1 GPa [10, 34] and to have high thermal and chemical stability. They also exhibit outstanding porosity with values between 70 and 91 % reported [8, 9, 35] and have high specific surface areas up to 2,200 $m^2/g$ [10, 36]. They are also simple and cheap to fabricate

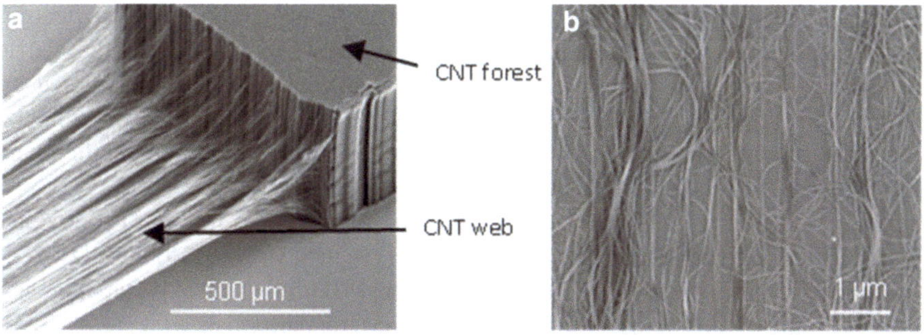


**Fig. 4.1** SEM images showing a web of CNTs being drawn from the edge of a "spinnable" forest (**a**), reprinted from ref. 40 with permission from Elsevier, and (**b**) a densified CNT web on a glass substrate

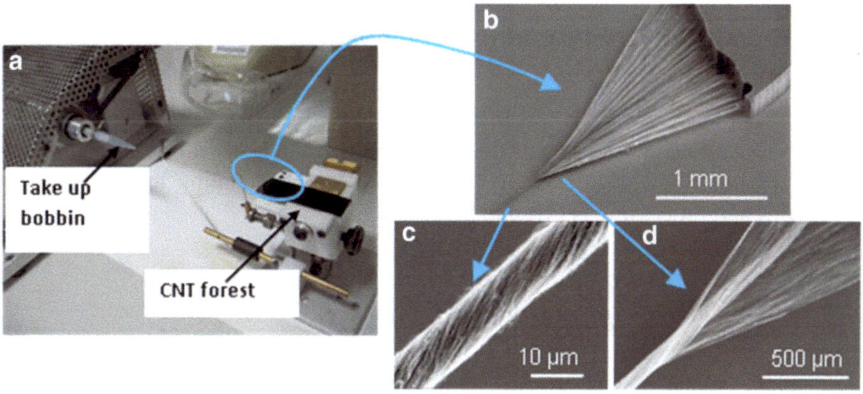


**Fig. 4.2** (**a**) Photo of a flyer spinner used to fabricate CNT yarns. (**b**)–(**d**) SEM images of the spinning process showing (**b**) the entire process and the spinning wedge/triangle that forms, (**c**) the resulting CNT yarn and (**d**) the convergence point of the spinning wedge. Reprinted from ref. 40 with permission from Elsevier

and offer an ideal platform for the incorporation of other additives and polymers. These properties make bucky-papers a promising material for sensors [7, 8], electrodes [37], separation [9–11] and composites [13, 38].

The properties of both CNT yarns and bucky-papers are highly dependent on their internal structure. Therefore techniques to better visualise or infer details about the CNT arrangement are desperately needed. However, there has, until recently, been a general lack of understanding regarding the effect of processing conditions and additives on the internal CNT arrangement and morphology. The traditional sample preparation techniques of microtoming and freeze fracturing used to prepare samples for scanning electron microscopy (SEM) and transmission electron microscopy (TEM) are not suitable for CNT yarn and bucky-papers as they typically cause severe and permanent structural deformation. Prior to FIB milling, information about the arrangement of nanotubes was typically inferred from SEM images of the sample surface or from X-ray diffraction which can give an indication of the degree of their alignment [17, 39]. More recently FIB milling has emerged as an effective tool for forming cross sections which can then be further imaged by either SEM [40] or TEM [24]. FIB milling is hence a powerful tool for examining the internal structure and enables the development of methods to manipulate the internal structure for specific properties.

In this chapter we describe in detail the procedure used to mill CNT yarns and bucky-papers with gallium ions in an FIB system (Sect. 4.3). This is followed by specific examples from our own work and others where FIB milling has led to an improved understanding of the internal structure and its impact on performance (Sects. 4 and 5). Section 2 also gives further background details for the interested reader regarding the CNT yarns and bucky-papers relevant to this work. However this section is not critical for the understanding of Sects. 4.3–4.6 and can be skipped if desired.

## 4.2 Background: Spinnable CNT Forests, CNT Yarn and Bucky-Paper

The CNT yarns and bucky-papers discussed in this chapter are primarily fabricated from "spinnable" CNT forests, and further background relating to their growth, properties and processing is given in this section.

### *4.2.1 Growth of Spinnable CNT Forests*

"Spinnable" CNT forests are grown by CVD. A silicon wafer bearing a thin (~50 nm) layer of thermal oxide and an iron catalyst is used as the growth substrate. Acetylene or ethylene diluted by helium is used as the carbon source and flows through a quartz tube heated to ~700 °C (Fig. 4.3) [21]. During synthesis the iron catalyst remains fixed to the substrate and CNTs grow vertically forming a forest-like structure (Fig. 4.3b). The individual CNTs tend to consist of 7–9 concentric walls with an inner diameter of ~5 nm and length of ~300–500 μm.

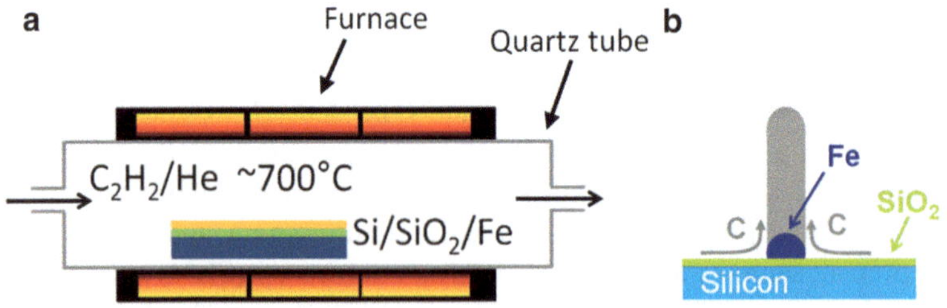


**Fig. 4.3** Schematics showing (**a**) the chemical vapour deposition process used to grow spinnable CNTs and (**b**) growth mechanism of the CNTs

While CNT forests can be grown over a wide parameter range (temperature, reaction time, heating/cooling rates, hydrocarbon source/concentration, catalyst type), the characteristic of "spinnability" is only achieved within a very narrow subset of these. "Spinnability" is not yet fully understood, but it is known to be related to the properties of the CNT forest such as [21]:

- The absence of free catalyst and amorphous carbon which is believed to lead to enhanced van der Waals interactions (i.e. high-purity CNTs)
- A suitably high forest (i.e. CNTs typically >200 μm in length)
- A degree of entanglement
- CNT areal density
- Adhesion to the substrate

### 4.2.2 CNT Sheets and Spinning into Yarns

Spinnable CNTs can be pulled away from the forest edge as shown in Fig. 4.1 to form a web. The as-drawn webs are like an aerogel with a low areal density of 2.7 mg/cm$^2$ and thickness of ~20 μm [41]. The webs can be used directly by placing it onto a desired substrate, where they can be fixed in place and densified using a solvent such as acetone or *iso*-propanol. As the solvent evaporates, surface tension effects draw the CNTs together, reducing the web thickness to ~50 nm (Fig. 4.1b) [41]. While the CNTs in the web show a preferential orientation due to the drawing process, there is still considerable branching and interconnection between them as shown in Fig. 4.1b.

Alternatively, the as-drawn web can be twisted into a yarn, in much the same way as conventional textile fibres. The application of twist provides radial compressive stresses to pack the CNTs into a denser structure, improving van der Waals forces between them and imparting tensile strength to the structure. Figure 4.2 shows the use of a flyer spinner to continuously draw CNT web from the forest edge and twist it into yarn, which is then collected on a rotating bobbin. Unless specified otherwise the CNT yarns discussed in this chapter have been processed using this "standard" method. The amount of twist applied to the yarns can be accurately controlled, and typically twist densities above 5,000 turns/m (5 mm$^{-1}$) are needed to form a coherent, uniform, load-bearing yarn structure.

In order to cater for the addition of various nanoparticles and nanofibres into the CNT yarn structure, Lima et al. developed an alternative method for the formation of CNT yarns as illustrated in Fig. 4.4 and designated as "twist-spun yarn" [15]. In this process CNT web layers are first drawn across filter paper. Once on the filter paper, the webs are densified by wetting them with *iso*-propanol or another solvent. Various other treatments or additives can then be applied to the web. To insert twist the filter paper with attached CNT web is suspended at one end from a rigid support while a metal rod is attached to the other end. The filter paper/CNT web is immersed in acetone to dissolve away the filter paper, and then the metal rod is rotated by placing it over a magnetic stirrer, and hence the web is twisted into a yarn. This method of twisting can also be applied in air.

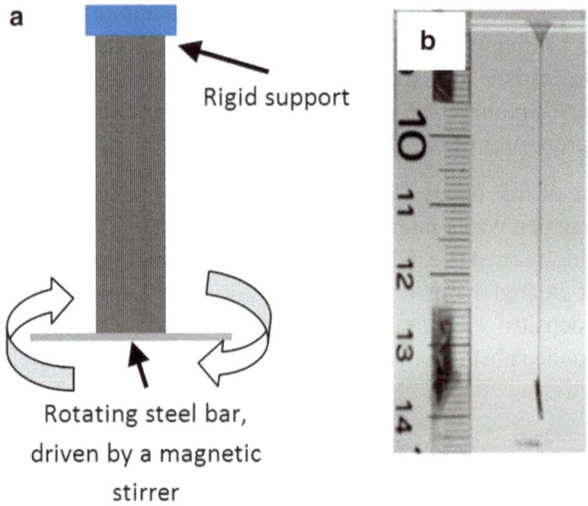


**Fig. 4.4** (**a**) Schematic showing the spin-twist process for forming CNT yarn. The top end of a CNT sheet is fixed to a rigid support, while the bottom is attached to a metal bar and twisted using a magnetic stirrer. (**b**) Photo of the process—reprinted from ref. 42 © 2012 WILEY-VCH Verlag GmbH & Co. KGaA, Weinheim

### *4.2.3 Bucky-Paper Processing*

The most commonly reported wet route to form bucky-papers is illustrated in Fig. 4.5. It involves first purifying and dispersing the CNTs into solution followed by vacuum filtration through a porous filter paper, which results in a mat of randomly entangled CNTs (Fig. 4.5b). Depending on bucky-paper thickness, the type of CNTs and porous filter used, it is sometimes possible to peel the bucky-paper away from the filter paper. Alternatively, the bucky-paper can be separated from the filter using a polydimethylsiloxane (PDMS) stamp or by dissolving away the porous support membrane. Bucky-papers can be highly flexible and mechanically robust as demonstrated by the origami plane in Fig. 4.5c.

The final properties of the bucky-paper are highly dependent on the type of CNTs used (their length, purity and diameter) and the method used to purify and disperse them [10]. The dispersion process for example often involves sonication [43, 44], the use of surfactants [45, 46] or functionalisation of the CNT outer walls [38, 47], all of which can alter the properties of the individual CNTs. Furthermore, a well-dispersed solution is desirable to achieve a uniform and defect-free bucky-paper structure. Hence it is critical to carefully optimise each of these steps in order to reproducibly manufacture mechanically strong bucky-papers.

The bucky-papers discussed in this chapter were formed by dispersing CNTs into *iso*-propanol by mild sonication (15-min intervals at a power of 150 W [9]). Spinnable forests were used with the CNTs being first scrapped off the silicon substrates and then dispersed. The spinnable CNTs are highly pure being free of other carbonaceous materials and catalyst [21], which avoids lengthy purifying steps which can alter the CNT properties.

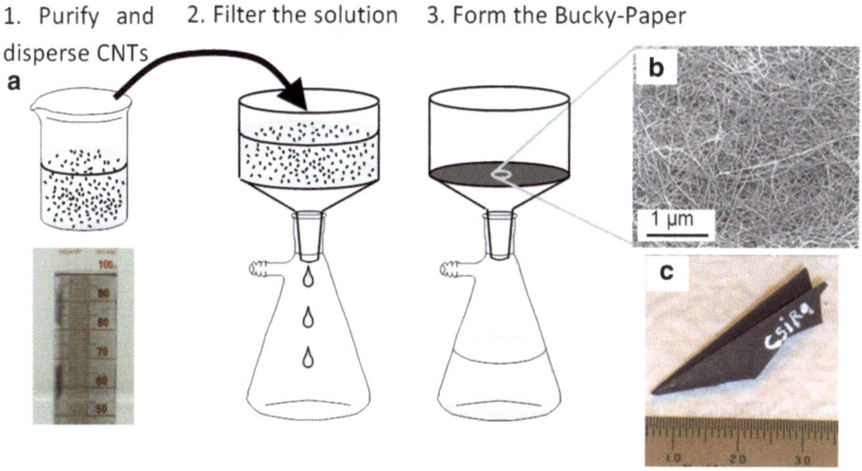


**Fig. 4.5** (**a**) Process for manufacturing bucky-papers, (**b**) SEM image showing the bucky-paper surface and (**c**) bucky-paper origami aeroplane demonstrating their flexibility and mechanical robustness. Reprinted from ref. 10

In addition to wet routes, bucky-papers can also be processed via two solid-state routes which enable CNT alignment within the structure. In the first approach a vertically aligned nanotube forest, generally grown on a silicon wafer by CVD, is flattened through mechanical rolling [48]. This technique generates very thin, semi-aligned bucky-papers and can be applied to any type of CNT forest, regardless of its CNT surface density, diameter or length or spinnability. The second route involves drawing webs from spinnable CNT forests [49]. Multiple layers can be overlapped and densified to form well-aligned bucky-papers, referred to as bucky-ribbons. This technique has the advantage over the previous rolling method of allowing different thicknesses to be prepared from the same CNT batch by varying the number of overlapped layers.

## 4.3 FIB Milling Procedure for CNT Yarns and Bucky-Papers

Gallium ion milling in a FIB system is an excellent technique to obtain cross sections and hence study the interior morphology of CNT yarns and bucky-papers [15, 40]. However the milling strategy needs to be carefully designed to minimise (1) redeposition of material onto the cross section of interest and (2) crystalline damage and implantation by the gallium ions.

Sample preparation for the yarns involved suspending a yarn across a single slot copper TEM grid, which was then fixed to a pin head SEM stub using carbon tape (Fig. 4.6). This procedure avoids carbon tape contamination of the yarn and also minimises the redeposition of sputtered material from the SEM stub or the support

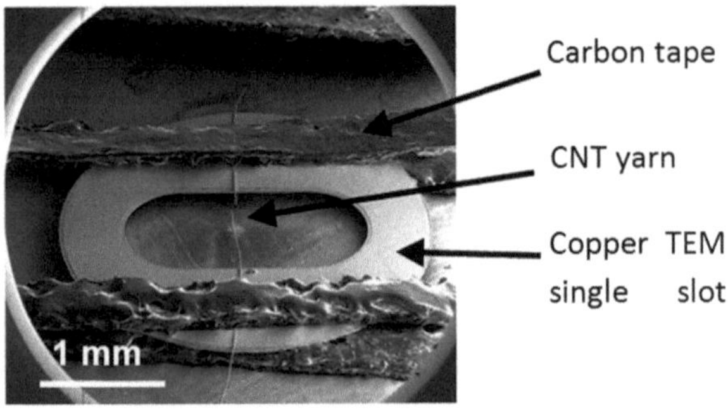


**Fig. 4.6** Low-magnification SEM image showing the sample set-up for gallium ion milling of a CNT yarn (52° tilt)

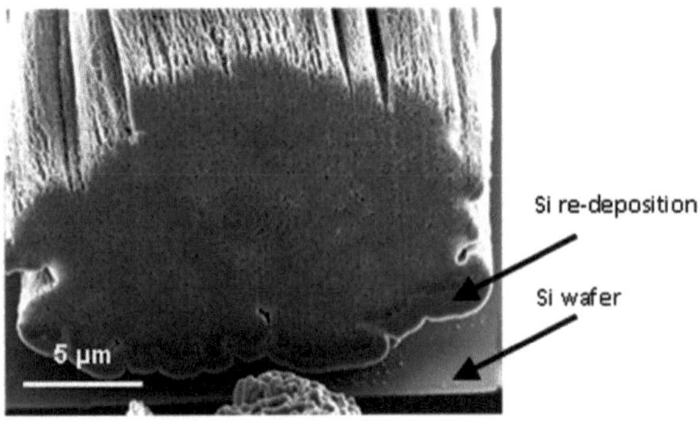


**Fig. 4.7** SEM image of a cross section milled through a CNT yarn supported on a Si-wafer (52° tilt). The gallium ions also penetrated into the Si wafer leading to Si redeposition onto the CNT yarn section

substrate onto the yarn cross section. Figure 4.7 shows a poor example where the gallium ions have also milled the supporting substrate (a silicon wafer) with subsequent redeposition onto the yarn cross section.

Gallium ion milling typically proceeds with an initial "quick and rough" cut using high beam currents of 5–7 nA, at a standard voltage of 30 kV. Figure 4.8a–d shows the progression of this initial rough cut. Once cut, the two yarn segments tend to separate due to tension in the initial suspended yarn. This allows direct observation of the yarn cross section (at a 52° angle due to the FIB geometry [50], see Fig. 4.9). As can be seen in the higher magnification image (Fig. 4.8d), while milling at high currents is fast, it also causes considerable damage so that little detail is discernable. The beam damage can be minimised through a series of final "polishing" slices at successively lower beam currents from 5 to 0.5 nA (Figs. 4.8e and 4.10). This fine milling results in a relatively clean cross section like that shown in Fig. 4.10 for a high-twist (25 $mm^{-1}$) yarn. A core–sheath arrangement is observed for this yarn, with a lower CNT packing density at the surface than at the yarn centre, and clearly illustrates the power of FIB milling to elicit details of the internal yarn structure. Close to the surface, the CNT bundles are easily resolved. However close to the yarn centre, where the density is higher, the bundles merge into one another forming an apparently solid block, but resolution may be ultimately limited by ion beam damage.

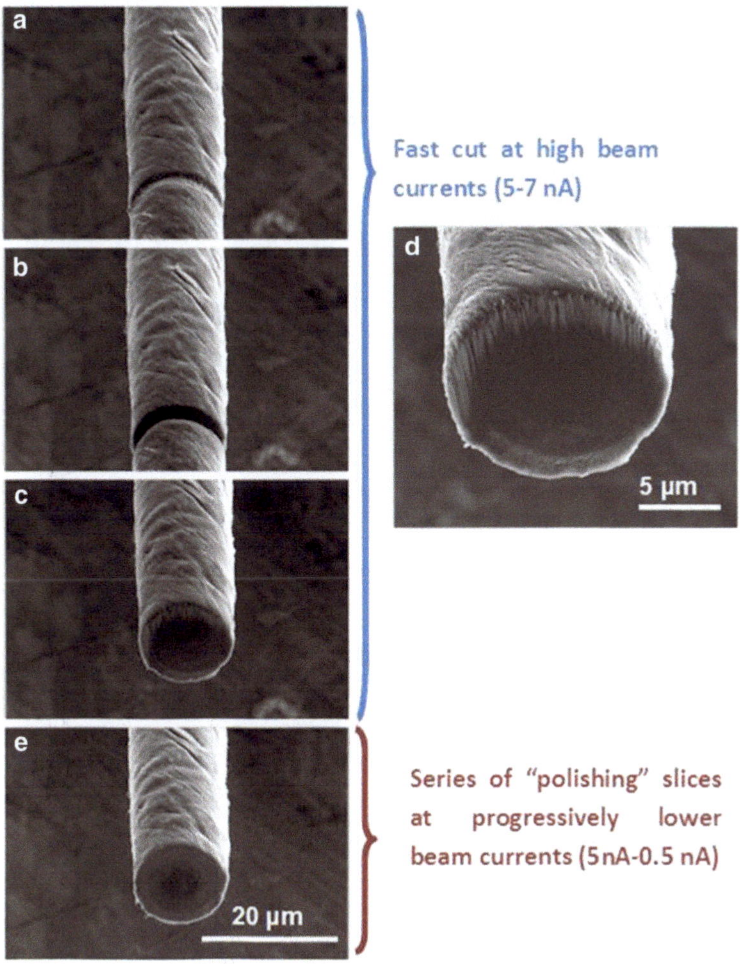


**Fig. 4.8** SEM images (52° tilt) showing progressive milling through a CNT yarn. Initially a fast cut at high beam currents of 5–7 nA is made (**a**–**d**). (**d**) is a higher magnification image of (**c**) and illustrates the large amount of damage incurred. This is followed by a series of polishing slices at progressively lower beam currents between 5 and 0.5 nA (**e**)

A similar procedure of rough milling followed by fine milling is used for the CNT bucky-papers. However due to the large, planar geometry of the bucky-papers, a trench is typically milled as shown in Fig. 4.11. It is particularly important to mill a wide enough trench to allow space for the sputtered material to escape and be either pulled away by the vacuum or deposit well away from your region of interest. If the trench is too narrow then material can redeposit on the cross section of interest and substantially affect the observations. The bucky-papers were typically placed onto pin head SEM stubs with carbon tape, and only a partial cross section made through the structure.

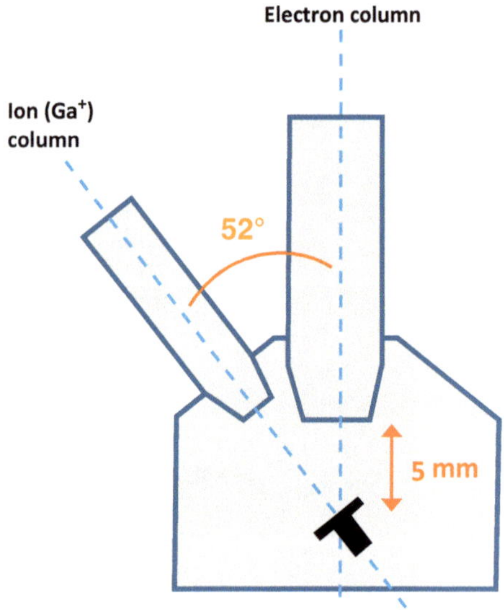


**Fig. 4.9** Schematic showing the typical arrangement of an FIB system. During milling, the sample is tilted 52° so that the gallium ions are incident normal. Consequently SEM imaging is performed with a 52° tilt

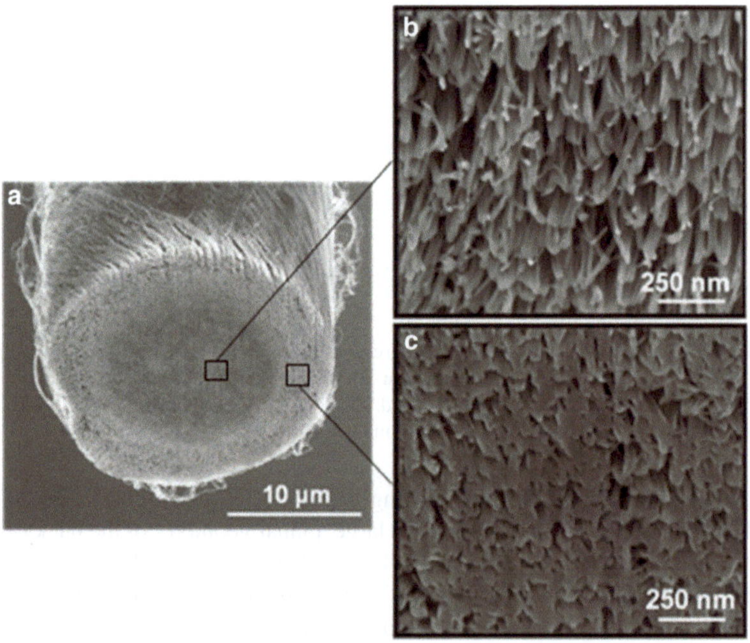


**Fig. 4.10** (**a**) SEM images (52° tilt) showing the internal structure of a 25 $mm^{-1}$ yarn after a series of polishing slices at progressively lower beam currents. A core–sheath structure is clearly evident. (**b**) and (**c**) are higher magnification images of the yarn edge and centre, respectively. This is the same yarn that is shown in Fig. 4.8

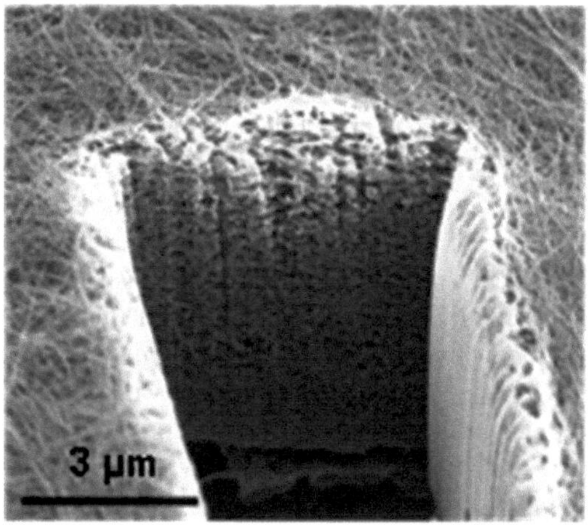


**Fig. 4.11** Trench milled into a CNT bucky-paper using gallium ions

## 4.4 Examples: CNT Yarns

FIB milling of cross sections followed by SEM imaging has provided insights into the internal yarn structure not previously possible. A strong emphasis of recent research has been towards improving yarn strength with a number of approaches investigated including:

1. The effect of fundamental spinning conditions (such as twist) [40]
2. Densification with solvents [22, 27, 40]
3. Modification of the spinning process itself [32, 51]
4. Functionalisation of the CNTs or the addition of polymers/fillers [31, 33, 52]

Researchers have also investigated the addition of various guest materials into the yarn structure to create multifunctional yarns [15]. While these studies have led to significant improvement in yarn strength and wider application of CNTs, until recently there has been little understanding of the effects on the internal yarn structure. In this section we give various examples from our work and others where FIB milling/SEM imaging has helped gain a better understanding of the how the above processes affect the internal structure and hence the properties of the yarn.

### *4.4.1 Effect of Fundamental Spinning Parameters on Internal CNT Yarn Structure: Twist and Solvent Densification*

Sears et al. performed one of the first comprehensive studies in which FIB milling was used to monitor changes in CNT yarn structure with varying spinning parameters [40]. In this study, FIB milling was used to make sections both parallel and perpendicular to the yarn axis and in this way monitor the internal structure of yarns with a range of twist densities and acetone densification. Furthermore, these changes were correlated with tensile strength measurements allowing a better understanding of the influence of key spinning parameters.

#### 4.4.1.1 FIB Milling and SEM Imaging

Figure 4.12 shows FIB-milled cross sections for CNT yarns with twist densities between 5 and 25 turns/mm (or $mm^{-1}$). This series of images reveals three main trends of interest. Firstly, at twist densities below ~5 $mm^{-1}$(Fig. 4.12a), the cross section has an irregular shape with a number of striations along the yarn axis. However, these striations do not penetrate deeply into the yarn interior. As the twist is increased beyond 5 $mm^{-1}$, these striations diminish and a circular cross section is observed.

Secondly, the packing density at the yarn centre continuously increases with twist density as shown by the higher magnification images in Fig. 4.12. This increase in CNT packing density correlates well with the reduced yarn diameter, both due to the increased radial compressive force generated by greater inserted twist.

Finally, and most importantly, at twist densities above ~20 $mm^{-1}$ a prominent core–sheath structure develops. The origin of this core–sheath structure is unclear. However it is suspected that the high compressive forces introduced during spinning and twist insertion lead to strong compaction of CNTs in the yarn centre. Consequently, when spinning is complete and the yarn allowed to relax to a stable configuration (torque stabilisation), these interior CNTs remain tightly packed due to strong van der Waals interactions between them. In contrast, the outer CNTs are under less pressure and are able to relax to form a lower density sheath. Also, as the yarn is twisted the outer layers are extended into longer helical paths than the inner layers. At high twists the outer nanotube bundles may be extended beyond their elastic limit and drafted permanently so that when the yarn relaxes and contracts longitudinally they take up a larger volume.

FIB milling was also used to mill sections parallel to the yarn axis for a low- (15 $mm^{-1}$) and high-twist (35 $mm^{-1}$) yarn as shown in Fig. 4.13. These cross sections show more clearly the variation in CNT orientation throughout the yarn interior. At the yarn centre the CNTs are aligned predominantly along the yarn axis and very few CNT ends are observed, while closer to the surface, the CNTs are increasingly oriented out of the page.

Another technique commonly used to densify CNT webs and yarns is to immerse them in a solvent and allow them to evaporate. As the solvent evaporates, surface tension effects pull the nanotubes and bundles together leading to densification. This has been reported to improve the CNT yarn strength [26, 29]. Figure 4.14 shows FIB sections made through an acetone-treated, low-twist (7.5 $mm^{-1}$) yarn. During spinning the CNT yarn was brought in contact with acetone at the point of twist insertion. This process resulted in a smaller diameter yarn (see Fig. 4.14c, d) with a correspondingly higher packing density (see Fig. 4.14a, b).

#### 4.4.1.2 Correlation of CNT Yarn Structure with Tensile Strength Measurements

The relationship between structure and strength was investigated by performing tensile testing on the same yarns used for FIB analysis. Figure 4.15a, b shows the dependence of breaking stress (MPa—breaking load/yarn area) and breaking load

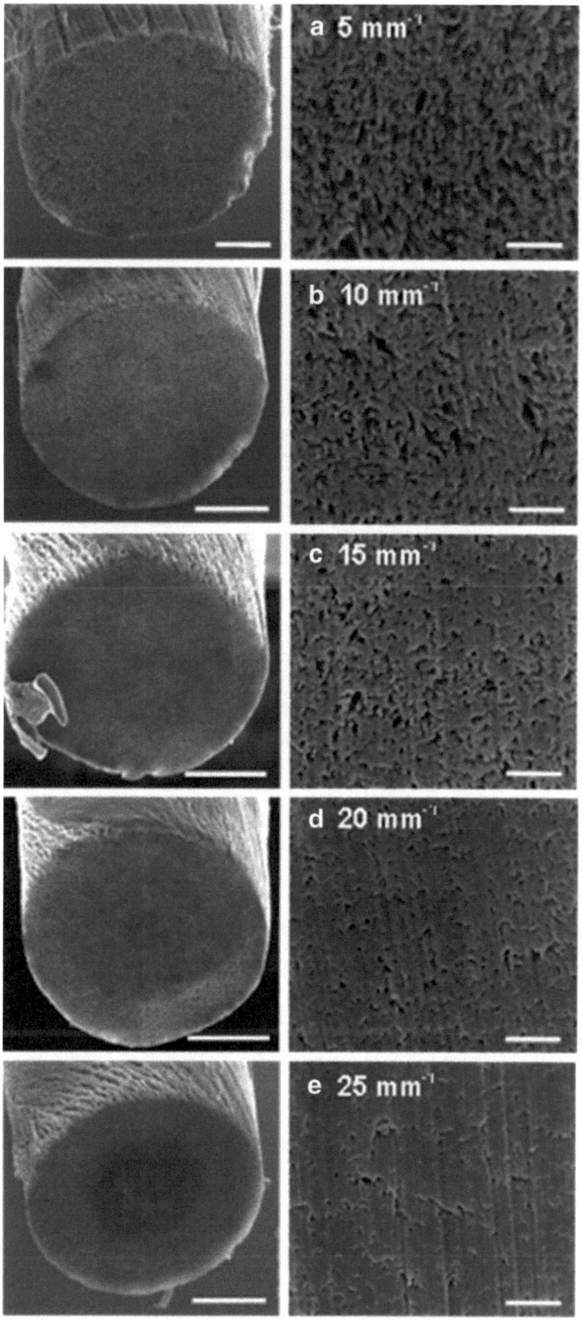


**Fig. 4.12** SEM images of FIB sections milled through CNT yarns with as-spun twist densities of (**a**) 5 mm$^{-1}$; (**b**) 10 mm$^{-1}$; (**c**) 15 mm$^{-1}$; (**d**) 20 mm$^{-1}$ and (**e**) 25 mm$^{-1}$. The left image shows the whole cross section (scale bar represents 5 μm), while the right image is a higher resolution image taken at the yarn centre (scale bar represents 500 nm). All images were taken with a sample tilt of 52°. Reprinted from ref. 40 with permission from Elsevier

(mN) as a function of the as-spun twist, respectively. A maximum breaking stress of ~580 MPa (breaking load of ~125 mN) was measured at a twist density of 15 mm$^{-1}$. A strong decline in tensile stress and breaking load is observed at higher twist

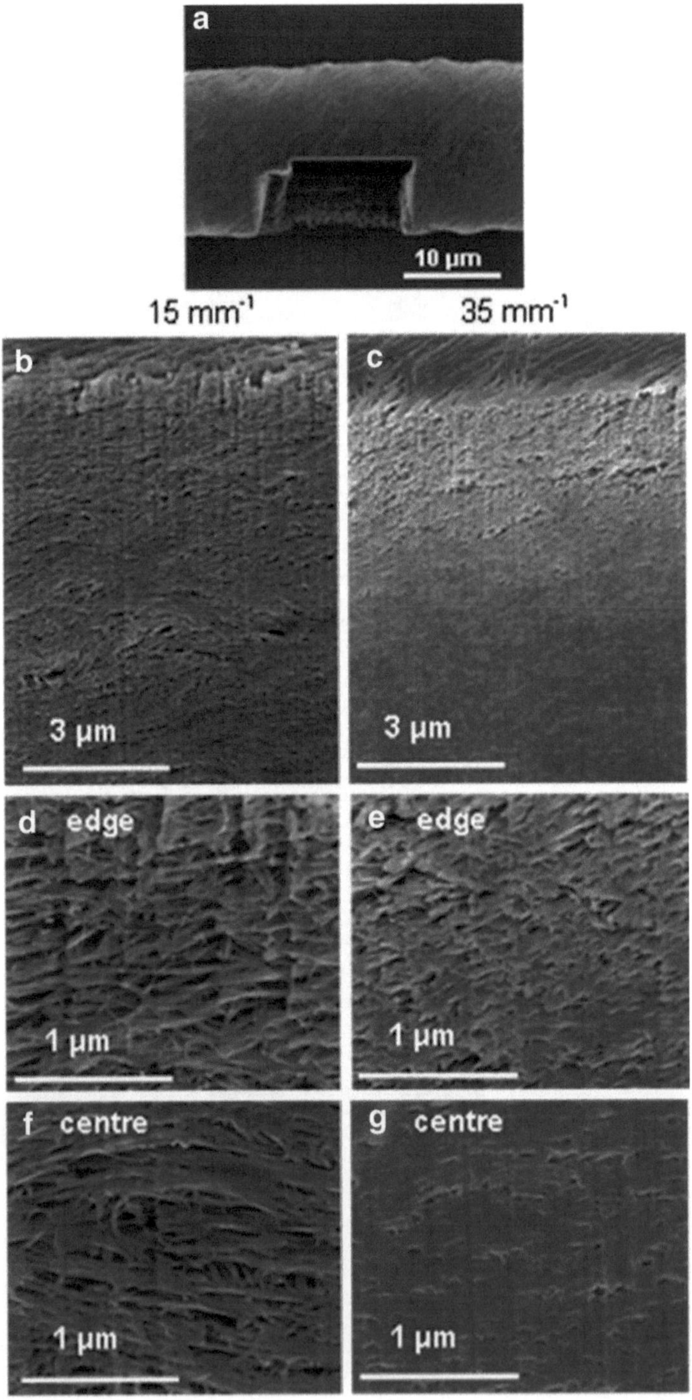


**Fig. 4.13** SEM images of sections milled parallel to the yarn axis for yarns with twist densities of 15 and 35 $mm^{-1}$: (**a**) Low-magnification view from the top of the yarn showing the milled section; (**b, c**) SEM images for the 15 and 35 $mm^{-1}$ twist yarns, respectively, showing a section taken from the centre of the yarn to the outer edge (52° tilt); (**d, e, f, g**) higher magnification SEM images for the 15 and 35 $mm^{-1}$ yarn showing regions from the centre and edge of the milled sections (52° tilt)

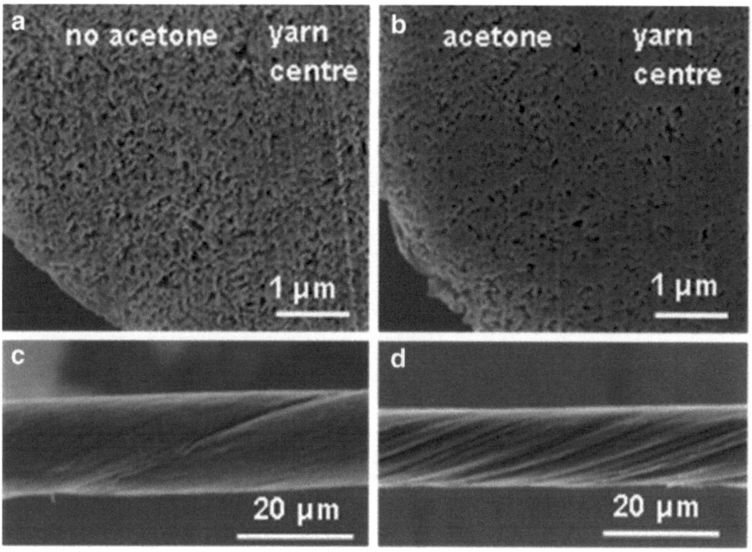


**Fig. 4.14** SEM images of a CNT yarn spun with a twist density of 7.5 $mm^{-1}$ (**a, c**) without acetone and (**b, d**) with acetone. The micrographs in (**a, b**) show one-quarter of the yarn cross section (52° tilt), while the images in (**c, d**) show the yarn surface (no tilt). Reprinted from ref. 40 with permission from Elsevier

densities and coincides with the onset of the core–sheath structure. This reduction with increasing twist is probably due to the oblique orientation of CNTs in the sheath away from the yarn axis and load direction and is similar to that observed for conventional yarns. The low-density sheath also implies that the different sections of the yarn will not simultaneously reach peak load when the yarn is extended, leading to a reduction in the peak load that the yarn can sustain.

An increase in the tensile breaking stress with twist up to 15 $mm^{-1}$ can be observed in Fig. 4.15a and is primarily due to the decreased yarn diameter. If breaking load is plotted instead of breaking stress as in Fig. 4.15b little dependence on twist is observed between 5 and 15 $mm^{-1}$. This is a somewhat surprising result as it indicates that the closer CNT packing and increased compressive stresses that accompany higher twist do not improve load transfer between the individual CNTs. A similar result is also observed for the acetone-treated yarns (solid circles in Fig. 4.15), where the acetone treatment leads to a denser, smaller diameter yarn but has little impact on the breaking load. These results highlight that the factors determining the strength in CNT yarns differ significantly from those of conventional yarns. This is not that surprising considering the large number of CNTs in a yarn, hundreds of thousands compared to forty to one hundred in a conventional yarn, along with the nanoscale nature of the CNTs and the existence of van der Waals forces between them.

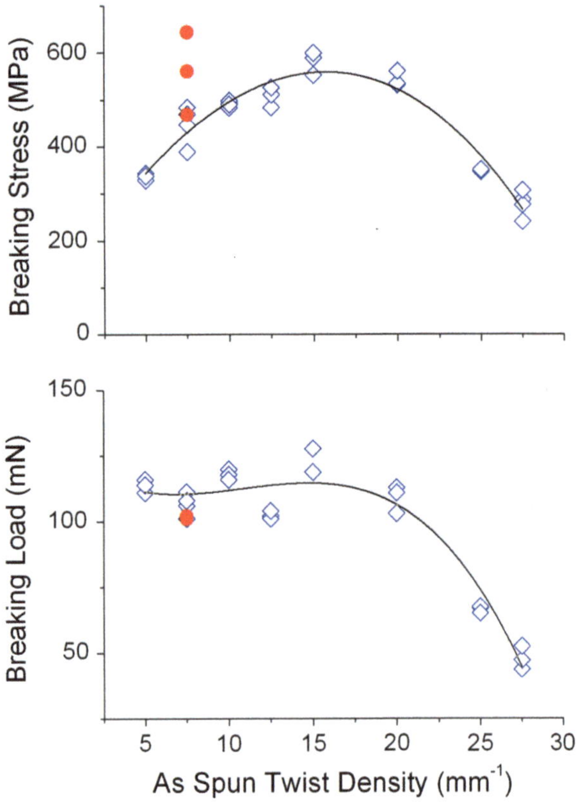


**Fig. 4.15** Dependence of breaking stress (MPa) (**a**) and breaking load (mN) (**b**) on as-spun twist density. The lines are a guide to the eye only and are based on a polynomial fit. The *solid circles* (*red* online) represent acetone-densified yarns

### *4.4.2 Improving CNT Yarn Strength: Modified Spinning Processes and Treatments*

Using the standard spinning process described in Sect. 4.2, CNT yarn strengths between 500 and 600 MPa are routinely achieved [23, 30, 40]. While impressive, this still falls considerably below expectations based on conventional yarns which typically achieve tensile strengths that are 60–70 % of their individual fibres. Assuming that a multi-walled CNT has a tensile strength of 14 GPa, current CNT yarns only achieve strengths around 4 % of their individual CNT strengths. Consequently there has been a drive to investigate methods to further improve CNT yarn strength, and FIB milling has been a key characterisation tool in this research.

For example Tran et al. developed a modified spinning process as shown in Fig. 4.16b where the web first goes through a tensioning/drafting zone before twisting into a yarn. The tensioning zone consists of a series of rollers with the aim of achieving better CNT alignment in the web and hence a more optimal arrangement in the final yarn. SEM imaging of cross sections formed by FIB milling seem to show an improved alignment of the CNTs in the web after the tensioning/drafting zone (see Fig. 4.17), and this correlated with a considerable increase in tensile strength of up to 1.4 GPa compared to the standard process [32, 33].

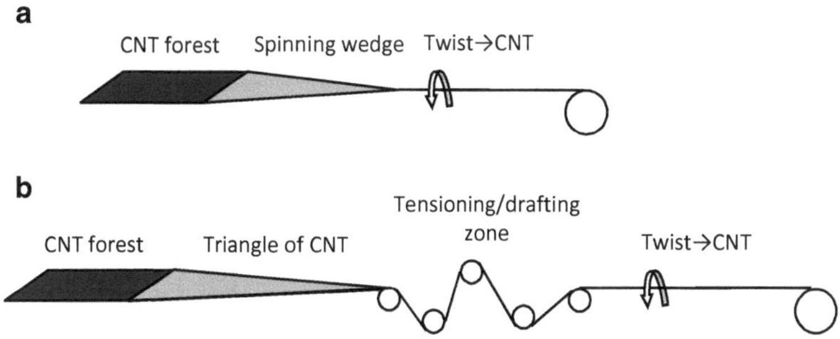


**Fig. 4.16** Schematics showing (**a**) the standard spinning process in which the web is directly spun into yarn forming a spinning wedge and (**b**) the different zones in the modified spinning process which separates the web drawing and spinning processes

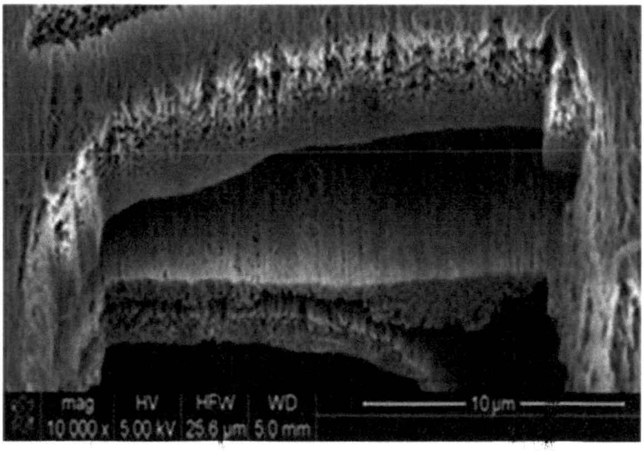


**Fig. 4.17** An SEM image of a cross section milled through three layers of CNT web formed using the modified process is shown in Fig. 4.16. Reprinted from ref. 33 doi:10.1088/0957-4484/22/14/145302

The modified spinning process also enables application of various treatments to the web prior to twisting, and FIB milling/SEM is an excellent tool for monitoring how these treatments affect the final yarn structure. For example, Min et al. pretreated the CNT web as shown by the scheme in Fig. 4.18 to create a highly cross-linked network within the yarn [31]. SEM of FIB-milled sections through untreated and treated yarns showed a significantly reduced porosity in the case of the treated yarn (Fig. 4.19), which was suggested to lead to greater interaction between CNTs contributing to the much improved yarn strength of 2.5 GPa and Young's modulus of 121 GPa.

A final example in which FIB milling is particularly relevant for evaluating a new process is the rub-densified CNT yarns reported by Miao et al. [51]. In the rub-densified yarns the so-called twistless or false twist yarn is formed by drawing the CNT web through two padded rollers which rotate and move laterally (Fig. 4.20). The rotation motion pulls the CNT web up through the rollers, while densification of the yarn is achieved by the lateral motion where the two rollers move in reciprocating directions creating a rubbing action. SEM imaging of cross

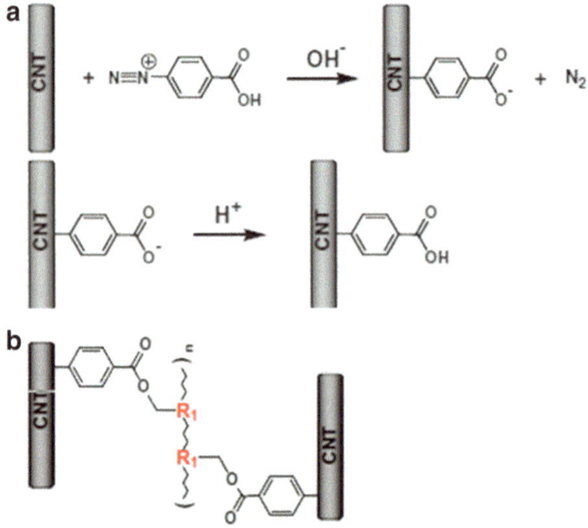


**Fig. 4.18** (**a**) The proposed chemical reactions involved in the sidewall modification of CNTs with 4-carboxybenzenediazonium salt. (**b**) Proposed cross-linked network structure after treatment of the functionalised CNT yarn with hexa(methoxymethyl) melamine (HMMM). Reprinted from ref. 31 with permission from Elsevier

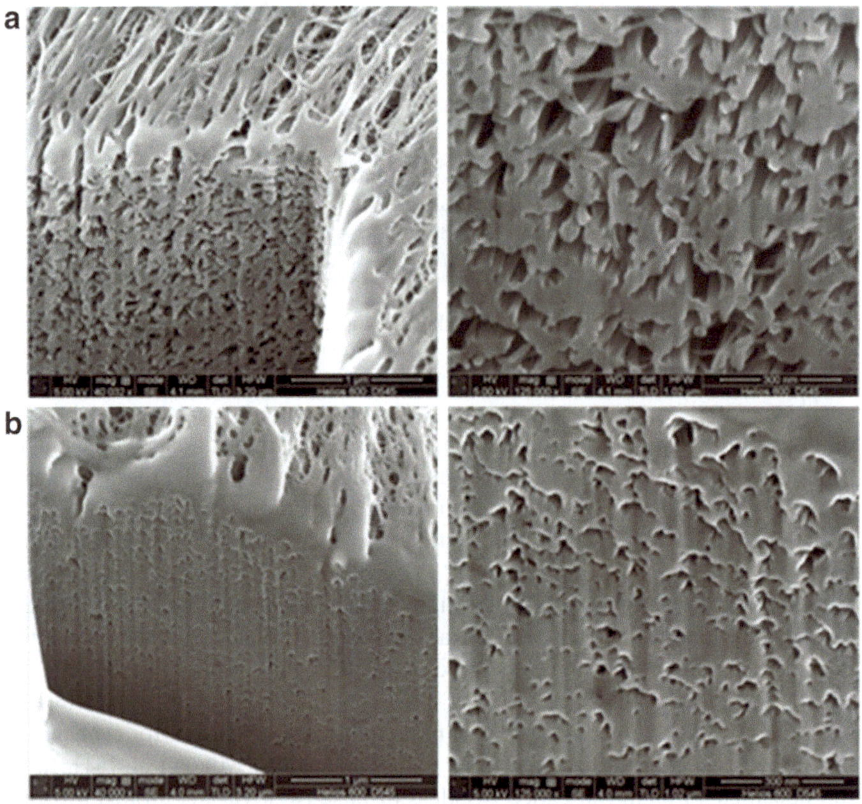


**Fig. 4.19** FIB–SEM images of untreated (**a**) and the cross-linked (**b**) CNT yarns. Reprinted from ref. 31 with permission from Elsevier

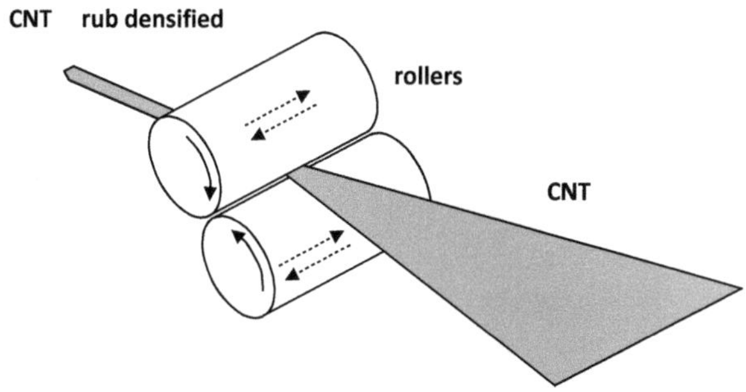


**Fig. 4.20** Schematic showing the action of the two rollers used to form the rub-densified yarns

sections made through the yarns using FIB milling revealed a core–sheath structure (Fig. 4.21). However unlike that formed using the standard spinning process, there is a high-packing-density sheath and a low-density core with many macroscopic voids. Furthermore the cross sections confirm that the CNTs within the sheath are lying straight and parallel to yarn axis due to the twistless process, again in contrast to that for yarns produced using the standard spinning process. The improved modulus of 41 GPa vs. 14 GPa for the standard process was attributed in part to the greater packing density and alignment observed within the FIB-milled sections of the sheath.

### *4.4.3 Multifunctional CNT Yarns*

Prof Baughman's group at the University of Texas at Dallas have also reported that CNT sheets and yarns are excellent host materials for heavy loading with various guest powders, nanofibres and other additives via a process they term biscrolling [15]. Using this biscrolling technique they demonstrate various functionalities such as superconducting yarn loaded with magnesium diboride, flexible yarn for Li-ion batteries loaded with graphene oxide nano-ribbons and photocatalytic yarn loaded with $TiO_2$ particles.

In their paper, FIB milling proved an excellent tool to elucidate details of how the guest materials are distributed within the CNT yarn and their effect on the CNT interior morphology [15]. Figure 4.22 shows SEM images of cross sections milled through four different CNT yarns loaded with either titanium by e-beam evaporation (Fig. 4.22a, b) or $TiO_2$ particles by liquid-based filtration (Fig. 4.22c, d) and aerosol deposition (Fig. 4.22e, f). The internal structures are highly complex, and different morphologies and porosities can be clearly observed depending on the guest materials and processing method used. For example Fig. 4.22a, b in which a layer of titanium is deposited onto the initial CNT sheet by electron beam evaporation shows a highly porous structure and sheet folding. In contrast a yarn formed with $TiO_2$ powder, Fig. 4.22c, d, shows a less porous structure, and evidence of a scroll formation can be discerned [15]. Lima et al. have also been able to engineer the yarn to form the core–sheath structure shown in Fig. 4.22g, h [15].

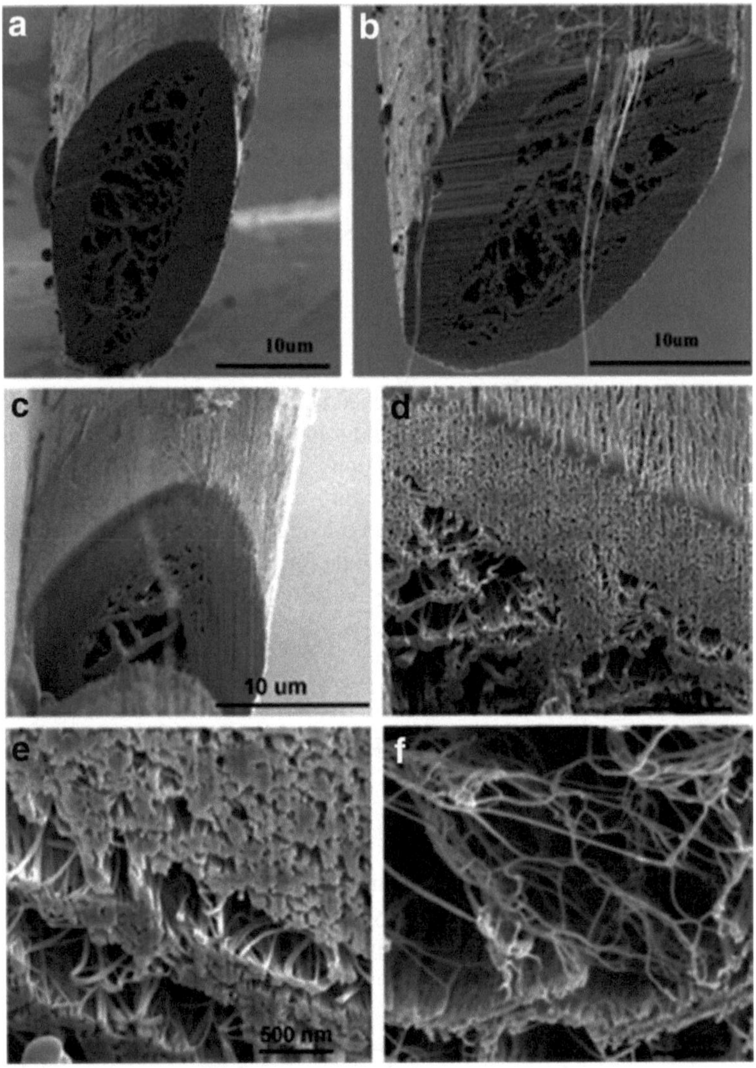


**Fig. 4.21** FIB-milled cross sections of rub-densified carbon nanotube yarns. (**a**–**c**) SEM images of FIB-milled yarn cross sections. (**d** and **e**) SEM images of high-CNT-density areas. (**f**) SEM images of large voids in the yarn core. Reprinted from ref. 51 with permission from Elsevier

## 4.5 Examples: Bucky-Papers

The internal morphology of CNT bucky-papers directly affects their porosity, specific surface area, degree of CNT alignment and interconnects and hence dictates their final properties such as catalytic activity, gas adsorption and electrical and thermal conductivities. Consequently, information regarding the bucky-paper morphology is critical to better understand their properties and performance. To this end FIB is now used as a routine tool to reveal the bucky-paper morphology, and this section gives a number of examples illustrating its applicability.

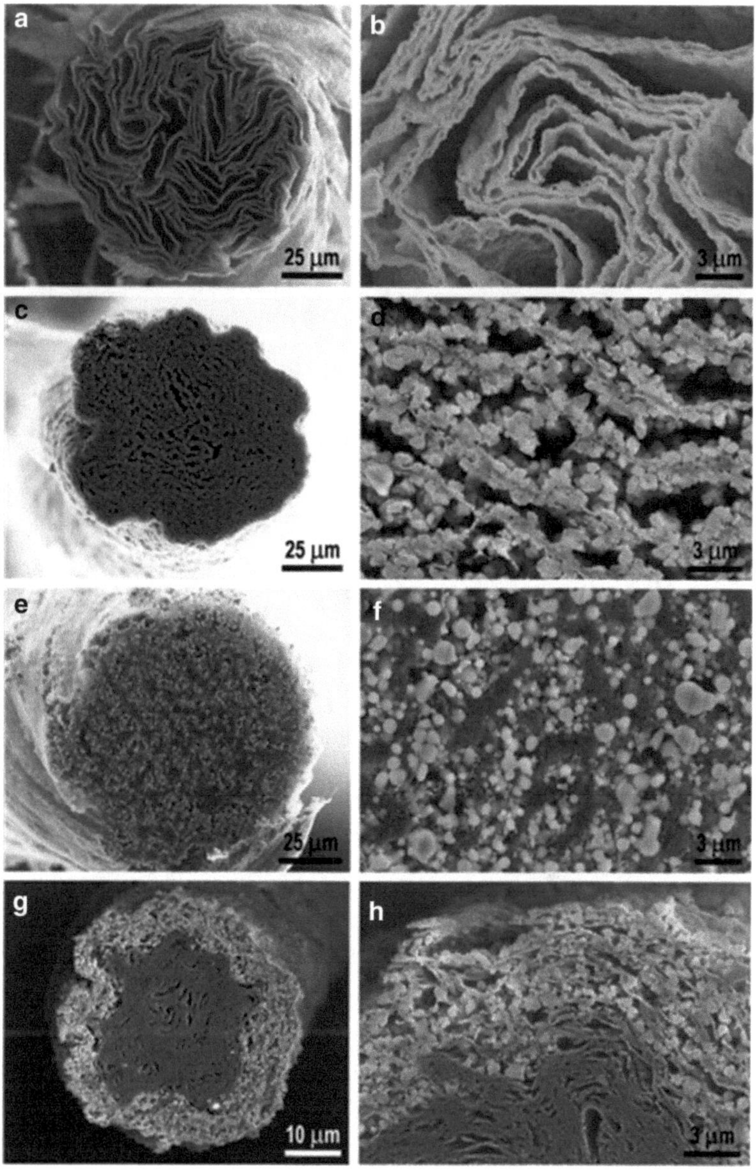


**Fig. 4.22** SEM images of cross sections milled through biscrolled yarns. (**a** and **b**) 70 wt% titanium deposited by electron beam evaporation onto two stacked web layers and then twisted into a yarn in liquid; (**c** and **d**) 93 wt% $TiO_2$ powder deposited by vacuum filtration onto two stacked web layers. Another layer of web was then placed on top and the stacked web structure twisted into yarn in liquid; (**e** and **f**) $TiO_2$ powder deposited by aerosol onto two stacked web layers and twisted into yarn in air; (**g** and **h**) $TiO_2$ yarn made by patterned filtration-based deposition of $TiO_2$ powder on MWNT sheets and asymmetrical twist insertion in liquid. The patterned deposition was limited to a strip along the sheet's edge spanning 15 % of its width. All of the yarns were formed by twist spinning a rectangular strip of CNT sheet in either air or liquid (see Sect. 4.2.2 "twist spun yarn" for further details). Reprinted from [15] http://www.sciencemag.org/content/331/6013/51.full. Reprinted with permission from AAAS

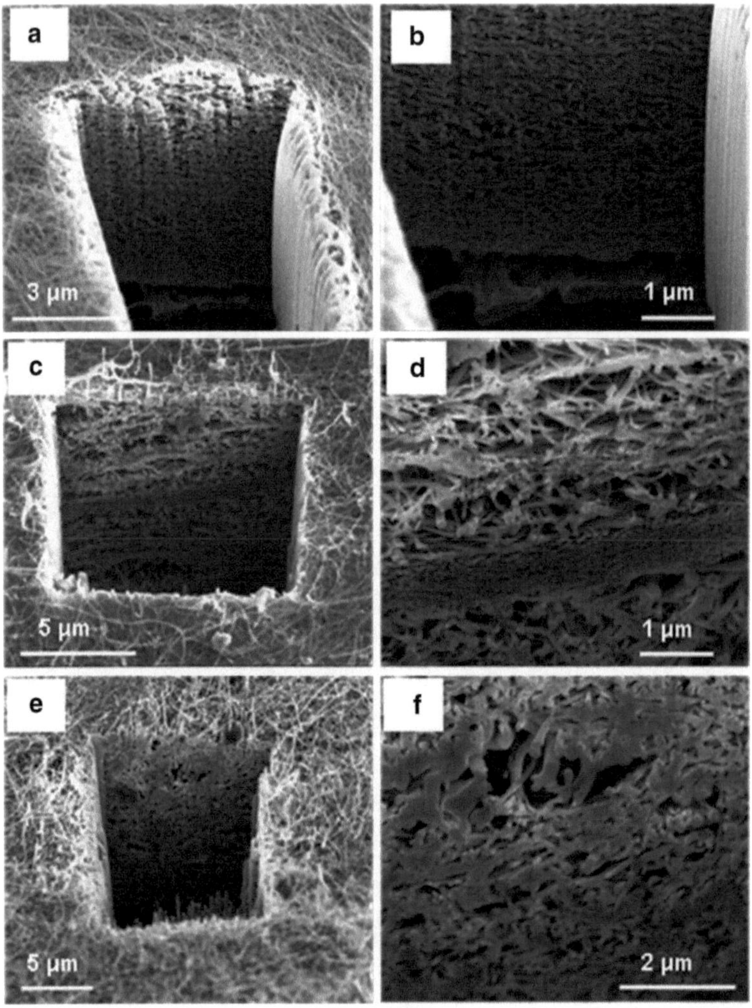


**Fig. 4.23** FIB milling reveals the porosity of bucky-papers processed from (**a**, **b**) thin (~10 nm diameter) CNTs, (**c**, **d**) a mixture of thin (~10 nm diameter) and thick (~50 nm diameter) CNTs and (**e**, **f**) thick (~50 nm diameter) CNTs. The degree of dispersion is clearly visible from the SEM cross section as a layered structure is formed within the mixture suggesting inhomogeneous mixing

### *4.5.1 Revealing the Internal Structure of Bucky-Papers*

As shown by Fig. 4.23, the porosity and CNT arrangement across bucky-papers prepared from different CNTs can be readily accessed by gallium ion-FIB milling of sections. The bucky-papers shown in Fig. 4.23 were processed by vacuum filtration from suspensions consisting of either thin (~10 nm diameter) or thick (~50 nm diameter) nanotubes or a mixture of the two (1:1 ratio). Clearly the type of CNTs

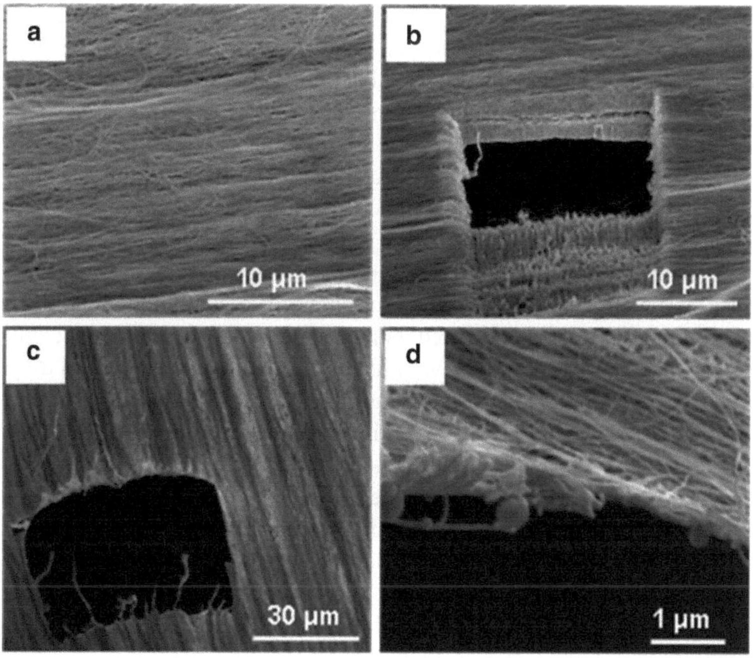


**Fig. 4.24** CNT alignment within bucky-ribbons formed from (**a**, **b**) 50 layers of stacked web densified with acetone and (**c**, **d**) 5 layers of web densified with acetone. The thickness of the bucky-ribbon in (**d**) is <100 nm thick corresponding to 8–10 CNTs in cross section. Some carbon redeposition is visible on the underside of the bucky-ribbon in (**d**)

used effects the bucky-paper porosity and structure. Interestingly, as seen in Fig. 4.23c, d, a layered structure was formed for the mixture and may indicate inhomogeneous mixing or different settling process during the filtration process.

FIB milling also facilitates thickness measurements on bucky-papers. FIB cross sections were used to confirm measurements made with a micrometer where compaction of the porous bucky-paper is possible when sandwiched within the grip of the micrometer [9, 14]. This technique was also used to determine the thickness and density of bucky-ribbon structures consisting of stacked layers of aligned CNT webs (see Sect. 4.2.3) (Fig. 4.24). It was possible to cross section bucky-ribbon structures as thin as ~100 nm, corresponding to less than ten CNTs in cross section (Fig. 4.24d).

### 4.5.2 *Revealing the Interior of Composite Bucky-Paper Structures*

Composite bucky-papers have also been investigated for various applications such as sensing and separation. FIB milling is particularly useful to gain a better understanding of how additives alter the final structure and hence their properties.

For example, metal nanoparticles were grown by electroless deposition within bucky-papers for targeted gas adsorption and catalytic activity [8]. The size of the gold nanoparticles as a function of the plating time and conditions was determined by analysis of FIB cross sections as shown in Fig. 4.25. These FIB cross sections show the formation of discrete nanoparticles at short plating times (1 h). With increasing plating time clustering of gold nanoparticles is observed (5 h) until at long plating times (20–26 h) the CNTs are uniformly coated with a continuous network of gold. This progression of gold plating was also used to confirm the porosity and pore size values experimentally obtained for the gold-plated bucky-papers by He-porosimetry and perm-porometry, respectively. The formation of a continuous gold network at long plating times was also correlated with an enhanced electrical conductivity.

Similarly, FIB was found to be very useful to evaluate the degree of dispersion of poly(styrene) (PS) beads within a bucky-paper structure (Fig. 4.26) and the effectiveness of the beads for tuning the bucky-paper porosity. The PS beads (100 nm and 1 μm) were dispersed into solution with the CNTs and then vacuum filtered in the usual manner to form the composite bucky-paper. The bead size clearly has a strong influence on the final structure. In the case of the larger beads, a layered structure consists of dense regions of CNTs and more porous regions filled with PS beads (Fig. 4.26a, b). In contrast a more uniform structure is obtained for the smaller bead size (Fig. 4.26e, f). Furthermore, FIB milling was used to reveal the cavities that formed after carbonisation of the beads (Fig. 4.26c, d) with the diameter of the cavities closely matching that of the original PS beads. Carbon shells, formed during the carbonisation of the beads, were also visible in the SEM images and explained the unexpected lower gas permeation measured across the carbonised samples compared to the uncarbonised ones [35]. No technique other than FIB milling is available to reveal these carbon shells and so explain the change in permeation.

### *4.5.3 Aging and Crack Formation Revealed by FIB Milling*

Bucky-papers were also used as membranes in a direct contact membrane distillation set-up to desalinate model sodium chloride solutions [9, 14]. In direct contact membrane distillation, the bucky-paper acts as a hydrophobic barrier separating a solution of cold freshwater from a solution of warm saltwater while allowing water vapour to cross through. Although the bucky-paper membranes were shown to perform better than commercial Teflon membranes of similar pore size, which are considered the benchmark in direct contact membrane distillation, their lifespan was found to be limited due to premature mechanical failure. This issue was attributed to progressive ageing of the bucky-paper membranes and to micro-crack formation across the structure. FIB was used to reveal the cracks as well as determine that the unwanted salt transport across the membranes was solely through these cracks and not by wicking within the bucky-paper pores (Fig. 4.27). Electron-dispersive spectroscopy was used to determine

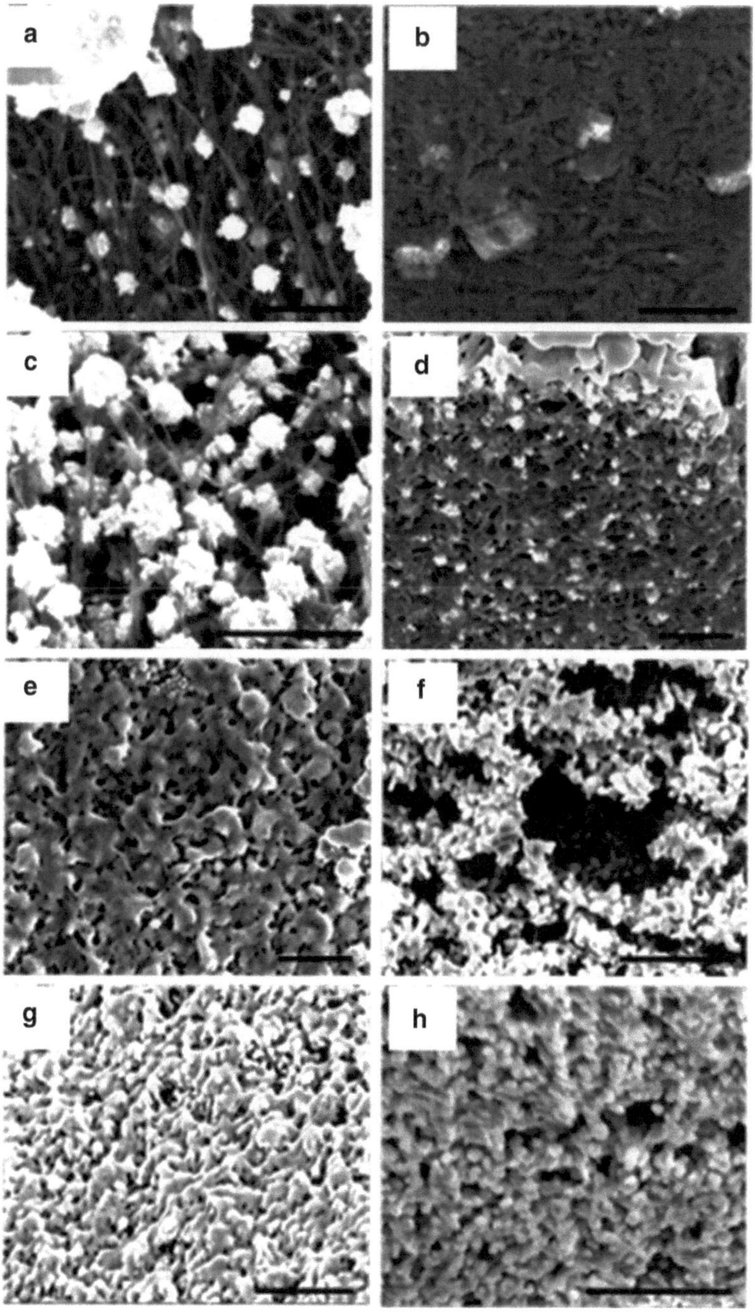


**Fig. 4.25** SEM images of bucky-paper–gold composites formed by electroless deposition of gold within the bucky-paper structure. The plating times are 1 (**a**, **b**), 5 (**c**, **d**), 20 (**e**, **f**) and 26 (**g**, **h**) [8]. The SEM images are of the surface (*left column*) and of cross sections (*right column*) milled with gallium ions in an FIB. All scale bars correspond to 500 nm

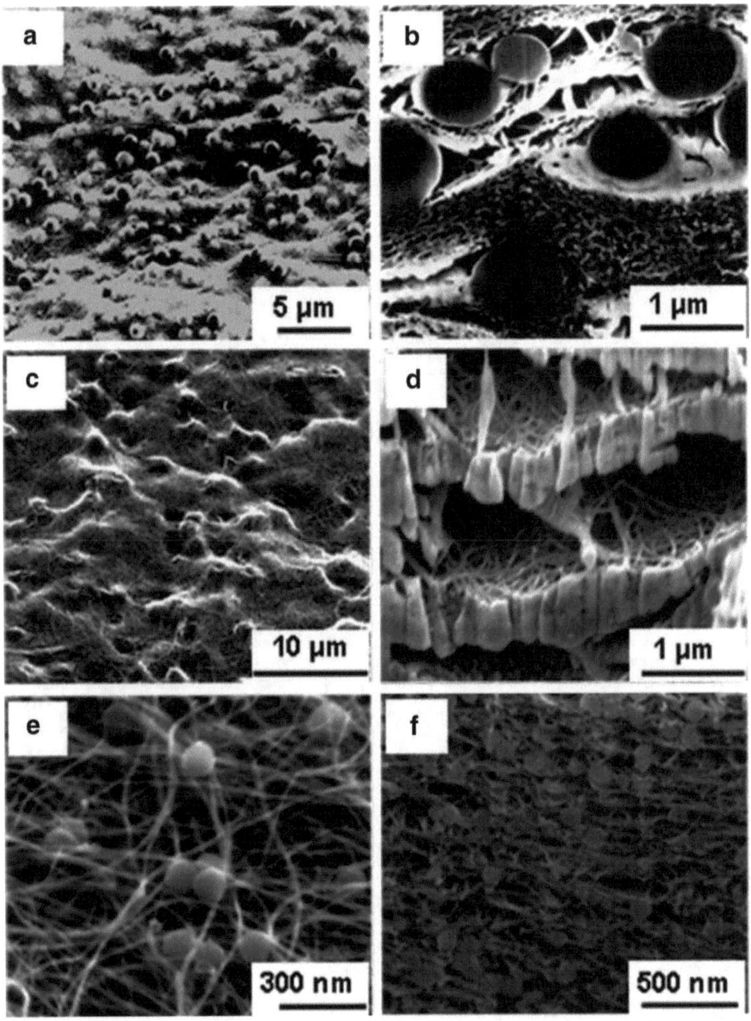


**Fig. 4.26** SEM images of the surface (*left*) and FIB-milled cross sections (*right*) through bucky-papers filled with (**a**, **b**) 1 µm diameter PS beads, (**c**, **d**) 1 µm diameter PS beads which were then carbonised at 150 °C for 2 h under $N_2$ and (**e**, **f**) 100 nm diameter PS beads [35]

the nature of the agglomerates found to form within these cracks, and only sodium and chloride corresponding to the salts used in the model seawater solutions were found, confirming the ageing theory [9, 14, 53]. No similar cracks were found when cross-sectioning unused membranes, also confirming mechanical ageing. The origin of the crack formation was assumed to be related to the presence of undispersed nanotube bundles within the bucky-paper structure.

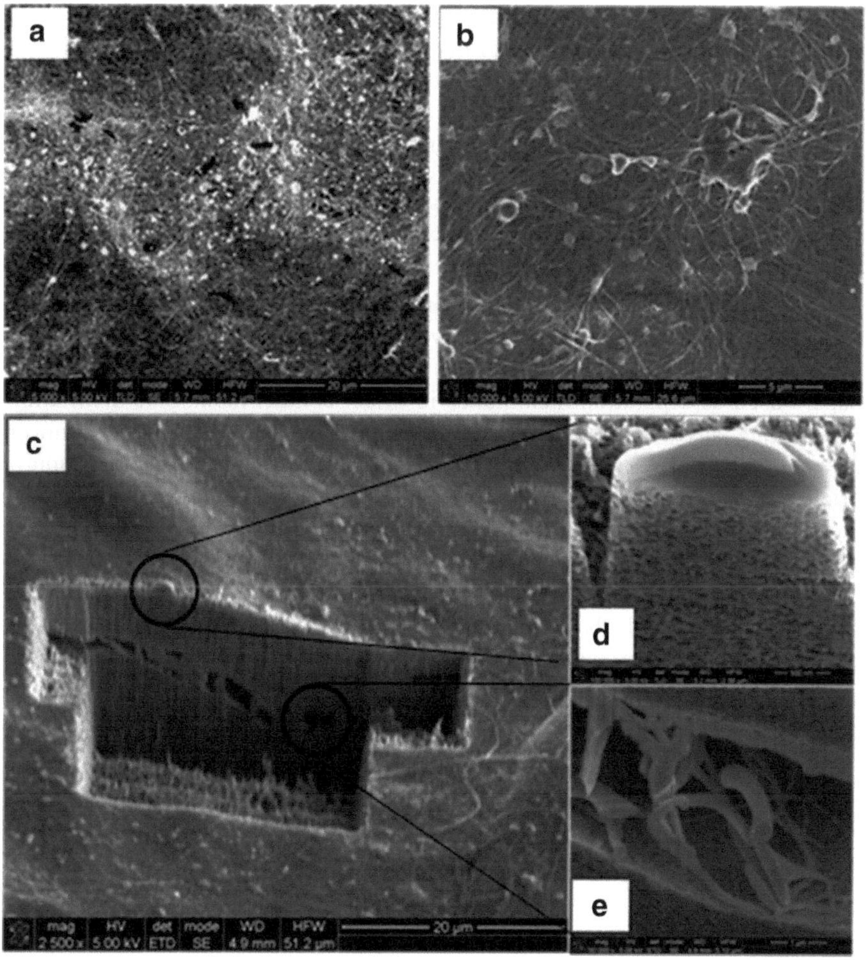

**Fig. 4.27** Surface and cross section analysis of bucky-papers used in direct contact membrane distillation for water desalination; if unwashed after use, NaCl salt crystals were found to be present at the surface (**a**, **b**) and within cracks (**c**) formed over time due to mechanical ageing across the bucky-papers; the two inserts (**d**, **e**) were taken from (**c**) and show more clearly the build-up of NaCl salts within cracks through the bucky-papers. Note that the surface in (**c**) has only undergone "rough" FIB milling, whereas further "polishing" milling was performed for the cross sections shown in (**d**) and (**e**)

## 4.6 Summary

FIB milling followed by SEM has emerged as an effective tool for revealing the internal structure and morphology of CNT yarns and bucky-papers. FIB milled cross sections using gallium ions avoid the severe and permanent deformation typical of the more conventional techniques of microtoming and freeze fracturing, allowing the internal structure to be clearly discerned. However the milling strategy needs to be carefully designed to minimise redeposition of material onto the cross section of interest and avoid crystalline damage associated with the gallium ions.

The properties of CNT bucky-papers and yarns are directly dependent on their internal structure. Hence FIB has been fundamental to gaining a better understanding of their internal structure and how this can be optimised for specific applications. For example, using FIB milling the effect of processing conditions such as twist and densification on the internal nanotube yarn structure has been investigated and correlated with changes in tensile strength. Similarly for bucky-papers, FIB milling has allowed the effect of different CNT types and dispersion conditions on the structure to be monitored. In the case of bucky-papers used as separation membranes in direct contact membrane distillation, FIB milling was able to reveal the formation of macro-cracks after use, confirming that their lifespan was limited by ageing. FIB milling is particularly appropriate when studying composite structures, and readily reveals any changes in the micro-structure due to the various additives.

In summary, FIB milling is a relatively accessible and straightforward method to form cross sections through CNT bucky-papers and yarns and hence investigate their internal structure. No doubt the use and scope of FIB milling to study such structures will broaden in years to come and be equally well applied to other nano-structures.

**Acknowledgements** We wish to acknowledge Dr. Stephen Hawkins and Chi Huynh who developed and grew the spinnable CNT forests used for the majority of the work presented in this chapter and also for fruitful discussions. Thanks also go to Chris Skourtis for expert technical advice regarding the spinning process and bucky-ribbon processing. We would like to thank Dr. Sergey Rubanov, Ken Goldie and Roger Curtain at the Bio21 Institute, University of Melbourne, for their expert advice with FIB milling and SEM.

## References

1. Baughman, R.H., Zakhidov, A.A., de Heer, W.A.: Carbon nanotubes-the route toward applications. Science **297**, 787–792 (2002)
2. Coleman, J.N., Khan, U., Blau, W.J., Gun'ko, Y.K.: Small but strong: a review of the mechanical properties of carbon nanotube-polymer composites. Carbon **44**, 1624–1652 (2006)
3. Dresselhaus, M.S., Dresselhaus, G., Avouris, P. (eds.): Carbon Nanotubes: Synthesis, Structure, Properties, and applications. Springer (2001)
4. O'Connell, M.J. (ed.): Carbon Nanotubes: Properties and Applications. Taylor & Francis Group (2006)
5. Yu, M.-F., Files, B.S., Arepalli, S., Ruoff, R.S.: Tensile loading of ropes of single wall carbon nanotubes and their mechanical properties. Phys. Rev. Lett. **84**(24), 5552–5555 (2000)
6. Zhao, H., Zhang, Y., Bradford, P.D., Zhou, Q., Jia, Q., Yuan, F.-G., Zhu, Y.: Carbon nanotube yarn strain sensors. Nanotechnology **21**, 305502 (2010)
7. Rein, M.D., Breuer, O., Wagner, H.D.: Sensors and sensitivity: carbon nanotube buckypaper films as strain sensing devices. Compos. Sci. Technol. **71**(3), 373–381 (2011). doi:http://dx.doi.org/10.1016/j.compscitech.2010.12.008

8. Dumee, L., Hill, M.R., Duke, M., Velleman, L., Sears, K., Schütz, J., Finn, N., Gray, S.: Activation of gold decorated carbon nanotube hybrids for targeted gas adsorption and enhanced catalytic oxidation. J. Mater. Chem. **22**(18), 9374–9378 (2012). doi:10.1039/c2jm16458b
9. Dumee, L.F., Sears, K., Schütz, J., Finn, N., Huynh, C., Hawkins, S., Duke, M., Gray, S.: Characterization and evaluation of carbon nanotube bucky-paper membranes for direct contact membrane distillation. J. Membr. Sci. **351**(1–2), 36–43 (2010). doi:10.1016/j.memsci.2010.01.025
10. Sears, K., Dumee, L., Schütz, J., She, M., Huynh, C., Hawkins, S., Duke, M., Gray, S.: Recent developments in carbon nanotube membranes for water purification and gas separation. Materials **3**(1), 127–149 (2010). doi:10.3390/ma3010127
11. Sweetman, L.J., Nghiem, L., Chironi, I., Triani, G., in het Panhuis, M., Ralph, S.F.: Synthesis, properties and water permeability of SWNT buckypapers. J. Mater. Chem. **22**(27), 13800–13810 (2012)
12. Foroughi, J., Spinks, G.M., Wallace, G.G., Oh, J., Kozlov, M.E., Fang, S., Mirfakhrai, T., Madden, J.D.W., Shin, M.K., Kim, S.J., Baughman, R.H.: Torsional carbon nanotube artificial muscles. Science **334**(6055), 494–497 (2011)
13. Gou, J.: Single-walled nanotube bucky paper and nanocomposite. Polym. Int. **55**(11), 1283–1288 (2006). doi:10.1002/pi.2079
14. Dumee, L., Sears, K., Schütz, J., Finn, N., Duke, M., Gray, S.: Carbon nanotube based composite membranes for water desalination by membrane distillation. Desalination Water Treat. **17**(1–3), 72–79 (2010). doi:10.5004/dwt.2010.1701
15. Lima, M.D., Fang, S., Lepró, X., Lewis, C., Ovalle-Robles, R., Carretero-González, J., Castillo-Martínez, E., Kozlov, M.E., Oh, J., Rawat, N., Haines, C.S., Haque, M.H., Aare, V., Stoughton, S., Zakhidov, A.A., Baughman, R.H.: Biscrolling nanotube sheets and functional guests into yarns. Science **331**, 51–55 (2011)
16. Edwards, S.L., Church, J.S., Werkmeister, J.A., Ramshaw, J.A.M.: Tubular micro-scale multiwalled carbon nanotube-based scaffolds for tissue engineering. Biomaterials **30**, 1725–1731 (2009)
17. Ericson, L.M., Fan, H., Peng, H., Davis, V.A., Zhou, W., Sulpizio, J., Wang, Y., Booker, R., Vavro, J., Guthy, C., Parra-Vasquez, A.N.G., Kim, M.J., Ramesh, S., Saini, R.K., Kittrell, C., Lavin, G., Schmidt, H., Adams, W.W., Billups, W.E., Pasquali, M., Hwang, W.-F., Hauge, R.H., Fischer, J.E., Smalley, R.E.: Macroscopic, neat, single-walled carbon nanotube fibers. Science **305**, 1447–1450 (2004)
18. Behabtu, N., Young, C.C., Tsentalovich, D.E., Kleinerman, O., Wang, X., Ma, A.W.K., Bengio, E.A., Waarbeek, R.F., de Jong, J.J., Hoogerwerf, R.E., Fairchild, S.B., Ferguson, J.B., Maruyama, B., Kono, J., Talmon, Y., Cohen, Y., Otto, M.J., Pasquali, M.: Strong, light, multifunctional fibers of carbon nanotubes with ultrahigh conductivity. Science **339**, 182–186 (2013)
19. Vigolo, B., Pénicaud, A., Coulon, C., Saunder, C., Pailler, R., Journet, C., Bernier, P., Poulin, P.: Macroscopic fibers and ribbons of oriented carbon nanotubes. Science **290**, 1331–1334 (2000)
20. Atkinson, K.R., Skourtis, C., Hutton, S.R.: Properties and applications of dry-spun carbon nanotube yarns. Adv. Sci. Technol. **60**, 11–20 (2008)
21. Huynh, C.P., Hawkins, S.C.: Understanding the synthesis of directly spinnable carbon nanotube forests. Carbon **48**, 1105–1115 (2010)
22. Jiang, J., Li, Q., Fan, S.: Spinning continuous carbon nanotube yarns. Nature **419**, 801 (2002)
23. Zhang, M., Atkinson, K.R., Baughman, R.H.: Multifunctional carbon nanotube yarns by downsizing an ancient technology. Science **306**(5700), 1358–1361 (2004)
24. Zhang, X., Li, Q., Holesinger, T.G., Arendt, P.N., Huang, J., Kirven, P.D., Clapp, T.G., DePaula, R.F., Liao, X., Zhao, Y., Zheng, L., Peterson, D.E., Zhu, Y.: Ultrastrong, stiff, and lightweight carbon-nanotube fibers. Adv. Mater. **19**(23), 4198–4201 (2007)
25. Kuznetsov, A.A., Fonseca, A.F., Baughman, R.H., Zakhidov, A.A.: Structural model for dry-drawing of sheets and yarns from carbon nanotube forests. ACS Nano **5**(2), 985–993 (2011)

26. Koziol, K., Vilatela, J., Moisala, A., Motta, M., Cunniff, P., Sennett, M., Windle, A.: High-performance carbon nanotube fiber. Science **318**, 1892–1895 (2009)
27. Li, Y.-L., Kinloch, I.A., Windle, A.H.: Direct spinning of carbon nanotube fibers from chemical vapor deposition synthesis. Science **304**, 276–278 (2004)
28. Motta, M., Li, Y.-L., Kinloch, I., Windle, A.: Mechanical properties of continuously spun fibers of carbon nanotubes. Nano Lett. **5**(8), 1529–1533 (2005)
29. Liu, K., Sun, Y., Zhou, R., Zhu, H., Wang, J., Liu, L., Fan, S., Jiang, K.: Carbon nanotube yarns with high tensile strength made by a twisting and shrinking method. Nanotechnology **21**, 045708 (2010)
30. Miao, M., McDonnell, J., Vuckovic, L., Hawkins, S.C.: Poisson's ratio and porosity of carbon nanotube dry-spun yarns. Carbon **48**, 2802–2811 (2010)
31. Min, J., Cai, J.Y., Sridhar, M., Easton, C.D., Gengenbach, T.R., McDonnell, J., Humphries, W., Lucas, S.: High performance carbon nanotube spun yarns from a crosslinked network. Carbon **52**, 520–527 (2013)
32. Tran, C.D., Humphries, W., Smith, S.M., Huynh, C., Lucas, S.: Improving the tensile strength of carbon nanotube spun yarns using a modified spinning process. Carbon **47**(11), 2662–2670 (2009)
33. Tran, C.D., Lucas, S., Phillips, D.G., Randeniya, L.K., Baughman, R.H., Tran-Cong, T.: Manufacturing polymer/carbon nanotube composite using a novel direct process. Nanotechnology **22**, 145302 (2011)
34. Zaeri, M.M., Ziaei-Rad, S., Vahedi, A., Karimzadeh, F.: Mechanical modelling of carbon nanomaterials from nanotubes to buckypaper. Carbon **48**(13), 3916–3930 (2010). doi:http://dx.doi.org/10.1016/j.carbon.2010.06.059
35. Dumee, L., Sears, K., Schütz, J., Finn, N., Duke, M., Gray, S.: A preliminary study on the effect of macro cavities formation on properties of carbon nanotube bucky-paper composites. Materials **4**(3), 553–561 (2011). doi:10.3390/ma4030553
36. AIST: Development of materials with large specific surface areas by using single-walled carbon nanotubes. http://www.aist.go.jp/aist_e/latest_research/2010/20100212/20100212.html (2010)
37. Roy, S., Bajpai, R., Jena, A.K., Kumar, P., Kulshrestha, N., Misra, D.S.: Plasma modified flexible bucky paper as an efficient counter electrode in dye sensitized solar cells. Energy Environ. Sci. **5**(5), 7001–7006 (2012)
38. Dumee, L.F., Sears, K., Marmiroli, B., Amenitsch, H., Duan, X., Lamb, R., Buso, D., Huynh, C., Hawkins, S., Kentish, S., Duke, M., Gray, S., Innocenzi, P., Hill, A.J., Falcaro, P.: A high volume and low damage route to hydroxyl functionalization of carbon nanotubes using hard X-ray lithography. Carbon **51**, 430–434 (2013). doi:10.1016/j.carbon.2012.08.065
39. Vilatela, J.J., Deng, L., Kinloch, I.A., Young, R.J., Windle, A.H.: Structure of and stress transfer in fibres spun from carbon nanotubes produced by chemical vapour deposition. Carbon **49**, 4149–4158 (2011)
40. Sears, K., Skourtis, C., Atkinson, K., Finn, N., Humphries, W.: Focused ion beam milling of carbon nanotube yarns to study the relationship between structure and strength. Carbon **48**, 4450–4456 (2010)
41. Zhang, M., Fang, S., Zakhidov, A.A., Lee, S.B., Aliev, A.E., Williams, C.D., Atkinson, K.R., Baughman, R.H.: Strong, transparent, multifunctional, carbon nanotube sheets. Science **309**, 1215–1219 (2005)
42. Lepró, X., Ovalle-Robles, R., Lima, M.D., Elías, A.L., Terrones, M., Baughman, R.H.: Catalytic twist-spun yarns of nitrogen-doped carbon nanotubes. Adv. Funct. Mater. **22**, 1069–1075 (2012)
43. Ham, H.T., Choi, Y.S., Chung, I.J.: An explanation of dispersion states of single-walled carbon nanotubes in solvents and aqueous surfactant solutions using solubility parameters. J. Colloid Interface Sci. **286**(1), 216–223 (2005). doi:http://dx.doi.org/10.1016/j.jcis.2005.01.002
44. Dumée, L., Sears, K., Schütz, J., Finn, N., Duke, M., Gray, S.: Influence of the sonication temperature on the debundling kinetics of carbon nanotubes in propan-2-ol. Nanomaterials **3**(1), 70–85 (2013)

45. Liu, Y., Yu, L., Zhang, S., Yuan, J., Shi, L., Zheng, L.: Dispersion of multiwalled carbon nanotubes by ionic liquid-type Gemini imidazolium surfactants in aqueous solution. Colloids Surf. A Physicochem. Eng. Asp. **359**(1–3), 66–70 (2010). doi:http://dx.doi.org/10.1016/j.colsurfa.2010.01.065
46. Vaisman, L., Wagner, H.D., Marom, G.: The role of surfactants in dispersion of carbon nanotubes. Adv. Colloid Interface Sci. **128–130**, 37–46 (2006). doi:http://dx.doi.org/10.1016/j.cis.2006.11.007
47. Michelis, P., Vlachopoulos, J.: Complete CNT disentanglement–dispersion–functionalisation in a pulsating micro-structured reactor. Chem. Eng. Sci. **90**, 10–16 (2013). doi:http://dx.doi.org/10.1016/j.ces.2012.12.003
48. Wang, D., Song, P., Liu, C., Wu, W., Fan, S.: Highly oriented carbon nanotube papers made of aligned carbon nanotubes. Nanotechnology **19**, 075609 (2008)
49. Musameh, M., Notivoli, M.R., Hickey, M., Kyratzis, I.L., Gao, Y., Huynh, C., Hawkins, S.C.: Carbon nanotube webs: a novel material for sensor applications. Adv. Mater. **23**(7), 906–910 (2011). doi:10.1002/adma.201003836
50. Gierak, J.: Focused ion beam technology and ultimate applications. Semicond. Sci. Technol. **24**, 043001 (2009)
51. Miao, M.: Production, structure and properties of twistless carbon nanotube yarns with a high density sheath. Carbon **50**, 4973–4983 (2012)
52. Miao, M., Hawkins, S.C., Cai, J.Y., Gengenbach, T.R., Knott, R., Huynh, C.P.: Effect of gamma-irradiation on the mechanical properties of carbon nanotube yarns. Carbon **49**, 4940–4947 (2011)
53. Dumee, L., Germain, V., Sears, K., Schütz, J., Finn, N., Duke, M., Cerneaux, S., Cornu, D., Gray, S.: Enhanced durability and hydrophobicity of carbon nanotube bucky paper membranes in membrane distillation. J. Membr. Sci. **376**(1–2), 241–246 (2011). doi:10.1016/j.memsci.2011.04.024

# Chapter 5
# Nanoscale Electrical Contacts Grown by Focused Ion Beam (FIB)-Induced Deposition

**J.M. De Teresa, R. Córdoba, A. Fernández-Pacheco, S. Sangiao, and M.R. Ibarra**

**Abstract** A detailed description of the use of the focused ion beam (FIB) to grow electrical contacts is presented. The combination of FIB with precursor compounds, which are delivered on the area of interest by means of gas-injection systems, allows the growth of electrical contacts with nanometric resolution on targeted places. The technique has been coined focused ion beam-induced deposition (FIBID). Pros and cons with respect to other existing techniques for contacting are discussed. The FIBID contacts reported in this chapter are based on the use of Pt and W precursors, which result in the growth of deposits with resistivities down to 100 μΩ cm without any post-treatment. A comparison of FIBID with focused electron beam-induced deposition, the sister technique that uses focused electrons instead of ions, is also presented. The steps to follow in order to be successful in the contacting process by means of FIBID are described. Examples of the contacting to individual nanowires and nanoparticles carried out in our laboratory are shown, together with the corresponding four-probe transport measurements. Below 5 K, W deposits are superconduct-

J.M. De Teresa (✉)
Departamento de Física de la Materia Condensada, Instituto de Ciencia de Materiales de Aragón (ICMA), Universidad de Zaragoza-CSIC, 50009 Zaragoza, Spain

Laboratorio de Microscopías Avanzadas (LMA), Instituto de Nanociencia de Aragón (INA), Universidad de Zaragoza, 50018, Zaragoza, Spain
e-mail: deteresa@unizar.es

R. Córdoba • S. Sangiao • M.R. Ibarra
Laboratorio de Microscopías Avanzadas (LMA), Instituto de Nanociencia de Aragón (INA), Universidad de Zaragoza, 50018 Zaragoza, Spain
e-mail: rocorcas@unizar.es; soraya.sangiao@gmail.com; ibarra@unizar.es

A. Fernández-Pacheco
Cavendish Laboratory, Department of Physics, University of Cambridge, Cambridge, UK
e-mail: af457@cam.ac.uk

Z.M. Wang (ed.), *FIB Nanostructures*, Lecture Notes in Nanoscale Science and Technology 20, DOI 10.1007/978-3-319-02874-3_5,

ing and can be therefore used for zero-resistance lead contacts, superconductor-based nanocontacts and probing of proximity effects, opening new perspectives as described here.

## 5.1 Introduction

Nanotechnology involves small dimensions, often calling for new tools and approaches. Whereas well-established techniques exist for metal contacting in micron-scale technology, the situation is different in the case of nanoscale technology. Thus, nanostructures such as nanowires, nanoparticles, carbon nanotubes, graphene and nanoconstrictions challenge the existing techniques for electrical contacting. Electron beam lithography (EBL) is one of the techniques available to perform electrical contacting to nanostructures [1, 2]. In EBL, several steps are involved such as resist spinning, electron beam exposure, resist development, metal evaporation and lift-off process. Some of these steps can damage the nanostructure to contact [3] or simply change its physical properties [4, 5]. Moreover, resist residues on the sample can be detrimental for subsequent experiments, such as those using scanning probe microscopes. On the other hand, one cannot forget that EBL is a multi-step time-consuming technique implying very tedious processing. Therefore, alternative techniques to produce electrical contacting at the nanoscale can be very useful, either solving the existing problems or just accelerating the processing. One of these alternative techniques is focused ion beam-induced deposition (FIBID), described in Sect. 5.1.1.

### *5.1.1 Description of the FIBID Technique*

FIBID and its sister technique focused electron beam-induced deposition (FEBID) are both single-step lithography techniques based on the dissociation of some precursor molecules by a focused ion beam (FIB) or a scanning electron microscope (SEM), respectively. These techniques can be considered as focused beam-assisted chemical vapour deposition (CVD) processes (Fig. 5.1). In this case, the energy required to break the precursor molecules is not provided thermally, as in CVD, but by the electron or the ion irradiation, whilst the substrate is maintained at room temperature. The precursor molecules are delivered onto the substrate surface by means of a nearby gas-injection system, and the focused beam is scanned on the substrate, creating a deposit with the same shape as the beam scan. Secondary electrons produced by the ion or the electron beam are thought to be the main responsible agent for the precursor molecule dissociation. Some review articles are recommended to the reader in order to go deeper in the FIBID technique [6–8].

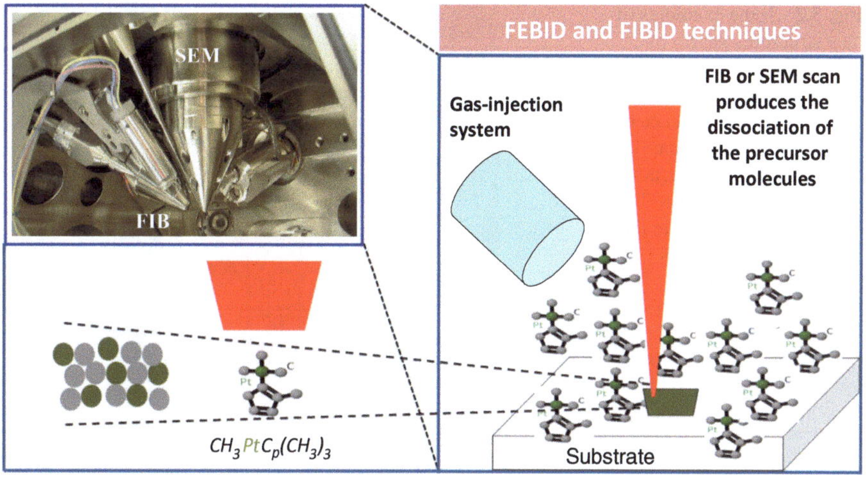


**Fig. 5.1** Sketch of focused electron beam-induced deposition (FEBID) and focused ion beam-induced deposition (FIBID). These are, respectively, electron- and ion-assisted chemical growth techniques giving rise to high-resolution patterning in a single step. The precursor molecules are locally dissociated by a scanning electron microscope (SEM) or a focused ion beam (FIB). The picture shows the interior part of a dual-beam equipment, with an SEM and an FIB placed at 52°

Two significant characteristics of FIBID are the flexibility in the shape of the grown nanostructures as well as the achievable resolution. This is illustrated in Fig. 5.2, where different nanostructures grown in our laboratory using the $W(CO)_6$ precursor are shown: nanowires, dots, stars and pillars. Pillars are an example of another advantage of FIBID: the possibility to grow three-dimensional structures [9]. Regarding resolution, sub-100 nm structures are routinely achieved with FIBID. Figure 5.2 provides one example, a 30-nm-wide wire, but even narrower structures might be obtained. Essential features to grow structures of that size and below are the quality of the FIB column, use of low beam currents and good beam focusing on the sample surface. State-of-the-art Ga-based FIB columns (based on liquid ion metal sources) [6, 10] as well as He-based FIB columns (based on gaseous sources) [11] provide, using low currents, the required beam size, profile and stability to obtain deposits with lateral resolution below 50 nm.

### *5.1.2 Deposition Versus Milling During FIBID*

One difficulty of FIBID compared to FEBID is that the deposition is competing with milling. The ions impacting on the sample surface are able to generate the precursor molecule dissociation but also sample milling. Sample milling is caused by the high energy and momentum of the incoming ions, which are able to produce

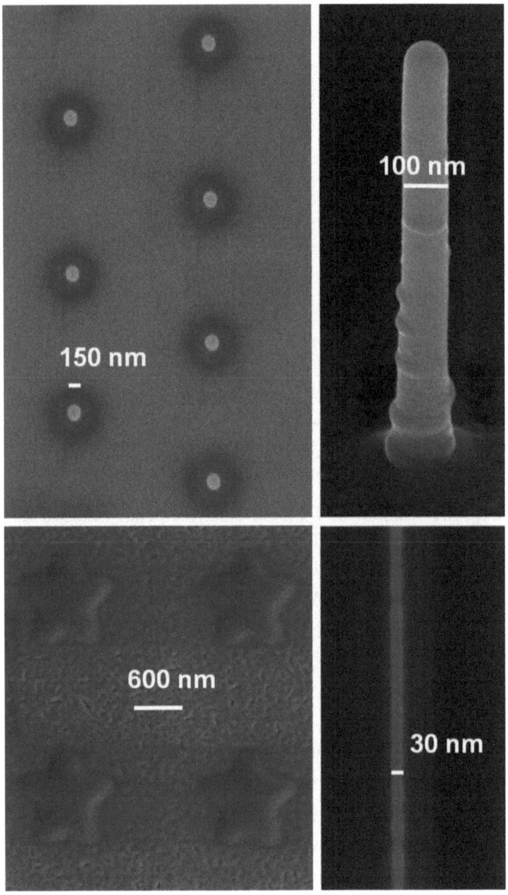


**Fig. 5.2** Some examples of nanostructures grown by FIBID in the Laboratory of Advanced Microscopies (Institute of Nanoscience of Aragón, Zaragoza, Spain) using the $W(CO)_6$ precursor: *dots, pillar, stars, wire*. The achievable resolution in the growth of nanowires is remarkable, 30 nm

surface-atom bond breaking. In order to avoid sample milling, sufficient precursor flux to the area of interest should be provided. In the deposition process, as the precursor molecule dissociation proceeds, new precursor molecules should replace the dissociated molecules. This can be jeopardised if the precursor flux, the molecule residence time or its diffusion on the sample surface are too low or if the beam current is too high. The precursor refreshment is thus crucial to avoid milling. As a consequence, some process parameters must be well controlled. For example, the precursor flux will be preferably high, whereas the ion beam current will be preferably low. The beam dwell time at each addressing point should be low enough to guarantee appropriate precursor replenishment. It can be also useful to increase the refresh time (time that the beam waits between consecutive passes) to allow full precursor replenishment. Moreover, each precursor molecule shows specific cross section for molecule dissociation as well as different stickiness coefficients to the

**Fig. 5.3** Deposition versus milling during FIBID process using the $(CH_3)_3Pt(CpCH_3)$ precursor. If the parameters used during deposition are not appropriate, instead of producing sample growth (*left image*), there is sample milling (*right image*). In this case the beam current was too high, 2.1 nA

substrate surface, which calls for detailed investigations of this effect in each case. Figure 5.3 illustrates this effect, where two sample depositions are compared with different process parameters. The used precursor was $(CH_3)_3Pt(CpCH_3)$. The deposit is a square of side 5 μm. On the left part of the image, one can observe that using the ion beam parameters 20 kV and 81 pA, there is appropriate growth of the sample. However, the right part of the image shows that the use of high beam current, 2.1 nA, and the same voltage produces milling instead of deposition.

### 5.1.3 Halo Effect

FIBID is accompanied by a side effect: the presence of a halo around the main deposit. One should learn to minimise the impact of this halo if its presence is detrimental for a particular application. The origin of this halo is mainly twofold. First, some ions are scattered inside the main deposit to high angles, dissociating precursor molecules up to a few microns away from the point where the FIB is focused. Second, some of the scattering processes induced by the primary ions on the substrate can generate secondary electrons escaping away from the area of the main deposit and contributing to the precursor dissociation out of the main deposit. Together with these two factors, the use of high beam currents (nA range) produces extended rounded beam spots as well as redeposition of material after being milled. The effect is illustrated in Fig. 5.4, where the halo around one deposit grown using the $W(CO)_6$ precursor is analysed. This deposit has width of 2 μm and thickness of 1 μm. It was grown using beam voltage of 30 kV and beam current of 1 nA. One can notice that very close but away from the main deposit, the halo is composed of a continuous film, with thickness about 10 % of the main deposit, evolving towards a more granular and thinner material as the distance from the main deposit increases.

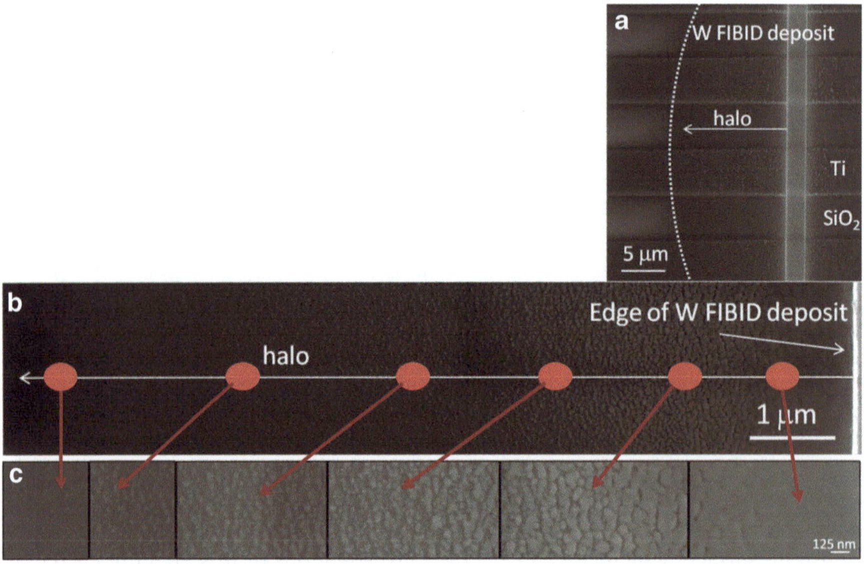


**Fig. 5.4** Illustration of the halo effect which is produced during FIBID deposition. Around the main deposit, an unwanted residual layer is grown, spanning a few microns away. (**a**) SEM image of one deposit using the $W(CO)_6$ precursor, marking with a *dashed line* the halo around the main deposit. (**b**) SEM images of the halo microstructure as one moves away from the main deposit. (**c**) Zoom-in into the microstructure of the halo as one moves away from the main deposit

After about 5 µm, the halo has almost disappeared. The composition of the halo is different from the main deposit as it becomes free from Ga. Moreover, as one moves away from the main deposit, the ratio W/C decreases. The halo thickness usually increases when several deposits are grown in the same area, which can be a serious problem in certain applications. As an example, let us consider those deposits grown to contact a nano-object. If the halo thickness is large and the contacts are close, electrical short circuits can be produced. The use of low beam currents and the reduction of the main deposit dimensions (area and thickness) can be useful to minimise the thickness and extension of the halo. The resistivity of the halo is expected to be higher than that of the main deposit due to the change in composition. The halo resistivity will increase as a function of the distance to the main deposit.

### *5.1.4 Precursors Available for Metallic Deposits*

The available precursors for FEBID and FIBID have been extensively reviewed by Utke et al. [6]. These precursors can be organic compounds, carbonyls, organometallic compounds, etc., and the electrical properties of the corresponding deposits

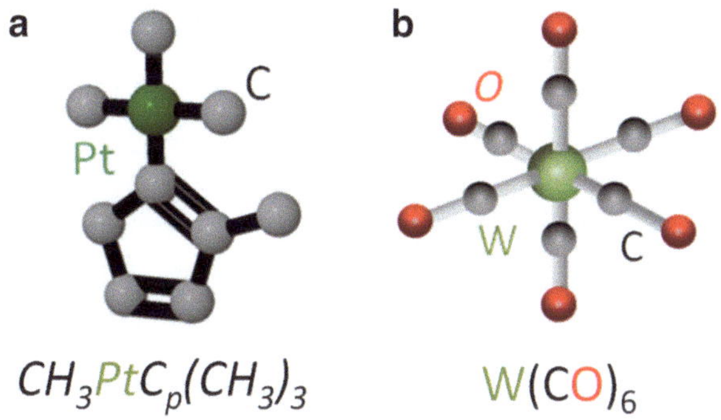


**Fig. 5.5** Precursor molecules that are used in the growth of metal contacts by FIBID reported in the present chapter

can be metallic, semiconducting or insulating. In this section we just focus on the most common precursors in order to create electrical contacts by FIBID, which is the topic of this chapter. We will refer to Pt-based and W-based precursors, which are installed in most of the commercial equipment used for FIBID. For Pt-based contacts we use the methylcyclopentadienyl-trimethyl platinum precursor, with formula $(CH_3)_3Pt(CpCH_3)$. This precursor is a solid compound in ambient conditions. In most of the published work on electrical contacting using Pt precursors, either this precursor or cyclopentadienyl-trimethyl platinum, $(CH_3)_3PtCp$, is used. For W-based contacts, we use the tungsten hexacarbonyl precursor, $W(CO)_6$. This precursor is also a solid compound in ambient conditions. The corresponding molecules are represented in Fig. 5.5.

### *5.1.5 Experimental Details*

The FIBID experiments reported along the next sections have been carried out using a dual-beam equipment (commercialised by FEI Company), which combines a 30 kV SEM and a 30 kV FIB placed inside the process chamber at 52° and having a coincidence point, which will be the working area. The presence of the SEM facilitates the sample inspection before and after the FIBID process. In addition, the SEM also allows the growth of deposits by FEBID. We have used two dual-beam models in this experiment, the Nova Nanolab 200 and the Helios Nanolab 600. In both cases the ion column is the Magnum column, the picture of which is shown in Fig. 5.6. This field-emission column is based on a Ga liquid ion metal source delivering beam currents in the range of 1 pA to 20 nA. The lowest current (1 pA) corresponds to a beam spot of 5–7 nm, which is one of the main limitations in the achievable resolution.

The precursor gas is delivered via an individual gas injector. This warrants the absence of crossed contamination between the different precursor chemistries. The

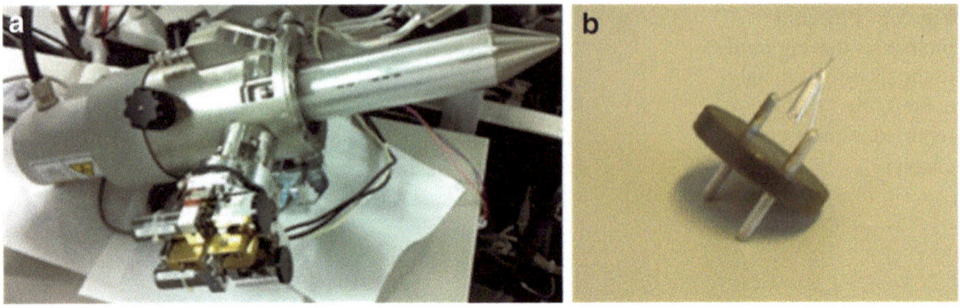


**Fig. 5.6** (**a**) Picture of the Magnum focused ion beam column used for the FIBID experiments reported here. (**b**) Picture of one Ga liquid ion metal source as that used in the experiments reported here

**Fig. 5.7** Picture plus sketch of a typical process performed inside our equipment to simultaneously grow a metallic deposit and investigate the electrical properties of the created device

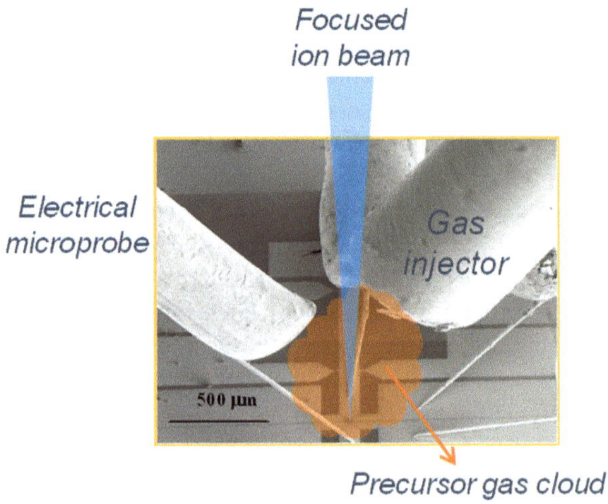


precursor is inside a crucible, and, for the FIBID process, it is heated to increase the vapour pressure, and a valve is open, allowing the flow of vapour from the crucible into the process chamber. In our case, the Pt-based precursor is heated typically to 45 °C and the W-based precursor to 55 °C. As shown in Fig. 5.7, the tip of the gas injector must be placed close to the area of interest, so that the local gas pressure in the patterning area is high. Typically, the tip of the injector is set 50 μm away from the central point in the *x*-direction and 150 μm away from the central point in the *z*-direction.

In the case of applications related to electrical contacts, the circuit resistance can be simultaneously monitored by means of electrical microprobes, as those shown in Fig. 5.7. This allows high control in the electrical properties of the created devices. In our equipment, we use a system provided by Kleindiek Company, with a suitable stage to hold four electrical microprobes.

The composition of deposits grown by FIBID was analysed by energy-dispersive X-ray spectroscopy (EDX). These measurements were performed in a commercial Oxford INCA 200 EDX set-up whose detector is driven in the vicinity of the sample. The selected energy for the microanalysis is 20 kV or below. Prior to each EDX experiment, energy calibration was done by means of a Co sample.

High-resolution transmission electron microscopy (HRTEM) experiments shown here were carried out using a Jeol 2010F equipment operated at 200 kV (point-to-point resolution 0.19 nm) on Pt and W deposits grown on Cu TEM grids with a supporting carbon membrane.

## 5.2 Pt-Based Metallic Deposits by FIBID

### *5.2.1 Introduction*

Pt-based deposits by FEBID and FIBID are commonly used to protect an area of interest that is subsequently sputtered by the ion beam in order to observe its cross-sectional view by SEM or TEM [10]. This is one of the most extended uses of Pt-based deposits but not the only one. When the Pt content in the deposits is high enough, its electrical resistivity is low enough to perform electrical contacts to micro- or nanostructures. This property has been used in the semiconductor industry for circuit repair. The advantages (compared to other techniques such as EBL) are the short processing time, absence of resists and chemical products on the sample and real-time inspection of the process by SEM. On the other hand, an important disadvantage of this technique compared to EBL is that the contact resistance can be very high, jeopardising the electrical measurements and/or the device functionality due to the series resistance or electrical noise introduced. Thus, a generous effort has been devoted to characterise the electrical contacts grown by FIBID and FEBID as well as to develop strategies to enhance the metal content in the deposits in order to decrease the deposit resistance. These include after-deposition annealing [12, 13], deposition on heated substrates [14], injection of a reactive gas during deposition [15], and use of carbon-free precursors [16]. A review article on the deposit purity has been published [17].

In order to contact a nano-object, it is important to take into account its resistivity and how different it is with respect to the material deposited by FIBID and FEBID. As a rule of thumb, if the resistivity of the micro/nano-object is in a range typical for a metal, $\rho < 100\ \mu\Omega$ cm, it will be difficult to have noise-free electrical contacts with these techniques. Thermal effects at the contact can also give rise to spurious results [18]. On the other hand, semimetal and semiconductor micro/nano-objects ($\rho > 1{,}000\ \mu\Omega$ cm) contacted using materials grown by FIBID and/or FEBID will be likely measured with low electrical noise.

One should also consider the different effects produced by the ion and the electron beams on the sample of interest. FIBID tends to sputter away a few nm of the sample simultaneously to the growth of the deposit and can produce Ga implantation and amorphisation on the first layers of the sample, as observed by Tham et al. [19]. This was the reason why these authors observed ohmic contacts between GaAs nanowires and Pt FIBID deposits [20].

There are several examples in literature where semimetal or semiconducting nanowires have been successfully contacted by Pt-based deposits. For example, Hernández-Ramírez et al. succeeded in the contacting of tin oxide ($SnO_2$) nanowires by Pt FEBID and FIBID [21]. These nanowires were applied for gas sensing as portable gas microsensors [22]. Cronin et al. contacted Bi nanowires with a thick oxide shell by first sputtering away the oxide layer by FIB and then growing Pt FIBID contacts [23]. Fabrega et al. contacted $TiO_2$ nanotubes, with application in photoconductive devices [24]. Some examples exist as well on contacting of carbon nanotubes [12, 25].

Although some examples have been published on the use of Pt FEBID and FIBID for contacting metallic wires [26, 27], the results are not systematic and show resistivity values of the wires much higher than the bulk values, casting doubts on the appropriateness of these techniques to contact low-resistivity nano-objects.

### *5.2.2 Thickness Dependence of the Resistivity of Pt Deposits by FIBID*

As noticed by Fernández-Pacheco et al. [28], several non-coincident results have been published regarding the resistivity of Pt deposits grown by FIBID. The deposit resistivity is crucial to understand and control the contacts formed between these deposits and the targeted micro/nano-object. In that reference, it is argued that the Pt precursor is a first important parameter to take into account. In literature, two main precursors have been used: $(CH_3)_3PtCp$ and $(CH_3)_3Pt(CpCH_3)$. Deposits using the first precursor (one carbon less present in the molecule) have the lowest resistivity reported (in some case only seven times higher than that of bulk Pt, 9.7 μΩ cm) [29, 30], whereas when using the second one [31–35], higher resistivity is usually found together with a non-metallic temperature dependence. A second important parameter is the deposit dimensions. Vaz et al. found that the resistivity significantly increased when the dimensions became small [36]. Although these are the most important parameters to consider, other growth parameters cannot be disregarded such as the presence of residual gases in the chamber, beam voltage and current used and precursor flux.

Hereafter, we show that under specific conditions, the deposit thickness is fundamental to determine the deposit resistivity, since the deposit composition varies as a function of thickness. Figure 5.8 shows the measurements of resistivity versus thickness performed in situ during the growth of a microwire, using $(CH_3)_3Pt(CpCH_3)$ as precursor. Those measurements were carried out using electrical microprobes compatible with the deposition process, combined with appropriate lithography designs [28, 37]. One can notice that for low thickness (smaller than ~50 nm) the Pt

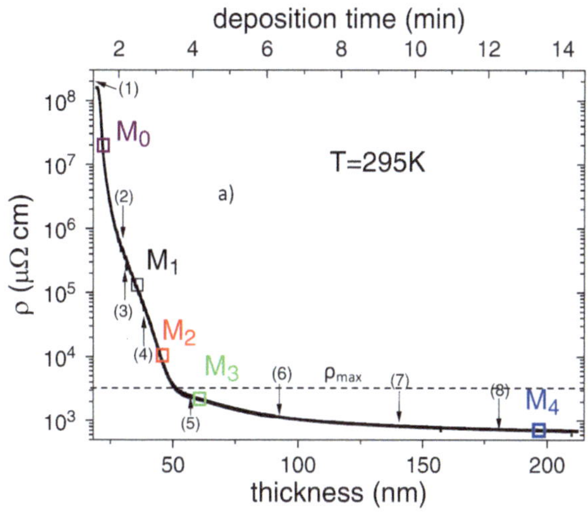


**Fig. 5.8** Resistivity as a function of deposition time and thickness for a typical wire grown by using the $(CH_3)_3Pt(CpCH_3)$ precursor. The resistivity varies by more than four orders of magnitude as a consequence of a change in composition. A thickness of 50 nm marks the transition between a non-metallic conduction ($t < 50$ nm) and metallic one ($t > 50$ nm). The resistivity saturates to a constant value for $t > 150$ nm, $\rho \approx 700$ μΩ cm (about 65 times higher than bulk Pt). The maximum resistivity for metallic conduction in non-crystalline materials ($\rho_{max}$), as calculated by Mott, is *marked*. This *line* separates the insulator NWs from the metallic ones. The numbers in *parentheses* correspond to different probe currents during the measurements: (*1*): 5 nA, (*2*): 10 nA, (*3*): 30 nA, (*4*): 100 nA, (*5*): 1 μA, (*6*): 3 μA, (*7*): 5 μA, (*8*): 10 μA. Reprinted with permission from [28]. Copyright (2009), APS

deposits have resistivity above the theoretical limit of metallic conduction (~3,000 μΩ cm) [38]. Above 50 nm, the resistivity is below this threshold value and eventually saturates at thicknesses above 150 nm to the value of 700 μΩ cm [28]. Current-versus-voltage measurements performed for certain values of thickness (denoted M0, M1, M2, M3, M4, marked in the figure) indicate that samples with thickness below 50 nm (M0, M1, M2) show non-linear features whereas linear behaviour is observed for samples with thickness above 50 nm (M3, M4) [28]. This is an important result, indicating that, under the conditions chosen for this study, Pt FIBID contacts to nano-objects should not have thickness lower than 50 nm to avoid elevated contact resistances and undesired non-metal features (such as non-linear current versus voltage).

The dependence of the Pt deposit resistivity with thickness has been explained by the changes in composition observed as a function of the deposit thickness [28]. This is illustrated in Fig. 5.9. As measured by EDX, the atomic C content in the deposits is about 70 % when the thickness is lower than 50 nm, resulting in a high resistivity. It evolves with thickness, decreasing to about 40 % for sample thickness beyond 200 nm. Simultaneously, the Pt and Ga contents increase with thickness to

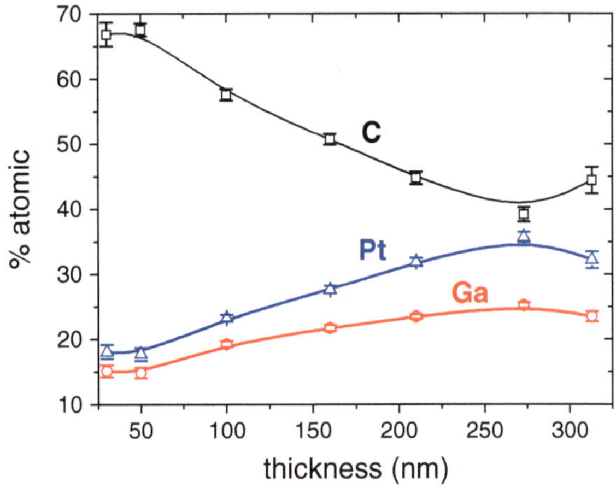


**Fig. 5.9** (**a**). Atomic percentage compositions of Pt–C nanowires as a function of their thickness, determined by EDX. We observe a clear difference in composition for the NWs smaller than 50 nm in comparison with the others, where a higher metallic content is present. Oxygen is always below 1 %. Reprinted with permission from [28]. Copyright (2009), APS

a saturated value of 60 % (35 % of Pt plus 25 % of Ga). These results indicate that in order to obtain a metallic resistivity in FIBID Pt, a high enough Pt plus Ga content is required. When this condition is fulfilled, metallic percolation is met and the material is suitable for good metal contacting to nano-objects. The resistivity of metallic Pt deposits varies little with temperature from 300 K down to 4 K [28] as expected for a dirty metal [38]; hence, they can be used for nanocontacting in a broad temperature range.

### *5.2.3 Comparison Between Pt Deposits by FIBID and FEBID*

It is illustrative to compare the behaviour of Pt deposits grown by FIBID and FEBID. FEBID deposits have the advantage that electrons do not produce milling. However, the metal content of FEBID deposits is in general low, lower than the percolation threshold values [6]. Some exceptions to this rule exist, such as Co FEBID deposits [39], but Pt deposits do not belong to this group. We have grown Pt deposits by FEBID, showing Pt content lower than 18 % at any tested beam energy and current [37]. This has a tremendous impact on the deposit resistivity. As shown in Fig. 5.10, the resistivity of Pt deposits by FEBID (~$10^7$ μΩ cm) is four orders of magnitude higher than that of Pt deposits by FIBID. Also, non-linear current-versus-voltage features are observed at any deposit thickness [37]. These findings indicate that Pt deposits by FEBID can be applied to contact nano-objects only if they show high resistivity (such as semiconductors) in order to avoid the problems previously mentioned (electrical noise, thermal effects, etc.).

In order to study the Pt deposit microstructure and understand its impact on the electrical properties, HRTEM experiments have been carried out. The HRTEM images shown in Fig. 5.11 allow the observation of atomic planes in the deposits. Both FEBID and FIBID Pt deposits consist of ellipsoidal crystalline Pt grains

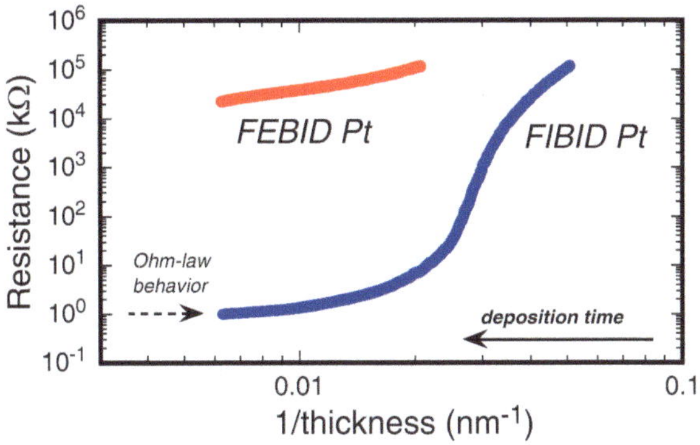


**Fig. 5.10** Representation of the resistance as a function of the inverse of thickness for one selected FEBID Pt nanowire and one selected FIBID Pt nanowire, which illustrates that the constant value expected if the Ohm law applies only occurs for the FIBID Pt deposits above ≈100 nm. Notice that the FEBID Pt deposits display resistivity values four orders of magnitude higher than the FIBID ones. Reprinted from [37]

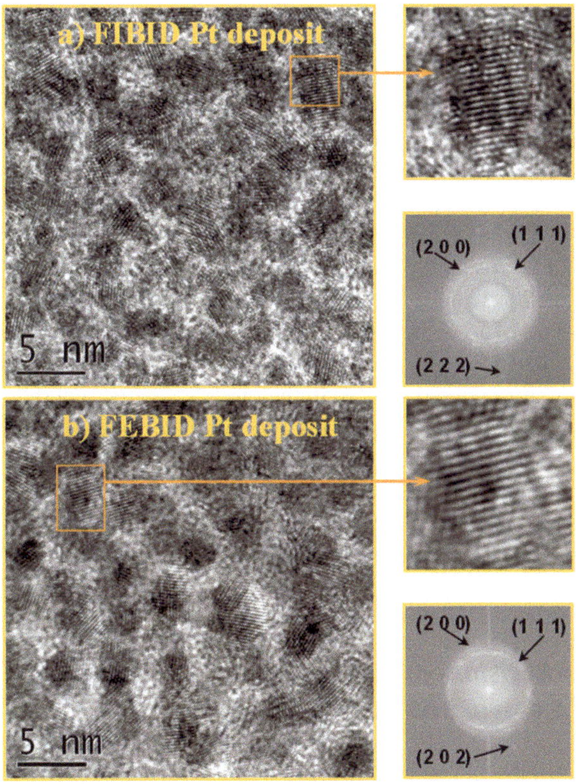


**Fig. 5.11** HRTEM images of a Pt nanodeposit by FIBID (**a**) and FEBID (**b**), in both cases grown at 30 kV beam energy on top of a TEM Cu grid covered with a thin supporting holey carbon membrane. One Pt grain has been selected in each case for magnification and clear observation of the corresponding atomic planes. The fast Fourier transform of the full image gives diffraction spots that correspond to the (200), (111), (222) and (202) atomic planes of fcc Pt. Reprinted from [37]

embedded in an amorphous carbonaceous matrix. The crystalline Pt grains are identified due to the clear observation of Pt atomic planes with the expected planar distances. Fast-Fourier transforms of the images show the presence of diffraction spots corresponding to different atomic planes of fcc Pt, with lattice parameter of 0.392 nm. An average size of Pt crystallites has been obtained from the measurement of about 50 individual grains in each image through the counting of the number of atomic planes. It is found that no matter what beam energy and current are used, in all FEBID and FIBID deposits the average Pt crystallite size is about $3.2 \pm 0.8$ nm. The differences between the transport properties of Pt FEBID and FIBID deposits will thus arise from the different amounts of Pt grains in each case, much more abundant in the case of Pt deposits by FIBID.

## 5.3 W-Based Metallic Deposits by FIBID

### *5.3.1 Introduction*

W wires are relevant in the semiconductor industry for integrated-circuit interconnects [40]. It was soon realised that W-based deposits by FIBID could be useful for circuit edit and mask repair [41]. W-based deposits by FIBID have high metallic content, as measured by EDX and X-ray photoelectron spectroscopy (XPS), with W content about or higher than 40 % [42]. In contrast with Pt FIBID deposits shown in the previous section, the content of the W deposit is quite homogeneous across the thickness [42]. The high metal content along the full thickness explains the low resistivity of the deposits, typically ranging between 100 and 600 μΩ cm [43–45], only 20 times higher than the bulk W resistivity, 5.6 μΩ cm. Post-annealing of these deposits by means of Joule heating was found to be efficient to reduce a wire resistivity from 300 to 55 μΩ cm [46]. In addition to the aforementioned properties, W FIBID deposits show superconductivity below 5 K, as first reported by Sadki et al. [47]. This material can therefore be used to fabricate superconducting contacts to nano-objects. Superconductivity at that temperature likely arises from the amorphous nature of the deposit [42, 48]. As shown in Fig. 5.12, HRTEM measurements do not give evidence for the presence of crystallites in the deposit, suggesting an amorphous structure, in sharp contrast with Pt deposits.

### *5.3.2 Low-Resistance Metallic Deposits by FIBID and Comparison with FEBID*

As mentioned above, the room-temperature resistivity of W FIBID deposits is low, typically ranging between 100 and 600 μΩ cm, and varies little with temperature between 300 and 6 K [44, 45]. This will allow efficient nanocontacting in a broad

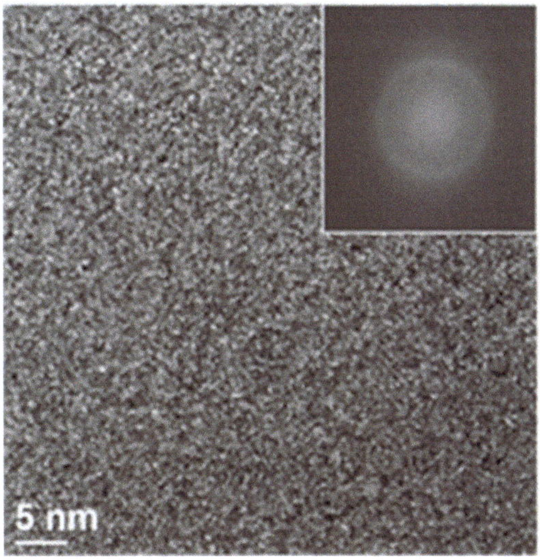


**Fig. 5.12** HRTEM image of a 20-nm-thick W FIBID deposit grown at 30 kV and 10 pA on a Cu TEM grid. The *inset* shows the absence of diffraction rings in the fast Fourier transform of the image

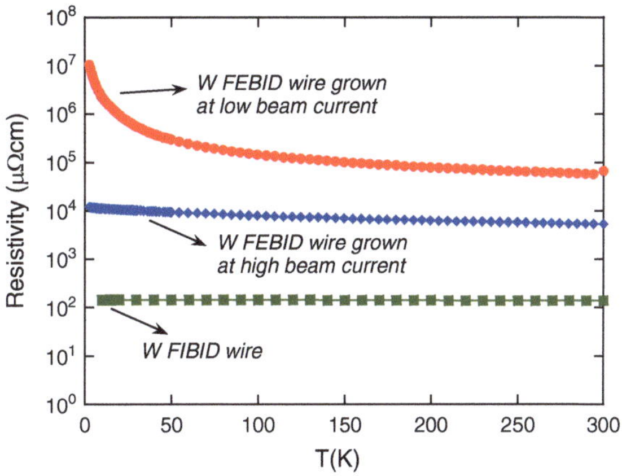


**Fig. 5.13** Temperature dependence of resistivity for two FEBID W wires grown, respectively, under high (6.3 nA) and low (0.4 nA) beam current and one FIBID W wire

temperature range. A comparison between the resistivity (and temperature dependence) of W FIBID and W FEBID deposits is shown in Fig. 5.13. In the case of W FIBID the room-temperature resistivity has a value of around 100 μΩ cm, with little temperature variation down to 10 K. In the case of W FEBID deposits, two samples grown under different electron beam current are shown. The sample grown using a beam current of 6.3 nA presents a W content of 42 %, with room-temperature resistivity of about $0.5 \times 10^4$ μΩ cm, which increases up to $1.2 \times 10^4$ μΩ cm at 3 K. On the contrary, the sample grown using a beam current of 0.4 nA presents a W content of 21 %, with room-temperature resistivity of about $6.5 \times 10^4$ μΩ cm, which becomes $85 \times 10^4$ μΩ cm at 3 K. These results indicate that, under appropriate growth conditions, W FEBID deposits can show resistivities about 10–20 times higher than

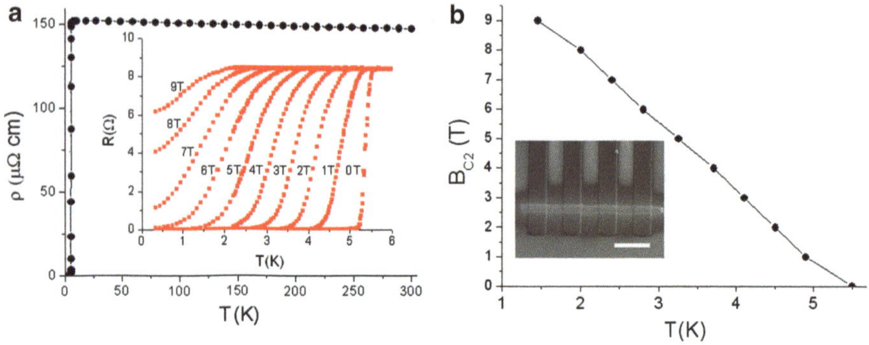


**Fig. 5.14** For a W FIBID microwire, (**a**) resistivity as a function of temperature. The *inset* shows measurements of the resistivity under various applied magnetic fields. (**b**) The critical field, $B_{C2}$ (defined at 90 % of the normal-state resistance, $\rho/\rho_n=0.9$), as a function of temperature. An SEM image with the four aluminium pads used for four-point measurements is shown in the *inset*, where the scale bar is 10 μm. Reprinted from [44]

FIBID ones and with little temperature dependence. Thus, W FEBID deposits can also be suitable for nanocontacting many nano-objects. However, the growth speed is significantly low (~$10^{-5}$ μm³/nC) compared to W FIBID deposits (~$10^{-2}$ μm³/nC) or Pt FEBID deposits (~$10^{-1}$ μm³/nC), which limits their applications.

Similar W FEBID resistivity values and temperature dependence to the ones presented here have been found by Porrati et al. [49]. They also reported an increase of the resistivity when these deposits are exposed to air. The same effect has been observed by us, with changes in the room-temperature resistivity by a factor of 2–10. In addition, they observed that electron-beam post-irradiation of the W FEBID samples makes the resistivity decrease [49].

### 5.3.3 Superconducting Metallic Deposits Below 5 K

Superconductivity is a special collective electronic state in condensed matter physics that allows free-dissipation current flow. The absence of resistance is very attractive for the contacting of nano-objects. Figure 5.14 shows the temperature dependence of resistivity of a W FIBID microwire, indicating a superconducting transition at $T_C \approx 5$ K [44, 47]. The application of an external magnetic field shifts this transition to lower temperatures, as expected.

Figure 5.15 displays the dependence of $T_C$ with the room-temperature resistivity of several micro- and nanowires grown in our laboratory. All the samples show $T_C$ in the range 4.5–5.5 K. Li et al. have found $T_C$ values up to 6.5 K [45]. We can notice in Fig. 5.15b that the superconducting transition is quite sharp, suggesting good sample homogeneity. The critical current that destroys superconductivity in these wires is of the order of $10^5$ A/cm² at 2 K.

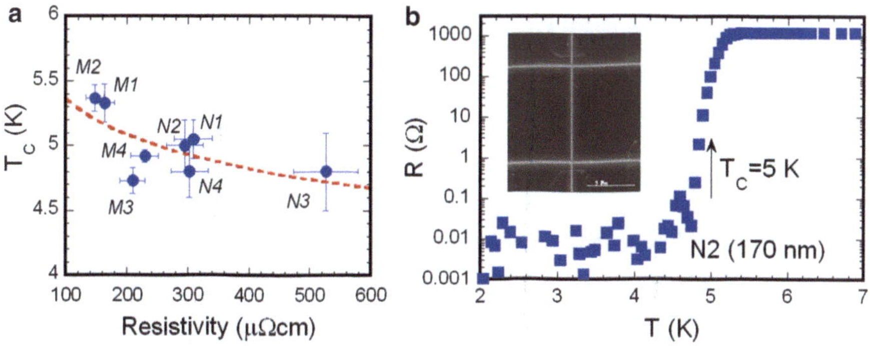


**Fig. 5.15** (**a**) $T_C$ versus resistivity of several microwires (M1, M2, M3, M4) and nanowires (N1, N2, N3, N4) studied in the present work. $T_C$ ranges between 4.7 and 5.4 K, with a slight tendency to decrease with the resistivity. Error bars come from the transition temperature width (*y*-axis) and in the determination of the wire section (*x*-axis). (**b**) Resistance (in log scale) versus temperature of N2 (width = 170 nm), indicating $T_C = 5.0 \pm 0.2$ K. The *inset* is an SEM image of this nanowire (*vertical line* with *white contrast*) and the two horizontal wires for the measurement of voltage during the resistance measurements, where the current flows through the nanowire. Reprinted from [44]

Apart from making contacts to nano-objects, which will be widely discussed in the following sections, W FIBID superconducting deposits have recently found additional applications. For example, basic studies of the behaviour of the superconducting vortex lattice as a function of magnetic field and temperature have been carried out [42, 50, 51]. A novel re-entrance of superconductivity as a function of the applied magnetic field has been observed in narrow wires due to size effects [52]. Also, W FIBID deposits have been used for the edit of superconducting circuits [53] and for the fabrication of three-dimensional superconducting devices [54].

## 5.4 Electrical Contacts to Nanowires and Nanoparticles by FIBID

A few examples exist in literature regarding FIBID contacting of nanowires and nanoparticles, as mentioned before. However, several points have to be addressed to be successful in this aim and avoid spurious results. From our experience, these are the points to take care of:

- The nanostructures should be thick enough because a FIBID process also mills part of them.
- Their properties should not be very sensitive to possible damage produced by the ion beam (amorphisation, implantation, milling, etc.). In any case, the exposure time to the ion beam should be minimised, and the use of SEM is encouraged for inspection.
- They should not be very conductive in order to avoid electrical noise and thermal effects. For those structures, it is more convenient to use EBL.

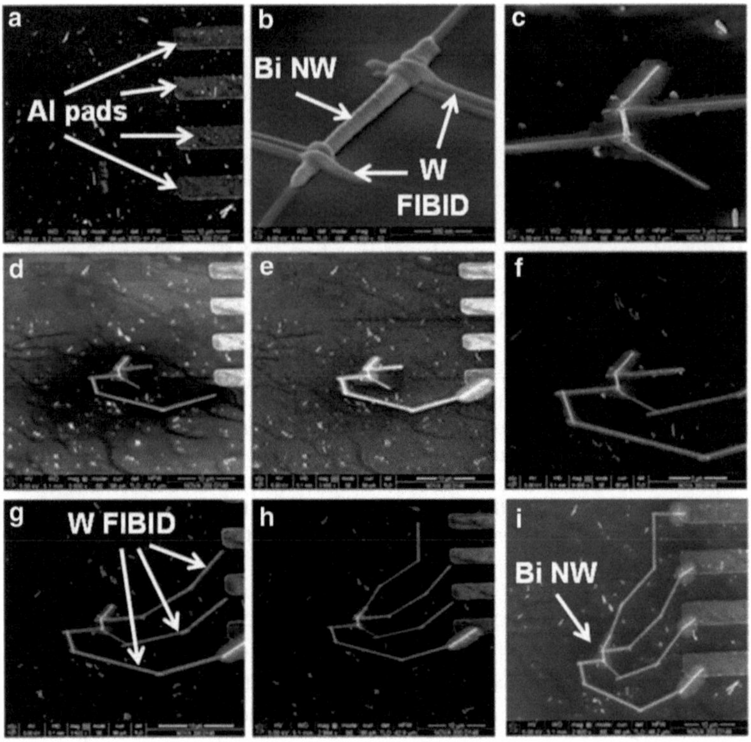


**Fig. 5.16** Different steps followed for the patterning of nanoelectrodes on an individual Bi nanowire. The droplet of the solution containing the nanowires has been placed on the surface of a thermally oxidised Si substrate (**a**) where four external metallic pads had been previously micropatterned by optical lithography. The part of the four nanoelectrodes closest to the nanowire is first grown (**b** and **c**). Finally, the four electrodes are connected to the four external metallic pads by means of W-based nanodeposits grown by FIBID (from (**d**) to (**i**))

- They should be long enough to avoid short circuits through the halo of the deposit. The contacts should be separate enough to avoid this problem or perform some cleaning process of the halo.
- Four-probe measurements are required to get reliable electrical measurements of a nano-object as two-probe measurements suffer from high contact resistance and lead resistance.

Hereafter, we present the successful process used to contact semimetal bismuth nanowires by means of W FIBID deposits [55]. To enable the patterning of nanoelectrodes on individual Bi nanowires they have to be deposited on a substrate. Thus, in the first step a membrane containing the nanowires was dissolved in dichloromethane and a droplet of the resulting solution was placed onto a thermally oxidised Si substrate where four external metallic pads of 5 μm in width had been previously micropatterned by optical lithography (Fig. 5.16a). After selecting a nanowire using SEM, the nanowire was contacted by four nanoelectrodes using FIBID: Two longitudinal electrodes were deposited at both ends of the nanowire to allow current injection, and

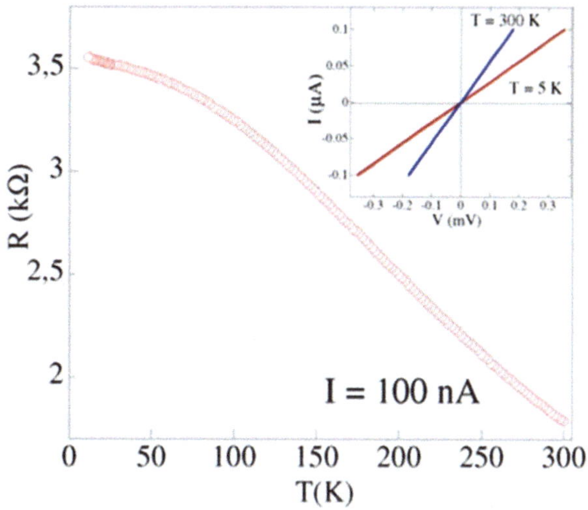


**Fig. 5.17** Temperature dependence of the resistance for the 100-nm-diameter Bi nanowire shown in Fig. 5.16. *Inset*: Current-versus-voltage curves taken at both 5 K and room temperature for this Bi nanowire

other two were deposited perpendicular to the nanowire to detect the voltage drop across it (Fig. 5.16b, c). Additional nanoelectrodes were subsequently grown to make electrical connections with the four external metallic pads (see Fig. 5.16d–i). All nanoelectrodes were made of W FIBID that results in low-resistive contacts.

After the nanocontacting process, the electrical resistance was first measured in situ to monitor any possible change in the sample resistance when exposed to ambient conditions. Afterwards, the sample was measured ex situ as a function of temperature and magnetic field. The current-versus-voltage measurements in the nanowire were found to be linear from 300 K down to 5 K, as expected for ohmic contacts (see the inset of Fig. 5.17). This indicates that the grown contacts allow reliable transport measurements despite a 5-nm-thick oxide shell layer around the wires, as observed in HRTEM images [55]. It is very likely that the FIBID process produces the milling of a few nanometres of the wire, thus eliminating this oxidised surface layer. The resistivity at the highest and lowest temperature measured was, respectively, $1.83 \times 10^{-3}$ μΩ cm at 300 K and $6 \times 10^{-4}$ μΩ cm at 5 K, which are reasonable values for a Bi nanowire. Magnetoresistance measurements below 10 K (not shown here) presented the signature of 1D weak antilocalisation effects in Bi, superimposed to the classical Lorentz magnetoresistance. From the corresponding fits, relevant transport parameters were obtained [55].

The investigation on magnetic micro/nanoparticles is a hot topic due to the large amount of their applications. In some cases, the properties of individual particles can be different from their collective behaviour, meriting additional individual characterisation. For example, the electrical resistivity is sensitive to grain boundaries and can mask the intrinsic electrical behaviour of a single nanoparticle. Also, in the measurement of transport properties, features related to discrete electronic states of nanoparticles can be smeared out when a large number of particles are probed.

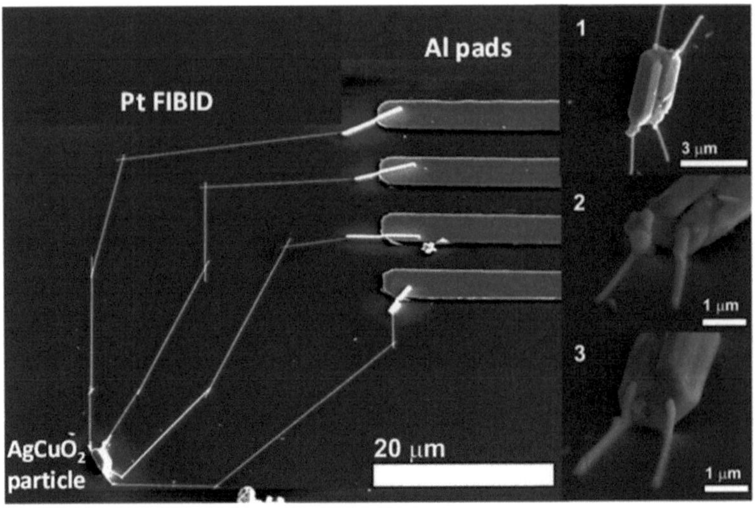


**Fig. 5.18** The *left* panel shows an SEM image of a $AgCuO_2$ particle contacted using Pt FIBID deposits. On the *right* panel, a close-up of the particle from different perspectives is shown. Reprinted with permission from [57]. Copyright (2010), ACS

Therefore, there exists the need to measure the electrical transport properties of individual micro/nanoparticles. The small size and the rounded shape of the particles is an additional difficulty to fabricate electrical contacts in comparison with the case of nanowires. Martínez-Boubeta et al. have reported the successful contacting of a core–shell Fe–MgO nanoparticle 250 nm in diameter by means of FIBID [56]. From the measurements, they were able to extract features related to the tunnelling mechanism governing the conductivity in the nanoparticle.

Here, we describe the experiments performed in our laboratory in order to contact a $AgCuO_2$ particle [57]. In this oxide compound, spectroscopy measurements had revealed the existence of delocalised electrons [58], but no microscopic evidence existed on the proposed metallic conduction. The particle shown in Fig. 5.18 had approximate dimensions of $1 \times 1 \times 3\ \mu m^3$ and did not show any observable damage caused by the ion beam used to make the contacts. The particle seemed to be the result of twinning, as is the case for many of the big particles obtained using a hydrothermal method. Nevertheless, it could be considered a single crystal since both twin crystals were clearly welded and aligned at the bottom face. EDX experiments also showed that the particle contains both Ag and Cu in a 1:1 ratio. Contacting the particles was indeed highly challenging due to their height (1 μm or over) and their low aspect ratio 3:1.

Once the particle was contacted, in situ four-probe current-versus-voltage measurements were performed. To check the quality of the contacts, initial two-probe measurements were made, using alternative couples of nanocontacts. In both cases, a linear dependence between applied current and voltage was obtained. Linear fitting of the lines obtained for the different set of nanocontacts gave comparable resistance values of 250.6 and 235.0 kΩ, respectively, indicating a good quality of

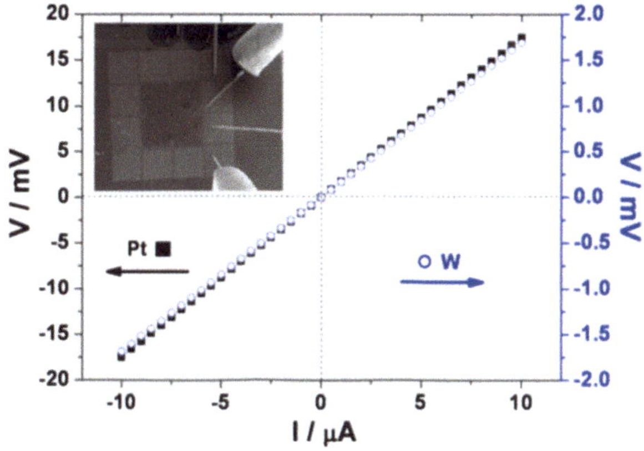


**Fig. 5.19** Current-versus-voltage plots obtained for contacted particles using Pt FIBID and W FIBID contacts. The *inset* shows the four probes contacting the patterned Al microelectrodes in the wafer, to which the metallic Pt FIBID nanowires are connected. The nanocontacted $AgCuO_2$ particle is in this case the one shown in Fig. 5.18. Reprinted with permission from [57]. Copyright (2010), ACS

the nanocontacts on the $AgCuO_2$ particle. Four-probe in situ measurements were then made using the four nanocontacts simultaneously. To avoid any risk of short circuit between the nanocontacts, the two intensity probes were connected to nano-contacts placed in opposite corners of the crystal. The measurements again showed a linear dependence between applied current and voltage, as shown in Fig. 5.19. The values obtained for the electrical resistivity are about 50–300 times lower than in the case of bulk non-sintered low-density pellets, where particle connections are expected to be quite resistive [59]. Thus, the transport results obtained on individual particles are strongly indicative of the metallic conduction inside the particles and thus of the electron delocalisation in the $AgCuO_2$ compound [57].

## 5.5 Superconducting Contacts to Nano-Objects by FIBID

The finding of superconductivity below $T_C = 5$ K in W FIBID deposits opened new perspectives to establish contacts to nano-objects. First, below $T_C$ the W FIBID leads will show zero resistance, avoiding the annoying lead resistance. The value of the contact resistance between a superconductor and a metal is normally low [60], but some physical phenomena occurring at the interface such as Andreev reflections can produce contact resistances that are not negligible [61]. This is why four-probe measurements are still preferred. Secondly, the superconducting contacts can be used to induce superconductivity in the nano-object due to the proximity effect, having access to a rich physics [62]. Thirdly, if the superconductivity is maintained at nanoscale dimensions, it will provide a method to probe spectroscopic information of another nano-object through point-contact spectroscopy [63].

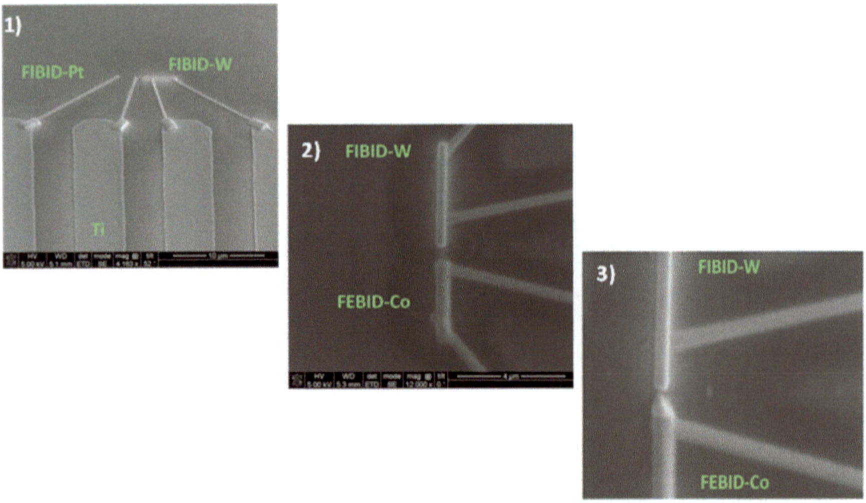


**Fig. 5.20** Fabrication of a controlled nanocontact between a superconducting W FIBID deposit and a ferromagnetic Co FEBID deposit. First, FIBID Pt wires are grown to perform the contact to Ti metallic pads fabricated by optical lithography. In a second step, the W FIBID deposit is grown. The last step is the growth of the Co FEBID deposit. The in situ monitorisation of the electrical resistance allows one to stop the process at the desired nanocontact resistance between the ferromagnet and the superconductor

In the following, we describe some experiments performed in our laboratory in which W FIBID deposits are used to investigate the electrical conduction between nanocontacts of size a few nanometres formed by superconducting W FIBID deposits and ferromagnetic Co FEBID deposits [64, 65]. From those experiments, we were able to extract useful information about the physical properties of the ferromagnet (its spin polarisation) and the superconductor (its superconducting gap). Figure 5.20 shows the process followed to grow the nanocontacts. First, Pt FIBID wires are contacted to Ti pads previously fabricated by means of optical lithography. Then, a W FIBID electrode is grown in the form of a wire of width equal to 500 nm. Afterwards, a Co FEBID wire is grown with a pointed shape so that a nanocontact of a few nanometres is established to the W wire. The size of this nanocontact can be controlled at a certain extent by in situ resistance measurements, using electrical microprobes placed on top of the Ti pads. We found that for our purpose of the study of Andreev reflection phenomena, nanocontact resistance values of a few hundreds of ohms were appropriate, which corresponds to W–Co contacts with sizes smaller than 5 nm.

The samples were studied ex situ as a function of temperature and magnetic field. The current-versus-voltage characteristics were measured at fixed temperatures, from where the differential conductance as a function of voltage, $G(V)=\mathrm{d}I/\mathrm{d}V$, can be extracted. From these results it was possible to obtain quantitative information on the superconductor gap and the ferromagnet spin polarisation. We analysed the differential conductance as a function of the bias voltage using the model proposed by Blonder, Tinkham and Klapwijk (BTK) [66] for a contact between a normal metal

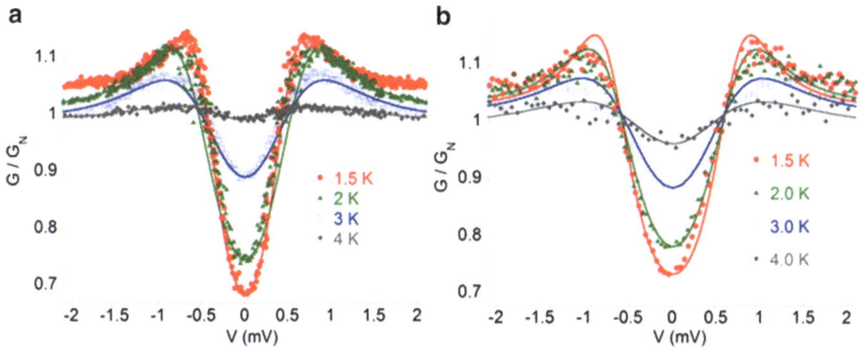


**Fig. 5.21** Differential conductivity measurements of two nanocontacts formed between a ferromagnetic (Co FEBID) and a superconducting (W FIBID) deposit. From fits of these curves, one can extract the ferromagnet spin polarisation and the superconductor superconducting gap. Reprinted with permission from [64]. Copyright (2011), Elsevier

and a superconductor but extended to include the spin polarisation *P* of the metal (reflecting the different spin-up and spin-down population of electrons at the Fermi level). The fittings indicated that the spin polarisation of the cobalt is 35 %, whereas the superconducting gap of the W FIBID wire is found to be 0.7 meV, in good agreement with scanning tunnelling spectroscopy measurements (Fig. 5.21) [42].

In the following, we discuss the application of W FIBID nanostructures to induce superconductivity in nano-objects and to measure their transport properties. Kasumov et al. deposited two suspended W FIBID electrodes and filled in the small gap (few-nanometre size) with metallofullerene molecules, observing proximity effects and multiple Andreev reflections [67]. HRTEM imaging previous to the molecule deposition allowed them to establish the lack of connection between the W electrodes. Another subsequent work used a different approach, where W FIBID wires were grown on the top of a nano-object [68–70]. This is an unreliable process because there is no way to verify the lack of electrical contact between the W FIBID wires, which present halo effects as we discussed in Sect. 5.1.3. This halo can extend up to several microns away from the main deposit, and, from our experience, deposition on top of nano-objects makes the halo bigger. Thus, control experiments where the W FIBID wires are grown at similar distances on flat substrates are not sufficient to guarantee the absence of electrical contact between the W FIBID wires if they are grown on top of the nano-objects. We propose a different approach, which is certainly more reliable, sketched in Fig. 5.22. In this approach, four W FIBID wires are first grown on a flat substrate. The electrical connection between these wires is then probed. If it results that the resistance between the contacts is finite, some cleaning of the halo has to be undertaken to guarantee the absence of connection. This can be done via oxygen plasma treatments or FIB cleaning. In the last step, the nano-object is placed on top of the W FIBID wires, which guarantees that the measured resistance comes exclusively from the nano-object. Following this approach, we have been able to reliably probe proximity effects on nano-objects with distance between the central W FIBID wires as short as 100 nm [71].

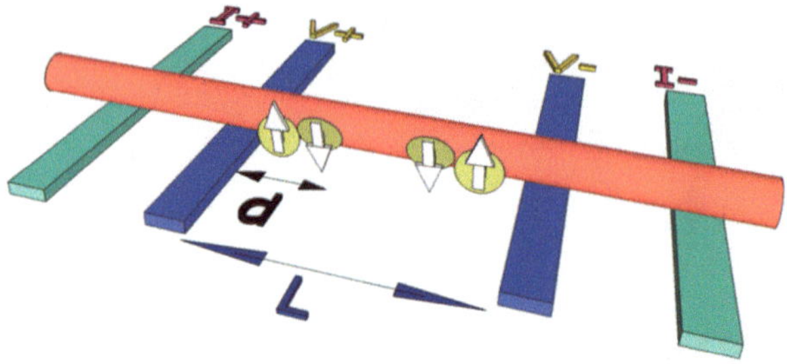


**Fig. 5.22** Sketch of the configuration appropriate to induce superconductivity by means of W FIBID deposits on a nano-object via the proximity effect, which is especially relevant when the superconducting carriers penetrate in the nano-object some distance (*d*) comparable to the distance between the central W FIBID wires (*L*). Moreover, this configuration allows the measurement of the transport properties across the nano-object

## 5.6 Summary

In this chapter, the FIBID process for nanoscale electrical contacting based on the use of FIB and precursor compounds has been described. Pt-based and W-based deposits with low resistance can be produced by FIBID and have been applied for the contacting of nano-objects such as nanowires and nanoparticles. This allows probing individual objects at the nanoscale, providing information that can be substantially different to that obtained when a collective group of nano-objects is probed. The range of applications of FIBID has been discussed, and comparisons with other techniques such as FEBID and EBL have been established. In addition, superconducting W FIBID contacts have been described and their applications discussed. FIBID is thus a powerful technique which can be used in a wide variety of applications in nanotechnology.

**Acknowledgements** We warmly acknowledge our close collaborators in the work presented in this chapter: J. Arbiol, L. Casado, I. Guillamón, N. Marcano, L. Morellón, D. Muñoz-Rojas, L. Pérez, M. Plaza, J. Sesé, H. Suderow and S. Vieira. Financial support by the Spanish Ministry of Economy (through project MAT2011-27553-C02, including FEDER funding) and the Aragón Regional Government is acknowledged.

## References

1. Xia, Y., Rogers, J.A., Paul, K.E., Whitesides, G.M.: Unconventional methods for fabricating and patterning nanostructures. Chem. Rev. **99**, 1823 (1999)
2. Martín, J.I., Nogués, J., Liu, K., Vicent, J.L., Schuller, I.K.: Ordered magnetic structures: fabrication and properties. J. Magn. Magn. Mater. **256**, 449 (2003)
3. Fan, J., Michalik, J., Casado, L., Roddaro, S., Ibarra, M.R., De Teresa, J.M.: Investigation of the influence on graphene by using electron-beam and photo-lithography. Solid State Commun. **151**, 1574–1578 (2011)

4. Bezryadin, A., Verschueren, A.R.M., Tans, S.J., Dekker, C.: Multiprobe transport experiments on individual single-wall carbon nanotubes. Phys. Rev. Lett. **80**, 4036–4039 (1998)
5. Gao, B., Chen, Y.F., Fuhrer, M.S., Glattli, D.C., Bachtold, A.: Four-point resistance of individual single-wall carbon nanotubes. Phys. Rev. Lett. **95**, 196802 (2005)
6. Utke, I., Hoffmann, P., Melngailis, J.: Gas-assisted focused electron beam and ion beam processing and fabrication. J. Vac. Sci. Tech. B **26**, 1197–1276 (2008)
7. Kim, C.-S., Ahn, S.-H., Yang, D.-Y.: Review: developments in micro/nanoscale fabrication by focused ion beams. Vacuum **86**, 1014–1035 (2012)
8. Russell, P.E., Utke, I., Moshkalev, S. (eds.): Nanofabrication using focused ion and electron beams: principles and applications. Oxford University Press, New York, NY (2012). ISBN 9780199734214
9. Li, W.X., Warburton, P.A.: Low-current focused-ion-beam induced deposition of three-dimensional tungsten nanoscale conductors. Nanotechnology **18**, 485305 (2007)
10. Giannuzzi, L.A., Stevie, F.A.: Introduction to Focused Ion Beams. Springer Science, Boston, MA (2005)
11. Chen, P., Van Veldhoven, E., Sanford, C.A., Salemink, H.W.M., Maas, D.J., Smith, D.A., Rack, P.D., Alkemade, P.F.A.: Nanopillar growth by focused ion helium ion-beam-induced deposition. Nanotechnology **21**, 455302 (2010)
12. Gopal, V., Radmilovic, V.R., Daraio, C., Jin, S., Yang, P., Stach, E.A.: Rapid prototyping of site-specific nanocontacts by electron and ion beam assisted direct-write nanolithography. Nanoletters **4**, 2059–2063 (2004)
13. Botman, A., Mulders, J.J.L., Weemaes, R., Mentink, S.: Purification of platinum and gold structures after electron-beam-induced deposition. Nanotechnology **17**, 3779–3785 (2006)
14. Córdoba, R., Sesé, J., De Teresa, J.M., Ibarra, M.R.: High-purity cobalt nanostructures grown by focused-electron-beam-induced deposition at low current. Microelectron. Eng. **87**, 1550–1553 (2010)
15. Langford, R.M., Ozkaya, D., Sheridan, J., Chater, R.: Effects of water vapour on electron and ion beam deposited platinum. Microsc. Microanal. **10**, 1122–1123 (2004)
16. Takeguchi, M., Shimojo, M., Furuya, K.: Post-deposition processes for nanostructures formed by electron beam induced deposition with $Pt(PF_3)_4$ precursor. Appl. Phys. A **93**, 439–442 (2008)
17. Botman, A., Mulders, J.J.L., Hagen, C.W.: Creating pure nanostructures from electron-beam-induced deposition using purification techniques: a technology perspective. Nanotechnology **20**, 372001 (2008)
18. Schwamb, T., Burg, B.R., Schirmer, N.C., Poulikakos, D.: On the effect of the electrical contact resistance in nanodevices. Appl. Phys. Lett. **92**, 243106 (2008)
19. Tham, D., Nam, C.-Y., Fischer, J.E.: Microstructure and composition of focused-ion-beam deposited Pt contacts to GaN nanowires. Adv. Mater. **18**, 290–294 (2006)
20. Nam, C.-Y., Tham, D., Fischer, J.E.: Disorder effects in focused-ion-beam deposited Pt contacts on GaN nanowire. Nanoletters **5**, 2029–2033 (2005)
21. Hernández-Ramírez, F., Tarancón, A., Casals, O., Rodríguez, J., Romano-Rodríguez, A., Morante, J.R., Barth, S., Mathur, S., Choi, T.Y., Poulikakos, D., Callegari, V., Nellen, P.M.: Fabrication and electrical characterization of circuits based on individual tin oxide nanowires. Nanotechnology **17**, 5577–5583 (2006)
22. Hernández-Ramírez, F., Prades, J.D., Tarancón, A., Barth, S., Casals, O., Jiménez-Díaz, R., Pellicer, E., Rodríguez, J., Juli, M.A., Romano-Rodríguez, A., Morante, J.R., Mathur, S., Helwig, A., Spannhake, J., Mueller, G.: Portable microsensors based on individual $SnO_2$ nanowires. Nanotechnology **18**, 495501 (2007)
23. Cronin, S.B., Lin, Y.-M., Rabin, O., Black, M.R., Ying, J.Y., Dresselhaus, M.S., Gai, P.L., Minet, J.-P., Issi, J.-P.: Making electrical contacts to nanowires with a thick oxide coating. Nanotechnology **13**, 653–658 (2002)
24. Fàbrega, C., Hernández-Ramírez, F., Prades, J.D., Jiménez-Díaz, R., Andreu, T., Morante, J.R.: On the photoconduction properties of low resistivity $TiO_2$ nanotubes. Nanotechnology **21**, 445703 (2010)

25. Moshkalev, S.A., León, J., Verissimo, C., Vaz, A.R., Flacker, A., de Moraes, M.B., Swart, J.W.: Controlled deposition and electrical characterization of multi-wall carbon nanotubes. J. Nano Res. **3**, 25–32 (2008)
26. De Marzi, G., Iacopino, D., Quinn, A.J., Redmond, G.: Probing intrinsic transport properties of single metal nanowires: direct-write contact formation using a focused ion beam. J. Appl. Phys. **96**, 3458 (2004)
27. Valizadeh, S., Abid, M., Hernández-Ramírez, F., Romano-Rodríguez, A., Hjort, K., Schweitz, J.A.: Template synthesis and forming electrical contacts to single Au nanowires by focused ion beam techniques. Nanotechnology **17**, 1134–1139 (2006)
28. Fernández-Pacheco, A., De Teresa, J.M., Córdoba, R., Ibarra, M.R.: Metal-insulator transition in Pt-C nanowires grown by focused-ion-beam-induced deposition. Phys. Rev. B **79**, 174204 (2009)
29. Lin, J.-F., Bird, J.P., Rotkina, L., Sergeev, A., Mitin, V.: Large effects due to electron-phonon-impurity interference in the resistivity of Pt/C-Ga composite nanowires. Appl. Phys. Lett. **84**, 3828–3830 (2004)
30. Tsukatani, Y., Yamasaki, N., Murakami, K., Wakaya, F., Takai, M.: Transport properties of Pt nanowires fabricated by beam-induced deposition. Jpn. J. Appl. Phys. **44**, 5683–5686 (2005)
31. Tao, T., Ro, J., Melngailis, J.: Focused ion beam deposition of Pt. J. Vac. Sci. Technol. B **8**, 1826–1829 (1990)
32. Puretz, J., Sawson, L.W.: Focused ion beam deposition of Pt containing films. J. Vac. Sci. Technol. B **10**, 2695–2698 (1992)
33. Telari, K.A., Rogers, B.R., Fang, H., Shen, L., Weller, R.A., Braski, D.N.: Characterization of Pt films deposited by focused ion beam-assisted chemical vapour deposition. J. Vac. Sci. Technol. B **20**, 590–595 (2002)
34. Langford, R.M., Wang, T.-X., Ozkaya, D.: Reducing the resistivity of electron and ion beam assisted deposited Pt. Microelectron. Eng. **84**, 784–788 (2007)
35. Peñate-Quesada, L., Mitra, J., Dawson, P.: Non-linear electronic transport in Pt nanowires deposited by focused ion beam. Nanotechnology **18**, 215203 (2007)
36. Vaz, A.R., Macchi, M., Leon, J., Moshkalev, S.A., Swart, J.W.: Platinum thin films deposited on silicon oxide by focused ion beam: characterization and application. J. Mater. Sci. **43**, 3429–3434 (2008)
37. De Teresa, J.M., Córdoba, R., Fernández-Pacheco, A., Montero, O., Strichovanec, P., Ibarra, M.R.: Origin of the difference in the resistivity of as-grown focused-ion and focused-electron-beam-induced Pt nanodeposits. J. Nanomater. **2009**, 936863 (2009)
38. Mott, N.F., Davis, E.A.: Electronic Processes in Non-crystalline Materials. Oxford University Press, New York, NY (1971)
39. Fernández-Pacheco, A., De Teresa, J.M., Córdoba, R., Ibarra, M.R.: High-quality magnetic and transport properties of cobalt nanowires grown by focused-electron-beam-induced deposition. J. Phys. D **42**, 055005 (2009)
40. Traving, M., Schindler, G., Engelhardt, M.: Damascene and subtractive processing of narrow W lines: resistivity and size effect. J. Appl. Phys. **100**, 094325 (2006)
41. Stewart, D.K., Stern, L.A., Morgan, J.C.: Focused-ion-beam induced deposition of metal for microcircuit modification. SPIE Proc. **1089**, 18 (1989)
42. Guillamón, I., Suderow, H., Vieira, S., Fernández-Pacheco, A., Sesé, J., Córdoba, R., De Teresa, J.M., Ibarra, M.R.: Nanoscale superconducting properties of amorphous W-based deposits grown with focused-ion-beam. New J. Phys. **10**, 093005 (2008)
43. Prestigiacomo, M., Roussel, L., Houe, A., Sudraud, P., Bedu, F., Tonneau, D., Safarov, V., Dallaporta, H.: Studies of structures elaborated by focused ion beam induced deposition. Microelectron. Eng. **76**, 175–181 (2004)
44. De Teresa, J.M., Fernández-Pacheco, A., Córdoba, R., Sesé, J., Ibarra, M.R., Guillamón, I., Suderow, H., Vieira, S.: Transport properties of superconducting amorphous W-based nanowires fabricated by focused-ion-beam-induced-deposition for applications in Nanotechnology. Mater. Res. Soc. Symp. Proc. **1180**, 1180-CC04-09 (2009)

45. Li, W., Fenton, J.C., Wang, Y., McComb, D.W., Warburton, P.A.: Tunability of the superconductivity of tungsten films grown by focused-ion-beam direct writing. J. Appl. Phys. **104**, 093913 (2008)
46. Prestigiacomo, M., Bedu, F., Jandar, F., Tonneau, D., Dallaporta, H., Roussel, L., Sudraud, P.: Purification and crystallization of W wires fabricated by focused ion beam induced deposition. Appl. Phys. Lett. **86**, 192112 (2005)
47. Sadki, E.S., Ooi, S., Hirata, K.: Focused-ion-beam-induced deposition of superconducting thin films. Appl. Phys. Lett. **85**, 6206–6208 (2004)
48. Collver, M.M., Hammond, R.H.: Superconductivity in "amorphous" transition metal alloy films. Phys. Rev. Lett. **30**, 92–95 (1972)
49. Porrati, F., Sachser, R., Huth, M.: The transient electrical conductivity of W-based electron-beam-induced deposits during growth, irradiation and exposure to air. Nanotechnology **20**, 195301 (2009)
50. Guillamón, I., Suderow, H., Fernández-Pacheco, A., Córdoba, R., Sesé, J., De Teresa, J.M., Ibarra, M.R., Vieira, S.: Direct observation of melting in a 2D superconducting vortex lattice. Nat. Phys. **5**, 651–655 (2009)
51. Guillamón, I., Suderow, H., Vieira, S., Córdoba, R., Sesé, J., De Teresa, J.M., Ibarra, M.R.: Direct observation of stress accumulation and relaxation in small superconducting vortex bundles. Phys. Rev. Lett. **106**, 077001 (2011)
52. Córdoba, R., Baturina, T.I., Sesé, J., Mironov, A.Y., De Teresa, J.M., Ibarra, M.R., Nasimov, D.A., Gutakovskii, A.K., Latyshev, A.V., Guillamón, I., Suderow, H., Vieira, S., Baklanov, M.R., Palacios, J.J., Vinokur, V.M.: Magnetic field induced dissipation free state in superconducting nanostructures. Nat. Commun. **4**, 1437 (2013)
53. Martínez-Pérez, M.J., Sesé, J., Córdoba, R., Luis, F., Drong, D., Shurig, T.: Circuit edit of superconducting microcircuits. Supercond. Sci. Tech. **22**, 125020 (2009)
54. Romans, E.J., Osley, E.J., Young, L., Warburton, P.A., Li, W.: Three dimensional nanoscale superconducting quantum interference device pickups coils. Appl. Phys. Lett. **97**, 22506 (2010)
55. Marcano, N., Sangiao, S., Plaza, M., Pérez, L., Fernández-Pacheco, A., Córdoba, R., Sánchez, M.C., Morellón, L., Ibarra, M.R., De Teresa, J.M.: Weak-antilocalization signatures in the magnetotransport properties of individual electrodeposited Bi nanowires. Appl. Phys. Lett. **96**, 082110 (2010)
56. Martínez-Boubeta, C., Balcells, L., Monty, C., Ordejón, P., Martínez, B.: Tunnelling spectroscopy in core/shell structured Fe/MgO nanospheres. Appl. Phys. Lett. **94**, 062507 (2009)
57. Muñoz-Rojas, D., Córdoba, R., Fernández-Pacheco, A., De Teresa, J.M., Sathier, G., Fraxedas, J., Walton, R.I., Casañ-Pastor, N.: High conductivity in hydrothermally grown $AgCuO_2$ single crystals verified using focused-ion-beam-deposited contacts. Inorg. Chem. **49**, 10977–10983 (2010)
58. Muñoz-Rojas, D., Subías, G., Fraxedas, J., Gómez-Romero, P., Casañ-Pastor, N.: Electronic structure of $Ag_2Cu_2O_4$. Evidence of oxidized silver and copper and internal charge delocalization. J. Phys. Chem. B **109**, 6193–6203 (2005)
59. Sauvage, F., Muñoz-Rojas, D., Poeppelmeier, K.R., Casañ-Pastor, N.: Transport properties and lithium insertion study in the p-type semiconductors $Ag_2Cu_2O_4$ and $Ag_2Cu_{0.5}Mn_{0.5}O_4$. J. Solid State Chem. **182**, 374–380 (2009)
60. Ekin, J.W., Larson, T.M., Bergren, N.F., Nelson, A.J., Swartzlander, A.B., Kazmerski, L.L., Panson, A.J., Blankenship, B.A.: High $T_C$ superconductor/noble-metal contacts with surface resistivities in the $10^{-10}$ $\Omega cm^2$ range. Appl. Phys. Lett. **52**, 1819–1821 (1988)
61. Van Son, P.C., Van Kempen, H., Wyder, P.: New method to study the proximity effect at the normal-metal-superconductor interface. Phys. Rev. Lett. **59**, 2226–2228 (1987)
62. Guéron, S., Pothier, H., Birge, N.O., Esteve, D., Devoret, M.H.: Superconducting proximity effect probed on a mesoscopic length scale. Phys. Rev. Lett. **77**, 3025–3028 (1996)
63. Soulen, R.J., et al.: Measuring the spin polarization of a metal with superconducting point contact. Science **282**, 85–88 (1998)

64. Sangiao, S., Morellón, L., Ibarra, M.R., De Teresa, J.M.: Ferromagnet-superconductor nanocontacts grown by focused electron/ion beam techniques for current-in-plane Andreev reflection measurements. Solid State Commun. **151**, 37–41 (2011)
65. Sangiao, S., De Teresa, J.M., Ibarra, M.R., Guillamón, H., Suderow, H., Vieira, S., Morellón, L.: Andreev reflections under high magnetic fields in ferromagnetic-superconductor nanocontacts. Phys. Rev. B **84**, 233402 (2011)
66. Blonder, G.E., Tinkham, M., Klapwijk, T.M.: Transition from metallic to tunneling regimes in superconducting microconstrictions: excess current, charge imbalance, and supercurrent conversion. Phys. Rev. B **25**, 4515–4532 (1982)
67. Kasumov, A., et al.: Proximity effect in a superconductor-metallofullerene-superconductor molecular junction. Phys. Rev. B **72**, 033414 (2005)
68. Shailos, A., Nativel, W., Kasumov, A., Collet, C., Ferrier, M., Guéron, S., Deblock, R., Bouchiat, H.: Proximity effect and multiple Andreev reflections in few-layer graphene. Europhys. Lett. **79**, 57008 (2007)
69. Wang, J., et al.: Proximity-induced superconductivity in nanowires: mini-gap state and differential magnetoresistance oscillations. Phys. Rev. Lett. **102**, 247003 (2009)
70. Wang, J., et al.: Interplay between superconductivity and ferromagnetism in crystalline nanowires. Nat. Phys. **6**, 389–394 (2010)
71. Sangiao, S., et al.: Manuscript in preparation

# Chapter 6
# Metal-Induced Crystallization of Focused Ion Beam-Induced Deposition for Functional Patterned Ultrathin Nanocarbon

**Gemma Rius, Xavier Borrisé, and Narcís Mestres**

**Abstract** This chapter recapitulates our work based on focused ion beam-induced deposition of ultrathin amorphous carbon (FIBID-C) patterns as a route towards the easy integration of carbon nanomaterials. The use of postdeposition thermal processing causes structural changes, chemical modification, and tuning of functional properties of the as-deposited FIBID-C. Particularly, in combination with metals our approach allows obtaining site-specific graphene-like features on insulators. We explore how to convert FIBID-C into functional materials, namely, graphene-like layers or diamond-like coatings, or combination with various metals.

First, we provide a general description of FIBID technique to frame the profuse detailed case of carbon deposition of ultrathin carbon micropatterns. Then, the study of the transformation of FIBID-C by thermal processing reveals the unique characteristics of the as-deposited materials and particularities of the metal-induced crystallization. A detailed section on Raman spectroscopy is included as it represents a powerful tool to identify any carbon form, from amorphous to perfectly crystalline,

G. Rius (✉)
Nagoya Institute of Technology, Gokiso cho Showa ku, 466-8555 Nagoya, Japan
e-mail: rius.gemma@nitech.ac.jp; gemrius@gmail.com

X. Borrisé
Institut Catala de Nanotecnologia, Campus UAB, 08193 Bellaterra, Spain

Institut de Microelectronica de Barcelona-CSIC, Campus UAB, 08193, Bellaterra, Spain
e-mail: xavier.borrise@imb-cnm.csic.es

N. Mestres
Institut de Ciencia dels Materials de Barcelona (ICMAB) Consejo Superior de Investigaciones Científicas (CSIC), Campus UAB, 08193 Bellaterra, Spain
e-mail: narcis.mestres@icmab.es

Z.M. Wang (ed.), *FIB Nanostructures*, Lecture Notes in Nanoscale Science and Technology 20, DOI 10.1007/978-3-319-02874-3_6, 

and other nanostructured carbon elements. We describe the Raman scattering in terms of both the basics of the technique and application to the structural determination of the FIBID-C layers in the different processing stages and material structural variants. In the last part, we provide additional demonstration of the technological capabilities of the FIBID-C approach by testing several processing options, such as resistance to etching, and integration routes for electronic devices.

## 6.1 Introduction: Integration of Carbon Nanomaterials

Focused ion beam-induced deposition (FIBID) is a direct patterning technique that allows versatile definition of features with diverse materials [1]. This chapter shows how to use the FIBID of ultrathin carbon patterns as an original technological platform to integrate several types of nanostructured carbon materials into functional structures.

The carbon family of nanometer-scale materials is large, as the allotropes of carbon range from fullerenes and carbon nanotubes (CNTs) and graphene to diamond nanoparticles and ultrananocrystalline diamond or low-crystalline carbon forms such as diamond-like carbon and amorphous or glassy carbon nanocoatings (or even organic conductive thin films) [2]. Relevantly, the properties of each of them do dramatically vary with their characteristics of morphology, atomic structure, doping, etc. making them potential elements as active and passive functional units for an important number of applications [3]. While $sp^2$ carbon networks, such as CNTs or graphene, endow special and exceptional mechanical, electrical, thermal, chemical, and optical properties [4], carbon $sp^3$ systems, where diamond is the record of the extremes, exhibit extraordinary hardness and thermal conductivity or transparency among other outstanding properties, alike in nanodiamond [5].

Aside from the inherent interest to use them to study their intrinsic fundamental properties, the group of nanocarbons is proposed as advanced materials impactful in the energy-, human health-, and environment-oriented developments, seen as substitute materials or added-value constituents derived from their dimensionality in the nanometric order. As an example, aspects like biocompatibility and chemical stability in physiological solutions make them suitable for sensors while especially promising because they could provide increased sensitivity, with additional possibility to be tuned or functionalized for optimal sensing selectivity [6].

A different approach is applied/implemented if products such as nanocomposites or powder materials are desired or, otherwise, nanometer-size defectless crystalline or nanostructured carbon units which could be useful, for instance, to build innovative electronic devices. Their preparation also involves a wide variety of techniques and experimental conditions. Synthesis and integration play a noteworthy role on carbon nanomaterial research, where flexibility and precise control for the production of the nanocarbon materials and capability to determine their properties are often required [3]. And achieving this is very often too challenging.

Chemistry-based synthesis is the leading method for nanocarbon production, in particular for CNTs and graphene, due to the relative simplicity of chemical vapor deposition apparatuses, relatively mild processing conditions, and commonly easily scalable systems and conditions (and thus, potential for mass production) [7]. However, it is not always capable of providing pursued architectures, such as vertical alignment in CNTs or high growth rate, the reason why enhancement by physical procedures, such as plasma or RF, is often introduced into CVD systems [8]. A typical example is the use of plasma reactors for nanodiamond deposition. Pure physical methods, bulk thermal treatment and surface modifications as in ion bombardment/deposition, such as used in the graphene growth on SiC substrates, imply either harsh or highly controlled contour conditions and defect-free supports or complex and expensive instrumentation (or a combination of these aspects). Higher control and structural determination of nanocarbon products often are disguised by impracticality for real and commercial applications.

An advantageous fact of the production of carbon materials is the availability of abundant forms of organic compounds and alternative solid carbon sources that can be used as precursors [9]. Ability to produce significant amount of and cost-effective nanomaterials is one of the issues to be able to move them massively to the market and accomplish an impactful societal benefit. Reproducibility of obtained products under equivalent conditions is also important. Being not surprising that less strict forms in crystalline structure, defect, and dimensionality, such as diamond-like carbon, glassy carbon nanofilms, or fullerenes, were the first to reach the applications market, a big challenge comes from the necessity to integrate nanosize carbon elements into existing technologies or establishing radically different approaches, i.e. (driven by), leading to the adoption of disruptive processing. The integration of carbon nanomaterials having very determined specifications and properties into desired configurations presents fabrication and control challenges. The present intensity on graphene research, as an effort to enable the full exploitation of its excellent properties, evidences the necessity to find alternative procedures, for example, to completely integrate graphene in the standard semiconductor planar technologies if this is the route for future (graphene-based) nanoelectronics. The bottleneck points are clearly identified in the use of metal as a catalyst for the crystallization and the need of transfer to insulating substrates and metal etching while preserving ideal graphene features. At times, growth temperature is also a concern.

Examples of difficulties and smart solutions to enable nanocarbon-based applications are abundant, but no final solution has been proposed or established yet. Hereby the present approach is another of those options and follows a processing route and processing methods inspired on the planar technology point of view and strategies. Merging the carbon nanomaterial preparation and lithography in a single-stage processing, in other words direct patterning, plus compatibility with Si-based materials, the aim is to introduce our concept as an example of a robust and flexible nanofabrication methodology.

In this chapter of the FIB nanostructures book comprised in the book series of Lecture Notes of Nanoscale Science and Technology of Springer Ed., we cover the description of our work based on FIBID of ultrathin amorphous carbon and its change, modification, or tuning by its thermal processing and combination with metals.

First, we review (1) the basics of FIBID technique, (2) the concept of local and controlled deposition techniques, and (3) the key aspects and elements of the instrumental. Examples of state-of-the-art applications in the literature are included. On this basis, we address a complete description of our processing platform based on FIBID of ultrathin carbon.

A short description on the platform design and its implementation is given. The patterning capabilities and limitations of our system are presented as they determine the characteristics of the patterned features that can be obtained, such as spatial resolution, minimum thickness, and control. Effect of thermal annealing is carefully examined, paying special attention to the metal-assisted treatment, which allows obtaining site-specific graphene-like features [10]. In this section, distinguishing aspects of the patterned material, chemical and atomic transformation, crystalline structure, and so on are provided in detail. Specifically, in-depth structural evaluation will be based on Raman spectroscopy for its tradition, abundant references, and therefore power to precisely describe the characteristics of carbon materials and as a confirmation of previously presented observations. Within the section on Raman spectroscopy for nanocarbon analysis, a short introduction to the basics of Raman scattering and instrumentation as well as review of the signatures for all the carbon allotropes are included. Following this basic analysis of the processing products from a materials point of view, a significant part is dedicated to illustrate the versatility and flexibility of the FIBID-C approach. Therefore, examples of process ability, such as etching or diverse substrates, other compatible metallic materials to assist heat treatment, and original configurations and integration methods for macroscopic interfacing of the nanostructured carbon features are shown.

The chapter concludes with a simple summary of the work presented here and a brief comment on the potential of mask-less ion beam-based deposition/patterning as well as manufacturing capability.

## 6.2 Focused Ion Beam-Induced Deposition

Most of the actual focused ion beam equipment is combined with an electron beam column to allow the observation of modified surfaces as well as a gas injection system (GIS) to allow the modification of the surface, either by ablation, enhanced etching or ion-induced deposition. This, in combination with electrical probes inside the chamber, and the usual SEM detectors convert the actual FIB systems in real "nanolaboratories," where prototyping is combined with in situ physical, chemical, and electrical characterization.

A key technology for such success is the introduction of a small injector (nozzle) that is able to locally deliver a gas over the surface. This is based normally on some kind of organometallic compound that the ions from the FIB conveniently dissociate in a metal part, deposited on the surface, and a volatile part evacuated through the normal vacuum system of the equipment. The resolution, then, depends primarily on the incident beam.

The possibility to deposit different kinds of metals or even insulators, combined with normal milling, opens a wide window of prototyping at the nanoscale. In the following we review the technique of FIBID, materials, limitations, and different applications.

### 6.2.1 Technique

Typically in SEM, when zooming out, the contamination of the surface due to hydrocarbons present in the chamber has been always observed. In this process, the hydrocarbons present in the chamber and the surface of the sample are fixed on the sample under the presence of an energetic charged beam, in this case, electrons. Many efforts have been made to solve such issue: actual systems are oil-free pumps, so the main source of contamination is the sample itself. Plasma cleaners or liquid nitrogen cold traps can be added today to any SEM chamber to avoid contamination [11].

The principle described to burn contamination on the sample is essentially the same used for FIBID and focused electron beam-induced deposition (FEBID). The physical processes involved are equivalent for both types of charges and rely on the ability of the primary beam and the generated secondary electrons (SE) to dissociate the molecules delivered by the nozzle and adsorbed onto the surface, by breaking some of the bonds of the molecule and creating volatile products (mainly H- and O-related compounds) while depositing the desired element [12].

Figure 6.1 depicts the principle for the deposition process. A nozzle is closely placed above the surface of the sample, typically few hundred μm, at a certain angle, in an arrangement that does not disturb the ion path to the sample (typically, 30–60°). The injector is connected to a gas reservoir (in gas, liquid, or solid phase) with some heater to achieve and control the ejection of the gas through the nozzle. In such situation the vacuum is reduced locally over the surface and must be monitored and dynamically controlled to avoid contamination of the ion column. Calibration of the distance of the nozzle to the surface, and adjustment of the temperature of the reservoir, is necessary to obtain a proper flux of the gas. To achieve this control, protocols for cleaning the nozzle from undesired outgassing or contamination are used. The nozzle is normally controlled by motors for finely adjusting its position. If the distance from the nozzle is too high the feed of gas component to the surface would be inefficient, ultimately leading to a milling dominant scheme, and, if too close, the probability to trap non-dissociated molecules on the deposited material increases, thus contaminating the deposited material. The temperature control is used to promote the gas flux towards the surface, but at excessive temperature the gas may become unstable due to thermal effects. The incidence angle also plays a role, and although in most commercial systems the primary position of the injectors is fixed, the angle of incidence must be carefully chosen to obtain good deposition efficiency, both in rate and composition of the deposited material.

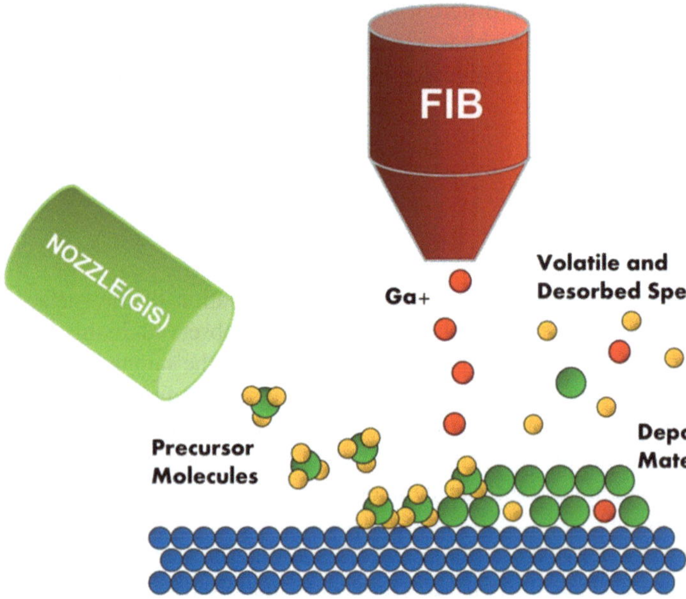


**Fig. 6.1** Principle of FIBID. The precursor molecules are adsorbed on the surface, and the ion beam dissociates them into a volatile part and the desired deposited material. Also impurities due to $Ga^+$ and volatile molecules are incorporated to the deposited material. Adapted from [1]

After (and then during) the gas delivery, the beam scans the surface in the presence of the adsorbed molecules and decomposes them into volatile and nonvolatile products. Simultaneously, the ion beam interacts with the surface, so there is always a balance between the sputtering yield and the deposition yield achieved by the chosen specific conditions.

Initially the beam impacts the bare surface of the substrate with some of the adsorbed molecules; therefore, some sputtering of the substrate is always occurring, which could be undesirable for some applications, as the surface undergoes little modification or damage. As the film starts to grow up, the sputter to deposition competition is related to the deposited material. This is an important difference with respect to FEBID, as electrons do not physically damage the surface. On the other hand, as ions do not penetrate to the substrate as electrons do, in some cases, the protection on inner parts of the substrate, which could allocate active devices, gives preference to the use of ions to avoid charge trapping caused during the FEBID process.

At the end of the process, and if the right conditions are selected, the residual injected gas together with the volatile part of the decomposition is evacuated through the normal vacuum system of the equipment, leaving the nonvolatile part as a layer of deposited material.

It is accepted that the SE are responsible for breaking the molecular bounds of the injected gas, typically an organometallic, as there is a correspondence on the energies, but this does not fully explain the differences in deposition rate between ion- and electron-induced deposition [1], so it seems that the primary beam also plays a role in the decomposition of the adsorbed molecules and fixing of the material. Finally, the best is always to calibrate the rates for each material to be deposited and for each specific support.

An important parameter to obtain good deposition conditions is the beam current of the incident ion beam. This is in close relation with the area to be deposited and the algorithm to perform the deposition. Several methods can be implemented, and all commercial apparatuses deliver automated routines for such purposes. The main notion is that the higher the current the faster the deposition, but this also increases the sputtering yield, so a compromise must be found. Beam diameter also correlates with the current so that higher current implies lower spatial resolution. For Pt, for example a good working point is around 2–6 pA/$\mu m^2$ [13]. As a rule of thumb, the current (in pA) to be used comes from multiplying by 5 the area to be scanned (in microns). For example, an area of $2\times2$ $\mu m^2$ can be scanned at a current of 20 pA (i.e., $2\times2\times5$). The routine for scanning is also important to minimize the dwell time at each point to avoid the dominance of the milling process, and related to this, the step size also must be selected to avoid incident beam overlap, which increases the milling probability. Depending on the software/equipment used, all these parameters can be selected directly (setting for example the step size and dwell time) or indirectly (for example, by setting the scanning frequencies over the area to be exposed).

In summary, once the material to be deposited is selected for the application, the important rules are to have clean samples to deposit the material, well-calibrated precursor deposition, clean nozzles, and reservoirs at a stable temperature. Then the area (pattern) and thus the current to be used must be selected.

### 6.2.2 *Precursor Materials and Deposited Materials*

Many materials can be deposited by FIBID, from gas, liquid, or solid phase, and even the same material can be deposited using different precursor chemistries. Utke et al. [1] made an excellent review for the focused electron beam- and ion beam-induced deposition, covering all aspects of the process for deposition, the chemistry research for new components, and the characteristics of many deposited materials.

However, most of the FIB users usually work with commercial equipment and use commercial precursors available from the same company, which leads to a few typical materials to be deposited, depending on the type of GIS (from gas or solid phase essentially) (Table 6.1).

Nevertheless, the increasing demand for nanotechnology platforms, like the FIB equipped with an electron column and a GIS, is pushing the research and development of new and safe chemistries to be used in normal working laboratories, giving the opportunity to fabricate new and more exotic devices at the nanoscale using all sorts of elemental materials [1].

**Table 6.1** Summary of the most common materials deposited and an example of some precursors used to deposit them

| Material | Precursor | Compound name | Reference |
|---|---|---|---|
| Carbon | $C_{14}H_{10}$ | Phenanthrene | [14] |
| Platinum | $(Me_3)MeCpPt$ | (Methylcyclopentadienyl) Trimethyl platinum | [15] |
| Silicon oxide | $CH_3Si(OC_2H_5)_3$ | Triethoxymethylsilane | [16] |
| Tungsten | $W(CO)_6$ | Tungsten hexacarbonyl | [17] |
| Gold | $C_7H_7O_2F_6Au$ | Dimethyl gold hexafluoro acetylacetonate | [18] |

**Table 6.2** Composition and resistivity of Pt, W, and C deposited by FIBID from [14, 15, 17]

| Material | Element composition (%) | Resistivity |
|---|---|---|
| Platinum | 45 Pt/45 C/5 Ga/5 O | 5–1 mΩ cm |
| Tungsten | 51 W/37 Pt/12 Ga/O Traces | 200 μΩ cm |
| Carbon | 90 C/10 Ga | 100 Ω cm |

Depending on the application, the purity of the deposited material is of highest importance. In other words, it is not the same to apply FIBID to protect a layer, e.g. when preparing a lamella for TEM inspection, than to prepare a device with an active layer deposited by FIBID. Hence, the determination of the properties of the deposited material is essential. Chemical composition, electrical conductance, and mechanical properties are among the interesting parameters to be determined depending on the application.

Table 6.2 summarizes some reported results for FIBID. In general the result of the deposition is a mixture of the desired compound material, with some residual carbon and gallium. The percentages vary upon precursor and conditions, and, in general, a better conductivity is achieved with ions rather than electrons, which gives deposited material with higher carbon concentration. However, sometimes it is likely that the surface is damaged with $Ga^+$ for the ion-induced deposition; thus, a compromise must arise between yield, properties, and $Ga^+$ damage or contamination. For W deposition, for example, Luxmoore et al. [17] found for FIBID a resistivity one order of magnitude lower than FEBID, and even at low temperature the FIBID material exhibited superconductor properties which was not the case for the FEBID tungsten. Deposition of Pt also gives two or three orders of magnitude higher resistivity for the FEBID, as seen by Langford et al. [15], where annealing is also a possibility to improve the conductivity, as they show for FEBID Pt layers.

### 6.2.3 *Applications and Limitations*

The most common use for deposition is the preparation of samples to be inspected by different high-resolution characterization techniques, TEM, SEM, auger, SIMS, etc. Albeit it is a very expensive technique compared to more conventional procedures, it has same exceptional advantages that make it very useful and more and more popular every day (indispensable at times). In TEM preparation, for example, the ability to prepare the sample with high-resolution positioning makes it practically the unique technique used in the semiconductor industry to prepare samples for atomic resolution inspection. The speed is also a factor to consider as with few hours the lamella is ready for inspection. Lastly, when the amount of available materials or the fragility of the sample must be considered, the FIB is the best choice for sample preparation [19]. As a drawback, the main considerations are the Ga contamination, amorphization of the surface (10–40 nm depending on the energy used for the final polishing), and size of the sample to be prepared, typically of few microns in size and depth.

The first step in lamella preparation by FIB is the deposition of Pt or any other Z material to make a homogeneous protecting layer to the surface to be milled. In this way the so-called waterfall or curtain effect is avoided [20]. This is also useful in normal cross-section preparation, and the deposition is typically of few hundred nanometers. Finally, another advantage of this method for preparation in the actual FIB/SEM platforms is the in situ characterization of the prepared sample, using any of the normal SEM detectors, EDS, EBSD, STEM, etc. before proceeding for TEM analysis.

In the semiconductor industry, FIB was used since the very beginning for mask repair [21] and also preparation of lamellas for TEM inspection, but the actual FIB/SEM platforms allow the device inspection and edition, by disconnecting lines, connection lines, vias preparation or milling and cleanups. All these processes deal with milling and deposition of Pt or W for metal contacts as well as insulating materials, which are used to either fill milled areas with the dielectric material during the edition or to protect the zones before milling edition, for example, vias definition, as the redeposition could affect the area nearby [22]. The success of this process relies on the ability for navigating over the sample and finding the structures, in many cases hidden few microns below the surface, so sometimes CAD navigation software is required. After the circuit is edited, the testing is straightforward, improving the cycle time for mask redesign.

More interesting from the nanofabrication point of view is the application of the FIBID technique in new architectures, exploiting the capabilities of resolution and localization. Matsui et al. [23] developed de FIBID technique to perform 3D structures that cannot be made by any other technique, based on CAD design and a very fine control of the positioning of the beam, combined with induced deposition of different materials like C, Pt, or W. Also functionalization of AFM tips by selective deposition of material is possible, changing the properties of the AFM tip for different purposes [24].

FIBID is also used to contact nano-objects [25] or in combination with FEBID when the device could be sensitive to ions [26]. Resolution below 100 nm is easily achieved by the FIBID technique. As a result, it allows, for example, the deposition of very fine lines of W with sub-20 nm resolution, which exhibit superconductor properties at low temperatures [27], or fabrication of metal heaters in bimorph cantilevers [28] proposed to be used as nanotweezers. Devices as FET transistors, fabricated in combination with FEBID [29], nanotemplates [28], or deposition of catalytic material for nanowire growth [30], are other applications for FIBID, enabling an increasing number of novel nanoarchitectures.

## 6.3 FIBID-C-Based Processing

### *6.3.1 Concept of the Platform: Methodological Approach*

As introduced, FIB is established as a key tool for the preparation of TEM sample specimens and used even commercially for photomask repair or restructuring of integrated circuits, generally as a combination of milling (controlled local material removal) and FIBID (controlled local material addition) [13].

Here we focus on an additional potential of FIB-based techniques by exploring some technological approach which demonstrates an alternative route for nanomaterial structuring taking the FIBID as the key element [1].

Following some of the concepts of the basic routines for nanofabrication in the planar technology style, FIBID can be thought as a mixed process of bottom-up and top-down approaches, as it takes notions from both methodologies. In our case, phenanthrene molecules are used as the building blocks of the ultrathin nanocarbon patterns (bottom-up) that are formed by their interaction with a nanosized Ga ion beam, which is a miniaturized version of patterning/deposition technique, as in stencil or ion implantation techniques (top-down). In essence, FIBID is a lithographic technique for direct (meso/nano) patterning.

Our work focuses in the nanostructuring of carbon, so it is a demonstrator for only this kind of material. Although it retains undeniably much interest of the scientific community for presenting special structural and functional features, the processing concept could also be eventually transferred to other instances of materials.

After patterning by FIBID-C, additional heat treatment is applied to control and diversify the possibilities of FIBID carbon layers. We investigate the tuning of nanopatterned carbon-based films, being particularly interested to determine the crystalline grade and then test and understand the formation of pure or compound materials, so that we are finally able to build a number of nanoarchitectures and form the diverse carbon allotropes.

Additionally, we propose the FIBID-C as an optional and creative way to address the integration of various C allotropes for device fabrication, such as electronic applications. Issues on FIBID-C integration and compatibility in terms of both the C nanomaterial and its processing (including alignment capability with other lithography levels and techniques) are also possible to be planned and overcome, respectively, as in part of a sequence or a complex system as presented in Sect. 6.3.5.

The design of the experimental procedure and implementation is quite simple, typically involving two main steps: direct patterning and subsequent annealing.

The simplest version consists in deposition of ultrathin carbon patterns onto an insulating silica support plus thermal treatment, typically at mid–high temperatures. More sophisticated heating treatments include the use of metals to catalyze crystallization or tailor the patterned carbon nanofilms. Metals (Ni, Cu, Fe, Pt, …) are introduced in the processing sequence in various forms, such as foils, thin-layer deposition, patterned thin layers, or nanoparticles (metals being added either before or after FIBID-C, as an additional preparation of heat treatment or supplementary step of the whole sequence).

### *6.3.2 Patterning: Properties and Limitations*

The FIB apparatus capabilities and specifications can differ from maker to maker and primarily as a function of the purpose application of the system. Common systems for TEM specimen preparation now combine both an electron beam and an ion beam column mainly to minimize the exposure to ions of the sample.

When ion probe system is planned for lithography of resists, fine surface 3D nanostructuring, or FIBID a dedicate attachment for precise beam control is desirable, which consists in additional electronic beam positioning control and ultrafast beam shutter and preferably together with a precise, flexible, and user-friendly design interface. For the present development of nanostructured carbon based on FIBID, we used a multibeam system JIB-4500 from JEOL. Designed for TEM transparent specimen preparation rather than for high-resolution processing, it does not include a lithography accessory and has several automated routines which implies that there will be intrinsic limitations in some aspects of the patterning, as we specify and describe next.

The gallium ion beam is operated at 30 kV and using the smallest probe that is allowed for FIBID by the software, which is the beam current of 300 pA (smaller apertures are available only to be used for bare ion exposure). In these conditions, focused beam has a diameter of ~42 nm, one of the aspects restricting the attainable spatial resolution. Related to this, most of the developments that we have been testing, and are presented here, are made of tens of micron size square or rectangular patterns, and the standard substrate is 100–500 μm thick $SiO_2$ thermally grown on Si wafer. Phenanthrene ($C_{14}H_{10}$) is the gas precursor used in our case for the FIBID-C, although other organic molecules such as styrene or naphthalene can also be used [1].

Performed at normal incidence angle, the experimental FIBID-C rate on $SiO_2$ follows the trend as shown in Fig. 6.2a. Actually, the software only allows the minimum thickness to be set to 80 nm. For the abovementioned beam conditions and to put it into numbers, the $40 \times 40$ μm$^2$ patterns for the thickness of 80 nm are deposited in nearly 9 min, based on instrument calibration. Experimentally, to achieve the thin layers of typically 10–20 nm that we present here, we interrupt manually the deposition. Considering the experimental calibration, ~5-s (or less) deposition time would be needed for $40 \times 40$ μm$^2$ patterns of thicknesses of about (or below) 5 nm. For this deposition time the deposited thickness precision/uncertainty is not only the same order of the intended layer but also dominated by the operator stoppage action. As a result, it is practically impossible for us to study high-spatial-resolution patterns or thinner layers either appropriately or robustly.

The AFM characterization, as used for deposited thickness determination, also allows us to observe a few features representative of FIBID. First, smoothness (or low RMS) increase is a characteristic of complete layer formation, a few nanometers thick. Second, typical patterning defects derive from the lateral diffusion of adsorbed species, beam electronic control limitations, etc., leading to notorious deformation of complex and dense patterns and eventually thickness nonuniformities in simple features (Fig. 6.2b). This fact is especially relevant if pattern stitching or 3D nanostructures are executed.

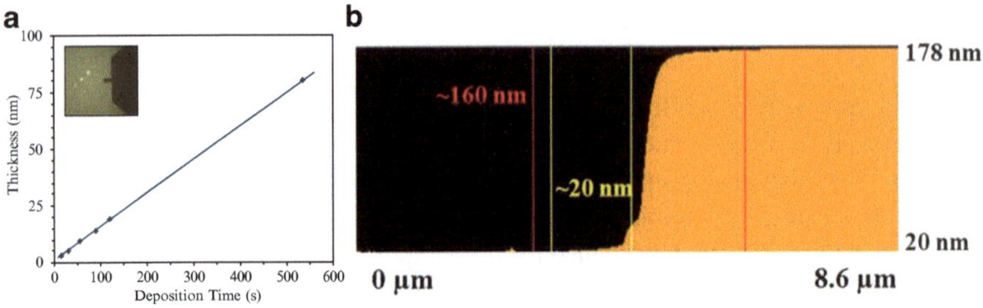


**Fig. 6.2** (**a**) Deposition rate for FIBID-C from a phenanthrene precursor onto $SiO_2$ and using 30 kV $Ga^+$; (**b**) AFM profile of a FIBID-C pattern, where layer smoothness and edge effects due to lateral diffusion are observed

One of the powerful capabilities of FIBID is not only to perform the direct patterning but also doing so onto a variety of substrates, including relatively odd samples, semiconducting, metallic, or even insulating materials. For the electrical insulators charging compensation by electron beam during FIBID dramatically enhances the pattern definition [31]. FIBID-C has been tested on glass samples, as shown later in Sect. 6.3.5, without charge compensation. As another example we have tested successfully FIBID-C directly on polycrystalline metal foils, such as Ni or Cu. A related use that would certainly benefit from the direct 3D nanopatterning flexibility would be, for example, the nanostructuring which is compatible with CMOS circuitry or preserves the integrity of nanosized materials, as in [32, 33].

The as-deposited processed material characteristics have been studied by TEM and XPS. In the TEM cross-sectional view, FIBID-C appears as an even, continuous, and uniform layer with apparently no atomic arrangement and no contamination (e.g., Ga from the beam). A basic point, intrinsically related to the deposition mechanism as explicated in previous Sect. 6.2, is that no clear interface between FIBID-C layer and the $SiO_2$ support can be observed. XPS informs on the existence of C–H and C–C $sp^2$ bonds accounting for the incomplete phenanthrene decomposition and (disordered) atomic C arrangement. Supplementary structural determination of as-deposited FIBID-C based on Raman spectroscopy is discussed in great detail in Sect. 6.3.4. Morphological, structural, and chemical characteristics will suffer a significant transformation as a result of the heat treatment, as explained next.

### 6.3.3 *Heat Transformation Versus Metal-Assisted Heat Transformation*

Our interest to study the heat treatment of FIBID-C is to explore the possibility to overcome some of the intrinsic properties of FIBID, such as its as-deposited high resistance (as explained in Sect. 6.3.5). In other words, we explore here how to convert FIBID-C into added-value crystalline materials, namely, graphene-like or diamond-like materials.

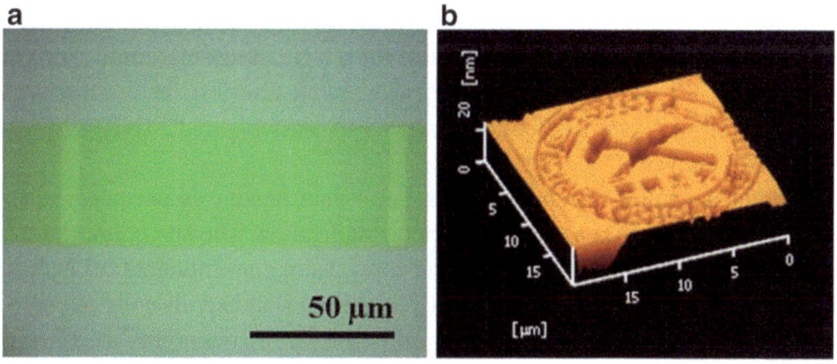


**Fig. 6.3** (**a**) Example of an annealed 10 nm thick pattern. *Long stripe* is done by pattern stitching (overlap seen as *brighter fringes*); (**b**) complex pattern representing TTI logo. Poor pattern definition is due to the density of the design in relation to limited spatial resolution of FIBID

The thermal process is performed in a simple infrared lamp furnace MILA 5000 from Ulvac Corp. in vacuum ambience, typically below 0.4 Pa. The system would allow gas flow during treatment, although we have not explored this option yet. Maximum operating temperature is ~1,100 °C, but we kept annealing temperatures below 1,000 °C. We obtained positive results of metal-assisted crystallization of FIBID-C in the ~900 °C range as well as a way to ensure sample support integrity and eventually reasons related with future processing compatibility. Due to comparison with the heating processes that include the accessory metal foil, a $SiO_2$ chip is always used to cover the FIBID-C-patterned sample during the thermal annealing [10]. Doing so also decreases the probability to absorb some residual elements from foreign materials unintentionally present in the chamber.

In general, we apply a fast heating rate from room temperature to the target temperature in 5 min and then steady for 30 min. Natural cooling down is typically used, which represents about 20 min in total with a characteristic rapid temperature decrease down to about 500 °C in 1 min and to 250 °C in 5 min, which might be an especially relevant point for the heat annealing processes involving metal catalyst.

En exemplary optical image of FIBID-C after annealed at 975 °C is shown in Fig. 6.3a. The patterned features look as if no ostensible morphological distortion has occurred both in terms of shape and thickness (AFM assessment) and clean of contamination. The TEM shows that atomic C arrangement has not suffered significant modification, and effectively XPS suggests that the FIBID-C is essentially constituted by a mixture of C $sp^2$ and $sp^3$ bonds and a small portion of C–O–C bonds. Two relevant aspects can be highlighted. FIBID-C interface with $SiO_2$ is now clear and diamond-like C–C bonds are found, and no obvious glassy-to-graphitic layer transition is obtained. Related to these distinctive features, we briefly refer to and illustrate the mechanical properties and resistance in Sect. 6.3.5. Parametric analyses of FIBID-C patterning and heat process are discussed in depth within the Raman spectroscopy in Sect. 6.3.4.

Other groups [34] had investigated the effect of thermal treatment on FIBID-C. In particular, their results quantify the decrease of H contents of their films beyond

773 K as well as gallium starting at 523 K, although the numbers and rate may differ as compared to our case primarily because they were studying significantly thicker FIBID-C films, 80–270 nm. Atomic fraction by Rutherford backscattering spectrometry and elastic recoil detection analysis of as-deposited FIBID-C was determined to be C:Ga:H = 87.4:3.6:9.0.

A different annealing mechanism is found when heating is executed including a metal foil, such as Ni or Cu as a catalyst for crystallization. The procedure follows well-known techniques, for example, mimicking concepts of conventional CVD nanocarbon synthesis, namely, graphene and CNTs [4], or akin to standard alloying-inspired methodologies [35]. The experiments not only demonstrate the possibility to convert amorphous FIBID-C to patterned graphene-like material but also reveal particularities intrinsic to the use of FIBID-patterned material and metal foil, both specifically. Remarkably perhaps it all is that especial just because we are dealing with carbon.

We use high-purity and commercially available Ni foils from Nilaco. After cutting them into pieces slightly bigger than the processed sample, Ni foil piece is placed between the patterned FIBID-C sample and the capping chip, which results doubly useful to sustain the sandwiched structure during pumping and to promote flattening of the easily bendable Ni foil. Then, the thermal treatment is applied, analogous to the process described above.

The better results for obtaining the transformation of amorphous FIBID-C into graphene-like patterns were achieved using 975 °C and 10 μm Ni. Thinner foils (5 μm) tended to melt, and thicker foils (50 μm) seemed less efficient to induce crystallization, probably strongly linked to the foil stiffness. Using higher temperature also promoted certain fusion of the metal and patterned sample, whereas lower temperatures appear insufficient to sponsor graphitization of the ultrathin FIBID-C. Further description of these parameters is discussed in terms of Raman scattering characterization in the next section. Optical inspection exposes that, as in previous case, wafer surface keeps most of its integrity and patterns of any size, shape, and thickness remain on the surface, but SEM and AFM inspection reveals that eventually deformation of the patterns may be found in thickness, and roughness, as well as impurities (Fig. 6.4).

The TEM inspection confirms the conversion of as-deposited compact FIBID-C into multiple carbon layers. The impurities are identified as compounds of Ni and silicides, which is considered to be the result of a eutectic formation, probably related to the escape of residual Ga. The XPS also supports qualitatively and quantitatively the predominance of C–C $sp^2$ bonds and existence of defects as clearly deduced from Raman spectrum signatures.

Again, other authors have also studied a similar idea and effectively combined the FIBID-C with some metallic catalyst to boost crystallization by thermal treatment [36]. In that case, they propose an "autarkic" approach where the very gallium, from the ion probe which is used for the FIBID, is implanted. Here, not only the crystalline grade is not better than our route for same processing temperatures, but also the patterns tend to lose most of their original shape.

The use of the metal catalyst in the specific form of foils, instead of other more conventional strategies, such as the use of thin metal layer deposition, appears to be

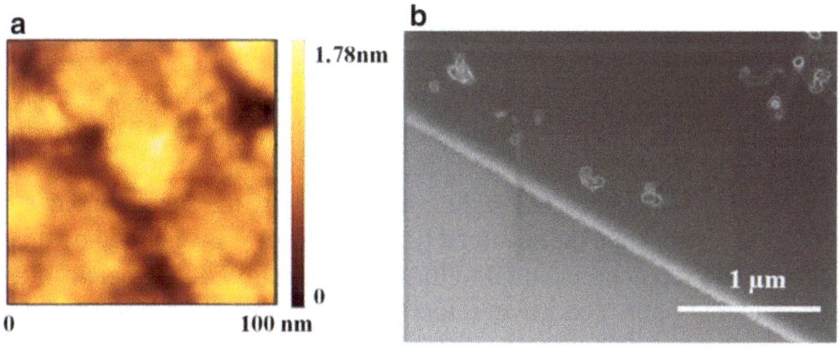


**Fig. 6.4** Examples of products of Ni-assisted thermal annealed of FIBID-C. (**a**) AFM image of a relatively low roughness area; (**b**) SEM image of another feature where some impurities, Ni precipitates, embedded in the FIBID-C pattern can be observed

multiplex advantageous. In addition to the convenience of allowing the synthesis directly on an insulating substrate such as relevant $SiO_2$, spontaneous detachment of metal and FIBID-C patterns after cooling makes a metal etching step unnecessary. In summary, it would be useful for the achievement of a larger crystal domain to have certain control and uniformity of the metal foil, in order to guarantee the planarity over the FIBID-C patterns and sample support. Besides, eventual rests of metal impurity and silicide particle formation limit the approach potential. Additionally, the minimum FIBID-C thickness available and compatible with the present graphitization mechanism probably discards its pertinence for strictly single-layer graphene applications.

### 6.3.4 *Raman Scattering in Carbons*

Raman spectroscopy is based on the inelastic scattering of light by matter, including from molecules to crystals [37]. The effect is highly sensitive to the physical and chemical properties of the scattering material as well as to any external perturbation that may modify these properties. Raman spectrometers are nowadays widely available; the technique is relatively simple to perform, is possible to carry out at room temperature, and, under ambient pressure, requires relatively simple or no specific sample preparation. For all these reasons and applied to the present study, Raman spectroscopy has historically played an important role in the structural characterization of carbon-based materials [38, 39] and, in particular, is commonly used to characterize all $sp^2$ carbons from bulk to low-dimensional objects, such as 3D graphite, 2D graphene, 1D CNTs, and 0D fullerenes [40–45].

In a Raman scattering process with incident photon energy $\hbar\omega_L$, the scattered photons, $\hbar\omega_s$, have lost (Stokes scattering) or gained (anti-Stokes scattering) the energy $\hbar\omega$ of an elementary excitation, for instance, a vibrational excitation in molecules or phonons in solids:

$$\hbar\omega_S = \hbar\omega_L \mp \hbar\omega \tag{6.1}$$

The Raman scattering is sensitive to probe the local atomic arrangement. Therefore, crystalline and amorphous phases as well as the influence of strain can be detected in the Raman vibrational spectra.

In a first-order Raman process, light is absorbed with the energy and momentum of the incident photon, and the scattered light is typically downshifted by the energy of a phonon with momentum equal to the momentum difference between the incident and scattered light. The wavevectors of the phonons thus produced are therefore very small compared to the size of the Brillouin zone, BZ ($q \sim 0$, $\vec{q}$ being the wavevector transferred to the excited phonon). These phonons near the Γ-point are Raman-active modes in the solid state.

In a second-order Raman process, the requirement of $q \sim 0$ is relaxed, and a pair of $q \neq 0$ phonons with wavevectors $q$ and $-q$ are relevant. The second (and higher)-order Raman processes generally give a broad background signal in the Raman spectra. Such higher order processes can have large intensities, comparable to the first-order process when two of the three intermediate states in the two-phonon scattering (or one-phonon and one-elastic scattering) process are real electronic states. Then, the double-resonance (DR) effect results in spectral features as intense as the first-order resonance Raman process. Because of the DR Raman process, $sp^2$ carbons show a strong Raman feature, and we are able to detect the Raman signal of a single monolayer graphene or of a single CNT.

#### 6.3.4.1 Instrumentation

The basic setup for the Raman scattering experiment consists of an excitation laser, a monochromator, and a detector. An interference filter is needed before the sample to remove plasma lines (or luminescence) from the lasers. For the excitation, typically both discrete excitation energies from Ar:Kr and He:Ne lasers and tunable Ti:sapphire and dye lasers are used. Both Ti:sapphire and dye lasers are usually pumped by an Ar ion laser with output power of ~6 W. The Ti:sapphire laser provides tunable output wavelengths in the near-infrared (NIR) range, and the dye lasers allow to change the output wavelengths in the visible region, and the specific wavelength range depends on the kind of dye solution used.

When we use one laser energy and observe the intensity of the scattered photons as a function of the shift of energy in $cm^{-1}$ from the incident light (the Raman shift, refer to expression (6.1)), this plot is called a Raman spectrum, from which we determine the phonon energy and its spectral width. The Raman spectrum is observed by passing the scattered light through a monochromator which disperses the light entering the instrument. At the zero energy shift a strong Rayleigh or elastically scattered signal exists which needs to be eliminated.

The separation of the Raman signals from the much stronger elastically scattered light has been achieved in the past by use of double or triple monochromators. The technical solution used nowadays is Raman spectrometers consisting of a holographic notch filter for stray light suppression and a single grating spectrograph. The notch filters suppress the incident laser light by a factor of more than $10^6$, while the rest of the spectrum is transmitted with a high efficiency (close to 100 %). In this way, a more intensive Raman signal is obtained with a single monochromator (usually 10–100 times stronger). For recording the scattered light, the charge-coupled devices (CCD) are nowadays used in the most Raman spectrometers. They have very high quantum efficiency over a broad spectral range and very low dark count level when cooled.

In micro-Raman measurements, an optical microscope is attached to the system so that we can get a spatial resolution of the order of the wavelength of light. The main additional option is the motorized stages for XY(Z) mappings. Raman maps are images generated from spectra recorded at discrete points of the sample (where the recording is automated). Recent developments in detector and computer technology as well as optimized instrument design have reduced integration times of Raman spectra by orders of magnitude, so that complete chemical mappings can be acquired in seconds or minutes with high spatial resolution. They show the variation of any fitted parameter (i.e., intensity, width, or position of one band) as a function of the point of analysis. In this way, it is possible to plot a "smart map" of the parameter (color or contrast scaling) superimposed with the optical image of the probed area [46].

#### 6.3.4.2 Raman Spectra of Diamond, Graphite, Graphene, Carbon Nanotubes, and Amorphous Carbons

The Raman Spectrum of $sp^3$ Bonded Carbon: Diamond

Diamond crystallizes in the cubic diamond structure, which consists of two interpenetrating fcc lattices, shifted from each other by one-quarter of the diagonal of the unit cell. Each $sp^3$ bonded carbon is surrounded by a regular tetrahedron formed by the nearest neighbors. The diamond crystal has only one triply degenerate optical phonon of $T_{2g}$ symmetry near the center of the Brillouin zone. This phonon is observed in the Raman spectrum at 1,332 $cm^{-1}$, as a sharp line (full width half maximum <2 $cm^{-1}$). A broad weak band appears in the frequency range of 2,100–2,700 $cm^{-1}$ associated with second-order Raman scattering.

The Raman Spectra of $sp^2$ Bonded Carbons: Graphene and Graphite

The Raman spectrum of diamond is easily distinguished from the spectrum of graphite even though both are composed entirely of C–C bonds. The graphite spectrum has several bands in the spectrum, and the main band has shifted from 1,332 $cm^{-1}$ in diamond to 1,582 $cm^{-1}$ in graphite as it is composed of $sp^2$ bonded carbon in planar sheets in which the bond energy of the $sp^2$ bonds is higher than the $sp^3$ bonds of diamond. The higher energy of the $sp^2$ bonds in graphite pushes the vibrational frequency of the bonds and hence the frequency of the band in the Raman spectrum to a higher frequency. The 1,582 $cm^{-1}$ band of graphite is known as the G band.

Graphene is the fundamental building block of many important carbon materials, including graphite. Graphite consists of stacks of planar graphene sheets. Monolayer graphite, graphene, is the simplest among the $sp^2$ carbon systems and has, consequently, the simplest Raman spectrum. Graphene is constituted by a single atom network of C atoms, perfectly ordered in a hexagonal arrangement. A large and ideal graphene sheet manifests two Raman phonon modes: the standard first-order Raman phonon, the G band at 1,580 $cm^{-1}$ arising from the in-plane bond-stretching vibration of pairs of $sp^2$ C atoms [43]. This mode does not require the presence of sixfold rings, and so it occurs at all $sp^2$ sites, not only those in rings. It always lays in the range of 1,500–1,630 $cm^{-1}$, as it does in aromatic (e.g., phenanthrene) and olefinic molecules [47]. The second band is a band at $<$2,700 $cm^{-1}$, historically named G′, since it is the second most prominent band always observed in graphite samples [48]. However, we now know that the G′ band is due to second-order Raman scattering. Thus, we refer to the G′ band as the 2D band. The evolution of the 2D band for different graphene sheets has been used for determining the number of graphene layers as well as for probing electronic structure through the double-resonance process. A symmetric and sharp 2D band (~30 $cm^{-1}$) can be used as a "detector" for monolayer graphene [41, 42].

In the presence of under-coordinated C atoms, structural defects, or edge atoms, the so-called defect-induced D band at frequencies around 1,345 $cm^{-1}$ can be resolved, having a variation in the intensity correlated with the edge conditions. The D band involving phonons near the K zone boundary is attributed to the breathing mode of $A_{1g}$ symmetry of aromatic rings [49–52]. This mode is forbidden in perfect graphene and involves an intervalley double-resonance process, so it only becomes active in the presence of disorder, i.e., activated by the relaxation of the $q=0$ selection rule (where $q$ is the phonon wavevector). The defect provides the missing momentum in order to satisfy momentum conservation in the Raman scattering process [53, 54]. Its intensity is strictly connected to the presence of sixfold aromatic rings. The D band grows in intensity with increasing disorder or decreasing crystal size.

Double resonances also can happen, as intravalley process, i.e., connecting two points belonging to the same cone around K or K′ [55]. This fact gives rise to the so-called D′ peak. Measured with the 514.5 nm excitation wavelength, D′ peak is centered at 1,620 $cm^{-1}$ in defective samples [49].

The 2D peak at ~2,700 $cm^{-1}$ is the second order of the D peak [41, 49], and the 2D′ peak at ~3,210 $cm^{-1}$ is the second order of D′. Since 2D and 2D′ originate from a process where momentum conservation is satisfied by two phonons with opposite wavevectors and no defects are required for their activation, they are always present. On the other hand, the D + D′ band (~2,940 $cm^{-1}$) is the combination of phonons with different momenta, around K and Γ, and thus requires a defect for its activation.

There are clear differences between the Raman spectra of graphene and graphite. A change in the shape and intensity of the 2D peak of graphene exists compared to bulk graphite. The 2D peak in bulk graphite consists of two components $2D_1$ and $2D_2$ [49, 50], roughly 1/4 and 1/2 the height of the G peak, respectively. On the contrary, the line shape of the 2D band of single-layer graphene is unique compared to multilayer graphene and graphite. It has a single Lorentzian line shape (~30 $cm^{-1}$)

and a high intensity roughly four times more intense than the G peak [41]. The 2D band can be explained with a double-resonance Raman process [53, 54] and has a close correlation with the electronic band structure of the graphitic materials [41, 42]. As the number of graphene layers increases, the electronic band structure varies and approaches that of graphite [54]. The line shape of the 2D band changes significantly as the thickness increases from 1 to 4 layers. Since the line shape of the 2D band reflects the electronic band structure, we can infer that the electronic band structure of five or six layers of graphene is very similar to that of graphite. There is also an overall blue shift of the 2D band with increasing number of layers. Moreover, the G band Raman frequency clearly downshifts with increasing number of layers (from ~1,585 to ~1,581 $cm^{-1}$ for graphite). The G band intensity increases linearly with the number of layers up to ten layers, and from the relative intensity of the G- and the 2D bands, one can reliably estimate the number of graphene layers.

### The Raman Spectrum of $sp^2$ Bonded Carbons: Carbon Nanotubes

CNTs are essentially rolled-up graphene sheets that have been sealed to form hollow tubes. Because of their unique mechanical, electrical, and thermal properties, CNTs are one of the most active areas in the field of modern carbon nanotechnology. Single-wall carbon nanotubes (SWCNTs) are cylindrical tubes with a single (outer) wall, the diameters of which are usually only 0.78–2 nm. There are also double-wall carbon nanotubes (DWCNTs) that have a second layer of graphene wrapped around a (inner) single wall. The latter are a subset of the larger category of multi-wall carbon nanotubes (MWCNTs) that have many layers of graphene wrapped concentrically around the core tube.

The Raman spectrum of an SWCNT bears similarity to graphene, which is not surprising as it is simply a rolled-up sheet of graphene. There are however two dominant signatures in the Raman spectra that distinguish SWNTs from other forms of $sp^2$ carbon.

The first relates to the low-frequency feature, usually in the range of 100–300 $cm^{-1}$, arising from scattering by the radial breathing modes (RBM). It corresponds to symmetric in-phase displacements of all the carbon atoms of the SWNT in the radial direction. The frequency of these bands can be correlated to the diameter of SWCNTs, and they can provide important information on their aggregation state.

The second signature relates to the multicomponent features associated with the tangential (G band) vibrational modes of SWNTs. The G band, ranging from 1,500 to 1,600 $cm^{-1}$ in SWNTs, is originated from the Raman active optical phonon mode $E_{2g}$ of 2D graphene by zone folding the 2D graphene Brillouin zone into the 1D SWNT Brillouin zone. Due to the phonon wavevector confinement along the SWNT circumferential direction and also due to symmetry-breaking effects associated with SWNT curvature, the G band is composed of several peaks, which can be assigned to different Raman modes by polarized Raman experiment. Based on a simple analysis, there are two intense peaks: $G^+$ and $G^-$. For semiconducting SWNTs, $G^+$ and $G^-$ stand for atomic displacements along the tube axis and the circumferential direction, respectively, which is opposite for metallic SWNTs [56, 57]. The G band can also be used to distinguish metallic SWNTs from semiconducting ones through strong differences in line shape and to characterize doping level and strains in SWNTs. Neither the RBM feature nor the multicomponent G band features are observed in any other $sp^2$ bonded carbon material.

MWCNTs have similar spectra to those of SWCNTs. The primary differences are the lack of RBM modes in MWCNTs and a much more prominent D band. The RBM modes are not present because the outer tubes restrict the breathing mode. Given the multilayer configuration a more prominent D band in MWCNTs is to be expected. It indicates more disorder and defects in the structure.

Fullerenes are molecules composed of $sp^2$ bonded carbon atoms that form hollow, cage-like structures. The arrangement of the atoms produces pentagonal and hexagonal (or sometimes heptagonal) rings that prevent the sheet from being planar. Fullerenes are highly stable chemically and have a variety of unusual properties. The most abundant fullerene is $C_{60}$, the so-called buckminsterfullerene, where the carbon atoms are located at the vertices of a truncated icosahedron. Since $C_{60}$ has the icosahedral $I_h$ symmetry (the highest possible molecular symmetry) its Raman spectrum is in principle fairly simple and shows only eight strong Raman lines [58, 59]. The most prominent one is the so-called pentagonal pinch mode (PPM) or $A_g(2)$ at ~1,469 $cm^{-1}$ mode that has been used repeatedly as an analytical probe for the structural and electronic properties of $C_{60}$ [60]. Besides investigation of pristine $C_{60}$ by Raman scattering, crystal field effects, photosensitivity, and breaking of selection rules by crystal defects or isotope contamination can make the detailed interpretation of the spectra of fullerenes not straightforward.

### The Raman Spectra of Mixed $sp^2$–$sp^3$ and $sp^1$ Bonded Carbons: Amorphous Carbons

Typical noncrystalline, amorphous carbon (a-C) shows a broad Raman band in the range between 1,000 and 1,800 $cm^{-1}$ (see Fig. 6.5, lower spectrum). It can be often fitted very well by two Gaussian curves on a linear background, centered at 1,300–1,380 $cm^{-1}$ and 1,520–1,580 $cm^{-1}$, respectively, if excited with $\lambda_L = 514.5$ nm. Commonly they are labeled D and G bands as in the case of disordered graphite. However, their frequencies and assignments differ from those observed in disordered graphite, since amorphous carbon consists of a mixture of $sp^2$, $sp^3$, and even $sp^1$ bonding structures. Accordingly, many vibrations other than graphitic ones can contribute to the two bands.

By excitation with visible light the contributions due to vibrations of $sp^2$ bonded carbon are resonantly enhanced, therefore dominating the Raman spectrum. With UV excitation, an additional peak, the T peak, at approximately 1,100 $cm^{-1}$ appears [52]. This T peak is due to the C–C $sp^3$ vibrations. If $sp^3$ carbon bonding is very high, then this a-C can be denoted as a tetrahedral amorphous carbon (ta-C) [61]. For very hard, diamond-like amorphous carbon with high amount of $sp^3$ bonded carbon, the G band dominates and can be shifted to frequencies above 1,580 $cm^{-1}$.

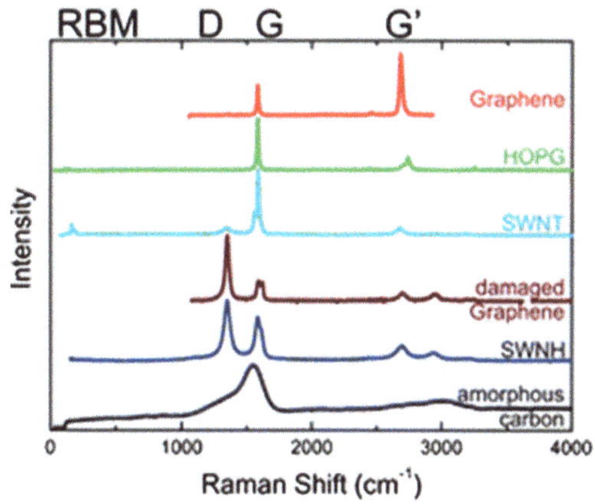


**Fig. 6.5** Summary of the exemplary Raman signatures of various graphene/graphite-like materials and nanostructured carbon. Reprinted with permission from Dresselhaus, M. S., Jorio, A., Hofmann, M., Dresselhaus, G., Saito, R.:. Nano Lett. 10, 751–758 (2010). Copyright 2010 American Chemical Society

Significant efforts have been devoted to quantify defects and disorder using Raman spectroscopy for nanographites [62–65], amorphous carbons [66], CNTs [67], and graphene [68, 69]. The first attempt was the pioneering work of Tuinstra and Koenig [70]. They reported the Raman spectrum of graphite and nanocrystalline graphite. They noted that the ratio of the D to G intensities varied inversely with the crystallite size, $L_a$. Later on, significant laser excitation energy, $E_L$, dependence of the intensity ratio was reported [51, 52]. Cançado et al. [64] measured this laser excitation energy dependency in the Raman spectra of nanographites, and the ratio between the D and G bands was shown to depend on $E_L{}^4$.

#### 6.3.4.3 Raman Characterization of Graphene-Like FIBID-C

Ni Foil Catalyst

The analysis of the synthesized products by Raman spectroscopy is based on this tradition and precision for the identification and characterization of the structural properties of C materials [40–45]. As stated before, the positions and the relative intensities of D and G Raman bands give information about the film's structural properties.

A Raman spectrum of an as-deposited carbon film (25 nm thick) is shown in Fig. 6.6, consisting of a broad and asymmetric Raman band centered at ~1,560 cm$^{-1}$. The broad band can be deconvoluted into two broad sub-bands, an intense G band at ~1,560 cm$^{-1}$ and a weak D band at ~1,350 cm$^{-1}$. This kind of Raman signature is typical of an amorphous and highly sp$^3$ bonded (ta-C) with only short olefinic sp$^2$ chains dispersed within this sp$^3$ matrix [51, 52]. Remarkably, due to the large optical absorption of FIBID-C, the prominent signal of the original Si substrate, the characteristic signature of which is a sharp peak at 520 cm$^{-1}$ and a second-order Raman band at 960 cm$^{-1}$, cannot be observed at those places where a FIBID-C has been deposited.

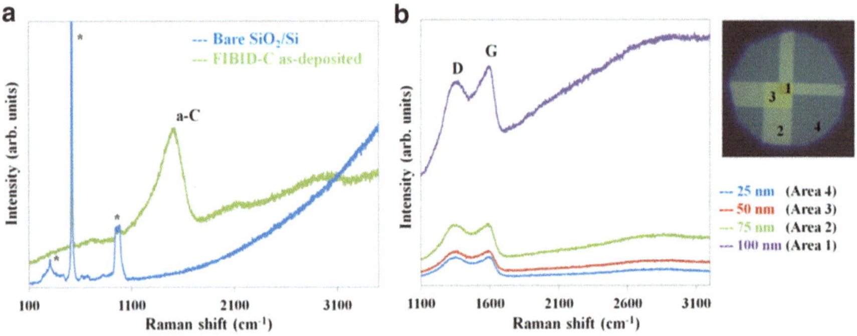


**Fig. 6.6** (**a**) Referential Raman spectra of the $SiO_2$/Si substrate used as support (*blue line*) (* Si bands) and of the as-deposited FIBID-C (before thermal treatment, i.e., amorphous), showing the typical Raman signature of a-C (*green line*); (**b**) FIBID-C patterns annealed without catalyst as a function of thickness. The broad D and G bands, as well as the absence of 2D band, indicate a low degree of graphitization. Reprinted with permission from Rius, G., Mestres, N., Yoshimura M.: J. Vac. Sci. Technol. B 30, 03D113-1 (2012). Copyright 2012 American Vacuum Society

In a vacuum-annealed sample (e.g., 30 min at 1,000 °C) the $sp^2$ sites begin to diffuse and condense into clusters of increasing size. Aromatic rings form and produce a clear D peak in the Raman spectrum. Decomposition of C–H bonds is an essential part of this transformation. As the $I$(D)/$I$(G) ratio is a measure of the ordering of the $sp^2$ phase, the $I$(D)/$I$(G) ratio increase, the shift towards higher frequencies of the G peak position (from 1,560 up to 1,590 $cm^{-1}$), and the shrinking of its linewidth in the annealed sample are correlated to an increase of the $sp^2$ phase and of growth in size of the graphitic domains. However, the graphitization is not complete or crystal relaxation is partial as understood from the broad D and G bands and the poor definition of the 2D and D′+D bands at high wave numbers (2,600–3,000 $cm^{-1}$) and supported by TEM observations.

Figure 6.7 illustrates the metal-induced heat transformation. Raman spectra as a function of deposited a-C thickness for 25, 50, and 100 nm measured after thermal annealing at 1,000 °C during 30 min using a Ni foil as a catalyst can be checked. The typical Raman signatures of graphene are observed at ~1,350 $cm^{-1}$ (D band), ~1,590 $cm^{-1}$ (G band), and ~2,695 $cm^{-1}$ (2D band). In addition, a small shoulder in the G band at ~1,620 $cm^{-1}$ assigned to the D′ band and the D+D′ band centered at 2,950 $cm^{-1}$ are observed.

The feature that becomes evident upon decreasing deposited a-C thickness, see Fig. 6.7, is that a larger intensity of the Raman peaks and a better resolution of the different Raman signatures are attained due to a reduction in the phonon bandwidth. Additionally, a decrease of the $I$(D)/$I$(G) intensity ratio occurs. Using the Ni foil catalysis, the $I$(D)/$I$(G) intensity ratio for annealed samples is typically around 0.9 or larger, which corresponds to a crystallite in-plane size ($L_a$) of about 20 nm or smaller. An in-depth limiting factor and having some relation with processing time, transformation rate, for the metal-assisted crystallization using the present foil-based method, can be derived from the experimental results, where the degree of

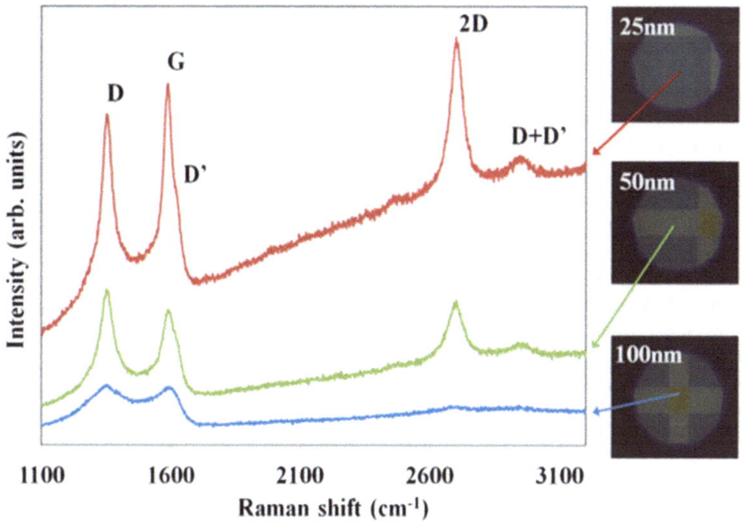


**Fig. 6.7** Raman spectra of 25 nm (*top*), 50 nm (*middle*), and 100 nm (*bottom*) thick C features after annealing at 1,000 °C during 30 min. The optical microscope images show the typical appearance of the patterned areas. Reprinted with permission from Rius, G., Mestres, N., Yoshimura M.: J. Vac. Sci. Technol. B 30, 03D113-1 (2012). Copyright 2012 American Vacuum Society

graphitization is inversely proportional to the layer thickness. The use of thinner C films (below 20 nm) did not lead to larger crystallite size of the produced graphene-like features. And then, the rate of graphene formation for ta-C features thicker than 100 nm is poor, as deduced from the Raman spectrum in Fig. 6.7.

Based on the correlation with previously reported results [43, 64], our FIBID carbon-based nanostructures annealed using a Ni foil as a catalyst can be identified as nanosized crystalline graphene. The AFM image in Fig. 6.4a gives an idea of the resulting surface morphology after metal-assisted annealing, corresponding to one of the Raman spectra sites in Fig. 6.7.

Raman mappings can be used to evaluate the uniformity of the resulting nanographene patterns. Figure 6.8 includes some mapping of the Raman intensity for a sample with an as-deposited ta-C thickness of 20 nm also processed at 1,000 °C for 30 min, in particular, for the G mode at 1,587 $cm^{-1}$ (left) and the 2D mode at 2,693 $cm^{-1}$ (right). The partial mapping reveals good uniformity of the nanographene sheet along the patterned area, although a slightly higher sensitivity of the 2D peak intensity as a function of position is observed.

### Cu Foil Catalyst

Complementarily, we have studied the effectiveness of Cu foil (10 μm thick) as a metal catalyst for graphene formation. Compared to Ni, we observe a less effective graphitization, indicated by a strong D mode and a broadened G mode

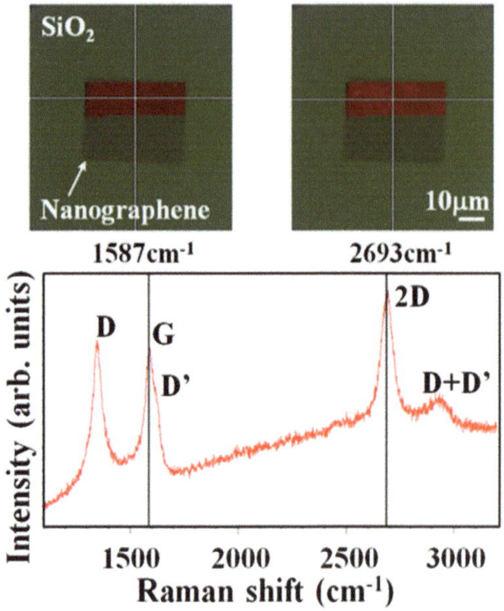


**Fig. 6.8** *Red*-colored areas correspond to the partial area Raman intensity mapping of the G mode at 1,587 cm$^{-1}$ (*left*) and the 2D mode at 2,693 cm$^{-1}$ (*right*) for a 20 nm thick FIBID-C layer (40×40 µm$^2$, seen as a *darker area*). Support is 500 nm thick $SiO_2$. Reprinted with permission from Rius, G., Mestres, N., Yoshimura M.: J. Vac. Sci. Technol. B 30, 03D113-1 (2012). Copyright 2012 American Vacuum Society

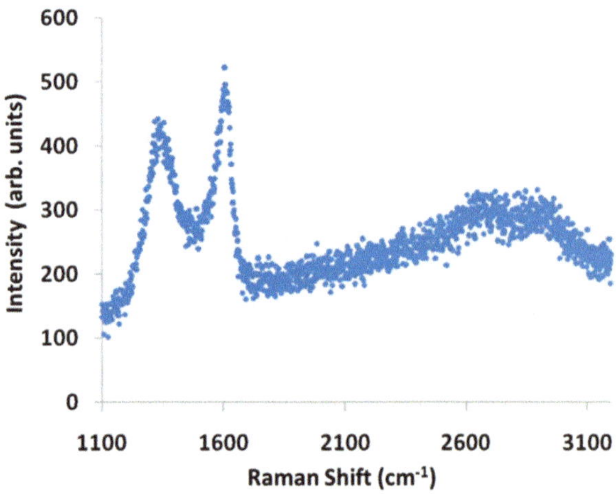


**Fig. 6.9** Raman spectra of ~4 nm thick C feature after annealing at 950 °C during 5 min using a 10 µm thick Cu foil catalyst

and a weak and broad 2D band. See Fig. 6.9 for a sample annealed at 950 °C for 5 min; the AFM estimated film thickness was ~3.5 nm. Longer thermal processing times and higher annealing temperature did not lead to a better graphitization. But more important, these parameters, both a prolonged processing time and a higher temperature, eventually were detrimental to sample integrity (sticking of Cu foil and Si support). The fact that the melting temperature of Cu is lower than for Ni together with the fact that carbon solubility in Cu is very low

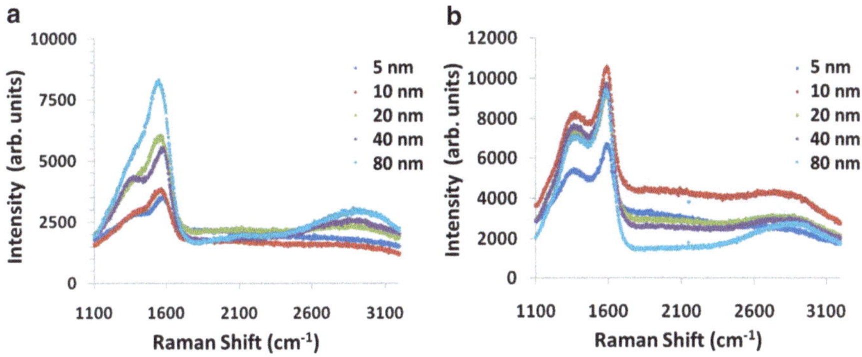


**Fig. 6.10** Evolution of the Raman spectra with the FIBID-C thickness for two different temperatures, (**a**) 500 °C and (**b**) 700 °C, using a Cu foil (10 μm) as catalyst

may explain the difference in the behavior and degree of the crystallization of FIBID-C patterns by Cu with respect to Ni.

Considering now the evolution of Raman scattering with processing temperature we can observe that there is a clear direct correlation of D band width and annealing temperature. As observed in Fig. 6.10, annealing at lower temperatures leads to broader Raman bands. The 2D band indicative of graphitization starts to appear on thicker patterns when the annealing is performed at 500 °C. This fact indicates that graphitization can take place at these lower temperatures. When annealing at 700 °C the splitting of D and G bands is more pronounced indicative of the reduction of their width. However, the broader D and G Raman bands observed in the spectra of samples annealed at low temperature still indicate smaller size of the graphite clusters and pronounced disorder of the carbon structure as compared to Ni results, indicating reduced efficiency of Cu metal thermal-induced crystallization catalysis.

## *6.3.5 Platform Versatility*

### 6.3.5.1 Combination of FIBID-C and Metals: Film and Particle Deposition

As it has been presented along the previous sections, a basic route to modify FIBID-C is a supplementary heat treatment. When crystalline carbon nanomembranes are pursued the metal-assisted treatment of FIBID-C boosts crystallization albeit keeping temperatures relatively low, because it lowers activation energy.

After seeing the practicality of using metal foils of Ni and Cu, we complement our knowledge on thermal transformation of FIBID-C by some other metallic materials and several methodologies to introduce the metal catalysis. As the exemplary cases, Pt thin-layer deposition, NiOx nanoparticles, and micron-patterned Fe thin films are used to illustrate the additional potential of our approach.

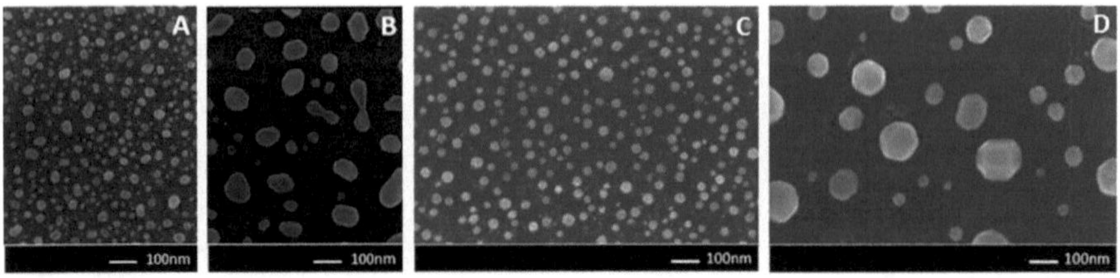


**Fig. 6.11** SEM top-view images of the effect of 975 °C annealing of ultrathin Pt layer deposited on $SiO_2$ (**a, b**) versus FIBID-C (**c, d**). Pt thicknesses are 5 nm for (**a**) and (**c**) and 10 nm for (**b**) and (**d**)

A simple way to study the effect of a metal on the thermal annealing of FIBID-C is depositing thin metal layers by sputtering or evaporation. Ion sputtering of ultrathin Pt films has been tested on our standard substrate ($SiO_2$), both before and after FIBID-C.

Figure 6.11c, d shows the typical morphology of Pt after 30-min heating at 975 °C, i.e., treatments are analogous to the foil case. Decomposition of Pt thin film is observed for all temperature range tested, which includes treatment temperatures above 900 °C. The as-deposited Pt films were inspected by AFM to corroborate the continuousness of the films ≥ 3.5 nm, where thicknesses of 5 and 10 nm are typically used. Slight roughness proportional correlation is observed for the Pt coating, as a function of deposited thickness. Although we have performed some more systematic trials, including the temperature and thickness tests and depositing Pt before and after FIBID-C, the discussion is here centered in the surface phenomena causing Pt nanostructuring.

On the one hand, it is clear that Pt film melts which would suggest that the interaction with the FIBID-C is likely. On the other hand, the contrast of, for example, the 10 nm thick FIBID-C pattern is rather homogeneous, which would indicate little metal-induced transformation. We cannot precisely and completely quantify the degree of graphitization by Raman spectroscopy because it may be preferentially circumscribed to the resulting interface between Pt and FIBID-C, but in general the degree of metal-induced crystallization on FIBID-C does not reach the results obtained using Ni or Cu foil. Even though Pt, as well as Ni, Cu, Co, etc., is known as an efficient metal catalyst for the synthesis of carbon nanomaterials [71], what can be found on C material resembles the thermal annealing performed without metal, as explained in Sects. 6.3.2 and 6.3.3.

The whole behavior is in accordance with the melting point depression (of Pt), the low solubility of C in Pt, and robustness of as-deposited FIBID-C (more details on this aspect are addressed later, Sect. 6.3.5.2). Remarkably, the formation of homogeneous shape Pt nanoparticles is not intrinsic to the high-temperature treatment, since random clustering is observed, for example, on the bare $SiO_2$ (Fig. 6.11a, b). This suggests that surface tension and FIBID-C roughness also play a role in the Pt NP formation.

Consequently, FIBID-C provides a convenient support to obtain controllable Pt particle size by adjusting the starting Pt film thickness, where faceted Pt can be clearly observed from the thicker Pt films (Fig. 6.11d). Rather than for the purpose

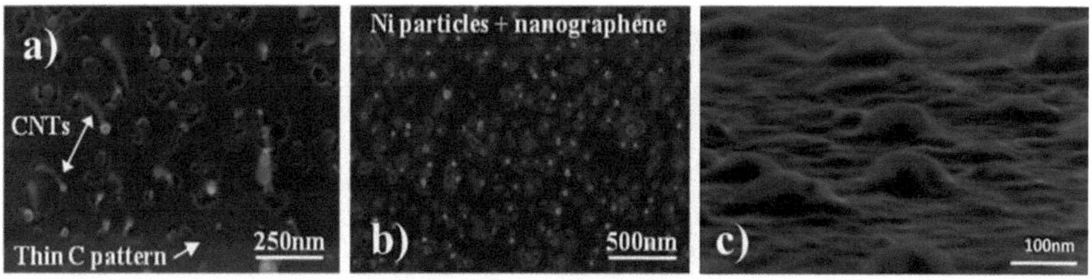


**Fig. 6.12** Crystallization of FIBID-C induced by heating process with $NiO_x$ nanoparticles: (**a**) Thin carbon precursor makes possible the formation of tubular structures, (**b**) while Ni particles embedded in nanographene are obtained for thicker FIBID-C

of graphitization catalysis, this method would represent an easy way to decorate carbon with metal nanoparticles for sensing or catalysis.

Using metal-containing nanoparticles has also been tested, specifically, in order to observe the direct and local interaction of FIBID-C with the metal (Ni) under the thermal treatment. In particular, crystalline Ni/$NiO_x$ NPs synthesized by sol–gel and precipitation method are used to observe the metal-induced crystallization. Clearer results are obtained when performing the FIBID-C after a low-density Ni/$NiO_x$ NPs have been dispersed; simple drop casting or spin coating onto the $SiO_2$ support is used (Fig. 6.12).

Combined with thin FIBID-C, below 5 nm, tubular carbon structures can be obtained. For thicker FIBID-C the Ni particles are found embedded (and even capped) inside the nanographene film. The crystalline degree of the latter has been confirmed by Raman spectroscopy, signature of local nanosized spots of few layer graphene surrounded by amorphous C (not for the tubular structures).

Briefly explaining the observed phenomena, if FIBID-C layers are significantly thinner than the average $NiO_x$ particle size FIBID-C acts as the precursor material for the formation of the CNTs. The formation of the CNTs proceeds by tip growth. For thicker FIBID-C, the particles graphitize the immediate carbon material in a thermally activated process, similarly to the one seen for the Ni foil. Although a number of the metal-containing nanoparticles are originally oxidized, their contact with carbon during low–middle stage of heating (>300 °C) transforms them into metallic nickel, thus enabling the high-temperature crystallization of adjacent FIBID-C.

As a last choice of metal to be exemplified, thin layers of Fe simply micropatterned by photolithography are used as the catalyst to convert FIBID-C into graphene-like material. Actually, related to this development, some group [72] has explored patterned Fe in order to obtain site-specific (transfer-free) graphene FETs, combined with microelectrodes for the external electrical test.

In this work, they used Fe patterns instead of a continuous Fe film which is the conventional catalyst form for growth of graphene by CVD. The idea allowed them to obtain a few layers to multilayer graphene in determined locations with reasonably large grain size, which can be routinely interfaced macroscopically by photolithography. The best results are obtained by rather thick, about 200–500 nm, iron features. Obviously a metal etching is still necessary, but transfer is eluded.

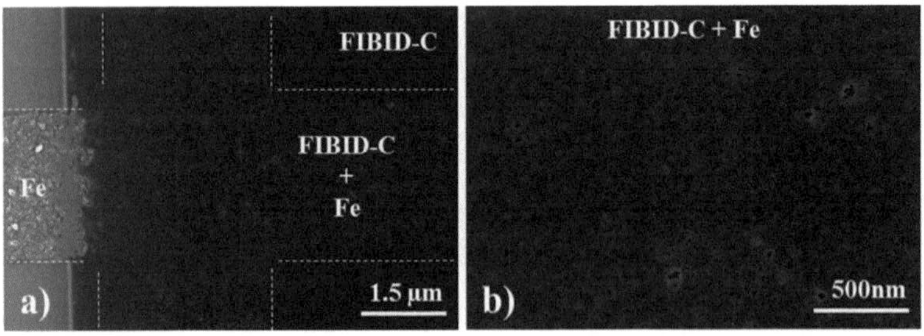


**Fig. 6.13** SEM images of ~8 nm thick Fe layers after thermal annealing: (**a**) General view (*dotted lines* are guide for the eye to recognize cross-like Fe pattern underneath the FIBID-C); (**b**) detailed image of the compound FIBID-C+Fe area. Formation of minute Fe particles beneath the C layer is apparent

The synthesis approach we have tested roots in a similar idea but substitutes the CVD stage by FIBID-C plus thermal annealing. Still in its preliminary stage of investigation, the temperature and processing time conditions have been similar to the previous cases, 975 °C and 30 min. Iron patterns are ~8, 18, and 92 nm thick, and FIBID-C-deposited thickness ranges from 10 to 80 nm.

The FIBID-C-patterned area is always designed to exceed the Fe pattern. Although a large number of structures can be observed as a result of thermal treatment, here only the most relevant aspects are presented. In all cases, the presence of iron promotes crystallization to a certain extent. Particularly there is a strong correlation of Fe and FIBID-C relative thicknesses.

For very thin Fe layers, ~8 nm, when deposited FIBID-C is also thin, same order ~10 nm, the decomposition into particles or cracked film of both FIBID-C and Fe is obtained. If covered by a thicker FIBID-C layer, Fe is transformed into spherical particles inside or underneath the FIBID-C (Fig. 6.13). Remarkably graphitization occurs both inside and outside the overlap area of FIBID-C and Fe patterns, suggesting some Fe migration to accomplish metal-induced crystallization.

For intermediate thickness of the Fe patterns, ~18 nm, severe decomposition of Fe features is also obtained. FIBID-C thickness test reveals that there is a threshold C thickness to obtain nanographene, similar to the Ni foil described in Sects. 6.3.3 and 6.3.4. Formation of a new compound, resembling carbide, and metal rearrangement are also apparent.

The best (strictly) graphene obtained from FIBID-C, in absolute terms of catalyst configuration method, metal material, etc. based on FIBID-C, is achieved using some ~92 nm thick Fe patterns. Control Fe features already account for the formation of big clusters due to the heat treatment, but the most relevant result is that the Fe pattern decomposition leads to the formation of graphene islands which always correlate with a terraced morphology underneath. It occurs when FIBID-C thickness is about half–one-third the Fe layer thickness. The heat process in this configuration leads to remote graphitization as well.

In brief, we hypothesize that the formation of graphene using FIBID-C and Fe catalyst thin patterns is a kind of carbon-derived carbide [7], i.e., graphene results from the precipitation of atomic C due to a saturation of the iron carbide (excess of C).

Summarizing this last instance, the results of Fe–FIBID-C show obvious differences in the graphene growth mechanism and conditions with respect to the Fe-CVD graphene. As learned from our experience, although as-deposited FIBID-C can be considered metastable atomically under temperature treatment it is actually a robust film. Its transformation to graphene-like materials has a lot to do with the particular characteristics of the FIBID-C film due precisely to its ion beam-based formation and molecular precursor.

#### 6.3.5.2 Processing Capability of FIBID-C: Support Materials and Chemical Etching

Within this section we relate a few aspects that have been explored in present FIBID-C-based approach as examples of its processing capability: universal substrate patterning, etching resistance, and mechanical properties.

As in the assortment of compatibility of FIBID-C with the diversity of metals and strategies in the previous section, we skip a profound description of, for example, the resulting interface and the deposition rate of FIBID-C for the collection of support materials that are tested and presented. Instead, only the materials resulting from heat treatment of FIBID-C are discussed.

One of the powerful characteristics of FIB, and particularly applicable to FIBID, is its potential to be applied over almost any substrate and material if the appropriate procedures are executed [32, 33], including odd surfaces and semiconducting or insulating materials.

Glass has been tested as the support for the experiments of FIBID-C plus thermal annealing. The main limitation in terms of FIBID may come from the fact that the glass has surface irregularities in the order or above of the typical FIBID thicknesses. As a result, deposited thickness precision decreases which roots on ion beam focus uncertainty. However, FIBID is a conformal technique; consequently, glass patterning by FIBID-C has been possible (Fig. 6.14a, b). The most evident observation is a certain distortion of the pattern design, attributed to the charging, which is common for insulating substrates. It can be completely compensated by simultaneous electron beam flow (although this option is not available in our system). Ni foil-assisted thermal annealing on glass works similar to $SiO_2$ supports as understood from Raman spectra while suggesting that increased glass roughness favors the contact of FIBID-C with the metal foil and thus crystallization (Fig. 6.14).

Testing HOPG as the FIBID-C support also provides valuable information on the as-deposited materials as well as the deposition process itself, for example, in terms of electrical conduction as it will be explained with more detail in the next Sect. 6.3.5.3. The deposition proceeds with no noticeable problem. Conductive AFM clearly shows that the as-deposited FIBID-C is basically insulating. Heat treatment restores the carbon surface so that electrical conduction is recovered but evidences that FIBID-C is not incorporated strictly ordered in HOPG or transforms into crystalline as a sort of epitaxy. More exhaustive and dedicated experiments would be needed to fully understand the complete nature of the damage and recovery process of HOPG and role of FIBID-C.

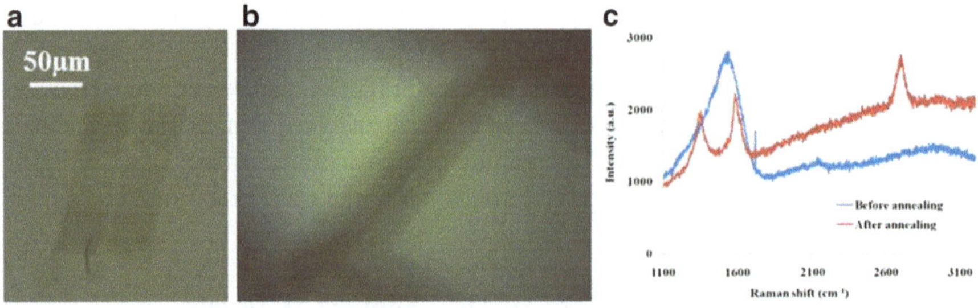


**Fig. 6.14** FIBID-C on glass (**a**) and (**b**) are optical microscope images of the patterns. Pattern distortion, attributed to support charging, is observed; (**c**) Raman spectra before and after the Ni-assisted thermal annealing, where nanographene formation is confirmed

Preliminary tests on SiC also confirmed that any epitaxial phenomena can be expected which can be easily explained as due to (1) the surface amorphization of the irradiated area and (2) the firmness of the as-deposited carbon layer, which requires higher temperatures, or otherwise metal catalysis, to liberate or decompose FIBID-C into atomic C that would eventually reconstruct into graphene-like material. However, SiC may be an attractive option to explore ion beam carbon deposition plus thermal annealing (for instance, for CNT growth due to material compatibility and combinability). What is more, some literature indicates that FIBID-C could be a powerful approach for selective graphene formation in SiC, as reported in [73] done by ion irradiation.

Moving to chemical etching, as metal catalysts are used to promote FIBID-C crystallization issues concerning metal contamination naturally arise. Rests and impurities, as well as reaction with other metals existing on the surface, may occur if metal-assisted thermal treatment of FIBID-C is combined with pre-patterned surfaces like those described in Sect. 6.3.5.3. As we have mentioned, the contamination by the very Ga ion beam can be neatly eliminated by some thermal annealing, but the presence of an additional metal may interfere.

Effectively, when large areas are patterned by FIBID-C and thermally treated with Ni foil, we observe some tendency to precipitate into the surface Ni-containing particles. The resistance of annealed FIBID-C to HCL:$HNO_3$ etching has been proven good and has allowed successful reduction of Ni remaining on surface (Fig. 6.15a).

Also concerning chemical etching, an interesting property of FIBID-C that has notorious implications for nanofabrication is the fact that FIBID-C is inert to HF solution. This allows envisioning a functional role of FIBID-C patterns, as it is masking for pattern transfer into $SiO_2$ or actually using FIBID-C films as an ultrathin structural layer for NEMS (Fig. 6.15b).

Intrinsically related to this latter possibility, the mechanical properties of FIBID-C are also outstanding. Again from the work of Kanda et al. [34], they have characterized the Young's modulus and hardness of FIBID-C films and monitored their evolution under thermal treatment. Their results show that Young's modules and hardness of ~12 and 23 GPa are characteristic of as-deposited FIBID-C and indicate how the thermal annealing causes their change (typically a degradation), associated to the H loss (and incomplete molecular bond or crystal annealing).

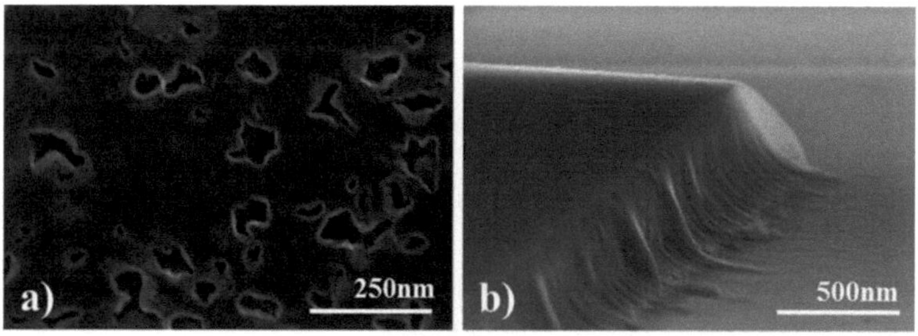


**Fig. 6.15** Testing FIBID-C behavior to etching processes: (**a**) Etching of Ni impurities; (**b**) FIBID-C used as a mask for pattern transfer during $SiO_2$ etching

#### 6.3.5.3 Proof of Technology via Electronic Devices: Integration and Compatibility Issues

Integration of FIBID-C plus heat treatment into processing sequences including other lithography levels are described here, in particular, for its application as a component of electronic devices.

Correlated to the idea of integrative approaches of FIBID-C, it is worth mentioning that studying the metal to nanostructured carbon boundary and how to obtain low contact resistance (Ohmic contacts) is intrinsically an important topic. Incorporation of thin carbon layers into the metal electrode has been proposed as an alternative to optimize nanocarbon electronics, for example, for the CNT–metal electrode interface [74].

Preliminary evaluation of the electrical conduction has been done by conductive AFM (an AFM option where one of the electrodes is the same tip which is also used as the topographical imaging probe). For the conduction tests on FIBID-C the second electrode has been manually placed using conductive carbon tape. Since manual preparation is used, a large area would be needed which can be achieved by pattern stitching (remember that typical pattern dimensions are $40 \times 40\ \mu m^2$). However, advanced beam control and precise motorized stage have not been available in our system, so pattern overlap resulted necessary. The important results of that experiment are that they confirmed the high resistivity of as-deposited FIBID-C, while it significantly reduced after thermal annealing.

Similarly, overlapping strategy has been used to produce large areas which, this time, had been combined with the definition of macroscopic contacts/electrodes by wire masking. Those results were already included within reported data in [10]. Still the method is not ideal to determine the electrical properties of single FIBID-C.

FIBID-C combined with the milling capabilities of FIB may also provide original architectures useful for the electrical characterization and device realization. As depicted in the sketch, $SiO_2$ removal by FIB can be used to drill a hole down to the doped Si, which is filled by the same FIBID-C, and then desired FIBID-C is produced on the surface (ready for thermal annealing). The convenience of the configuration has been demonstrated and the electrical conduction has been confirmed by, again, conductive AFM (Fig. 6.16). This strategy and sophisticated versions of it would be interesting for example for sensing or production of particular test templates.

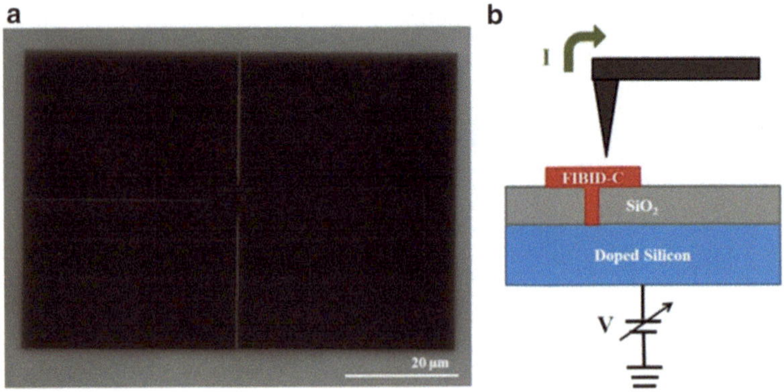


**Fig. 6.16** (**a**) Example of a structure produced by combining milling and FIBID capabilities; (**b**) proposed AFM-based methodology for testing the electrical performance of the nanoarchitecture

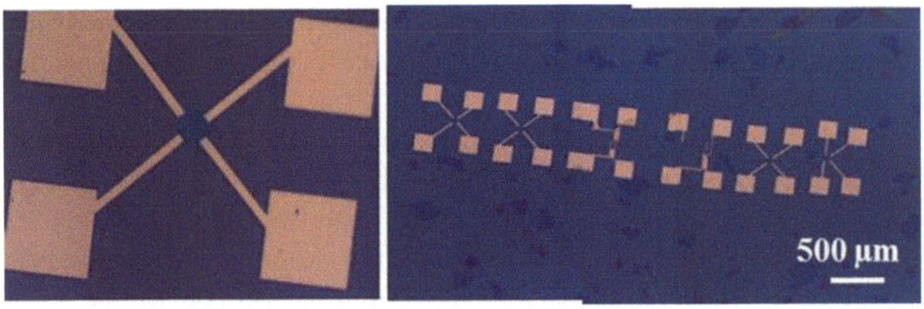


**Fig. 6.17** Optical microscope images of FIBID-C patterns combined with EBL, metal deposition, and resist lift-off for their robust electrical testing

Undoubtedly, the best option to provide electrical interface to FIBID-C for its macroscopic access is the "mix-and-match" routine, to combine FIBID with another precise lithography technique, as it could be photolithography or electron beam lithography.

Effectively, EBL has been the ideal tool to generate some good prototypes, hence allowing a more systematic study of FIBID-C variables, such as film thickness or degree of graphitization [75] (Fig. 6.17). The two key points that EBL provides for the present case are pattern flexibility and capability for precise positioning. Thanks to the precision of the EBL stage the use of marks for alignment could be avoided. An FIB system properly equipped would do the work similarly, or even simpler, since direct patterning of the metal contact could be possible.

Bringing the considerations about compatibility of the FIBID-C approach with other lithography techniques to a higher degree of development, such as photolithography, the obvious constraint is the thermal treatment. There are basically two issues for including FIBID-C plus heating as a part of multiple-step processing such as in microelectronic technology: (1) existing features can sustain the heat treatment with no or significant degradation and (2) materials do not participate through unwanted reactions. In the simplest scenario, integration means to make it compatible either with pre-patterned electrodes or alignment marks.

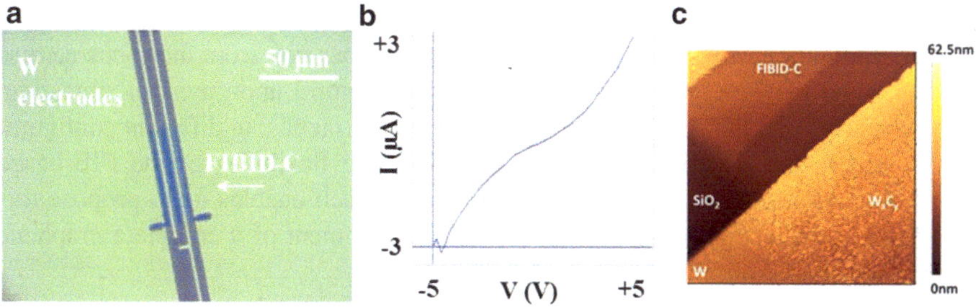


**Fig. 6.18** (**a**) Combination of FIBID-C (20 nm thick) with pre-patterned W electrodes, (**b**) electrical testing after FIBID-C thermal annealing (975 °C), (**c**) AFM image of the FIBID-C, W electrodes, and compound area, where reaction of FIBID-C is clear at the metal-C interface

In combination with photolithography, pre-patterned W electrodes had been tested (Fig. 6.18). In this instance, FIBID-C is performed onto the electrodes made of W that should sustain the thermal treatment. Effectively, no significant change is observed after heating the W layers, but the overlap of FIBID-C and W is a weak point, as seen in the electrical characteristics. Again a reaction between W and C occurs, probably creating some carbide which is manifested in the electrode boundary. Last option for compatibility with photolithography would be to perform FIBID-C on substrates containing alignment marks. The choice of the material has to be carefully studied to avoid incompatibilities and side effects (such as unwanted crossed reactions/alloying, degradation of the marks), especially, as already presented, if metal-induced crystallization of FIBID-C is used. Safe spacing should be used from each FIBID pattern to the alignment marks to avoid crossed effects during heating, which may lead to some sort of design rules similar to those characteristic of the resolution for each planar technology in microelectronics.

As a final remark, using topographical marks made by FIB milling, such as holes, not only may provide too low contrast for example for EBL alignment but also may expose materials during the heat treatment, like Si which is more sensitive than $SiO_2$ to high-temperature process, oxidizing, desorbing, or interacting with the metal catalyst.

Advanced device architectures exploiting the multiple materials are typically available in FIB systems, e.g., TEOS and Pt, such as in the case of FET configuration where top dielectric and gate metal contact could be also realized by FIBID, but have not been explored by us yet.

## 6.4 Summary and Outlook

This chapter has compiled our work on FIBID applied to the fabrication of patterned graphene-like materials. We have addressed a comprehensive description of the main technological points: FIBID patterning, metal-induced crystallization, processing compatibility, and integration aspects, towards device or specific template production, with an especial attention devoted to distinguish their atomic structure, mainly using Raman scattering technique.

Remarkably, the transition metal-assisted vacuum annealing of FIBID-C boosts the graphitization of the few-nm-thick patterns and benefits from the robustness of as-deposited FIBID-C to make possible a higher control upon thermal annealing. We have explored the capabilities of several kinds of metals, in different configurations, and tested electrical properties thanks to the flexibility of the FIB-based approach. As a prospect, the potential of the approach enables us to propose this platform as an inspirational concept for the development of a complete graphene technology based on ion beam-based deposition.

**Acknowledgements** The experimental work has been done during the postdoctoral fellowship of Dr. G. Rius at Toyota Technological Institute (TTI), Nagoya (Japan), and made possible thanks to the host, Professor M. Yoshimura, and the facilities of the Center for Local Structuring Control of TTI. The financial support of the Strategic Research Foundation Grant-aided project for Private Universities from MEXT (Japan) is acknowledged.

Financial support from Spanish MICINN (Consolider NANOSELECT (CSD2007-00041)) and Generalitat de Catalunya (2009 SGR 770 and Xarmae) are also acknowledged.

We are very thankful for the partnership with the Centro Nacional de Microelectronica (CNM), in Bellaterra (Spain), especially to Professor F. Pérez-Murano, Professor P. Godignon, M. Sansa, and L. Solé, for some microfabricated samples, electrical testing, and technical support. We wish to thank the GICSERV access program for granting access to the ICTS CNM Clean room.

XPS characterization and TEM inspection were done at Politecnico de Torino (Italy) and Nagoya University (Japan), respectively. We appreciate the collaboration of Prof. Tagliaferro and Prof. Tanji and their group members.

## References

1. Utke, I., Hoffmann, P., Melngailis, J.: Gas-assisted focused electron beam and ion beam processing and Fabrication. J. Vac. Sci. Technol. B **26**, 1197 (2008)
2. Dai, L.: Carbon Nanotechnology: Recent Developments in Chemistry, Physics, Materials Science and Device Applications. Elsevier Science, Amsterdam (2006)
3. Jariwala, D., Sangwan, V.K., Lauhon, L.J., Marks, T.J., Hersam, M.C.: Carbon nanomaterials for electronics, optoelectronics, photovoltaics, and sensing. Chem. Soc. Rev. **42**, 2824–2860 (2013)
4. Singh, V., Joung, D., Zhai, L., Das, S., Khondaker, S., Seal, S.: Graphene based materials: past, present and future. Progr. Mater. Sci. **56**, 1178–1271 (2011)
5. Williams, O.A., Nesladek, M., Daenen, M., Michaelson, S., Hoffman, A., Osawa, E., Haenen, K., Jackman, R.B.: Growth, electronic properties and applications of nanodiamond. Diam. Relat. Mater. **17**, 1080–1088 (2008)
6. Yamashita, T., Yamashita, K., Nabeshi, H., Yoshikawa, T., Yoshioka, Y., Tsunoda, S.-I., Tsutsumi, Y.: Carbon nanomaterials: efficacy and safety for nanomedicine. Materials **5**, 350–363 (2012)
7. Batzill, M.: The surface science of graphene: metal interfaces, CVD synthesis, nanoribbons, chemical modifications, and defects. Surf. Sci. Rep. **67**, 83–115 (2012)
8. Meyyappan, J.: A review of plasma enhanced chemical vapour deposition of carbon nanotubes. J. Phys. D Appl. Phys. **42**, 213001 (2009)
9. Li, Z., Wu, P., Wang, C., Fan, X., Zhang, W., Zhai, X., Zeng, C., Li, Z., Yang, J., Hou, J.: Low-temperature growth of graphene by chemical vapor deposition using solid and liquid carbon sources. ACS Nano **5**, 3385–3390 (2011)
10. Rius, G., Mestres, N., Yoshimura, M.: Synthesis of patterned nanographene on insulators from focused ion beam induced deposition of carbon. J. Vac. Sci. Technol. B **30**, 03D113-1 (2012)
11. Levesque, T.: Review of plasma cleaning technology for decontamination in FIB, SEM and TEM. Microsc. Anal. **25**, 11–14 (2011) (EU)

12. Giannuzzi, L.A., Stevie, F.A.: Introduction to Focused Ion Beams: Instrumentation, Theory, Techniques and Practice (Chapter 3). Springer, New York, NY (2004)
13. Kang, H.H., Chandler, C., Weschler, M.: Focused Ion Beam Systems: Basics and Applications (Chapter 3). Cambridge University Press, New York, NY (2007)
14. Matsui, S.: Focused-ion-beam deposition for 3-D nanostructure fabrication. Nuclear Instrum. Methods Phys. Res. B **257**, 758–764 (2007)
15. Langford, R.M., Wang, T.-X., Ozkaya, D.: Reducing the resistivity of electron and ion beam assisted deposited Pt. Microelectron. Eng. **84**, 784–788 (2007)
16. Young, R.J., Puretz, J.: Focused ion beam insulator deposition. J. Vac. Sci. Technol. B **13**, 2576–2579 (1995)
17. Luxmoore, I.J., Ross, I.M., Cullis, A.G., Fry, P.W., Orr, J., Buckle, P.D., Jefferson, J.H.: Low temperature electrical characterisation of tungsten nano-wires fabricated by electron and ion beam induced chemical vapour deposition. Thin Solid Films **515**, 6791–6797 (2007)
18. Sheed, G.M., Lezec, H., Dubner, A.D., Melngailis, J.: Focused ion-beam induced deposition of gold. Appl. Phys. Lett. **49**, 584–1586 (1986)
19. Langford, R.: Focused Ion Beam Systems: Basics and Applications (Chapter 8). Cambridge University Press, Cambridge (2007)
20. Phaneuf, M.W.: Introduction to Focused Ion Beams: Instrumentation, Theory, Techniques and Practice (Chapter 8). Springer, New York, NY (2004)
21. Orloff, J.: High-resolution focused ion-beams. Rev. Sci. Instrum. **64**, 1105–1130 (1993)
22. Hooghan, K.N.: Introduction to Focused Ion Beams: Instrumentation, Theory, Techniques and Practice (Chapter 5). Springer, New York, NY (2004)
23. Matsui, S., Kaito, T., Fujita, J., Domuro, M., Kanda, K., Haruyama, Y.: Three-dimensional nanostructure fabrication by focused-ion-beam chemical vapor deposition. J. Vac. Sci. Technol. B **18**, 3181–3184 (2000)
24. Menozzi, C., Gazzadi, G.C., Alessandrini, A., Facci, P.: Focused ion beam-nanomachined probes for improved electric force microscopy. Ultramicroscopy **104**, 220–225 (2005)
25. Motayed, A., Davydov, A.V., Vaudin, M.D., Levin, I., Melngailis, J., Mohammad, S.N.: Fabrication of GaN-based nanoscale device structures utilizing focused ion beam induced Pt deposition. J. Appl. Phys. **100**, 024306 (2006)
26. Vila, A., Hernandez-Ramirez, F., Rodriguez, J., Casals, O., Romano-Rodriguez, A., Morante, J.R., Abid, M.: Fabrication of metallic contacts to nanometre-sized materials using a focused ion beam (FIB). Mater. Sci. Eng. C Biomim. Supramol. Syst. **26**, SI1063–SI1066 (2006)
27. Li, W.X., Fenton, J.C., Gu, C.Z., Warburton, P.A.: Superconductivity of ultra-fine tungsten nanowires grown by focused-ion-beam direct-writing. Microelectron. Eng. **88**, 2636–2638 (2011)
28. Zakharov, D., Lebedev, G., Irzhak, A., Afonina, V., Mashirov, A., Kalashnikov, V., Koledov, V., Shelyakov, A., Podgorny, D., Tabachkova, N., Shavrov, V.: Submicron-sized actuators based on enhanced shape memory composite material fabricated by FIB-CV. Smart Mater. Struct. **21**, 052001 (2012)
29. Watanabe, K., Morita, T., Kometani, R., Hoshino, T., Kondo, K., Kanda, K., Haruyama, Y., Kaito, T., Fujita, J., Ishida, M., Ochiai, Y., Tajima, T., Matsui, S.: Nanoimprint using three-dimensional microlens mold made by focused-ion-beam chemical vapor deposition. J. Vac. Sci. Technol. B **22**, 22–26 (2004)
30. Hofmann, S., Cantoro, M., Kaempgen, M., Kang, D.J., Golovko, V.B., Li, H.W., Yang, Z., Geng, J., Huck, W.T.S., Johnson, B.F.G., Roth, S., Robertson, J.: Catalyst patterning methods for surface-bound chemical vapor deposition of carbon nanotubes. Appl. Phys. A Mater. Sci. Process. **81**, 1559–1567 (2005)
31. Stokes, D.J., Vystavel, T., Morrissey, F.: Focused ion beam (FIB) milling of electrically insulating specimens using simultaneous primary electron and ion beam irradiation. J. Phys. D Appl. Phys. **40**, 874 (2007)
32. Rius, G., Llobet, J., Arcamone, J., Borrise, X., Perez-Murano, F.: Electron and ion beam lithography for the fabrication of nanomechanical devices integrated on CMOS circuits. Microelectron. Eng. **86**, 1046–1049 (2009)

33. Rius, G., Llobet, J., Esplandiu, M.J., Sole, L., Borrise, X., Perz-Murano, F.: Using electron and ion beams on carbon nanotube-based devices. Effects and considerations for nanofabrication. Microelectron. Eng. **86**, 892–894 (2009)
34. Kanda, K., Igaki, J.-Y., Saikubo, A., Kometani, R., Susuki, T., Niihara, K., Saitoh, H., Shinji, A., Matsui, S.: Effects of annealing on material characteristics of diamond-like carbon film formed by focused ion beam chemical deposition. Jpn. J. Appl. Phys. **47**, 7464–7466 (2008)
35. Presser, V., Heon, M., Gogotsi, Y.: Carbide-derived carbons – from porous networks to nanotubes and graphene. Adv. Funct. Mater. **21**, 810–833 (2011)
36. Hatakeyama, T., Kometani, R., Warisawa, S., Ishihara, S.: Selective graphene growth from DLC thin film patterned by focused-ion-beam chemical vapor deposition. J. Vac. Sci. Technol. B **29**, 06FG04 (2011)
37. Raman, C.V., Krishnan, K.S.: A new type of secondary radiation. Nature **121**, 501 (1928)
38. Saito, R., Dresselhaus, G., Dresselhaus, M.S.: Physical Properties of Carbon Nanotubes. Imperial College Press, London (1998)
39. Robertson, J.: Diamond-like amorphous carbon. Mater. Sci. Eng. R **37**, 129–281 (2002)
40. Dresselhaus, M.S., Dresselhaus, G., Saito, R., Jorio, A.: Raman spectroscopy of carbon nanotubes. Phys. Rep. **409**, 47–99 (2005)
41. Ferrari, C., Meyer, J.C., Scardaci, V., Casiraghi, C., Lazzeri, M., Mauri, F., Piscanec, S., Jiang, D., Novoselov, K.S., Roth, S., Geim, A.K.: Raman spectrum of graphene and graphene layers. Phys. Rev. Lett. **97**, 187401 (2006)
42. Gupta, A., Chen, G., Joshi, P., Tadigadapa, S., Eklund, P.C.: Raman scattering from high-frequency phonons in supported n-graphene layer films. Nano Lett. **6**, 2667–2673 (2006)
43. Pimenta, M.A., Dresselhaus, G., Dresselhaus, M.S., Cançado, L.G., Jorio, A., Saito, R.: Studying disorder in graphite-based systems by Raman spectroscopy. Phys. Chem. Chem. Phys. **9**, 1276–1291 (2007)
44. Camara, N., Huntzinger, J.-R., Rius, G., Tiberj, A., Mestres, N., Pérez-Murano, F., Godignon, P., Camassel, J.: Anisotropic growth of long isolated graphene ribbons on the C face of graphite-capped 6H-SiC. Phys. Rev. B **80**(125410), 1–8 (2009)
45. Dresselhaus, M.S., Jorio, A., Hofmann, M., Dresselhaus, G., Saito, R.: Perspectives on carbon nanotubes and graphene Raman spectroscopy. Nano Lett. **10**, 751–758 (2010)
46. Colomban, P.: Raman analyses and "smart imaging" of nanophases and nanosized materials. Spectrosc. Eur. **15**, 8 (2003)
47. Lin-Vien, D., Colthurp, N.B., Fateley, W.G., Grasselli, J.G.: The Handbook of Infrared and Raman Characteristic Frequencies of Organic Molecules. Academic, New York, NY (1991)
48. Vidano, R.P., Fishbach, D.B., Willis, L.J., Loehr, T.M.: Observation of Raman band shifting with excitation wavelength for carbons and graphites. Solid State Commun. **39**, 341 (1981)
49. Nemanich, R.J., Solin, S.A.: First- and second-order Raman scattering from finite size crystals of graphite. Phys. Rev. B **20**, 392–401 (1979)
50. Lespade, P., Al-Jishi, R., Dresselhaus, M.S.: Model for Raman scattering from incompletely graphitized carbons. Carbon **20**, 427–431 (1982)
51. Ferrari, A.C., Robertson, J.: Interpretation of Raman spectra of disordered and amorphous carbon. Phys. Rev. B **61**, 14095–14107 (2000)
52. Ferrari, A.C., Robertson, J.: Resonant Raman spectroscopy of disordered, amorphous, and diamondlike carbon. Phys. Rev. B **64**, 075414–075513 (2001)
53. Saito, R., Jorio, A., Souza Filho, A.G., Dresselhaus, G., Dresselhaus, M.S., Pimenta, M.A.: Probing phonon dispersion relations of graphite by double resonance Raman scattering. Phys. Rev. B **88**, 027401 (2002)
54. Thomsen, C., Reich, S.: Double resonant Raman scattering in graphite. Phys. Rev. Lett. **85**, 5214 (2000)
55. Saito, R., Jorio, A., Souza Filho, A.G., Dresselhaus, G., Dresselhaus, M.S., Pimenta, M.A.: Probing phonon dispersion relations of graphite by double resonance Raman scattering. Phys. Rev. Lett. **88**, 027401–027404 (2001)
56. Piscanec, S., Lazzeri, M., Robertson, J., Ferrari, A.C., Mauri, F.: Optical phonons in carbon nanotubes: Kohn anomalies, Peierls distortions, and dynamic effects. Phys. Rev. B **75**, 035427 (2007)

57. Farhat, H., Son, H., Samsonidze, G.G., Reich, S., Dresselhaus, M.S., Kong, J.: Phonon softening in individual metallic carbon nanotubes due to the Kohn anomaly. Phys. Rev. Lett. **99**, 145506 (2007)
58. Kuzmany, H., Matus, M., Burger, B., Winter, J.: Raman scattering in C60 fullerenes and fullerides. Adv. Mater. **6**, 731 (1994)
59. Kuzmany, H., Pfeiffer, R., Hulman, M., Kramberger, C.: Raman spectroscopy of fullerenes and fullerene–nanotube composites. Phil. Trans. Roy. Soc. Lond. A **362**, 2375–2406 (2004)
60. Lorentzen, J.D., Guha, S., Menéndez, J., Giannozzi, P., Baroni, S.: Raman cross section for the pentagonal-pinch mode in buckminsterfullerene C60. Chem. Phys. Lett. **270**, 129–134 (1997)
61. Mekenzie, D.R.: Tetrahedral bonding in amorphous carbon. Rep. Prog. Phys. **59**, 1611 (1996)
62. Cançado, L.G., Jorio, A., Pimenta, M.A.: Measuring the absolute Raman cross section of nanographites as a function of laser energy and crystallite size. Phys. Rev. B **76**, 064304–064310 (2007)
63. Ferrari, A.C.: Raman spectroscopy of graphene and graphite: disorder, electron-phonon coupling, doping and nonadiabatic effects. Solid State Commun. **143**, 47–57 (2007)
64. Cançado, L.G., Takai, K., Enoki, T., Endo, M., Kim, Y.A., Mizusaki, H., Jorio, A., Coelho, L.N., Magalhaes-Paniago, R., Pimenta, M.A.: General equation for the determination of the crystallite size La of nanographite by Raman spectroscopy. Appl. Phys. Lett. **88**, 3106–3109 (2006)
65. Camara, N., Rius, G., Huntzinger, J.-R., Tiberj, A., Mestres, N., Godignon, P., Camassel, J.: Selective epitaxial growth of graphene on SiC. Appl. Phys. Lett. **93**(123503–1), 123503–3 (2008)
66. Casiraghi, C., Ferrari, A.C., Robertson, J.: Raman spectroscopy of hydrogenated amorphous carbon. Phys. Rev. B **72**, 085401–085414 (2005)
67. Hulman, M., Skakalova, V., Roth, S., Kuzmany, H.J.: Raman spectroscopy of single-wall carbon nanotubes and graphite irradiated by $\gamma$ rays. J. Appl. Phys. **98**, 024311–024315 (2005)
68. Lucchese, M.M., Stavale, F., Ferriera, E.H., Vilane, C., Moutinho, M.V.O., Capaz, R.B., Achete, C.A., Jorio, A.: Quantifying ion-induced defects and Raman relaxation length in graphene. Carbon **48**, 1592–1597 (2010)
69. Cançado, L.G., Jorio, A., Martins Ferreira, E.H., Stavale, F., Achete, C.A., Capaz, R.B., Moutinho, M.V.O., Lombardo, A., Kulmala, T.S., Ferrari, A.C.: Quantifying defects in graphene via Raman spectroscopy at different excitation energies. Nano Lett. **11**, 3190–3196 (2011)
70. Tuinstra, F., Koenig, J.L.: Raman spectrum of graphite. J. Phys. Chem. **53**, 1126–1130 (1970)
71. Martin-Fernandez, I., Gabriel, G., Rius, G., Villa, R., Perez-Murano, F., Lora-Tamayo, E., Godignon, P.: Vertically aligned multi-walled carbon nanotube growth on platinum electrodes for bio-impedance applications. Microelectron. Eng. **86**, 806–808 (2009)
72. Kondo, D., Sato, S., Yagi, K., Harada, N., Sato, M., Nihei, M., Yokohama, N.: Low temperature synthesis of graphene and fabrication of top-gated field effect transistors without using transfer process. APEX **3**, 025102 (2010)
73. Tongay, S., Lemaitre, M., Fridmann, J., Hebard, A.F., Gila, B.P., Appleton, B.R.: Drawing graphene nanoribbons on SiC by ion implantation. Appl. Phys. Lett. **100**, 073501-1–073501-3 (2012)
74. Nagashio, K., Nishimura, T., Kita, K., Toriumi, A.: Metal/graphene contact as a performance Killer of ultra-high mobility graphene analysis of intrinsic mobility and contact resistance. Electron Devices Meeting (IEDM), IEEE International, Baltimore, MD, 7–9 December, pp. 1–4 (2009)
75. Micaela, C., Gemma, R., Alessandro, V., Alberto, T. Nanographene patterns from focused ion beam induced deposition. Structural characterization of graphene materials by XPS and Raman scattering. Handbook of Graphene Science, Taylor and Francis Ed. (Under review, 2013)

# Chapter 7
# Deterministic Fabrication of Micro- and Nanostructures by Focused Ion Beam

**Jining Sun and Xichun Luo**

**Abstract** Focused ion beam (FIB) machining is an ideal technique for both micro- and nanofabrication. This chapter describes a deterministic fabrication approach to accurately obtain micro- and nanostructures. It starts with a detailed introduction of FIB machining mechanism, followed by a surface topography simulation method and a divergence compensation method. The effectiveness of this fabrication approach has been fully demonstrated through a number of experiments which show that the surface topography of the machined material can be precisely predicted; the divergence compensation method can reduce the machining error, and the machined surface form accuracy can be dramatically improved. Combining with other techniques, some hybrid manufacturing techniques based on FIB machining are also introduced.

## 7.1 Introduction

Due to the broad applications of micro- and nanostructures in multiple disciplines, the capability of fabrication of highly accurate micro- and nanostructures with great repeatability on different materials is essential for modern science and engineering research. Innovative techniques and processing methods are continually developed to meet the demand of current research and industrial progress. One of the representative

J. Sun
Institute of Mechanical, Process and Energy Engineering, School of Engineering and Physical Sciences, Heriot-Watt University, Riccarton, Edinburgh EH14 4AS, UK
e-mail: Jining.Sun@hw.ac.uk

X. Luo (✉)
Department of Design, Manufacture and Management,
Faculty of Engineering, University of Strathclyde, Glasgow, GI 1XQ, UK
e-mail: Xichun.Luo@strath.ac.uk

Z.M. Wang (ed.), *FIB Nanostructures*, Lecture Notes in Nanoscale Science and Technology 20, DOI 10.1007/978-3-319-02874-3_7,

techniques is focused ion beam (FIB) machining. Fundamentally, an FIB system produces and directs a beam of high-energy ions (5–50 keV), focusing them onto a sample for the purpose of both machining and imaging. It is a direct writing technique that requires no mask during the fabrication process. Even more important, FIB machining enables the fabrication of both two-dimensional and three-dimensional micro- and nanostructures.

The primary use of FIB is for the fabrication, analysis or repair of a sample at the micro- and nanoscale. The FIB technique is capable of building two-dimensional and three-dimensional structures on almost any kind of solid material. Combined with other techniques such as nanoimprint lithography (NIL) and single-point diamond turning (SPDT), these structures can be duplicated in a cost-effective way. However, it is a great challenge in the FIB machining process to precisely (nanometre level) control the shape of the machined structures, especially for the fabrication of nanostructures. This is mainly due to the redeposition effect, variation of sputter yield and limitation of ion beam spot size. Theoretical and technical measures need to be developed to overcome the complications of FIB machining. The achievement of this goal will provide the whole area of research a totally new perspective.

## 7.2 FIB Machining Mechanism

When a beam of ions is incident on a solid material, several ion–matter interaction mechanisms take place simultaneously below, on and above the target surface, such as scattering, sputtering and redeposition.

### *7.2.1 Below the Surface*

Below the substrate surface, the ion–solid collision cascades are the principal interaction mechanism. According to the mass ratio of the particles involved in the scattering process, the ion–solid collision cascades can be divided into three regimes, including the single knock-on regime where the ion mass is a lot less than the mass of the particle, the linear cascade regime where the masses of the incident ion and the target particle are around the same and the spike regime where the ion mass is much more than the mass of the particle. The FIB sputtering process generally operates in the linear cascade regime where the incident ion beam collides with the substrate atoms. The ion–solid scattering theory can provide valuable analytical expressions of the important physical mechanisms and dependencies. However, the solutions are often complicated and require simplifying assumptions [40]. The first assumption is that the particles do not move in any preferred directions, which exist in, e.g., a crystalline target. The second assumption is that only elastic collisions occur between the incident ions and the target atoms. The third assumption is that the scattering is made up of hard-sphere collisions, that is to say the particles are

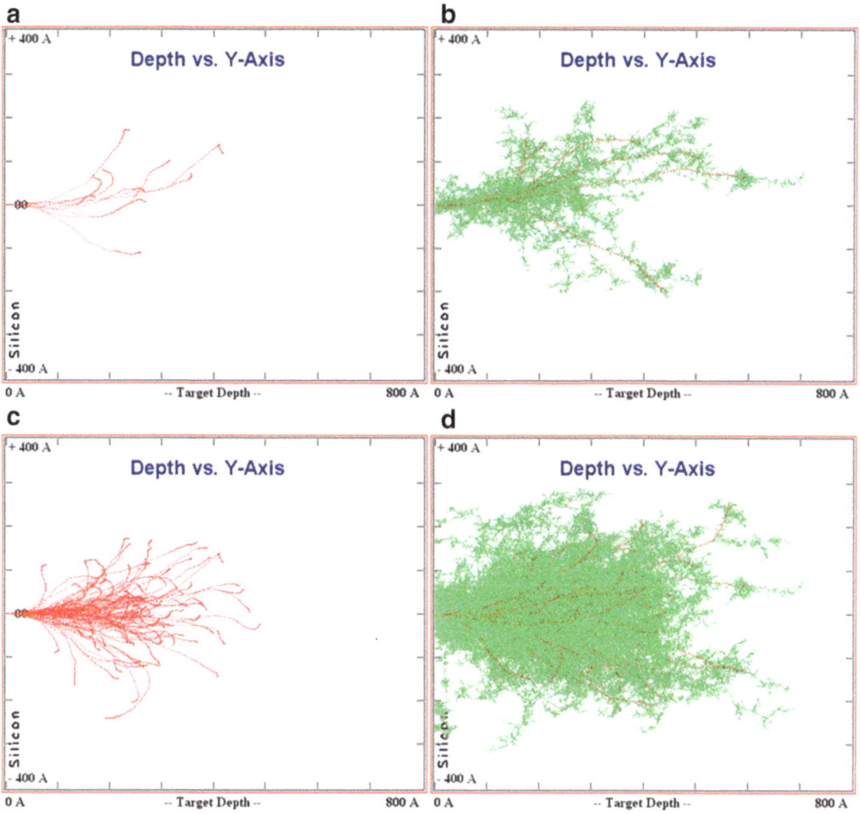


**Fig. 7.1** Ion only (**a**, **c**) and ion plus recoil (**b**, **d**) trajectories for 30 keV gallium ions incident on silicon, for 10 (**a**, **b**) and 100 (**c**, **d**) incident ions, as obtained from TRIM

assumed to be classical perfect spheres doing the collisions. Binary collision approximation (BCA) is a simplified model used to calculate the paths of the particles in the cascade based on the above assumptions. In this model, the interaction of an incident ion with atomic nucleus in the target sample is regarded as an elastic collision process, while the interaction with electrons is regarded as an inelastic process without any scattering effects. Only two charged particles are involved in one scattering process, i.e. the incident ion and the target atom. The BCA model is fundamental for investigating complex situations that depend on individual ion–solid combinations. This model has been widely applied in many simulation programmes based on the Monte Carlo method. The trajectories of each incident ion and all associated recoil atoms are traced until the kinetic energy has fallen below the cut-off energy or the traced particles have escaped from the defined region. Figure 7.1 shows the spatial distributions of the ions and the recoil trajectories for 30 keV gallium ions in silicon as obtained from simulations using a Monte Carlo simulation code called TRIM. The left column of the figure (a, c) indicates the ion trajectories in the

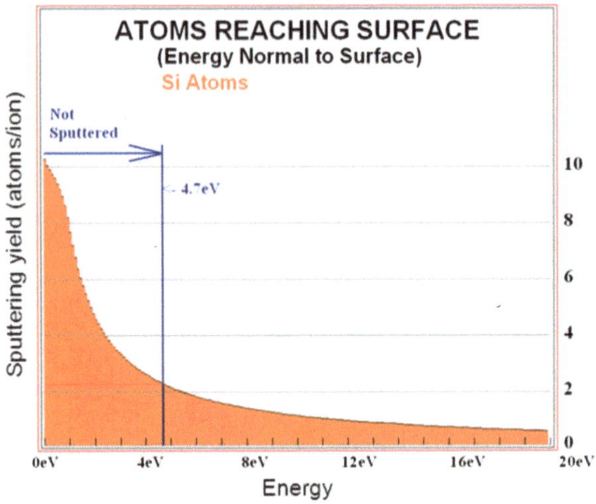


**Fig. 7.2** Simulation result of the energy distribution of the recoil atoms in a silicon substrate under 30 keV gallium ion beam radiation

substrate, and the range of implanted ions is predicted from this simulation. The right column (b, d) indicates the moving and stopped recoiling atoms of the target substrate. In Fig. 7.1a, it is evident that two of the ten ions are backscattered. Figure 7.1b indicates that, for the present ion–target combination, an individual ion creates several smaller sub-cascades with little overlap. The overlap of many incident ions (Fig. 7.1d) forms a cascade region which is similar to the region of the ion tracks (Fig. 7.1c).

## 7.2.2 *On the Surface*

On the substrate surface, ion milling takes place as a result of physical sputtering of the substrate atoms. The ion beam sputtering is a very special scattering process which takes place near the surface of the target material. It is the most pronounced ion–solid interaction that occurs in the FIB milling process. A recoil atom near the surface can obtain enough energy in the collision cascade induced by the ion bombardment to be knocked out as a sputtered particle if it receives sufficient kinetic energy to overcome the surface-binding energy. Figure 7.2 shows the energy distribution of recoil atoms in a silicon substrate under 30 keV gallium ion beam radiation calculated by TRIM. Since the surface-binding energy for silicon is 4.7 eV, any recoil atoms with kinetic energy below this threshold cannot escape from the surface of the silicon substrate.

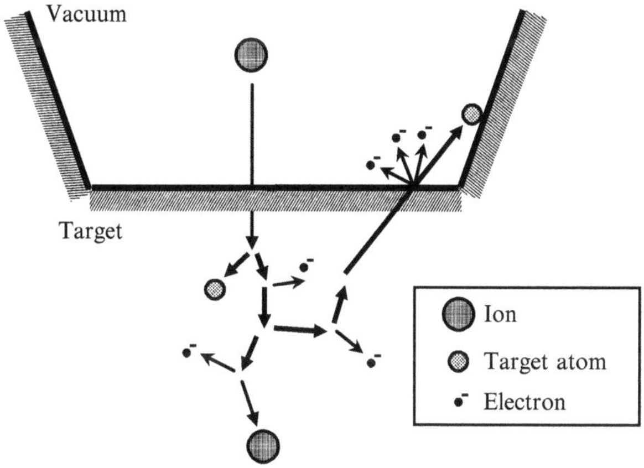


**Fig. 7.3** Sch. of redeposition effect

## 7.2.3 *Above the Surface*

Above the substrate surface, the redeposition effect takes place accompanied with the ion sputtering process. Some of the sputtered atoms induced by the incident ion beam will deposit at the site of interest, reducing the net removal rate and creating unwanted features. As illustrated in Fig. 7.3, when the sputtered atom leaves the target material, it is ejected with a finite kinetic energy. Therefore, it can be considered as a projectile that can exert interactions on its trajectory. Depending on the sticking coefficient of the material, the sputtered atom may attach to the sidewall of the structure. The sticking coefficient is defined as the ratio of atoms that properly react and attach on the surface to the total sputtered atoms. It is a statistical measure of the material's affinity to adhere to a surface. It depends on the chemistry of the material being deposited and on the substrate temperature. For FIB sputtered atoms, the sticking coefficient is assumed to be constant for different materials [15].

The area that can be affected by the redeposited atoms depends on the surface topography, angular distribution of the sputtered atoms and surface local pressure. Due to the cosine rule of the angular distribution of the sputtered atoms the redeposition effect is more pronounced in structures with a high aspect ratio, in which case, the projectile is more liable to collide with the sidewall and therefore be trapped into the structure. The surface local pressure also has a profound effect on the degree of redeposition [14, 36]. Due to the confining geometry restricting the escape of the sputtered material, the local pressure at the sputtering point is much higher than that in other parts in the vacuum chamber; therefore, the collision mean free path is decreased, which leads to a further increase in the probability of the sputtered atom collisions and redeposition rate.

## 7.3 Key Fabrication Parameters in FIB Machining

The FIB machining process is a combination of physical sputtering and material redeposition. The material removal process depends on many processing parameters including the sputter yield, beam scanning strategy, scanning passes, dwell time and beam overlap. Understanding these processing parameters can lead to an enhancement of the sputtering rate and further improvement of machining accuracy.

### 7.3.1 *Sputter Yield*

The sputter yield is an elementary parameter which characterises the basic sputtering phenomena essential to the milling process in making microstructures from simplified two-dimensional or three-dimensional structures. The accurate determination of the sputter yield is an essential requirement in FIB machining.

The sputter yield is defined as the average number of target atoms ejected from the sample per incident ion. Correspondingly, the sputter rate (or material removal rate) is the number of atoms being sputtered from the target per unit time. A reasonable estimation of the sputter rate by sputter yield is

$$Y_{\mathrm{t}} = Y \times \frac{I}{e} \tag{7.1}$$

where $Y_{\mathrm{t}}$ and $Y$ are the sputter rate and the sputter yield, respectively; $I$ is the ion beam current, and $e$ is the elementary charge.

Experimentally, the sputter yield can be measured at specific sputtering conditions. Through measuring the volume of the ion caved cavity, the average sputter yield can be evaluated. Theoretically, the sputter yield can be calculated by Sigmund's linear collision cascade model [41]:

$$Y(E,\theta) = \frac{0.042}{U_{\mathrm{s}}} \alpha S_{\mathrm{n}}(E) \cos^{-f_{mr}}(\theta) \tag{7.2}$$

where $U_{\mathrm{s}}$ is the atom-binding energy; $\alpha$ is a factor that depends only on the efficiency of energy transfer; $S_{\mathrm{n}}(E)$ is the elastic stopping power; $f_{\mathrm{mr}}$ is a function of the mass ratio of the incident ion and the target atom and $\theta$ is the incident angle. This indicates that the sputter yield depends not only on the substrate material but also on many processing parameters, including the ion energy, angle of incidence and milling conditions. Alternatively, Yamamura proposed a semi-empirical formula that can analytically describe the angle dependency of the sputter yield over a wide range of the incident angles [54]:

$$Y(\theta) = Y(0)\cos^{-\alpha}(\theta)\exp\left(-\beta\left(\cos^{-1}(\theta)-1\right)\right) \tag{7.3}$$

where $\alpha$ and $\beta$ are adjustable parameters.

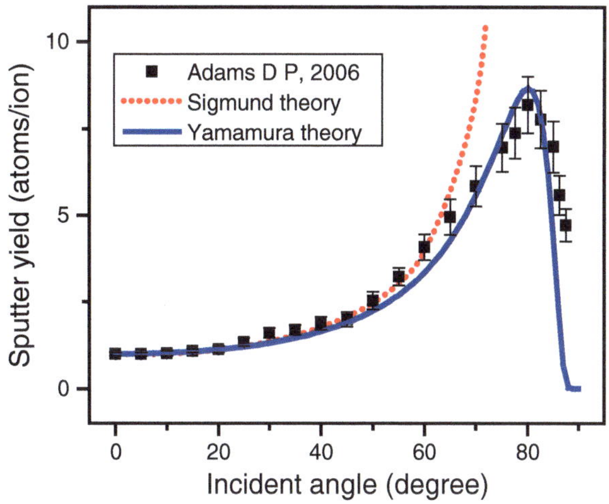


**Fig. 7.4** Sputter yield of gallium ion incidents on a silicon substrate at 20 keV

In most FIB machining experiments, the material of the substrate, ion energy and milling conditions are kept constant. The only variable during the milling process is the ion beam incident angle. The theoretical results of sputter yield angular dependence are compared with the experimental results [1] in Fig. 7.4. This illustrates a beam of gallium ions with an energy level of 20 keV bombarded on a silicon substrate with different incident angles. It indicates that the Sigmund theory coincides with the experimental data only for incident angles smaller than 65°, while Yamamura's semi-empirical expression is consistent with the experimental data for all incident angles. The sputter yield reaches a maximum at about 80° and then decreases very rapidly to zero as the incident angle approaches 90°.

The mechanism of this angular dependence can be explained by Fig. 7.5, which shows the simulations of the collision cascades under different incident angles. From 0° to 80° the ion trajectories and recoiling atoms generated by the higher incident angle are observed to be closer to the surface. From a statistical perspective the more collisions occur proximal to the surface, the higher the sputter yield. Therefore, the sputter yield keeps increasing with the ion incident angle. For large incident angles, ion reflection plays an important role at glancing angles and results in a rapid drop in the sputter yield.

### *7.3.2 Scanning Strategy*

In a typical FIB system, the machining process is performed by placing the beam sequentially at predefined positions to form an area of quadrilateral or more general shape. The scanning strategy produces the trajectories of the scanning beam in the ion beam machining process. The trajectories of the scanning beam spot are the

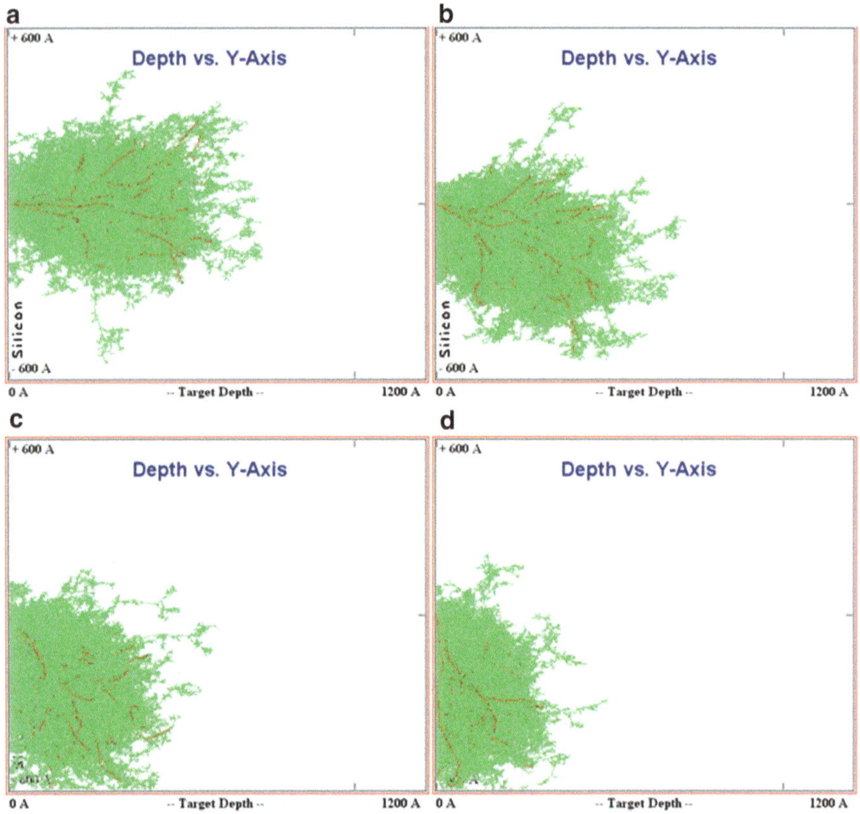


**Fig. 7.5** 20 keV gallium ion sputtering trajectories under (**a**) 0°, (**b**) 30°, (**c**) 60° and (**d**) 80°

beam scanning lines. Each of the scanning lines constituting the scanning path is in fact a linear array of discrete points to which the beam is rapidly deflected and at which it pauses for a predetermined period (dwell time) to remove the required amount of material. There are two kinds of scanning path that are commonly used in FIB machining: raster scan and serpentine scan, both of which have been illustrated in Fig. 7.6. In the raster scan, the beam proceeds along each scanning line in the same direction. The scanning beam is blanked at the end of each scanning line and retraced to the opposite end of the next scanning line. In serpentine scan, the beam moves in opposite directions between each pair of adjacent scanning lines. Therefore, the retrace is eliminated, and the beam does not need to be blanked during the scanning process. Apart from the raster and serpentine scan, self-defined scanning paths are also available. This kind of scanning path is essential for machining complex structures where high precision is required. The beam blanking tails generated in the ion column can also be minimised by optimising the scanning path. In the beam scanning process, the pattern is represented by a series of milling points, the coordinates of which are stored sequentially in the computer memory.

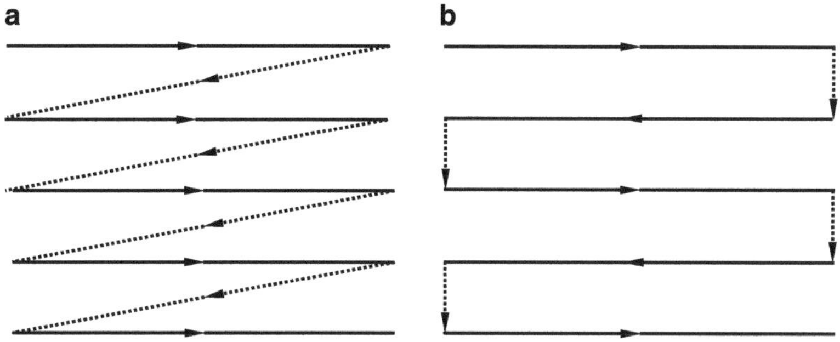


**Fig. 7.6** Scanning strategies in FIB machining process with *arrows* indicating scanning direction: (**a**) raster scan; (**b**) serpentine scan

Whatever scanning mode is chosen, the beam is always blanked after the last scanning line of each scanning pass while being deflected back to the start point on the first scanning line of the next pass.

### *7.3.3 Beam Diameter*

The analysis of beam profile is one of the fundamental issues in FIB machining. The beam profile is usually considered as a Gaussian distribution with a circular cross section, even if the beam had a wide beam skirt following a Holtsmark distribution [16, 52]. The intensity distribution is Gaussian with the long-ranging tails satisfying the following equation:

$$J(x,y) = \frac{I}{2\pi\sigma^2}\exp\left[-\frac{(x-x_p)^2+(y-y_p)^2}{2\sigma^2}\right] \tag{7.4}$$

where *J*(*x*,*y*) is the ion flux density at a point (*x*,*y*); *I* is the ion current and $\sigma$ is the standard deviation of the Gaussian distribution; the centre of the beam is located at point $(x_p, y_p)$.

The commonly used characteristic value to describe the beam spot size is the beam diameter which is the full width at half maximum (FWHM) of the intensity or $e^{-1}$ (36.7 %) of the maximum intensity. The beam diameter *d* can be estimated from the Gaussian distribution:

$$d = \sqrt{8\ln 2}\,\sigma \tag{7.5}$$

Therefore, the ion flux density can be expressed as a function of the beam diameter:

$$J(x,y) = \frac{4I\ln 2}{\pi d^2}\times 16^{-\frac{(x-x_p)^2+(y-y_p)^2}{d^2}} \tag{7.6}$$

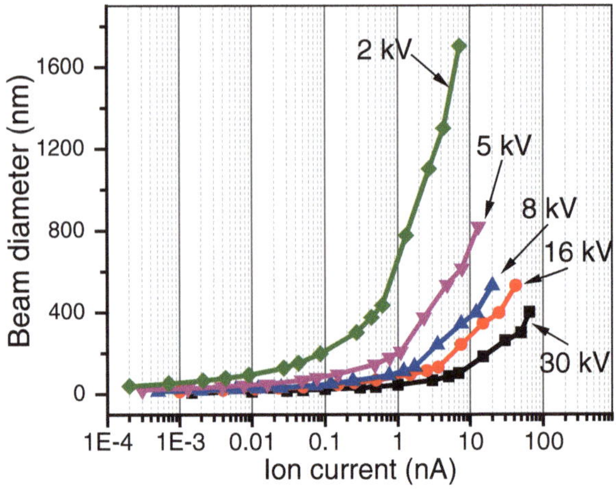


**Fig. 7.7** The relationship between the ion current and the beam diameter on an FEI Quanta 3D FEG FIB system

Generally the beam diameter grows as the ion current increases. An illustration of this relationship is shown in Fig. 7.7.

Under different acceleration voltages ranging from 2 to 30 kV, the beam diameter is reduced proportionally in relation to the ion current. A small beam diameter results in a high fabrication resolution, and this constraint relationship reveals that in order to fabricate structures with a high resolution, only small current can be used in the FIB machining process. The inversely proportional relationship between the ion current and the spot size also reveals the contradiction between the throughput and the resolution. High throughput and high resolution cannot be achieved simultaneously based on the current settings of the FIB system. Revolutionary design focusing on improving the performance of the ion column needs to be carried out. One solution of this contradiction is to introduce parallel ion beam machining to the ion beam system, which is the Charged Particle Nanotech (CHARPAN) project [28, 34, 47, 49]. It can generate hundred thousands of ion beams working in parallel structured by a programmable aperture plate system. This unique design enables the system with a direct structuring power at high throughput and high resolution.

## *7.3.4 Pixel Spacing*

As the ion beam scans across the surface, FIB milling is performed by a precise pixel-by-pixel movement. After scanning all the points on a particular line in the sputtering area the beam moves to the next line. After scanning the whole sputtering area, the beam restarts scanning until it delivers the fully assigned ion dose. This is also known as a digital scan and is schematically shown in Fig. 7.8. The distance between

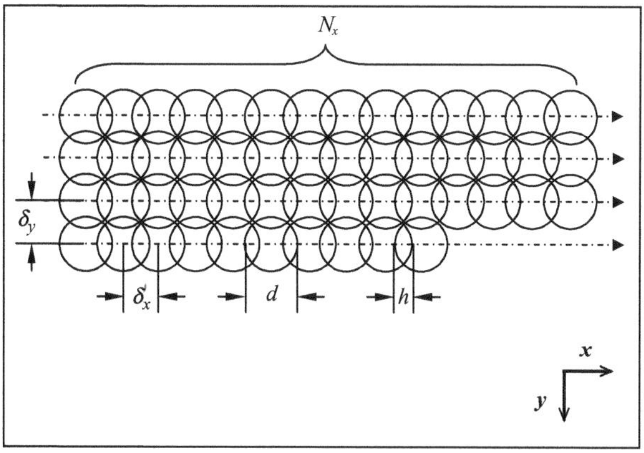


**Fig. 7.8** Schematic diagram of digital scan. Each *circle* represents a beam spot

two adjacent pixels $\delta$ is called the pixel distance. The overlap of the two adjacent beam spots is $h$. To fabricate a smooth profile with a constant rate of material removal the pixel spacing must be small enough to allow a proper overlap between the adjacent pixels along both the $x$- and $y$-directions [48].

As the intensity profile of the FIB can be defined as a Gaussian distribution [16], the ion flux distribution during the scanning process can be calculated as the sum of every beam profile:

$$J(x,y) = \frac{I}{2\pi\sigma^2} \sum_{n_x=0}^{N_x} \sum_{n_y=0}^{N_y} \exp\left[ -\frac{\left(x - x_p - n_x\delta_x\right)^2 + \left(y - y_p - n_y\delta_y\right)^2}{2\sigma^2} \right] \tag{7.7}$$

where $N_x$ and $N_y$ are the total pixel numbers along the $x$- and $y$-directions, respectively.

In order to obtain a steady and unwavering ion flux, the threshold value for the minimum amount of the pixel spacing can be determined mathematically. Figure 7.9 shows the normalised cumulative ion flux distribution in a digital scanning process for a beam with 50 nm diameter. The pixel spacing along the $x$-direction changes from $8\sigma$ to $1.5\sigma$. It is found that when the pixel spacing is smaller than $1.5\sigma$, a constant flux field can be obtained. This uniformity condition of the ion flux with respect to the scanning direction can be directly extended to satisfy the uniformity condition between the scan lines.

### 7.3.5 *Scanning Passes and Dwell Time*

In FIB machining, the incident ions are to repetitively scan with the same scanning pattern on a sample many times. This is called the multi-pass scanning method. The iteration of each scanning pattern is called a scanning pass. The multi-pass

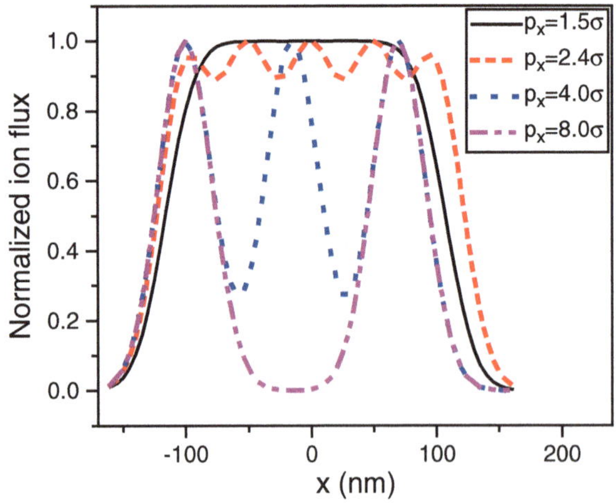


**Fig. 7.9** Normalised ion flux distribution in multiple scan-line FIB milling

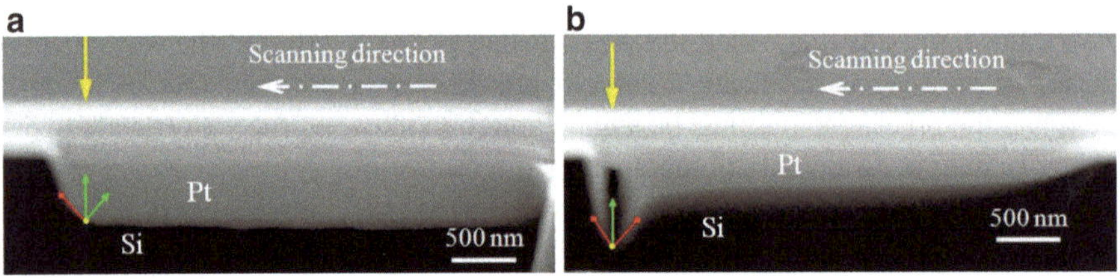


**Fig. 7.10** Cross sections of micro-grooves on silicon (100) surface (coated with 500 nm platinum) fabricated using (**a**) a single-pass scanning process and (**b**) a multi-pass scanning process (20 passes in total)

scanning process is helpful in forming a uniform bottom surface [53]. Figure 7.10 shows a comparison of two micro-grooves which were fabricated on a silicon substrate (100) surface by FIB machining using a single scanning pass method and a multi-pass scanning method. Under the same ion dose of $1.2 \times 10^{18}$ ions/cm$^2$ the number of scanning passes was inversely proportional to the dwell time. The dwell time is defined as the stationary period that a beam of ions stays at a spatial point. It is also a measure of the scanning speed. The surface topography of the scanned area could be very different under the same incident ion dose but with different dwell time. Compared to using multiple scanning passes with a smaller dwell time, the single scanning pass had a longer dwell time which led to a deeper and narrower scanning channel. Due to the cosine rule of the angular distribution of the sputtered atoms, the sputtered atoms generated in the single scanning pass have more difficulty in escaping from the micro-channel which led to a higher redeposition rate as shown in Fig. 7.10a. The longer dwell time can also result in a higher local pressure at the sputtering point. This phenomenon will further lead to a reduction of the collision mean free path of the sputtered atoms; therefore, the probability of redeposition increases. However, when using the multi-pass scanning method some deposited atoms can subsequently be removed by the next scanning pass. The multi-pass

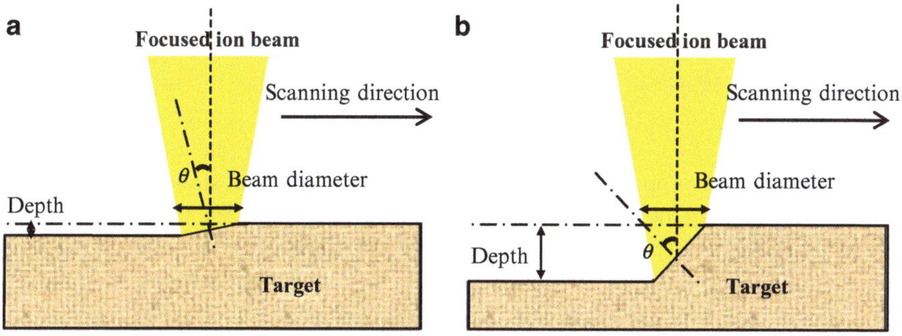


**Fig. 7.11** Schematic diagram of a one-dimensional sputter model at steady state: (**a**) Material removed in each scanning pass is much smaller than the beam diameter; (**b**) material removed in each scanning pass is at the same level as the beam diameter

scanning process is also helpful in forming a uniform bottom surface in the groove as shown in Fig. 7.10b. A similar phenomenon has also been observed by Yamaguchi et al. [53].

A longer dwell time not only leads to a boost in redeposition but also results in a higher local sputter yield. Santamore et al. found that in the FIB machining process, the scanning speed has a significant effect on the sputter yield [38]. For very rapid scanning with small dwell time at normal incidence, each pass of the beam removes a thickness of material that is much smaller than the beam diameter. In this case, the sputter yield corresponds to the yield at normal incidence. If the scan speed is slowed down, the thickness removed in each scanning pass is at the same size as the beam diameter. The actual incident angle at the sputtering point is no longer zero even though the beam is still perpendicular to the surface. A schematic diagram of this phenomenon is illustrated in Fig. 7.11. This illustrates that, under normal incident condition, the effective ion beam incident angle $\theta$ at the local milling point is widely divergent from the normal incident angle when the material removed in each scanning pass is at the same level as the beam diameter. The increase in the local incident angle leads to a variation in the sputter yield and thus causes the fabrication process to be unpredictable. Therefore, the material removed in each scanning pass should be much less than the beam diameter at the selected scanning speed in order to confine the divergence between the practical sputter yield and the theoretical sputter yield applied in the machining model. The lower limit of number of the scanning passes should be controlled under this constraint.

## 7.4 Surface Topography Simulation in FIB Machining

Surface topography simulation is essential for FIB manufacturers to evaluate the sputtering result and provide references for FIB operators. The cornerstone of surface topography simulation is the evaluation of the sputter depth. The depth of microstructures with simple geometries fabricated by FIB machining is usually guided by the assumption of the existence of a proportional relationship with the local ion dose. However, this assumption becomes invalid for high-aspect-ratio structures or curved surfaces, where angular dependence of sputter yield as well as the redeposition effect cannot be ignored. The redeposition phenomenon depends on the local surface topography, material properties of the substrate and ion–target interaction between the sample and the ion source used. Besides, the distribution of the redeposited atoms highly depends on the local surface topography which can only be predicted by the use of adequate, theoretically motivated numerical methods. A sophisticated ion–solid interaction model combined with surface topography simulation may contribute to the understanding of these effects.

The theoretical description of the changes on surface topography under a given process has long been of great scientific interest. To the authors' knowledge, the earliest research on surface topography evolution under FIB radiation was carried out by Mueller et al. [29] in 1986. Two-dimensional simulation program Complete Modelling Program of Silicon Technology (COMPOSITE) was modified for this process with the beam confined in one direction and given a Gaussian distribution. This work provided an insight into the interplay of the angular dependent sputter yield and redeposition effect. Topography simulation was implemented by a string-based algorithm in COMPOSITE. In 1989, Katardjiev et al. presented a generalised kinematic theory of surface evolution to simulate the topography development of three-dimensional surfaces during growth and erosion based on the Huygens' principle [21]. A three-dimensional simulation code direct numerical evaluation of surface evolution (DINESE) [19, 20, 22] was developed to simulate the evolution of real three-dimensional structures during milling and deposition. The TRIDYN code was used to obtain angle-dependent sputter yields, but the redeposition effect is not considered in DINESE code which means that it is only applicable to simulate the machining of low-aspect-ratio structures. Biedermann and Platzgummer developed a software package called Ionshaper [33] which included first- and second-order sputtering as well as the first- and second-order redeposition of sputtered atoms. The natural erosion process was characterised by a surface velocity vector normal to the surface and a surface shape with a fully continuous derivative. A novel ion erosion model was presented in this work to produce a good simulation of sharp edges where the surface slope is ill defined. Boxleitner and Hobler developed a dynamic Monte Carlo simulation code FIBSIM [4] to evaluate the surface topography evolution and damage formation during TEM sample preparation [5]. FIBSIM combined the simulation of binary collision cascades with two-dimensional cell-based topography simulation model. Analogously, Kim et al. developed a two-dimensional [24] and three-dimensional [25] simulation code named AMADEUS 2D/3D (Advanced Modelling and Design Environment for Sputter Processes) to predict the surface topography evolution under ion beam radiation. The surface was represented by a structured or an unstructured grid. Each node on the grid moved according to the calculated sputtered and redeposited fluxes. The node velocity along the surface normal direction is proportional to the fluxes received. Therefore, the surface evolution can be simulated by this string/cell-based tracking technique.

**Table 7.1** Topography simulation codes for FIB machining

| Simulation code | Dimension | Topography model | Redeposition | Reference |
|---|---|---|---|---|
| COMPOSITE | 2D | String based | Yes | Muller et al. [29] |
| DINESE | 3D | Huygens' principle | No | Katardjiev et al. [22] |
| FIBSIM | 2D | Cell based | Yes | Boxleitner and Hobler [4] |
| Ionshaper | 2D, 3D | Huygens' principle | Yes | Platzgummer et al. [33] |
| AMADEUS-2D | 2D | String based | Yes | Kim et al. [24] |
| AMADEUS-3D | 3D | Cell based | Yes | Kim et al. [25] |
| AMADEUS-Level set | 2D | Level set method | Yes | Kim et al. [26] |

This method works well for the simulation of machining low-aspect-ratio structures. For structures with a high aspect ratio, nonetheless the ion–solid interaction model is still applicable, but the string-based topography tracking technique would crash due to the sharp gradients and cusps. A more robust surface tracking technique needs to be developed. Based on the same ion–solid interaction model, Kim et al. developed a two-dimensional topography simulation implemented using the level set method [26]. Introduced by Osher and Sethian [31], it is a highly robust and accurate computational technique for tracking moving interfaces. Via this method the surface topological merging and breaking, sharp gradients and cusps can form naturally with the help of an upwind difference scheme. The effects of curvature can also be easily incorporated, which makes it a more robust method.

A summary of the reviewed simulation codes is listed in Table 7.1. Both two-dimensional and three-dimensional models have been developed to predict the surface topography evolution in the FIB machining. The redeposition effect has been considered in most of the models. However, for structures with a high aspect ratio, more robust simulation techniques need to be developed. A three-dimensional topography simulation based on level set method may be competent to give a solution for this issue.

### 7.4.1 Ion Sputtering Model

Figure 7.12 illustrates a surface sputtering model for FIB machining process. When a beam of ions bombards a work piece, some atoms will be sputtered out of the work piece surface. Some of the sputtered atoms will adhere to the machined surface due to the redeposition effect which strongly depends on the work piece material and ion beam properties. Moreover, the local surface topography also has an important role in the redeposition effect; this effect can only take place when the motional trajectories of the sputtered atoms interfere with an existing boundary of the machined structure (see Fig. 7.12, point d$A'$). Since the redeposition effect is random, the actual tracking of the fabrication process cannot be achieved analytically.

The prediction of this dynamic process can be achieved by using a numerical simulation method. The surface topography can be represented by a number of

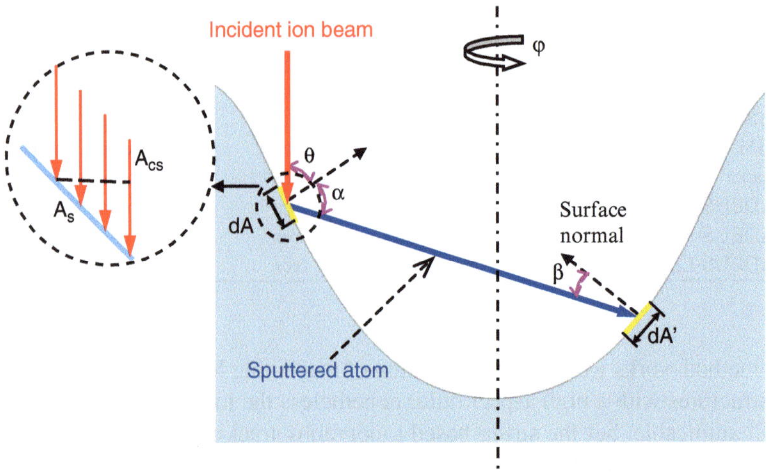


**Fig. 7.12** Schematic illustration of the generation mechanism for a machined surface fabricated by FIB

nodes whose trajectories evolve as a function of time during the FIB fabrication process. The normal velocity $v_\perp$ of each node is proportional to the total flux at the corresponding node. It can be described as

$$v_\perp = \frac{F_{total}}{N} \tag{7.8}$$

where $F_{total}$ is the total flux (atoms/μm²/s) and $N$ is the atom density of target material (atoms/μm³). The total flux at each node on the target surface consists of two parts, i.e. the flux of sputtered atoms ($F_{direct}$) caused by the incident ion beam directly and the flux of redeposition atoms ($F_{indirect}$) which is contributed to by $F_{direct}$. They can be described as

$$F_{total} = F_{direct} + F_{indirect} \tag{7.9}$$

$$F_{direct} = \frac{F_{incident} Y(\theta) A_{cs}}{A_s} = F_{incident} Y(\theta)\cos(\theta) \tag{7.10}$$

$$F_{indirect} = -S_c \int \frac{F_{direct} f(\alpha) d\Omega}{dA' \iint f(\alpha) d\alpha d\varphi} dA = -S_c \int \frac{F_{direct} f(\alpha)\cos(\beta)}{d^2 \iint f(\alpha) d\alpha d\varphi} dA \tag{7.11}$$

As shown in Fig. 7.12, $A_{cs}$ and $A_s$ are the cross-sectional area and the actual area on the target surface, respectively. $Y(\theta)$ is the sputter yield at an incident angle $\theta$. $S_c$ and $f(\alpha)$ are the sticking coefficient and the angular distribution of the sputtered atoms. dΩ is the infinitesimal emission solid angle of the sputtered atoms; d$A$ and d$A'$ are both the infinitesimal areas on the target surface. For symmetrical structures, a rotational angle $\varphi$ is needed to reconstruct three-dimensional surfaces.

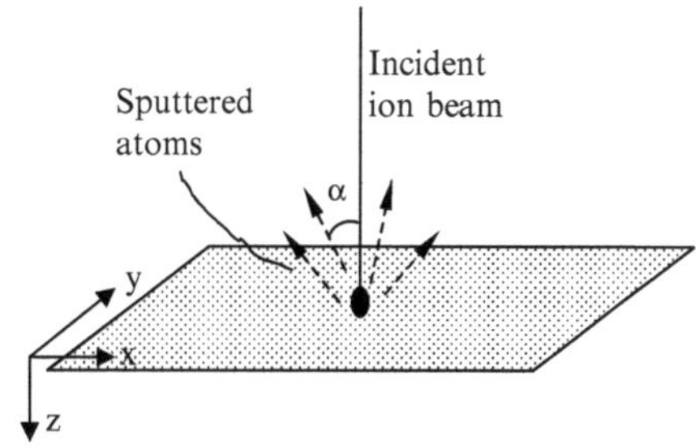


**Fig. 7.13** Scheme of the simulation plan

## *7.4.2 Determination of Angular Distribution of Sputtered Atoms*

The angular distribution of sputtered atoms $f(\alpha)$ generally follows a cosine distribution for normal incidence ion bombardment [9, 42]. For varying ion incident angles, the angular distribution of the sputtered atoms is slightly different, and the deviations from the cosine rule can be corrected by a power of $n$ [35]:

$$f(\alpha) = \cos^n(\alpha) \tag{7.12}$$

In order to determine this correction factor in (7.12), the angular distribution function was simulated by using the Monte Carlo method. This work simulated a beam of 30 keV gallium ions bombarding a silicon substrate in the $x$–$y$-plane with an incident angle of 0° along the $z$-axis as shown in Fig. 7.13. The total history number was set to be 10,000 to guarantee the simulation accuracy. The heat of sublimation, 4.7 eV, was taken as the surface-binding energy of silicon. The lattice-binding energy[1] and displacement energy[2] were assumed to be 2 and 15 eV, respectively.

The simulated distribution of sputtered atoms in the $x$–$y$-plane is shown in Fig. 7.14. The coordinate of each point in Fig. 7.14 is represented by the cosine of the final trajectory of sputtered atom. The angular distribution in the whole scattering space was analysed from two orthogonal cross sections, the $x$–$z$-plane and the $y$–$z$-plane. As shown in Fig. 7.15, in each plane the total scattering angle ($\pi$) was discrete with an angle step of $\Delta\alpha$. In this work $\Delta\alpha$ was set to be 4° to meet the statistical requirement that there were at least ten particles falling in each angular region. All of the sputtered atoms in the whole scattering space were projected onto this plane. The number of the atoms that fall into the region $[\alpha - \Delta\alpha/2, \alpha + \Delta\alpha/2]$ was counted as $\Delta n$. The angular distribution was the sputtered atoms per unit area:

$$f(\alpha) = \frac{\Delta n}{2\pi\left(\cos(\alpha - \Delta\alpha/2) - \cos(\alpha + \Delta\alpha/2)\right)} \tag{7.13}$$

[1] The minimum energy needed to remove an atom from a lattice site.

[2] The minimum energy required to knock a target atom far enough away from its lattice site so that it will not immediately return.

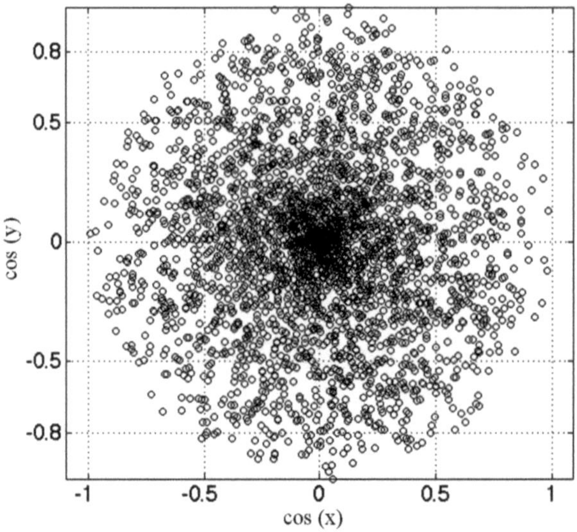


**Fig. 7.14** Simulated distribution of sputtered atoms in the *x*–*y*-plane

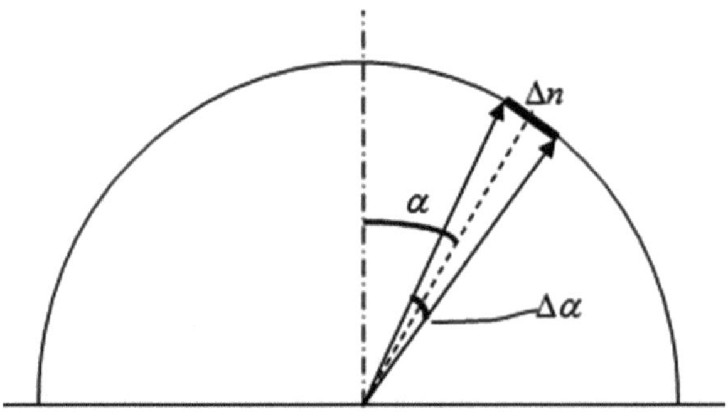


**Fig. 7.15** Schematic of the *x*–*z* (or *y*–*z*)-plane in the scattering space

Equation (7.13) can be further simplified and normalised as

$$f(\alpha) = \frac{\Delta n}{2\sin\alpha \sin(\Delta\alpha / 2)} \tag{7.14}$$

Based on (7.14) and the simulation results, the angular distributions of the sputtered atoms in the *x*–*z*-plane and *y*–*z*-plane were obtained. The two distributions are shown in Fig. 7.16a, b, respectively. Both of the distributions are fitted tightly to the cosine rule in the form of equation (7.12). In the *x*–*z*-plane and the *y*–*z*-plane the correction factor *n* was found to be 1.6.

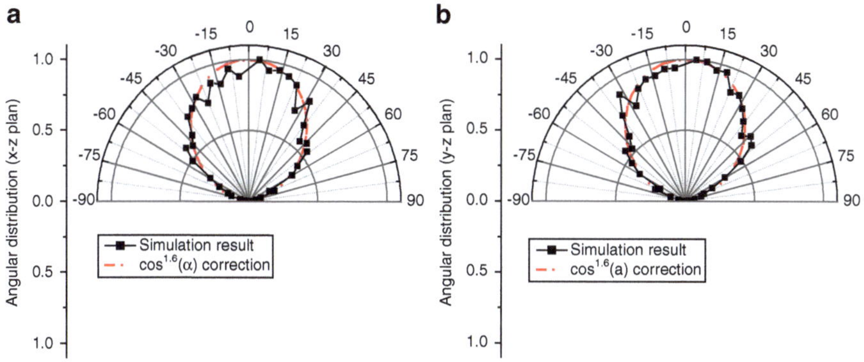


**Fig. 7.16** Monte Carlo simulation of the angular distribution of the sputtered atoms when 30 keV gallium ions strike a silicon substrate. (**a**) Angular distribution of sputtered atoms in *x*–*z*-plane; (**b**) angular distribution of sputtered atoms in *y*–*z*-plane

In this work, the simulation of FIB fabrication is assumed to start from a flat surface. However, from the start of the simulation, the machined surface will move and then becomes rough at the atomic level which results in a variation of the ion incident angles from −90° to 90°. This is expected to spread the calculated angular distribution and finally give a result not very different from a cosine function. For simplicity a cosine function with correction factor 1.6 is applied as the angular distribution function in this model and is used for further calculations.

### 7.4.3 Determination of Sticking Coefficient

The sticking coefficient $S_c$ is defined as the ratio of the flux caused by redeposition to the flux of sputtered atoms generated by the incident ion beam. It stands for the absorption degree between the redeposited atoms and the machined surface. In this model $S_c = 1$ was adopted since no chemical gas assistance is involved in the machining process.

### 7.4.4 Simulation of Surface Topography Evolution

The simulation of surface topography was realised by using the level set method [39]. The level set method is a numerical technique designed to track the evolution of interfaces and shapes between two different regions. This method makes it very easy to follow irregular and dynamic shapes that change along with the variation of time. A "level set" of a function $\phi$ with $n$ variables is a set of the form

$$L_c(\phi) = \{(x_1,,\cdots,,x_n) \mid \phi(x_1,,\cdots,,x_n) = c\} \tag{7.15}$$

where $x_1$–$x_n$ are the $n$ variables of function $\phi$; $c$ is a given constant value and $L_c(\phi)$ represents a set of such variables that meet equation (7.15).

The idea of the level set method is to take the propagating interfaces ($L_c(\phi)$) as a certain level set ($c$) of a higher dimensional function which is called a level set function ($\phi$). The advantage is that merging and breaking, sharp gradients and cusps on the interfaces can be handled easily. Therefore it is suitable for tracking freeform interfaces. As with the evolution of the interfaces, the level set function also evolves under certain restraint conditions. The moving state of the interfaces can always be represented by the level set $c$ of the level set function. Tracing the evolution of interfaces by the level set method is actually a combination of solving level set equations and constructing the restraint conditions.

Considering a vector $x$ evolving in the time $t$ with velocity $v_\perp$,[3] the level set function can be expressed as $\phi(x,t)$. At any time $t$, the moving front of vector $x$ can be embedded in the level set function in the form of

$$\phi(x,t) = c \tag{7.16}$$

where $c$ is a given constant. By applying the chain rule on both sides of (7.16)

$$\frac{\partial \phi(x,t)}{\partial x}\mathrm{d}x + \frac{\partial \phi(x,t)}{\partial t}\mathrm{d}t = 0 \tag{7.17}$$

$$\frac{\partial \phi(x,t)}{\partial x} \cdot \frac{\mathrm{d}x}{\mathrm{d}t} + \frac{\partial \phi(x,t)}{\partial t} = 0 \tag{7.18}$$

The moving front of the vector $x$ at any time $t$ is then determined by solving time-dependent Hamilton–Jacobi equation in the form of equation (7.19) with an initial condition expressed in (7.20):

$$v_\perp(x,t)\left|\nabla\phi(x,t)\right| + \phi_t(x,t) = 0 \tag{7.19}$$

$$\phi(x,t=0) = \Gamma \tag{7.20}$$

where

$$\nabla\phi(x,t) = \vec{n} \cdot \left|\nabla\phi(x,t)\right| = \vec{n} \cdot \left|\frac{\partial \phi(x,t)}{\partial x}\right| \tag{7.21}$$

and

$$v_\perp(x,t) = \vec{n} \cdot \frac{\mathrm{d}x}{\mathrm{d}t} \tag{7.22}$$

$\vec{n}$ is the normal direction of the moving front. $\Gamma$ is the initial level set function. $\Gamma - c$ is a signed distance function, which represents the distance from the point $x$ to the

[3] The subscript "⊥" indicates that the direction of the velocity is perpendicular to the moving front of vector $x$. That is to say, the velocity $v_\perp$ is along the gradient direction.

moving front where a positive sign means that the point is outside the moving front and vice versa. Therefore, the moving front is the data set of the points corresponding to $\Gamma - c = 0$. Particularly, when $c$ is set to be zero, the signed distance function becomes $\Gamma$; the moving front is the zero level set of the solutions derived from (7.19) and (7.20).

Here, a hidden point is that the velocity $v_{\perp}$ is defined not only for the level set corresponding to the moving front of the vector $x$ but also for the other level sets which do not have physical meanings. In order to solve (7.19), $v_{\perp}$ is assumed to be defined for all the level sets. Therefore, both the interface and the velocity are embedded in a higher dimensional function. The newly constructed velocity is called extension velocity ($v_{\text{ext}}$). The extension velocity should, in the limit as one approaches the level set $c$, yield the velocity of the level set $c$, i.e.

$$\lim_{x \to a} v_{ext}(x) = v_{\perp}(a) \tag{7.23}$$

where $a$ is a point on the moving front.

All in all, it can be concluded that in order to track the surface topography evolution of the propagating interfaces by the level set method, one could follow the following procedures:

1. Determination of the velocity of the propagating interfaces ($v_{\perp}$) during the evolution process.
2. Construction of the initial level set function ($\Gamma$) where the initial state of the propagating interfaces is embedded into certain level set $c$ (see footnote 2).[4]
3. Construction of extension velocity ($v_{\text{ext}}$).
4. Solving level set functions (7.19).
5. Inferring the level set $c$ from the solution of the level set function at time $t$: The level set $c$ is actually the state of the propagating interfaces at time $t$.

The simulation started from a beam of gallium ions with Gaussian distribution bombarding a flat silicon substrate perpendicularly. The flat plane was embedded into a level set function as the zero level set. In the sputtering process, the dwell time was set at discrete time steps of 1 μs. At each consecutive step, the simulation sets the zero level set as the current surface topography of the nanodot. Based on the current surface topography, the sputter yield at each pixel for the next time step was then obtained. Therefore, the direct sputtering flux $F_{\text{direct}}$ was achieved.

The calculation of $F_{\text{indirect}}$ is more complex. The value of $F_{\text{indirect}}$ is built on $F_{\text{direct}}$; it can only be obtained when $F_{\text{direct}}$ is determined. In addition, not every point on the surface can receive $F_{\text{indirect}}$ from other points; $F_{\text{indirect}}$ only worked when the two points can "see" each other. Therefore a visibility test was needed in every time step. In level set method, the visibility test has been contained in the level set function. This can be clearly observed in Fig. 7.17. If the level set function $\phi$ is always positive between two points on the surface (point O and point A), these two points are visible to each other which means that the redeposition atoms from one point

[4] For the sake of convenience, $c$ is generally set to be 0.

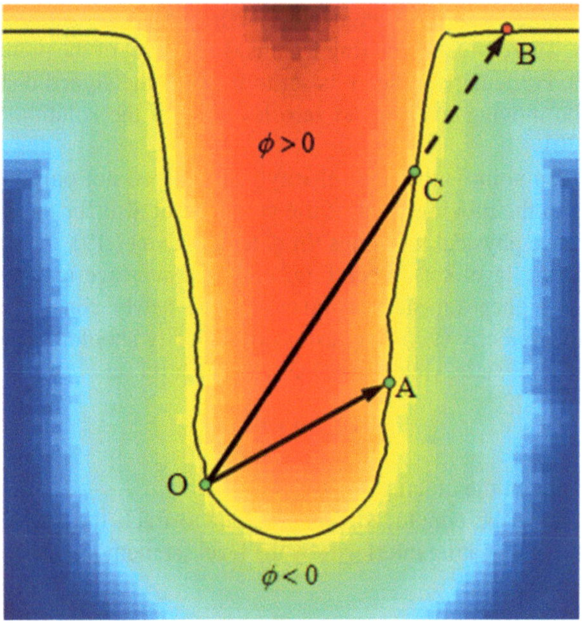


**Fig. 7.17** Visibility test on level set function

can reach the other point. Between the points O and B, $\phi$ is only positive in the OC segment. After passing through the moving front (point C where $\phi=0$), $\phi$ changes to negative in CB. This means that point O and B cannot see each other. Sputtered atoms from O to B are blocked at point C. Thus the redeposition will not occur between point O and point B.

All the calculations and analysis above led to the possibility of determining the surface velocity evolution with redeposition effect. However, in order to solve the level set function, the velocity was defined not only on the machined surface but also over the whole domain of the level set function. Therefore, extension velocities in the whole domain had to be constructed. Extension velocities are needed to maintain the propagation of the level set function. In this work, a simple copy of the velocity from the nearest point on the machined surface was applied to construct the extension velocities.

Forward Euler time discretisation and upwind spatial differencing were used as a robust way to solve the hyperbolic level set equation. The approximate solutions of the level set function at different times are shown in Fig. 7.18. It indicates that the level set function evolves from a plane ($t=0$ ms) to a curved surface ($t=5.5$ ms). Meanwhile, the cross section of the machined structure is naturally formed by tracking the zero level set of the level set function. Due to the spatial symmetry of the nanodot, three-dimensional re-construction relative to the symmetry axis was applied as shown in Fig. 7.19. Finally a quasi three-dimensional simulation program based on the level set method was developed for FIB nanofabrication.

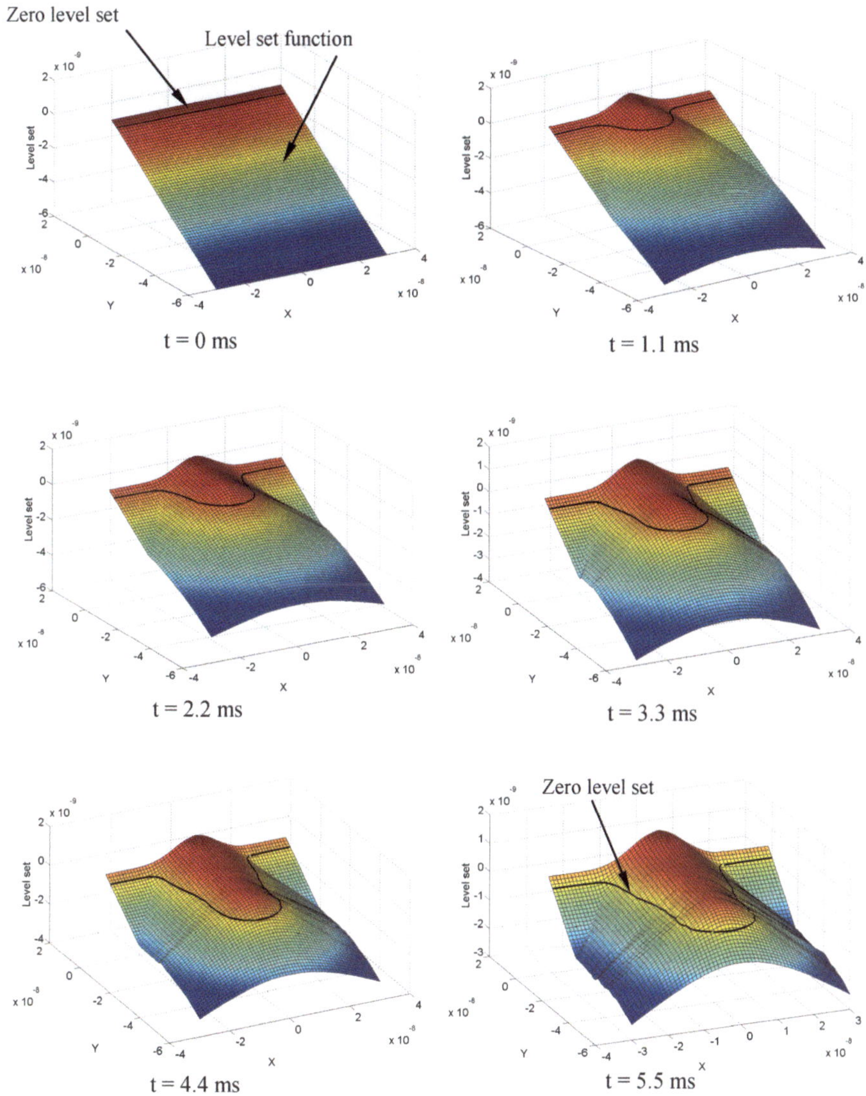


**Fig. 7.18** Evolution of nanodot profile in the simulation process

## *7.4.5 Experimental Validation*

Experimental work on the fabrication of nanodot arrays on a silicon substrate was carried out by using a dual-beam FIB system (FEI Quanta3D FEG) with gallium ion source. As a comparison test to evaluate the simulation results, a 30 keV gallium beam with current of 10 pA was used to drill two series of nanodots on (100) surface

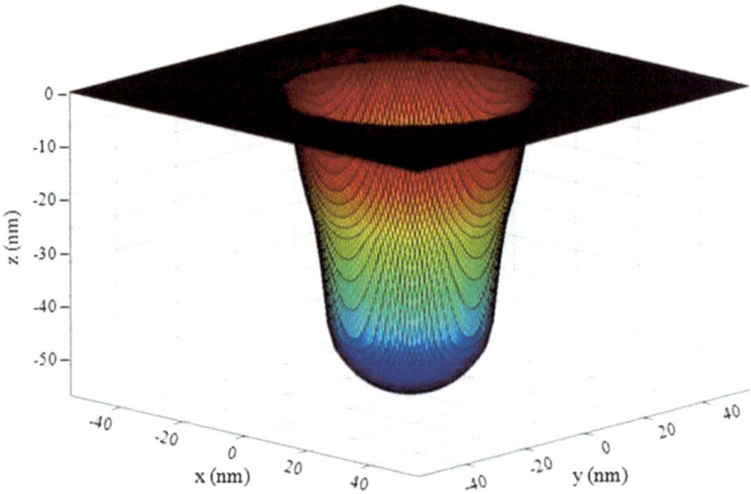


**Fig. 7.19** Surface topography of a nanodot after three-dimensional re-construction

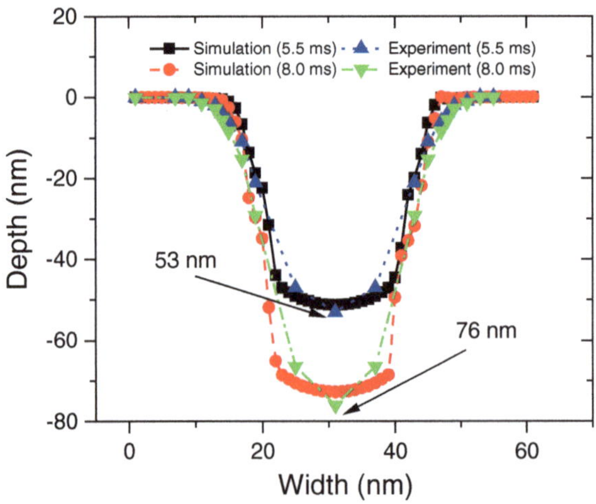


**Fig. 7.20** Comparison of simulation and experimental results of two nanodots fabricated by FIB

of a silicon substrate. The dwell times were fixed at 5.5 and 8.0 ms, respectively. According to the simulation results, nanodot arrays with depth of 51 and 71 nm can be obtained in these situations. The FIB-fabricated nanodot arrays were measured by an atomic force microscope (AFM, HYSITRON TI900 TriboIndenter). The maximum measured depths were 53 and 76 nm, respectively. A comparison of the simulation and experimental results is shown in Fig. 7.20. It can be seen that the simulation results agreed well with the experimental results in terms of maximum fabrication depth with errors less than 7 %. The maximum divergence takes place at the bottom part of the nanodots, where a relatively flat bottom is predicted by the simulation but

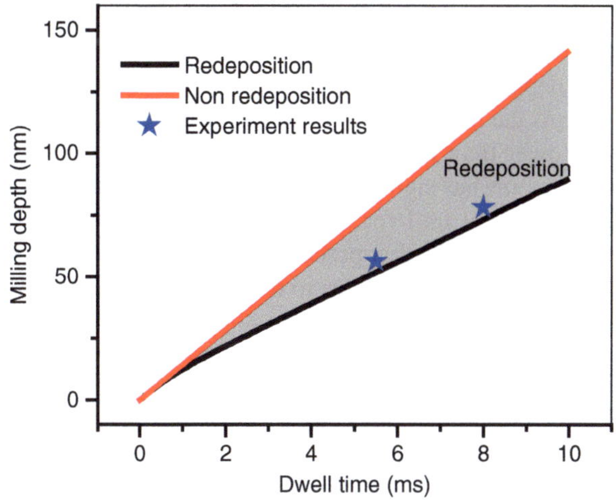


**Fig. 7.21** Effect of redeposition on the nanodot fabrication process

a sharp bottom is observed in the measurement results. The reason for these discrepancies is the limitation of the geometry of the AFM tip which is not sharp enough to reach the sidewall of the nanodot.

This validation indicates that a good agreement between the simulation and experimental results had been obtained with simulation errors less than 10 %. The modelling and simulation programme, therefore, could be used to determine machining parameters to achieve deterministic FIB nanofabrication.

### *7.4.6 Investigation of the Redeposition Effect*

The difficulty in predicting the depth of nanodot is due to the redeposition effect. If no redeposition occurs, the milling depth of the nanodot would be proportional to the ion fluence and can be easily determined. Therefore, it is necessary to study the contribution of the redeposition effect during the sputtering process. An accurate evaluation of the redeposition effect enables the prediction of milling depth without the simulation.

In order to study this effect, two sets of simulations were carried out using a 30 keV gallium beam with 10 pA ion beam current to fabricate nanodot arrays on a silicon substrate under dwell times ranging from 0 to 10 ms. In the first set of simulation the sticking coefficient $S_c$ was set to 0 in order to emulate an ideal machining condition without redeposition. In the other simulation set the sticking coefficient $S_c$ was set to 1 so as to take fully into account the influence of atom redeposition. Figure 7.21 illustrates the variation of the depth of nanodot against dwell time from both the simulation test and experiment results under a dwell time of 5.5 and 8 ms.

It was found that the milling depth increased linearly with the increase in dwell time when ion redeposition was ignored. On the other hand, there was a nonlinear increase of the FIB milling depth against dwell time, especially when the dwell time was less than 4 ms. After 4 ms the increase in milling depth could be regarded as linear again but with a slope lower than that without atom redeposition. This confirms that the redeposition of atoms has significant effects in the FIB nanofabrication process. The simulation is an ideal tool for the quantitative evaluation of the effects of redeposition on the nanofabrication process. As illustrated in Fig. 7.21, more than 1/3 of the milling depth was reduced due to the redeposition effect.

## 7.5 Determination of Fabrication Parameters in FIB Machining

The fabrication of customised three-dimensional structures is the reverse of the surface topography prediction. Several methods have been reported to realise three-dimensional structures using FIB machining. In the 1990s, Ishitani et al. first demonstrated the feasibility of FIB in two-dimensional fabrication by cutting a 45 μm diameter micro-gear in stainless steel foil with the aid of a sample rotator [18]. Nellen et al. [30] introduced a two-dimensional simulation model to estimate the sputter depth of simple geometric structures. Fabrication parameters, including ion fluence, beam diameter and scanning strategy were taken into consideration.

Several fabrication methods for the fabrication of three-dimensional structures were proposed by former researchers. Vasile et al. introduced a depth control method by controlling the dwell time on each milling pixel [50, 51]. In this method, a general mathematical model was developed which directs the ion beam material removal process and has the capability of milling a three-dimensional cavity from a given initial geometry into a pre-specified final geometry. The factors that affect the ion milling process, such as sputter yield and ion beam intensity distribution, were considered. Redeposition effect and the influence of the scanning path were ignored. Fu et al. [13] employed a two-dimensional slice-by-slice method to fabricate three-dimensional structures by FIB. Sequential two-dimensional slices with small thicknesses were used to approach the desired three-dimensional structures. The number of the slices can be determined according to the maximum depth of the three-dimensional microstructures. The milling depth for each slice depends on the slice thickness. Analogously, Lalev et al. [27] developed a data preparation program for FIB machining of complex three-dimensional structures utilising this slice-by-slice method. Complex surfaces can be easily designed in any three-dimensional CAD package and then converted into Graphic Database System II (GDSII) streams for FIB sputtering or deposition. Compared with the depth control method, this method has the advantages of a simple mathematical model, less memory space used and fast milling speed. The sputter yield and the dwell time are constant during the machining of flat slices. However, the fabricated structures have stair-step sidewalls due to the limitation of the number of finite slices. Regarding the surface quality of

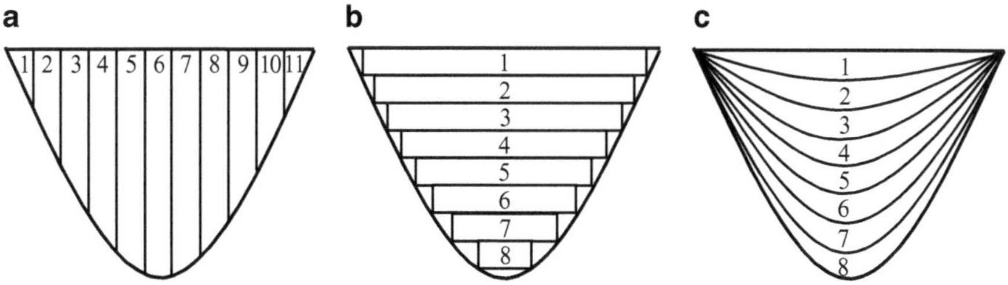


**Fig. 7.22** Three different methodologies for producing a 3D shape: (**a**) Depth control method: sputtering by changing the dwell time on the pixel; (**b**) slice-by-slice method: sputtering the sequential 2D slice by slice; and (**c**) continuous slicing method: modified by repeating Fu's method by continuously reducing the ion dose for each slice, whereas the total ion dose is the same as Fu's method

the sidewall, Hopman et al. [17] proposed a "trial-and-error" fabrication method to optimise surface quality. It was found that the scanning routine of the ion beam can be used as a sidewall optimisation parameter. For circular patterns, a spiral scanning routine can significantly reduce the amount of redeposition. Based on Fu and Hopman's methods, Kim et al. [23] chose a continuous slicing method which used spiral scanning as a part of the vector scan instead of using a raster scan to produce accurate circular patterns. Rather than using a discrete slice-by-slice process, the circular structure was fabricated in a contiguous way where the ion beam followed a spiral routine; therefore, stair-step structures on the sidewall were avoided. However, this is not a general method for all kinds of three-dimensional structures since only circular structures were studied in this work. The differences between the above fabrication methods are illustrated schematically in Fig. 7.22.

There are other methods developed for only fabricating specific structures. A quasi-direct writing method [11] and a self-organised formation [12] method were introduced to fabricate diffractive structures on silicon substrate. In these the dwell time was kept at a constant during the whole process. Diffractive structures with an amplitude of less than 1 μm are generated by changing the pixel spacing along the direction of the cross scan with zero overlap and keeping the pixel spacing constant along the other scanning direction with a normal overlap of 50–60 %. The advantage of this method is that there is no need to program and it can be applied on almost any material. However, this is not a general method for micro-fabrication; only blazed grating-like structures can be machined in this way. Svintsov et al. developed their IonRevSim software [46] specifically for data preparation and prediction of the shape of FIB-machined structures. It demonstrated a quantitative description of FIB machining for three-dimensional structures by means of an isotropic local etching model which assumes that the sputter yield is proportional to $\frac{1}{\cos(\theta)}$, where $\theta$ is the incident angle. This assumption works under a low incident angle; therefore, this model can be used for the prediction of FIB-machined structures with a low aspect ratio and with an inclination not greater than 45°.

It can be concluded that FIB is a competent technology for fabricating three-dimensional microstructures by the aid of sophisticated data processing or data preparation. However, the limitations are also very obvious. The depth control method is able to fabricate complex structures precisely, but sophisticated data processing is needed before the fabrication. That is to say that user-specified program is required and the calculation of the dwell time matrix is a time-consuming work especially for microstructures over large area. The slice-by-slice method is an efficient way for three-dimensional microstructure fabrication, but it will result in a stair-step structure on the sidewall which degrades the machining accuracy. The continuous slicing method and the quasi-direct writing method are only applicable to circular structures and blazed grating-like structures, respectively. The IonRevSim software is only used for the prediction and fabrication of structures with low aspect ratio. Redeposition effect is ignored in all of the above methods. Novel data processing and data preparation techniques that are based on the current FIB control software need to be developed for the fabrication of complex structures with high accuracy and high usability.

### *7.5.1 The Divergence Compensation Method*

In this section, the development of a divergence compensation method is presented in order to facilitate the production of ultra-precision three-dimensional structures. Key fabrication parameters, such as dwell time distribution, scanning passes, scanning pitch and scanning strategy, can be determined by this method.

When FIB is applied for the fabrication of three-dimensional structures, the machining accuracy can be degraded due to the finite spot size of the incident ion beam and sophisticated ion–solid interactions. There are three main aspects which have an impact on this:

- Firstly, the beam intensity profile is usually considered as Gaussian distribution with circular cross section [16]. The tail of the Gaussian distribution broadens the beam profile and contributes extra ion flux to the adjacent area. For each milling point, the accumulation of all the extra fluxes contributed by the adjacent points results in an unintentional increase of the local ion dose.
- Secondly, the variation of the sputter yield under different ion incident angles also degrades the machining accuracy of FIB milling. The prediction of the milling depth based on the incident ion dose is hampered by the variation of the sputter yield.
- Thirdly, the redeposition effect, as well as the scanning strategy, can further affect the machining accuracy.

Proper divergence compensation techniques are therefore needed to overcome the above issues in order to fabricate complex structures with high accuracy using FIB. The predictive divergence compensation approach can be implemented through the development of a three-dimensional surface topography generation model.

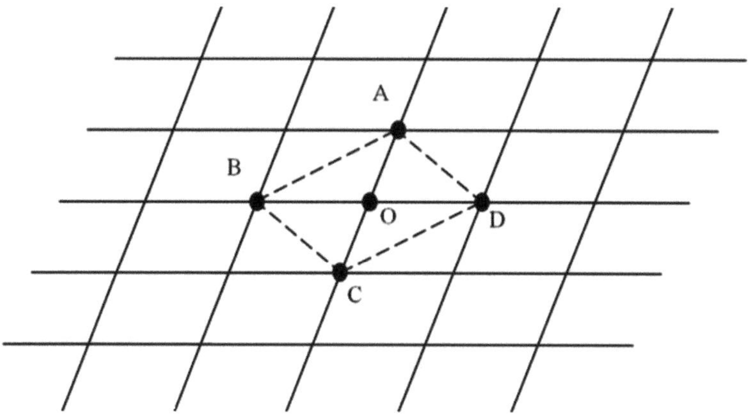


**Fig. 7.23** Schematic illustration of surface topography model

In this model, the machined surface is described as a grid in the $x$–$y$-plane composed of arrays of equally spaced intervals, where the gridlines intersect at the nodes (Fig. 7.23). The numerical value of each node represents the elevation of this surface point above a fiducial plane. Each node is surrounded by four other adjacent nodes, which can be combined into patches to investigate the surface normal at the target node. The surface normal direction at the target node is determined by the relative positions of the four adjacent nodes. As illustrated in Fig. 7.23, the surface normal at the target node O is defined as the average value of the normal directions of pitches OAB, OBC, OCD and ODA. Once the direction of a node is defined, the separation angle between the incident ion beam and the machined surface is specified. Therefore, the sputter yield at each node can be obtained.

## 7.5.2 Dwell Time Distribution

The FIB milling depth is derived from the ion incident flux, atom density of the target material and sputter yield. The total incident flux at a certain node is contributed not just from the local pixel area but also from all of the adjacent areas due to the overlap effect. Therefore, the dwell time for each node can be obtained from a matrix relationship as showed in (7.24), which includes the ion beam distribution, ion beam overlap and angular dependent sputter yield [51]:

$$\sum_{k,l=1}^{n} C_{ijkl} sp_{kl} t_{kl} p_x p_y = \Delta Z_{ij} \tag{7.24}$$

where $C_{ijkl}$ is the coefficient matrix which describes the flux contribution from the adjacent node $(k, l)$ to the local node $(i, j)$. $sp_{kl}$ and $t_{kl}$ are the scanning passes and the dwell time at node $(k, l)$, respectively. $p_x$ and $p_y$ are pitches along the pixel address scheme, and $\Delta Z_{ij}$ is the milling depth increment at node $(i, j)$.

The incident ion beam current density needs to be corrected when the ion beam is presented to inclined surfaces. In this case, the number of ions per area decreases as $\cos(\theta)$, where $\theta$ is the separation angle between the incident ion beam and the surface normal. Therefore, the corrected milling depth at node $(i, j)$ can be expressed as

$$\Delta Z_{i,j} = \sum_{k,l=1}^{n} \frac{\Phi(k,l)}{\eta} J_{kl}(i,j) Y(\theta_{ij}) \cos(\theta_{ij}) s p_{kl} t_{kl} p_x p_y \tag{7.25}$$

where $\Phi(k,l)$ is the ion flux density at node $(k, l)$. $\eta$ and $J_{kl}$ are the atom density of the solid and the ion beam intensity distribution function at node $(k, l)$, respectively. Generally $J_{kl}$ follows Gaussian distribution. A three-dimensional ion beam intensity distribution is shown in (7.6). $Y(\theta)$ is the angular dependent sputter yield. The summation over the indices $k$ and $l$ accounts for total dose received at node $(i, j)$ from all the pixels in the address scheme.

Equation (7.25) can be applied to each node on the machined surface to obtain a corrected FIB milling depth. In an area comprising $N \times N$ nodes, the total milling time $p_{kl}t_{kl}$ at each node is derived from an $N \times N$ equation set composed of (7.25). Therefore, the dwell time distribution is calculated by (7.26)

$$t = \frac{Z}{C} \tag{7.26}$$

where $t$ and $Z$ are dwell time matrix and milling depth matrix, respectively. $C$ is the coefficient matrix of the equation set (7.25).

### 7.5.3 The Implementation of the Divergence Compensation Method in FIB Machining

In a modern FIB system, pattern generation in conjunction with graphic bitmap files as inputs enables the achievement of complex two-dimensional structures. Each pixel in the bitmap represents an incident beam spot. The local colour value of each pixel in the bitmap delineates the dwell time and beam blanking status. In a 24-bit RGB bitmap each pixel consists of:

- A red component—not used.
- A green component—determines if the beam is blanked, any other value other than 0 activates the beam.
- A blue component—determines the dwell time per pixel, the blue ranges from 0 to 255, which corresponds to an 8-bit colour depth.

Figure 7.24 shows an example of an FIB-fabricated two-dimensional structure based on the corresponding bitmap image. It reveals the transformation from the colour scale level in the bitmap to the milling depth by varying the dwell time for each

**Fig. 7.24** FIB milling of a two-dimensional structure. (**a**) The bitmap image applied; (**b**) the corresponding structures obtained from the bitmap

pixel. The brightest part in the bitmap corresponds to the longest dwell time, leading to a concave structure, while the darkest part corresponds to the shortest dwell time. The flexibility in controlling the dwell time at each pixel enables a precise method for manufacturing complex three-dimensional structures by FIB milling.

However, the structures obtained using such a bitmap always diverge from the original design due to the intrinsic problems associated with the FIB machining mechanism, including atom redeposition, overlap effect of the ion beam and variation of the sputter yield under different incident angles. This divergence can be compensated for by the application of the divergence compensation method; based on this, the dwell time distribution is corrected layer by layer and a series of bitmaps can be generated. In each scanning pass, the colour spread of each bitmap is identical to the dwell time distribution calculated by (7.26). Therefore, the corrected bitmaps with a specified number of pixels ($M \times N$) for the intended structures can be generated. These bitmaps were in 24-bit RGB scale, in which the local colour of each pixel represents the dwell time. The pixel number in the bitmap was determined by the following equation (7.27):

$$\begin{cases} M = \dfrac{W}{p_x} \\ N = \dfrac{H}{p_y} \end{cases} \tag{7.27}$$

where $p_x$ and $p_y$ are the pitches between adjacent pixels along the $X$- and $Y$-directions; $W$ and $H$ are the width and height of the designed pattern.

In three-dimensional FIB machining, the milling depth at each pixel is determined by the sputter yield and dwell time. The dependence between sputter yield and ion incident angles was determined as the first step. The divergence compensation method was then applied to calculate the corrected dwell time matrix based on the sputter yield. In the following step, the dwell time matrix was input into the FIB system in the form of bitmaps.

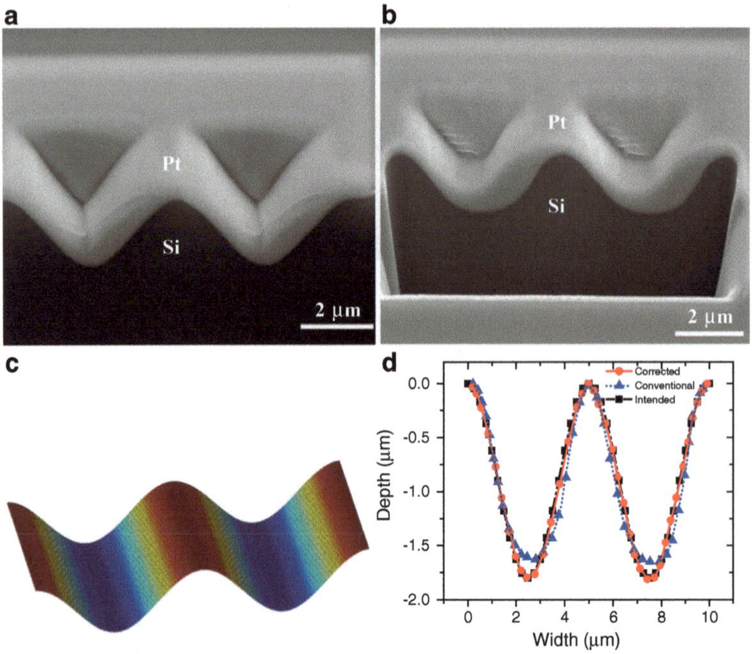


**Fig. 7.25** Comparison of sinusoidal structure fabricated by the conventional and predictive divergence compensation method: (**a**) Cross section of the corrected sinusoidal structure; (**b**) cross section of the sinusoidal structure fabricated by conventional bitmap milling; (**c**) intended sinusoidal shape and (**d**) comparison of the sinusoidal structures fabricated by both methods

One assumption in the divergence compensation method is that the incident ions bombard the substrate simultaneously in each scanning pass—the scanning strategy is not taken into consideration. However, experimental results show that the scanning strategy does impact on the machined surface topography because of the redeposition effect [17]. A comparison of the structures fabricated by different scanning strategies will be carried out as a supplement of this research.

### 7.5.4 *Experimental Validation*

The FIB machining experiments were carried out on a dual-beam FIB system (FEI Quanta 3D FEG). The system was operated at an acceleration voltage of 30 kV. The beam diameter (FWHM) and the ion current are 50 nm and 1 nA, respectively. Three-dimensional structures, including a parabolic structure (amplitude of 2.5 μm, width of 10 μm), two hemispherical structures (radius of 5 μm and 500 nm) and a sinusoidal structure (period 5 μm, amplitude of 0.9 μm), were fabricated on a P-type silicon substrate (100) surface; the surface roughness of the silicon substrate is less than 5 Å.

The sinusoidal microstructures fabricated by the conventional method and the predictive divergence compensation method are shown in Fig. 7.25. With the

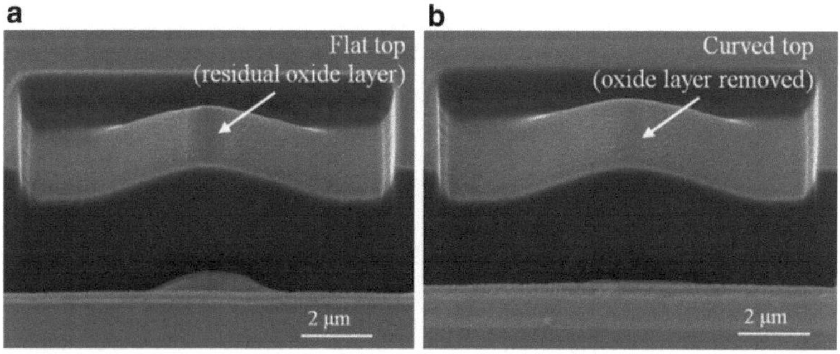


**Fig. 7.26** Comparison of sinusoidal structures fabricated on (**a**) silicon substrate with a thin oxide layer and (**b**) silicon substrate after removing the oxide layer

predictive divergence compensation method the peak-to-valley form error of the machined structure has been reduced using the conventional bitmap milling approach from 200 to 30 nm. The machining accuracy was much improved, especially at the bottom part of the structure which was ameliorated to a curved profile instead of the flat one obtained using the conventional bitmap milling approach (see Fig. 7.25b). Further FIB machining experiments indicated that a flat top/bottom could always be found on the top of the convex- or the bottom of the concave-shaped structures using conventional bitmap milling. The reason for this phenomenon is mainly related to the aspect ratio of the machined structure. Due to the low aspect ratio near the base, the differences in the ion dose used to fabricate the neighbouring points in the apex area of these structures are very tiny. An overlap of the Gaussian beam skirt in this region eliminates these tiny ion dose differences and results in the formation of a flat top/bottom. Since the overlap effect had been taken into consideration in the predictive divergence correction method, extra ion doses caused by beam overlap were compensated; therefore, more accurate structures were formed.

In addition, the existence of an oxide layer on the silicon substrate also contributes to the formation of a flat top/bottom on the low-aspect-ratio structures because the oxide layer is less sensitive to gallium ions than silicon. Figure 7.26a shows a low-aspect-ratio sinusoidal structure fabricated on a silicon substrate covered by a thin oxide layer; a flat part on the top of the structure was observed. Figure 7.26b shows that a precise sinusoidal feature was obtained using a silicon substrate after removing the oxide layer. Figures 7.27 and 7.28 show fabricated parabolic and hemispherical microstructures on silicon substrate on which the oxide layer was removed before FIB machining. Form accuracies (*p*–*v*) of 20 and 120 nm, respectively, were achieved for these structures.

The effectiveness of this divergence compensation method at the micron scale has been demonstrated through comparisons between the experimental results and the intended structures. This methodology is also effective for FIB milling at the submicron scale. A hemispherical structure with 480 nm radius (500 nm intended) was fabricated by using the divergence compensation approach as shown in Fig. 7.29. Along the circumference a reduction of 20 nm was found. Such relative divergence is primarily due to the redeposition effect which is more pronounced in the nanometre range.

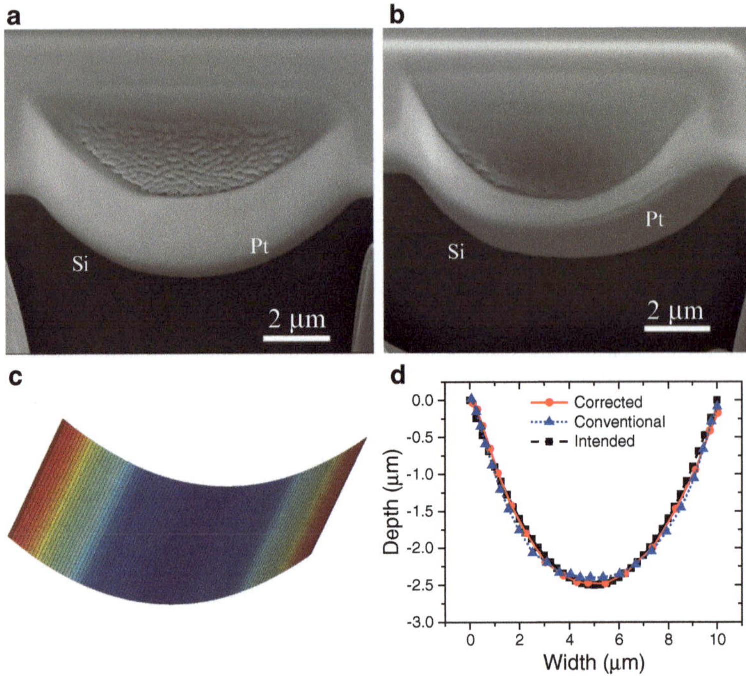


**Fig. 7.27** Comparison of parabolic microstructures fabricated by the conventional and predictive divergence compensation method: (**a**) Cross section of the corrected parabolic structure; (**b**) cross section of the parabolic structure fabricated by conventional bitmap milling; (**c**) intended parabolic shape and (**d**) comparison of the parabolic structures fabricated by both methods

It is worth mentioning that although the fabricated structures were very close to the designed geometries, there were still divergences along the curved surface. Such divergences were mainly caused by atom redeposition and the limited resolution of the dwell time. The redeposition process is inevitable within the FIB fabrication process due to its randomness. Although redeposition can be reduced by using the multi-pass scanning method, there is still certain amount of sputtered atoms deposited on the machined structure, especially for high-aspect-ratio and nanostructures. Besides, the limited resolution of dwell time contributes to the fabrication divergence of structures with a low aspect ratio. A user-specified maximum dwell time is assigned to each bitmap. In terms of the local colour, the dwell time for each pixel in the bitmap is linearly interpolated between zero and the maximum dwell and rounded to the value from a fixed dwell time table. For low-aspect-ratio structures, the dwell times are very close between adjacent pixels. Differences between dwell times may be omitted due to rounding error. This phenomenon also caused divergence of the fabricated structures from the designed shape.

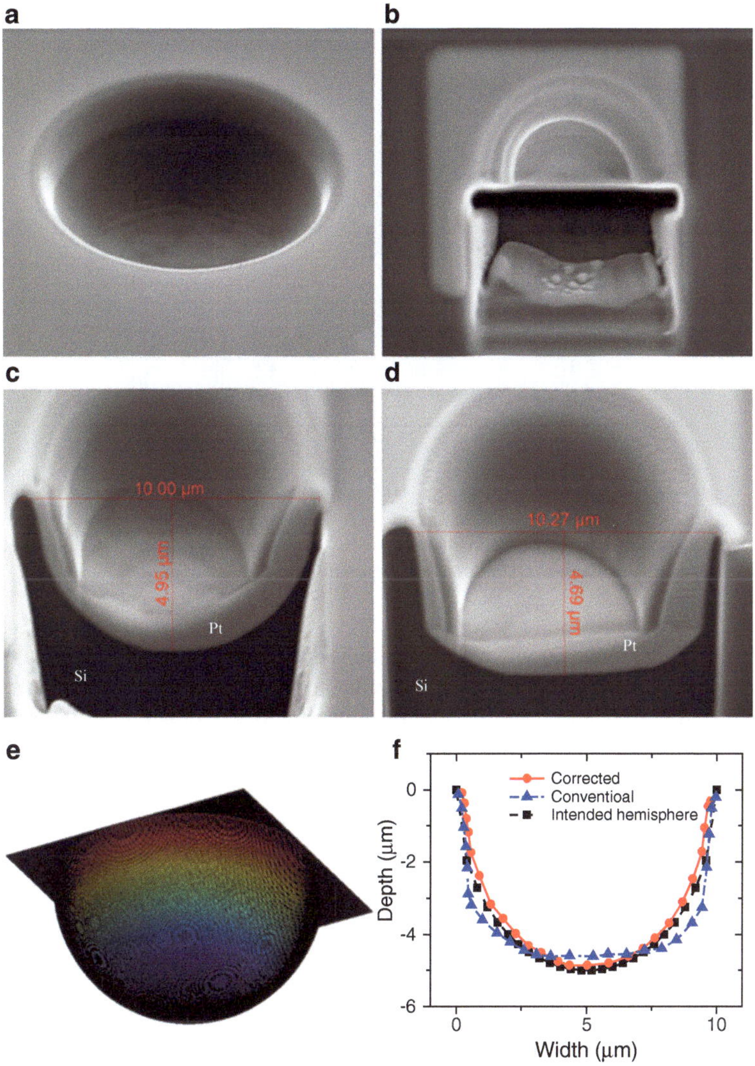


**Fig. 7.28** SEM images of hemisphere structures with 5 μm radius fabricated by the predictive divergence compensation method and conventional method: (**a**) 52° tilt view; (**b**) cross section cutting through the diameter; (**c**) cross section of the hemisphere fabricated by the predictive divergence compensation method; (**d**) cross section of the hemisphere fabricated by the conventional method; (**e**) sketch of the designed hemisphere and (**f**) comparison of hemispherical structures

### *7.5.5 Optimisation of the Divergence Compensation Method*

The limitation of the divergence compensation method is that due to the computing of dwell time at each pixel it can be very time consuming. For a milling area comprising $M \times N$ pixels, the equation set with a coefficient matrix of dimensions of $M^2 \times N^2$ needs

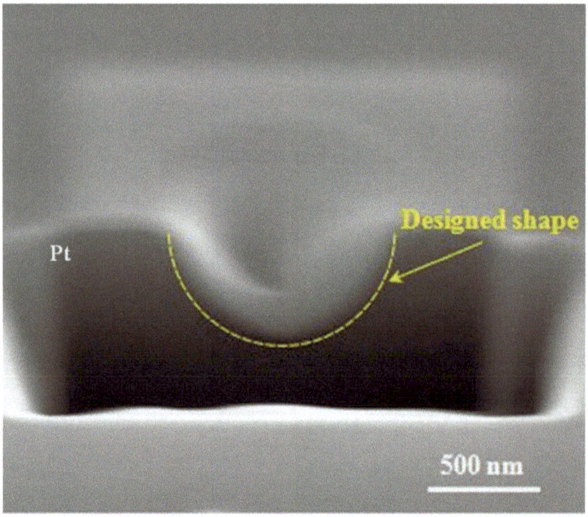


**Fig. 7.29** Cross section of nano-hemisphere with 500 nm radius fabricated on silicon; a reduction of 20 nm along the radius was found due to redeposition

to be solved. The huge amount of calculation is due to the consideration of the overlap effect where the ion dose at each pixel is influenced by the ion dose from all the other pixels. In order to integrate this correction method with a commercial FIB system and enable it to be applied online, an optimisation of the divergence compensation approach is carried out through the theoretical and experimental evaluation of the overlap effect in order to reduce the processing time required.

The overlap effect is characterised by normalised pixel spacing $p_s/d_f$, where $p_s$ is the pixel spacing along the *X*- and *Y*-directions and $d_f$ is the beam diameter. It was found that $p_s/d_f$ should be equal to or smaller than 0.673 in order to have a uniform scanning ion flux in channel milling [48]. In this case, the extra dose from the adjacent pixel contributes almost 33 % of the peak dose shown in Fig. 7.30a. The overlap here can be reduced by increasing the pixel spacing; indeed, when $p_s/d_f$ equals to 1.0 only an extra dose of 7 % is contributed. To evaluate the influence of the overlap effect under such condition the ion current was kept at 1 nA as that used in the above experiments. The variation rate of the ion dose is around $1.4\times10^{13}$ ions/ms as shown in Fig. 7.30b. The variation of milling depth can be estimated based on the ion dose variation. Taking the fabrication of the hemispherical structure as an example, there are around 50,000 pixels in the bitmap and the total fabrication time is 10 min. The average machining time on each pixel is 12 ms. Calculation results show that $1.4\times10^{13}$ ions/ms result in a depth variation of 160 nm; however, because of the redeposition effect, the actual depth variation is actually less than this. Such depth variation is in the same order as the divergence between the fabricated structure and the intended structure. Validation tests on hemispherical structures are shown in Fig. 7.30c, and these indicate that, by applying $p_s/d_f=1$ without the overlap effect correction, the machined profile coincides with the intended structure very well with a maximum divergence of only 150 nm. Further increases in the normalised

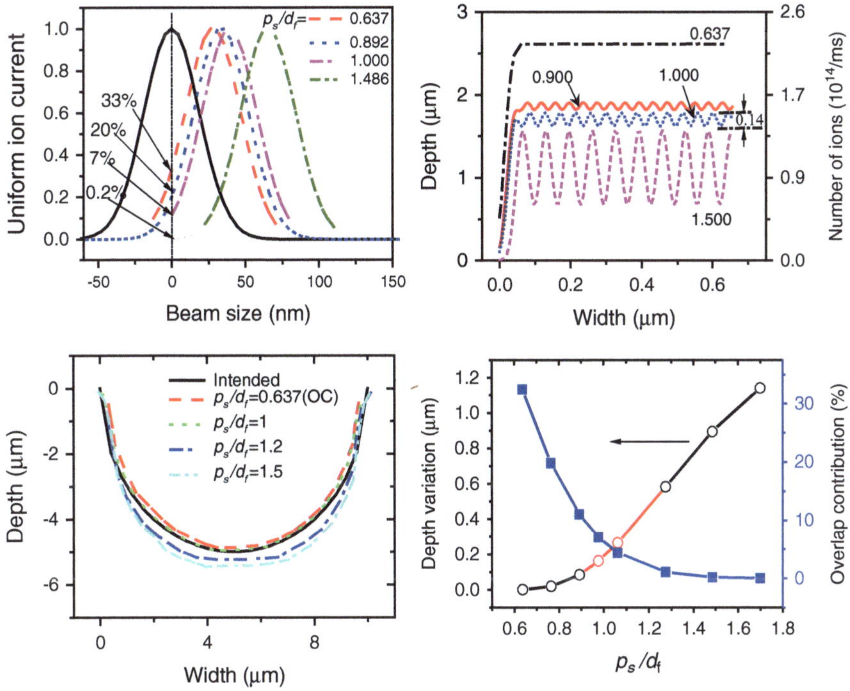


**Fig. 7.30** Influence from normalised pixel spacing in the machining process: (**a**) Overlap of ion current under different pixel spacing; (**b**) overlap effect on depth distribution for Gaussian FIB milling (uniformed pixel spacing $p_s/d_f$ is indicated); (**c**) comparison of hemisphere structures fabricated under different normalised pixel spacing without consideration of the overlap effect ("OC" indicates the overlap correction) and (**d**) depth variation and the contribution of overlap effect under different normalised pixel spacing

pixel spacing will result in a deeper milling cavity and a flat base due to a bigger depth variation. This can be clearly observed in Fig. 7.30c, d with $p_s/d_f = 1.2$ and $p_s/d_f = 1.5$ present. Therefore, it is recommended that the overlap effect can be suppressed by choosing the pixel spacing equal to the beam diameter. The same result has also been verified on the sinusoidal and parabolic structures.

## 7.6 FIB Hybrid Machining

### *7.6.1 FIB Machining and Nanoimprint Lithography*

FIB machining is an ideal technology for obtaining micro- and nanostructures. However, the low material removal rate as well as the gallium ion implantation effect limit the applications of FIB technology. Over the years there have been demonstrations of other methods which are capable of forming sub-micron

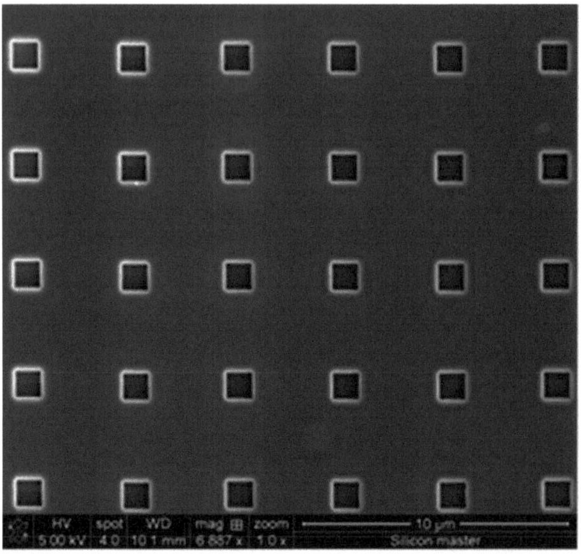

**Fig. 7.31** Hard silicon master fabricated by FIB

structures, such as electron beam lithography, dip-pen lithography and laser direct writing. However, there are still no universal machining techniques in existence. None of the machining techniques can dominate the manufacturing at nano-, micro-, human and large scales. The combination of different techniques is the norm especially when fabricating complex structures on special materials. One important machining method that can be easily combined with other micro- and nano-manufacturing process is NIL. NIL was initially proposed and developed by Prof. Chou's group [7, 8] since 1990s. They used a hard-structured mould to emboss into polymer material cast on the wafer substrate under controlled temperature and pressure conditions to create a thickness contrast pattern in a thin resist film carried on a substrate. Patterns were then transferred through the entire resist layer by an anisotropic etching process. Recently it has been reported that FIB machining technique can be used as a high-resolution patterning method suitable for NIL [6, 37, 43, 45]. An FIB-machined silicon substrate patterned with 1 μm × 1 μm cavity arrays which can be used as a hard mould in NIL is shown in Fig. 7.31. Once the silicon mould is completed it is reusable in the fabrication of PDMS mould. Therefore, the machining efficiency of FIB technique can be boosted when combining with NIL. However, NIL only works on a small scale. Meanwhile, both the master and the work piece must be flat. The replication of nanostructures on curved surfaces is still a challenging task. For nanostructures to become ubiquitous they will have to be mass produced at low cost.

### 7.6.2 FIB Machining and Single-Point Diamond Turning

In order to overcome this limitation, another hybrid machining approach based on FIB machining and SPDT has been developed. SPDT has been exploited successfully to generate a wide range of optical structures from discrete optical components to large-area micro-optic based films such as brightness enhancement films for LCD displays [34]. It is suitable at the industrial scale production with a high throughput. However, the size of the cutting tools limits the application of SCD turning in nanotechnologies. Because of the extreme hardness of diamond, it is very difficult to machine SCD tools by routine micromachining techniques. The choice of shaping techniques is very limited. The literature reports that laser [49] and FIB methods [32] can be used to shape diamond. By applying advanced ion milling techniques to trim diamond tools, it is possible to increase the complexity and ultimate dimensional resolution of diamond-machined products.

In the last few years, the feasibility of fabricating micro-tools by FIB sputtering has been successfully demonstrated by many researchers. Micro-cutting and micro-milling tools made of diamond [10], cobalt M42 high-speed steel and C2 tungsten carbide [2, 3] were successfully developed. These tools' dimensions range from 15 to 100 μm. In order to extend the application of this hybrid machining method to nanofabrication, Sun et al. developed diamond-cutting tools with nanoscale periodical chisel structures on the tip as shown in Fig. 7.32 [44]. For the sake of contrast, the original SCD tool is shown in Fig. 7.32a. Figure 7.32b, c shows SEM images of an FIB-fabricated nano-SCD tool. An ion-induced secondary electron image (taken from the clearance facet) of the diamond tip with a magnification $\times 20,000$ is shown in Fig. 7.32d. Periodical chisel structures with finely finished edges are successfully obtained. The ripple-free sidewalls on the diamond tip are clearly observed in Fig. 7.32c. A surface roughness Ra of 1 nm was obtained on both of the clearance facet and the rake facet.

The cutting performance of the nano-SCD tool was tested on a Precitech Freeform 700 ultra-precision diamond-turning machine. The FIB-machined nano-SCD tool was mounted on this machine to fabricate nano-grating array on an electroless nickel substrate as shown in Fig. 7.33a. The fabricated micro-grating array is shown in Fig. 7.33b, where colourful interference fringe can be clearly observed. An SEM image of the nano-grating array with a magnification of 15,000 is shown in Fig. 7.33c. Figure 7.33d shows the cross-sectional view of the nano-grating array. Nano-grating array with good form coherence was successfully obtained. Using such a nanoscale diamond tool in diamond turning process to fabricate nanostructures, the material removal rate is 30,000 times higher than purely FIB machining with ion current of 1 nA [44]. The machining efficiency can be further improved by fabricating more nanostructures on the diamond tool tip.

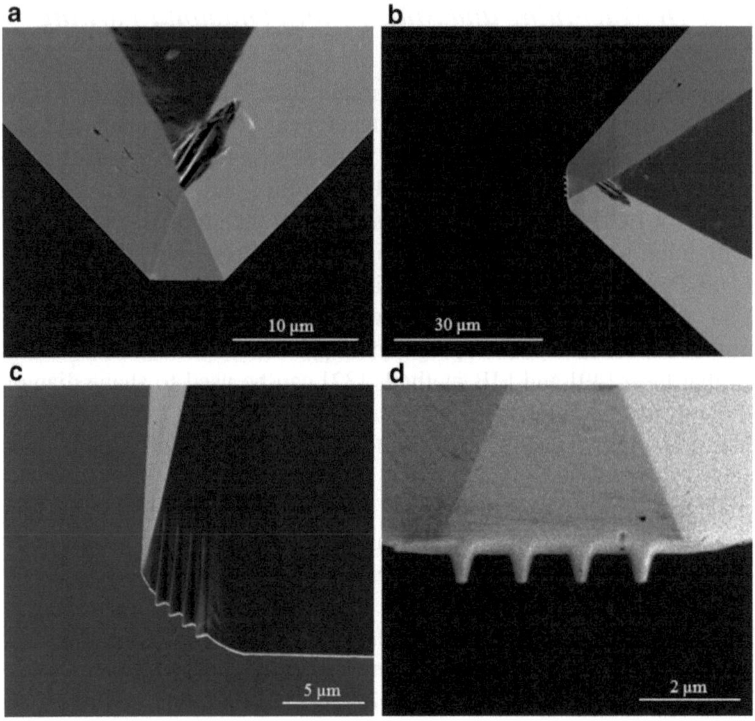


**Fig. 7.32** A nanoscale SCD tool fabricated by FIB machining. (**a**) Original SCD tool; (**b**) a lower magnification view of the same tool trimmed by FIB; (**c** and **d**) lateral and top view of the nanoscale SCD tool

### *7.6.3 FIB Machining and Laser/Plasma Machining*

In addition to the above hybrid machining techniques, laser and plasma sources have also been integrated with the FIB machining system. This combination allows the FIB machining system to satisfy high-resolution requirements and ultra-fast material removal rates at the same time. There have already been several such kinds of commercial FIB systems successfully developed and are available on the market. For example, the Zeiss AURIGA Laser system integrates a laser source, an ion source and an electron source into the same workstation which enables an extremely fast material removal and precise fine removal machining capability. The Tescan FERA3 XMH Plasma FIB-FESEM workstation integrates a xenon plasma source into the system to remove bulk material effectively.

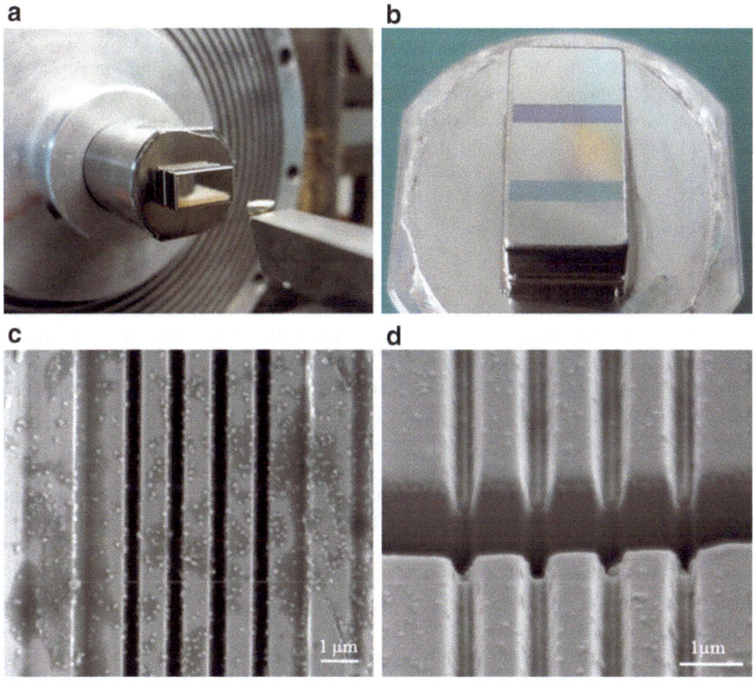


**Fig. 7.33** Nano-grating array diamond turned by the nano-SCD diamond tool: (**a**); (**b**) Electroless nickel; (**c**) the SEM image of the nano-grating array; (**d**) cross-sectional view of the nano-grating array

## 7.7 Conclusions

In this chapter a deterministic fabrication of micro- and nanostructures by FIB has been introduced. The sputter yield and the distribution of sputtered atoms are calculated using the Monte Carlo method. The FIB machining process is modelled and simulated based on an ion–solid sputtering model combined with the level set method. The machined surface topography generation is accurately predicted.

The divergence compensation method can reduce the divergence caused by the variation of sputter yield, overlap effect and redeposition effect. The machined surface form accuracy can be dramatically improved by using this method. A less than 3 % relative divergence was consistently achieved for the basic microstructures. The divergence compensation method can also be implemented in FIB machining to obtain fabrication parameters for accurately machining micro- and nanostructures.

By combining with other manufacturing techniques such as NIL and SPDT, the authors believe that it is very promising to develop high-throughput manufacturing techniques which can meet the industrial requirement to fabricate micro- and nanostructures over extremely large area cost-effectively.

**Acknowledgments** The authors wish to thank Mr. J Chien, Dr. A Lee, Mr. C Wu and Mr. H Huang (Metal Industries Research and Development Centre); Dr. Tomas Hrncir (TESCAN); Mr. David Beamer (FEI); Mr. R Green, Mr. M Norris and Mr. J Bryden (Spanoptic) and Mr. A Cox (Contour Fine Tooling) for collaborations and support. This study has also received financial support from EPSRC's IMRC at Heriot-Watt University (Project no. 113955) and Royal Society of Edinburgh (Project no. 113666) whom the authors also gratefully acknowledge.

## References

1. Adams, D., Vasile, M.: Accurate focused ion beam sculpting of silicon using a variable pixel dwell time approach. J. Vac. Sci. Technol. B Microelectron. Nanometer Struct. **24**, 836 (2006)
2. Adams, D.P., Vasile, M.J., Benavides, G., Campbell, A.N.: Micromilling of metal alloys with focused ion beam-fabricated tools. Precis. Eng. **25**, 107–113 (2001)
3. Adams, D.P., Vasile, M.J., Krishnan, A.S.M.: Microgrooving and microthreading tools for fabricating curvilinear. J. Int. Soc. Precis. Eng. Nanotechnol. **24**, 347–356 (2000)
4. Boxleitner, W., Hobler, G.: FIBSIM—dynamic Monte Carlo simulation of compositional and topography changes caused by focused ion beam milling. Nucl. Inst. Methods Phys. Res. B **180**, 125–129 (2001)
5. Boxleitner, W., Hobler, G., Kluppel, V., Cerva, H.: Simulation of topography evolution and damage formation during TEM sample preparation using focused ion beams. Nucl. Inst. Methods Phys. Res. B **175**, 102–107 (2001)
6. Bruinink, C.M., Burresi, M., De Boer, M.J., Segerink, F.B., Jansen, H.V., Berenschot, E., et al.: Nanoimprint lithography for nanophotonics in silicon. Nano Lett. **8**, 2872–2877 (2008)
7. Chou, S.Y., Krauss, P.R., Renstrom, P.J.: Imprint of sub 25 nm vias and trenches in polymers. Appl. Phys. Lett. **67**, 3114 (1995)
8. Chou, S.Y., Krauss, P.R., Renstrom, P.J.: Nanoimprint Lithography. J. Vac. Sci. Technol. B Microelectron. Nanometer Struct. **14**, 4129–4133 (1996)
9. de Winter, D.A.M., Mulders, J.J.L.: Redeposition characteristics of focused ion beam milling for nanofabrication. J. Vac. Sci. Technol. B Microelectron. Nanometer Struct. **25**, 2215 (2007)
10. Ding, X., Lim, G.C., Cheng, C.K., Butler, D.L., Shaw, K.C., Liu, K., et al.: Fabrication of a micro-size diamond tool using a focused ion beam. J. Micromech. Microeng. **18**, 075017 (2008)
11. Fu, Y., Bryan, N., Zhou, W.: Quasi-direct writing of diffractive structures with a focused ion beam. Opt. Express **12**, 1803–1809 (2004)
12. Fu, Y., Bryan, N., Zhou, W.: Self-organized formation of a Blazed-grating-like structure on Si (100) induced by focused ion-beam. Opt. Express **12**, 227–233 (2004)
13. Fu, Y., Bryan, N.K.A.: Fabrication of three-dimensional microstructures by two-dimensional slice by slice approaching via focused ion beam milling. J. Vac. Sci. Technol. B Microelectron. Nanometer Struct. **22**, 1672 (2004)
14. Giannuzzi, L.A., Stevie, F.A.: Introduction to Focused Ion Beams: Instrumentation, Theory, Techniques and Practice. Springer, New York (2005)
15. Anderson, H.H., Bay, H.L.: Sputtering by particle bombardment I. In: Behrisch, R. (ed.) Topics in Applied Physics. Sputtering by Particle Bombardment I. Springer, Berlin, New York (1981)
16. Harriott, L.: Beam size measurements in focused ion beam systems. J. Vac. Sci. Technol. A Vac. Surf. Films **8**, 899–901 (1990)
17. Hopman, W.C.L., Ay, F., Hu, W., Gadgil, V.J., Kuipers, L., Pollnau, M., et al.: Focused ion beam scan routine, dwell time and dose optimizations for submicrometre period planar photonic crystal components and stamps in silicon. Nanotechnology **18**, 195305 (2007)
18. Ishitani, T., Ohnishi, T., Kawanami, Y.: Micromachining and device transplantation using focused ion beam. Jpn. J. Appl. Phys. **29**, 2283–2287 (1990)
19. Katardjiev, I.: A kinematic model of surface evolution during growth and erosion: Numerical analysis. J. Vac. Sci. Technol. A Vac. Surf. Films **7**, 3222–3232 (1989)

20. Katardjiev, I.: Simulation of surface evolution during ion bombardment. J. Vac. Sci. Technol. A Vac. Surf. Films **6**, 2434–2442 (1988)
21. Katardjiev, I., Carter, G., Nobes, M.: The application of the Huygens principle to surface evolution in inhomogeneous, anisotropic and time-dependent systems. J. Phys. D Appl. Phys. **22**, 1813 (1989)
22. Katardjiev, I., Carter, G., Nobes, M., Berg, S., Blom, H.: Three dimensional simulation of surface evolution during growth and erosion. J. Vac. Sci. Technol. A Vac. Surf. Films **12**, 61–68 (1994)
23. Kim, C., Park, J., Chu, W., Jang, D., Kim, S., Ahn, S.: Fabrication of silicon micro-mould for polymer replication using focused ion beam. Microelectron. Eng. **86**, 556–560 (2009)
24. Kim, H.B., Hobler, G., Lugstein, A., Bertagnolli, E.: Simulation of ion beam induced micro/nano fabrication. J. Micromech. Microeng. **17**, 1178–1183 (2007)
25. Kim, H.B., Hobler, G., Steiger, A., Lugstein, A., Bertagnolli, E.: Full three-dimensional simulation of focused ion beam micro/nanofabrication. Nanotechnology **18**, 245303 (2007)
26. Kim, H.B., Hobler, G., Steiger, A., Lugstein, A., Bertagnolli, E.: Level set approach for the simulation of focused ion beam processing on the micro/nano scale. Nanotechnology **18**, 265307 (2007)
27. Lalev, G., Dimov, S., Kettle, J., Van Delft, F., Minev, R.: Data preparation for focused ion beam machining of complex three-dimensional structures. Proc. Inst. Mech. Eng. B J. Eng. Manuf. **222**, 67–76 (2008)
28. Muehlberger, M., Boehm, M., Bergmair, I., Chouiki, M., Schoeftner, R., Kreindl, G., et al.: Nanoimprint lithography from CHARPAN Tool exposed master stamps with 12.5 nmhp. Microelectron. Eng. **88**, 2070–2073 (2011)
29. Muller, K., Weigmann, U., Burghause, H.: Simulation of focused ion beam milling. Microelectron. Eng. **5**, 481–489 (1986)
30. Nellen, P.M., Callegari, V., Bronimann, R.: FIB milling of photonic structures and sputtering simulation. Microelectron. Eng. **83**, 1805–1808 (2006)
31. Osher, S., Sethian, J.: Fronts propagating with curvature-dependent speed: algorithms based on Hamilton-Jacobi formulations. J. Comput. Phys. **79**, 12–49 (1988)
32. Picard, Y.N., Adams, D., Vasile, M., Ritchey, M.: Focused ion beam-shaped microtools for ultra-precision machining of cylindrical components. Precis. Eng. **27**, 59–69 (2003)
33. Platzgummer, E., Biedermann, A., Langfischer, H., Eder-Kapl, S., Kuemmel, M., Cernusca, S., et al.: Simulation of ion beam direct structuring for 3D nanoimprint template fabrication. Microelectron. Eng. **83**, 936–939 (2006)
34. Platzgummer, E., Loeschner, H., Gross, G.: Results obtained with the CHARPAN engineering tool and prospects of the ion mask exposure tool (iMET). Photomask Technology, 06 Oct 2008, vol. 7122
35. Plummer, J.D., Deal, M.D., Griffin, P.B.: Silicon VLSI Technology: fundamentals, practice, and modeling. In: Behrisch, R. (ed.) Topics in Applied Physics. Sputtering by Particle Bombardment I, 1st edn. Prentice Hall, Upper Saddle River, NJ (2000)
36. Prenitzer, B.I.: Investigation of variables affecting focused ion beam milling as applied to specimen preparation for electron microscopy: a correlation between Monte Carlo based simulation and empirical observation [Thesis], University of Central Florida (1999)
37. Robinson, M.T., Torrens, I.M.: Computer simulation of atomic-displacement cascades in solids in the binary-collision approximation. Phys. Rev. B **9**, 5008 (1974)
38. Santamore, D., Edinger, K., Orloff, J., Melngailis, J.: Focused ion beam sputter yield change as a function of scan speed. J. Vac. Sci. Technol. B Microelectron. Nanometer Struct. **15**, 2346 (1997)
39. Sethian, J.A.: Level Set Methods and Fast Marching Methods: Evolving Interfaces in Computational Geometry, Fluid Mechanics, Computer Vision, and Materials Science, 2nd edn. Cambridge University Press, Cambridge (1999)
40. Sigmund, P.: On the number of atoms displaced by implanted ions or energetic recoil atoms. Appl. Phys. Lett. **14**, 114–117 (1969)
41. Sigmund, P.: Theory of sputtering. I. Sputtering yield of amorphous and polycrystalline targets. Phys. Rev. **184**, 383–416 (1969)

42. Springer: In: Sputtering by Particle Bombardment I: Physical Sputtering of Single-Element Solids. Springer (1981)
43. Sun, H.-W., Liu, J.-Q., Chen, D., Gu, P.: Optimization and experimentation of nanoimprint lithography based on FIB fabricated stamp. Microelectron. Eng. **82**, 175–179 (2005)
44. Sun, J., Luo, X., Chang, W., Ritchie, J., Chien, J., Lee, A.: Fabrication of periodic nanostructures by single-point diamond turning with focused ion beam built tool tips. J. Micromech. Microeng. **22**, 115014 (2012)
45. Sun, J., Shu, W., Luo, X., Ritchie, J.: Fabrication of Periodic Metal Nanodots Arrays. In: 25th Annual American Society of Precision Engineering Meeting, Vol. 1, pp. 1–4. Atlanta, USA (2010)
46. Svintsov, A., Zaitsev, S., Lalev, G., Dimov, S., Velkova, V., Hirshy, H.: FIB sputtering optimization using Ion Reverse Software. Microelectron. Eng. **86**, 544–547 (2009)
47. Tian, Z., Yam, S.-H., Barnes, J., Bock, W., Greig, P., Fraser, J.M., et al.: Refractive index sensing with Mach–Zehnder interferometer based on concatenating two single-mode fiber tapers. IEEE Phot. Technol. Lett. **20**, 626–628 (2008)
48. Tseng, A.A.: Recent developments in micromilling using focused ion beam technology. J. Micromech. Microeng. **14**, 15–34 (2004)
49. van Delft, F., van de Laar, R., Verschuuren, M., Platzgummer, E., Loeschner, H.: Charged Particle Nanopatterning (CHARPAN) of 2D and 3D masters for flexible replication in Substrate Conformal Imprint Lithography (SCIL). Microelectron. Eng. **87**, 1062–1065 (2010)
50. Vasile, M.J., Niu, Z., Nassar, R., Zhang, W., Liu, S.: Focused ion beam milling: depth control for three-dimensional microfabrication. J. Vac. Sci. Technol. B Microelectron. Nanometer Struct. **15**, 2350 (1997)
51. Vasile, M.J., Xie, J., Nassar, R.: Depth control of focused ion-beam milling from a numerical model of the sputter process. J. Vac. Sci. Technol. B Microelectron. Nanometer Struct. **17**, 3085 (1999)
52. Ward, J., Kubena, R., Utlaut, M.: Transverse thermal velocity broadening of focused beams from liquid metal ion sources. J. Vac. Sci. Technol. B Microelectron. Nanometer Struct. **6**, 2090–2094 (1988)
53. Yamaguchi, H., Shimase, A., Haraichi, S., Miyauchi, T.: Characteristics of silicon removal by fine focused gallium ion beam. J. Vac. Sci. Technol. B Microelectron. Nanometer Struct. **3**, 71–74 (1985)
54. Yamamura, Y., Itikawa, Y., Itoh, N.: Angular dependence of sputtering yields of monatomic solids. IPPJ-AM reports, Nagoya, Japan Institute of Plasma Physics, Nagoya Univ (1983)

# Chapter 8
# Application of Ion Beam Processes to Scanning Probe Microscopy

**Ashley D. Slattery, Christopher T. Gibson, and Jamie S. Quinton**

**Abstract** Focused ion beam (FIB) techniques have played an important role in nanoscience over the past ~10 years, providing a versatile method of manipulating matter on the nanoscale. During this time, FIB techniques have been used to modify and enhance probes for atomic force microscopy (AFM) techniques, enabling a variety of ingenious and exciting AFM applications with specialized probes. This chapter highlights novel applications of FIB processing, where ion beam milling has been used to quantitatively and accurately determine the mechanical properties of atomic force microscope cantilevers. Several other novel contributions of FIB methods towards the field of AFM are also reviewed.

## 8.1 Introduction

Over the past 10 years, focused ion beam (FIB) microscopy has developed into an extremely versatile technique for manipulating materials in nanoscale science. The ability to remove and deposit material with nanometer resolution provides a direct method of manipulating nanoscale systems and fabricating unique structures. Another relatively new method in nanoscale science is scanning probe microscopy (SPM), a term which encompasses a collection of techniques which

A.D. Slattery • C.T. Gibson • J.S. Quinton (✉)
Flinders Centre for NanoScale Science and Technology, School of Chemical and Physical Sciences, Flinders University of South Australia,
GPO Box 2100, Adelaide 5001, SA, Australia
e-mail: Ashley.Slattery@flinders.edu.au; Christopher.Gibson@flinders.edu.au

Z.M. Wang (ed.), *FIB Nanostructures*, Lecture Notes in Nanoscale Science and Technology 20, DOI 10.1007/978-3-319-02874-3_8,

all utilize a nanoscale probe to sense surface forces and interactions. The design of SPM probes to sense specific interactions has resulted in a host of microscopy techniques capable of mapping a variety of forces in three dimensions [1–4]. This chapter provides a brief background on FIB processing and on SPM methods (primarily the atomic force microscope). The application of FIB processes to the field of atomic force microscopy (AFM) is reviewed by highlighting two recent publications by the authors and several other interesting applications reported in the literature.

## 8.2 Background

### *8.2.1 Focused Ion Beam Processes*

At the fundamental level, FIB techniques rely on the interaction between a beam of ions with kiloelectron volt energies and the surface of interest. There are several processes that can occur in this situation; these are implantation of the ions in the sample, sputtering of the sample atoms, emission of electrons, and possible chemical reactions. These effects provide several ways to interact with and characterize a sample; using a narrowly focused beam of ions, these tasks can be performed on the nanoscale. The generation of secondary electrons means that FIB microscopes operate in a similar manner to scanning electron microscopes (SEMs). The basic design of an FIB column is shown in Fig. 8.1, and the operation and development of these columns are briefly introduced.

There were several important advancements which enabled the field of FIB processing to emerge, one of which is the development of the liquid metal ion source (LMIS) [6–8]. These consist of a tungsten tip (with end diameter as small as 5 nm) and a reservoir of liquid metal (usually Ga) which wets the tip with a thin metallic film. High potential applied to the tip results in the formation of a Taylor cone [9], which forms a tip size of approximately 10 nm diameter. Field emission is used to extract the Ga ions from the tip of the Taylor cone, and the ions are accelerated down the column. The application of ion focusing columns with these LMIS sources enabled the development of the first scanning ion microscope [10].

Initially, FIB instruments were applied extensively in the microelectronics industry, where the ability to remove and deposit material was invaluable, specifically for repairing photomasks, doping, and performing direct lithography. Over the past decade, FIB techniques have been adopted extensively in materials science, as the ion beam can be used to image samples and also to sputter material with high precision. One of the disadvantages of imaging with an ion beam is that the imaging process is destructive, and care must be taken to ensure that sample damage is minimal. To avoid this problem, there are a number of "dual-beam" systems available commercially which incorporate an electron beam for imaging and an ion beam for

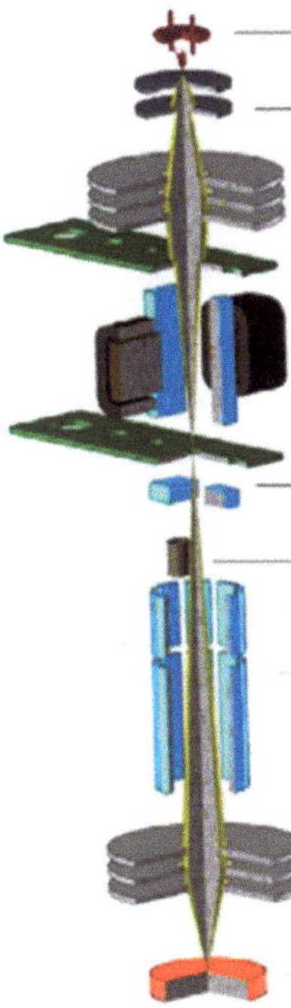


**Fig. 8.1** Schematic of an FIB column, showing the various components and geometry. Reprinted with permission from [5]. Copyright 2013, Elsevier

milling. The ability to nondestructively image a sample with the electron beam and mill features with the ion beam makes dual-beam instruments incredibly useful for applications in materials science.

### *8.2.2 Scanning Probe Microscopy*

SPM has contributed substantially to the advancement of nanoscience. The concept of a scanning probe microscope was developed by Young in 1966 [11]. Young et al. developed a device they called the "topografiner," and its application in profiling surfaces was reported in 1972 [12]. The pioneering work by Young et al. led to the invention of the scanning tunneling microscope (STM) by Binnig and Rohrer in 1983, which enabled microscopists to visualize individual atoms for the first time [13]. The atomic force microscope (AFM) was developed by Binnig, Gerber, and Quate shortly after in 1986 [14], and the family of SPMs has been developing ever since. The versatility and high resolution of the AFM in particular have resulted in widespread adoption and extensive use in materials science, biology, nanotechnology,

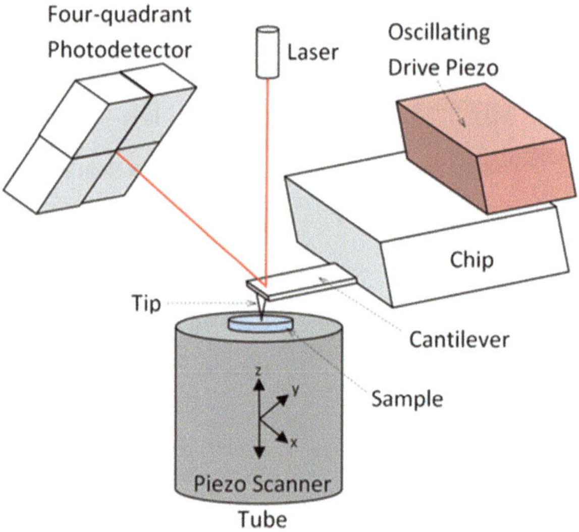


**Fig. 8.2** Schematic of the basic components of a sample-scanning atomic force microscope

and industry. This review focuses mainly on the operation of the AFM; however, there are many excellent, detailed introductions to the field of SPM available to the interested reader [15, 16].

There are a wide variety of commercial AFM systems available; however, the majority operates on the same basic principles which are depicted in Fig. 8.2.

At the heart of the AFM is the force-sensing probe, which interacts with the sample surface. The probe is generally a sharp tip attached to the end of a flexible cantilever, and by measuring the deflection of the cantilever when the tip interacts with the surface, the force applied can be determined.

The probe consists of the sharp imaging tip, the cantilever, and the comparatively large piece of material on which the cantilever is mounted, which is often called the chip and is typically silicon or glass. Cantilevers can be manufactured with many geometries; however, the most common are the so-called beam-shaped and V-shaped types which are shown in Fig. 8.3.

In practice, the deflection of the cantilever is usually measured using a laser beam reflected off of the back of the cantilever into a four-quadrant photodiode. As the cantilever deflects, the position of the reflected laser spot is then detected on the photodiode. This setup is commonly referred to as the "optical lever" or "beam bounce" method and provides a low-cost, simple method to measure the cantilever deflection with sub-nanometer precision. Other methods for measuring the force interaction include laser Doppler vibrometry (LDV) [17] and probes with in-built sensors for direct force readout [18].

Imaging of the sample is achieved by rastering the tip across the surface, using piezoelectric movement such that the probe can be moved in the *x*-, *y*-, and *z-directions* as shown earlier in Fig. 8.2. A feedback loop is employed to maintain constant force interaction as the tip is scanned over the surface, and so the resulting

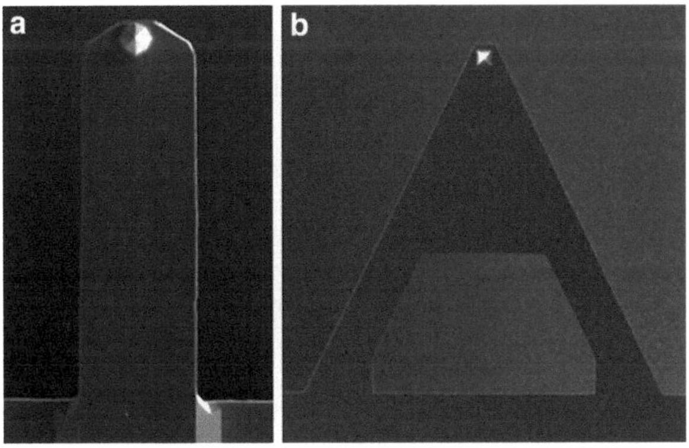


**Fig. 8.3** SEM images of a beam-shaped cantilever (**a**) and a V-shaped cantilever (**b**)

image is a map of constant force over the sample surface which generally reflects the sample topography. This describes the simplest imaging mode available, which is termed "contact mode" where the static deflection of the cantilever is used as the feedback signal. There are a variety of modes developed, but certainly the most widespread is tapping or intermittent contact mode. Tapping mode is a dynamic imaging mode where the cantilever is oscillated sinusoidally at its resonant frequency. As the probe interacts more strongly with the surface, the amplitude of oscillation is damped and so in most applications it is the amplitude of the cantilever that is used as the feedback signal to maintain a constant tapping force as the probe scans.

In addition to imaging, AFM can be used to directly measure forces on the nanoscale. This is commonly achieved through force spectroscopy or force–distance curves, an example of which is given in Fig. 8.4.

Figure 8.4 shows the tip approaching the surface in region A, until B where attractive forces cause the cantilever to bend towards the surface and the tip makes contact. Further extending the tip in region C causes the cantilever to deflect away from the surface as repulsive forces are experienced and the tip pushes into the surface. The tip is then retracted in region D and the cantilever unloads as the pressure on the tip and surface decreases. Adhesive forces result in the tip adhering to the surface during retraction, until these forces are overcome at position E when the tip detaches from the surface and the cantilever returns to the equilibrium position.

In order for such tip–surface interaction measurements to be quantitative, critical properties of the cantilever such as deflection sensitivity and spring constant must be well understood. Here we provide an overview of these properties as well as describe several methods that are used to determine these quantities.

#### 8.2.2.1 Deflection Sensitivity

As discussed previously, the AFM is an ideal instrument for measuring forces on the nanoscale and has found many applications in this area. Despite its name, AFM does not measure force directly. The deflection of the cantilever (a photodiode signal often

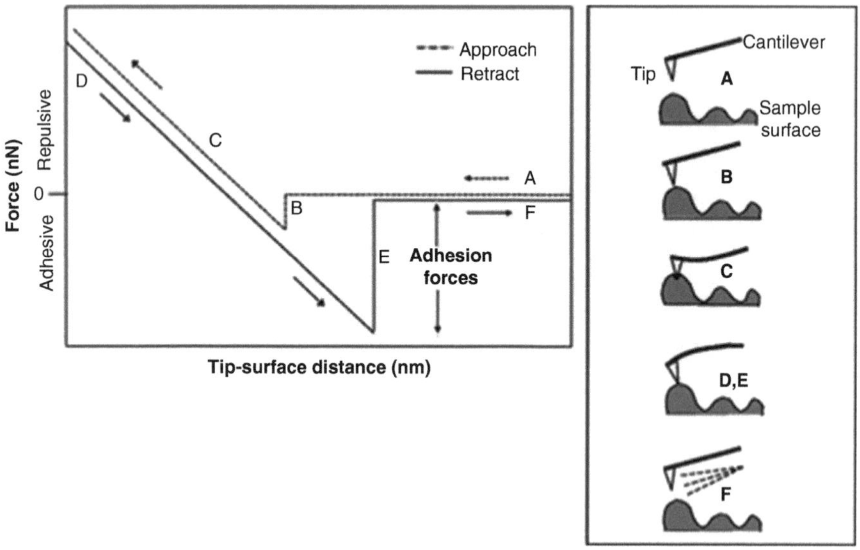


**Fig. 8.4** Depiction of a basic force–distance curve, showing the different regimes and corresponding position of the cantilever relative to the surface. Reproduced/adapted with permission from [19]

in volts or milliamps) is the raw data that is obtained when the cantilever experiences a force. This signal must be converted to a deflection (often in nanometers) and then to a force (often in nanonewtons). The conversion factor that is used to obtain deflection (distance) is known as the tip's deflection sensitivity and typically has units of nm $V^{-1}$. The deflection sensitivity is unique to each individual cantilever and can even change if a probe is repositioned or re-loaded into the instrument, as the geometry of the optical lever only needs to be altered slightly to effect a change in the deflection sensitivity.

The sensitivity is most commonly measured by performing a force–distance curve on a hard surface such as silicon or sapphire, so that there is no indentation or flexing of the surface. In this case, any extension of the scanner results in an equal amount of cantilever deflection. Dividing the known cantilever deflection by the change in photodiode signal gives the deflection sensitivity in nm $V^{-1}$. Practically, the sensitivity is obtained by measuring the slope of the force curve in the contact regime (i.e., during phases denoted C and D shown in Fig. 8.4). The main disadvantage to this approach is that the tip must contact a hard surface, which often results in tip damage to some extent. There have been several methods reported in literature to eliminate the problem of tip–surface contact during deflection sensitivity measurements, although often at the cost of additional uncertainty [17, 20–22].

#### 8.2.2.2 Spring Constant

Treating the cantilever as a first-order spring (which is a good approximation for small deflections), Hooke's law can be used to determine the force acting on the cantilever for a given deflection. Shown in (8.1), $F$ is the force in Newtons acting on the cantilever, resulting in a normal displacement of $x$ meters and $k$ is the cantilevers' spring constant:

$$F = -kx \tag{8.1}$$

In order to quantitatively measure force using the AFM, the cantilevers' spring constant and (8.1) are used. Often the spring constant must be determined as a function of position along the cantilever, one example is when the tip is not located exactly at the free end of the cantilever (where the spring constant is often specified). The spring constant of the cantilever increases with position from the free end to the fixed end, and Sader et al. derived an expression for this which is valid for beam-shaped cantilevers [23]. This expression shown in (8.2) describes the spring constant ($k\Delta_L$) at a position $\Delta L$ from the cantilevers' free end as a function of the spring constant at the cantilevers' free end ($k_{\text{end}}$) where $L$ is the length of the cantilever:

$$k_{\Delta L} = k_{end}\left(\frac{L}{L-\Delta L}\right)^3 \tag{8.2}$$

Manufacturers of AFM probes commonly supply nominal values for the spring constant (amongst other cantilever properties), which are specified for each type of probe. Unfortunately there is significant variation in cantilever parameters, even amongst a single batch. This is due largely to poor control over cantilever thickness which leads to a wide variation in spring constant [24]. For this reason, quantitative force measurements necessitate the accurate determination of each individual cantilevers' spring constant before use.

Determining the spring constant of AFM cantilevers is usually referred to as "cantilever calibration" and can be a very difficult procedure to perform accurately; this is demonstrated by the wide variety of methods developed specifically for this purpose. The following sections provide a brief overview of some of the more commonly used cantilever calibration methods.

### Theoretical Methods

Basic beam theory is one of the most fundamental methods of determining the spring constant of the AFM cantilever. Applying the Euler–Bernoulli beam theory to an ideal-shaped rectangular beam with uniform material properties, the spring constant can be determined using the width ($w$), length ($L$), thickness ($t$), and Young's modulus ($E$) of the cantilever, using the following formula [25]:

$$k = \frac{Et^3w}{4L^3} \tag{8.3}$$

This approach is often referred to as the "simple beam" method, as it is most accurate when applied to rectangular cantilevers that have perfect beam-shaped geometry and well-known Young's modulus such as single-crystal silicon.

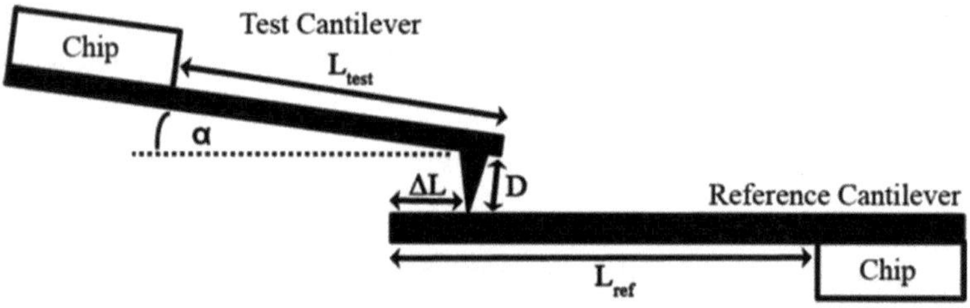


**Fig. 8.5** A schematic of the reference cantilever method showing important measured quantities. Reprinted with permission from [29]. Copyright 2013, IOP Publishing

Significant uncertainty arises for cantilevers with metallic coatings, cantilevers with poorly defined material properties such as silicon nitride [26], non-rectangular cross sections, and V-shaped cantilevers. The majority of cantilevers used in practice are often affected by one or several of these factors, and as a result there have been many corrections developed for the simple beam method which reduce the uncertainty somewhat [27, 28].

Static Methods

Perhaps the most common static cantilever calibration method is the reference cantilever technique. This method is one of the simplest and direct techniques for calibrating an AFM cantilever. This method generally involves pressing a test cantilever of unknown spring constant against a pre-calibrated reference cantilever or standard with known spring constant. The main advantage of this method is that the spring constant is measured directly and at the tip of the test cantilever, and the procedure is also quite simple to implement.

This process is shown in Fig. 8.5, along with the important measured quantities. After performing the required measurements, the spring constant of the test cantilever can be determined by the following equation, where $k_{\text{test}}$ and $k_{\text{ref}}$ are the test and reference cantilever spring constants, respectively; $S_{\text{C}}$ and $S_{\text{H}}$ are the deflection sensitivity measured on the reference cantilever and a hard surface, respectively; and the remaining quantities are as defined in Fig. 8.5:

$$k_{test} = k_{ref}\left(\frac{S_{\text{C}}}{S_{\text{H}}} - 1\right)\cos^2\alpha \times \left(1 - \tan\alpha\frac{3D}{2L_{test}}\right)\left(\frac{L_{ref}}{L_{ref} - \Delta L}\right)^3 \tag{8.4}$$

Unfortunately there are also several disadvantages which limit the accuracy of the technique. The uncertainty inherent in the technique is largely due to the uncertainty in the spring constant of the reference cantilever, which is a function of how accurately the reference cantilever is calibrated and precise knowledge of the location that the test cantilever is pressing ($\Delta L$).

It is important when applying the reference cantilever method that the spring constant of the test and reference cantilevers be as close as possible. A Large mismatch between the two cantilevers can result in a significant increase in the uncertainty of the method, which limits the range of cantilevers that a particular reference cantilever can calibrate.

### Dynamic Methods

There are many calibration methods which determine the spring constant as a function of the cantilevers' dynamic properties; certainly, the most widely adopted of these is the thermal noise method [30]. Thermal energy results in the cantilever oscillating with very small amplitude at its resonant frequency. Hutter and Bechhoefer used the equipartition theorem to relate the cantilevers' spring constant to the mean square of its displacement $\langle d_c^2 \rangle$ due to thermal motion, Boltzmann's constant ($k_B$), and ambient temperature ($T$) to obtain the following equation:

$$k = \frac{k_B T}{\chi^2 d_c^2} \tag{8.5}$$

This method has been implemented widely and is available on most commercial AFMs due to its ease of use. Once the deflection sensitivity of the cantilever is measured, the deflection of the cantilever is monitored for a period of time and the power-density spectrum obtained. Using the relation shown above, the spring constant can be determined easily. The thermal method is applicable to a wide range of cantilevers as long as the appropriate corrections are applied. These corrections are grouped into a single term called the Chi factor ($\chi$) and are determined using finite element analysis with a value of 1.106 for beam-shaped cantilevers [31] and 1.144 for V-shaped cantilevers [32]. There are also corrections which must be made for laser spot position [33], tip height, and approach angle of the AFM to the surface [34].

The thermal noise method has been shown to provide very accurate results under certain applications; however, there are several factors (including the Chi factor corrections) which can introduce uncertainty. The thermal fluctuations measured are generally very small for cantilevers with high spring constant or very low mass; as a result, the signal measured approaches the noise level of the optical lever detection system. It is possible to use LDV to increase the accuracy of the measurement [17]; however, these detection systems are uncommon on commercial AFMs.

The Sader hydrodynamic method is another technique that has been widely implemented and determines the spring constant by analyzing the hydrodynamic drag on the cantilever while oscillating. The cantilevers' spring constant is related to its plan view dimensions ($L$ and $w$), resonant frequency ($\nu_f$), $Q$ factor ($Q_f$), Reynolds number ($\Gamma_i$), and density of the medium ($\rho_f$) by the following equation:

$$k = 0.1906 L w^2 \rho_f Q_f \Gamma_i \left(2\pi\nu_f\right)^2 \tag{8.6}$$

One of the reasons this technique has been so widely applied is due to its nondestructive nature and ease of use. Simply by knowing the length, width, and resonant properties of the cantilever, the spring constant can be determined to good accuracy. Originally this method was only applicable to beam-shaped cantilevers; however, recent work by Sader et al. has extended the technique to cantilevers with arbitrary geometries [35].

Sader et al. also developed a method which determines the spring constant based on the resonant frequency and effective mass of the cantilever [23]. Here, $\rho_{\text{lever}}$ is the density of the cantilever material and $\nu_0$ is the resonant frequency, while $L$, $w$, and $t$ are the cantilever dimensions defined previously:

$$k = \left(nLwt\rho_{lever}\right)\left(2\pi v_0\right)^2 \tag{8.7}$$

The effective mass is related to the actual mass of the cantilever by the factor $n$, which can be determined using tables provided in the original publication. This method is also affected by the non-ideal properties described for the simple beam method, although the effect of uncertainty in thickness is significantly reduced in comparison to this method as the thickness term is only to the first power.

Cleveland et al. also developed a technique to determine the spring constant based on the cantilevers' resonant frequency but without the requirement to measure the cantilevers' thickness [25]. This technique is a result of combining (8.3) and (8.7) and is usually termed the Cleveland formula, which is given in (8.8) with the quantities as defined previously:

$$k = 2\pi^3 wL^3 \left(\frac{\rho_{lever}^3}{E}\right)^{1/2} v^3 \tag{8.8}$$

Deviation from non-ideal cantilever geometry can introduce significant uncertainty in this method. The application of the technique however is very simple and non-destructive and offers a reasonable estimate of the spring constant which is a significant advantage over other more complicated approaches. It should be noted again that for the methods described previously which incorporate a term for cantilever density, the application to silicon nitride cantilevers introduces significant uncertainty due to a wide variation in the material properties of the cantilever.

The Cleveland added mass method is the last method that is introduced and is most relevant to the work presented in the following section. This method involves damping of the cantilevers' oscillation by loading it with a known mass and measuring the reduction in resonant frequency as a result [25]. Treating the cantilever as a simple harmonic oscillator, Cleveland et al. determined that the relationship between the mass added, shift in resonant frequency, and spring constant would be given by the following equation. Here, $\Delta M$ represents the addition of mass to the cantilever and $\nu_0$ and $\nu_1$ are the cantilevers' resonant frequency before and after the mass change, respectively:

$$k = \frac{4\pi^2 \Delta M}{\left(1/v_0^2\right) - \left(1/v_1^2\right)} \tag{8.9}$$

The method has been most commonly applied by loading the cantilever with a microsphere. The Cleveland method is one of the first spring constant calibration techniques developed; however, it suffers from relatively low precision due to two main factors. The first is the uncertainty in the mass with which the cantilever is loaded, both due to the difficulty in measuring the microsphere volume (Au, W, Si, etc.) and also due to the glue required to attach the sphere. The second source of uncertainty is in knowing the loading position of the mass on the sphere, which is critical for determining the spring constant. These sources of uncertainty provide the motivation for the work presented in the following section.

Calibration methods can also be categorized based on whether they are potentially destructive to the cantilever or not. Many techniques require the probe to be engaged on a hard surface in order to measure the deflection sensitivity (such as the thermal and reference cantilever methods) which can cause damage to the tip. Methods such as the Cleveland added mass technique require the cantilever to be modified (by attaching microspheres) which often requires the spheres to be fixed with adhesive and also has the potential to significantly damage the cantilever. There are significant benefits for techniques which are able to calibrate the cantilever not only with reduced risk of damage but also without tip–surface contact. The following techniques presented incorporate both of these advantages.

## 8.3 Cantilever Calibration Using FIB Methods

### *8.3.1 FIB Mass Removal*

The use of FIB processing in the calibration of AFM cantilevers offers the unique ability to remove material on a very small scale with extremely high precision. The precision of FIB processing was utilized as a powerful tool to reduce the main sources of uncertainty identified previously in the "Cleveland added mass method" [36]. The aim of this application was to significantly improve the accuracy of the technique, providing the AFM community with a method of calibrating a wide range of cantilevers with extremely high accuracy and without tip damage.

The ion beam was used to remove a complete section from the cantilever; a circle was generally chosen and was milled through the cantilever just behind the tip. The milled regions' shape and position on the cantilever were precisely defined, and several examples are given in Fig. 8.6. Knowing the thickness of the cantilever and the density of the material removed, the precise quantity of mass removed and its location on the cantilever can be determined. Instead of measuring a decrease in resonant frequency as with the Cleveland added mass method, an increase in resonant frequency was observed as expected.

The spring constant is determined at the position of the mass removed and is given by (8.9). This expression determines the spring constant at the position of the mass removed, and so in order to correctly determine the force sensed by the tip,

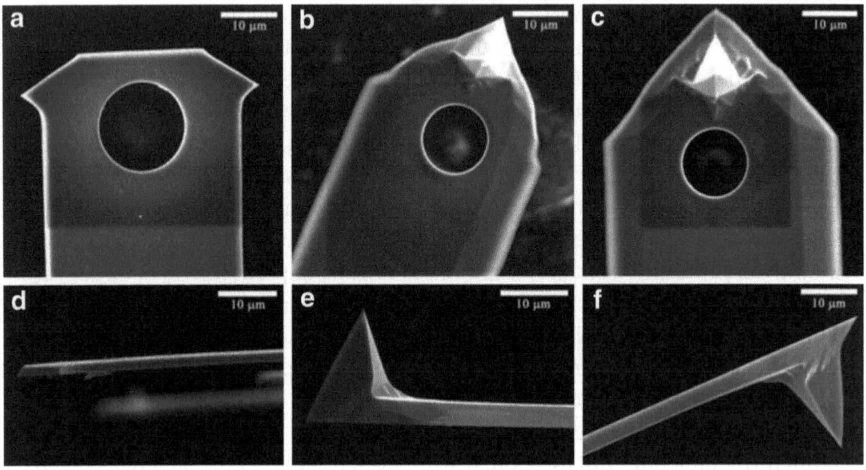


**Fig. 8.6** Plan-view and corresponding side-view SEM images (*directly below*) of cantilevers C1 (**a**), F1 (**b**), and F2 (**c**). Reprinted with permission from [36]. Copyright 2013, IOP Publishing

the spring constant must be corrected using (8.2). Incorporating (8.2) into (8.9), the following equation is obtained where $L$ represents the length from the cantilevers' fixed end to the tip in this case, and $\Delta L$ is the distance from the removed mass to the tip:

$$k_{tip} = \frac{4\pi^2 \Delta M}{\left(1/v_0^2\right)-\left(1/v_1^2\right)}\left(\frac{L}{L-\Delta L}\right)^{-3} \tag{8.10}$$

Equation (8.10) allows the spring constant to be determined at the tip for a single mass removal step. It is also possible to remove mass in several steps and measure the resonant frequency shift after each milling process, which is similar to the approach that Cleveland et al. used in their original paper [25]. This is effectively a multiple point measurement and can be described by the following equation where $M_{\text{effective}}$ is the effective mass added to the cantilever in each step and $\nu$ is the corresponding resonant frequency, $k_{\text{effective}}$ is the spring constant at the position of the added mass, and $M^*$ is the effective mass of the entire cantilever prior to milling:

$$M_{effective} = k_{effective}\left(2\pi v\right)^{-2} - M^* \tag{8.11}$$

In this expression, $M_{\text{effective}}$ is to correct for a mass applied (or removed) that is not located at the free end of the cantilever. If the mass is set back by a distance $\Delta L$, then it can be corrected using (8.2) in the same manner as the spring constant.

Performing multiple point measurements and testing them against this relation can be used to test the validity of the assumptions made in (8.2) and (8.9). A plot of the effective mass removed versus the resonant frequency should produce a linear relationship where the slope is equal to the spring constant and the intercept is the effective mass of the cantilever before milling.

**Table 8.1** Spring constant values determined using the single-point FIB method, compared to results from established methods

| | Spring constant (N m$^{-1}$) | | | | | |
|---|---|---|---|---|---|---|
| Cantilever | FIB (7–10 %) | Thermal (10–20 %) | Sader hydrodynamic (5–10 %) | Sader dimensional (5–10 %) | Cleveland dimensional (5–10 %) | Euler beam formula (10 %) |
| [a]R1 | 0.0950 ± 0.021 | – | 0.0842 | 0.0811 | 0.0820 | 0.0860 |
| [a]R2 | 0.746 ± 0.060 | – | 0.727 | 0.690 | 0.701 | 0.726 |
| [a]R3 | 6.40 ± 0.45 | – | 6.51 | 6.22 | 6.22 | 6.77 |
| M1 | 44.3 ± 2.8 | 42.1 | 46.8 | 44.1 | 36.70 | 42.3 |

[a]These cantilevers possess ideal beam-shaped geometry, and as such the estimated uncertainty range of the established methods is low

#### 8.3.1.1 Single-Point Calibration

FIB milling was used to remove mass from a variety of different cantilevers, initially in a single-point measurement approach. Three tipless silicon CLFC reference cantilevers (R1–3) from Bruker and a silicon tapping mode probe (M1) from Mikromasch were calibrated in this manner. Careful measurement of the dimensions using SEM allows the mass to be removed and its location to be accurately determined. After measuring the shift in the resonant frequency, (8.10) was applied to determine the spring constant at the cantilevers' tip or at the free end for tipless types (Table 8.1).

The agreement between the FIB method and the others is quite good for cantilevers R2 and R3, with the results agreeing to within 2–8 %. We expect that most of the other methods will also be accurate, as the geometry of these cantilevers is close to ideal beam shaped and the cantilevers' properties have been measured very carefully. Conversely, the spring constant is overestimated by 10–15 % by the FIB method for cantilever R1, and we believe that this is a result of the error of small shift in resonant frequency observed.

This result identifies a limitation in the FIB method where a small shift in resonant frequency increases the uncertainty in this measured quantity. In order to keep the uncertainty in the resonant frequency low, a shift of approximately 2 kHz is required for the measurement method used. The minimum quantity of material that must be removed to generate such a resonance shift with a cantilever of spring constant 0.7 N m$^{-1}$ and thickness 1 μm is approximately 1–5 % of the cantilevers' mass.

For cantilevers where the resonance shift is large enough, the uncertainty observed is very low. Cantilever R3 possesses an uncertainty of only 7 %, which is primarily due to the uncertainty in their thickness. This level of uncertainty is comparable with the most accurate calibration methods currently available and is expected to be reduced to less than 5 % for certain types of cantilevers and with the ability to measure cantilever thickness with lower uncertainty.

#### 8.3.1.2 Multiple Point Calibration

Three cantilevers (C1, Mikromasch CSC12) and (F1-2, Bruker FMV) were calibrated using the multiple-point measurement approach introduced previously. This was performed to validate the method itself and would also be expected to reduce the uncertainty of the method at the cost of increased calibration time. Mass was removed incrementally by enlarging the hole size in steps, and the resonance shift after each mass removal was measured; SEM images of these cantilevers are shown in Fig. 8.6.

By plotting the data to satisfy (8.11), a linear relation is observed which is shown in Fig. 8.7 for each cantilever. The spring constant can be read directly from these plots as the slope, and these results are shown in Table 8.2. The excellent fit between the data and the linear regression suggests that the use of a single-point measurement is accurate and that the inherent assumptions in the technique are valid. The validation of the technique and the assumptions made is discussed in more detail in the following section.

Comparing these results with other established methods in Table 8.2, it can be seen that the agreement between results is generally very good. The uncertainty quoted on the FIB results is determined from the propagation of error in measurement, while the uncertainty given for the other techniques is from their reported uncertainty in literature. In accordance with the sources of uncertainty identified in the introduction of the various established techniques, the highest uncertainty is often observed for cantilevers with complex geometry and other non-ideal factors. However, we expect that the uncertainty using the multiple-point measurements will be lower than that associated with the single-point results shown previously, making this procedure useful for highly accurate spring constant calibration.

#### 8.3.1.3 Validation of the Method

A potential disadvantage of the FIB method is that the cantilever is altered quite significantly. It is reasonable to propose that the milling process may damage or alter the cantilevers' properties in some way, thus affecting the operation of the cantilever or introducing uncertainty into the method. A number of tests were performed to determine whether this was the case or not.

In order to determine the effect of FIB milling on the spring constant of the cantilever, the reference cantilever method was used to directly measure any changes. The spring constant of cantilever C1 was measured at the end of the cantilever during the multiple-point measurement process. After the hole was made wider in each step, the spring constant (relative to the initial value) was measured and plotted against the ratio of the hole diameter to the width of the cantilever. The result of this experiment is shown in Fig. 8.8, and it is evident that the change in spring constant as a result of mass removal at the end of the cantilever is very minor. Almost no change is observed for hole diameters up to 40 % of the cantilevers' width, and even when the hole reaches 85 % of the width of the cantilever the reduction in spring constant is approximately 10 %.

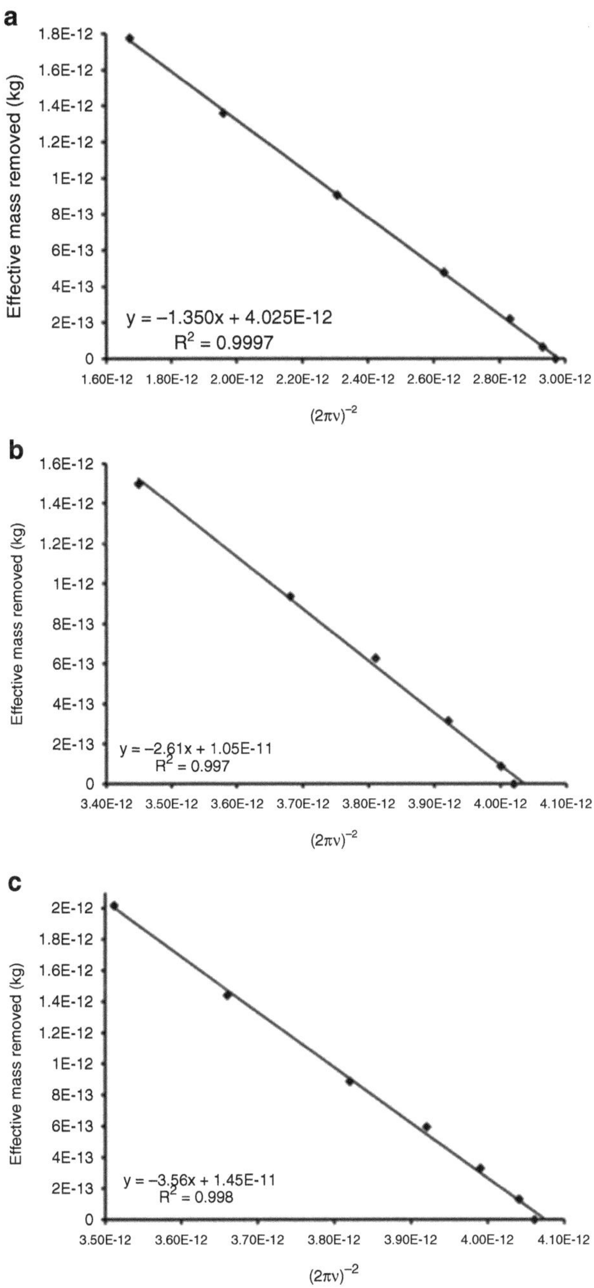


**Fig. 8.7** The multiple-point measurement method applied to cantilevers C1 (**a**), F1 (**b**), and F2 (**c**). Reprinted with permission from [36]. Copyright 2013, IOP Publishing

**Table 8.2** Results for the multiple-point calibration method compared to several other established calibration methods

| | Spring constant ($N\ m^{-1}$) | | | | | | | |
|---|---|---|---|---|---|---|---|---|
| Cantilever | FIB (±7–10 %) | Thermal[a] [45] (±10–20 %) | Sader hydrodynamic [23] (±5–15 %) | Sader dimensional [22] (±10–20 %) | Cleveland dimensional [24] (±10–15 %) | Euler beam theory (±10–20 %) [24] | Reference lever before milling [6] (±10–20 %) | Reference lever after milling (±10–20 %) |
| C1 | 1.36±0.15 | – | 1.41 | 1.29 | 1.29 | 1.52 | 1.45 | 1.45 |
| F1 | 2.53±0.18 | 2.50 | 2.50 | 2.55 | 2.50 | 2.67 | – | 2.40 |
| F2 | 3.40±0.25 | 3.11 | 3.27 | 3.07 | 2.91 | 3.40 | – | 3.37 |

Reprinted with permission from [36]. Copyright 2013, IOP Publishing

[a]The thermal technique was not applied to the tipless cantilevers (type C), as high adhesion between the cantilever and surface prevented calibration of the deflection sensitivity

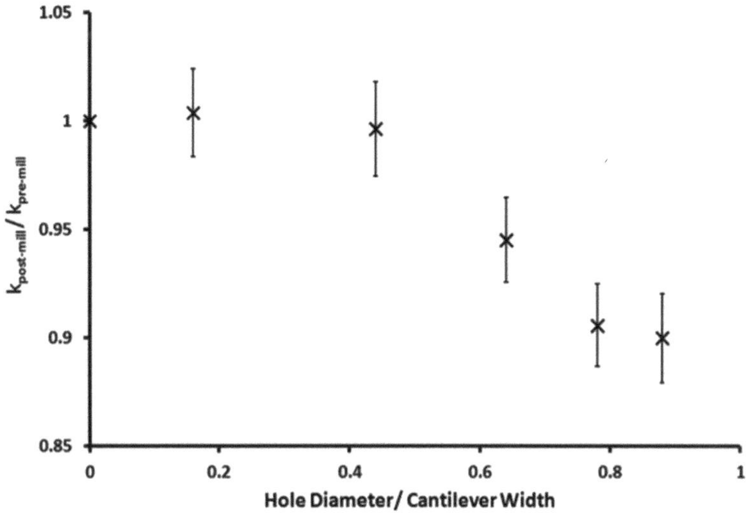


**Fig. 8.8** Variation of the spring constant at the end of cantilever C1 after each sequential milling step, measured by the reference lever method. Reprinted with permission from [36]. Copyright 2013, IOP Publishing

It is proposed that this result is due to the location of the removed mass. The hole is located as close as possible towards the free end of the cantilever, while the mechanical properties of the cantilever are more strongly defined by the material closer to the fixed end of the beam. This is where the majority of the strain occurs under flexing, and this may explain why the milling process has almost no effect on the cantilevers' spring constant. Further investigation showed that the milling location is an important factor in whether the mechanical properties of the cantilever are affected; these results are reported in Table 2 of [36].

It is also possible to theoretically determine the resonant frequency of the cantilever after mass removal. The derivation for this equation can be obtained in the supplementary material of the original publication [36] and assumes that the dimensions of the cantilever are uniform, that the effective mass removed has no effect on the spring constant, and that the effective mass removed is accurately described by (8.9). By comparing the theoretical values to the ones observed experimentally, all of these assumptions are tested. The terms in the following equation are as defined previously:

$$\nu_{FIB} = \left[ \frac{0.24wL}{0.24wL - 0.25\pi D^2 \left( (L - \Delta L)/L \right)^3} \right]^{1/2} \nu_0 \tag{8.12}$$

These results are shown in Fig. 8.9, where the resonant frequency is plotted against the ratio of the hole diameter to the width of the cantilever. Very good agreement is observed over the entire range of measurements, which further supports the

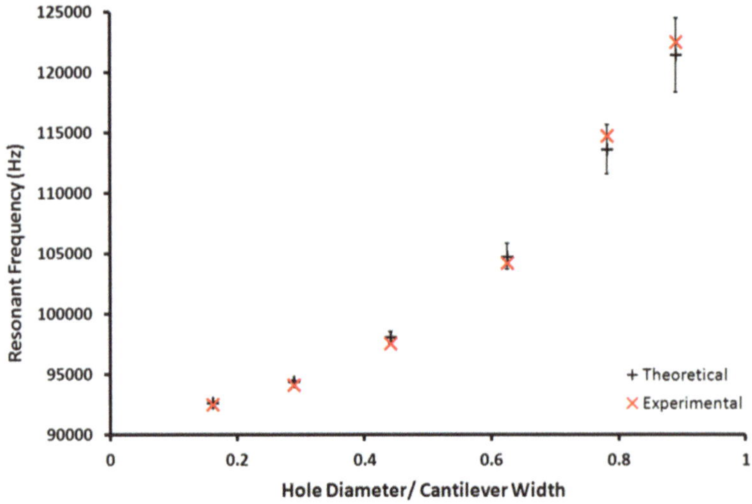


**Fig. 8.9** Comparison between theoretically derived and experimentally determined resonant frequencies plotted as *D*/*W* increases for cantilever C1. Reprinted with permission from [36]. Copyright 2013, IOP Publishing

assumption that the milling process has little effect on the cantilevers' mechanical properties and that the assumptions made in the technique are valid.

The data presented in Figs. 8.8 and 8.9 appear to be inconsistent, as they show the same cantilever observed using two different approaches to measure the mechanical properties as the hole size increases. Monitoring the spring constant directly with the reference cantilever method, a 10 % decrease in spring constant at the largest hole is observed. Conversely, the data presented in Fig. 8.9 shows insignificant difference between the resonant frequency expected theoretically and the one observed experimentally. This apparent inconsistency is attributed to a change in the relationship between the static and the dynamic spring constant. The static and dynamic spring constants of a rectangular cantilever are not equal for the first resonance mode; these quantities differ by a few percent and are related by $k_d/k_s = 1.03$. Hahner and Melcher et al. recently demonstrated that cantilevers with non-rectangular geometries can have dynamic spring constants increased relative to their static spring constants [37, 38]. Even small changes in the geometry at the end of the cantilever were shown to increase this ratio by approximately 5 %. The removal of a circular section behind the tip is also expected to increase the ratio between the static and dynamic spring constant. The data in Fig. 8.8 shows the static spring constant of the cantilever, while the data in Fig. 8.9 measures the dynamic property of resonant frequency; hence, it is proposed that the milling process does not change the dynamic spring constant. However when the diameter of the hole is approximately 50 % of the cantilever width the ratio of the dynamic to static spring constant increases due to the changing shape of the cantilever. This must result in a decrease in the static spring constant. Essentially a hinge is formed when *D*/*W* is 0.50 or greater.

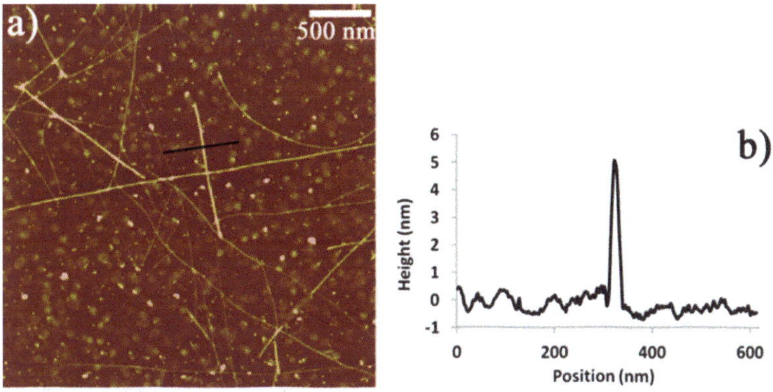


**Fig. 8.10** AFM image of a silicon surface covered with CVD-grown carbon nanotubes, acquired using cantilever F1 after the milling procedure. Reprinted with permission from [36]. Copyright 2013, IOP Publishing

Therefore to avoid this issue it is recommended to keep the diameter of the milled hole less than 50 % of the cantilever width.

A significant advantage of the FIB calibration method is that the cantilever can be calibrated without the tip ever touching a surface. This preserves the sharpness of the tip for accurate measurements and is especially important for tips that have been modified/functionalized. An AFM image of a silicon surface with CVD-grown carbon nanotubes was acquired using a FIB-milled cantilever (Fig. 8.10); the image demonstrates that the FIB-milling process preserves the sharpness of the imaging tip and that the cantilever is capable of imaging. This result was expected, as the results presented previously show negligible change in the cantilevers' mechanical properties, and the tip is never contacted to a surface during the calibration procedure.

### *8.3.2 FIB-Milled Positional Markers*

In addition to the mass removal method, FIB milling was also applied to improve the reference cantilever technique [29]. Although the mass removal method is able to calibrate a wide range of cantilevers with very high accuracy, there are certain types of probes which it is unsuitable for. These are primarily cantilevers fabricated from silicon nitride, as the density of this material can vary substantially which introduces significant uncertainty into the technique. The presence of metallic films (such as Au, Ti, and Cr) also introduces uncertainty in the density of the removed mass, as the thickness of the films is not well known with only nominal values available from the manufacturer. The reference cantilever technique is very powerful in that it measures the spring constant of an unknown cantilever directly and is thus unaffected by these factors.

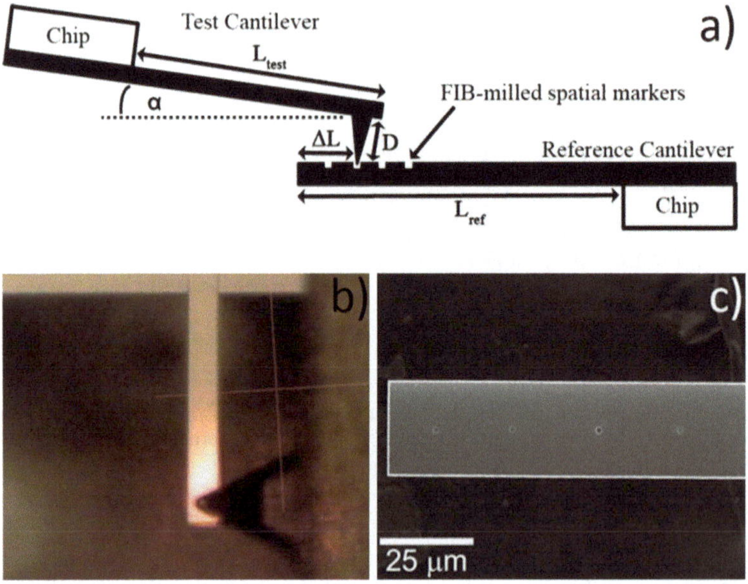


**Fig. 8.11** Schematic of the reference lever method applied with FIB-milled spatial markers (**a**), an optical image of the reference lever measurement (**b**), and an SEM image of the FIB-milled reference cantilever (**c**). Reprinted with permission from [29]. Copyright 2013, IOP Publishing

#### 8.3.2.1 Standard Geometry

In the introduction of the reference cantilever method, the main source of uncertainty was identified as knowing the value of $\Delta L$ accurately. To address this, FIB milling was used to produce small spatial markers along the length of several reference cantilevers as shown in Fig. 8.11b, c.

The test cantilever can be located on a marker using the AFM optics, the probe is then engaged on the reference cantilever, and the marker is located by AFM imaging as shown in Fig. 8.11a. The exact location on the cantilever is known by comparing with SEM images of the markers and their location on the reference cantilever, and calculation of the spring constant is identical to the procedure used in the standard reference lever technique with (8.4). An AFM image of the spatial markers is shown in Fig. 8.12, demonstrating the ease with which these features can be located.

This approach reduces the uncertainty in the loading position ($\Delta L$) to the order of 100 nm or less, reducing the uncertainty in $\Delta L$ substantially. The presence of spatial markers along the reference cantilever has the added benefit of providing a wide range of available spring constants along a single cantilever, as the spring constant scales to the cubed power with position ($\Delta L$).

The reference cantilevers used were Bruker CLFC type, which consist of three ideal beam-shaped cantilevers of different spring constants on a single chip. The ideal beam shape and uncoated nature of these cantilevers means that they can be calibrated very accurately using established methods such as those introduced in

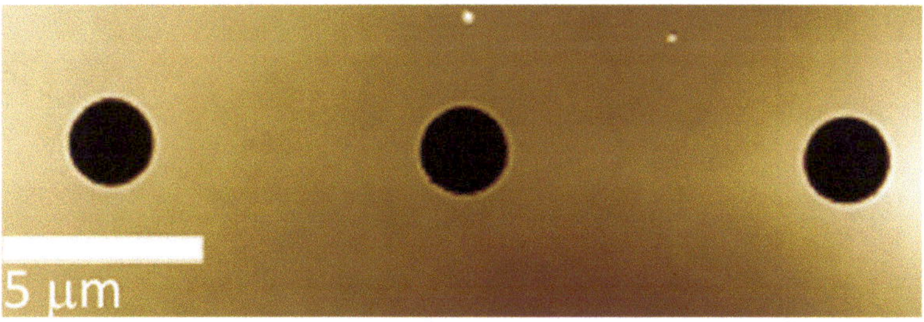


**Fig. 8.12** AFM image of FIB-milled spatial markers in the reference cantilever, acquired during the reference cantilever calibration procedure. Reprinted with permission from [29]. Copyright 2013, IOP Publishing

**Table 8.3** Spring constants determined using the single-point calibration technique for several cantilevers and compared to established calibration methods

| | Spring constant (N m$^{-1}$) | | | |
|---|---|---|---|---|
| Cantilever | Spatial marker standard reference lever method | Thermal noise (±10–20 %) | Sader hydrodynamic (±10–15 %) | Cleveland formula (±10–15 %) |
| V3 | 0.167±0.012 | 0.150±0.023 | – | – |
| V4 | 0.176±0.011 | 0.155±0.023 | – | – |
| V5 | 0.298±0.019 | 0.292±0.044 | – | – |
| | 0.312±0.024 | | | |
| V6 | 0.625±0.039 | 0.600±0.090 | – | – |
| | 0.633±0.043 | | | |
| F2 | 2.42±0.14 | 2.38±0.24 | 2.53±0.32 | 2.33±0.29 |
| M1 | 37.4±2.3 | 37.5±4.7 | 37.5±4.7 | – |
| M2 | 40.7±2.5 | 44.6±6.7 | 42±6.3 | – |

Reference lever methods were performed such that the spring constants of the two cantilevers matched as closely as possible. The cantilever types are the V-shaped (V, Bruker SNL) and the beam-shaped (F, Bruker FMV, and M, Mikromasch NSC15). Reprinted with permission from [29]. Copyright 2013, IOP Publishing

Sect. 8.2.2.1. These cantilevers were calibrated, and then FIB milling was used to produce spatial markers in 20 μm intervals along their length, which is shown in the figure. Using these three cantilevers and the range of markers milled into them, a spring constant range from approximately 0.1 to 30 N m$^{-1}$ was available with the upper limit being defined by how close the markers were positioned to the base of the cantilever.

### Single-Point Measurements

With the wide range of reference lever spring constants available, it was possible to calibrate a number of test cantilevers with equally diverse spring constants. These results are presented in Table 8.3, for beam-shaped and V-shaped cantilevers, and compared to several well-established calibration methods.

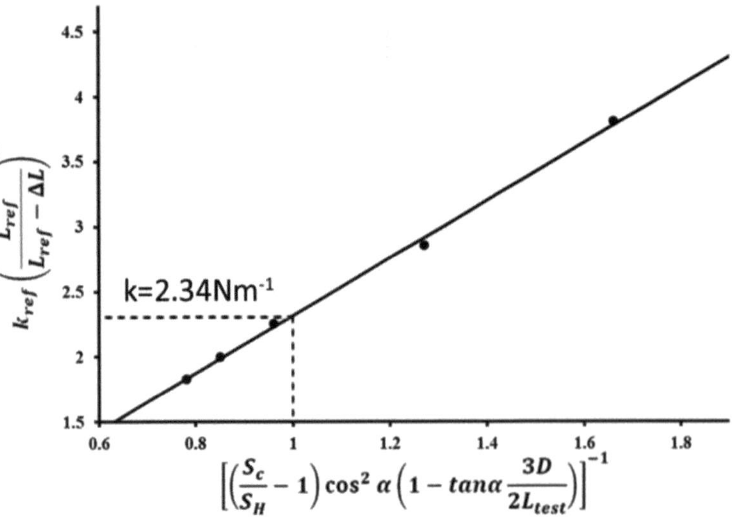


**Fig. 8.13** The multiple-point measurement method applied to cantilever F1 in the standard configuration. Reprinted with permission from [29]. Copyright 2013, IOP Publishing

The results presented above show good agreement between all methods, which demonstrates that all of the methods used can be quite accurate when applied carefully and only where they are applicable. An absence of results for a particular method indicates that the technique was not suitable for calibrating the cantilever. It is evident that the V-shaped cantilevers in particular are quite limited in the techniques available to calibrate them. An immediate benefit of the reference cantilever method is that it is applicable to any cantilever and is a direct measurement technique, requiring no correction for different cantilever geometry.

The results for the reference lever method using spatial markers in particular agree with the other methods within experimental uncertainty, which is between 6 and 10 %. A detailed uncertainty analysis is provided in the supplementary material of the original publication [29].

### Multiple-Point Measurements

The reference cantilever method can be applied at multiple points along the length of the cantilever; multiple-point measurements were used by Gates et al. and Clifford et al. to improve the accuracy of the reference lever technique [39, 40]. In an approach similar to that of Gates et al., single-point measurements were performed at several locations along the length of the cantilever, and the terms in (8.4) were plotted as shown in Fig. 8.13 to yield a linear relationship. The spring constant of the test cantilever ($k_{test}$) is obtained at the point where the $x$-axis is equal to 1, as shown in Fig. 8.13.

**Table 8.4** Spring constant measurements determined using the multiple-point reference lever method, compared to other calibration methods

| | Spring constant (N $m^{-1}$) | | | |
|---|---|---|---|---|
| Cantilever | Spatial marker standard reference lever method | Thermal noise (±10–20 %) | Sader hydrodynamic (±10–15 %) | Cleveland formula (±10–15 %) |
| V1 | 0.088±0.005 | 0.086±0.013 | – | – |
| V2 | 0.114±0.006 | 0.109±0.016 | – | – |
| V7 | 0.680±0.052 | 0.635±0.010 | – | – |
| F1 | 2.34±0.13 | 2.33±0.23 | 2.48±0.31 | 2.30±0.29 |

Cantilever types are as described in Table 8.3. Reprinted with permission from [29]. 

This analysis was performed on four cantilevers, and the results are shown in Table 8.4. The uncertainty introduced by the reference lever method in this case is expected to be very low; Clifford et al. determined that the potential uncertainty in a multiple-point calibration method could be as low as 0.6 % [40]. Given the increased accuracy in the measurement afforded by the use of the spatial markers, we believe that an uncertainty of 0.6 % is a reasonable estimate for the multiple-point measurement results presented in Table 8.4.

The results presented for the multiple-point reference lever method all agree with the established techniques as expected, given the results for the single-point measurements and the expectation that the uncertainty should be lower for this method. It is interesting to note that in several cases the uncertainty on the single-point measurements is comparable to the multiple-point technique. This is likely due to the use of FIB spatial markers, which demonstrates the benefit of these positional features when applying the reference lever method.

#### 8.3.2.2 Inverted Geometry

A significant disadvantage of the reference cantilever method is that the technique requires force curves to be performed both on a hard surface (silicon or sapphire) and on the reference cantilever itself. This almost certainly results in some degree of damage to the tip of the probe such as that shown in Fig. 8.14, which is a result of performing many static force curves on hard surfaces. Obviously this is undesirable, in particular for applications where the probe has been functionalized or where the tip is extremely delicate (i.e., carbon nanotube probes).

Using a similar approach as described previously, the reference cantilever method can be applied without potential damage to the test cantilevers' tip. FIB milling was again used to mill spatial markers, but now they are located along the length of the test cantilever. The reference cantilever method was then inverted as shown in Fig. 8.15a, allowing the test cantilever to be calibrated accurately without the tip ever contacting a surface.

When calibrating a V-shaped cantilever, it is also possible to apply the inverted reference cantilever method without the use of spatial markers. This is achieved by

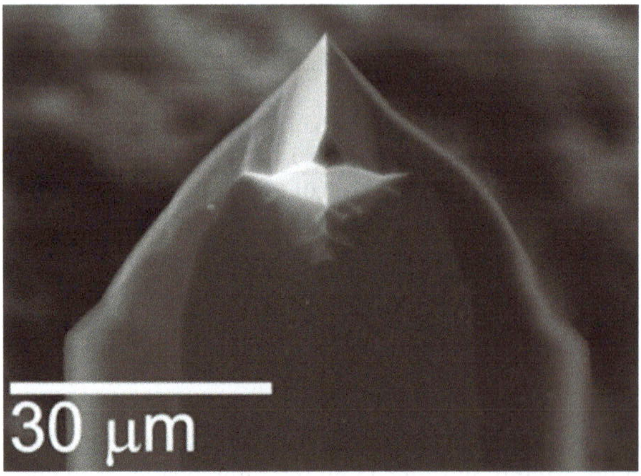


**Fig. 8.14** An SEM image of a Bruker FMV probe that has been blunted by repeated force curve measurements typical of those performed during reference cantilever calibration. Reprinted with permission from [29]. Copyright 2013, IOP Publishing

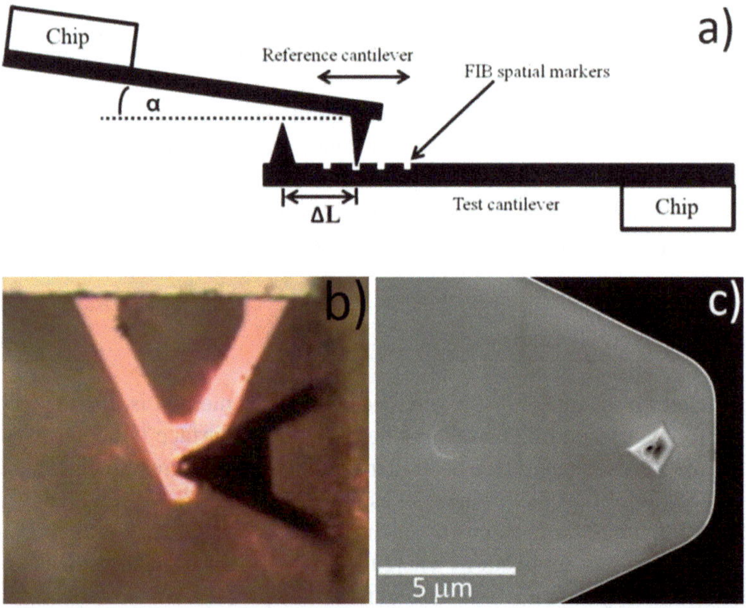


**Fig. 8.15** Schematic of the inverted reference lever method applied with FIB-milled spatial markers (**a**), an optical image of the reference lever measurement (**b**), and an SEM image of the FIB-milled test cantilever (**c**). Reprinted with permission from [29]. Copyright 2013, IOP Publishing

reverse imaging the V-arm intersection of the test cantilever with the AFM prior to performing the reference lever measurement. The distance from the V-arm intersection to the probe tip on the test cantilever can then be determined exactly by SEM

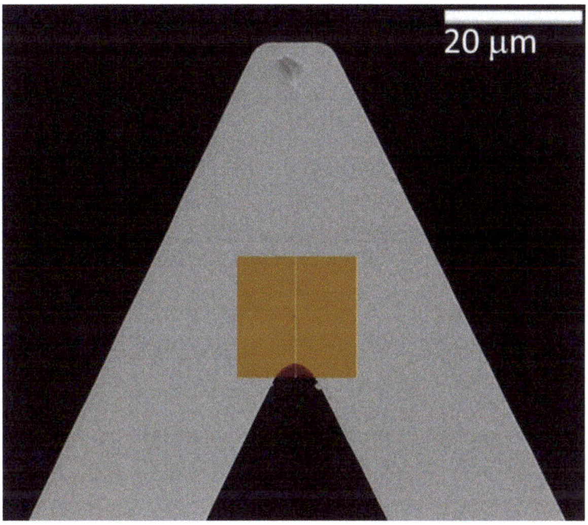


**Fig. 8.16** An AFM image of the V-arm intersection, overlaid on an SEM image of the cantilever. Reprinted with permission from [29]. Copyright 2013, IOP Publishing

imaging to precisely determine the loading position. This is shown in Fig. 8.16, where the AFM image obtained is overlaid on an SEM image of the probe; the location of the reference lever measurement is located in the centre of the AFM image.

To determine the spring constant, (8.4) must be rearranged in the case of the inverted method, which gives (8.13). In this expression, the terms after $k_{\text{ref}}$ have been inverted, and $L$ is now the length of the test cantilever ($L_{\text{test}}$), while the length of the reference cantilever is $L_{\text{ref}}$. In addition, $\Delta L$ is now the distance from the loading point of the reference cantilever tip to the test cantilevers' tip, and $D_{\text{ref}}$ is the height of the reference cantilevers' tip. These quantities are shown clearly in Fig. 8.15a:

$$k_{test} = k_{ref}\left[\left(\frac{S_{\text{C}}}{S_{\text{H}}}-1\right)\cos^2\alpha \times \left(1-\tan\alpha\,\frac{3D}{2L_{test}}\right)\right]^{-1}\left(\frac{L_{ref}}{L_{ref}-\Delta L}\right)^{-3} \tag{8.13}$$

The inverted reference lever method was applied to a number of different cantilevers with both beam- and V-shaped geometry. The spring constant was determined using (8.13), and the results obtained are shown in Table 8.5.

Comparing the inverted reference lever method with the standard method, good agreement is observed for all of the cantilevers studied with essentially the same level of uncertainty that was observed with the standard method. This indicates that in cases where preserving the imaging tip is important, the inverted reference lever method is a useful approach for accurately measuring the spring constant.

**Table 8.5** Spring constants determined using the inverted reference lever method compared to established techniques

| Cantilever | Spring constant ($N\ m^{-1}$) Inverted reference lever | Thermal noise (±10–20 %) | Standard reference lever (data from Tables 8.3 and 8.4) | Sader hydrodynamic (±10–15 %) | Cleveland formula (±10–15 %) |
|---|---|---|---|---|---|
| V1 (calibrated using cantilever V4) | $0.084^{a}_{s}\pm0.006$ ($\Delta L/L=0.19$) $0.088^{b}\pm0.006$ | $0.086\pm0.013$ | $0.088\pm0.006$ | – | – |
| V1 (calibrated using V7) | $0.085^{a}_{s}\pm0.008$ ($\Delta L/L=0.17$) $0.089^{b}\pm0.008$ | $0.086\pm0.013$ | $0.088\pm0.006$ | – | – |
| V2 | $0.117_{m}\pm0.010$ | $0.109\pm0.016$ | $0.114\pm0.006$ | | |
| V3 | $0.150_{m}\pm0.013$ | $0.150\pm0.023$ | $0.162\pm0.012^{c}$ | – | – |
| V5 | $0.281_{s}\pm0.020$ marker 1 ($\Delta L/L=0.12$) | $0.292\pm0.044$ | $0.305\pm0.024^{c}$ | – | – |
| F2 | $2.44_{m}\pm0.15$ | $2.38\pm0.24$ | $2.42\pm0.14$ | $2.53\pm0.32$ | $2.33\pm0.29$ |



The superscripts a and b represent slightly different ΔL offset corrections, and are described in more detail in reference [29]. The subscripts s and m refer to values obtained from single and multiple point measurements respectively

## 8.4 FIB Modification of AFM Probes

The ability to remove and deposit material on the nanoscale has resulted in many applications where FIB processing has been used to modify AFM probes. Some recent applications reported in literature are highlighted in this section.

### *8.4.1 Cantilever Property Enhancement*

The cantilever and the probe are at the heart of the AFM, and as such the properties of these components largely determine the performance of the instrument. It is no surprise that there have been many reports of using FIB processing to alter and enhance the way in which the cantilever behaves.

#### 8.4.1.1 Improving Cantilever Performance

Altering the shape of the cantilever can have significant effects on its properties. One important property for AFM measurements is the cantilevers' resonant frequency. Increasing the resonant frequency allows for a higher imaging speed [41], greater temporal resolution [42], and reduced noise due to thermal vibration [31]. Hodges et al. used FIB milling to modify the geometry and surface area of beam-shaped and

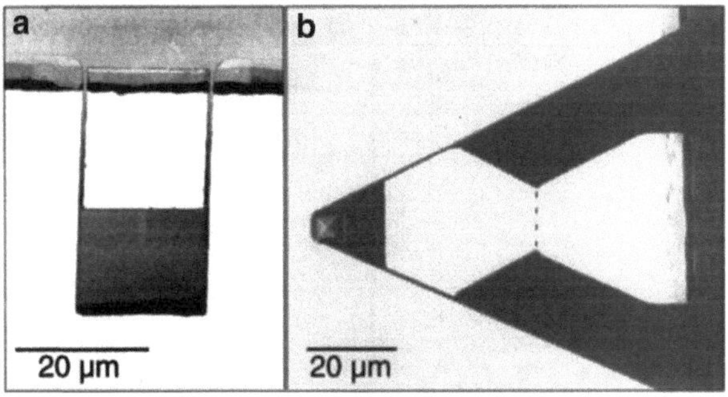


**Fig. 8.17** Modified AFM cantilevers with reduced mass to improve performance. Reprinted with permission from [43]. Copyright 2013, American Institute of Physics

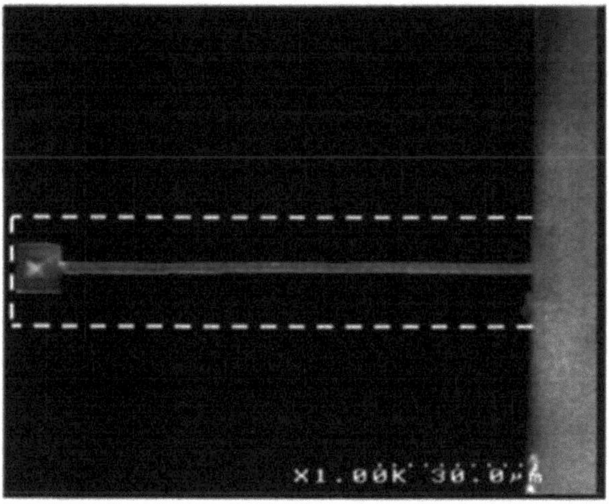


**Fig. 8.18** FIB-modified AFM cantilever to reduce the effects of hydrodynamic drag and increase resonant frequency. Reproduced with permission from [44]. Copyright, American Institute of Physics

V-shaped AFM cantilevers in order to increase their performance by increasing their resonant frequency [43]. These cantilevers were designed to function optimally in fluid, such that the reduction in resonant frequency due to fluid damping would be minimal. Hodges et al. observed a two- to threefold increase in the performance of the modified cantilevers, which was limited mainly by the requirement for a large reflective surface on which to align the AFM laser spot (Fig. 8.17).

FIB modification of cantilevers has also been applied to reduce the hydrodynamic drag during dynamic operation modes (Fig. 8.18). Dynamic AFM was developed in liquid mode to reduce the force applied to the soft, delicate samples often encountered in this mode. A factor which can significantly affect the performance of a cantilever in dynamic mode is hydrodynamic damping, which is strongly dependent on the cantilevers' size and shape. Typical cantilevers oscillating close to the surface in fluid can experience hydrodynamic damping forces on the order of several nanonewtons, which can have adverse effects on force measurements and the phase signal.

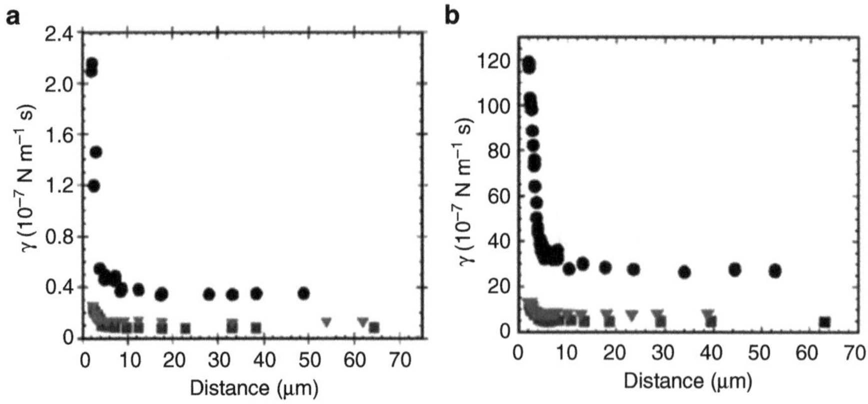


**Fig. 8.19** Damping coefficient of the cantilever as a function of distance from the surface in air (**a**) and fluid (**b**). *Circles* show cantilever 1 (standard), *triangles* show cantilever 2 ($w=4$ μm), *squares* show cantilever 3 ($w=2$ μm). Reproduced with permission from [44]. Copyright, American Institute of Physics

Maali et al. used FIB milling to reduce the width of a standard beam-shaped AFM cantilever in order to reduce its hydrodynamic drag [44]. Using this approach, the hydrodynamic damping was observed to decrease by a factor of 10 in air and a factor of 7 in water (Fig. 8.19). This provides an effective method to reduce hydrodynamic forces that may hinder delicate measurements, using FIB processing to modify the probe.

#### 8.4.1.2 Application to Sensors

In addition to the use of cantilevers for AFM applications, there is often overlap into the field of micromechanical sensors. Cantilever sensors are especially susceptible to thermal fluctuations, making them excellent thermal detectors. Lavrik et al. used FIB milling to optimize two commercially available cantilevers for calorimetric spectroscopy [45]. The FIB was used to remove material from the cantilever to reduce the thermal conductance between the sensing region and the cantilever support, as can be seen in Fig. 9 of [45].

In the field of micromechanical sensors, the FIB offers a powerful tool for rapidly producing prototype sensors [46]. This allows complicated sensor designs to be tested and optimized before considerable resources are invested in lithographic techniques and the production of specific masks.

### *8.4.2 AFM Tip Modification*

In addition to modification of the cantilever, the FIB can be used to modify the imaging tip of the AFM probe. The function of these modified probes ranges from

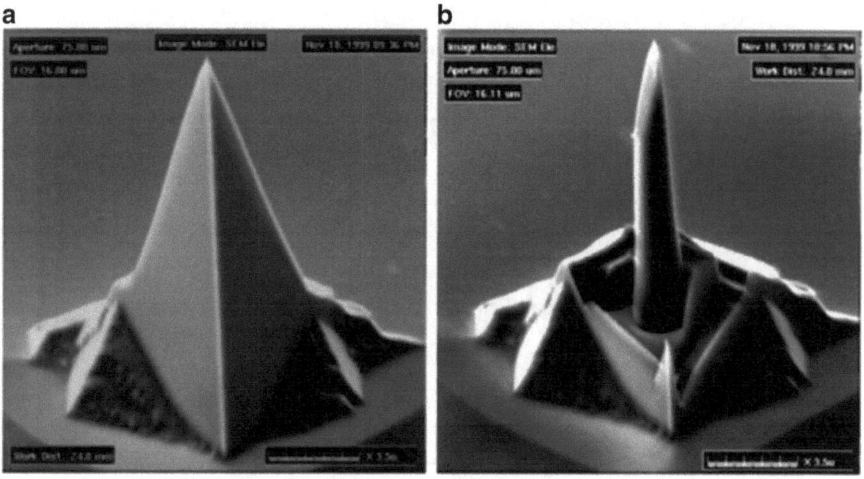


**Fig. 8.20** SEM images of a silicon nitride AFM probe tip before and after milling to improve the aspect ratio of the probe for characterizing deep trenches. Reproduced with permission from [47]

simple reshaping of the tip [47–49] to the formation of electrodes [50] and plasmonic structures [51, 52].

#### 8.4.2.1 Improving Aspect Ratio

The AFM is used extensively to characterize deep trenches and vertical steps on samples such as patterned photoresists and plasmonic structures like waveguides. Unfortunately the tip manufacturing process can only generate AFM tips with pyramidal geometry. These tips have a very sharp apex but often have a broad cone angle of approximately 34° [47], which results in significant tip convolution when imaging steep features. Increasing the aspect ratio of the tip allows the probe to image these structures effectively, and FIB milling has enabled researchers to achieve this.

Fu et al. used FIB milling to increase the aspect ratio of a tapping mode AFM probe in order to profile plasmonic structures consisting of deep trenches [47]. The modified tip is shown in Fig. 8.20 and has an aspect ratio of 24:1, which allowed the probe to penetrate the deep structures and resolve the base of the trench.

This application is so developed in fact that probes modified in this manner are available commercially from manufacturers such as the probe shown in Fig. 8.21 produced by Nanosensors. Even though the first work of this type was reported almost 10 years ago, FIB processing still provides the most effective means of producing high-aspect-ratio AFM probes with precise orientation.

#### 8.4.2.2 Modifying Tip Function

In addition to improving resolution, the probe tip can be modified to alter the way in which the probe interacts with the surface. Menozzi et al. shaped the tip of a

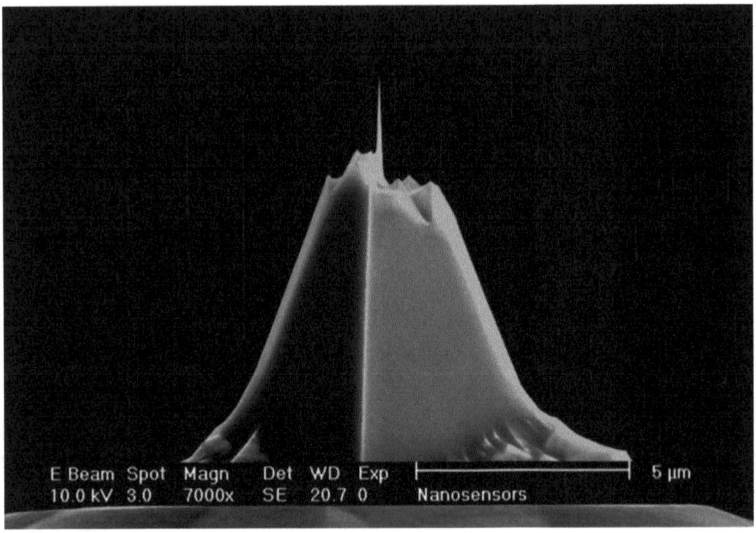


**Fig. 8.21** SEM image of an FIB-modified AR10-NCH probe from Nanosensors. Reprinted with permission from [53]. Copyright 2013, Nanosensors

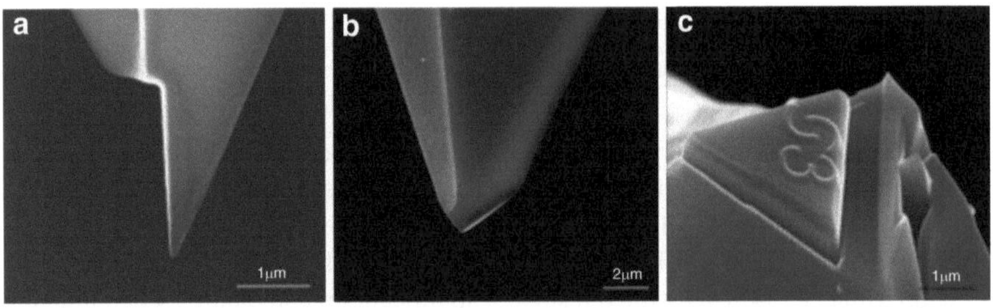


**Fig. 8.22** AFM probe tips modified for indentation experiments with specific tip geometry (**a**) and (**b**); and (**c**) a micro-stamp milled into the flattened tip of an AFM probe for micro-stamping. Reprinted with permission from [54]

silicon probe to perform indentation experiments with the desired probe geometry [54]. The tip (which was initially a quadratic pyramid) was milled such that its geometry was that of a triangular pyramid and the angle of the indentation region precisely defined (Fig. 8.22). The hardness measured using the two different tip geometries was compared with a theoretical model previously reported [55], and the results were found to be in good agreement.

Dinelli et al. and Menozzi et al. also fabricated a stamp feature on the end of a flattened AFM probe to perform nanoimprint lithography (NIL) [54, 56]. This NIL tip consisted of a sharp tip for imaging and a stamp for imprinting the surface. These probes were used to imprint polymer films and monitor the indentation parameters in real time.

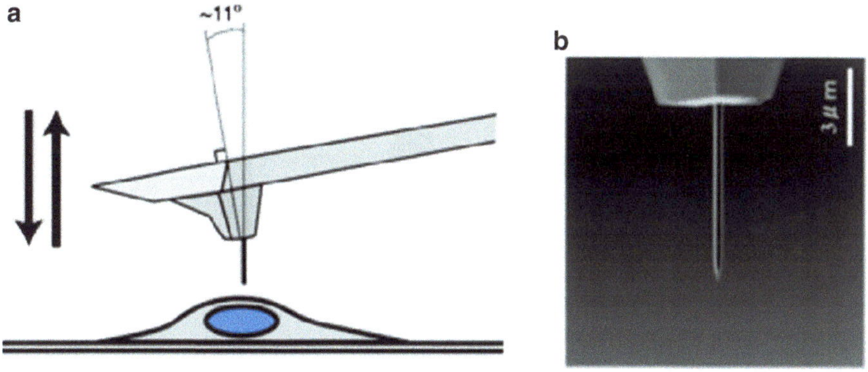


**Fig. 8.23** Schematic of the nanoneedle AFM probe experiment on the living cell (**a**), and an SEM image of the nanoneedle tip (**b**). Reprinted with permission from [57]. Copyright (2013) American Chemical Society

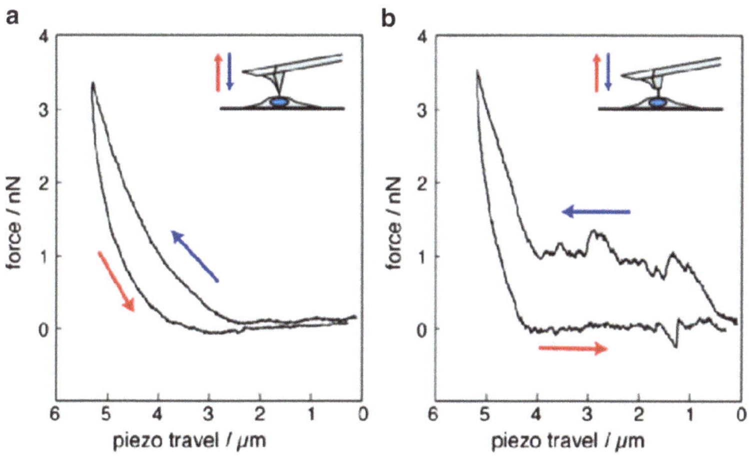


**Fig. 8.24** Force–distance curves on a living cell using the unmodified (**a**) and nanoneedle (**b**) AFM probes. Reprinted with permission from [57]. Copyright (2013) American Chemical Society

Obataya et al. reported the operation on a living cell with the use of a nanoneedle AFM probe [57]. A combination of FIB milling and etching processes was used to produce a needle 6–8 μm in length and 200–300 nm in diameter. A schematic showing the indentation of the cell and an SEM image of one of the nanoneedle probes fabricated is shown in Fig. 8.23a, b, respectively.

These probes were then used to puncture living cells and observe the force characteristics as the needle penetrated the cell membrane and nucleus. The same experiment was performed using a standard AFM probe, and the comparison of the force curves obtained is shown in Fig. 8.24a, b. The two force curves show very different behavior of the probes; sharp steps observed with the nanoneedle suggest penetration

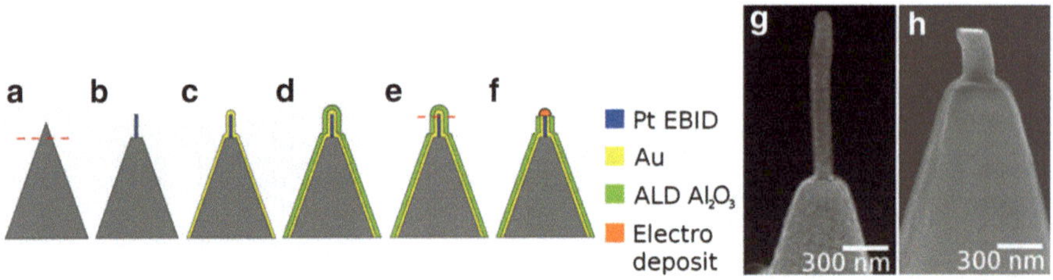


**Fig. 8.25** Fabrication of an electrode at the AFM tip apex using FIB milling. Reprinted (adapted) with permission from [50]. Copyright 2013, American Institute of Physics

through the various barriers in the cell. The blunt probe however was unable to penetrate the cell with the force applied, which is evident by the constant increase in cantilever deflection.

The FIB-modified probes penetration of the cell by exerting much lower force than would be required with standard probes and thus less potential for damage to the cell. Obataya et al. suggested that the nanoneedle could be loaded with molecules such as nucleic acids and proteins to deliver them directly to the cell.

#### 8.4.2.3 Nano-electrode Probes

There are an increasing number of applications where functional probes have been used to extend the capability of the AFM. These include magnetic force microscopy (MFM), scanning electrochemical microscopy (SECM-AFM), and tip-enhanced Raman spectroscopy (TERS). Comstock et al. reported the fabrication of a nanoneedle SECM-AFM probe using several deposition processes to form the nanoneedle and FIB milling to expose the nanoscale electrode at the end of the probe [50].

This process is shown in Fig. 8.25; FIB milling is initially used to flatten a region of the tip where electron beam-induced deposition (EBID) is used to deposit a long platinum spike. A series of atomic layer deposition (ALD) steps are used to deposit a conductive gold layer with an insulating layer of alumina, and the FIB is then used to expose the gold electrode at the end of the probe. In addition to demonstrating that these probes are capable of imaging normally, electrodeposition was used to form metallic clusters of silver or nickel at the end of the nanoneedle. The ability to form nanostructures at the tip of the nanoneedle would allow these probes to be applied in areas such as MFM and apertureless near-field optical microscopy.

#### 8.4.2.4 Nanopipette Probes

FIB milling has also been used to fabricate a "nanopipette" AFM probe, which is able to deposit droplets of attoliter volume [58]. A standard AFM probe was modified by milling a pore through the probe tip and forming a liquid-loading area on the back of the cantilever. SEM images of the modified probe are shown in Fig. 8.26.

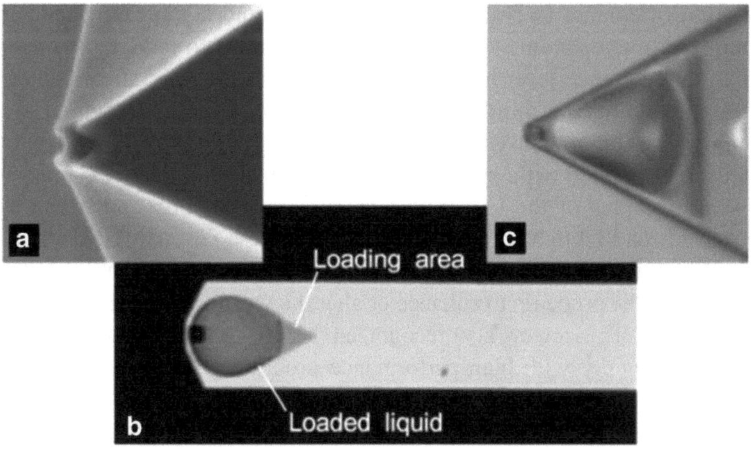


**Fig. 8.26** AFM probe modified to deposit attoliter volumes of liquid. Reprinted with permission from [58]. Copyright, American Institute of Physics

The nanopipette was used to deposit arrays of droplets consisting of glycerol and glycerol-containing polystyrene nanoparticles. Meister et al. determined that the volume of the droplets deposited was as low as 5 aL and could be deposited with droplet spacing below 500 nm.

## 8.5 Conclusion

FIB processes have enabled researchers in materials science to sculpt material on the nanoscale, with a speed and precision that were previously unachievable. Applying FIB techniques to the field of AFM, researchers have been able to customize the imaging probe to improve performance and perform specific functions. We have highlighted two recent publications which describe the application of FIB milling to improve the accuracy of two spring constant calibration methods. The Cleveland added mass method was modified, whereby instead of adding mass to the cantilever, a section was milled using the FIB and the resonant frequency shift used to determine the spring constant. Utilizing FIB milling allows mass to be removed with a high level of precision, significantly reducing the sources of uncertainty in the original method. The use of FIB milling allowed us to demonstrate an uncertainty as low as 7 %, with uncertainties below 5 % achievable for certain cantilevers. This level of uncertainty is comparable with the most accurate calibration methods currently available and is applicable to a wide range of cantilevers.

FIB milling was also used to improve the precision of the reference cantilever calibration method. The reference lever technique is very powerful in that the spring constant is measured directly and that it can be applied to any cantilever. FIB milling was used to produce positional markers along the length of the cantilever, allowing the position of contact to be determined precisely and significantly reducing the uncertainty in the technique. This approach was shown to reduce the uncertainty added by the reference lever method to below 1 %, which is comparable or better than the uncertainty of multiple-point measurement methods, which are time consuming. The ability to invert the method was also demonstrated, which allows the technique to be applied without damaging the imaging tip of the test cantilever.

Several examples of the use of FIB processing to modify and enhance AFM probes have been highlighted from the literature. These covered two distinct areas: modification of the cantilever to improve the performance of the cantilever itself and modification of the probe tip to enhance or alter the way in which the probe interacts with the sample. This review also recognized the use of FIB in industry to provide the AFM community with high-performance probes for specific application. FIB techniques enable researchers to perform modifications to AFM probes quickly and with extremely high precision, enabling the fabrication of highly specialized probes to perform specific tasks. The large number of publications reporting the FIB modification of AFM probes highlights the significant contribution of FIB processing to enabling diverse experiments in the field of AFM.

**Acknowledgments** The authors would like to acknowledge the Australian Microscopy and Microanalysis Research Facility (AMMRF) network and Adelaide Microscopy for providing access to the Helios dual-beam instrument and the AMMRF network for providing access to the atomic force microscopes. The authors wish to thank Mr. Len Green of Adelaide Microscopy in particular for his assistance with the FIB microscope and Dr. Adam Blanch from Flinders Centre for NanoScale Science and Technology for assistance with SEM imaging.

## References

1. Martin, Y., Wickramasinghe, H.K.: Magnetic imaging by "force microscopy" with 1000 Å resolution. Appl. Phys. Lett. **50**(20), 1455 (1987)
2. Betzig, E., Trautman, J.K., Harris, T.D., Weiner, J.S., Kostelak, R.L.: Breaking the diffraction barrier: optical microscopy on a nanometric scale. Science **251**(5000), 1468–1470 (1991)
3. Anderson, M.S.: Locally enhanced Raman spectroscopy with an atomic force microscope. Appl. Phys. Lett. **76**(21), 3130 (2000)
4. Stöckle, R.M., Suh, Y.D., Deckert, V., Zenobi, R.: Nanoscale chemical analysis by tip-enhanced Raman spectroscopy. Chem. Phys. Lett. **318**(1–3), 131–136 (2000)
5. BenAssayag, G., Salles, P., Bertin, F., Cooper, D.: TEM specimen preparation by Au+, Ga+, Si+ and Si++ focused ion beams for off-axis electron holography. Microelectron. Eng. **87**(5–8), 1579–1582 (2010)
6. Krohn, V.E., Ringo, G.R.: Ion source of high brightness using liquid metal. Appl. Phys. Lett. **27**(9), 479 (1975)
7. Clampitt, R.: Intense field-emission ion source of liquid metals. J. Vac. Sci. Technol. **12**(6), 1208 (1975)
8. Sudrad, P.: Thèse d'état Université d'Orsay (1979)
9. Taylor, G.: Disintegration of water drops in an electric field. Proc. Roy. Soc. A. Math. Phys. Eng. Sci. **280**(1382), 383–397 (1964)
10. Seliger, R.L.: High-resolution, ion-beam processes for microstructure fabrication. J. Vac. Sci. Technol. **16**(6), 1610 (1979)
11. Young, R.D.: Field emission ultramicrometer. Rev. Sci. Instrum. **37**(3), 275 (1966)

12. Young, R.D.: The Topografiner: an instrument for measuring surface microtopography. Rev. Sci. Instrum. **43**(7), 999 (1972)
13. Binnig, G., Rohrer, H.: Scanning tunneling microscopy. Surf. Sci. **126**(1–3), 236–244 (1983)
14. Binnig, G., Quate, C.F., Gerber, C.: Atomic force microscope. Phys. Rev. Lett. **56**(9), 930–933 (1986)
15. Butt, H., Cappella, B., Kappl, M.: Force measurements with the atomic force microscope: technique, interpretation and applications. Surf. Sci. Rep. **59**(1–6), 1–152 (2005)
16. Wiesendanger, R.: Scanning Probe Microscopy and Spectroscopy: Methods and Applications. Cambridge University Press, Cambridge (1994)
17. Ohler, B.: Cantilever spring constant calibration using laser Doppler vibrometry. Rev. Sci. Instrum. **78**(6), 063701 (2007)
18. Chui, B.W., Kenny, T.W., Mamin, H.J., Terris, B.D., Rugar, D.: Independent detection of vertical and lateral forces with a sidewall-implanted dual-axis piezoresistive cantilever. Appl. Phys. Lett. **72**(11), 1388 (1998)
19. Shahin, V., Ludwig, Y., Schafer, C., Nikova, D., Oberleithner, H.: Glucocorticoids remodel nuclear envelope structure and permeability. J. Cell Sci. **118**(Pt 13), 2881–2889 (2005)
20. Higgins, M.J., Proksch, R., Sader, J.E., Polcik, M., Mc Endoo, S., Cleveland, J.P., Jarvis, S.P.: Noninvasive determination of optical lever sensitivity in atomic force microscopy. Rev. Sci. Instrum. **77**(1), 013701 (2006)
21. D'Costa, N.P., Hoh, J.H.: Calibration of optical lever sensitivity for atomic force microscopy. Rev. Sci. Instrum. **66**(10), 5096 (1995)
22. Tourek, C.J., Sundararajan, S.: An alternative method to determining optical lever sensitivity in atomic force microscopy without tip-sample contact. Rev. Sci. Instrum. **81**(7), 073711 (2010)
23. Sader, J.E., Larson, I., Mulvaney, P., White, L.R.: Method for the calibration of atomic force microscope cantilevers. Rev. Sci. Instrum. **66**(7), 3789 (1995)
24. Cumpson, P.J., Clifford, C.A., Hedley, J.: Quantitative analytical atomic force microscopy: a cantilever reference device for easy and accurate AFM spring-constant calibration. Meas. Sci. Technol. **15**(7), 1337–1346 (2004)
25. Cleveland, J.P., Manne, S., Bocek, D., Hansma, P.K.: A nondestructive method for determining the spring constant of cantilevers for scanning force microscopy. Rev. Sci. Instrum. **64**(2), 403 (1993)
26. Drummond, C.J., Senden, T.J.: Characterisation of the mechanical properties of thin film cantilevers with the atomic force microscope. Mater. Sci. Forum **189–190**, 107–114 (1995)
27. Poggi, M.A., McFarland, A.W., Colton, J.S., Bottomley, L.A.: A method for calculating the spring constant of atomic force microscopy cantilevers with a nonrectangular cross section. Anal Chem **77**(4), 1192–1195 (2005)
28. Sader, J.E.: Parallel beam approximation for V-shaped atomic force microscope cantilevers. Rev. Sci. Instrum. **66**(9), 4583 (1995)
29. Slattery, A.D., Blanch, A.J., Quinton, J.S., Gibson, C.T.: Calibration of atomic force microscope cantilevers using standard and inverted static methods assisted by FIB-milled spatial markers. Nanotechnology **24**(1), 015710 (2013)
30. Hutter, J.L., Bechhoefer, J.: Calibration of atomic-force microscope tips. Rev. Sci. Instrum. **64**(7), 1868 (1993)
31. Butt, H.-J., Jaschke, M.: Calculation of thermal noise in atomic force microscopy. Nanotechnology **6**(1), 1–7 (1995)
32. Stark, R.W., Drobek, T., Heckl, W.M.: Thermomechanical noise of a free v-shaped cantilever for atomic-force microscopy. Ultramicroscopy **86**(1–2), 207–215 (2001)
33. Proksch, R., Schäffer, T.E., Cleveland, J.P., Callahan, R.C., Viani, M.B.: Finite optical spot size and position corrections in thermal spring constant calibration. Nanotechnology **15**(9), 1344–1350 (2004)
34. Hutter, J.L.: Comment on tilt of atomic force microscope cantilevers: effect on spring constant and adhesion measurements. Langmuir **21**(6), 2630–2632 (2005)
35. Sader, J.E., Sanelli, J.A., Adamson, B.D., Monty, J.P., Wei, X., Crawford, S.A., Friend, J.R., Marusic, I., Mulvaney, P., Bieske, E.J.: Spring constant calibration of atomic force microscope cantilevers of arbitrary shape. Rev. Sci. Instrum. **83**(10), 103705 (2012)

36. Slattery, A.D., Quinton, J.S., Gibson, C.T.: Atomic force microscope cantilever calibration using a focused ion beam. Nanotechnology **23**(28), 285704 (2012)
37. Hähner, G.: Dynamic spring constants for higher flexural modes of cantilever plates with applications to atomic force microscopy. Ultramicroscopy **110**(7), 801–806 (2010)
38. Melcher, J., Hu, S., Raman, A.: Equivalent point-mass models of continuous atomic force microscope probes. Appl. Phys. Lett. **91**(5), 053101 (2007)
39. Gates, R.S., Reitsma, M.G.: Precise atomic force microscope cantilever spring constant calibration using a reference cantilever array. Rev. Sci. Instrum. **78**(8), 086101 (2007)
40. Clifford, C.A., Seah, M.P.: Improved methods and uncertainty analysis in the calibration of the spring constant of an atomic force microscope cantilever using static experimental methods. Meas. Sci. Technol. **20**(12), 125501 (2009)
41. Butt, H., Siedle, P., Seifert, K., Fendler, K., Seeger, T., Bamberg, E., Weisenhorn, A.L., Goldie, K., Engel, A.: Scan speed limit in atomic force microscopy. J. Microsc. **169**(1), 75–84 (1993)
42. Viani, M.B., Schäffer, T.E., Paloczi, G.T., Pietrasanta, L.I., Smith, B.L., Thompson, J.B., Richter, M., Rief, M., Gaub, H.E., Plaxco, K.W., Cleland, A.N., Hansma, H.G., Hansma, P.K.: Fast imaging and fast force spectroscopy of single biopolymers with a new atomic force microscope designed for small cantilevers. Rev. Sci. Instrum. **70**(11), 4300 (1999)
43. Hodges, A.R., Bussmann, K.M., Hoh, J.H.: Improved atomic force microscope cantilever performance by ion beam modification. Rev. Sci. Instrum. **72**(10), 3880 (2001)
44. Maali, A., Cohen-Bouhacina, T., Jai, C., Hurth, C., Boisgard, R., Aimé, J.P., Mariolle, D., Bertin, F.: Reduction of the cantilever hydrodynamic damping near a surface by ion-beam milling. J. Appl. Phys. **99**(2), 024908 (2006)
45. Lavrik, N.V., Sepaniak, M.J., Datskos, P.G.: Cantilever transducers as a platform for chemical and biological sensors. Rev. Sci. Instrum. **75**(7), 2229 (2004)
46. Ghatnekar-Nilsson, S., Graham, J., Hull, R., Montelius, L.: Multi-frequency response from a designed array of micromechanical cantilevers fabricated using a focused ion beam. Nanotechnology **17**(20), 5233–5237 (2006)
47. Fu, Y., Zhou, W., Lim, L.: Geometrical characterization issues of plasmonic nanostructures with depth-tuned grooves for beam shaping. Opt. Eng. **45**(10), 108001 (2006)
48. Ximen, H., Russell, P.E.: Microfabrication of AFM tips using focused ion and electron beam techniques. Ultramicroscopy **42–44**, 1526–1532 (1992)
49. Akiyama, K., Eguchi, T., An, T., Fujikawa, Y., Yamada-Takamura, Y., Sakurai, T., Hasegawa, Y.: Development of a metal–tip cantilever for noncontact atomic force microscopy. Rev. Sci. Instrum. **76**(3), 033705 (2005)
50. Comstock, D.J., Elam, J.W., Pellin, M.J., Hersam, M.C.: High aspect ratio nanoneedle probes with an integrated electrode at the tip apex. Rev. Sci. Instrum. **83**(11), 113704 (2012)
51. Jin, E.X., Xu, X.: Focussed ion beam machined cantilever aperture probes for near-field optical imaging. J. Microsc. **229**(Pt 3), 503–511 (2008)
52. Huth, F., Chuvilin, A., Schnell, M., Amenabar, I., Krutokhvostov, R., Lopatin, S., Hillenbrand, R.: Resonant antenna probes for tip-enhanced infrared near-field microscopy. Nano Lett. **13**(3), 1065–1072 (2013)
53. Nanosensors: Trench measurement probe series. www.nanosensors.com/prod_cat_ar.html. Accessed 29 Jan 2013
54. Menozzi, C., Calabri, L., Facci, P., Pingue, P., Dinelli, F., Baschieri, P.: Focused ion beam as tool for atomic force microscope (AFM) probes sculpturing. J. Phys. Conf. Ser. **126**, 012070 (2008)
55. Pugno, N.M.: A general shape/size-effect law for nanoindentation. Acta Mater. **55**(6), 1947–1953 (2007)
56. Dinelli, F., Menozzi, C., Baschieri, P., Facci, P., Pingue, P.: Scanning probe nanoimprint lithography. Nanotechnology **21**(7), 75305 (2010)
57. Obataya, I., Nakamura, C., Han, S., Nakamura, N., Miyake, J.: Nanoscale operation of a living cell using an atomic force microscope with a nanoneedle. Nano Lett. **5**(1), 27–30 (2005)
58. Meister, A., Liley, M., Brugger, J., Pugin, R., Heinzelmann, H.: Nanodispenser for attoliter volume deposition using atomic force microscopy probes modified by focused-ion-beam milling. Appl. Phys. Lett. **85**(25), 6260 (2004)

# Chapter 9
# Fabrication of Needle-Shaped Specimens Containing Subsurface Nanostructures for Electron Tomography

**Jesús Hernández-Saz, Miriam Herrera, and Sergio I. Molina**

**Abstract** This chapter describes the use of the focused ion beam instrument for the fabrication of needle-shaped specimens containing specific subsurface features of a sample. These specimens are useful for the analysis of isolated features of a material by different characterization techniques, among other applications. We show the application of this methodology in the field of semiconductor materials, in particular for the analysis by electron tomography of InAs/InP- and InAs/GaAs-capped quantum dots. The proposed methodology allows the correlation of the structural properties of isolated objects located inside the material with their functional properties, avoiding the interference with other features in the material.

## 9.1 Introduction

Nowadays, the development of advanced materials with applications in nanotechnology relies not only on the ability to design and fabricate these materials but also on the capability to fully characterize their structural and functional properties. Generally, these structures are characterized in bulk, where information from the material as a whole is obtained. However, often it is useful to investigate isolated objects in the material, avoiding the interference from other features of the structure. As an example, the experimental analysis of the effect of size, shape, and arrangement of silver nanoparticles on the electromagnetic fields induced by their localized

J. Hernández-Saz • M. Herrera • S.I. Molina (✉)
INNANOMAT Group, Departamento de Ciencia de los Materiales e I.M. y Q.I., Facultad de Ciencias, Universidad de Cádiz, Campus Río San Pedro, s/n, Puerto Real 11510 Cádiz, Spain
e-mail: jesus.hernandez@uca.es; miriam.herrera@uca.es; sergio.molina@uca.es

Z.M. Wang (ed.), *FIB Nanostructures*, Lecture Notes in Nanoscale Science and Technology 20, DOI 10.1007/978-3-319-02874-3_9,

surface plasmon resonances studied theoretically in [1] requires isolating those nanoparticles; the electronic properties of single-walled carbon nanotubes are altered by intertube coupling [2]; therefore, their analysis needs avoiding the interference from other nanotubes.

Other example of the importance of isolating particular objects in a material for their characterization can be found in the semiconductor field, where the electronic properties of semiconductor quantum dots (QDs) are highly dependent on their individual size and shape [3–5]. Often, the conventional growth of these nanostructures produces a broad size and shape distribution of the QDs [6–8], making the correlation between their functional properties measured in bulk and their individual structural properties unfeasible. In order to solve this, the fabrication by focused ion beam (FIB) of specimens in the form of nano-needles containing only the QDs of interest has been proposed [9, 10]. In this way, the QDs can be characterized independently, which represents a breakthrough in the understanding of the behavior of these complex nanostructures.

The fabrication of needles containing a single QD by FIB is a challenging task. First, in most cases, the QDs are capped by a layer of a different semiconductor material; therefore, the location of the QDs cannot be directly inferred from the observation of the surface of the sample in the FIB. Secondly, it should be considered that depending on the application that the needle-shaped specimen will be used for, it should meet additional requirements. For example, a complete structural characterization of semiconductor QDs implies obtaining 3D information of the specimen, which can be obtained by electron tomography (ET). However, acquiring good quality transmission electron microscopy (TEM) images from a needle requires that it has a small diameter and that the needle surface is free of damage from the fabrication process.

In this chapter we show a methodology for the fabrication by FIB of needle-shaped specimens containing specific QDs located below the sample surface that meet the requirements for a good-quality ET analysis. This methodology can be adapted to a wide variety of materials and applications, including the fabrication of nano-tips for techniques such as near-field scanning optical microscopy (SNOM) and ultrasonic force microscopy (UFM) as well as the analysis of isolated objects by atom probe tomography (APT).

## 9.2 Methodology

We have developed two different methods for the fabrication of needle-shaped specimens containing QDs located below the sample surface. The first one is for samples with a low density of QDs, where a random needle milled in the material is not likely to contain one of the QDs. To exemplify this methodology, we have used a sample of InAs/InP QDs grown by molecular beam epitaxy (details of the sample fabrication process can be found elsewhere [11]), where the density of QDs is as low as $40\times10^{8}$ dots $cm^{-2}$. This kind of samples receive attention for their applications in optical devices [12], as they can be used as active elements in lasers [13], frequently in optical fiber communication technology. We have also developed a simplified method for samples with very high density of nanostructures. In this case, we have used a sample consisting of 50 layers of stacked InAs QDs grown on a GaAs substrate (additional information about the sample growth can be found in [14]). The stacking of QD layers is a common practice to increment the photocurrent in optoelectronic devices; these InAs/GaAs structures have been designed for intermediate-band solar cells [15, 16].

The two proposed needle fabrication methods start with the extraction of a thin lamella from the bulk material in the FIB and its deposition on a TEM grid. For this, we use the in situ lift-out method, widely reported in the literature [17–20]. Because of this, we explain this method briefly in Sect. 9.2.1, including useful advices for the most critical steps. After this, we show the general procedure for milling a needle from a lamella using the annular milling method (AMM) in Sect. 9.2.2. In this section, we also specify the procedure and parameters to obtain a good-quality needle-shaped specimen for ET.

The additional steps required for the fabrication of needle-shaped specimens containing QDs located below the sample surface are explained in Sects. 9.2.3 and 9.2.4, for structures with low and high density of QDs, respectively. In the first case, the QDs initially need to be found in order to mill a needle containing one of them. Here, it should be mentioned that the QDs in a lamella are not visible in the FIB with the secondary electron detector or with the backscattered electron detector. This difficulty has been overcome by developing a procedure of marking the lamella in the FIB with the ion beam and observing the location of the QDs with regard to these marks by TEM. In the case of samples with a high density of QDs, an easier method of fabricating multiple needles from the lamella and then choosing the needle with better characteristics is preferred. Finally, the analysis by ET of the needle described in Sect. 9.2.4 is presented in Sect. 9.3 to demonstrate the applicability of the described method.

The proposed methodology has been carried out using a dual-beam FEI Quanta200 3D FIB instrument equipped with an Omniprobe micromanipulator. For the TEM analysis, a JEOL 2010F microscope operating at 200 kV and a JEOL 1200 microscope working at 120 kV have been used.

### *9.2.1 Fabrication of a Lamella from the Bulk Material*

As explained above, the first step of the developed methodology consists of extracting a thin lamella of material from the bulk sample and attaching it to a grid. For this, several methods have been published in the literature with the objective of obtaining specimens for TEM [21–23]. Initially, specimens were prepared cutting and polishing a piece of material, attaching it to a grid outside the FIB and then thinning it to electron transparency in the FIB (this method is known as the H-bar technique [22]). However, the development of micromanipulators has allowed shortening the preparation time. Thus, the lift-out technique has become one of the most widely used methods to fabricate electron-transparent specimens [19, 21]. The lift-out method can be in situ (where the micromanipulator is located inside the FIB) or ex situ (the micromanipulator is outside the FIB), the first one being more reliable. Because of this, we have chosen the in situ lift-out method to fabricate the initial thin lamella, and this method is briefly explained in this section.

Before milling a lamella from the bulk material, the surface of the sample needs to be protected from possible damage caused by the interaction with the ion beam. This protection is normally made by depositing a rectangular metal layer (often Pt) on the surface of the sample to prevent undesired $Ga^+$ ions from reaching the sample surface. The deposition of a Pt layer in the FIB is normally made by an electron- or ion-assisted chemical vapor deposition (CVD) process [24, 25]. Depositing Pt with the ion beam has the advantage that it is a faster process but the disadvantage that it can cause damage to a depth of several tens of nm on a semiconductor specimen surface. The electron-assisted deposition is much slower, but normally it only affects few nm on the sample surface. Therefore, the best option is reaching a compromise between the electron- and ion-assisted deposition methods where initially an electron-assisted CVD process is applied for 30 min, and, after that, the ion-assisted method is used to deposit several microns of Pt. For special cases where the feature of interest is located very close to the sample surface and it can be damaged depositing Pt even with the electron beam, other protection methods are recommended such as covering the surface with a plasma-polymerized organic film [26]. Regarding the thickness of the Pt layer, it should be chosen considering that after all the milling processes that will take place to fabricate the needle, there should always be some Pt remaining. We have confirmed that a Pt layer of 2 μm thick is normally adequate to accomplish the whole fabrication process. Figure 9.1a shows a scanning electron microscopy (SEM) image of a Pt deposition on an InP surface obtained following the steps explained above.

Once the surface of the sample has been protected, stair-step trenches are milled on both sides of the Pt layer, forming a lamella in the bulk material. This lamella needs to be thinned to 1.5–2 μm in several steps (Fig. 9.1b). Table 9.1 [9] shows the parameters used in the FIB for the different steps of the process for semiconductors such as GaAs or InP. After the trenches have been milled, two of the three connections between this lamella and the bulk sample need to be cut, as shown in Fig. 9.1c.

At this point, the lamella is ready to be lifted out and taken to a TEM grid. For this, it needs to be attached to the micromanipulator by depositing a C layer (Fig. 9.1d), freed from the bulk material, then attached to the TEM grid, and freed from the micromanipulator. These are the most complex steps of the lift-out method, as the attachment of the lamella to the micromanipulator or to the grid requires approaching them to a critical distance. If the gap is too wide it would prevent the lamella from attaching firmly to the micromanipulator or to the grid, but the physical contact may break the lamella. In order to avoid the physical touch, the ion beam image should be continuously monitored, as there is a change in contrast that takes place due to the electrical contact that occurs right before the physical touch. This is the optimum moment to start the C deposition. The correct attachment between the lamella and

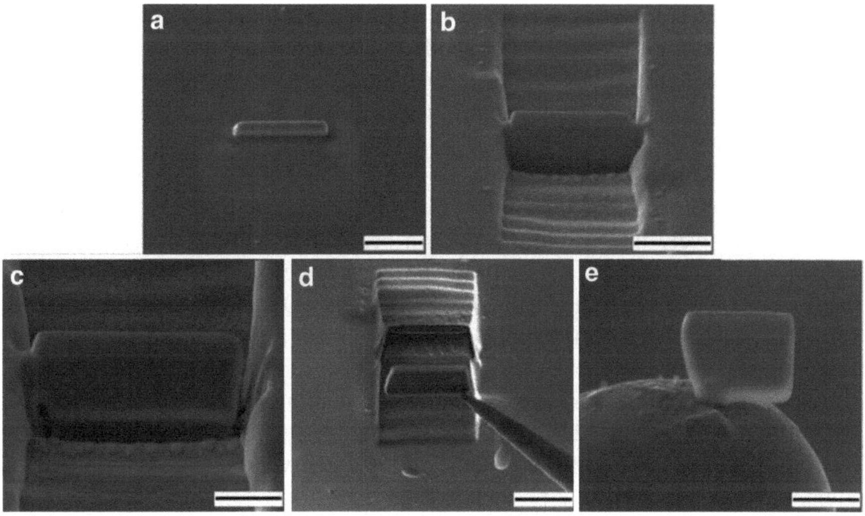


**Fig. 9.1** SEM images of the different steps of the in situ lift-out method for the fabrication of a lamella from the bulk material: (**a**) Pt deposition, (**b**) milling of the stair-step trenches, (**c**) three of the connections between the lamella and the bulk material are milled, (**d**) lamella attached to the micromanipulator, and (**e**) lamella attached to the TEM grid. The scale bars are 10 μm for (**a**), (**b**), and (**d**) and 5 μm for (**c**) and (**e**) (reproduced from [9])

**Table 9.1** Summary of the parameters used in the FIB for the in situ lift-out method for the fabrication of a lamella in InP or GaAs samples

| Step | Angle (°) | Voltage (kV) | Current (nA) | Thickness (μm) |
|---|---|---|---|---|
| Pt deposition (electron assisted) | 0 | 15 | 2 | 2 |
| Pt deposition (ion assisted) | 52 | 30 | 0.3 | 2 |
| Trench | 52 | 30 | 20 | 4 |
| Thinning | 52 | 30 | 7 | 2 |
| Thinning | 52 | 30 | 1 | 1–1.5 |
| Precut | 7 | 30 | 1 | 1–1.5 |
| Thinning in grid | 52 | 30 | 0.1 | 0.4 |
| Thinning in grid | 52 | 30 | 0.05 | 0.25 |
| Thinning in grid | 52 | 30 | 0.01 | 0.10–0.15 |

(Reproduced from [9])

the micromanipulator or the grid can be checked monitoring the grounding current during the C deposition, as once they are electrically connected through the C, the magnitude of the grounding current changes sharply.

When the lamella is already attached to the TEM grid (Fig. 9.1e), it needs to be further thinned with the ion beam. In this process, care should be taken to keep as much of the Pt protective layer as possible, as this protection will be needed for the needle preparation. For this, it is very helpful to follow the milling process by imaging with the electron beam during the process. The current of the ion beam needs to be carefully selected to avoid redeposition of the milled material: sequential steps lowering the $Ga^+$ current is recommended. The details of the parameters used for thinning the lamella down to electron transparency are included in Table 9.1. The final thickness of the sample in this step will depend on the needle fabrication process that will be used. For low-density QD samples and as explained before, we will mark the lamella in the FIB and take it to the TEM to check the position of the QDs with regard to the marks. For this, the lamella needs to be electron transparent but not excessively thin to prevent the ion beam from drilling the sample when introducing the marks. We find appropriate an average thickness for the electron-transparent lamella of about 150 nm. For high-density QD samples, the sample does not need to be electron transparent, and in this case a final thickness of around 1–1.5 μm is reasonable.

With regard to the GaAs sample, it is worth mentioning that GaAs has a lower sputtering yield than InP [27], which means that for a given ion beam current and voltage, the time required for a specific milling step will be higher. In this case, attention should be paid to a possible spatial drift of the sample with time, as its effects on the final geometry of the specimen will be more pronounced.

### 9.2.2 *Production of Needle-Shaped Specimens for Electron Tomography*

Frequently, the procedures for fabricating needle-shaped specimens have been linked to the development of the APT technique [28], given that for this technique the specimens need to have this geometry. These fabrication methods can be as well used for obtaining needles for other applications, although they need to be adapted to meet the specifications required in each case. Electro-polishing has been one of the most used techniques for preparing needles for APT [28], although this technique is not appropriate for some materials and structures. Other methods are based in selective chemical etching processes [29], sometimes applied after lithography methods [30] or after the mechanical cut of the sample [31].

In the last years, the FIB instrument has been widely used for fabricating needle-shaped specimens from different materials [32–34] as it provides advantages such as speed and reliability. For this, normally the AMM [35, 36] is used, where a bulk sample is milled with an annular pattern of variable diameter. This method has been applied to different materials such as metals [37], polymers [38, 39], carbon nanotubes [40], or semiconductors [41]. In the literature, maintaining the voltage and current of the ion beam fixed while varying the inner diameter of the annular pattern [30] is a common practice when using the AMM for fabricating a needle. However, when the needle-shaped specimen will be used for ET, the fabrication method should be adapted to obtain needles optimized for this application. For ET, the needles need to meet two basic requirements. First, they need to have a small diameter in the area of interest (in our case, where the QDs are) in order to be electron transparent. Secondly, the surface of the needle should be as clean as possible to avoid artifacts in the electron images. This means minimizing the amorphous layer that is generally created on the material surface by the high-energy ion beam. In this work, we have used the AMM to obtain the needle-shaped specimens from our samples. In the following, we explain the conditions for fabricating these needle-shaped specimens from a lamella of the material of interest with enough quality for ET.

#### 9.2.2.1 Reduction of the Needle-Shaped Specimen Diameter

Generally, when using the AMM for fabricating needle-shaped specimens, the final needle has a diameter of about 150–200 nm [30, 42]. However, in order to obtain the best resolution in TEM images it is preferably that the thickness of the needle in the area of interest is smaller, ideally lower than 100 nm. In order to achieve this reduction in diameter, the conditions of the ion beam current and voltage as well as the inner and outer diameters of the annular milling pattern should be carefully chosen.

To fabricate a needle with a reduced diameter from a bulk material or from a lamella, generally successive annular patterns are used, where the inner and outer diameters are progressively decreased. As mentioned before, generally these steps are carried out using the same ion beam current. However, if a needle with a small diameter is needed, this current should be adjusted. Here, the main limitation is the thickness of the Pt protective layer, given that when this layer is milled away the process should be stopped to avoid damaging the needle. The Pt layer can be preserved for a longer time by milling the needle with an ion beam of small diameter (with a small current), as the precision of the milling process improves [43]. However, the use of small currents increases the milling time, and a small drift of the sample during the process could damage the needle. Optimizing the ion beam current and the inner and outer diameters of the annular pattern, the minimum diameter that we have been able to obtain at 30 kV is 100 nm, slightly higher than expected.

Additional parameters that can be modified to try to reduce the needle diameter include the ion beam acceleration voltage. Working at very low voltages (5–10 kV), the image quality is not satisfactory and the spatial resolution for imaging and milling is not adequate to mill a thin needle. However, using an acceleration voltage of 20 kV and the appropriate ion beam current (that will be shown later) we have been able to reduce the needle diameter to 70 nm.

With regard to the ion beam current and the inner and outer radius of the annular milling process, as explained above the AMM is carried out in a number of steps, where these two parameters are progressively reduced. In order to select these parameters and the number of steps, the structure of the sample should be taken into consideration. The sample with InAs/InP QDs only has one layer of QDs; therefore, the area of interest has a small length along the needle axis. In this case, a small number of steps are required, which are included in Table 9.2. On the other hand, the InAs/GaAs sample has 50 layers of stacked QDs; therefore, the region of interest along the needle axis is much longer, about 900 nm for this structure. In this case, care should be taken to sculpt a needle with reduced diameter along this larger distance in the needle axis. In soft materials such as III–V semiconductors, milling

**Table 9.2** Parameters used in each step of the AMM for the fabrication of semiconductor needles with a reduced diameter

| | Step | Inner diameter (nm) | Outer diameter (nm) | Current (pA) | Voltage (kV) |
|---|---|---|---|---|---|
| InAs/InP sample | 1 | 400 | 1,000 | 10 | 30 |
| | 2 | 200 | 450 | 23 | 20 |
| | 3 | 70 | 250 | 23 | 20 |
| | 4 | – | – | 70 | 5 |
| InAs/GaAs sample | 1 | 1,000 | 1,500 | 100 | 30 |
| | 2 | 800 | 1,400 | 81 | 20 |
| | 3 | 700 | 1,200 | 23 | 20 |
| | 4 | 600 | 1,000 | 23 | 20 |
| | 5 | 500 | 850 | 23 | 20 |
| | 6 | 400 | 700 | 4 | 20 |
| | 7 | 300 | 600 | 4 | 20 |
| | 8 | 150 | 400 | 4 | 20 |
| | 9 | – | – | 70 | 5 |

a needle with the ion beam following an annular pattern normally produces a typical conical shape where the diameter increases rapidly as the distance from the top of the needle is raised. To avoid this, an increase in the annular milling steps has been introduced in the procedure, which allows obtaining a needle with nearly cylindrical shape. This also helps avoiding the effect of the drift mentioned before. Table 9.2 includes the steps followed for milling a needle from the InAs/GaAs lamella. In general, it is worth mentioning that the annular pattern is normally etched from the external surface of the needle inwards with dwell time of 1 μs.

Finally, it should be considered that after the last step of the AMM, there still should be at least 300 nm of Pt on the surface of the needle to allow for the cleaning step that will reduce the thickness of the amorphous surface layer.

#### 9.2.2.2 Reduction of the Surface Damage Caused by the Milling Process

When the ion beam of the FIB is focused on a semiconductor material in order to remove some parts of the structure, the sample can be damaged because of an increase in the material temperature due to beam heating, because of ion implantation, surface amorphization, etc. However, some of these factors may have a stronger effect in the final quality of the sample than others. For example, it has been reported that FIB milling does not produce the beam heating typical in broad argon beam milling because the angle between the ion beam and the prepared section is normally very small [44]. The temperature rise has actually been calculated for broad argon beam milling as few hundred K, whereas in FIB it is less than 10 K [45]. Therefore, damage from beam heating is not expected to reduce drastically the quality of the prepared sample.

Regarding ion implantation, several parameters should be considered in order to understand its effect on a sample. First, the incidence angle of the ion beam on the sample should be considered, as the implantation has been found to be reduced when the ion beam is parallel to the sample surface [46]. When milling a needle with the AMM, the ion beam is perpendicular to the Pt protective layer but parallel to the final needle surface; therefore, the implantation should not be very deep. Secondly, the acceleration voltage of the ion beam is also an important parameter. It has been reported that $Ga^+$ implantation is around 20 % at a depth of 20 nm when milling with the ion beam at 30 kV perpendicularly to a Cu/Co sample but almost zero when reducing the acceleration voltage to 5 kV [47, 48]. This shows that by decreasing the ion voltage in the last step of sample preparation, less damage is produced in the sample. Finally, it has been found that high-dose ion imaging at 30 kV even with low beam currents can produce a strong ion implantation on the sample during the fabrication process [49]. This should definitely be avoided by imaging the sample with the electron beam instead of with the ion beam whenever possible.

One of the main problems of working with semiconductor samples in the FIB is surface amorphization because of the high-energy ion beam. This process is probably related to ion implantation in that it has also been found to cause amorphization of the specimen surface [18, 50]. The thickness of the amorphous layer can be modulated considering parameters such as the beam energy or the angle of incidence. However, often it is easier to mill the sample with the optimum parameters for the fabrication process and then remove the amorphized surface layer. For this, several approaches have been proposed, such as gas-assisted, wet, or chemical etching [44]; broad argon ion milling [51]; or FIB milling with a lower ion beam energy [52, 53], where the effect of the orientation of the $Ga^+$ beam with regard to the sample surface has also been studied [54].

Scanning the sample with a low-energy ion beam in the FIB is one of the techniques used more often to eliminate the damaged area in a sample, as it is relatively fast and effective. Because of this, the effect of a low-voltage cleaning step has been studied for different materials such as Si [44], stainless steel [55], or AlGaAs [56]. Working at 5 kV in the FIB is complicated given that the quality of the ion beam image is very poor; therefore, it does not have enough precision for milling a specific pattern. In order to eliminate the amorphized area, imaging the sample with the low-voltage beam for some time is normally enough.

We have investigated the efficiency of milling at low voltage in the FIB for reducing the thickness of the surface amorphous layer in semiconductors such as GaAs (which has similar behavior as InP [27]). In order to compare the behavior of the material for different ion acceleration voltages, we have fabricated needles using 10, 20, and 30 kV and an additional needle milled at 30 kV but scanned with the ion beam at 5 kV and 70 pA for 30 s. Figure 9.2 shows a plot of the thickness of the amorphous layer measured in each needle-shaped specimen by TEM *vs.* the ion beam voltage used for its fabrication. This plot evidences that the reduction in the ion beam voltage in the last step of the fabrication process from 30 to 5 kV clearly reduces the damaged surface layer, from 16 nm to less than 5 nm.

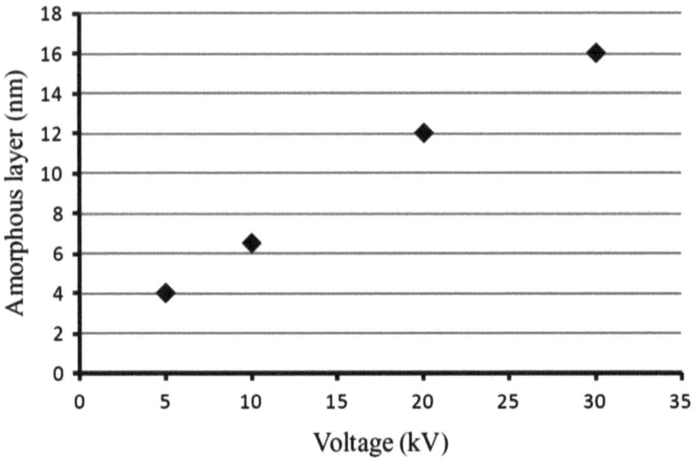


**Fig. 9.2** Plot of the thickness of the GaAs amorphous layer *vs.* ion beam voltage, where a reduction of the amorphous layer down to <5 nm has been achieved at small voltages (5 kV) (reproduced from [9])

The reduction in the surface amorphous layer produces a considerable improvement in the quality of electron microscopy images and, consequently, in the ET analysis, given that the proportion of crystalline material with respect to the total specimen (crystalline plus amorphous) is enhanced. For example, if we compare a needle-shaped specimen with 100 nm of diameter milled at 30 kV with the same needle but after a cleaning step at 5 kV, the relationship crystal/amorphous goes from 68 to 90 %; therefore, this step of a 5 kV is a key point for a good-quality ET analysis.

Figure 9.3 shows two bright field (BF) TEM images of the same needle (a) before the cleaning step and (b) after the cleaning step, where the reduction in the surface amorphous layer in the second micrograph is clear. The image in Fig. 9.3b shows a needle with a diameter smaller than 100 nm in the area of interest and an amorphous surface layer below 5 nm, demonstrating the success of the proposed method in the fabrication of needle-shaped specimens that fulfill the requisites for ET analysis.

The effect of the ion voltage in the specimen quality has also been analyzed by high-resolution (HR) TEM images. Figure 9.4 shows HRTEM images acquired from the edge of needles of GaAs milled/cleaned at different ion voltages, from left to right 5, 10, 20, and 30 kV. As it can be clearly observed, the noncrystalline region at the edge of the material is progressively increased when increasing the ion voltage. A reduction in the quality of the images has also been found, as the atomic columns become blurred for samples prepared at higher voltages. In addition, the increase in the rate of amorphous material in the needles is revealed by the fast Fourier transform (FFT) of each image included as an inset. For higher voltages, the intensification of the bright ring in the electron diffraction patterns clearly shows an increase in the amorphization degree.

In order to understand the relation between the ion voltages and the thickness of the amorphous surface layer, we have simulated the average penetration depth

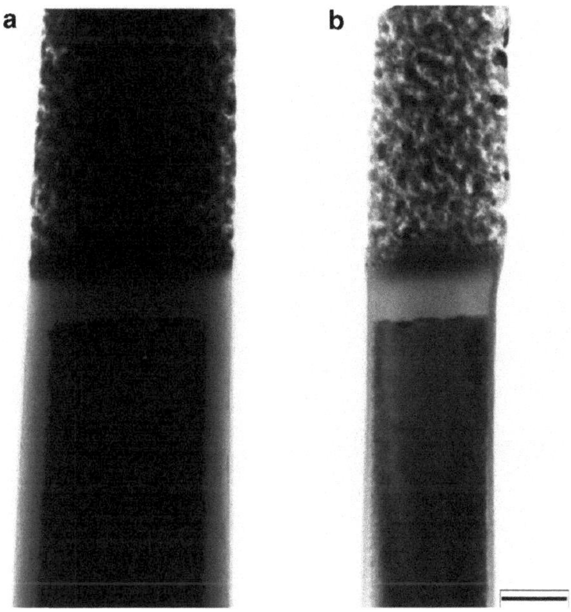


**Fig. 9.3** BF TEM images of a needle-shaped specimen (**a**) before the cleaning step and (**b**) after the cleaning step. The needle in (**b**) has a diameter <100 nm in the area of interest and has a surface amorphous layer <5 nm, fulfilling the requisites for ET. The images were taken at the same magnification, and the scale bar is 50 nm

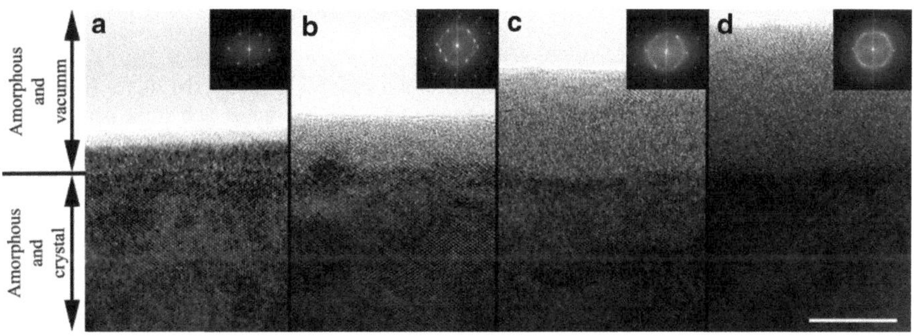


**Fig. 9.4** HRTEM images of the edge of needles milled/cleaned at different ion voltages: (**a**) 5, (**b**) 10, (**c**) 20, and (**d**) 30 kV. An increase in the surface amorphous layer for higher voltages can be clearly observed in the images as well as more intense amorphous rings in the FFT included as *insets*. The scale bar represents 10 nm

(usually called *the ion range*) of $Ga^+$ ions in a structure of GaAs for different acceleration voltages (5, 10, 20, and 30 kV). The simulations were carried out using the Monte Carlo method, with the free software SRIM (*stopping and range ions in matter*) [57]. During the preparation process, the needle-shaped specimens are milled from the upper part, where the incidence of the $Ga^+$ ions is almost parallel to the

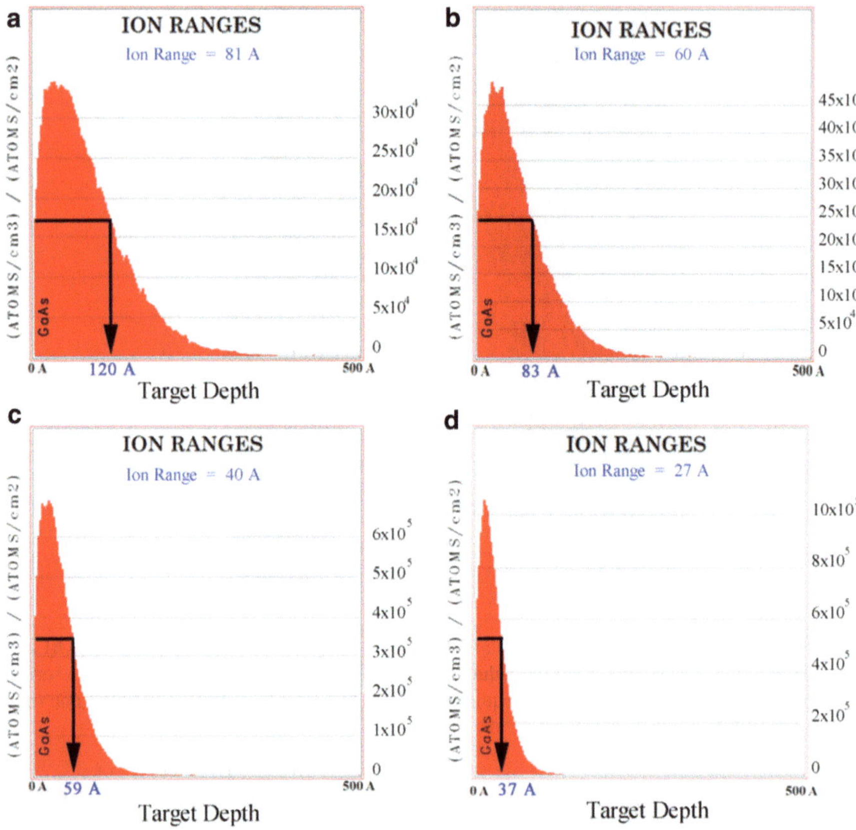


**Fig. 9.5** SRIM calculation of the $Ga^+$ ion ranges for ion voltages of (**a**) 30, (**b**) 20, (**c**) 10, and (**d**) 5 kV on a GaAs surface with an angle of incidence of 5°. The ion ranges at half the peak heights are marked with *arrows*

needle surface. Because of this, we have considered a value of 5° for the angle of incidence of the ions on the material.

Figure 9.5 shows the results of the simulations of the ion range for the different ion voltages. As it can be observed, the average penetration depth of the ions is reduced when reducing the ions energy, from 8 nm at 30 kV to 3 nm at 5 kV, confirming that the effect of the incident ions on the specimen surface can be decreased by reducing the ion voltage. In order to compare our experimental results shown in Fig. 9.4 with the simulations, we have calculated the penetration depth at half the peak height of the ion distribution curve for each ion voltage. We have found that for 30 kV this simulated penetration depth is around 12 nm, where the experimental value found is slightly higher, around 16 nm. For 20 kV, the simulated penetration depth (and the experimental value) is 8 nm (12 nm), for 10 kV is 6 nm (6.5 nm), and for 5 kV is 4 nm (4 nm). As it can be observed, there is a reasonable agreement between the thickness of the amorphous layer found experimentally and the

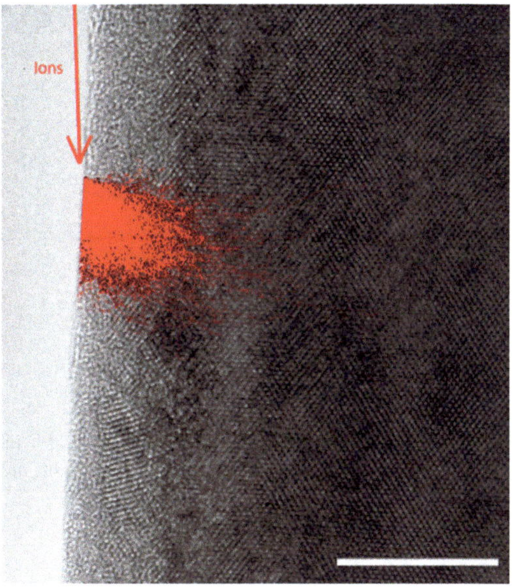


**Fig. 9.6** HREM image of a GaAs needle-shaped specimen prepared at 10 kV, where the simulated trajectories of 2,000 $Ga^+$ ions when interacting with the GaAs structure have been superimposed. The correlation between the ion penetration depth and the thickness of the amorphous surface layer is evidenced. The scale bar is 10 nm

penetration depth of $Ga^+$ ions in GaAs, showing that this penetration depth could be responsible for the damage in the needle surface. In Fig. 9.6 we have superimposed an experimental HREM image of the needle milled at 10 kV with the simulation of the $Ga^+$ ion trajectories (for 2,000 ions) when interacting with the GaAs structure. This image clearly shows the concordance between the amorphized region of the sample and the incident ion trajectories, evidencing the close relation between both effects.

### *9.2.3 Fabrication of Needles Containing Specific Features Located Below the Sample Surface: Samples with Low Density of QDs*

The fabrication of needle-shaped specimens that contain features of the material that are not visible from the surface of the structure requires further steps in the milling process. Also, additional difficulties arise when the density of those features is very small, in such a way that the fabrication of random needles from the material is not likely to include the desired object. We have developed a procedure to ensure that the feature of interest is included in the fabricated needle-shaped specimen. This procedure includes the design and introduction of some marks in the initial lamella that will help us locate the objects of interest in the FIB. In the following, we explain the steps that are necessary to reach our objective. In particular, we exemplify the methodology with a semiconductor sample of InAs/InP QDs, where the density of QDs is around $40 \times 10^8$ dots $cm^{-2}$, as mentioned earlier.

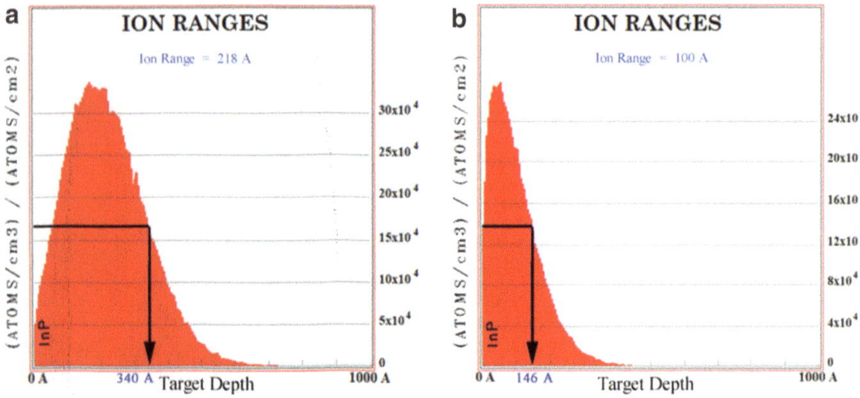


**Fig. 9.7** SRIM calculation of the $Ga^+$ ion ranges for incidence angles of (**a**) 90° and (**b**) 5° on a InP structure. The ion ranges at half the peak heights are marked with *arrows*

#### 9.2.3.1 Marks and QD Localization

Once a lamella of the InAs/InP sample has been milled and taken to a grid, the major difficulty for milling a needle that contains a QD is that these QDs are not visible with the secondary electron detector or the backscattered detector in the FIB. To solve this, the electron-transparent lamella can be taken to the TEM, where the QDs are clearly observable. However, the position of the QDs found by TEM should be recognizable once we are back to the FIB to allow milling a needle in that position. For this, we have designed some marks that will be introduced in the lamella with the ion beam in the FIB and which should be visible both in the FIB and in the TEM. In this way, the location of the QDs will be found in the FIB from their position with regard to the marks observed by TEM. Care should be taken when marking a thin lamella with the ion beam, as these marks should not affect the structural quality of the features to study. Also, the marks should be designed in such a way that the location of the QDs can be deduced from the position of the marks with a small spatial error. For this, we have designed the marks as a line of circles patterned on the surface of the lamella in the FIB with 50 nm of diameter and <50 nm of depth, whose centers are separated 150 nm. These marks should not be patterned on the Pt protective layer as in this case they would not be clearly visible in the TEM.

In order to make sure that the structural quality of the QDs will not be affected by these marks, we have simulated the ion range of $Ga^+$ ions at 30 kV when they are incident on InP. In order to choose the incident angle for these simulations, it is worth considering that the marks are expected to have a shape close to cylindrical, where in the sides of the cylinder the angle between the incident $Ga^+$ ions and the InP is very small (5° for the simulation), and in the base of the cylinder this angle is 90°. Figure 9.7 shows the results of these simulations. As it can be observed, the ion range in the sides of the marks is around 10 nm, where in the base of the marks it is close to 20 nm. We have also calculated the ion penetration depth at half the peak

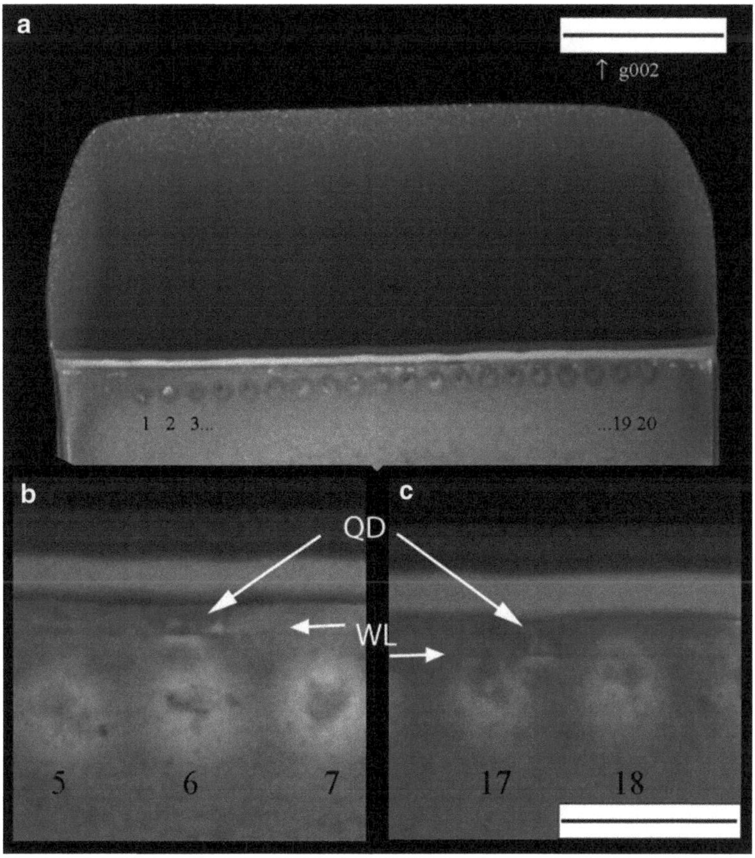


**Fig. 9.8** 002DF TEM image showing the relative position between the marks and the QDs: There is a QD above mark number 6 and another one between mark numbers 17 and 18. The wetting layers (WL) are also marked with *arrows*. The scale bars are 1 μm for (**a**) and 200 nm for (**b**) and (**c**) (reproduced from [9])

height of the ion distribution curve in order to estimate the surface damaged area caused by these marks. We have obtained a value of 15 nm for an incidence angle of 5° and 34 nm for 90°. This means that the marks should not be made on the QDs and closer than 15 nm from the QD layer to avoid possible ion implantation or amorphization of the QDs.

Once the electron-transparent lamella is marked, it is ready to be observed in the TEM. Figure 9.8 shows 002DF TEM images of a lamella marked following the procedure above. As it can be seen in Fig. 9.8a, 20 marks have been introduced in a 4 μm long lamella. Taking a closer look (Fig 9.8b, c), it can be observed that there are two QDs in this lamella, one right above the mark number 6 and another one between marks 17 and 18. This information will be used in the FIB in order to find the locations of the QDs in the lamella to mill the needles in those positions.

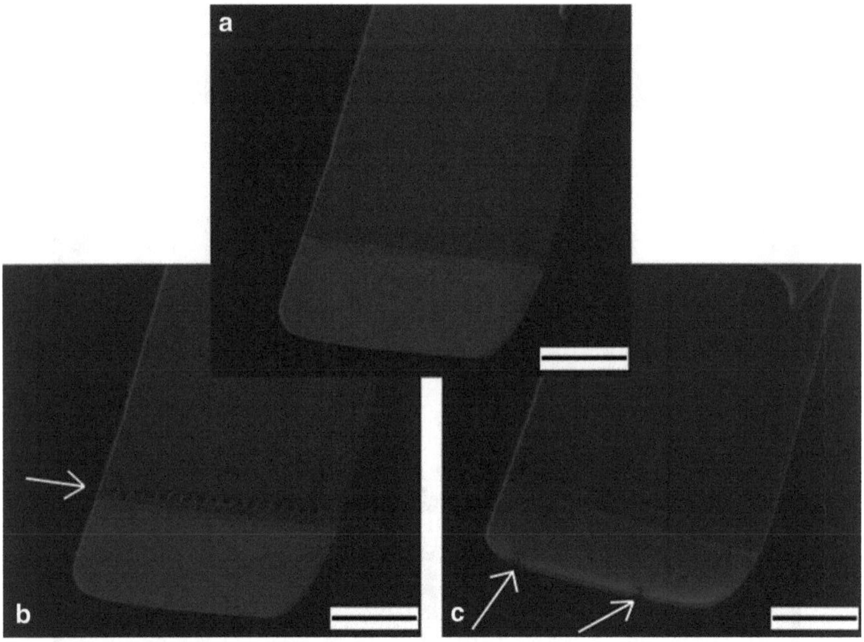


**Fig. 9.9** SEM images showing the lamella: (**a**) Without marks, (**b**) with marks to locate the QD, and (**c**) with additional marks milled in the top of the Pt layer, where the AMM will be applied. All the scale bars are 1 μm (reproduced from [9])

#### 9.2.3.2 Milling a Needle from an Electron-Transparent Marked Lamella

Once the lamella has been observed by TEM, it is time to fabricate a needle in the position indicated by the marks. For milling the needle, the lamella needs to be parallel to the ion beam, and in this configuration the marks are not visible in the FIB. In order to solve this, an additional mark needs to be introduced in the upper part of the Pt layer and right above the position of the QDs. This mark will be used as the center of the annular pattern. Figure 9.9 [9] shows SEM images of the complete process of marking the lamella. Initially, we have an electron-transparent lamella as shown in Fig. 9.9a. Here, a line of marks is introduced with the ion beam, as shown with an arrow in Fig. 9.9b; now the lamella is observed in the TEM to check the position of the QDs are with respect to the marks. Finally, a small mark is introduced in the surface of the Pt layer in the position where the QDs are supposed to be, as can be seen in Fig. 9.9c. When the center of the annular pattern is placed right over this last mark, the fabricated needle will contain the desired QD.

With the aim of obtaining an estimation of the spatial error in the fabrication of a needle in the desired location, we have measured the distance between the center of the needle and the position of the QD for needles with different diameters. For this, we have fabricated needles with annular pattern inner radiuses of 400, 200, and 70 nm.

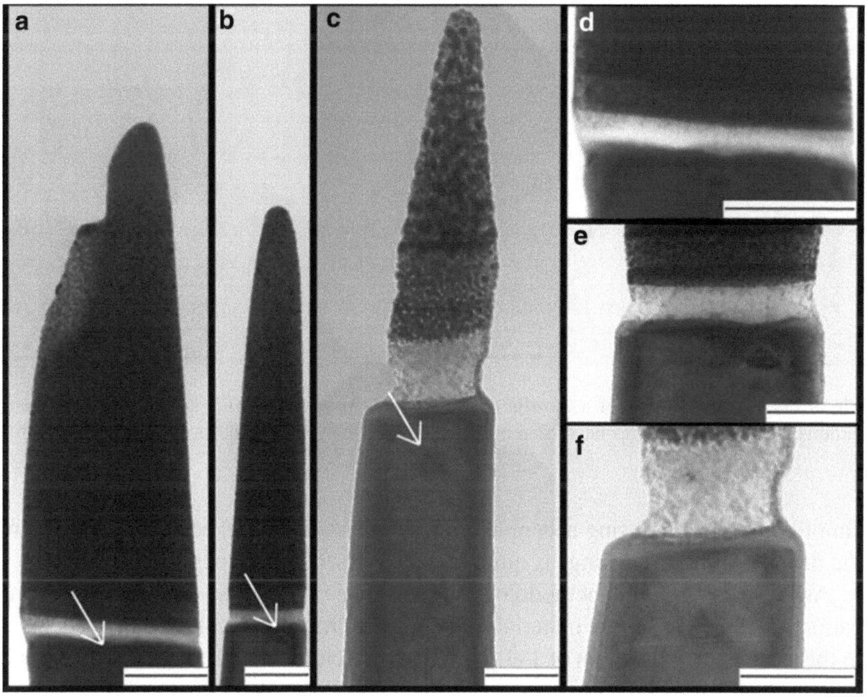


**Fig. 9.10** 002BF TEM images showing needles sculpted around single QDs (indicated by the *arrows*). Three needles were milled using annular pattern inner radius of (**a**) 400, (**b**) 200, and (**c**) 70 nm. Details of the same needles at the zones of interest are shown in (**d**–**f**). The scale bars are 200 nm for (**a**), (**b**), and (**d**), 100 nm for (**e**), and 50 nm for (**c**) and (**f**) (reproduced from [9])

Figure 9.10 [9] shows BF TEM images of some of these needles, where details of the zones of interest are presented in Fig. 9.10d–f. In these images, it can be observed that the average spatial error in the application of our methodology is around 25–30 nm. For cases where a better precision is required, the needle can be milled in two steps. In the first step the needle is milled down to a diameter of around 200 nm, then the position of the QD is observed in the TEM to correct possible deviations, and finally the needle is milled to get the final diameter.

### *9.2.4 Fabrication of Needles Containing Specific Features Located Below the Sample Surface: Samples with High Density of QDs*

When the sample contains a very high density of features of interest, milling a needle containing one of these features is quite straightforward and it is not necessary to follow the abovementioned steps to mark the lamella. We will show this

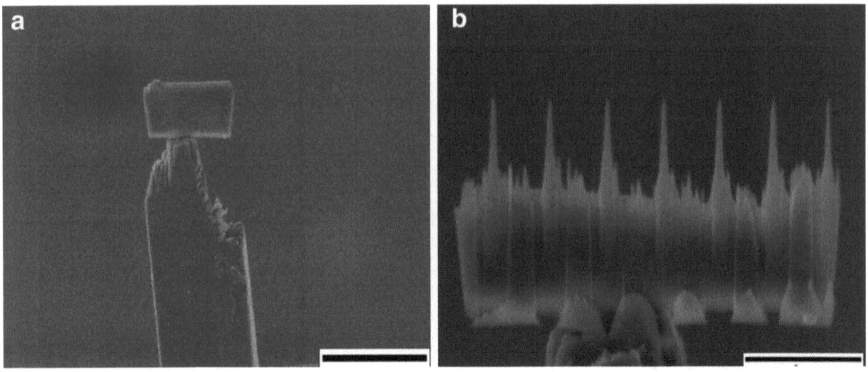


**Fig. 9.11** (**a**) SEM image of a lamella of the InAs/GaAs sample. (**b**) SEM image of the same lamella where seven needles have been milled. The scale bars are 20 μm for (**a**) and 5 μm for (**b**)

simplified procedure using a sample with 50 InAs/GaAs-stacked QD layers, where the density of these stackings is quite high, as shown in previous studies [14].

As in Sect. 9.2.3, the methodology for this type of samples starts with the fabrication of a lamella from the material following the procedure explained in Sect. 9.2.1, in this case with a thickness of 1–1.5 μm. It is preferable that the lamella has a good length of about 10 μm, for the reasons that will be explained later. Figure 9.11a shows a lamella from the InAs/GaAs sample attached to a TEM grid. As the density of QD stacks in this sample is high, it is very likely that a needle milled in a random position in the lamella using the AMM contains one of these QD stacks. However, these stacks occupy a large distance along the needle axis, and sometimes the needle may not contain the whole stack. In order to being able to obtain an optimized needle, where a large area of the QD stack is included, a good solution would be marking the lamella as shown in Sect. 9.2.3.1 and selecting the position where the needle should be milled. However, this is a time-consuming process, and in this case a different alternative can be taken in order to save time. In this way, after milling a lamella with a considerable length as said before, it is possible milling several needles in the lamella using the parameters shown in Table 9.2. Figure 9.11b shows a lamella where seven needles have been sequentially milled following this procedure. Here, it has to be considered that redeposition on previous needles of the material sputtered while milling a new needle should be avoided. For this, the distance between subsequent needles should not be smaller than 1.5 μm. Also, a final cleaning step at 5 kV is useful to remove possible contamination.

Once the needles have been milled, the specimen is taken to the TEM to observe their quality. Desired characteristics are that a QD stack is fully included in the needle, that the surface amorphous layer is smaller than 5 nm, and that the needle has a reduced diameter. Figure 9.12 shows a high-angle annular dark field (HAADF) image of one of the needles in Fig. 9.11b. In this technique, the intensity in the image is roughly proportional to the squared average Z number of the atoms in the material; therefore, the InAs QDs show brighter contrast than the GaAs matrix.

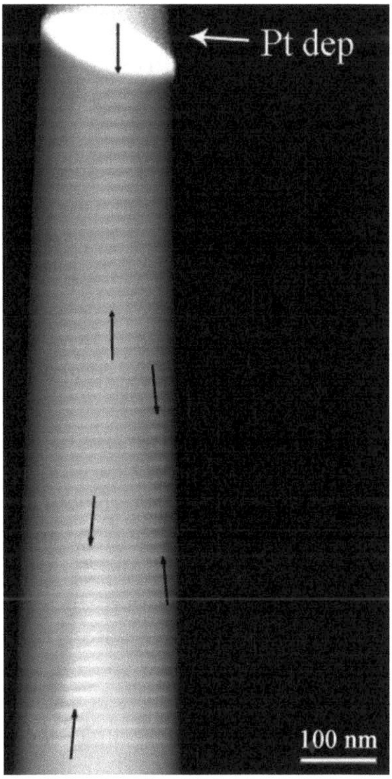


**Fig. 9.12** HAADF-STEM image of the QD stacks (marked with *arrows*) inside a needle-shaped specimen

As it can be observed in the image of Fig. 9.12, there are several QD stacks in the needle (marked with black arrows in the image), and it is worth mentioning that they do not follow a vertical alignment, but they are tilted with regard to the growth direction. This makes difficult having a full QD stack inside the needle. In this case, a compromise should be reached between the needle diameter and the amount of QDs included in the needle: for smaller diameters (preferred for ET) a reduced number of QDs will be found in the needle. In this case, a diameter of 150 nm has been found as the optimum one. As it can be observed in this example, milling several needles in the same lamella has the advantage that it allows choosing the most appropriate needle considering the specific characteristics of each material system.

Once the needle has been selected and depending on the application for which the needle is being fabricated, the rest of the needles can be milled away in the FIB in order to leave only the one of interest. For example, for ATP analysis there should be only one needle on the specimen holder; therefore, the rest of them should be removed. In our case, our objective is the study by ET of the stacking of QDs. In case the tomography holder allowed tilting the sample to high degrees (>70°), the contiguous needles may interfere in the electron beam path for high tilting angles. This is not our case; therefore, the ET analysis can be carried out in the sample containing all the seven needles, as it will be shown next.

## 9.3 Electron Tomography of InAs/GaAs Needle-Shaped Specimens

First images of the InAs/GaAs needle-shaped specimens fabricated following the proposed method have shown that the InAs/GaAs QD stackings do not follow a vertical alignment parallel to the growth direction, but they are tilted away with respect to this direction. The vertical distribution of the QDs has a direct effect in its electronic structure due to a possible electron tunnelling between layers [14], and it has also been found to influence optical properties such as the photoluminescence emission of the structure [58]. Because of this, understanding the 3D distribution of stacked QDs is essential to understand and optimize the functional properties of a wide range of devices. However, the vertical alignment of the stacking of QDs is often analyzed by TEM from 2D projections of the volume of the sample in one or several directions [59, 60], losing 3D information and making the complete correlation with the optical characteristics unfeasible.

In this section, we show the applicability of the fabricated needle-shaped specimen for the ET analysis of the 3D vertical alignment of the QD stack contained in the needle. In ET, a series of images are taken in the microscope while tilting the sample over an angular range with a tilting step of few degrees, and with these images the 3D volume of the specimen is reconstructed. For this, it is required that the electron beam crosses a constant thickness of the electron-transparent foil when travelling through the sample. This is not accomplished by the thin foils prepared by the conventional method of specimen preparation, and only cylindrical or conical shaped specimens with the symmetry axis parallel to the tilting axis as the needle-shaped specimens fabricated here meet this requirement. For the ET analysis, we have acquired 61 HAADF-STEM images over an angular range of 120° with a tilting step of 2° in a Jeol JEM 2010F electron microscope with a field emission gun working at 200 kV using a Fischione tomography holder (model 2030). The tilt series has been accurately aligned using the Inspect 3D software of FEI Company with the cross-correlation method in combination with the least-squares alignment mode with the AMIRA software. The 3D reconstruction has been carried out using the simultaneous iterative reconstruction technique and is visualized with the software AMIRA. A manual segmentation of the tomogram was carried out in order to locate the QDs. The position of the QDs has been considered as the geometric center of the QDs in the tomogram.

Figure 9.13 [10] shows some HAADF-STEM images of an InAs/GaAs QD stack that is contained in the prepared needle-shaped specimen, obtained at different tilting angles. About the vertical alignment of the QDs, in Fig. 9.13a it seems to be parallel to the growth direction. However, as mentioned before, TEM cross section images are 2D projections of the sample, and therefore, the volume information is lost; this should be taken into account to avoid the misinterpretation of the images. In this regard, Fig. 9.13b, c shows HAADF images of the same InAs/GaAs QD stack as in Fig. 9.13a but taken at different tilting angles, 90° apart from each other, and −10° and 80° from the micrograph in Fig. 9.13a, respectively. As it can be

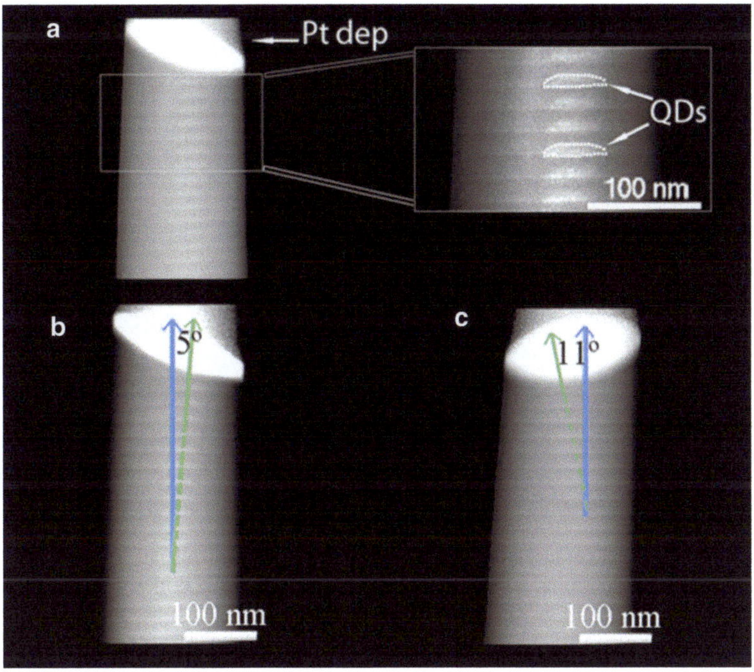


**Fig. 9.13** Cross-sectional HAADF-STEM images of the needle-shaped specimen taken at different tilting angles. Note that angles between the stackings of QDs and the growth direction are different for the three images: (**a**) 0°, (**b**) 5°, and (**c**) 11° (reproduced from [10])

observed, at these rotation angles, the stacking of QDs is not vertically aligned anymore. Instead, deviation angles of 5° and 11° with respect to the growth direction have been measured. Other values for the vertical alignment of the QDs have been measured from different tilting angles. These experimental results evidence that the conclusions obtained from the conventional 2D analysis of the stacking of QDs often found in the literature are not reliable, the 3D analysis of the sample being an essential step.

In order to obtain 3D information from the sample, we have computed the tomogram using the 61 images acquired in the tilting series. The results are shown in Fig. 9.14 [10]. Figure 9.14a shows a general view of the needle, including the upper QD stack and the Pt deposition. For the analysis of the distribution of the QDs, a segmentation of the reconstructed structure was carried out, as shown in Fig. 9.14b. This figure reveals that the real distribution of the QDs consists of a stacking that follows a straight line that deviates $10° \pm 1°$ from the growth direction Z, which is quite different from the results obtained from Fig. 9.13. The observed deviation is likely caused by the elastic interactions with the buried dots and by chemical composition fluctuations [7, 61]. From this analysis, we have also observed that there is an asymmetry in the size of the QDs, being around 30 % smaller in one direction than in the perpendicular one in the growth plane.

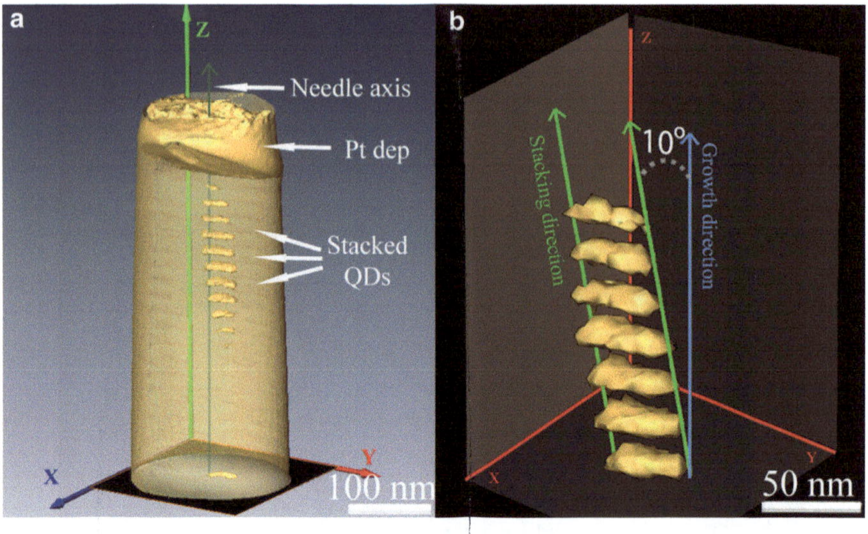


**Fig. 9.14** (**a**) Semitransparent external surface of the tomogram with opaque surfaces for the QDs below the platinum deposition. A projection of one of the central QDs is included in the base of the needle. (**b**) Segmentation of the QDs in the tomogram, showing that the stacking of QDs follows a *straight line* that deviates $10° \pm 1°$ from the growth direction (reproduced from [10])

The fabrication of needle-shaped specimens has allowed obtaining precise experimental data about the location of InAs QDs in the structure. These data can be correlated with the growth design and with the functional properties of the structure, allowing the optimization of the performance of devices based in stacking of QDs.

## 9.4 Conclusions

The fabrication of needle-shaped specimens by FIB allows the characterization of isolated objects in a material by different techniques such as ET, microPL, or APT. It is also an attractive procedure for preparing SNOM nano-tips with controlled optical properties and nano-tips for other techniques such as UFM.

We propose a methodology for fabricating needle-shaped specimens containing specific subsurface features of a material and for their optimization for ET analysis. For this, a complex process of marking a lamella of material and observing it in the TEM in order to infer the position of the feature of interest with regard to the marks has been developed and applied to a structure of low-density InAs/InP QDs. We also propose a simplified method for material systems where there is a high density of objects of interest. In this case, milling multiple needles in a lamella and choosing the optimum one is a time-saving process. We have applied this last method to a sample of InAs/GaAs QD stacks. The applicability of these needle-shaped specimens for the analysis by ET has been demonstrated by the characterization in 3D of this InAs/GaAs structure.

The developed process to isolate nanostructures such as QDs opens the way for the correlation of the structural and functional properties of individual objects, allowing a more precise characterization of the material than in bulk studies.

**Acknowledgments** The authors wish to thank D. Fuster, A. Rivera, Y. González, and L. González from IMM (CNM-CSIC) for the InAs/InP sample fabrication. The authors also thank D. Alonso-Álvarez for the stacked InAs/GaAs sample preparation. This work was supported by the Spanish MICINN (project TEC2011-29120-C05-03 and CONSOLIDER INGENIO 2010 CSD2009-00013) and the Junta de Andalucía (PAI TEP-946 INNANOMAT research group). Co-financing from UE-FEDER is also acknowledged. TEM measurements were carried out at DME-SCCYT-UCA.

## References

1. Hao, E., Schatz, G.C.: Electromagnetic fields around silver nanoparticles and dimers. J. Chem. Phys. **120**, 357–366 (2004)
2. Wang, F., Sfeir, M.Y., Huang, L.M., Huang, X.M.H., Wu, Y., Kim, J.H., Hone, J., O'Brien, S., Brus, L.E., Heinz, T.F.: Interactions between individual carbon nanotubes studied by Rayleigh scattering spectroscopy. Phys. Rev. Lett. **96**, 167401 (2006)
3. Mlinar, V., Bozkurt, M., Ulloa, J.M., Ediger, M., Bester, G., Badolato, A., Koenraad, P.M., Warburton, R.J., Zunger, A.: Structure of quantum dots as seen by excitonic spectroscopy versus structural characterization: using theory to close the loop. Phys. Rev. B **80**, 165425 (2009)
4. Sadi, I., Sellami, K., Yahyaoui, M., Testelin, C., Boujdaria, K.: Electron and hole energy levels in InAs/GaAs quantum dots: size and magnetic field effects. J. Appl. Phys. **109**, 033703 (2011)
5. Usman, M., Tasco, V., Todaro, M.T., De Giorgi, M., Oreilly, E.P., Klimeck, G., Passaseo, A.: The polarization response in InAs quantum dots: theoretical correlation between composition and electronic properties. Nanotechnology **23**, 165202 (2012)
6. Stangl, J., Holy, V., Bauer, G.: Structural properties of self-organized semiconductor nano-structures. Rev. Mod. Phys. **76**, 725–783 (2004)
7. Springholz, G.: Three-dimensional stacking of self-assembled quantum dots in multilayer structures. C. R. Physique. **6**, 89–103 (2005)
8. Muñoz-Matutano, G., Suárez, I., Canet-Ferrer, J., Alén, B., Rivas, D., Seravalli, L., Trevisi, G., Frigeri, P., Martínez-Pastor, J.: Size dependent carrier thermal escape and transfer in bimodally distributed self assembled InAs/GaAs quantum dots. J. Appl. Phys. **111**, 123522 (2012)
9. Hernandez-Saz, J., Herrera, M., Molina, S.I.: A methodology for the fabrication by FIB of needle-shape specimens around sub-surface features at the nanometre scale. Micron **43**, 643–650 (2012)
10. Hernandez-Saz, J., Herrera, M., Alonso-Alvarez, D., Molina, S.I.: Analysis of the 3D distribution of stacked self-assembled quantum dots by electron tomography. Nanoscale. Res. lett. **7**, 681 (2012)
11. Fuster, D., Rivera, A., Alen, B., Alonso-Gonzalez, P., Gonzalez, Y., Gonzalez, L.: Direct formation of InAs quantum dots grown on InP (001) by solid-source molecular beam epitaxy. Appl. Phys. Lett. **94**, 133106 (2009)
12. Shields, A.J.: Semiconductor quantum light sources. Nat. Photonics **1**, 215–223 (2007)
13. Li, S.G., Gong, Q., Cao, C.F., Wang, X.Z., Yue, L., Yan, J.Y., Wang, H.L.: Morphology and shape dependent characteristics of InAs/InP(1 0 0) quantum dot laser grown by gas source molecular beam epitaxy. Physica. E. Low. Dimens. Syst. Nonostruct. **44**, 1983–1987 (2012)
14. Alonso-Alvarez, D., Taboada, A.G., Ripalda, J.M., Alen, B., Gonzalez, Y., Gonzalez, L., Garcia, J.M., Briones, F., Marti, A., Luque, A., Sanchez, A.M., Molina, S.I.: Carrier recombination effects in strain compensated quantum dot stacks embedded in solar cells. Appl. Phys. Lett. **93**, 123114 (2008)

15. Luque, A., Marti, A.: Increasing the efficiency of ideal solar cells by photon induced transitions at intermediate levels. Phys. Rev. Lett. **78**, 5014–5017 (1997)
16. Nozik, A.J.: Quantum dot solar cells. Phys. E. **14**, 115–120 (2002)
17. Bicais-Lepinay, N., Andre, F., Pantel, R., Jullian, S., Margain, A., Kwakman, L.F.T.: Lift-out techniques coupled with advanced TEM characterization methods for electrical failure analysis. Microelectron. Reliab. **42**, 1747–1752 (2002)
18. Mayer, J., Giannuzzi, L.A., Kamino, T., Michael, J.: TEM sample preparation and FIB-induced damage. MRS Bull. **32**, 400–407 (2007)
19. Langford, R.M., Rogers, M.: In situ lift-out: steps to improve yield and a comparison with other FIB TEM sample preparation techniques. Micron **39**, 1325–1330 (2008)
20. Tomus, D., Ng, H.P.: In situ lift-out dedicated techniques using FIB-SEM system for TEM specimen preparation. Micron **44**, 115–119 (2013)
21. Langford, R.M., Clinton, C.: In situ lift-out using a FIB-SEM system. Micron **35**, 607–611 (2004)
22. Li, J., Malis, T., Dionne, S.: Recent advances in FIB-TEM specimen preparation techniques. Mater. Charact. **57**, 64–70 (2006)
23. Floresca, H.C., Jeon, J., Wang, J.G.G., Kim, M.J.: The focused ion beam fold-out: sample preparation method for transmission electron microscopy. Microsc. Microanal. **15**, 558–563 (2009)
24. Van Dorp, W.F., Hagen, C.W.: A critical literature review of focused electron beam induced deposition. J. Appl. Phys. **104**, 081301 (2008)
25. Park, B.C., Park, Y.C., Lee, H.J., Kim, Y.H.: Transmission electron microscopy study of damage layer formed through ion beam induced deposition of platinum on silicon substrate. J. Vac. Sci. Technol. B. Microelectron. Nanometer. Struct. Process. Meas. Phenom. **28**, C6F31–C36F37 (2010)
26. Kato, N.I., Miura, N., Tsutsui, N.: A plasma-polymerized protective film for transmission electron microscopy specimen preparation by focused ion beam etching. J. Vac. Sci. Technol. A **16**, 1127–1130 (1998)
27. Menzel, R., Bachmann, T., Wesch, W.: Physical sputtering of III-V-semiconductors with a focused Ga+-beam. Nucl. Instrum. Methods. Phys. Res. Sect. B-Beam Interact. Mater. Atoms. **148**, 450–453 (1999)
28. Melmed, A.J.: The art and science and other aspects of making sharp tips. J. Vac. Sci. Technol. B **9**, 601–608 (1991)
29. Hagedorn, T., Ouali, M.E., Paul, W., Oliver, D., Miyahara, Y., Grütter, P.: Refined tip preparation by electrochemical etching and ultrahigh vacuum treatment to obtain atomically sharp tips for scanning tunneling microscope and atomic force microscope. Rev. Sci. Instrum. **82**, 113903 (2011)
30. Larson, D.J., Wissman, B.D., Martens, R.L., Viellieux, R.J., Kelly, T.F., Gribb, T.T., Erskine, H.F., Tabat, N.: Advances in atom probe specimen fabrication from planar multilayer thin film structures. Microsc. Microanal. **7**, 24–31 (2001)
31. Morris, R.A., Martens, R.L., Zana, I., Thompson, G.B.: Fabrication of high-aspect ratio Si pillars for atom probe 'lift-out' and field ionization tips. Ultramicroscopy **109**, 492–496 (2009)
32. Duguay, S., Colin, A., Mathiot, D., Morin, P., Blavette, D.: Atomic-scale redistribution of dopants in polycrystalline silicon layers. J. Appl. Phys. **108**, 034911 (2010)
33. Gordon, L.M., Tran, L., Joester, D.: Atom probe tomography of apatites and bone-type mineralized tissues. ACS. Nano. **6**, 10667–10675 (2012)
34. Narayan, K., Prosa, T.J., Fu, J., Kelly, T.F., Subramaniam, S.: Chemical mapping of mammalian cells by atom probe tomography. J. Struct. Biol. **178**, 98–107 (2012)
35. Larson, D.J., Foord, D.T., Petford-Long, A.K., Anthony, T.C., Rozdilsky, I.M., Cerezo, A., Smith, G.W.D.: Focused ion-beam milling for field-ion specimen preparation: preliminary investigations. Ultramicroscopy **75**, 147–159 (1998)
36. Miller, M.K., Russell, K.F., Thompson, G.B.: Strategies for fabricating atom probe specimens with a dual beam FIB. Ultramicroscopy **102**, 287–298 (2005)
37. Saghi, Z., Gnanavel, T., Peng, Y., Inkson, B.J., Cullis, A.G., Gibbs, M.R., Mobus, G.: Tomographic nanofabrication of ultrasharp three-dimensional nanostructures. Appl. Phys. Lett. **93**, 153102 (2008)

38. Kawase, N., Kato, M., Nishioka, H., Jinnai, H.: Transmission electron microtomography without the "missing wedge" for quantitative structural analysis. Ultramicroscopy **107**, 8–15 (2007)
39. Kato, M., Kawase, N., Kaneko, T., Toh, S., Matsumura, S., Jinnai, H.: Maximum diameter of the rod-shaped specimen for transmission electron microtomography without the "missing wedge". Ultramicroscopy **108**, 221–229 (2008)
40. Ke, X.X., Bals, S., Cott, D., Hantschel, T., Bender, H., Van Tendeloo, G.: Three-dimensional analysis of carbon nanotube networks in interconnects by electron tomography without missing wedge artifacts. Microsc. Microanal. **16**, 210–217 (2010)
41. Yaguchi, T., Konno, M., Kamino, T., Watanabe, M.: Observation of three-dimensional elemental distributions of a Si device using a 360 degrees-tilt FIB and the cold field-emission STEM system. Ultramicroscopy **108**, 1603–1615 (2008)
42. Jarausch, K., Thomas, P., Leonard, D.N., Twesten, R., Booth, C.R.: Four-dimensional STEM-EELS: enabling nano-scale chemical tomography. Ultramicroscopy **109**, 326–337 (2009)
43. Orloff, J.: High-resolution focused ion-beams. Rev. Sci. Instrum. **64**, 1105–1130 (1993)
44. Kato, N.I.: Reducing focused ion beam damage to transmission electron microscopy samples. J. Electron Microsc. **53**, 451–458 (2004)
45. Ishitani, T., Yaguchi, T.: Cross-sectional sample preparation by focused ion beam: a review of ion-sample interaction. Microsc. Res. Tech. **35**, 320–333 (1996)
46. Rubanov, S., Munroe, P.R.: Damage in III-V compounds during focused ion beam milling. Microsc. Microanal. **11**, 446–455 (2005)
47. McKenzie, W.R., Quadir, M.Z., Gass, M.H., Munroe, P.R.: Focused ion beam implantation of diamond. Diamond. Relat. Mater. **20**, 1125–1128 (2011)
48. Roediger, P., Wanzenboeck, H.D., Waid, S., Hochleitner, G., Bertagnolli, E.: Focused-ion-beam-inflicted surface amorphization and gallium implantation-new insights and removal by focused-electron-beam-induced etching. Nanotechnology **22**, 235302 (2011)
49. Larson, D.J., Foord, D.T., Petford-Long, A.K., Liew, H., Blamire, M.G., Cerezo, A., Smith, G.D.W.: Field-ion specimen preparation using focused ion-beam milling. Ultramicroscopy **79**, 287–293 (1999)
50. Cooper, D., Ailliot, C., Barnes, J.P., Hartmann, J.M., Salles, P., Benassayag, G., Dunin-Borkowski, R.E.: Dopant profiling of focused ion beam milled semiconductors using off-axis electron holography; reducing artifacts, extending detection limits and reducing the effects of gallium implantation. Ultramicroscopy **110**, 383–389 (2010)
51. Jin, H.H., Shin, C., Kwon, J.: Fabrication of a TEM sample of ion-irradiated material using focused ion beam microprocessing and low-energy Ar ion milling. J. Electron Microsc. **59**, 463–468 (2010)
52. Miyajima, N., Holzapfel, C., Asahara, Y., Dubrovinsky, L., Frost, D.J., Rubie, D.C., Drechsler, M., Niwa, K., Ichihara, M., Yagi, T.: Combining FIB milling and conventional Argon ion milling techniques to prepare high-quality site-specific TEM samples for quantitative EELS analysis of oxygen in molten iron. J. Microsc. **238**, 200–209 (2010)
53. Schaffer, M., Schaffer, B., Ramasse, Q.: Sample preparation for atomic-resolution STEM at low voltages by FIB. Ultramicroscopy **114**, 62–71 (2012)
54. Montoya, E., Bals, S., Rossell, M.D., Schryvers, D., van Tendeloo, G.: Evaluation of top, angle, and side cleaned FIB samples for TEM analysis. Microsc. Res. Tech. **70**, 1060–1071 (2007)
55. Lozano-Perez, S.: A guide on FIB preparation of samples containing stress corrosion crack tips for TEM and atom-probe analysis. Micron **39**, 320–328 (2008)
56. Yabuuchi, Y., Tametou, S., Okano, T., Inazato, S., Sadayama, S., Yamamoto, Y., Iwasaki, K., Sugiyama, Y.: A study of the damage on FIB-prepared TEM samples of AlxGa1-xAs. J. Electron Microsc. **53**, 471–477 (2004)
57. Ziegler, J.F., Biersack, J.P.: SRIM the stopping and range of ions in matter. http://www.srim.org
58. Jin-Phillipp, N.Y., Phillipp, F.: Strain distribution in self-assembled InP/GaInP quantum dots. J. Appl. Phys. **88**, 710–715 (2000)

59. Gutierrez, M., Herrera, M., Gonzalez, D., Garcia, R., Hopkinson, M.: Role of elastic anisotropy in the vertical alignment of In(Ga)As quantum dot superlattices. Appl. Phys. Lett. **88**, 193118 (2006)
60. Shoji, Y., Oshima, R., Takata, A., Okada, Y.: The effect of spacer layer thickness on vertical alignment of InGaAs/GaNAs quantum dots grown on GaAs(3 1 1)B substrate. Phys. E. **42**, 2768–2771 (2010)
61. Molina, S.I., Ben, T., Sales, D.L., Pizarro, J., Galindo, P.L., Varela, M., Pennycook, S.J., Fuster, D., Gonzalez, Y., Gonzalez, L.: Determination of the strain generated in InAs/InP quantum wires: prediction of nucleation sites. Nanotechnology **17**, 5652–5658 (2006)

# Chapter 10
# Fabrication Technique of Deformation Carriers (Gratings and Speckle Patterns) with FIB for Microscale/Nanoscale Deformation Measurement

**Yanjie Li, Huimin M. Xie, Qinghua H. Wang, and Zhanwei W. Liu**

**Abstract** The fabrication technique of deformation carriers with focused ion beam (FIB) milling for microscale/nanoscale deformation measurement has been studied in this chapter. The deformation carriers refer to microscale/nanoscale gratings and speckle patterns, which are indispensable elements for moiré and digital image correlation (DIC) method under scanning electron microscope (SEM) respectively. The fabrication principle, the design, and the influencing factors of these two kinds of deformation carriers are studied respectively. Their successful applications to microscale/nanoscale deformation measurement are also presented, which demonstrate that the deformation carriers combined with photo-mechanics techniques under SEM are effective tools for microscale/nanoscale deformation measurement.

Y. Li
School of Civil Engineering and Architecture, University of Jinan, Jinan 250022, China

AML, Department of Engineering Mechanics, Tsinghua University, Beijing 100084, China
e-mail: yanjie.li.01@gmail.com

H.M. Xie (✉)
AML, Department of Engineering Mechanics, Tsinghua University, Beijing 100084, China
e-mail: xiehm@mail.tsinghua.edu.cn

Q.H. Wang
National Institute for Materials Science, 1-2-1 Sengen, Tsukuba, Ibaraki 305-0047, Japan
e-mail: WANG.Qinghua@nims.go.jp

Z.W. Liu
School of Aerospace Engineering, Beijing Institute of Technology, Beijing 100081, China
e-mail: liuzw@bit.edu.cn

Z.M. Wang (ed.), *FIB Nanostructures*, Lecture Notes in Nanoscale Science and Technology 20, DOI 10.1007/978-3-319-02874-3_10,

## 10.1 Introduction

Micro-electro-mechanical systems (MEMS) have drawn great attention in recent decades due to their promising applications in many industrial fields, including information and communication technologies, automotive industry, biomedicine, electronics, aerospace, and so on. It is interdisciplinary involving optics, materials, microelectronics, and mechanics. As mechanics researchers, we focus on the characterization of their mechanical properties, which are important factors affecting not only their performance but also the basis to analyze their reliabilities. Estimation of fundamental mechanical property requires that displacements and strains be measured at micron and submicron length scale [1–5], which means traditional macroscopic experimental testing methods like moiré method and digital image correlation (DIC) method are no longer applicable and it is urgent to develop novel microscopic testing methods.

Microscopic deformation measurement lies in improving the spatial resolution and displacement sensitivity. Fortunately, the development of high-resolution microscopes, especially scanning electron microscope (SEM), provides a good platform for microscopic deformation measurement. As widely applied experimental techniques, moiré method and DIC method can be effectively combined with SEM, which not only improves the spatial resolution and displacement sensitivity but also maintains the advantage of noncontact full field measurement. However, both methods need deformation carriers fabricated on specimen surfaces. For moiré method, gratings are indispensable and for DIC method, random speckle patterns are necessary.

1. In order to measure microscopic deformation with SEM moiré method, high frequency gratings are required to be fabricated on specimen surface. To the best of the authors' knowledge, three options are available so far to fabricate gratings with pitches at submicron scale, including holographic photolithography [6, 7], electron beam lithography (EBL) [8–10], and focused ion beam (FIB) milling [11–14]. Holographic photolithography consists of coating photoresist, exposure, developing, depositing metal film, and removing the rest of photoresist, which is a little cumbersome. This technique can produce large area gratings, usually dozens of millimeters, with high efficiency, but also demands high precision optics setup and good performance of vibration isolation. EBL is almost the same as photolithography except two differences. First, the exposure source is electron beam instead of laser beam. Second, the grating pattern is formed by pattern generator instead of the interference of two laser beams. Compared to the above two methods, FIB milling is a more straightforward technique (maskless and resistless), which directly writes gratings on a specific region of the specimen surface. Both FIB and EBL allow producing gratings with frequencies up to 10,000 lines/mm, but neither method is suited for large scale patterning since they are time-consuming and of low throughput due to their serial nature. Therefore, the grating made by EBL and FIB is more suitable to measure microscopic deformation. For microstructures commonly used in MEMS like free standing films, FIB is no doubt a better choice as photoresist coating and developing probably destroy the device if EBL is used.

Some successful applications of FIB milling to fabricate high frequency gratings have been reported. Xie and Li [11] fabricated gratings of 5,000 lines/mm on polysilicon micro-cantilever as deformation carrier to measure its residual deformation. Yan et al. [12] reported the fabrication of nano-gratings on silicon-on-insulator based cantilever. Shang and Xie et al. [13] realized the real-time observation of the grating fabricated on polysilicon cantilever by FIB as the temperature elevated from room temperature to 300 °C at intervals of 50 °C. Du and Xie et al. [14] applied this technique to porous Ti-Ni shape memory alloy. However, this fabrication technique has not been studied thoroughly and comprehensively. How to design a grating with a proper frequency and an appropriate size has not been clarified yet.

2. SEM-DIC requires the presence of a randomly patterned material surface and the random pattern must have a recognizable and resolvable structure at an appropriate scale for the interested problem. Although the fabrication of random patterns encounters a great challenge due to the tiny size of material at reduced length scale, some different patterning techniques have been developed for each length scale. Sutton et al. [15] developed chemical vapor exposure and UV lithography to fabricate random patterns for nanoscale and microscale measurement respectively. Berfield et al. [16] created random speckle patterns with a fine point air-brush for microscale measurement and fluorescent nanoparticles for nanoscale measurement. Tanaka et al. [17] utilized EBL to make random patterns for microscale/nanoscale measurement by backscattered electron imaging (BSEI) under field emission SEM. Liu et al. [18] made use of FIB milling for fabrication of periodic micro-markers on a MEMS device. Korsunsky and Sebastiani et al. [19–21] applied FIB milling to fabricate periodic holes as speckle patterns for nanoscale measurement. Sabate et al. [1, 22–24] acquired speckle patterns by deposition of 20 nm pt layer with FIB for nanoscale measurement. With the use of FIB, Kregting et al. [25] first defined a rectangular grid in spacing of 250 nm and then randomly positioned markers within a square of 150 nm size around each grid point. Kammers and Daly [26] provided a survey of small-scale patterning methods for SEM-DIC.

   Each patterning technique has its own advantages and limitations. For the microscale/nanoscale measurement of micro-devices like MEMS, FIB is an ideal choice for its direct milling or depositing capability. It should be noted that due to the disadvantage of time consumption, large-area pattern fabrication is impossible for FIB. However, this disadvantage is negligible for micro-device and micro-structure. Although FIB milling has been used to fabricate speckle patterns at reduced length scale, this fabrication technique itself has not been studied thoroughly. How to design an optimized template and an appropriate pattern size is not clear yet.

This chapter aims at providing guidance on the fabrication and the application of the deformation carriers including gratings and speckle patterns fabricated with FIB milling. The principle of fabricating deformation carriers with FIB milling is introduced. The design of deformation carriers is emphasized. The fabrication influencing factors are analyzed. To check out their high temperature performances, high temperature tests using fabricated deformation carriers are performed. The deformation carriers have been applied successfully to measure microscopic deformation, which demonstrate that the deformation carriers fabricated with FIB milling combined with SEM moiré method or SEM-DIC method are effective tools for microscopic deformation measurement.

## 10.2 Fabrication Technique of Gratings with FIB

### *10.2.1 Principle of Grating Fabrication with FIB*

FIB technologies have been widely applied as powerful tools in the semiconductor industry mainly for micro/nano-structuring, mask repairing, device modification, failure analysis, and integrated circuit debugging [27, 28]. Here we extend the application of FIB to fabricate specimen grating, which offers a new tool to fabricate gratings for moiré method. By accelerating concentrated ions to a specific site, FIB milling can etch off any exposed material, leaving a very clean hole, as shown in Fig. 10.1.

Dual beam systems (SEM and FIB) are used in our study. The flowchart of grating fabrication is shown in Fig. 10.2. The first step is to design grating patterns and the details are described in Sect. 10.2.2. Parallel gratings are for one dimensional

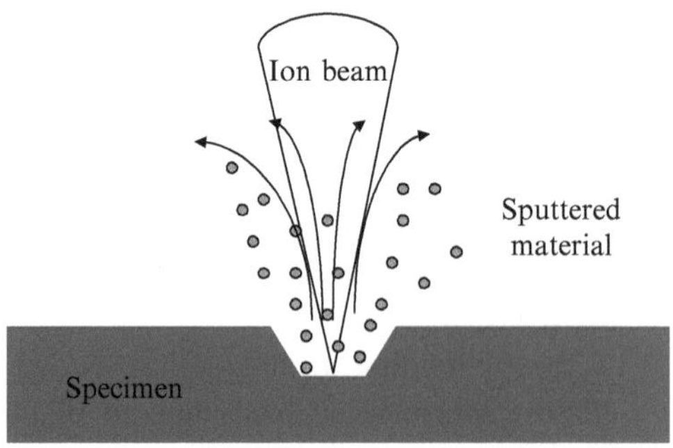


**Fig. 10.1** The diagram of FIB milling [29]

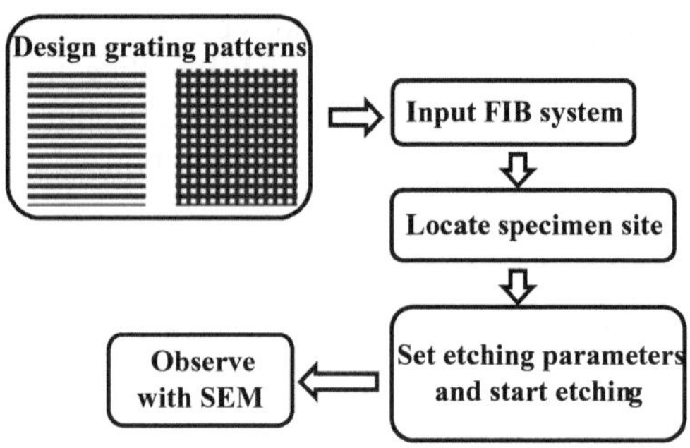


**Fig. 10.2** The flowchart of grating fabrication [29]

deformation measurement and cross gratings are for two dimensional deformation. Then input the designed grating patterns into the FIB system and locate the site of interest on the specimen. Set etching parameters, such as the ion beam current and etching depth, and start etching. Finally, observe the fabricated grating with SEM.

## 10.2.2 Design of FIB Gratings

The design of the grating patterns is the first and most critical step during the process of grating fabrication, including designing the grating frequency and the grating area at the same time. It overcomes the shortcoming of the traditional practice that only takes into account the grating frequency.

Denote $L_{vx}$ and $L_{vy}$ as the dimensions of the view field along axes $x$ and $y$, $L_{sx}$ and $L_{sy}$ as the dimensions of the sample along axes $x$ and $y$, $L_{gx}$ and $L_{gy}$ as the dimensions of the grating area along axes $x$ and $y$. In order to design a specimen grating with a proper frequency and a proper size, a process flow is devised as follows [30]:

1. Design the frequency of the specimen grating $f_s$ according to the displacement measurement accuracy;
2. Calculate the matched magnification $M_0$;
3. Calculate $L_{vx}$ and $L_{vy}$;
4. Compare the size of the sample to the size of the view field

   (a) If $L_{sx} \ll L_{vx}$ or $L_{sy} \ll L_{vy}$, enhance the magnification until $L_{sx} \geq L_{vx}/10$ and $L_{sy} \geq L_{vy}/10$ and recalculate $f_s$ according to the magnification, then the size of grating falls into three categories:

   If $L_{sx} \leq L_{vx}$ and $L_{sy} \leq L_{vy}$, $L_{gx} = L_{sx}$, $L_{gy} = L_{sy}$
   If $L_{sx} \leq L_{vx}$ and $L_{sy} > L_{vy}$, $L_{gx} = L_{sx}$, $L_{gy} = L_{vy}$
   If $L_{sx} > L_{vx}$ and $L_{sy} \leq L_{vy}$, $L_{gx} = L_{vx}$, $L_{gy} = L_{sy}$

   (b) If $L_{sx} \gg L_{vx}$ and $L_{sy} \gg L_{vy}$, then the size of the grating should satisfy that $L_{vy}/10 \leq L_{gy} \leq L_{vy}$ and $L_{vx}/10 \leq L_{gx} \leq L_{vx}$.
   (c) If $L_{sx} \approx L_{vx}$ and $L_{sy} \approx L_{vy}$, $L_{gx} = \min(L_{sx}, L_{vx})$, $L_{gy} = \min(L_{sy}, L_{vy})$

## 10.2.3 Influencing Factors of Grating Fabrication with FIB

### 10.2.3.1 Roughness of Specimen Surface

A flat plane is indispensable to fabricate high quality specimen gratings. Generally, except silicon, silicon-based materials, MEMS devices and thin films, most of the specimen surface especially metal materials need polishing. First polish the specimen surface with sand papers (400#, 600#, 800#, 1,000#, 1,500# and 2,000# in sequence) along two perpendicular directions alternatively and then with w0.5

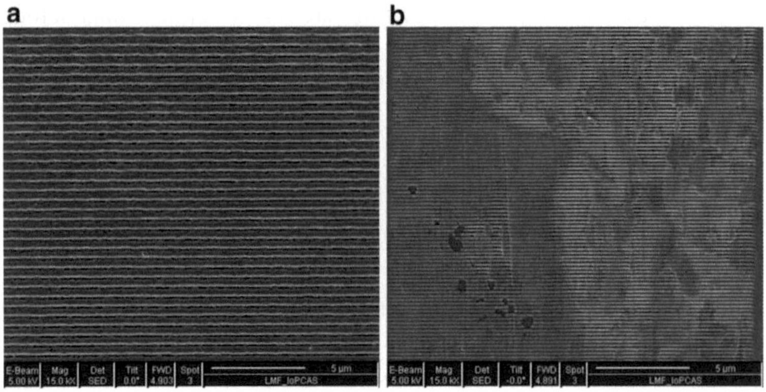


**Fig. 10.3** Gratings fabricated on the surface of stainless steel with different levels of roughness (**a**) low roughness, Ra=0.89 nm [30] and (**b**) high roughness, Ra=1.29 nm [31]

**Fig. 10.4** Diagram of grating structure

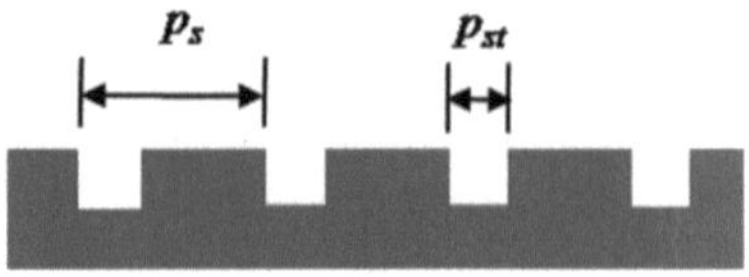


diamond compounds. Remember to wash the specimen surface with water before each replacement. Finally observe the specimen surface under microscope to ensure there is little or no scratch. Only if the specimen is polished to a mirror-like plane, can top quality gratings be made. As shown in Fig. 10.3, the grating on stainless steel with low level surface roughness is obviously better than that on the other specimen with high level surface roughness.

#### 10.2.3.2 Opening Ratio of Specimen Grating

The opening ratio of a grating is the ratio between the width of the grating trench ($p_{st}$) and the grating pitch ($p_s$), see Fig. 10.4, which directly influences the morphology of SEM moiré fringes. A group of gratings with different opening ratios have been fabricated on a silicon wafer, as shown in Fig. 10.5. The grating frequency is 2,000 lines/mm, which means the grating pitch is 0.5 μm. The setting depth is 0.2 μm, the width of grating area is 15 μm, the ion beam current is 500 pA, and the magnification of the FIB system is 5,000×. The corresponding SEM moiré fringes are presented in Fig. 10.6. It can be found that with a predefined frequency, the smaller the opening ratio is, the thinner dark fringes are. In general, thinner fringes are favorable. Therefore, a grating of smaller opening ratio will result in moiré fringes with higher quality. The smallest opening ratio corresponds to no-setting case, when the grating trench width is only determined by the width of ion beam current. Furthermore, it also consumes the least time, see Table 10.1.

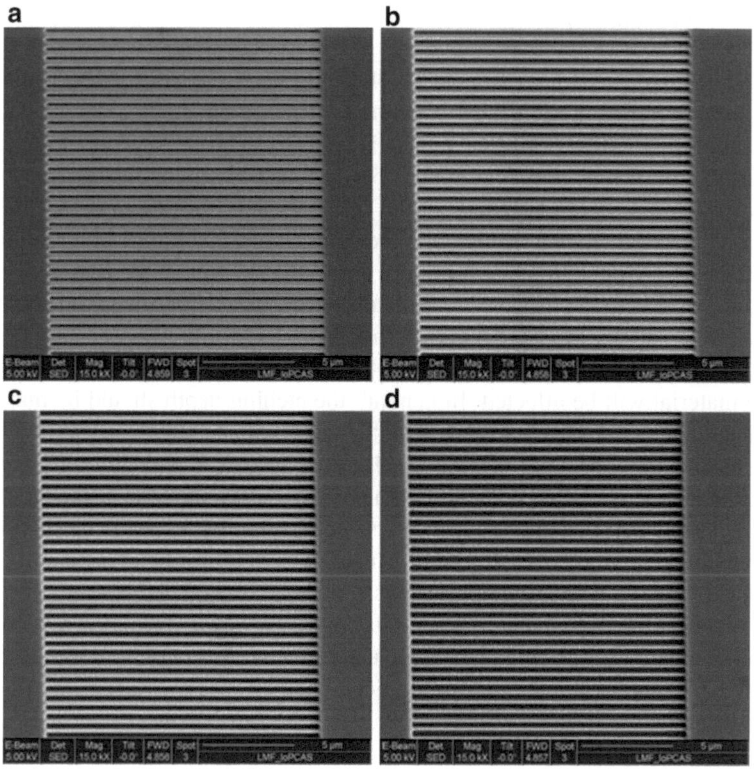


**Fig. 10.5** Gratings with different opening ratios (**a**)1/5, (**b**) 2/5, (**c**) 1/2, and (**d**) 3/5 [31]

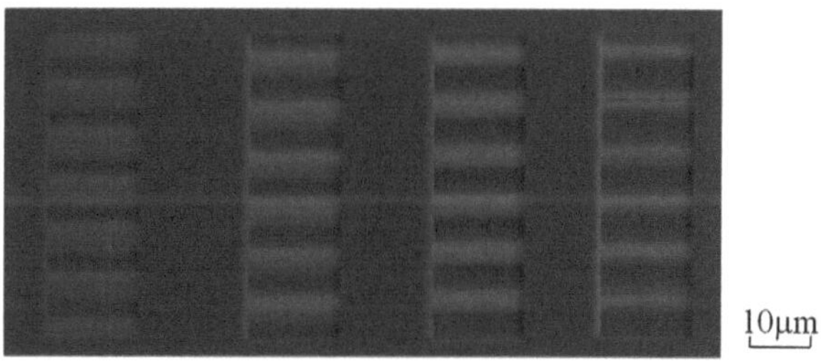


**Fig. 10.6** SEM moiré fringes generated by specimen gratings in Fig. 10.5, whose opening ratios are 1/5, 2/5, 1/2, and 3/5 respectively from *left* to *right* [31]

**Table 10.1** Time consumed by gratings with different opening ratios [31]

| Opening ratio | Time (s) |
|---|---|
| No setting | 220 |
| 1/5 | 446 |
| 2/5 | 834 |
| 1/2 | 1,057 |
| 3/5 | 1,246 |

#### 10.2.3.3 Etching Depth

The etching depth influences the contrast of the grating itself, which further influences the contrast of SEM moiré fringes. Gratings with different etching depths are illustrated in Fig. 10.7. The grating frequency is 2,000 lines/mm, the width of the grating area is 15 μm, the ion beam current is 500 pA and the magnification of the FIB system is 2,500×. It can be seen that the contrast of the grating increases with the increase of setting etching depth while the contrast of SEM moiré fringes increases with the increase of setting etching depth until 0.4 μm (Fig. 10.8). This implies that high quality SEM moiré fringes can be obtained when the etching depth is appropriate, neither too small nor too large. On the other hand, FIB milling itself is a destructive fabrication method. If the etching depth is too large, the mechanical property of the tested material will be affected. In general, the etching depth should be much less than the thickness of material, to be specific, their ratio should be smaller than 1/20. At the same time, a larger etching depth will consume longer time as shown in Fig. 10.9 and the cost will be increased. In order to reduce the cost, a smaller etching depth is preferable. In summary, the optimized etching depth should be shallow but ensure enough contrast of SEM moiré fringes. The optimized etching depth can be obtained by repeated trials and it is always varied with materials.

In practice, the etching depth is set in advance. However, due to the redeposition of sputtered atoms, the actual depth is usually much lower than the assumed value. Figure 10.10 shows the atomic force microscope (AFM) measurement results of gratings in Fig. 10.7.

#### 10.2.3.4 Ion Beam Current

The ion beam current is an important parameter that determines processing accuracy and rate [28]. If a higher milling depth or a fast milling rate is required, a higher-current ion beam is preferred. However, the higher-current ion beam results in a coarser definition of the milled structure and a larger amount of debris. Experiments demonstrate that higher quality of gratings can be fabricated when the ion beam current is between 100 and 800 pA.

As a result of the primarily Gaussian distribution of the current density of the ion beam, the grating patterns naturally have sinusoidal profiles, as shown in Fig. 10.11.

#### 10.2.3.5 Scan Mode

There are two types of scan modes for FIB milling: parallel mode and serial mode. In parallel mode, all periodic spots are etched in the whole expected region during each pass, and this procedure is repetitively carried out until the expected depth is realized, as shown in Fig. 10.12a, therefore the redeposition is proportionally reduced in each pass and a portion of the redeposition from the earlier passes is removed by the subsequent milling. In serial mode, one periodic spot is repetitively

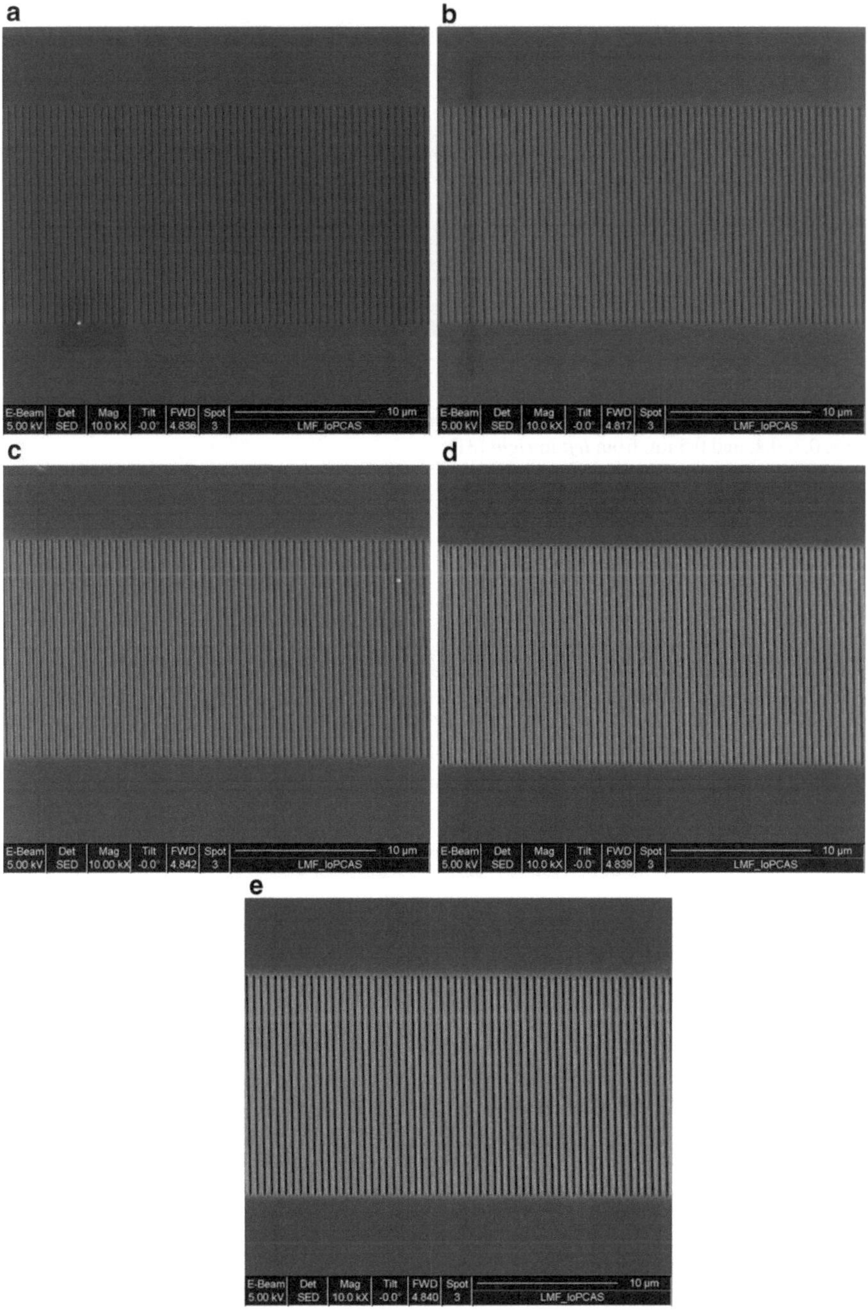


**Fig. 10.7** Gratings with different setting etching depths (**a**) 0.1 μm, (**b**) 0.2 μm, (**c**) 0.3 μm, (**d**) 0.4 μm, and (**e**) 0.5 μm [31]

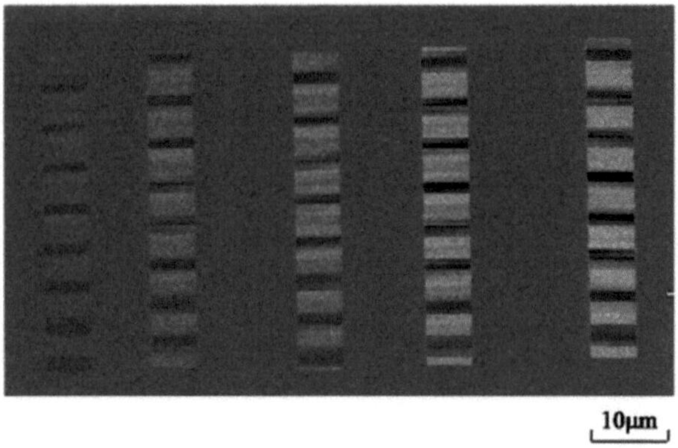


**Fig. 10.8** SEM moiré fringes generated by gratings in Fig. 10.7, the setting etching depths are 0.1, 0.2, 0.3, 0.4, and 0.5 μm from *left* to *right* [31]

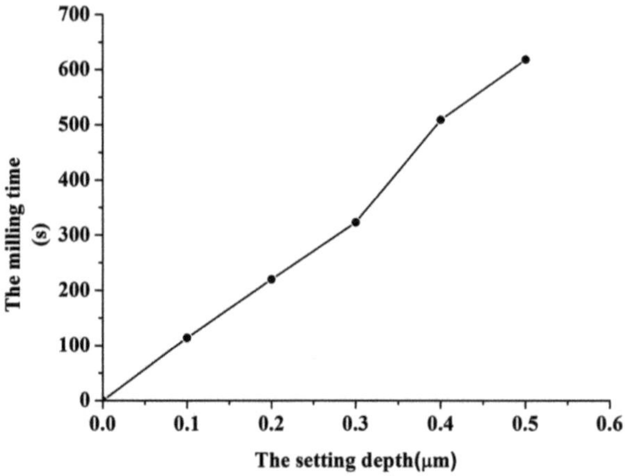


**Fig. 10.9** Milling time vs. setting depth [31]

etched until the expected depth is realized, and then the next periodic spot, as shown in Fig. 10.12b; therefore, sputtered materials are continuously redeposited into the region milled earlier, and the redeposited materials accumulate more on the regions milled earlier. In order to obtain flat and vertical channels, parallel mode is recommended.

#### 10.2.3.6 Magnification of the FIB System

In the fabrication process, the maximum grating area is inversely proportional to the magnification of the FIB system. Since the grating length is usually designed to be larger than or equal to its width, here the relation curve between the maximum grating length and the magnification of the FIB system of FEI DB 235 is presented in Fig. 10.13. However, one frequency can be manufactured under different magnifications.

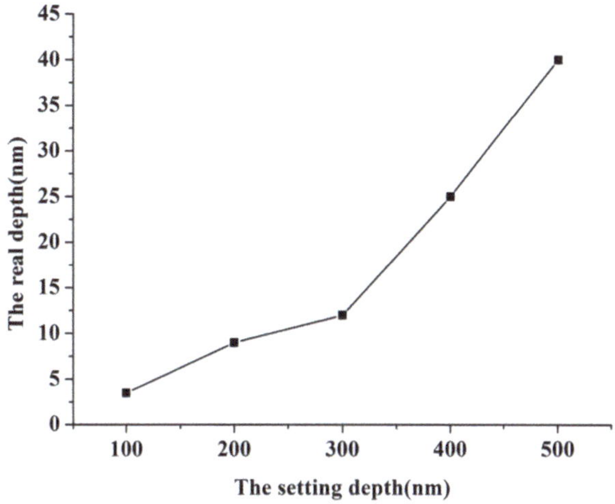


**Fig. 10.10** Real depth measured by AFM vs. setting depth [31]

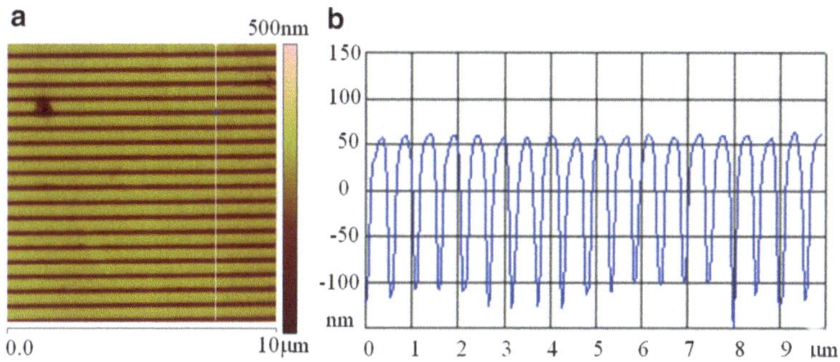


**Fig. 10.11** The morphology of a grating under AFM (**a**) the whole area and (**b**) the depth along the *white line*, the grating area is 10×10 μm [31]

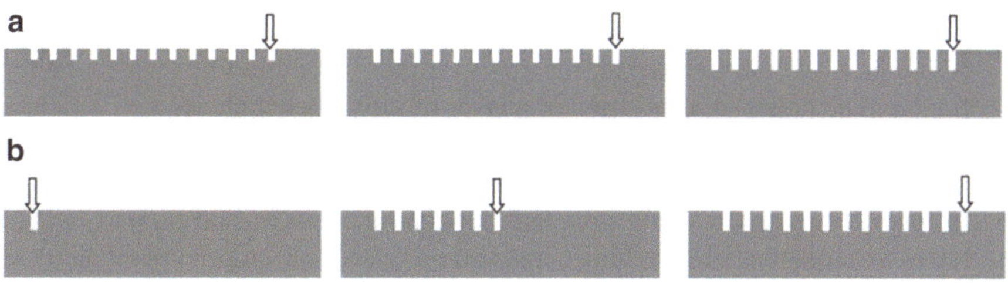


**Fig. 10.12** Scan modes of FIB milling (**a**) parallel mode and (**b**) serial mode [31]

For example, to fabricate a grating in length of 150 μm, any magnification less than 2,000× can be chosen. However, the higher the magnification is, the higher the resolution is and the better the quality of the grating is. Therefore, the highest magnification is the best choice.

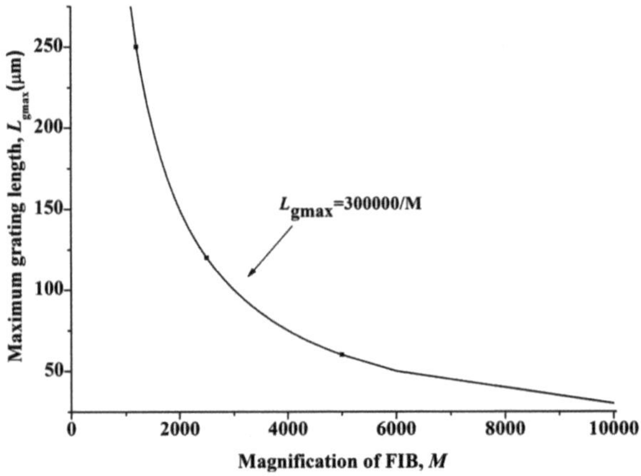


**Fig. 10.13** The maximum length of a grating vs. magnification of the FIB system of FEI DB235 [31]

#### 10.2.3.7 Mosaic

Mosaic is one of schemes to get large area gratings. However, we do not advocate mosaic gratings since imperfect alignment will lead to a nonuniform initial moiré and make the subsequent calculation complicated. As shown in Fig. 10.14, the initial moiré is nonuniform and the mosaic position can be observed clearly for the mosaic grating while the moiré is uniform for the other one without mosaic. By increasing the number of the grating lines and etching gratings under a lower magnification, mosaic can be avoided.

#### 10.2.3.8 Discussions on the Influencing Factors

1. A flat plane with a surface roughness of low level is indispensable to fabricate high quality specimen gratings. Fortunately, this requirement can be met by polishing in most cases. However, for tiny specimens like thin films which are too fragile to be polished, deposition of a layer of Pt is a better choice.
2. A smaller opening ratio of gratings results in a better quality of SEM moiré fringes, and the opening ratio is smallest in non-setting case.
3. In order to save the etching time and maintain the mechanical property of the tested material itself, the etching depth should be shallow on condition that it can ensure enough contrast of SEM moiré fringes. The optimized etching depth is usually obtained by repeated trials and varied with materials.
4. Repeated experiments demonstrate that higher quality gratings can be fabricated when the ion beam current is between 100 and 800 pA. Higher-current ion beam results in a coarser definition of the milled structure and lower-current will reduce the etching rate.

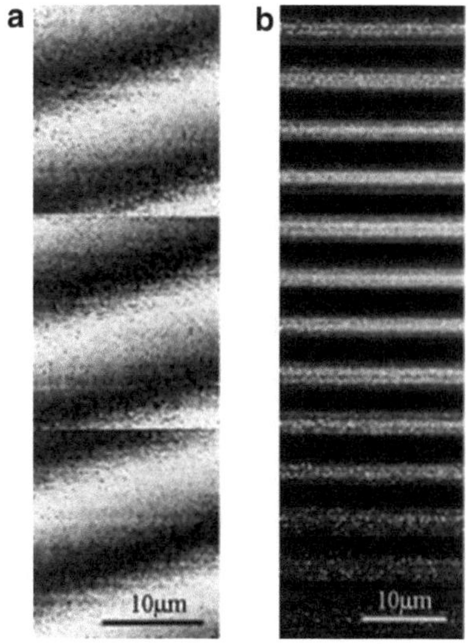


**Fig. 10.14** Moiré fringes generated by gratings with (**a**) and without mosaic (**b**) [30]

**Table 10.2** Influences of the ion beam current and the etching depth on the grating fabrication [31]

| Parameter | Magnitude | Etching time | Grating quality |
|---|---|---|---|
| Ion beam current | Large | Short | Bad |
| | Small | Long | Good |
| Etching depth | Large | Long | Good |
| | Small | Short | Bad |

5. Parallel mode is recommended to reduce the influence of material redeposition.
6. The magnification of the FIB system is inversely proportional to the maximum grating area. Therefore, the former is derived according to the designed value of the latter.
7. Mosaic can be avoided by increasing the number of the grating lines and etching gratings under a lower magnification.

It can be found that some parameters have been determined through our above mentioned analysis or during the design process. Thus, in practical fabrication process, only the ion beam current and the etching depth need to be optimized through repeated trials. The influences of the ion beam current and the etching depth to the grating fabrication are listed in Table 10.2.

### *10.2.4 Examples of Gratings Fabricated with FIB*

FIB is a widely used microscale/nanoscale machining tool, which can fabricate high quality gratings on the surface of various materials, as shown in Fig. 10.15. The grating fabrication parameters are listed in Table 10.3.

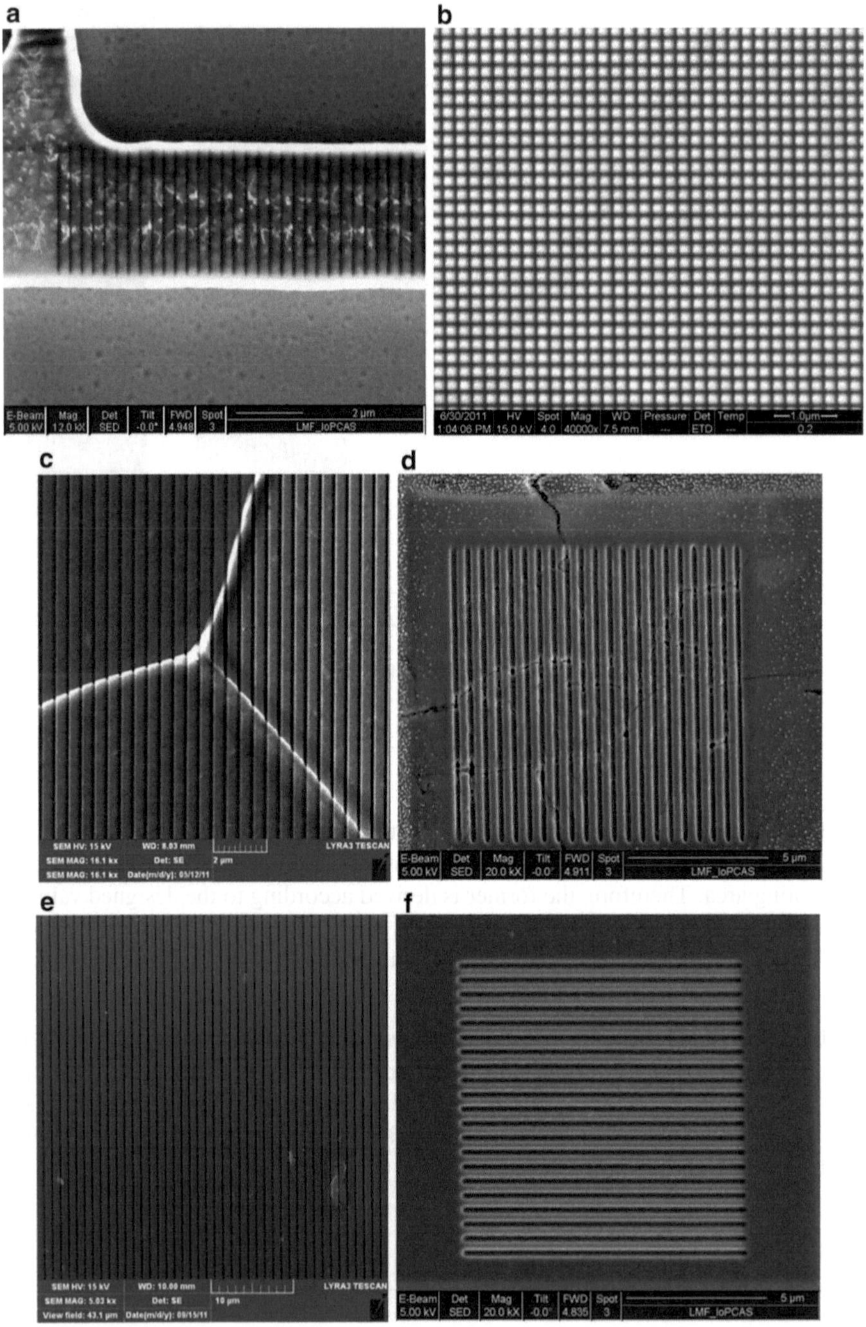


**Fig. 10.15** Gratings fabricated with FIB on different materials (**a**) amorphous SiC [31], (**b**) low-carbon steel [31], (**c**) low carbon steel [29], (**d**) thermal barrier coating [31], (**e**) optical fiber [31], and (**f**) silicon wafer [31]

**Table 10.3** The fabrication parameters of gratings in Fig. 10.15 [31]

| Number | Material | Grating type | Grating frequency (lines/mm) | Ion beam current (pA) | Etching depth (μm) |
|---|---|---|---|---|---|
| (a) | Amorphous SiC | Parallel | 3,000 | 138 | 0.2 |
| (b) | Low-carbon steel | Cross | 5,000 | 119 | 0.2 |
| (c) | Low-carbon steel | Parallel | 2,000 | 139 | 1 |
| (d) | Thermal barrier coating | Parallel | 2,000 | 739 | 1 |
| (e) | Optical fiber | Parallel | 1,000 | 123 | 0.1 |
| (f) | Silicon | Parallel | 2,000 | 460 | 0.4 |

## *10.2.5 High Temperature Tests*

Since the grating is by directly fabricated on the specimen surface with FIB, it is natural to conclude that if the specimen is stable under high temperature, then the grating is reliable to be used to measure high temperature deformation. To check out the performance of the grating under high temperature, the verification experiments are carried out with amorphous SiC micro-beam and thermal barrier coating.

### 10.2.5.1 Amorphous SiC Micro-beam [30]

A 2,000 lines/mm parallel grating is fabricated on the amorphous silicon carbide specimen surface with FEI DB235, as shown in Fig. 10.16a. The fabrication parameters are listed in Table 10.4.

Put the sample on the SEM stage (Shimadzu S-550) to record the gratings and the moiré fringes before and after high temperature, as shown in Figs. 10.16 and 10.17. It can be seen that although the morphology of the grating is not very clear after high temperature of 350 and 550 °C, the moiré is still distinguishable with a contrast degradation, which indicates that the FIB milled grating can suffer high temperature at least up to 550 °C. However, the pitch of the moiré fringes is altered after high temperature, which means that residual strain exists. The residual strain is probably attributed to the thermal mismatch between the micro-beam, the sacrificial layer and the Si substrate. The dirty things in Fig. 10.17c are caused by careless scratch. The new material in Fig. 10.16b, c is probably the product of the high temperature chemical reaction of sputtered gold with atmosphere.

### 10.2.5.2 Thermal Barrier Coating

The tested thermal barrier coating is zirconia thin film in thickness of 500 μm coated on the substrate of stainless steel, which is produced through a thermal spraying process. The specimen surface is polished to meet the roughness requirement. Since zirconia is not conductive, a thin layer of gold is coated on the specimen surface. A 2,000 lines/mm cross grating is fabricated on the interface of substrate and the adhesive layer with FEI DB235 and the fabrication parameters are listed in Table 10.5.

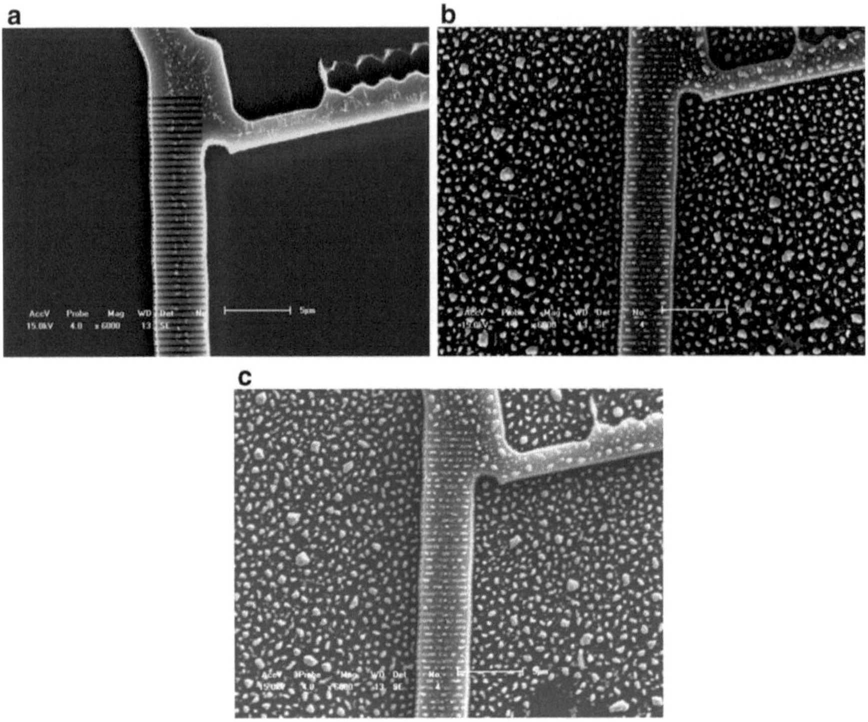

**Fig. 10.16** Gratings on amorphous SiC micro-beam (**a**) before high temperature, (**b**) after high temperature of 350 °C, and (**c**) after high temperature of 550 °C [30]

**Table 10.4** The grating fabrication parameters on the amorphous silicon carbide specimen surface

| Ion beam current (pA) | Setting depth (μm) | Dimensions of grating area (μm$^2$) | Milling time (min) | Magnification of FIB system |
|---|---|---|---|---|
| 164 | 0.3 | 75×4 | 4 | 4,000× |

Put the sample on the SEM stage (FEI Siron 400 NC) to record the moiré fringes before and after high temperature of 600 °C, as shown in Fig. 10.18. The fringes are still clear, which implies that the FIB milling grating on the surface of a thermal barrier coating specimen can bear high temperature of 600 °C. It can be observed that there is a relative displacement between both sides of the interface, which means that the thermal deformations of these two materials are different.

### *10.2.6 Application to Measure the Deformation Around Grain Boundary [29]*

Parallel gratings of 5,000 lines/mm are fabricated on low-carbon steel with FEI DB 235. In order to study the influence of grain boundaries on the plastic deformation,

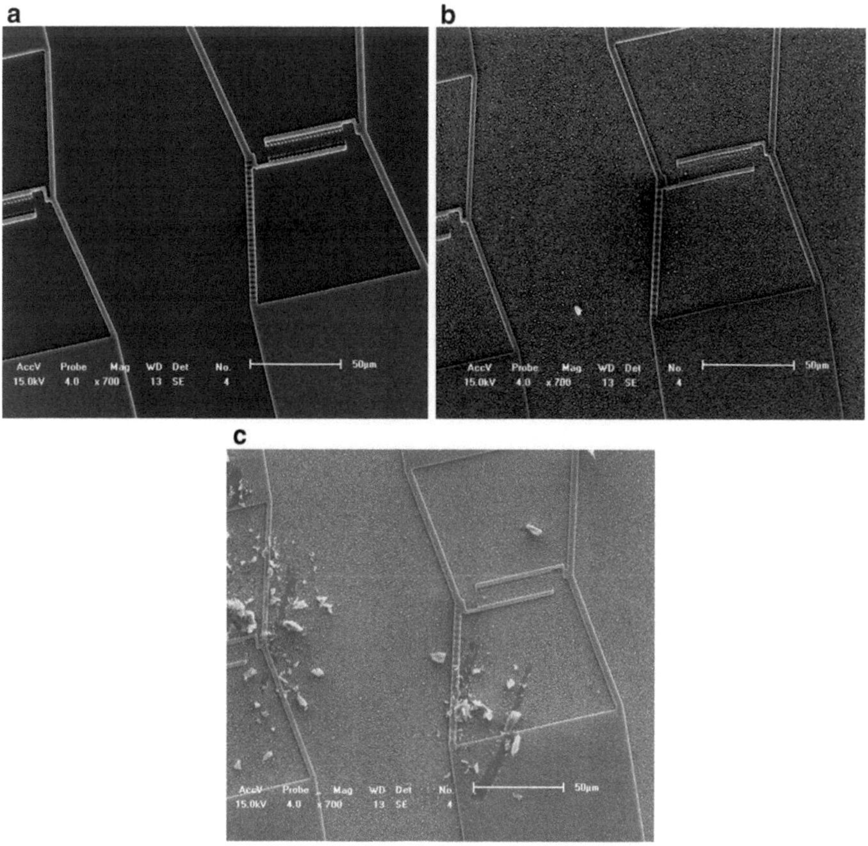


**Fig. 10.17** Moiré fringes (**a**) before high temperature, (**b**) after high temperature of 350 °C, and (**c**) after high temperature of 550 °C [30]

**Table 10.5** The grating fabrication parameters on the thermal barrier coating specimen surface

| Ion beam current (pA) | Setting depth (μm) | Dimensions of grating area (μm$^2$) | Milling time (min) | Magnification of FIB system |
|---|---|---|---|---|
| 700 | 0.2 | 100×100 | 37.6 | 3,000× |

the deformation characteristics of five different types of grain boundaries are investigated. Moiré fringes around each grain boundary are captured and their strain distributions are derived, as shown in Fig. 10.19, where the dashed lines are grain boundaries. The strain fields around these grain boundaries are inhomogeneous and increase with stress. When the direction of the grain boundary is almost parallel to the loading direction as Fig. 10.19a, the grain boundary has no obvious influence on the strain distribution. When the grain boundary is almost perpendicular to the loading direction as Fig. 10.19b, the strain along the grain boundary is minimum. When the angle between the direction of the grain boundary and the loading

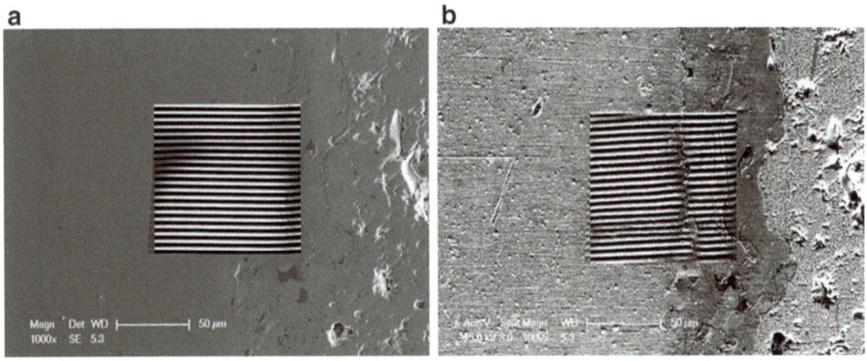

**Fig. 10.18** Moiré fringes (**a**) before high temperature and (**b**) after high temperature of 600 °C [31]

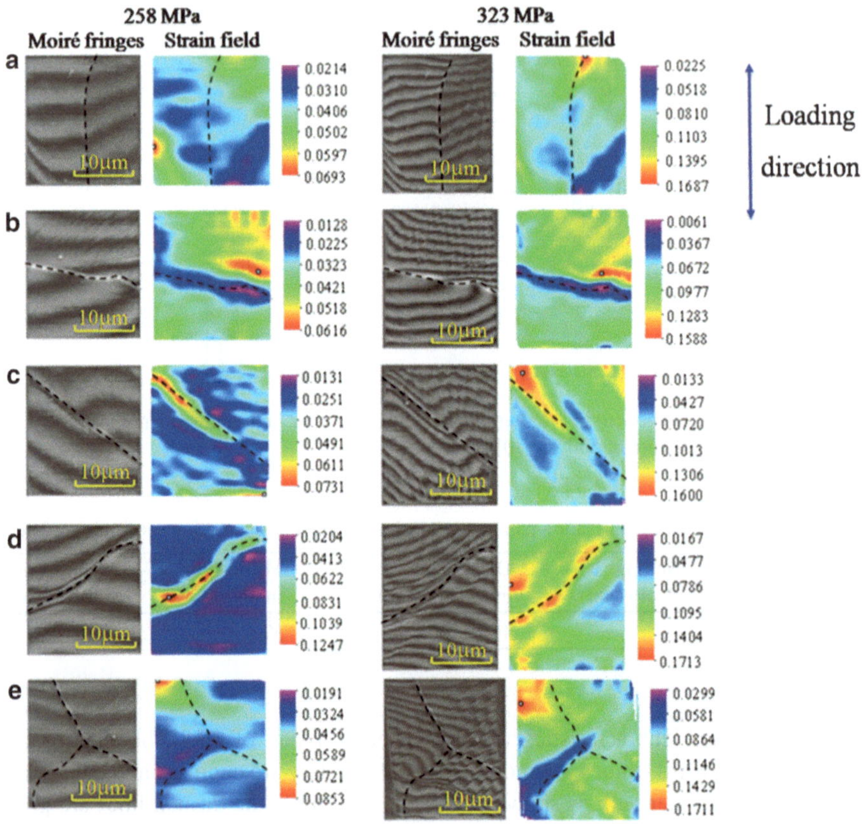


**Fig. 10.19** Moiré fringes and strain fields around different grain boundaries, each row corresponds to one type of grain boundary, where the first two images are the moiré fringes and strain field under the stress of 258 MPa, while the second two images are the moiré fringes and strain field under the stress of 323 MPa, the *dashed lines* are grain boundaries, the *dots* are the locations of the maximum strain [29]

direction is about 45° as Fig. 10.19c, d, the strain along the grain boundary is maximum. For a tricrystal grain boundary as Fig. 10.19e, only the strain around the left grain boundary is minimum and the other two grain boundaries have no obvious influence on the strain distribution. These different distributions are probably attributed to the differences of grain orientations around grain boundaries and the different angles between the direction of grain boundaries and the loading direction.

The moiré fringes and the enlarged grating when the specimen is subjected to tensile stress of 323 MPa are shown in Fig. 10.20. It is found that a slip band appears in the grating region and the bending direction of moiré fringes is along the direction of slip band. Furthermore, the moiré fringes become denser around the slip band, which indicates that the inhomogeneous strain is caused by slip.

## 10.3 Fabrication Technique of Speckle Patterns with FIB

### *10.3.1 Principle of Fabricating Speckle Pattern with FIB Milling*

Various microstructures can be fabricated with FIB system based on different input pattern images. This characteristic makes it possible to fabricate random speckle patterns with FIB milling. Using FIB milling to fabricate speckle patterns not only extends the application of FIB but also offers a new patterning technique for DIC method.

Dual systems combining FIB and SEM are used in our study. The flowchart of fabricating speckle pattern with FIB is shown in Fig. 10.21. A template (a binary bmp image, as seen in Fig. 10.22) of speckle pattern is first generated by software, converted with Patterns 1.04 (software package), and inputted into the pattern generator of the FIB system. Then an area of interest on the specimen is chosen and the speckle patterns are etched according to the template once the parameters including the magnification, the ion beam current, and the etching time are selected. Finally, speckle patterns are formed on the specimen surface and observed with SEM.

### *10.3.2 Design of the Speckle Pattern*

The design of the speckle pattern involves two aspects: the pattern morphology and the pattern size. The former directly influences the quality of the speckle pattern and depends on the template. The latter determines the displacement sensitivity and depends on the magnification of the FIB system. Therefore, the design of the speckle pattern includes choosing an optimized template and a proper magnification of the FIB system to meet the displacement measurement requirements.

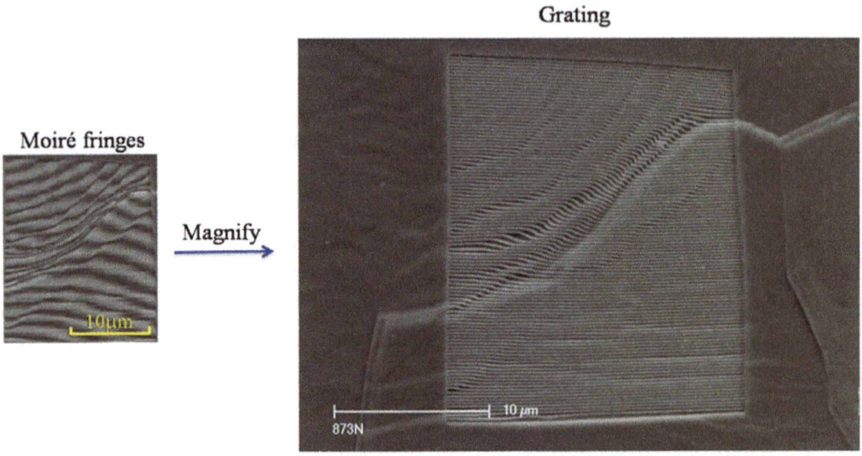


Fig. 10.20 Moiré fringes and grating corresponding to Fig. 10.19d when the specimen is subjected to tensile stress of 323 MPa [29]

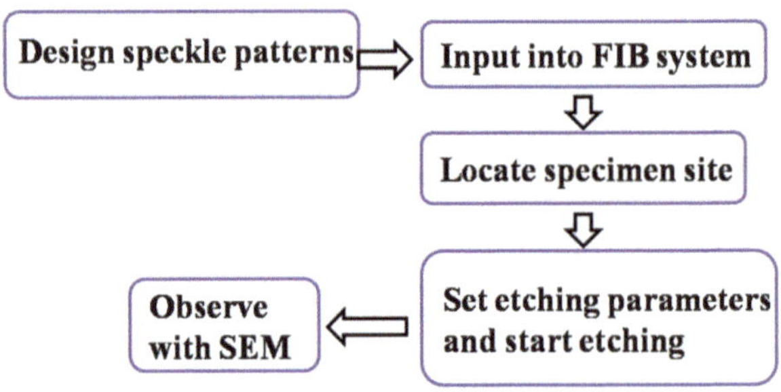


Fig. 10.21 Flowchart of speckle fabrication process with FIB [32]

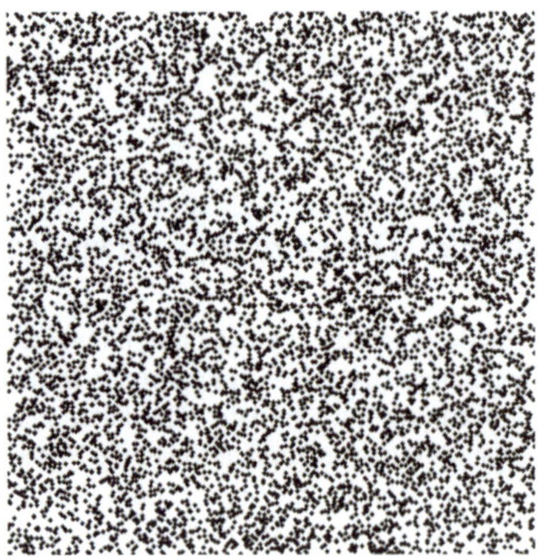

Fig. 10.22 A template of speckle pattern. The size is 512 pixels × 512 pixels, the number of the speckles is 9,000 and the radius of a single speckle is 2 pixels. During the fabrication process, only the counterpart of the specimen to the black part is etched. (Copyright by Science China Physics Mechanics Astronomy) [33]

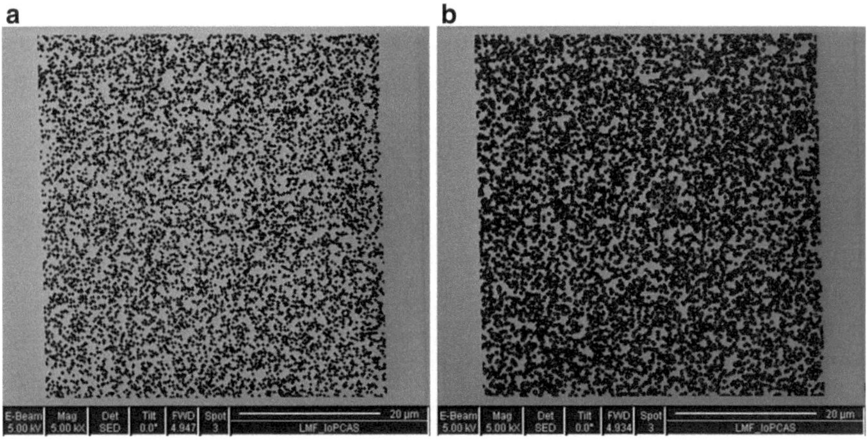


**Fig. 10.23** Fabricated speckle patterns based on two different templates (**a**) radius: 2 pixels, number: 9,000 and (**b**) radius: 3 pixels, number: 9,000 (Copyright by Science China Physics Mechanics Astronomy) [33]

#### 10.3.2.1 The Template

The morphology of the template is determined by two parameters: the radius and the number of the speckle. As reported in previous study [34], the optimized radius of a single speckle is 2–3 pixels, thus only 2 pixels and 3 pixels are taken into account. Templates with speckle amounts of 4,000, 5,000, 6,000, 7,000, 8,000, 9,000, and 10,000 are generated respectively for radiuses of 2 pixels and 3 pixels so as to choose a good one. The speckle patterns are fabricated on a silicon wafer and two typical SEM images of the fabricated speckle patterns are shown in Fig. 10.23, which are based on the templates with 9,000 speckles in radiuses of 2 pixels and 3 pixels. The etching time is 5 min, the ion beam current is 84 pA, and the magnification is 5,000×.

Since the quality of the speckle pattern directly affects the accuracy of the subsequent correlation calculation and a good template is the foundation to get a high quality speckle pattern, it is essential to choose a best template with a unified evaluation criterion. In order to assess the quality of a speckle pattern, some methods or parameters have been developed, including mean intensity gradient [35], mean subset fluctuation [36], image morphology method [37], subset entropy [38], and sum of squares of subset intensity gradient [39]. Here mean intensity gradient [35] is adopted as it is an effective global parameter.

The expression of mean intensity gradient is [35]

$$\delta_f = \sum_{i=1}^{W}\sum_{j=1}^{H}\left|\nabla f\left(\mathrm{x}_{ij}\right)\right| / \left(W \times H\right) \tag{10.1}$$

where $W$ and $H$ (in unit of pixels) are image width and height respectively, $\nabla f\left(\mathrm{x}_{ij}\right) = \sqrt{f_x\left(\mathrm{x}_{ij}\right)^2 + f_y\left(\mathrm{x}_{ij}\right)^2}$ is the local intensity gradient vector with $f_x(\mathbf{x}_{ij})$,

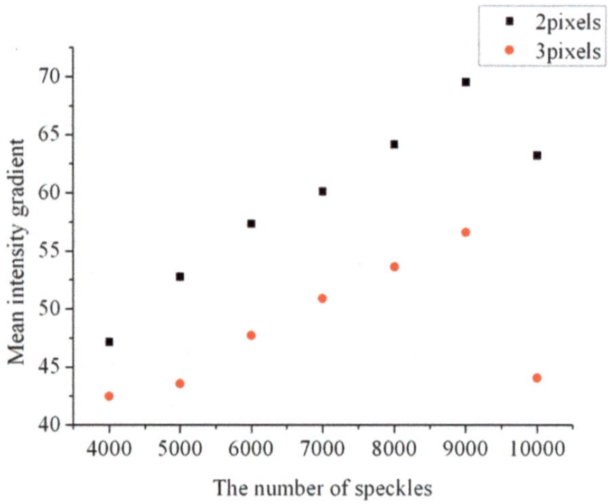


**Fig. 10.24** Mean intensity gradient of the fabricated speckle patterns (Copyright by Science China Physics Mechanics Astronomy) [33]

$f_y(\mathbf{x}_{ij})$ being the $x$- and $y$-directional intensity derivatives at pixel ($\mathbf{x}_{ij}$), which is calculated using a commonly used gradient operator. The greater the mean intensity gradient is, the better the quality of the fabricated speckle pattern is.

The calculated results of mean intensity gradient are shown in Fig. 10.24. It can be found that the qualities of the speckle patterns fabricated based on the templates composed of speckles of 2 pixels in radius are better than those fabricated based on the templates composed of speckles of 3 pixels in radius. Mean intensity gradient keeps growing with the increase of the number of speckles within 9,000, which implies that the quality of the speckle pattern is improved with the increase of the number of speckles until the number reaches 9,000. Therefore the template with 9,000 speckles of 2 pixels in radius is optimum for silicon. For other materials, an optimized template can be determined in the same way.

#### 10.3.2.2 Magnification of the FIB System

It is well recognized that the displacement sensitivity of DIC method is 0.01 pixel. Since the pixel size of the recorded SEM image is usually constant, the displacement sensitivity is directly related to its physical size, which is approximate to the pattern size. Therefore, the pattern size is designed according to the required displacement measurement sensitivity. During the fabrication process the pattern size is adjusted by altering the magnification of the FIB system. Once their relationship is determined, it is convenient to obtain a desired pattern size by altering the magnification.

Using the same template, the etched area is inversely proportional to the magnification of the FIB system. For simplicity, the template is generated as a square so that the relationship between the etched area and the magnification is transformed into the relationship between the length of the square and the magnification.

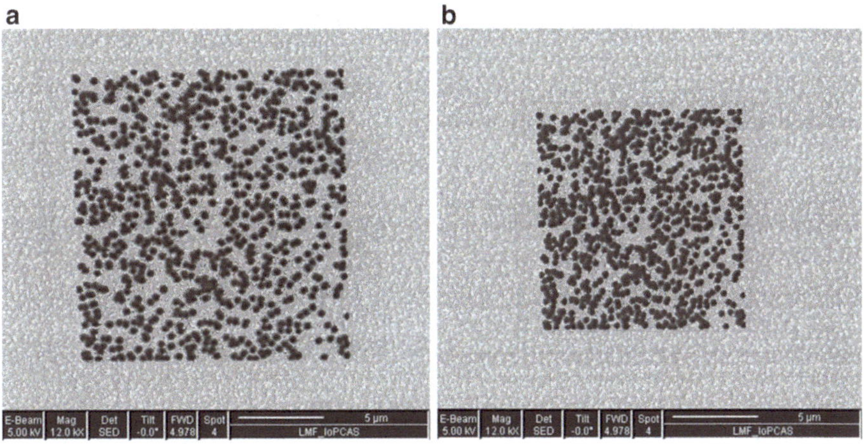


**Fig. 10.25** SEM images of speckle patterns fabricated at two different magnifications of the FIB system (**a**) 15,000× and (**b**) 20,000× (Copyright by Science China Physics Mechanics Astronomy) [33]

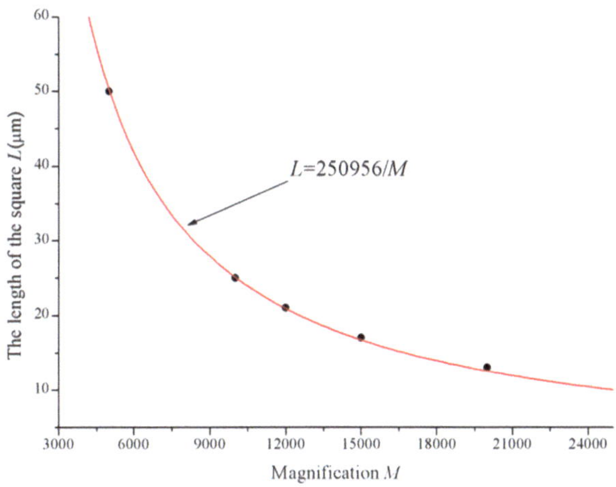


**Fig. 10.26** Relation curve of the length of the square versus magnification (Copyright by Science China Physics Mechanics Astronomy) [33]

To calibrate their relationship, the speckle patterns are fabricated on amorphous silicon carbide under the magnification of 5,000×, 10,000×, 12,000×, 15,000×, and 20,000× respectively with the same template and two of them are shown in Fig. 10.25. The etching time is 5 min and the ion beam current is 164 pA. It can be seen that the morphologies of the speckle patterns are identical but the size varies with the magnification of the FIB system. By measuring the lengths of the squares at different magnifications, the relation curve of the length versus the magnification is obtained, as shown in Fig. 10.26, and the best fitted expression is derived. Therefore, the magnification of FIB can be determined according to the designed pattern size.

## 10.3.3 Influencing Factors of FIB Patterning Technique

The etching time and the ion beam current are two major parameters that influence the quality of the speckle pattern.

### 10.3.3.1 Etching Time

In order to investigate the influence of the etching time on the fabricated pattern, five groups of speckle patterns are produced on silicon substrate with different etching times, as shown in Fig. 10.27. It can be found that the contrast of the speckle image is improved with the growth of etching time, which is attributed to a larger depth of the speckle as a result of a longer etching time. However, a longer etching time will lead to a higher cost and a larger etching depth will affect the mechanical properties of materials. Therefore, it is necessary to obtain a reasonable time by repeated trials to ensure enough contrast and the mechanical property simultaneously. For silicon, 5 min is optimum to fabricate 50 μm×50 μm speckle patterns.

### 10.3.3.2 Ion Beam Current

To investigate the effect of the ion beam current, the speckle patterns are fabricated on silicon with different ion beam currents, including 28, 55, 222, 589, and 983 pA, and three of them are presented in Fig. 10.28. The quality of speckle patterns is still assessed with mean intensity gradient and the results are plotted in Fig. 10.29, which suggests that 589 pA is optimum. If the ion beam current is lower, the contrast of the speckle pattern will be poor; otherwise the noise level will be higher.

## 10.3.4 Applications of FIB Milling Speckle Patterns

The speckle pattern fabricated on silicon based on the template consisting of 9,000 speckles of 2 pixels in radius is applied to rigid body translation tests and rotation tests under SEM (FEI Siron 400). All the displacement calculations are fulfilled by self-developed DIC software. The subset size is 41×41 pixels$^2$ and the sub-pixel displacement is calculated by Newton–Raphson method.

### 10.3.4.1 Translation Tests

A piezoelectric transducer (PZT) calibrated with a grating of 2,000 lines/mm is used to translate the specimen by increasing the voltage from 0 to 80 V with a step of 10 V. The speckle patterns are captured before and after the translation. The displacements

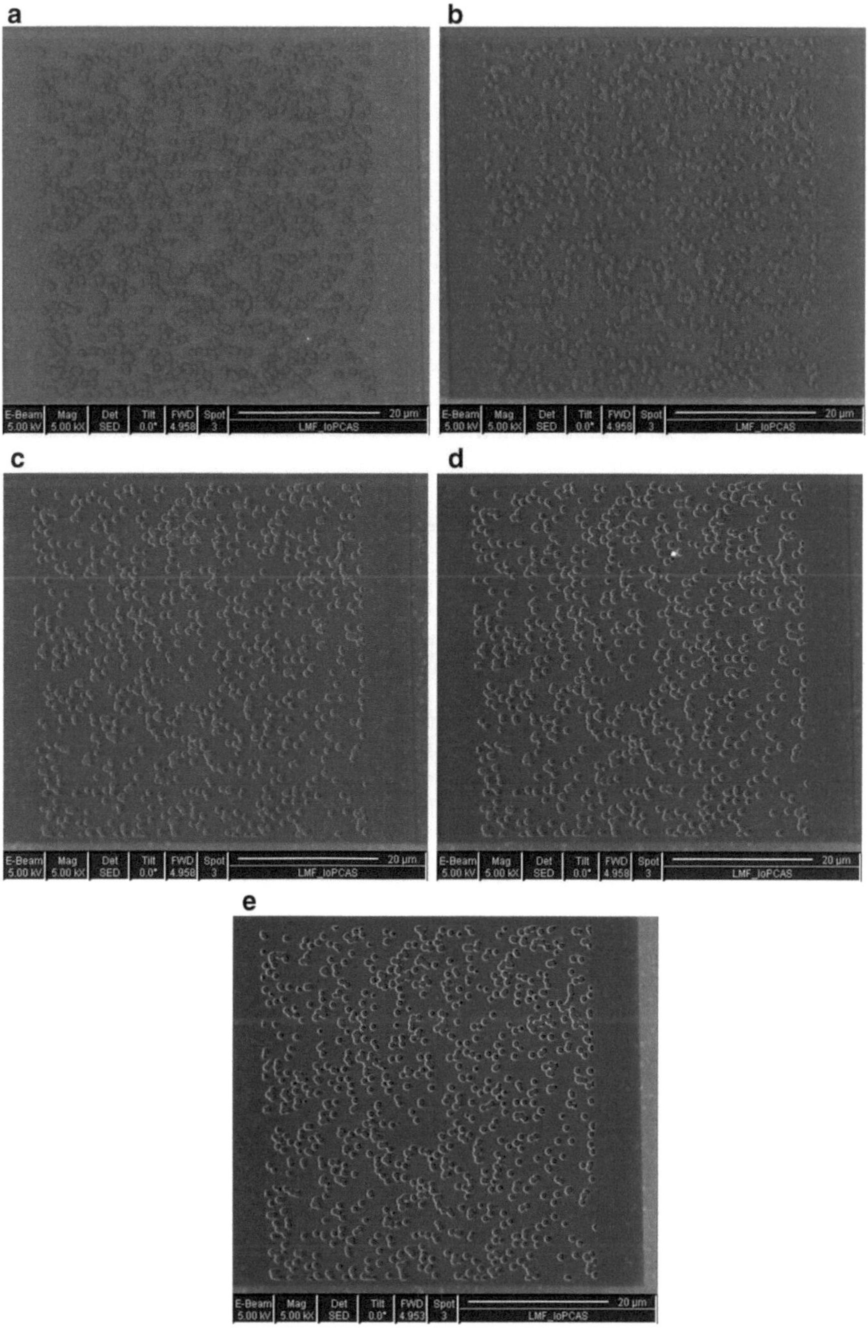


**Fig. 10.27** Speckle patterns with different etching times (**a**) 60 s, (**b**) 120 s, (**c**) 180 s, (**d**) 240 s, and (**e**) 300 s. The ion beam current is 986 pA and the magnification of the FIB system is 5,000×. (Copyright by Science China Physics Mechanics Astronomy) [33]

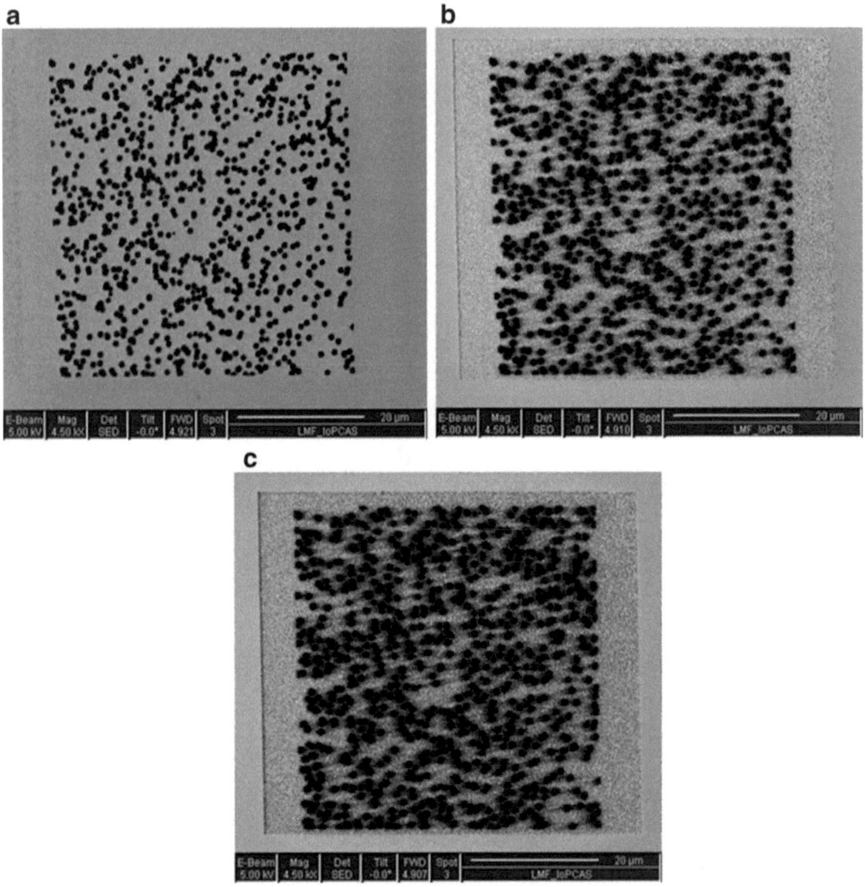


**Fig. 10.28** SEM images of speckle patterns fabricated with three different ion beam currents (**a**) 28 pA, (**b**) 589 pA, and (**c**) 983 pA. The magnification is 5,000× and the etching time is 5 min. (Copyright by Science China Physics Mechanics Astronomy) [33]

are calculated with DIC method and the results are listed in Table 10.6. The calculated average displacements are in good agreement with the imposed ones and the relative errors are within 5 %. To view the results more clearly, the results are illustrated in Fig. 10.30. As seen from Table 10.6, the standard errors are relatively small compared to the average value, which is the reason why standard errors are not displayed in the graph.

#### 10.3.4.2 Rotation Tests

Rotation tests are realized by rotating the scanning lines of the SEM. The speckle patterns are captured before rotation and after rotation of 1°, 2°, and 3°. The displacements

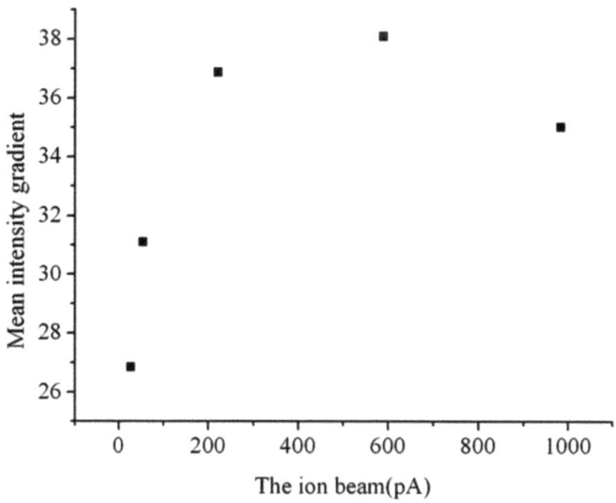


**Fig. 10.29** Mean intensity gradient corresponding to Fig. 10.28 (Copyright by Science China Physics Mechanics Astronomy) [33]

**Table 10.6** Comparison between the calculated displacement and the imposed one (Copyright by Science China Physics Mechanics Astronomy) [33]

| Imposed displacement (μm) | Calculated displacement (μm) | Standard error (μm) | Absolute relative error (%) |
|---|---|---|---|
| 0.909 | 0.929 | 0.0027 | 2.2 |
| 2.000 | 1.973 | 0.0017 | 1.4 |
| 3.212 | 3.116 | 0.0020 | 3.0 |
| 4.525 | 4.359 | 0.0028 | 3.7 |
| 5.859 | 5.598 | 0.0026 | 4.4 |
| 7.212 | 6.851 | 0.0041 | 5.0 |
| 8.525 | 8.098 | 0.0038 | 5.0 |
| 9.434 | 9.319 | 0.0032 | 1.2 |

are calculated with DIC method and the displacement vector after rotation of 3° is shown in Fig. 10.31. The rotation angle is expressed as

$$\theta = \frac{1}{2}\left(\frac{\partial u}{\partial y} - \frac{\partial v}{\partial x}\right) \tag{10.2}$$

The calculated results are listed in Table 10.7. Both the standard errors and the relative errors are very small, which confirms the reliability of the fabricated speckle pattern.

### *10.3.5 High Temperature Tests*

High temperature in situ tests are performed under SEM (Shimadzu S-550) using speckle patterns fabricated on silicon. Here are the fabrication parameters: the ion

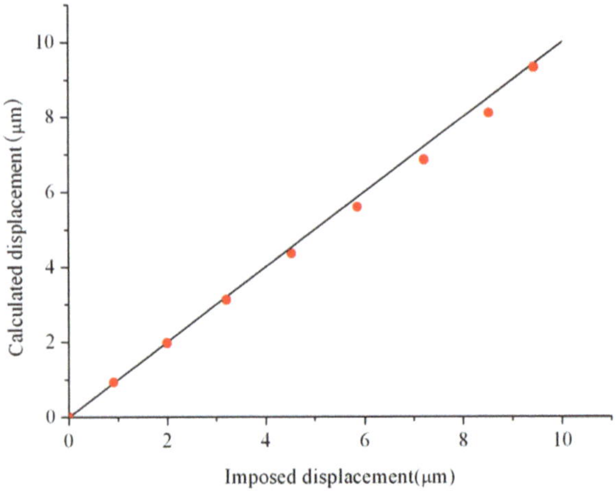


**Fig. 10.30** Calculated average displacement versus imposed displacement. The *solid line* is a reference line where the calculated displacement is equal to the imposed displacement and the *red points* are the experimental data. (Copyright by Science China Physics Mechanics Astronomy) [33]

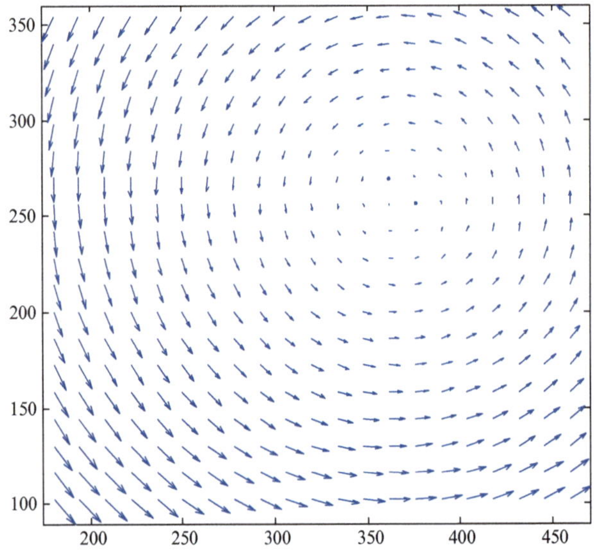


**Fig. 10.31** Displacement vector when scanning lines are rotated 3° (Copyright by Science China Physics Mechanics Astronomy) [33]

beam current is 627 pA, the magnification of FIB is 2,500, and the milling time is 5 min. In order to ensure good conductivity, a thin layer of gold is deposited on the specimen surface before experiment.

Record the speckle patterns at room temperature. Increase the temperature to 50, 100, 150, and 200 °C, keep for 5 min and record the corresponding speckle patterns respectively, as shown in Fig. 10.32. The speckle patterns at room temperature have a

**Table 10.7** Comparison between the calculated rotation angle and the real one (Copyright by Science China Physics Mechanics Astronomy) [33]

| Real angle (°) | Calculated angle (°) | Standard error (°) | Absolute relative error (%) |
|---|---|---|---|
| 1 | 0.998 | 0.007 | 0.2 |
| 2 | 2.003 | 0.015 | 0.15 |
| 3 | 2.983 | 0.020 | 0.57 |

good contrast. At 50 °C, the speckle patterns remain unchanged but the background becomes white due to the heating of filament. At 100 °C, speckle patterns cluster and several small bright points appear around. At 150 °C, the speckle patterns become white or gray, bright large particles appear in the periphery and the bright points become denser and bigger and far away from the speckle patterns. At 200 °C, the bright particles continue growing. Great changes have taken place with the increase of temperature, which will bring great error to the correlation calculation. The changes are perhaps caused by gold film and the $Ga^+$ left on the specimen surface during the fabrication process.

## 10.4 Conclusions

1. The fabrication technique of high frequency gratings with FIB milling is studied for use of SEM moiré method thoroughly and comprehensively. The concept is proposed that the frequency and the size of the grating must be designed to meet the requirements of displacement measurement accuracy and the view field at the same time. To get high quality gratings, the main influencing factors, including the surface roughness, the opening ratio of gratings, the scanning mode of the FIB system, the etching depth, the magnification of the FIB system and the mosaic of gratings, are investigated. Some technique skills are summarized to provide a guide for this grating fabrication method. To verify the capability of the fabricated gratings for high temperature use, high temperature tests are carried out, which will provide experiment evidences for high temperature deformation measurement. The successful application of gratings fabricated with FIB to measure the deformation around the grain boundary is presented, which proves FIB milled grating combined with SEM moiré method is effective for microscopic deformation measurement.
2. The fabrication technique of microscale/nanoscale speckle patterns with FIB milling is developed. The speckle patterns are fabricated by directly etching the black part of a software-generated binary template of random speckle patterns. Mean intensity gradient is used to evaluate the quality of these SEM images of speckle patterns fabricated with different templates so that a best template is selected. The size of the etching area depends on the displacement measurement sensitivity and can be adjusted by altering the magnification of FIB. The influencing factors including etching time and ion beam current are also analyzed. Rigid body translation tests and rotation tests are carried out under SEM with the patterns fabricated by FIB. The results demonstrate that FIB patterning technique coupled with SEM-DIC is well suited to measure the microscale/nanoscale deformation of macro-specimens and microscale/nanoscale devices like MEMS.

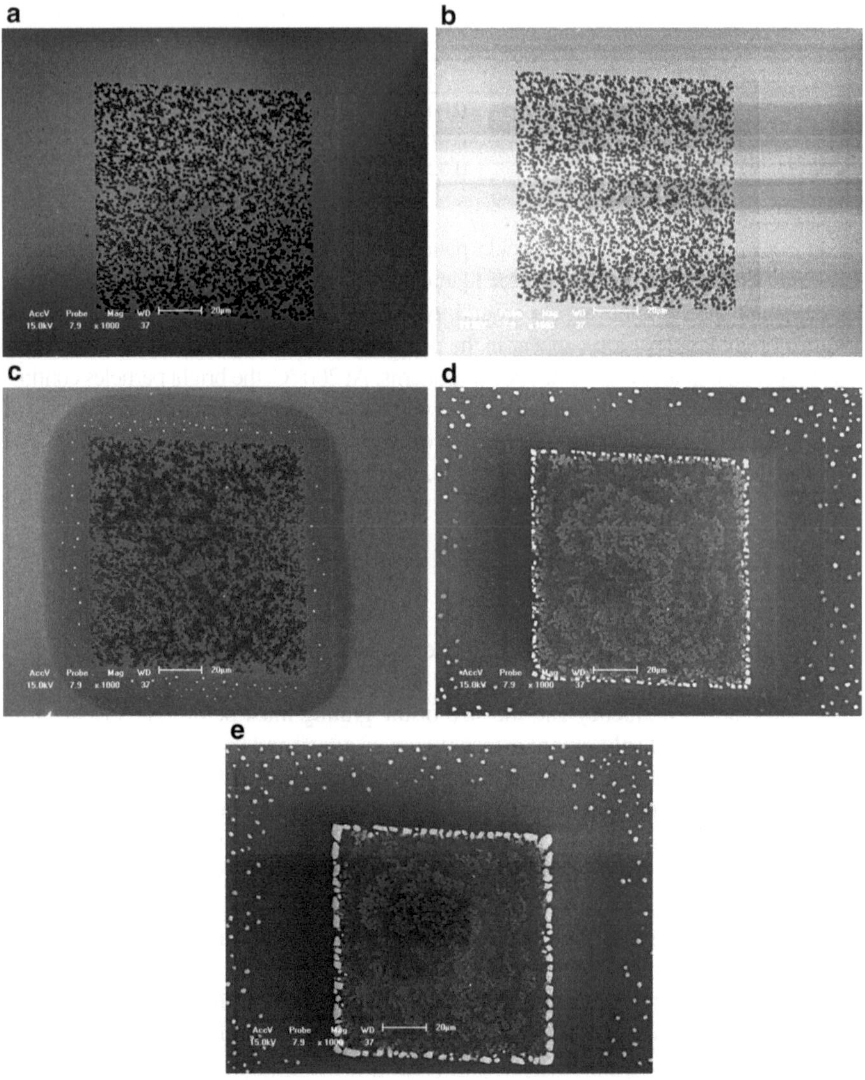

Fig. 10.32 Speckle patterns at different temperatures (a) room temperature, (b) 50 °C, (c) 100 °C, (d) 150 °C, and (e) 200 °C [32]

High temperature in situ tests are conducted with speckle patterns fabricated with FIB, which demonstrate that the speckle patterns fabricated by direct milling is not suited for high temperature in situ experiments.

**Acknowledgements** The authors are grateful to the financial supported by the National Basic Research Program of China ("973" Project) (Grant Nos. 2010CB631005, 2011CB606105), the National Natural Science Foundation of China (Grant Nos. 11232008, 91216301, 11227801, 11172151), Specialized Research Fund for the Doctoral Program of Higher Education (Grant No.20090002110048), Tsinghua University Initiative Scientific Research Program, the Initiative Scientific Research Program, and the Doctoral Program of University of Jinan (Grant No. XBS 1307).

## References


1. Sabate, N., Vogel, D., Gollhardt, A., Keller, J., Cane, C., Gracia, I., Morante, J., Michel, B.: Residual stress measurement on a MEMS structure with high-spatial resolution. J. Microelectromech. Syst. **16**, 365–372 (2007)
2. Zhou, W., Yang, J.L., Sun, G.S., Liu, X.F., Yang, F.H., Li, J.M.: Fracture properties of silicon carbide thin films by bulge test of long rectangular membrane. J. Microelectromech. Syst. **17**, 453–461 (2008)
3. Tong, C.J., Lin, M.T.: Design and development of a novel paddle test structure for the mechanical behavior measurement of thin films application for MEMS. Microsyst. Technol. **15**, 1207–1216 (2009)
4. Jonnalagadda, K.N., Chasiotis, I., Yagnamurthy, S., Lambros, J., Pulskamp, J., Polcawich, R., Dubey, M.: Experimental investigation of strain rate dependence of nanocrystalline Pt films. Exp. Mech. **50**, 25–35 (2010)
5. Kang, D.J., Park, J.H., Shin, M.S., Ha, J.E., Lee, H.J.: Specimen alignment in an axial tensile test of thin films using direct imaging and its influence on the mechanical properties of BeCu. J. Micromech. Microeng. **20**, 085001 (2010)
6. Post, D., McKelvie, J., Tu, M., Dai, F.: Fabrication of holographic gratings using a moving point source. Appl. Opt. **28**, 3494–3497 (1989)
7. Moulart, R., Rotinat, R., Pierron, F., Lerondel, G.: On the realization of microscopic grids for local strain measurement by direct interferometric photolithography. Opt. Lasers Eng. **45**, 1131–1147 (2007)
8. Kishimoto, S., Egashira, M., Shinya, N.: Microcreep deformation measurements by a moire method using electron beam lithography and electron beam scan. Opt. Eng. **32**, 522–526 (1993)
9. Xie, H., Kishimoto, S., Shinya, N.: Fabrication of high-frequency electron beam moire grating using multi-deposited layer techniques. Opt. Laser Technol. **32**, 361–367 (2000)
10. Xie, H., Kishimoto, S., Li, Y., Liu, Q., Zhao, Y.: Fabrication of micro-moire gratings on a strain sensor structure for deformation analysis with micro-moire technique. Microelectron. Reliab. **49**, 727–733 (2009)
11. Xie, H., Li, B., Geer, R., Xu, B., Castracane, J.: Focused ion beam Moire method. Opt. Lasers Eng. **40**, 163–177 (2003)
12. Yan, D., Cheng, J., Apsel, A.: Fabrication of SOI-based nano-gratings for Moire measurement using focused ion beam. Sensor. Actuator. Phys. **115**, 60–66 (2004)
13. Shang, H., Xie, H., Wang, X., Jiang, S., Dai, F., Wang, W., Fang, Y., Zhao, Y.: Thermal properties measurement of micro-electromechanical system sensors by digital Moiré method. Strain **41**, 157–162 (2005)
14. Hua, D., Hui-Min, X., Zhi-Qiang, G., Qiang, L., Chang-Zhi, G., Hai-Chang, Q., Li-Jian, R.: Development of the technique for fabricating submicron Moire gratings on metal materials using focused ion beam milling. Chin. Phys. Lett. **24**, 2521–2524 (2007)
15. Scrivens, W., Luo, Y., Sutton, M., Collette, S., Myrick, M., Miney, P., Colavita, P., Reynolds, A., Li, X.: Development of patterns for digital image correlation measurements at reduced length scales. Exp. Mech. **47**, 63–77 (2007)
16. Berfield, T., Patel, J., Shimmin, R., Braun, P., Lambros, J., Sottos, N.: Micro-and nanoscale deformation measurement of surface and internal planes via digital image correlation. Exp. Mech. **47**, 51–62 (2007)
17. Tanaka, Y., Naito, K., Kishimoto, S., Kagawa, Y.: Development of a pattern to measure multiscale deformation and strain distribution via in situ FE-SEM observations. Nanotechnology **22**, 115704 (2011)
18. Liu, Z., Xie, H., Fang, D., Gu, C., Meng, Y., Wang, W., Fang, Y., Miao, J.: Deformation analysis in microstructures and micro-devices. Microelectron. Reliab. **47**, 2226–2230 (2007)
19. Korsunsky, A., Sebastiani, M., Bemporad, E.: Focused ion beam ring drilling for residual stress evaluation. Mater. Lett. **63**, 1961–1963 (2009)
20. Korsunsky, A., Sebastiani, M., Bemporad, E.: Residual stress evaluation at the micrometer scale: analysis of thin coatings by FIB milling and digital image correlation. Surf. Coat. Technol. **205**, 2393–2403 (2010)

21. Sebastiani, M., Eberl, C., Bemporad, E., Pharr, G.M.: Depth-resolved residual stress analysis of thin coatings by a new FIB-DIC method. Mater. Sci. Eng. A **528**, 7901–7908 (2011)
22. Sabate, N., Vogel, D., Gollhardt, A., Marcos, J., Gracia, I., Cane, C., Michel, B.: Digital image correlation of nanoscale deformation fields for local stress measurement in thin films. Nanotechnology **17**, 5264–5270 (2006)
23. Sabate, N., Vogel, D., Gollhardt, A., Keller, J., Michel, B., Cane, C., Gracia, I., Morante, J.: Measurement of residual stresses in micromachined structures in a microregion. Appl. Phys. Lett. **88**, 071910 (2006)
24. Sabate, N., Vogel, D., Gollhardt, A., Keller, J., Cane, C., Gracia, I., Morante, J., Michel, B.: Measurement of residual stress by slot milling with focused ion-beam equipment. J. Micromech. Microeng. **16**, 254–259 (2006)
25. Kregting, R., Gielen, S., Driel, W., Alkemade, P., Miro, H., Kamminga, J.D.: Local stress analysis on semiconductor devices by combined experimental-numerical procedure. Microelectron. Reliab. **51**, 1092–1096 (2011)
26. Kammers, A.D., Daly, S.: Small-scale patterning methods for digital image correlation under scanning electron microscopy. Meas. Sci. Technol. **22**, 125501 (2011)
27. Tseng, A.: Recent developments in micromilling using focused ion beam technology. J. Micromech. Microeng. **14**, R15–R34 (2004)
28. Reyntjens, S., Puers, R.: A review of focused ion beam applications in microsystem technology. J. Micromech. Microeng. **11**, 287–300 (2001)
29. Li, Y., Xie, H., Tang, M., Zhu, J., Luo, Q., Gu, C.: The study on microscopic mechanical property of polycrystalline with SEM moiré method. Opt. Lasers Eng. **50**, 1757–1764 (2012)
30. Li, Y., Xie, H., Guo, B., Luo, Q., Gu, C., Xu, M.: Fabrication of high-frequency moiré gratings for microscopic deformation measurement using focused ion beam milling. J. Micromech. Microeng. **20**, 055037 (2010)
31. Li, Y.J.: Theoretical and experimental study of SEM moiré method and its applications. Ph.D. thesis, Department of engineering mechanics, Beijing Tsinghua University (2011)
32. Li, Y.J.: The measurement technique of digital image correlation and its applications. Master thesis, Department of engineering mechanics, Beijing Tsinghua University (2008)
33. Li, Y.J., Xie, H.M., Luo, Q., Gu, C.Z., Hu, Z.X., Chen, P.W., Zhang, Q.M.: Fabrication technique of micro/nano-scale speckle patterns with focused ion beam. Sci. China Phys. Mech. Astron. **55**, 1037–1044 (2012)
34. Zhou, P., Goodson, K.E.: Subpixel displacement and deformation gradient measurement using digital image/speckle correlation (DISC). Opt. Eng. **40**, 1613–1620 (2001)
35. Pan, B., Lu, Z.X., Xie, H.M.: Mean intensity gradient: an effective global parameter for quality assessment of the speckle patterns used in digital image correlation. Opt. Lasers Eng. **48**, 469–477 (2010)
36. Hua, T., Xie, H.M., Wang, S., Hu, Z.X., Chen, P.W., Zhang, Q.M.: Evaluation of the quality of a speckle pattern in the digital image correlation method by mean subset fluctuation. Opt. Laser Technol. **43**, 9–13 (2011)
37. Lecompte, D., Smits, A., Bossuyt, S., Sol, H., Vantomme, J., Van Hemelrijck, D., Habraken, A.: Quality assessment of speckle patterns for digital image correlation. Opt. Lasers Eng. **44**, 1132–1145 (2006)
38. Yaofeng, S., Pang, J.H.L.: Study of optimal subset size in digital image correlation of speckle pattern images. Opt. Lasers Eng. **45**, 967–974 (2007)
39. Pan, B., Xie, H., Wang, Z., Qian, K.: Study on subset size selection in digital image correlation for speckle patterns. Opt. Express **16**, 7037–7048 (2008)

# Chapter 11
# Controlled Quantum Dot Formation on Focused Ion Beam-Patterned GaAs Substrates

**Haoyu Zhang and Thomas Walther**

**Abstract** We combine focused ion beam patterning (FIB) and self-assembly of epitaxial quantum dots in order to produce regular quantum dot arrays. Island-shaped quantum dots are expected to nucleate at specific locations where the ion beam has formerly patterned the surface. Different ion beam patterning parameters are compared, including accelerating voltage, probe current, dwell time and pitch. Three different types of ion beam damage are explained. Secondary electron imaging (SEI) in a scanning electronic microscope (SEM) and atomic force microscopy (AFM) have been performed before and after overgrowth of the patterned areas. We investigate how the patterning parameters influence nucleation and overgrowth, including the diameter, height and distribution of three forms of structures formed, i.e. islands, rings and holes. Islands are preferably formed at arrays patterned at low acceleration voltage with low ion dose. Also, the relationship between patterning parameters and island sizes has been studied. Micro-photoluminescence was performed in and outside patterned regions, in order to compare their optoelectronic properties.

## 11.1 Introduction

In the last few decades, fabrication of quantum dots (QDs) has experienced rapid progress. Quantum dots are small artificially engineered crystal structures which have electronic properties that are characteristic of atomic dimensions [1]. The smaller the quantum dots are the larger their bandgap is. Quantum dots have been widely studied and are used in optoelectronic engineering for lasers, light-emitting diodes (LEDs) and solar cells. One of the dominating fabrication techniques is

H. Zhang • T. Walther (✉)
Department of Electronic & Electrical Engineering, University of Sheffield,
Mappin Building, Sheffield, S1 3JD, UK
e-mail: elp08hz@sheffield.ac.uk; t.walther@sheffield.ac.uk

Z.M. Wang (ed.), *FIB Nanostructures*, Lecture Notes in Nanoscale Science and Technology 20, DOI 10.1007/978-3-319-02874-3_11,

Stranski-Krastanov growth by molecular beam epitaxy (MBE). It is a top-down approach and a quick way to grow quantum dots on large areas of a wafer or substrate; however, these self-assembled quantum dots have a certain spread in their size and hence of their emission energies [2]. In addition, the inherently random position of nucleation puts limits on downsizing devices further. If only a few quantum dots are included in a device, the location and uniformity of quantum dots will have a strong influence on the reproducibility of the electronic and optoelectronic performance across multiple devices. The bottom-up growth by patterning the substrate before overgrowth is an alternative approach which can produce regular QD arrays on patterned substrates with better position control and homogeneity [3]. Photoluminescence spectra of QD arrays can exhibit significant inhomogeneous broadening if their shapes, sizes or chemical composition differ laterally [4–6]. Periodic arrays of regular QDs may have improved optical properties and are thus interesting for a number of applications. However, it is a challenge to produce highly ordered QD arrays.

Bottom-up QD growth can be achieved by using optical-, electron beam- or ion beam-based lithography [4, 7], nanoimprint techniques [8] or growth on cleaved edges or misoriented surfaces [9]. Focused ion beam (FIB) technology can be used to assist in the fabrication of regular QD arrays by patterning the substrate before overgrowth. The principle is similar to other forms of lithographic or chemical etching-induced quantum dot array fabrication. Molecular beam epitaxial quantum dots are expected to nucleate preferentially at specified locations previously scanned by the ion beam, avoiding undisturbed regions of the surface where uniform quantum dot nucleation, typical of self-assembly processes, will continue to be observed. As this technology has no chemical etching of a resist or mechanical imprint by a template involved, there are hardly any restrictions on the sample material [10] which is considered an advantage compared with other fabrication methods. A disadvantage of FIB patterning is that the procedure is serial and thus slow, so that only a limited field of view can be patterned in reasonable time, and if the field of view becomes very large, distortions of the patterns due to optical aberrations of the beam-forming optics may become relevant. Regular QD arrays have been shown to have better photoelectric properties (Purcell effect [11]) and sharper spectral peaks compared with self-assembled quantum dots [12, 13].

## 11.2 Regular Quantum Dot Array Fabrication by FIB

The fabrication of regular quantum dot arrays by FIB can be separated into two main stages: the FIB patterning stage and the MBE overgrowth stage. We have investigated InGaAs quantum dot growth on GaAs (001) substrates patterned in a JEOL 6500 F Fabrika SEM/FIB dual-beam instrument equipped with an Orsay Physics $Ga^+$ ion column and a Raith lithography system that can control both the ion and the electron beam. The ion beam that scans the surface of the substrate disturbs both the native surface oxide ($Ga_2O_3$) and the underlying crystalline GaAs, leading to regular arrays of protrusions, rings or holes, depending on the ion beam

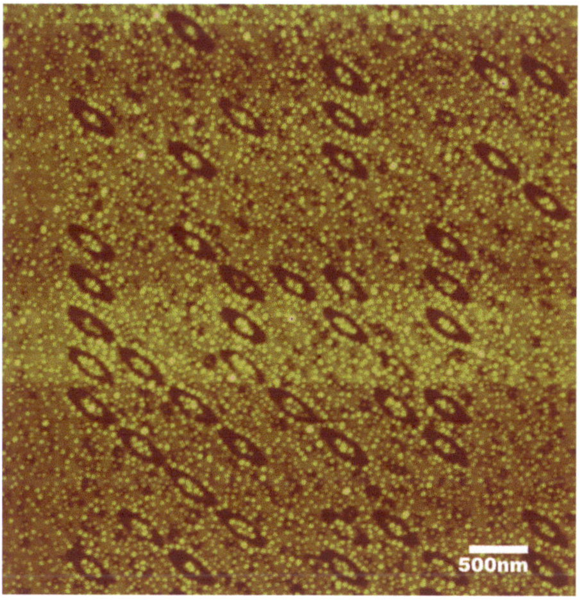


**Fig. 11.1** Atomic force microscope (AFM) image of an overgrown pattern produced with a stigmatic FIB, demonstrating the material deposited subsequently is more likely to form islands near the edges and on the valleys of deep holes

parameters. This will change the surface morphology and hence the surface energy near to the drilled features. This surface energy is expected to influence the quantum dot nucleation. A series of patterns have been milled with systematically varied ion beam parameters. The different parameters included accelerating voltage, ion probe current (aperture size), dwell time and the pitch size of the arrays. Quantum dots have been observed to nucleate more likely at the bottom of patterned holes, if these are sufficiently large, or in the vicinity of their edges during MBE overgrowth, as shown in Fig. 11.1.

In the second overgrowth stage, the patterned substrate is moved into the MBE chamber. First the sample is degassed at 400 °C for 1 h to remove carbon surface contamination. Prior to the overgrowth, the top native oxide layer is removed under an arsenic flux at 580 °C for 30 min, during which the surface is monitored using reflection high-energy electron diffraction (RHEED). A clear (2×4) reconstruction is indicative of a smooth and relative oxide-free surface. Higher temperature annealing should be avoided, in order to maintain the underlying FIB pattern [14]. The first deposited layer is a 20 nm GaAs buffer layer at 580 °C, followed by a 1.5 ML InAs layer at 520 °C. The growth rates of GaAs and InAs are 0.5 ML/s and 0.02 ML/s, respectively. The buffer layer thickness is well controlled so that it reduces the effect of surface ion damage but yet does not fill in the FIB fabricated patterns. During growth, the thickness of InAs is very important. The amount of material deposited ensures InGaAs quantum dots will nucleate based on the Stranski-Krastanov growth transition, which for InAs/GaAs is known to occur at 1.8 ML InAs thickness [14, 15] and trigger the formation of indium-rich islands connected by an indium-depleted wetting layer [16].

Before and after MBE growth, the sample has been investigated with atomic force microscopy (AFM) and scanning electronic microscopy (SEM). AFM is very sensitive to surface roughness changes, which allows us to understand the

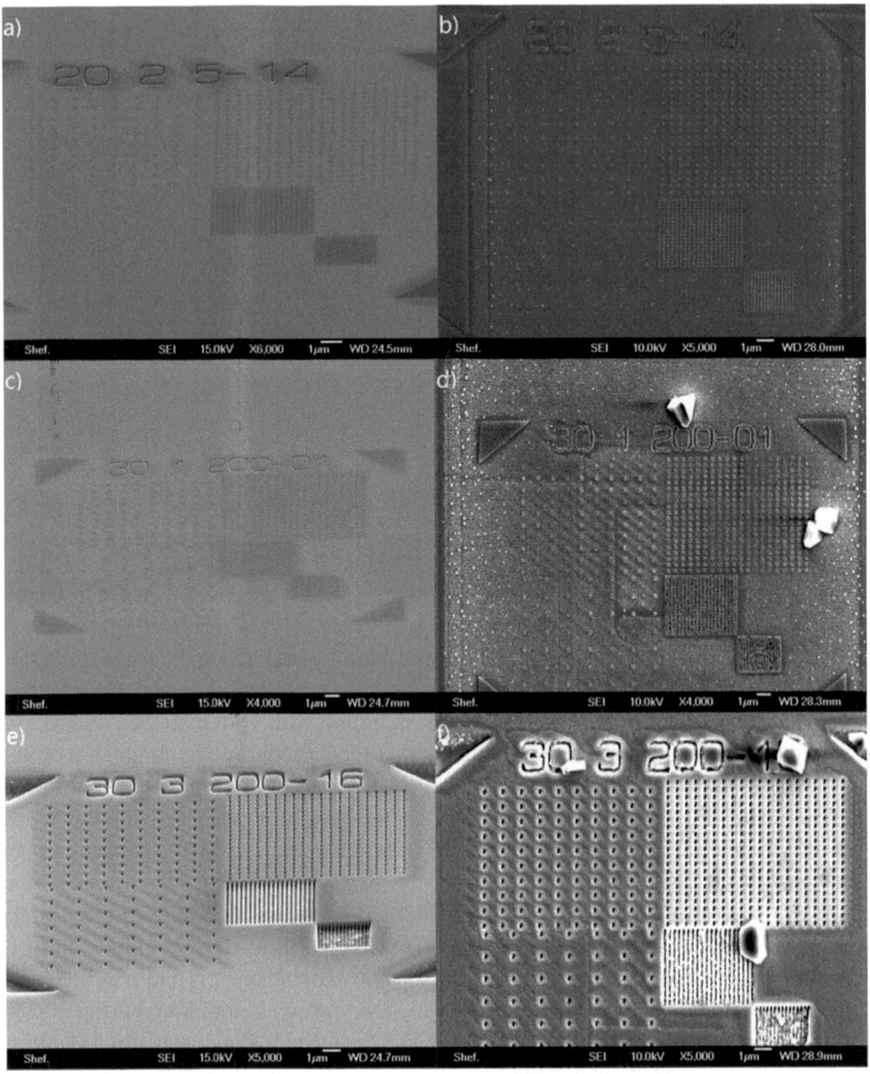


**Fig. 11.2** SEM secondary electron (SE) images captured before (**a**, **c**, **e**) and after (**b**, **d**, **f**) overgrowth. When comparing SE images directly after patterning but before overgrowth with those after overgrowth, the 55° sample tilt for FIB milling gives an apparent foreshortening of the vertical due to the projection, which has not been corrected here. The ion dose increases from *top* to *bottom*. Patterning parameters are (**a**, **b**) 20 keV, 5pA, 5 ms dwell time; (**c**, **d**) 30 keV, 2pA, 200 ms dwell time; (**e**, **f**) 30 keV, 40pA, 200 ms dwell time

relationship between the FIB milled pattern and the quantum dots grown later. AFM images have been captured by a Veeco Nanoscope V instrument operated in tapping (non-contact) mode. SEM secondary electron imaging (SEI) is an essential tool to examine the patterned region with sufficient lateral resolution. Some typical AFM and SEM images are shown in the following (Figs. 11.2 and 11.3).

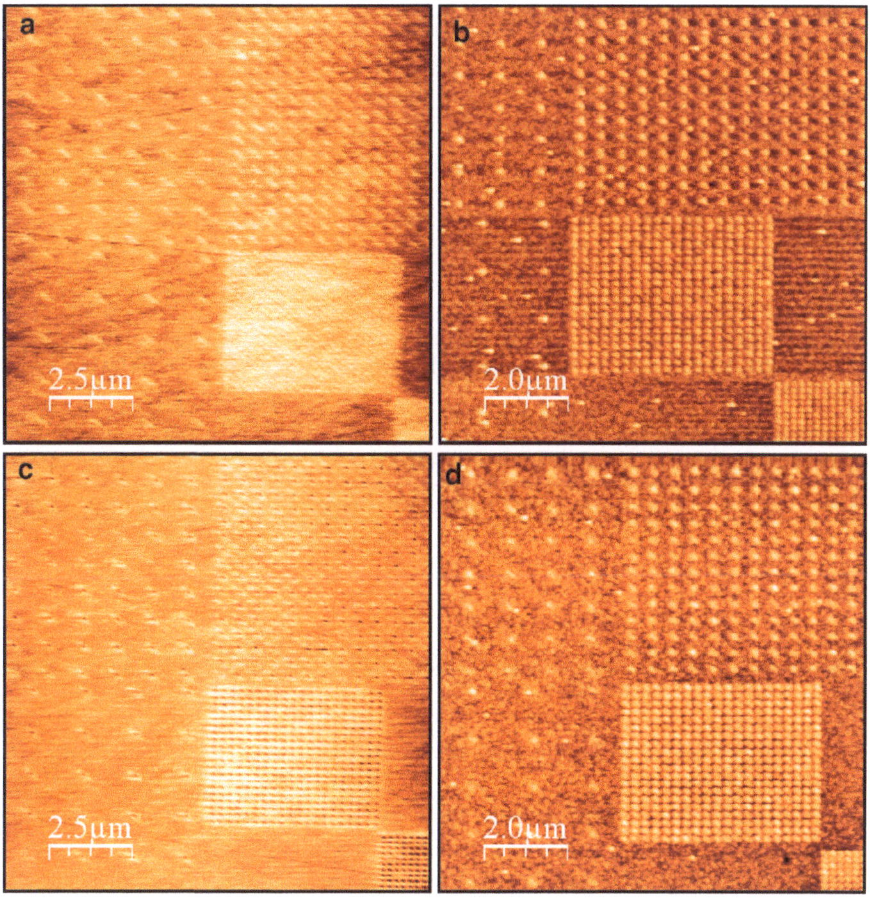


**Fig. 11.3** AFM images captured in tapping mode before (**a**, **c**) and after (**b**, **d**) overgrowth. The field of view is 10 μm and the sampling resolution is 10 nm, (**a**) shows the $Ga^+$ implantation created hillock features, while in (**c**) it created tiny holes. Both patterns successfully triggered nucleation of QDs, as (**b**) and (**d**) demonstrate. The patterning parameters are (**a**), (**b**) 20 keV ion energy, 5pA ion probe current, 5 ms dwell time and (**c**, **d**) 30 keV ion energy, 20pA ion probe current, 5 ms dwell time

## 11.3 Investigating the Role of Different Focused Ion Beam Parameters

The damage from ion beam implantation into the substrate is controlled by accelerating voltage, ion probe current, ion dose and angle of incidence. The crystallographic structure of the implanted substrate will affect the damage as well. We used the Monte Carlo ion beam implantation simulator code *TRIM* (*Transport of Ion in Matter*) [17] to simulate $Ga^+$ beam implantation into GaAs for different accelerating voltages (10 kV, 20 kV and 30 kV). The angle of incidence was 0°, so that the ion beam was perpendicular to the surface. Each simulation was for an ion dose of

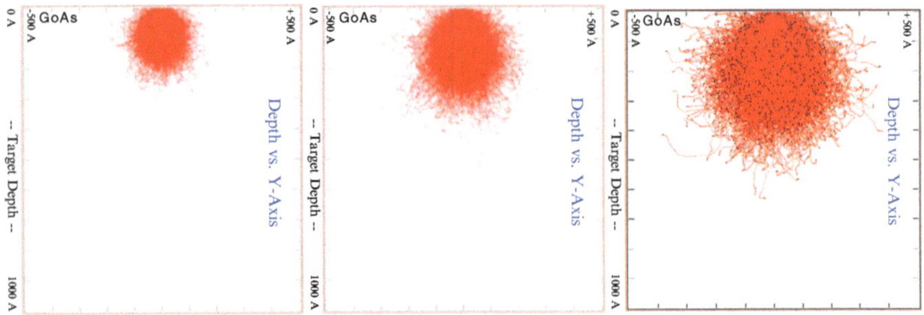


**Fig. 11.4** TRIM simulation for $Ga^+$ implantation into GaAs with different acceleration voltages (10 keV, 20 keV and 30 keV, from *left* to *right*). The FIB strikes from above. The fields of view are 100 nm × 100 nm

**Table 11.1** Ion energy vs. damage area size from SRIM

| Ion energy (keV) | Projected range (nm) | Longitudinal straggling (nm) | Lateral straggling (nm) |
|---|---|---|---|
| 10 | 8.0 | 5.1 | 3.6 |
| 15 | 10.5 | 6.4 | 4.6 |
| 20 | 12.9 | 7.6 | 7.6 |
| 25 | 15.2 | 8.8 | 8.8 |
| 30 | 17.4 | 9.9 | 7.2 |

10,000. Material re-sputtering was not considered. The simulation results are shown in Fig. 11.4.

The aim of this simulation is to show the influence of different acceleration voltages.The simulation indicates that a higher accelerating voltage will produce a larger damage volume inside the substrate, with the ions spreading wider laterally and penetrating deeper into GaAs. The *SRIM* (*Stopping Range of Ion in Matter*) simulation yields the relationship between the projected range and the ion energy from 10 keV to 30 keV, as listed in Table 11.1. This simulation is useful for choosing the later annealing and overgrowth process parameters, as it indicates how much surface layer can be removed without eliminating ion implantation-affected areas.

Corresponding to different combinations of accelerating voltage and ion dose, there are three types of damage at impact which we can distinguish according to their influence on later overgrowth, as sketched in Fig. 11.5. From left to right, the focused ion beam-induced damage is increasing. If the ion beam energy or dose is relatively low, the implanted $Ga^+$ ions create very low sputtering, and most of them will be implanted underneath the top surface, leading to hillock formation at the implanted region due to swelling of the lattice. Later annealing may completely repair the surface and flatten it, but residual influence from $Ga^+$ may still remain in the form of point and lattice defects formed at a certain depth beneath the surface, as in the case of ion implantation for doping. Slightly increasing the ion beam energy or dose, the ion beam will significantly damage the exposed region, first creating an amorphous surface layer. The damage region will be partially repaired

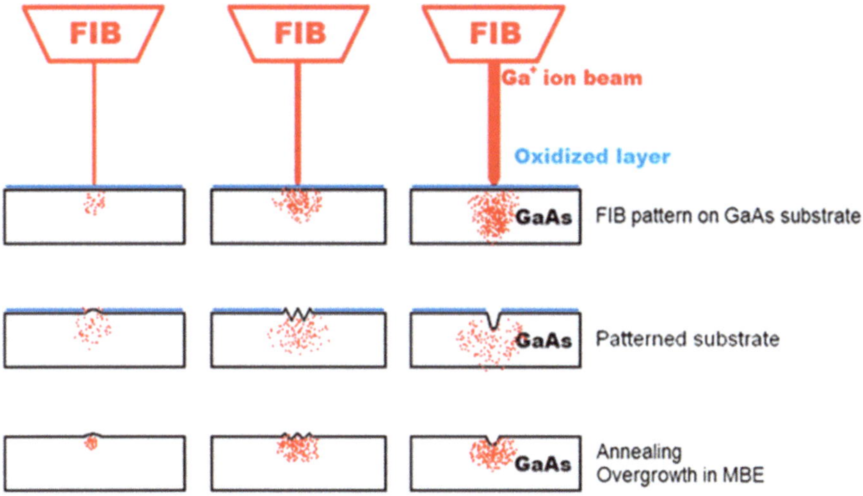


**Fig. 11.5** Patterning a GaAs substrate by FIB under different conditions, with ion beam energy and/or dose increasing from *left* to *right*, leading to lattice swelling by ion implantation (*left*), amorphisation (*centre*) and finally sputtering (*right*)

by subsequent annealing, but will not completely heal. A rough surface is expected to influence later overgrowth significantly, as shown in the middle part of Fig. 11.5. When the ion beam has both high energy and ion dose, it will sputter out material rapidly and eventually form a large and deep hole (diameter > 30 nm, depth >> 50 nm) at the patterned position. It should be noted here that once a hemispherical hole has been sputtered out, $Ga^+$ ions will hit its sides under a shallower angle than the bottom, which means the sputtering rate at the side walls will be much lower than at the bottom of the holes, and the holes will thus extend preferentially in depth, creating holes of very high aspect ratios and almost vertical side walls [18]. Such holes remain almost the same after annealing and will affect the overgrowth in ways different from the other two before-mentioned mechanisms, as the surface is strongly corrugated and the side walls of the holes can form various facets with respect to the [001] surface normal.

Different types of damage result in different nucleation geometries of QDs, as captured by both AFM and SEM and shown in Fig. 11.6. The fabricated patterns have distinctly different forms.

After growth, from the images captured, we clearly found InGaAs was more likely to nucleate within the ion beam-patterned area. Structures formed at the patterned locations fell qualitatively into three classes: islands, rings or deep holes. Islands were mainly found in the regions previously scanned by FIB with short dwell times and low probe currents ($<10^3$ ions/spot [19, 20]) where the ion beam implantation produced only very shallow holes or even hillocks rather than holes. When the dwell times and probe currents were increased, the ion beam produced deeper holes during patterning, and the material deposited during overgrowth

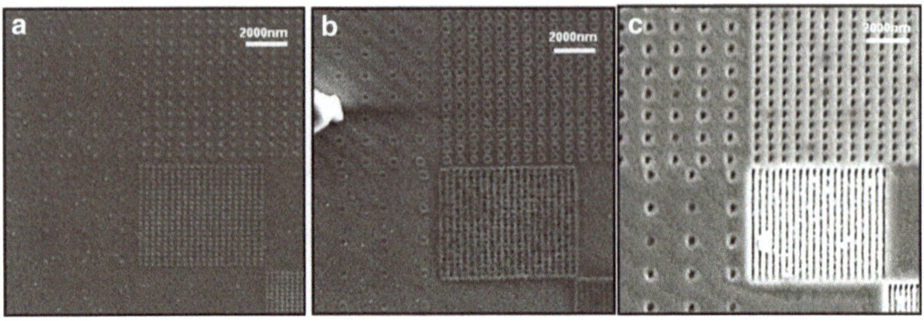


**Fig. 11.6** Three distinct types of In(Ga)As nucleation geometries with increasing ion dose from *left* to *right*: islands (**a**), rings (**b**) and deep holes (**c**)

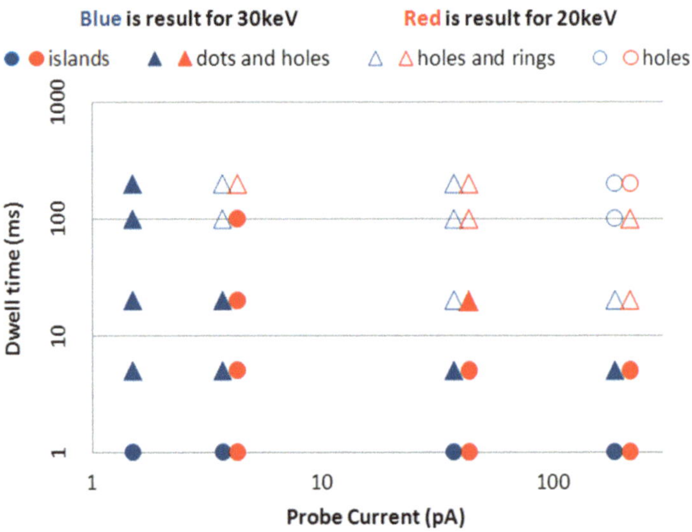


**Fig. 11.7** FIB patterns milled with short dwell times and small aperture sizes (low ion dose) produced islands. On the other hand, long dwell times and big aperture sizes (high ion dose) produced In(Ga)As rings or left only holes with some InGaAs nucleating preferentially at their edges. Most of the information has been collected from SEM

nucleated preferentially in the form of rings surrounding the edges of the patterned rather shallow and amorphous depressions. When dwell times and probe currents increased further, the drilled holes became too deep to be repaired by annealing and remained after overgrowth, surrounded by mounds that could form rings only if clearly separated laterally, as shown below in Fig. 11.6c. In regions not patterned by FIB, QD nucleation was random and generally less likely (Fig. 11.7 and Table 11.2).

The following investigation is focused only on islands, fabricated by MBE overgrowth of low-dose FIB-patterned GaAs. We have determined the relationship between the dimensions of islands and the FIB parameters. The data include the diameter of the drilled holes before overgrowth, the heights and diameters of the

**Table 11.2** Parameters for producing InGaAs quantum dots

| | Aperture (Ion current) | #1 (1.5 pA) | #2 (4.0 pA) | #3 (40 pA) | #5 (40 pA) |
|---|---|---|---|---|---|
| 20 keV | | | | | |
| Dwell time (ms) | 200 | N/A | Dots | Holes/rings | Holes |
| | 100 | N/A | Dots | Holes/rings | Holes/rings |
| | 20 | N/A | Dots | Dots/holes | Dots/rings |
| | 5 | N/A | Dots | Dots/holes | Dots |
| | 1 | N/A | Dots | Dots/holes | Dots |
| 30 keV | | | | | |
| Dwell time (ms) | 200 | Dots/rings | Holes/rings | Holes/rings | Holes |
| | 100 | Dots/holes | Holes/rings | Holes/rings | Holes |
| | 20 | Dots/holes | Dots/holes | Holes/rings | Holes/rings |
| | 5 | Dots/holes | Dots/holes | Dots/holes | Dots/holes |
| | 1 | Dots | Dots | Dots | Dots |

The 20 keV FIB with probe current 1.5pA produced a signal-to-noise ratio too low for evaluation, and this column is marked as N/A in the table

InGaAs islands, the pitch size of the arrays and the distribution homogeneity of the structures formed. The height and diameter of islands are measured by analysing AFM images as shown schematically in Fig. 11.8. The measurements were collected from four patterned areas and comprise data from three patterns formed at 20 keV $Ga^{+}$ ion energy and 5pA probe current with three different dwell times (1 ms, 5 ms and 20 ms), as well as from one pattern formed at 20 keV ion energy but with the probe current increased to 40pA at a dwell time of 1 ms. From Fig. 11.9a, it can be seen that the island heights are almost constant, covering a narrow range from 12 nm to 25 nm, with values slightly increasing with pitch. The diameter scales with the square root of the pitch, as can be seen from Fig. 11.9b, and is also increasing with the dwell time and probe current, which is not so obvious from the diagram. The result confirms the possibility of controlling the nucleated quantum dot size. However, it is clearly difficult to create islands under 100 nm in diameter and below 10 nm in height, which will be a prerequisite for strong quantum confinement effects in optical spectra. As the amount of deposited material per unit area is fixed (1.5 ML InAs) and the volume of each island can be estimated as shown in Fig. 11.8c, the total volume of material being converted into dots per unit area ($T = V_{QD}/\text{pitch}^2$) can be used to determine the fraction of InAs located in QDs rather than in the wetting layer. More material being converted into dots means the FIB pattern has a stronger influence on the nucleation of the islands.

For each test pattern, the ion beam parameter and gun alignment were kept fixed. Each pattern contained 5 sub-patterns with different pitch sizes of 200, 250, 500, 1,000 and 1,500 nm. This design helped us to analyse the relationship between the pitch sizes of the arrays and the distribution homogeneity of the structures formed, as shown in Table 11.3. We evaluated each pattern of sufficient signal-to-noise ratio with respect to two characteristics: the probability, $P^{+}$, of islands forming in unwanted positions (‘false positives’: islands formed outside the patterned sites)

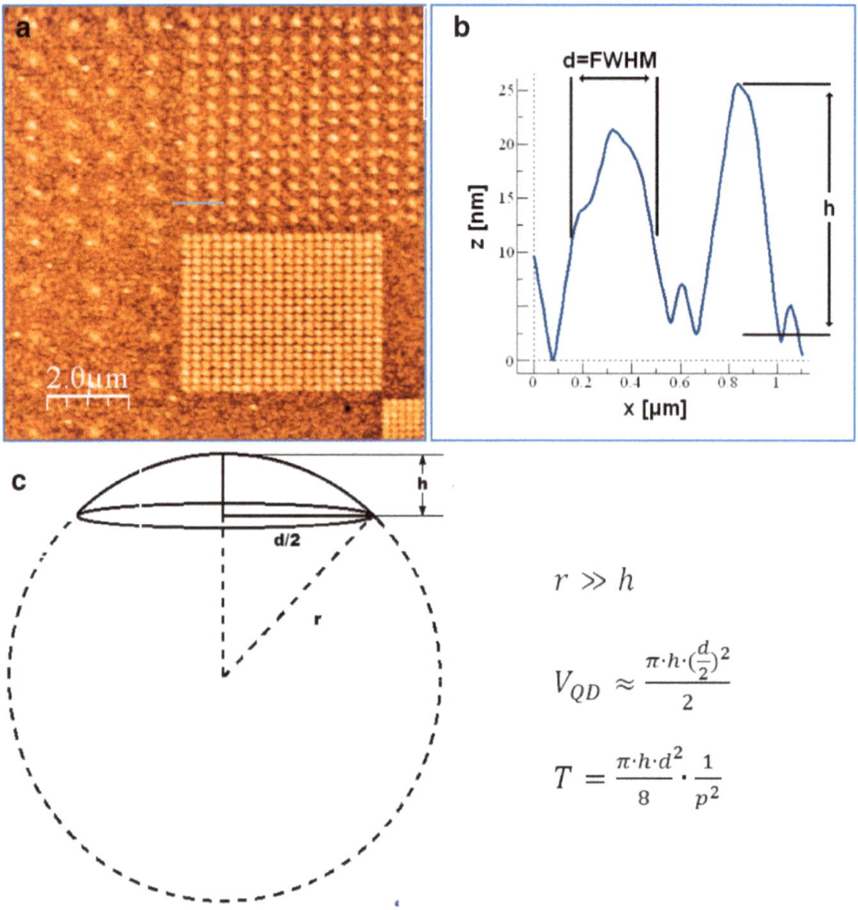


**Fig. 11.8** The size of quantum dots is measured by analysing AFM images. The heights are measured as peak heights above background and the diameters are measured as the full widths at half maximum (FWHM). The volume, $V_{QD}$, of quantum dots is then calculated from their diameter, d, and height, h, assuming the geometry of a spherical cap as sketched in **c**). $T = V_{QD}/p^2$, where p is the pitch of the array (i.e. distance between the QDs), then describes the average effective thickness of the QD material deposited

and the probability, $P^-$, of islands not having formed at patterned sites ('false negatives': islands missing at patterned sites). The results show that only patterns with a pitch of ≤500 nm create acceptable QD arrays with less than 3 % nucleation of islands outside the pattern and less than 3 % of islands missing at the patterned sites. This result agrees with Hull's study on the distribution control of SiGe QDs grown on patterned silicon [20]. In general, the failure rates $P^+$ and $P^-$ increase with ion dose, as shown in Fig. 11.10, which again confirms that low ion dose patterns create more perfect arrays of smaller islands that are more appropriate for optoelectronic applications. Generally, $P^+ > P^-$, i.e. it is much more likely that additional islands

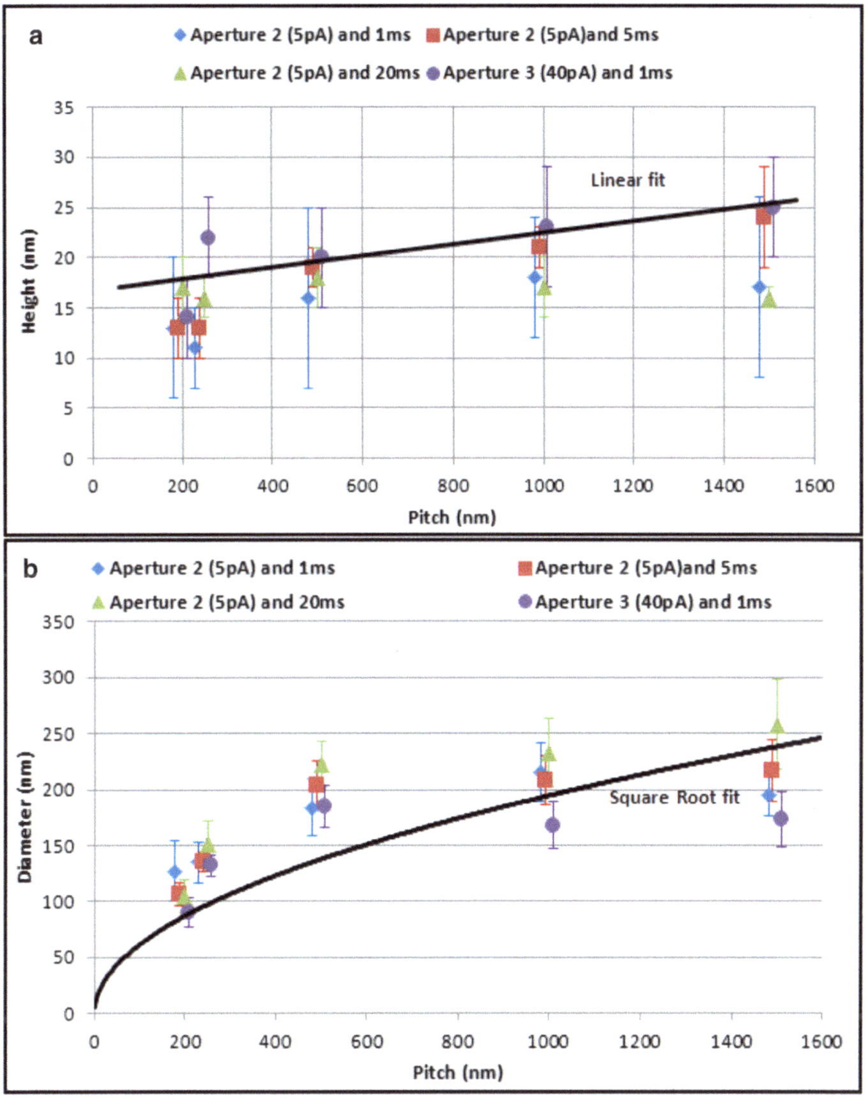


**Fig. 11.9** Plots of height (**a**) and diameter (**b**) of islands nucleated from four different FIB patterns at 20 keV

**Table 11.3** Statistical distribution of islands in patterned areas for one set of parameters

| 20 keV ion energy, 4pA probe current and 5 ms dwell time | | | | | |
|---|---|---|---|---|---|
| Pitch [nm] | 1,500 | 1,000 | 500 | 250 | 200 |
| $P^-$ [%] | 6 | 7 | 3 | <0.1 | <0.1 |
| $P^+$ [%] | 67 | 18 | 2 | 0.3 | 0.4 |

This pattern was created with fixed ion beam and column alignment, and it contained five arrays with different pitches

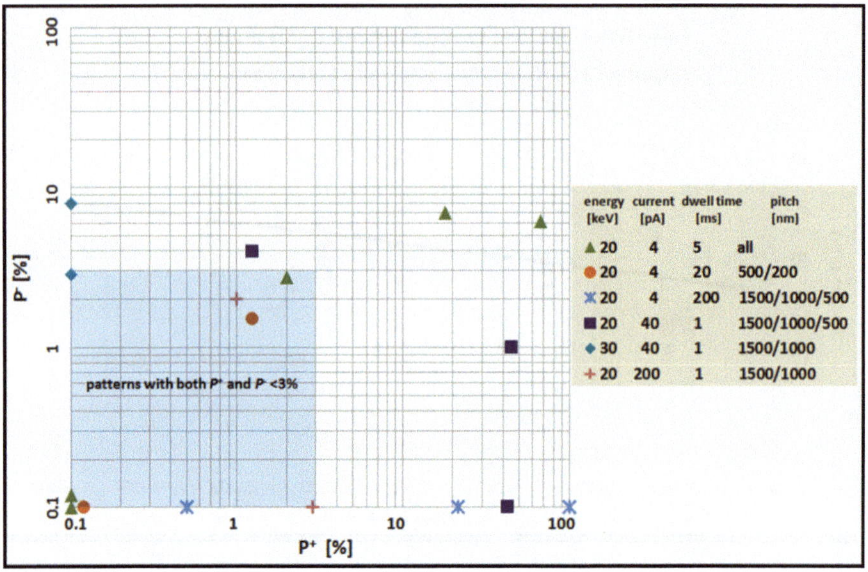


**Fig. 11.10** Comparison of two types of island nucleation defects. $P^+$ describes the fraction of islands nucleated off patterned sites, $P^-$ the fraction of islands having failed to nucleate at patterned sites

form off patterned sites than islands are missing at patterned sites. There is also the possibility of nucleating multiple QDs on the same specific site. The number of QDs formed per site depends on the geometry of the patterned feature and the overgrowth temperature [12, 21].

## 11.4 μ-Photoluminescence Investigation

Micro-photoluminescence (μ-PL) studies were performed at room temperature after overgrowth and SEM and AFM imaging. The exciting laser (green) had a diameter of ~5 μm, a wavelength of 532 nm and 1–2 mW power. μ-PL measurements were performed scanning the laser beam along a straight line from positions outside the patterns, where QDs had randomly nucleated on the surface, across the entire patterned area with a step size of 5 μm, as shown in Fig. 11.11a. The spectra are not as sharp as expected [22]. There are two rather broad peaks in the spectra at ~878 nm (1.41 eV) and ~1,152 nm (1.08 eV). They are related to the underlying GaAs substrate and the InGaAs QD signal. From the latter peak wavelength, we can estimate the compositional ratio of Ga:In atoms as 74:26 (i.e. $In_{0.26}Ga_{0.74}As$), assuming Vegard's law for bulk materials can be applied here as well [2]. The FIB-patterned area shows higher signals for both peaks, where the signal from the GaAs substrate increases more rapidly. The enhanced optical contributions from both the pure

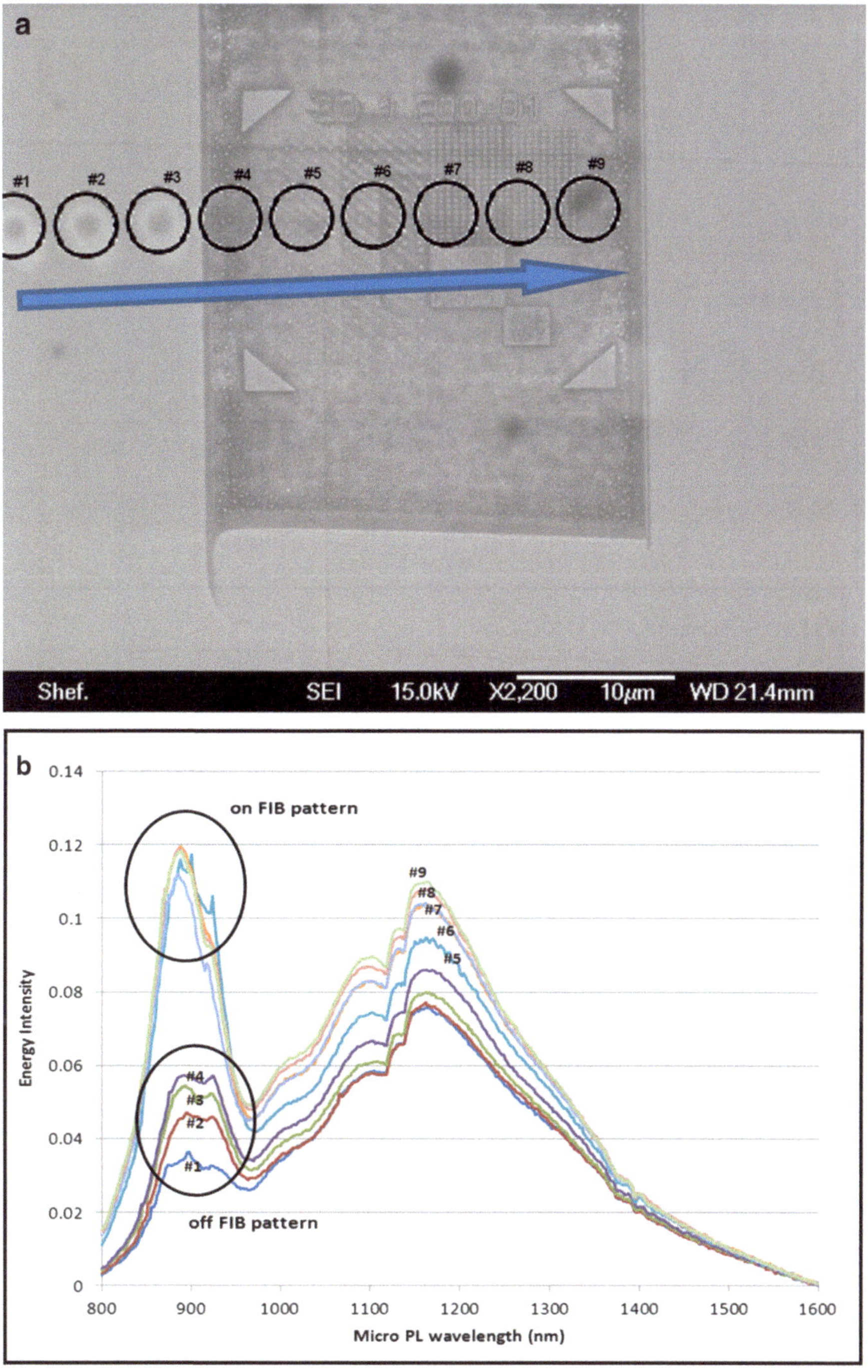


**Fig. 11.11** SEM image of array investigated by μ-PL. Scan started ~15 μm away from patterned regions and moved 40 μm across the patterned area (**a**). Set of experimental μ-PL spectra (**b**). The intensities at two wavelengths, 878 nm (GaAs) and 1,152 nm (InGaAs), were extracted at each scan position (**c**)

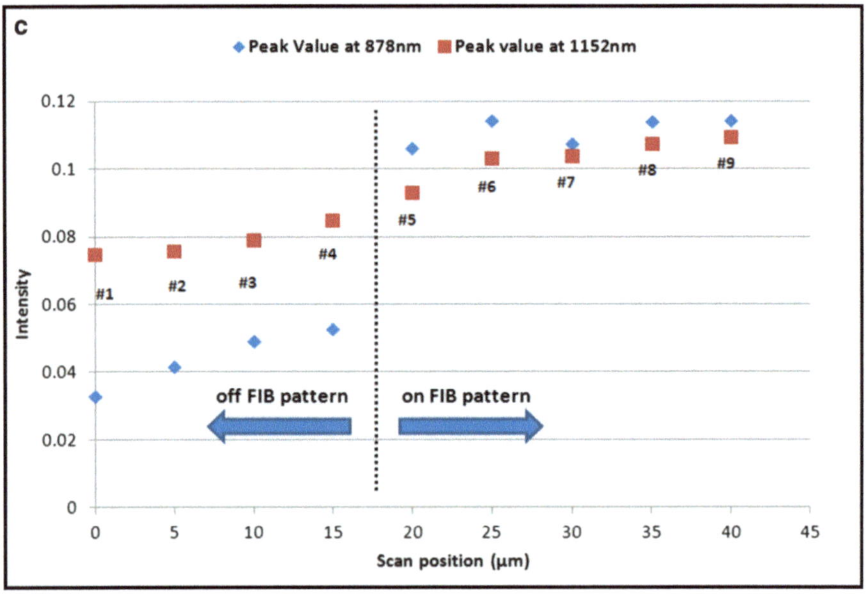


**Fig 11.11** (continued)

GaAs and the InGaAs islands may be explained by indium atoms from the wetting layer formed upon epitaxy having been more efficiently integrated into the islands when the surface has previously been FIB patterned, confirming the formation of QDs.

## 11.5 Conclusion

It has been clearly proven that FIB patterning of a GaAs substrate influences the nucleation of InGaAs islands which form QDs. The technique can be used to fabricate regular QD arrays if the patterning ion beam can be adjusted to a low ion dose. With this technique we gain control of the location, size and distribution of QDs formed.

**Acknowledgements** The authors thank Dr Ian M Ross of the department for assistance with the dual-beam FIB instrument and Dr David DT Childs of the National Centre for III/V Technologies for the μ-PL measurements.

## References


1. Franchi, S., Trevisi, G., Seravalli, L., Frigeri, P.: Quantum dot nanostructures and molecular beam epitaxy. Prog. Cryst. Growth Charact. Mater. **47**(2–3), 166–195 (2003)
2. Bhattacharya, P., Ghosh, S., Stiff-Roberts, A.D.: Quantum dot opto-electronic devices. Ann. Rev. Mater. Res. **34**, 1–40 (2004)
3. Gray, J.L., Hull, R., Floro, J.A.: Periodic arrays of epitaxial self-assembled SiGe quantum dot molecules grown on patterned Si substrates. J. Appl. Phys. **100**(8), 084312 (2006)
4. Dias, N.L., Garg, A., Reddy, U., Young, J.D., Verma, V.B., Mirin, R.P., et al.: Directed self-assembly of InAs quantum dots on nano-oxide templates. Appl. Phys. Lett. **98**(14), 141112 (2011)
5. Lee, J., Saucer, T.W., Martin, A.J., Tien, D., Millunchick, J.M., Sih, V.: Photoluminescence Imaging of Focused Ion Beam Induced Individual Quantum Dots. Nano Lett. **11**(3), 1040–1043 (2011)
6. Wang, J., Yuan, J., Otten, F.W.M.V., Notzel, R.: Combining selective area growth and self-organized strain engineering for site-controlled local InAs/InP quantum dot arrays. J. Cryst. Growth. **335**, 25–27 (2011)
7. Kalliakos, S., Garcia, C.P., Pellegrini, V., Zamfirescu, M., Cavigli, L., Gurioli, M., et al.: Photoluminescence of individual doped GaAs/AlGaAs nanofabricated quantum dots. Appl. Phys. Lett. **90**(18), 181902 (2007)
8. Chou, S.Y., Krauss, P.R., Renstrom, P.J.: Imprint of sub-25nm vias and trenches in polymers. Appl. Phys. Lett. **67**(21), 3114–3116 (1995)
9. Ben Naceur, H., Mzoughi, T., Moussa, I., Rebey, A., El Jani, B.: Properties of InAs grown on misoriented GaAs substrates by atmospheric pressure metal-organic vapor phase epitaxy. Nucl. Instrum. Meth B **268**(3–4), 236–240 (2010)
10. Hiroshima, H., Komuro, M.: Control of bubble defects in UV nanoimprint. Jpn. J. Appl. Phys. **46**(9B), 6391–6394 (2007)
11. Gerard, J.M., Sermage, B., Gayral, B., Legrand, B., Costard, E., Thierry-Mieg, V.: Enhanced spontaneous emission by quantum boxes in a monolithic optical microcavity. Phys. Rev. Lett. **81**(5), 1110–1113 (1998)
12. Cheng, C.-C., Meneou, K., Cheng, K.Y.: Effects of nano-pattern size on the property of InAs site-controlled quantum dots. J. Cryst. Growth **323**(1), 180–182 (2011)
13. Mehta, M., Reuter, D., Melnikov, A., Wieck, A.D., Remhof, A.: Focused ion beam implantation induced site-selective growth of InAs quantum dots. Appl. Phys. Lett. **91**(12), 123108 (2007)
14. Kiravittaya, S., Heidemeyer, H., Schmidt, O.G.: Growth of three-dimensional quantum dot crystals on patterned GaAs (001) substrates. Phys. E. **23**(3–4), 253–259 (2004)
15. Leonard, D., Pond, K., Petroff, P.M.: Critical layer thickness for self-assembled InAs island on GaAs. Phys. Rev. B **50**(16), 11687–11692 (1994)
16. Walther, T., Cullis, A.G., Norris, D.J., Hopkinson, M.: Nature of the Stranski-Krastanow transition during epitaxy of InGaAs on GaAs. Phys. Rev. Lett. **86**(11), 2381–2384 (2001)
17. Liu, F.Q., Wang, Z.G., Wu, J., Xu, B., Gong, Q., Liang, J.B.: Structure and photoluminescence of InGaAs self-assembled quantum dots grown on InP(001). J. Appl. Phys. **85**(1), 619–621 (1999)
18. Walther, T., Sehrbrock, A., Quandt, E.: Analysis of focused ion beam cut profiles by transmission electron microscopy. Proc ICM-16; Sapporo, Japan **2**, 1112 (2006)
19. Hull, R., Floro, J., Graham, J., Gray, J., Gherasimova, M., Portavoce, A., et al.: Synthesis and functionalization of epitaxial quantum dot nanostructures for nanoelectronic architectures. Mater. Sci. Semicond. Process. **11**(5–6), 160–168 (2008)
20. Hull, R., Floro, J.A., Gherasimova, M., Graham. J.F., Gray. J.L., Portavoce. A, et al.: Bridging the length scales between lithographic patterning and self assembly mechanisms in fabrication of semiconductor nanostructure arrays. In: Walther T, Nellist PD, Hutchison JL, Cullis AG (editors) J Phys Conf Ser. **209**, 012003 (2010)

21. Lee, J.Y., Noordhoek, M.J., Smereka, P., McKay, H., Millunchick, J.M.: Filling of hole arrays with InAs quantum dots. Nanotechnology **20**(28), 285305 (2009)
22. Srujan, M., Ghosh, K., Sengupta, S., Chakrabarti, S.: Presentation and experimental validation of a model for the effect of thermal annealing on the photoluminescence of self-assembled InAs/GaAs quantum dots. J. Appl. Phys. **107**(12), 123107 (2010)

# Chapter 12
# Development of Functional Metallic Glassy Materials by FIB and Nanoimprint Technologies

**A. Inoue, D.V. Louzguine-Luzgin, and Fahad Al-Marzouki**

Patterning techniques are very important in many areas of modern science and technology including applications. Production of micro- and nano-sized periodical structures has been intensively studied in the last years in order to generate one-, two-, and three-dimensional periodical structures in metals, polymers, and ceramics. The process of patterning commonly involves many individual steps before obtaining the desired structure. In general, a pattering method consists of the following elements: formation of a mask or muster, choice of transfer medium or replications of patterns, and choice of resist, which is usually a functional material capable of serving as the resist. Metallic glass is one of the promising materials for such a purpose. Recently, glassy alloys attracted much attention as imprinting materials because they soften on heating similar to polymers, but on subsequent cooling retain high hardness at room temperature and demonstrate good corrosion resistance [1]. It has been reported that a three-dimensional structure with several tens or hundreds of nanometer in characteristic length scale could be prepared using Pt-, Zr-, Au-based glassy alloy ribbon and Zr-based glassy alloy in bulk form [2–5] or in a thin film form [6]. In the present chapter we discuss usage of metallic glasses for nanoimprinting and other similar processes.

A. Inoue (✉)
Tohoku University, Sendai, 980-8577, Japan

Physics Department, King Abdulaziz University, Jeddah, Saudi Arabia
e-mail: ainouebmg@yahoo.co.jp

D.V. Louzguine-Luzgin
WPI Advanced Institute for Materials Research, Tohoku University, Sendai, 980-8577, Japan
e-mail: dml@wpi-aimr.tohoku.ac.jp

F. Al-Marzouki
Physics Department, King Abdulaziz University, Jeddah, Saudi Arabia
e-mail: fmphy@kau.edu.sa

Z.M. Wang (ed.), *FIB Nanostructures*, Lecture Notes in Nanoscale Science and Technology 20, DOI 10.1007/978-3-319-02874-3_12,

## 12.1 Metallic Glasses as Materials for Nanoimprinting

Nanoimprint technology is a promising method of mass production of micro- and nanodevices producing a high-quality product at relatively low cost [7]. Metallic glasses are the most promising metallic materials used for thermal nanoimprinting because of their amorphous structure and nanometer-scale homogeneity. They have excellent mechanical properties and material functionalities such as high corrosion resistance and desirable magnetic properties. Thermoplastic materials, especially metallic glasses, are used for nanoimprinting owing to their rapid decrease in viscosity above the glass transition temperature ($T_g$). For example, Fig. 12.1 shows the DSC traces of the $Cu_{36}Zr_{48}Al_8Ag_8$ bulk glassy sample and the viscosity data calculated using thermomechanical analysis (TMA) traces from the deformation of the sample as a function of temperature. The viscosity ($\eta$) was calculated from normal stress ($\sigma_n$) and strain rate using the following equation: $\eta = \sigma_n/3$ [8]. The viscosity changes about 1.5 orders of magnitude within the supercooled liquid region when the glass transition is complete. The dotted line is an extrapolation which predicts that the viscosity should reach about $10^{12}$ Pa·s in the beginning of the glass transition on heating.

Metallic glasses exhibit perfect Newtonian viscous flow in the supercooled liquid temperature region above $T_g$ and below the crystallization temperature ($T_x$) (Fig. 12.2) and demonstrate excellent micro- and nanoformability under low stress [1, 10, 11]. Here, it should be pointed out that both temperatures $T_g$ and $T_x$ are not constant but depend on the heating rate and keeping time [12, 13] (Fig. 12.2).

Bulk metallic glasses can be shaped by application of electrical current [14]. Shaping of bulk metallic glasses (BMG) and BMG-based composites into various complex forms has been achieved by an electromechanical process. Bulk metallic glasses having large supercooled region between the glass transition temperature $T_g$ and the crystallization temperature $T_x$ up to 1–1.5 hundred degrees were utilized for this purpose. In this range, the supercooled liquid in principal deforms in a Newtonian way, allowing thermomechanical shaping in the low viscosity range. Electromechanical shaping technology allows rapid shaping of BMGs at low applied stresses. It allows eliminating the effect of the thermal mass of the furnace and not heating the deformation dies. Joule heating is efficiently used owing to the high electrical resistivity of bulk metallic glasses. The crosses and other mechanically resistant complex forms were achieved from two amorphous rods or an amorphous rod and a crystalline bar.

Various physical and chemical coating techniques have been investigated for the deposition of thin films. The physical methods include ion beam sputtering [15], RF magnetron sputtering [16], DC magnetron sputtering [17], and pulsed laser deposition (PLD) [18]. Laser interference metallurgy is a new method for periodic surface microstructure design on multilayered metallic thin films [19]. Chemical methods include chemical vapor deposition (CVD) [20] and chemical solution deposition (CSD) [21]. Metallic glasses can be produced by sputtering, melt spinning, casting, etc.

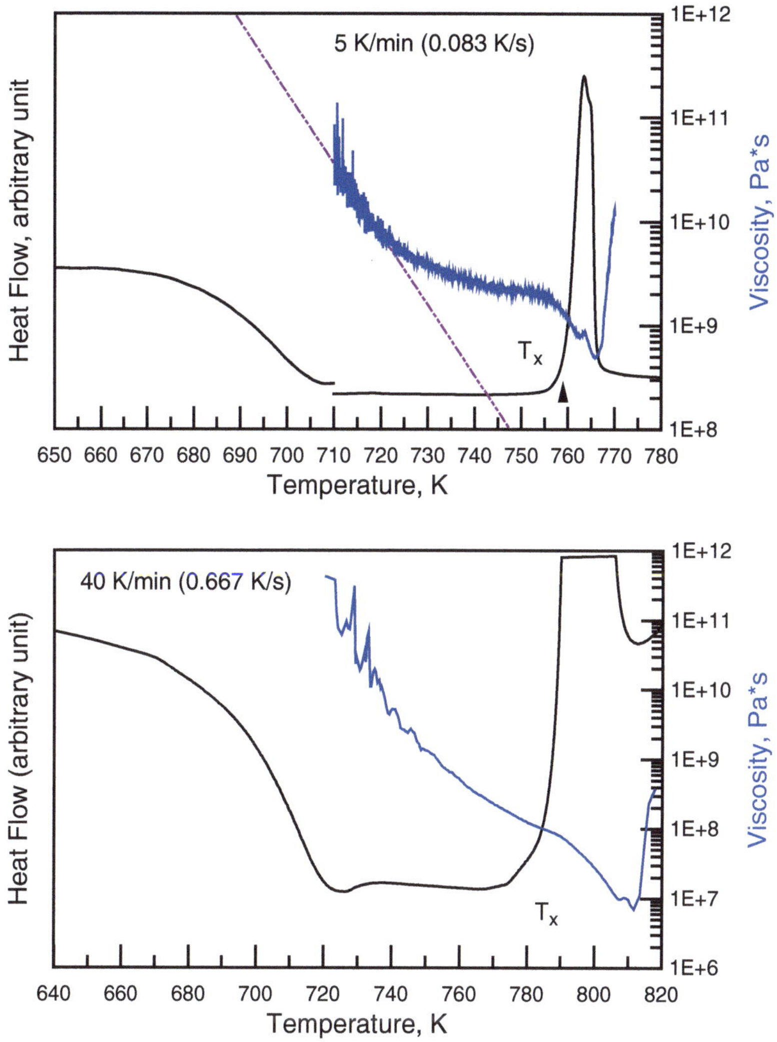


**Fig. 12.1** Viscosity of a $Cu_{36}Zr_{48}Al_8Ag_8$ glassy alloy as a function of temperature derived from the sample length change at the heating rate of 5 K/min (0.083 K/s) and 40 K/min (0.667 K/s) (*blue curve*). DSC trace is given for comparison (*black curve*). The DSC data below 710 K at 5 K/min were multiplied by 10 and shifted up for better visibility of the glass transition region. Reprinted from [9] with permission of Elsevier. © (2010) Elsevier

## 12.2 Fabrication of the Molds and Nanoimprinting

A pressing die must be prepared if a nanoscale pattern is going to be made on the surface of a metallic glassy material by an imprinting technology. Fabrication of micro- and nanometer-scale dies was carried out by electron beam lithography,

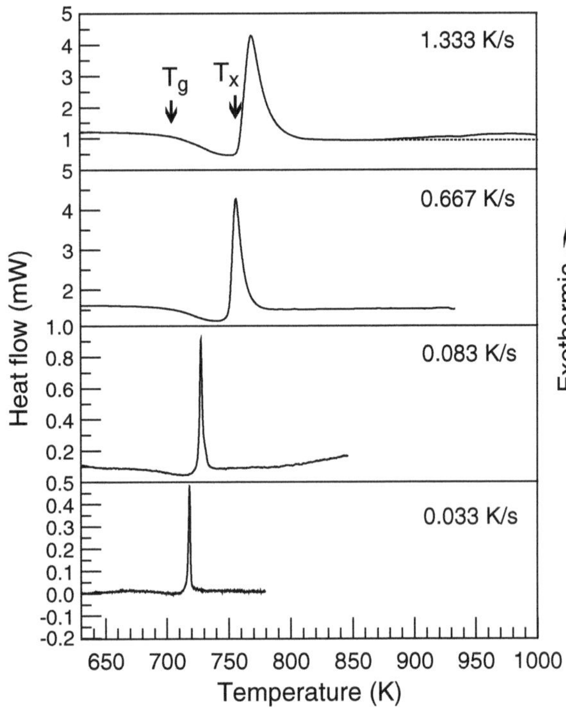


**Fig. 12.2** DSC curves of $Ti_{50}Ni_{25}Cu_{25}$ glassy alloy taken at different heating rates. Reprinted from [13] with permission of Springer. © (2000) Springer

electrolytic etching of silicon [22], UV-LIGA process [23], laser machining of polyimide film [24], and milling of various materials with a focused ion beam (FIB) [25]. For example, nano-dies having periodic structures, with dot pitches of 50 nm and 40 nm fabricated by FIB-assisted chemical vapor deposition (CVD) and reactive ion etching (RIE) of $SiO_2$, were used to perform nanoimprinting on Pt-based metallic glass [26]. Superior formability and nano-imprintability were shown using metallic glasses.

Nanoimprint lithography process consists of the following steps (Fig. 12.3). The first stage is preparation of the mold. It could be a Si wafer mold patterned using lithography and ion etching process. The second is the imprint process in which a mold with a nanostructure on its surface is pressed into a metallic glassy material, followed by removal of the mold. This step duplicates the nanostructures on the mold in the metallic glass. The metallic glass viscosity decreases as the temperature increases. During the imprint step, the metallic glass is heated to a temperature above its glass transition temperature but below $T_x$, the glass becomes a viscous liquid, can flow and, therefore, can be readily deformed into the shape of the mold. The mold creates a thickness variation pattern on the metallic glassy surface. In some cases reactive ion etching is used to remove the residual material in the compressed area. This process transfers the thickness variation pattern into the entire sample. Unlike conventional lithography methods, imprint lithography itself does not utilize expensive beam-current techniques. On cooling metallic glass becomes

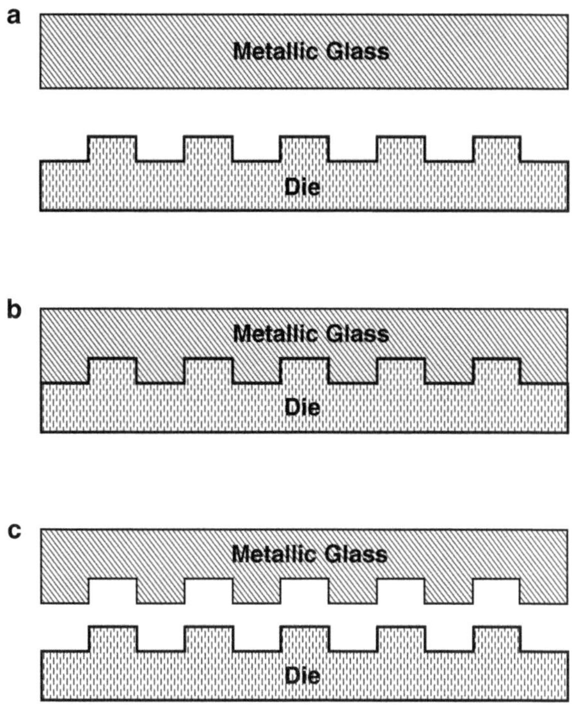


**Fig. 12.3** Schematic illustration of the imprinting process. (**a**) initial stage, (**b**) imprinting, (**c**) removal of the die

strong and solid again and can further be used as a die for other materials which soften at lower temperature.

In order to fabricate molds with concave nanohole arrays, Pt masks of crossed-line (lattice) patterns were deposited on $SiO_2/Si$ substrates by a dual-beam FIB-CVD system with a Ga- ion beam at an accelerating voltage of 30 kV, and nanohole arrays were fabricated by RIE of the $SiO_2$ layer with $CHF_3$ plasma [27]. The precursor gas [methylcyclopentadienyl trimethyl platinum, $(CH_3)_3(CH_3C_5H_4)Pt$] was preheated at 325 K and sprayed on the substrate through a fine nozzle heated at 403 K. The Pt was deposited on the substrate by decomposition of the gas molecules during ion beam irradiation in a vacuum atmosphere. The main Pt-deposition parameters are the ion beam current (IBC), dwell/irradiation time ($t_d$), and the number of scans (Ns). The beam diameter, deposition rate, and deposition area increase with increasing IBC. Therefore, a very low IBC of 0.3 pA corresponding to 10 nm in theoretical beam diameter was used for patterning the Pt mask on a nanometer scale. The deposition parameters of td and Ns influenced the cross-sectional profile of the mask patterns. The resulted line profiles were observed by SEM and measured by atomic force microscopy. Figure 12.4 shows an SEM microphotograph of a periodic line pattern of deposited Pt.

$SiO_2$ layer was treated by using a reactive ion etching apparatus. Figure 12.4b–f present the results of reactive ion etching of the periodic line pattern with a 25-nm pitch at radio frequency (RF) power values of (b) 5, (c) 50, (d) 100, and (e) and (f) 300 W under a pressure of 0.1 Pa. The etching depth of $SiO_2$ was set at 50 nm for

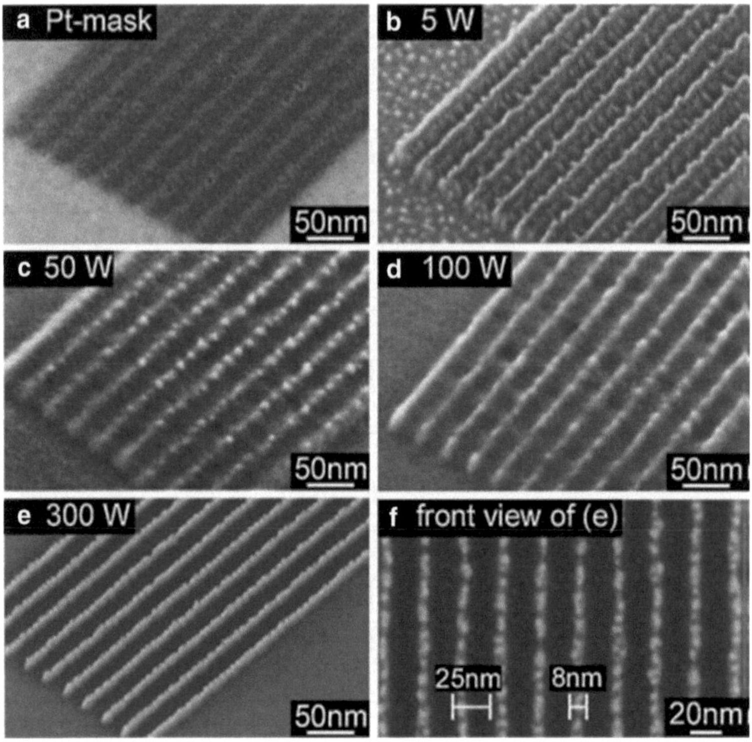


**Fig. 12.4** SEM microphotographs of (**a**) the deposited Pt mask of periodic line pattern with a 25-nm pitch on a $SiO_2$/Si substrate formed by FIB-CVD, and RIE of $SiO_2$ layer at RF powers of (**b**) 5, (**c**) 50, (**d**) 100, and (**e**, **f**) 300 W under a pressure of 0.1 Pa. Reproduced from [27] with permission of Japan Society of Applied Physics. © (2012) Japan Society of Applied Physics

each etching condition. The gaps between adjacent lines were not etched at low RF power conditions (Fig. 12.4b), whereas the clearest etching result was obtained at the highest RF power of 300 W (Fig. 12.4e). The minimum line width was 8 nm (Fig. 12.4f). The ion sputtering effects must be increased to fabricate periodic line pattern with a 25-nm pitch. The width and thickness of the line mask were increased by stacking at the crossing points and that nanohole arrays with pitches of 25 nm or less are therefore difficult to fabricate.

The surface quality of the metallic glassy deposited samples is an important feature and has to be taken into consideration if the sample is going to be applied as a nanoimprinting material. A successful preparation of Pd–Cu–Ni–P glassy alloy thin film using the pulsed laser deposition (PLD) method was reported recently [28]. The surface is relatively flat and smooth with an exception of some droplets and no distinct pore is seen. At high magnification, the surface structure appears to be granular. The average ($R_a$) and maximum ($R_y$) roughness values for such regions evaluated by AFM measurements are $0.11 \pm 0.02$ nm and $1.3 \pm 0.4$ nm, respectively. To confirm the oxidization resistance, the surface roughness was measured after 1 year

**Fig. 12.5** FE–SEM surface image of the Pd–Cu–Ni–P glassy alloy thin film in the as-imprinted state. Reproduced from [28] J. Non-Cryst. Solids with permission of Elsevier. © (2010) Elsevier

of deposition of the thin film. Even being exposed to the air, the surface roughness was not changed. It indicates that the Pd–Cu–Ni–P glassy alloy exhibits a high oxidization resistance even in the thin film state. The results on thermal imprinting test presented as shown in Fig. 12.5 revealed that the obtained Pd–Cu–Ni–P glassy alloy thin film has excellent nano-imprintability. After 10 s, the periodic dot pattern of the mold was precisely imprinted on the surface of the as-deposited film. Such a short production time is valuable for mass production. In addition, high oxidization resistance and high hardness lead to stable surface morphology. Consequently, it is concluded that the Pd-based glassy alloy thin film has excellent fundamental properties and nano-imprintability which exhibits potential for application of imprinting materials for fabricating bit patterned media.

Other metallic glasses also have a smooth surface and soften on heating about $T_g$ which allows creation of micron and nanoscale patterns on their surfaces by molding [2, 29, 30]. Nanoimprinting on the surface of the $Pt_{60}Ni_{15}P_{25}$ metallic glass was carried out by using the DLC/$Al_2O_3$ molds. This Pt-based metallic glass exhibits Newtonian viscous flow in the supercooled liquid temperature range of $\Delta T_x = 63$ K between $T_g = 482$ K and $T_x = 545$ K [31]. The Pt-based metallic glass specimen, of about 30 μm in thickness, was fabricated by a melt spinning method. Two kinds of straining behavior occur under compressive stress: specimen thickness reduction as a macroscopic deformation and local deformation near the specimen surface associated with nanoimprinting behavior. The $Al_2O_3$ substrate was mechanically and chemically polished and then subjected to a superplastic forming at 520 K for 30 s under a compressive stress of 140 MPa. Nanoimprinting was performed at 530 K for 15 s under a compressive stress of 50 MPa. After cooling down to room temperature, the mold could be released easily from the metallic glass. Figure 12.6 presents SEM microphotographs of the nanoimprinted surface of a metallic glass specimen obtained by using a mold with nanodot arrays with pitches of 33 and 25 nm. Concave arrays of nanoscale hole are observed on the surface of the specimen, and the replicated hole arrays precisely correspond to the nanodot shape of the mold. The depth of the nanoholes is below 10 nm. The wetting behavior, surface tension, and

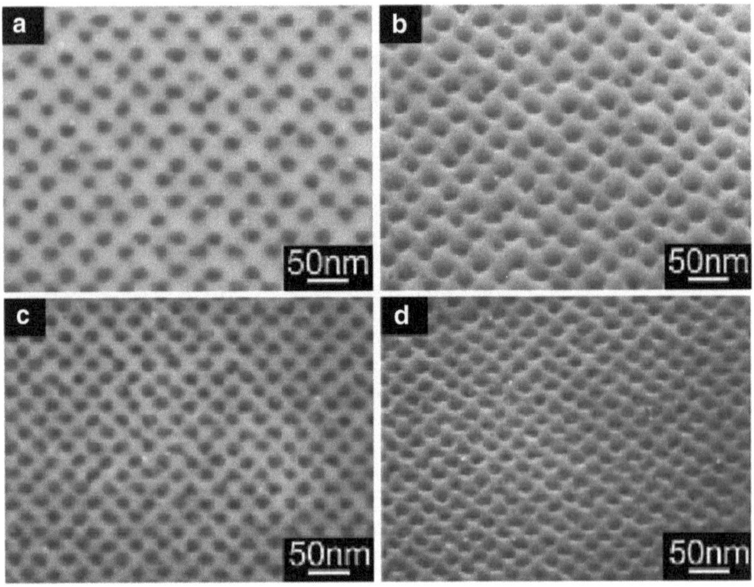


**Fig. 12.6** SEM microphotographs of nanoimprinted surfaces of $Pt_{60}Ni_{15}P_{25}$ metallic glass by using DLC molds with dot pitches of (**a, b**) 33 and (**c, d**) 25 nm. (**a**) and (**c**) are front views; (**b**) and (**d**) are tilted views by 45°. Reprinted from [27] with permission of Japan Society of Applied Physics. © (2012) Japan Society of Applied Physics

viscosity significantly influence the nanoformability of metallic glass at an ultrafine scale.

A technique for fabricating nanostructures by nanoforging using super fluidity of metallic glass in the supercooled liquid state was created by using nanoscale dies that are fabricated by a focused ion beam (FIB) technique [32]. FIB machining of Zr-based metallic glass is useful for fabricating nanoscale dies because of the isotropic homogeneity in the amorphous phase. These dies were used to nano-forge the samples of $Pt_{48.75}Pd_{9.75}Cu_{19.5}P_{22}$ metallic glass (Fig. 12.7). The thin foil specimens were heated in a small furnace and compressively loaded in a small vacuum chamber. A die-forged microgear and periodic nanostructures for optical applications were produced [32].

By using the low viscosity and good wetting properties of the supercooled liquid, different structures were obtained on the surface of $Pd_{40}Ni_{40}Si_4P_{16}$ metallic glass as presented in Fig. 12.8. The structures shown in Fig. 12.8c and d were formed using the Si molds shown in Fig. 12.8a, b. Complicated structures could also be formed as shown in Fig. 12.8e,f, indicating that this metallic glass is another good candidate material for imprinting. Figure 12.8g, h shows the porous alumina and the transferred nanoscale rods with diameters of about 100 nm on the surface of the $Pd_{40}Ni_{40}Si_4P_{16}$ alloy, respectively [34].

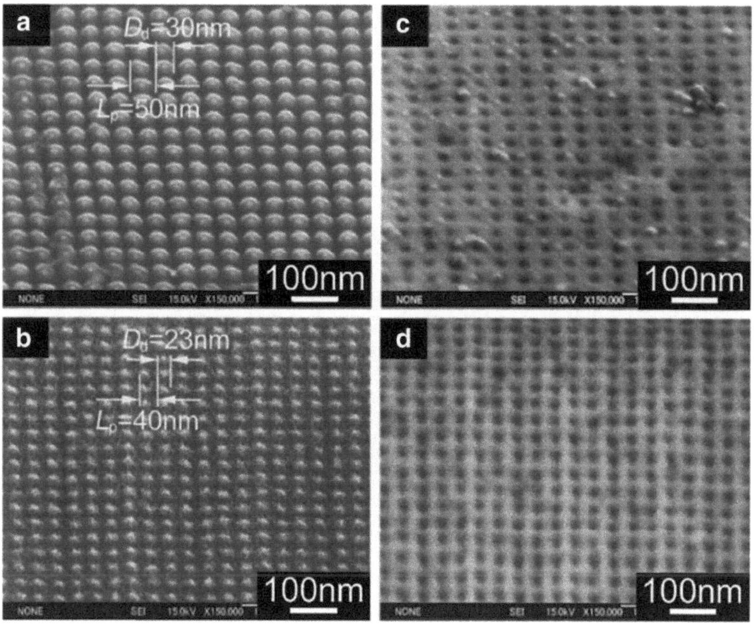


**Fig. 12.7** SEM microphotographs of dies having nanodot arrays with pitches of (**a**) 50 nm and (**b**) 40 nm. (**c**) and (**d**) are the nanoimprinted surfaces of the $Pt_{48.75}Pd_{9.75}Cu_{19.5}P_{22}$ metallic glass using the dies of (**a**) and (**b**), respectively. Reprinted from [33] with permission of Japan Institute of Metals and Materials. © (2011) Japan Institute of Metals and Materials

## 12.3 Near-Net-Shaped Casting

Near-net-shaped casting is used to fabricate castings of BMGs [35]. For example, die-cast BMG was used to fabricate a watch case [36]. A three-dimensional net-shaped microgear was successfully prepared by the precision die-casting technique (Fig. 12.9a). The alloy was melted in the ceramic sleeve using a high-frequency induction heating and injected into the mold by moving both the plunger and the sleeve. The temperature of the molten alloy was monitored using a radiation thermometer and fed back by a PID controller. The process was performed in a vacuum of $10^{-2}$ Pa. The maximum plunger speed and injection pressure were 3.0 m/s and 30 MPa, respectively. The as-cast microgear made of BMG has a glassy state in the injection temperature range from 1,300 to 1,690 K. Owing to good surface quality and high wear resistance [37], $Ni_{53}Nb_{20}Zr_8Ti_{10}Co_6Cu_3$ bulk glassy alloy was also used to cast a microgear (Fig. 12.9b). The glass transition temperature, crystallization temperature, and supercooled liquid region ($\Delta T_x$) are 842 K, 893 K, and 51 K, respectively. The heat of crystallization for a Ni-based microgear was found to be equal to that for the melt-spun ribbon, and the DSC traces were consistent with the XRD data. A microscale bearing made of $Ni_{53}Nb_{20}Zr_8Ti_{10}Co_6Cu_3$ glassy alloy showed an excellent durability and wear resistance [38]. The durability of Ni-based glassy alloy micro gears is about 300 times longer compared to a steel gear [39].

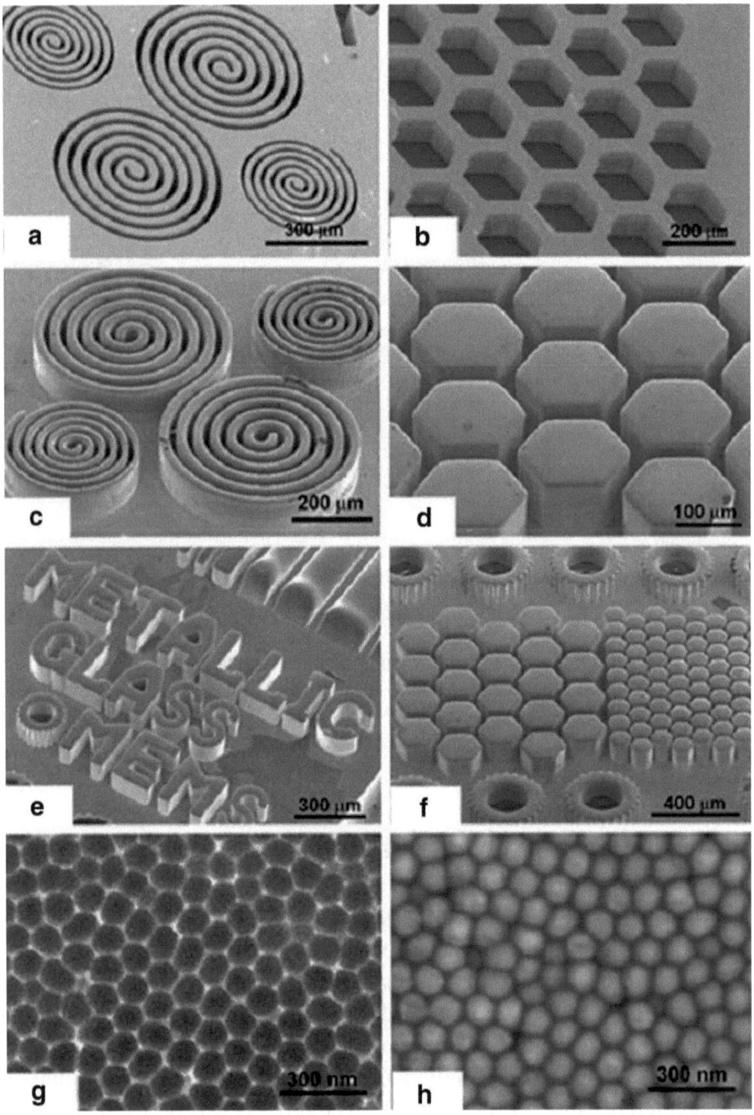


**Fig. 12.8** (**a** and **b**) are Si molds, (**c**–**f**) are the patterns fabricated with Si molds, (**g**) is porous alumina, and (**h**) is the corresponding nano rods fabricated with the porous alumina (**g**). All the above structures are obtained using molding on the $Pd_{40}Ni_{40}Si_4P_{16}$ glassy ribbons. Reprinted from [34] with permission of Springer © (2011) Springer

A Zr-based BMG was also prepared by the precision die-casting technique and exhibited excellent dimensional accuracy on the micrometer scale. BMGs were also applied for optical precision devices [40].

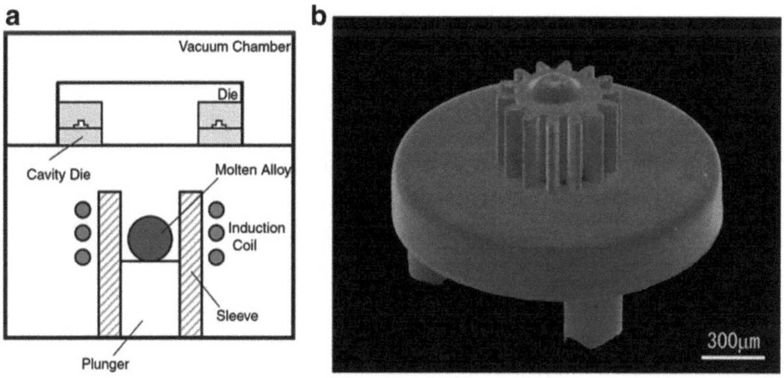


**Fig. 12.9** (**a**) Schematic illustration of the precision die-casting equipment. (**b**) SEM microphotographs of the extruded Ni-based glassy microgear prepared by the precision die-casting technique. Reprinted from [41] with permission of Japan Institute of Metals and Materials. © (2004) Japan Institute of Metals and Materials

## 12.4 Thermoplastic Forming for MEMs

Microelectromechanical systems (MEMS) devices become more and more incorporated in humans daily life. MEMS have found their application in many diverse areas. Injection molding is a long-established process for the manufacture of components, of sizes in a wide range. There is a growing market for components for micro-engineering applications or for multiscale components. MEMS materials have two major requirements. The material characteristics must be exploitable on a microscopic scale instead of the conventional macroscopic scale, and the forming, processing, and machining methods should be available for these materials. The micro-shape creation method includes complex processes. To make MEMS elements, especially structural elements, suitable materials and micro-machining technology must be developed. From the viewpoint of production engineering, plastic forming processes offer a significant advantage in productivity and enable mass production with controlled quality and low cost. In the forming process, the relationship between the die and the material is based on the difference in the strength of the materials. From this point of view, superplastic materials and glassy alloys in the supercooled liquid state have a great advantage in achieving deformation under very low stresses compared to conventional plastic deformation.

A backward extrusion machine was developed and employed to fabricate microgears. It is known that superplasticity allows achieving large elongation under relatively low stresses. The material exhibits a strain rate dependence on applied stresses, and the relationship between stress and strain rate is represented by equation $\sigma = \mathrm{ke}'^{m}$ of deformation under high temperatures. For superplastic alloys the strain rate sensitivity exponent $m \geq 0.5$. Similar characteristic deformation behavior is found in La-Al-Ni glassy alloy showed $m = 1$. This alloy was extruded at 490 K and microgear parts with high surface quality were successfully produced (Fig. 12.10).

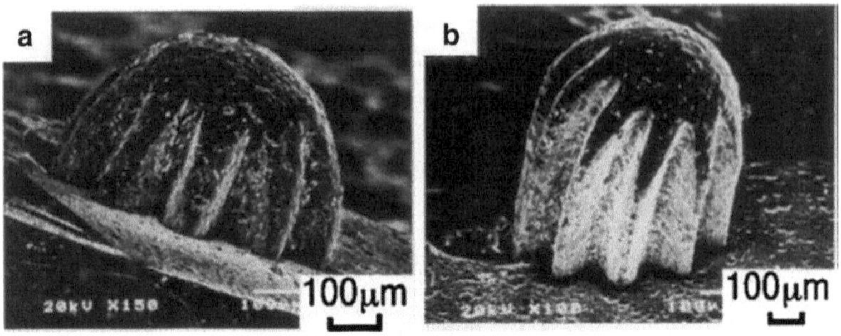


**Fig. 12.10** SEM microphotographs of the extruded microgear shafts made of $La_{55}Al_{25}Ni_{20}$ amorphous alloy. Reprinted from [42] with permission of Elsevier. © (2001) Elsevier

Thus, high-strength microgear shafts of 10–50 μm in size were extruded. It is found that upon the micro-extrusion, the tool surface roughness and lubrication conditions influence the forming behavior.

## 12.5 Magnetic Storage Devices

The capacity of data storage devices has increased greatly in recent years together with society's increased demand on storage of digital information. It is necessary to enhance the storage density of the magnetic storage devices used in hard disk drives (HDDs) as well. Current commercialized HDDs with a perpendicular magnetic recording method have areal recording densities of several hundred Gbit/in$^2$. However, the current recording method apparently cannot exceed an areal recording density of 1 Tbit/in$^2$ because of thermal instability phenomena connected with superparamagnetism [43]. The bit patterned media may help to overcome this obstacle [44, 45]. In bit patterned media, informational data are recorded in single isolated hard magnetic bits (islands) placed in a nonmagnetic materials. At an areal recording density of 2 Tbit/in$^2$, the dot pitch ($L_p$) of the magnetic bits is 18 nm. In order to produce bit patterned media, the ultrafine dot array is determined by nanopatterning techniques such as self-assembly with a block copolymer [46], electron beam (EB) lithography [47, 48], extreme ultra violet lithography, [49], and nanoimprint lithography [50]. Nanoimprinting is also a promising fabrication method for high efficiency and low-cost mass production of bit patterned media [51].

Recently, substantial research activities took place towards fabrication of molds for producing bit patterned media. The self-assembly techniques produce nanodot arrays with a sufficient resolution for bit patterned media. However, the position accuracy and dot shape uniformity issues remain unsolved [52]. Conventional electron beam lithography is often used for fabricating molds with ultrafine nanodot arrays. This method produces highly accurate dot arrays of $L_p = 18$ nm on resist materials owing to beam scanning controllability [53]. However, it is difficult to transfer the pattern to mold materials such as Si and $SiO_2$ because of the low etching stability on resist materials [54].

In Ref. [55] a focused ion beam-assisted chemical vapor deposition (FIB-CVD) technique was adopted [56] for patterning a platinum (Pt) etching mask with nanodot arrays. A mold with $L_p = 25$ nm was produced by diamond-like carbon by reactive ion etching (RIE) with oxygen plasma, confirming that FIB-CVD enables direct patterning of a metallic etching mask on the mold material. FIB-CVD and RIE offer a much simpler nanofabrication process than EB lithography. The deposited metallic mask achieves a higher selectivity than polymer resist materials produced by the reactive ion etching process. The resolution of FIB mask patterning is determined primarily by the distribution of the Pt etching mask, which is deposited by dissolving a precursor gas by secondary electrons emitted by ion irradiation. For a Ga+ion beam with an acceleration voltage of 30 kV, the areas of Pt deposition are 30–40 nm in radius. Although, a 25-nm pitch pattern was successfully formed in our previous study, an 18-nm pitch pattern was difficult to fabricate because adjacent dots in the Pt mask connected with each other. Thus, one of the purposes of this study is to investigate the feasibility of fabricating a mold at $L_p = 18$ nm by FIB-CVD and reactive ion etching.

A dual-beam FIB apparatus with a Ga ion beam at an accelerating voltage of 30 kV and a minimum ion beam diameter of 5 nm was used for the FIB-CVD experiments. The precursor gas [methylcyclopentadienyl trimethyl platinum, $(CH_3)_3(CH_3C_5H_4)Pt$] was preheated at 325 K in the reservoir and sprayed on the substrate through a fine nozzle heated to 403 K. The adsorbed precursor molecules on the substrate were dissolved by emitted secondary electrons, and only Pt was deposited on the surface. Pt mask patterning was performed on 50-nm-thick silicon dioxide on a single-crystal silicon ($SiO_2$/Si) substrate. FIB irradiation was conducted at a beam current of 0.3 pA. The spot drawing mode was used on FIB scanning. The number of beam scans Ns was fixed. The surface morphology of the deposited Pt mask was observed by an atomic force microscopy (AFM) and a field emission scanning electron microscopy (SEM) with a low accelerating voltage.

Figure 12.11a, b show a cross-sectional microphotograph of the Pt-deposited $SiO_2$/Si substrate with $L_p = 33$ nm [57]. Pt was found to exist at a depth of about 20 nm under the $SiO_2$ surface near the center of the irradiation point. In the spot drawing mode in FIB-CVD, a nanoscale hole structure is generated at the central part of the irradiation point by the FIB sputtering effect. Therefore, Pt was deposited in the concave nanoscale holes generated by FIB sputtering at almost the same time as the nanoscale hole was produced. Figure 12.11c shows an AFM image and the profile of the deposited Pt on the $SiO_2$ surface at $L_p = 200$ nm. A concave profile was also observed at the central irradiation point, as in Fig. 12.11a. The areas of deposited Pt are 30–40 nm in radius.

For the pattern with $L_p = 18$ nm, the adjacent dots are expected to connect. Figure 12.12a shows a secondary electrons SE image of a top view of the Pt deposited on $SiO_2$/Si substrate at $L_p = 18$ nm. A nanoscale dot array was not observed on the $SiO_2$ surface. However, the patterned nanoscale dot array can be seen in the

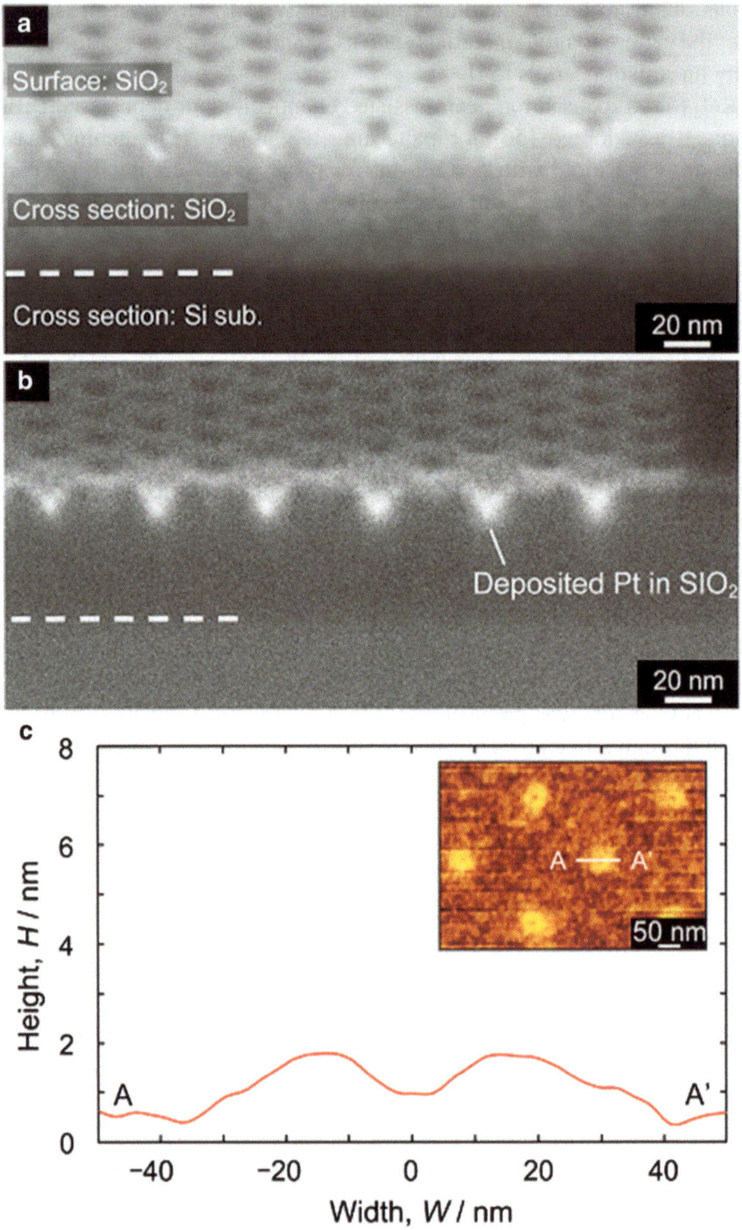


**Fig. 12.11** (**a** and **b**) SEM microphotographs of the cross section of Pt-deposited $SiO_2$/Si substrate with a dot pitch ($L_p$) of 33 nm. (**c**) AFM image and cross-sectional profile of the deposited Pt on $SiO_2$. Reprinted from [57] with permission of American Vacuum Society. © (2012) American Vacuum Society

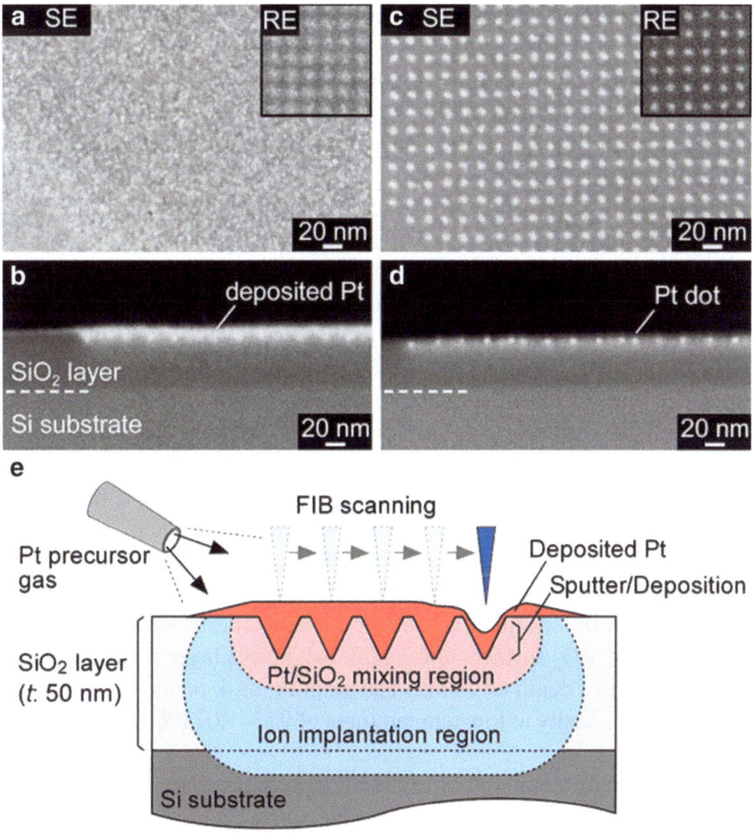


**Fig. 12.12** Pt mask patterning of nanodot array at $L_p = 18$ nm. (**a**) SE and RE images of top view of patterned specimen. (**b**) Cross-sectional RE image of Pt deposited on $SiO_2$/Si substrate in (**a**). (**c**) Result of Ar plasma etching of Pt-deposited $SiO_2$/Si substrate at $L_p = 18$ nm. (**d**) Cross-sectional RE image of substrate in (**c**). (**e**) Schematic illustration of Pt mask-patterned $SiO_2$/Si substrate at $L_p = 18$ nm in spot mode of FIB. Reprinted from [57] with permission of American Vacuum Society. © (2012) American Vacuum Society

backscattered/reflected electrons RE image, as shown on the upper right of Fig. 12.12a. Figure 12.12b shows an RE image of the cross section of the specimen; the nanoscale dots were observed in the superficial layer of deposited Pt.

Figure 12.12e shows a schematic illustration of the Pt deposited on $SiO_2$/Si substrate structure. An adjacent nanohole becomes filled by deposited Pt during the subsequent beam drawing and is covered with overspread Pt; thus, we could not obtain an 18-nm pitch resolution. However, Pt was also deposited in each nanohole generated by FIB sputtering. It seems that isolated nanodots should be obtained by removing only the superficial layer of the deposited Pt (Fig. 12.12e). Figure 12.12c shows a SEM microphotograph of the etched surface of $SiO_2$; an isolated Pt dot array at $L_p = 18$ nm was obviously observed on the specimen's surface. Pt dots appeared in each nanohole in the $SiO_2$ in a cross-sectional image (Fig. 12.12d), which suggests that the obtained Pt nanodot array can be used as an etching mask to develop a mold.

Figure 12.13a–c show SEM microphotographs of patterned substrates at $L_p = 18$ nm after heat treatment at temperatures of 973, 1,073, and 1,173 K. Each Pt particle aggregated into a more spherical shape than before heat treatment (Fig. 12.13c).

In addition, histograms of $L_p$ are depicted at the right side of each SEM image, and the number of data (N) and standard deviation ($\sigma_p$) in the $L_p$ are also presented. $L_p$ was measured from the oval center coordinates of each Pt particle using image analysis software and the SEM images in Fig. 12.13a–c. $L_p$ exhibited a normal distribution around the value of 18 nm, but $\sigma_p$ increased with increasing temperature. Figure 12.13d presents the effects of the temperature on $\sigma_p$ and the Pt dot circularity ($D_b/D_a$), which was estimated from the ratio of the major ($D_a$) and minor ($D_b$) axes of an elliptical shape fitted by the image analysis software. The measured values before heating (after Ar sputter etching) are also plotted for comparison. The circularity improved with increasing heat treatment temperature, but $\sigma_p$ increased exponentially. There was a trade-off relationship between $D_b/D_a$ and $\sigma_p$.

Figure 12.14a, b show cross-sectional RE images of patterned substrates at $L_p = 18$ nm that were heat treated at temperatures of 1,073 and 1,173 K, respectively. After treatment at 1,073 K, the Pt mask exhibited an orderly pattern located at regular intervals on the $SiO_2$ surface, and a high-brightness layer was observed underneath the Pt mask to a depth of 20 nm. The main Pt mask on the $SiO_2$ surface could be aggregated more easily at low temperatures of 973–1,073 K than the mixed Pt in the $SiO_2$ layer.

The aggregation of the mixed Pt in the $SiO_2$ progressed at a high temperature of 1,173 K. At the same time, the Pt mask on the surface and particulate mixed Pt in $SiO_2$ would be bound together. Consequently, the heating temperature is assumed to influence the three-dimensional position accuracy of the Pt mask. The heat treatment temperature at 973 K produced a $\sigma_p$ of 0.82 nm. Some Pt particles with a 5-nm diameter existed on the surface as well as underneath the main Pt mask. These Pt particles on the surface would impede the etching process in the early phase and increase the variation in the Pt mask diameter by clustering. An etch-back process with an Ar plasma to isolate Pt mask before heat treatment is essential for achieving a highly uniform mask. Figure 12.15a shows the fabricated results of Pt masks at $L_p = 16$–14 nm by FIB. These arrays were patterned on the same substrate. Ar plasma etching was conducted to remove the Pt surface layer. A small amount of Pt mask particles with diameters of 3–4 nm were observed on the surface in each $L_p$. The minimum obtained $L_p$ was 14 nm. Figure 12.15b shows relation between $L_p$ and the maximum dwell/irradiation time ($t_{d1}$) during formation of an isolated Pt mask. The limit of $t_{d1}$ decreased as $L_p$ was reduced.

The relationship between $L_p$ and the diameter ($D_d$) of Pt masks fabricated at $t_{d1}$ is shown in Fig. 12.15b; the standard deviation ($\sigma_p$) in $D_d$ is also presented as error range. $D_d$ and $\sigma_p$ were measured from the SEM images (Fig. 12.15a). Figure 12.15b, c reveal that $D_d$ is associated with $t_{d1}$ and decreased with decreasing $L_p$. At $L_p = 14$ nm, the $D_d$ values were less than 5 nm, which is equivalent to the minimum

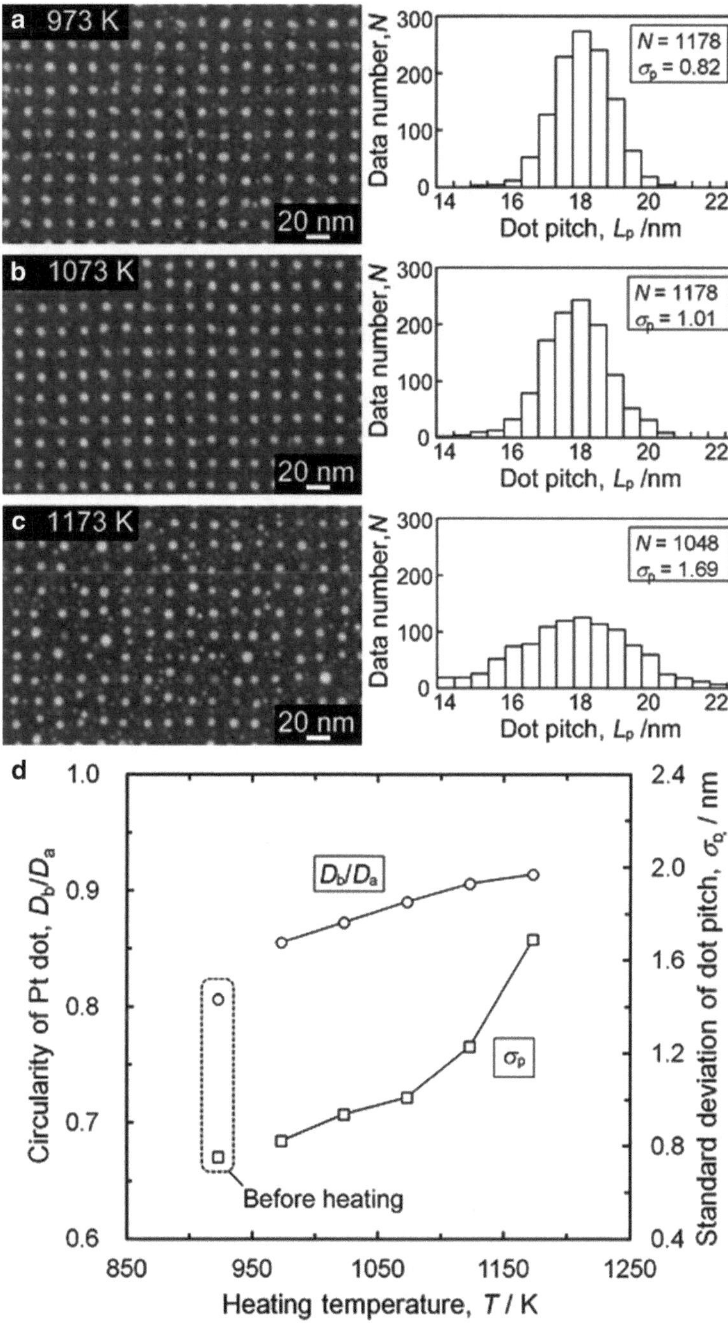


**Fig. 12.13** (**a–c**) SEM microphotographs and histograms of results of heat treatment of the Pt mask at $L_p = 18$ nm. Heating temperatures were (**a**) 973, (**b**) 1,073, and (**c**) 1,173 K. (**d**) Effects of heating temperature on circularity ($D_b/D_a$) and standard deviation of dot pitch ($\sigma_p$). Reprinted from [57] with permission of American Vacuum Society. © (2012) American Vacuum Society

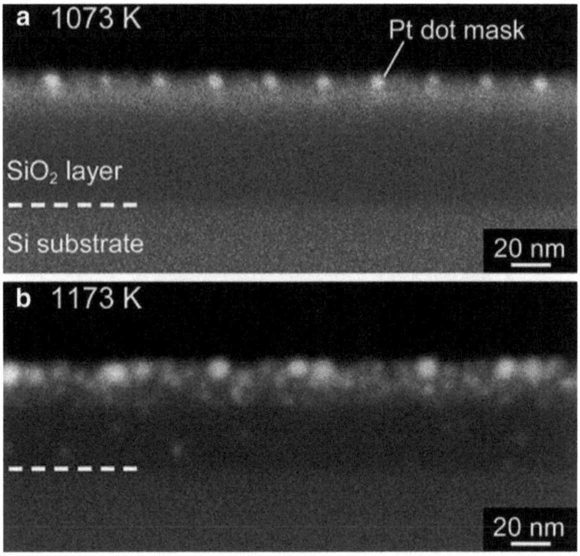


**Fig. 12.14** Cross-sectional RE images of the heated substrates at (**a**) 1,073 K and (**b**) 1,173 K after forming Pt mask at $L_p = 18$ nm. Reprinted from [57] with permission of American Vacuum Society. © (2012) American Vacuum Society

beam diameter in FIB. The values of $\sigma_p$ were 1–2 nm regardless of $L_p$ and $D_d$. The $\sigma_p/D_d$ ratio is plotted on the secondary axis in Fig. 12.15c. $\sigma_p/D_d$ ratio is inversely proportional to $L_p$.

The resolution limit depends primarily on the beam diameter associated with $I_b$, and finer resolution can be obtained by obtaining the balance between Pt deposition and $SiO_2$ sputtering in spot mode of FIB. The reactive ion etching (RIE) experiment was performed using a mixed gas of trifluoromethane ($CHF_3$) and oxygen. Figure 12.16a, b show SEM microphotographs of the mold with convex conical dots at $L_p = 18$ nm. The diameter of the retained Pt mask was 8 nm, and the dot height was 34 nm. The fabricated dot shapes had a high aspect ratio of about 2 in this nanometer-scale structure. In the cross-sectional image (Fig. 12.16b), the patterned area was etched more deeply than the non-patterned area, and a gradual slope shape appeared in boundary of the patterned area. A slight irregularity in the shape of the $SiO_2$ dot on the mold appears in Fig. 12.16b. The irregularity of the dot shape is expected to affect the quality of BPM associated with the imprinted hole shape. To fabricate BPM, hard magnetic materials, such as a Co/Pd multilayer, must be embedded in the imprinted holes on a nonmagnetic metallic glass. Therefore, the irregularity of the dot shape would ultimately affect the perpendicular magnetization of hard magnetic bits associated with the morphology of the Co/Pd multilayer. Suppressing the aggregation of Pt is preferable for obtaining highly uniform dot shapes. The aggregating temperature of the main Pt masks differs from that of the mixed Pt particles owing to differences in the Pt concentration distribution and surrounding conditions. In the experiment Pt was distributed more uniformly at 1,073 K than at 1,173 K. Therefore, the irregularity of the etched dot shape would also increase with increasing heating temperature as well as with the position accuracy of the main Pt mask.

Figure 12.16c–e show the results of etching with $L_p = 16$–12 nm. The RIE conditions were the same at the case of $L_p = 18$ nm. No traces of the Pt mask remained on

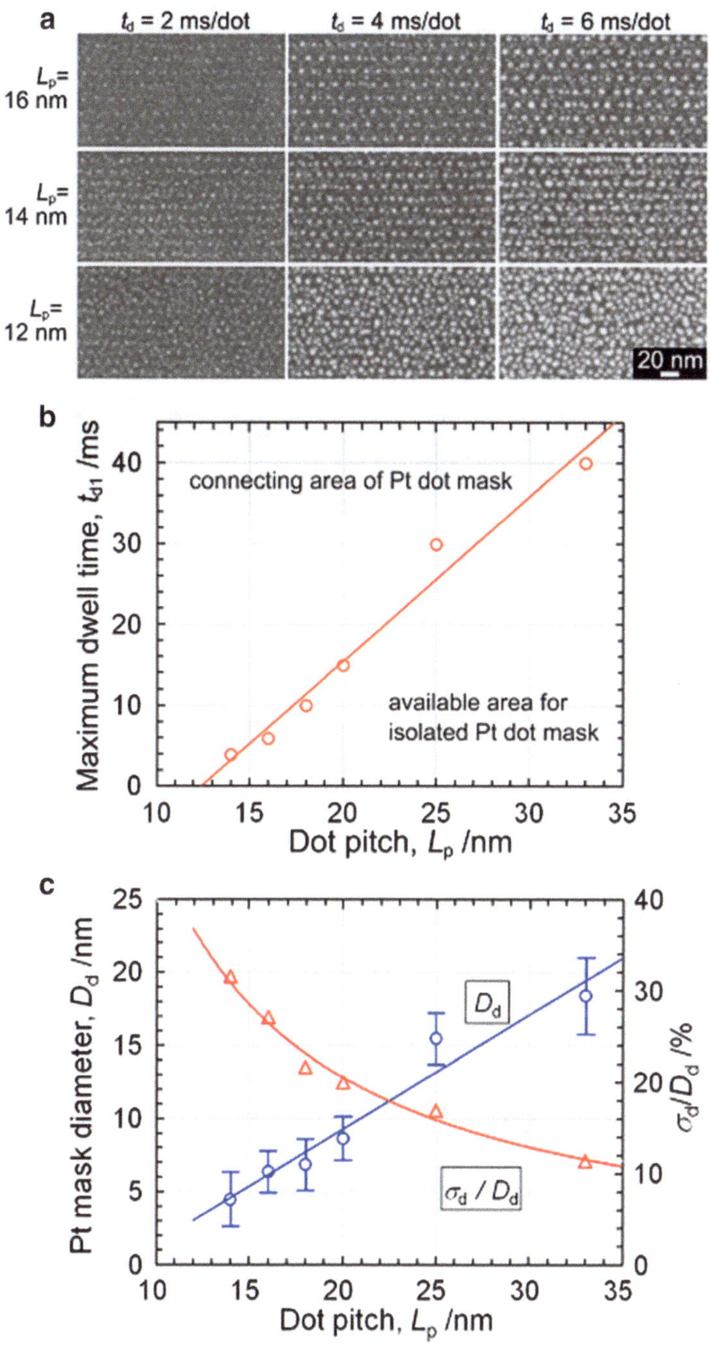


**Fig. 12.15** Investigation of resolution limit of Pt mask patterning. (**a**) SEM microphotographs of fabricated Pt masks with $L_p$=16–14 nm at $t_d$=2–6 ms/dot. (**b**) Relationship between $L_p$ and maximum dwell time ($t_{d1}$) during isolated Pt mask formation. (**c**) Dependence of $L_p$ on Pt mask diameter ($D_d$) and its variation ($\sigma_p$). Reprinted from [57] with permission of American Vacuum Society. © (2012) American Vacuum Society

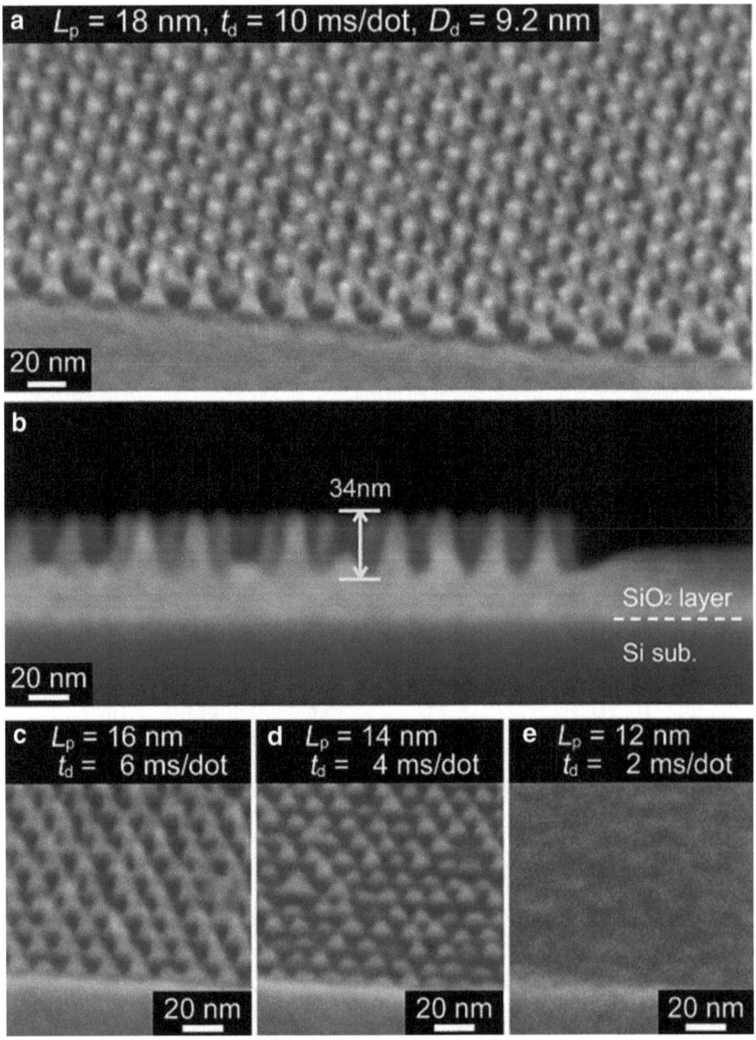


**Fig. 12.16** SEM microphotographs of fabricated $SiO_2$/Si mold of nanodot array at $L_p$ = (**a**) and (**b**) 18, (**c**) 16, (**d**) 14, and (**e**) 12 nm. (**a**) and (**c–e**) are tilted views, and (**b**) is a cross-sectional SE image of the specimen in (**a**). Reprinted from [57] with permission of American Vacuum Society. © (2012) American Vacuum Society

top of the convex dots at $L_p = 16$ and 14 nm, and the dot shape was conical with a low aspect ratio compared to that at $L_p = 18$ nm. At $L_p = 12$ nm, no dot arrays were observed on the patterned area. The scale effect with increasing specific surface seemed to affect the etching stability of the Pt mask at $D_d < 10$ nm. The mixing density of Pt increased with decreasing $L_p$ because the volume of beam irradiation per unit area also increased.

In comparison with other glassy alloys, Pd- and Pt-based glassy alloys [58] have attractive advantages such as extremely high thermal stability against

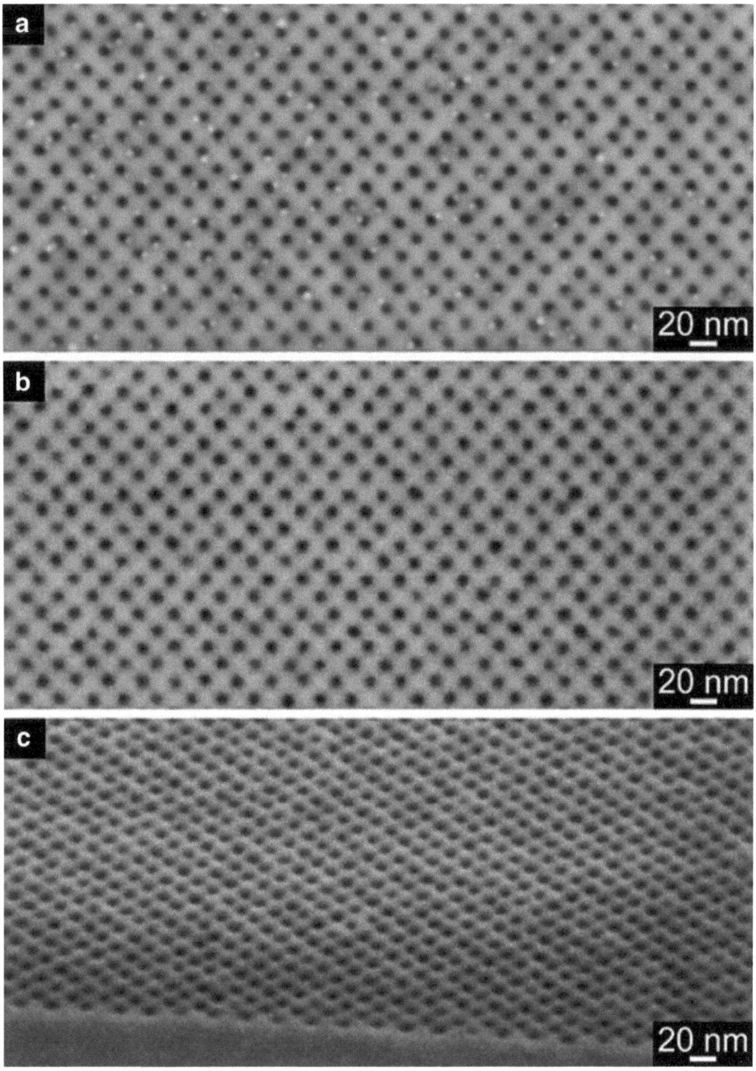


**Fig. 12.17** SEM microphotographs of surface of Pd-based metallic glass nanoimprinted by using the mold with nanodot array having 18-nm pitch: (**a**) first impression, (**b**) and (**c**) second impression. Reprinted from [57] with permission of American Vacuum Society. © (2012) American Vacuum Society

crystallization, high oxidization resistance, and low chemical reactivity with imprinting molds made of $SiO_2$, favorable for easy and precise pattern imprinting. Therefore, Pd-based glassy alloy thin film is a promising candidate to fabricate nanostructured surface pattern. Nanoimprinting was performed on the metallic glass using the fabricated mold. The prepared $Pd_{39}Cu_{29}Ni_{13}P_{19}$ metallic glass exhibited Newtonian viscosity flow in the supercooled liquid temperature range

($\Delta T_x = 83$ K) between $T_g = 575$ K and $T_x = 658$ K. The material was prepared by magnetron sputtering on a Si substrate with a thickness of 20 nm and surface roughness $R_a = 0.18$ nm. Figure 12.17a shows an SEM microphotograph of the result of the initial nanoimprinting of Pd-based metallic glass at 603 K for 10 s under a compressive stress of 10 MPa. The aggregated spherical Pt mask of the mold was observed around the replicated hole structure. Subsequently, nanoimprinting was performed for the second time at 603 K under a compressive stress of 40 MPa for 10 s. Figures 12.17b, c show SEM microphotographs of the nanoimprinted surface of the Pd-based metallic glass. The fine nanoscale hole arrays with $L_p = 18$ nm were clearly replicated. The height of the $SiO_2$ dot decreased naturally after the first imprinting because the Pt spherical mask was removed. The dot height would change to approximately 25 nm from 34 nm because of the loss of the Pt mask with a diameter of 8 nm. However, the depth of the imprinted hole was estimated to be less than 10 nm on the basis of the tilted SEM photograph in Fig. 12.17c. The final imprinted depth should be determined mostly by the imprinting conditions. We assumed that the wetting behavior and surface tension of the metallic glass would significantly affect the nanoscale formability at the ultrafine scale. In terms of fabricating BPM, the shape uniformity of $SiO_2$ dots on the mold is the major factor determining the magnetic recording properties. Therefore, evaluation of the uniformity of the dot shapes on the mold after removal of the Pt mask is a subject for further study. It is also important to investigate the nanoformability of metallic glass in terms of the wetting behavior between the mold material and the metallic glass, surface tension, viscosity, applied stress, and material thickness in order to control the depth of the imprinted holes at the ultrafine scale. Consequently, the excellent nanoscale formability and homogeneity of metallic glass were successfully demonstrated by nanoimprinting with the fabricated mold.

Thus in the spot drawing mode in FIB-CVD, Pt was deposited in the nanoscale holes generated by the sputtering effects of FIB and was also distributed near the beam irradiation point associated with the region of SE emission. A connected Pt dot array with an 18-nm pitch was observed on the $SiO_2$ surface. Heat treatment was applied to enhance the uniformity of the Pt dots' shape by thermal aggregation, and a Pt etching mask with spherical dots was obtained. RIE of the $SiO_2$ layer by using a mixed gas of $CHF_3$ and $O_2$ produced a mold having a convex nanodot array with an 18-nm dot pitch. Successful nanoimprinting of $Pd_{39}Cu_{29}Ni_{13}P_{19}$ metallic glass using the fabricated mold led to the formation of fine concave nanohole arrays. Thus, metallic glass is a nano-imprintable material for mass-producing nanodevices such as BPM with high areal recording densities of 2 Tbit/in$^2$.

## 12.6 Direct Formation of a Nanostructure on Deposition

Growing number of research activities are now focused on metallic glassy nano-objects, for example, nanowires [59, 60]. Nanowires have been used as barcodes for biological multiplexing, which could be visualized by an optical microscope [61]. Metallic nanowires may have analytical applications ranging from interconnects to

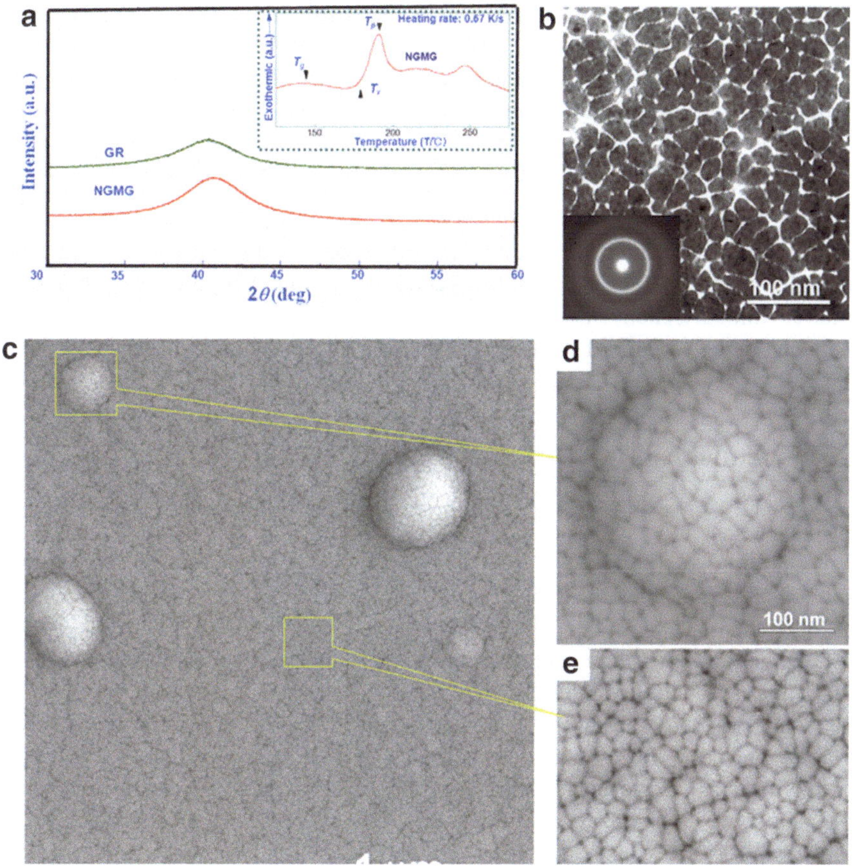


**Fig. 12.18** (**a**) XRD patterns of the $Au_{49}Ag_{5.5}Pd_{2.3}Cu_{26.9}Si_{16.3}$ GRs and the as-prepared Au-based nanostructured metallic glass. The inset of (**a**) is the DSC curve of the nanostructured metallic glassy sample. (**b**) Bright-field TEM image. The inset of (**b**) is the SAED pattern. (**c**) SEM images of the surface morphology of nanostructured metallic glassy sample, consisting of islands (**d**) and glassy grains (**e**), respectively. Reprinted from [65] with permission of Elsevier © (2011) Elsevier

sensors [62]. The micrometer long nickel nanowires could be internalized by cells allowing the manipulation of living cells through magnetic field [63].

Metallic glasses are suitable materials for scanner micro-mirrors. In order to achieve the large scanning angle without mechanical failure during actuation, the micro-mirror structure was fabricated using Fe-based metallic glass [64]. High values of mechanical strength and elastic strain limit are desired for the torsion bar for providing high performance of the mirror, including large tilting angle and good stability. Millimeter-size mirror device was successfully fabricated from Fe-based MG ribbon. Extremely large optical tilting angle of exceeding 100° was obtained on activation of the mirror by magnetic force when electrical current of 100 mA was applied. Large tilting angle of the mirror is due to the torsion bar which was fabricated with Fe-based MG material that has large elastic strain limit and high fracture toughness.

Recently developed Au-based metallic nanostructured metallic glasses with large surface area produced from a BMG forming alloy (Fig. 12.18) [65] opened up a new application area of such a material as catalyst. This material showed a good catalytic activity for the following reaction:

$$\underset{[1]}{PhMe_2Si-\mathrm{H}} + \mathrm{H_2O} \xrightarrow[Au\ based NGMG]{acetone, rt, 24\mathrm{h}} \underset{[2]}{PhMe_2Si-OH} + \mathrm{H_2}$$

of dimethylphenylsilane ($PhMe_2Si\text{–}H$) with water in the presence of Au-based metallic nanoglass. The reaction proceeded at room temperature for 24 h and the desired dimethylphenylsilanol was obtained. As similar granular structure was obtained in the case of $Pd_{78}Si_{22}$ nanoglass other catalysts can be prepared soon.

## 12.7 Conclusion

In the present work we summarized recent findings in the field of micro- and nanoscale processing and pattering of metallic glasses. Nanoimprinting was performed on various metallic glasses using the molds fabricated by focused ion beam-assisted chemical vapor deposition and reactive ion etching. The glassy alloys have a homogeneous structure and exhibit viscous flow at much lower temperatures in comparison with crystalline alloys. The high-quality imprint patterns and near-net-shape cast products were obtained, which indicates applicability of metallic glasses for these purposes. On the other hand metallic glasses themselves can be readily used as dies for nanoimprinting lithography replacing expensive nanoscale dies made of silicon which have lower durability. Direct formation of a nanostructure on deposition is also discussed.

## References

1. Inoue, A.: Acta Mater. **48**, 279–306 (2000)
2. Saotome, Y., Imai, K., Shioda, S., Shimizu, S., Zhang, T., Inoue, A.: Intermetallics **10**, 1241–1247 (2002)
3. Jeong, H.W., Hata, S., Shimokohbe, A.: J. Microelectromech. Syst. **12**, 42–52 (2003)
4. Schroers, J., Pham, Q., Desai, A.: J. Microelectromech. Syst. **16**, 240–247 (2007)
5. Kumar, G., Tang, H.X.: J. Schroers Nature **457**, 868–872 (2009)
6. Sharma, P., Zhang, W., Amiya, K., Kimura, H.M., Inoue, A.: J. Nanosci. Nanotechnol. **5**, 416–420 (2005)
7. Chou, S.Y., Krauss, P.R., Zhang, W., Guo, L., Zhuang, L.: J. Vac. Sci. Technol. B **15**, 2897–2901 (1997)
8. Spaepen, F.: Acta Metall. **25**, 407–415 (1977)
9. Louzguine-Luzgin, D.V., Wada, T., Kato, H., Perepezko, J., Inoue, A.: Intermetallics **18**(6), 1235–1239 (2010)
10. Y. Saotome and A. Inoue: Proc. 7th IEEE Workshop on Micro Electro Mechanical Systems. 343–348 (1994)

11. Fukuda, Y., Saotome, Y., Kimura, H., Inoue, A.: Mater. Trans. **52**, 239–242 (2011)
12. Louzguine-Luzgin, D.V., Inoue, A.: J. Nanosci. Nanotechnol. **5**(7), 999–1014 (2005)
13. Louzguine, D.V., Inoue, A.: J. Mater. Sci. **35**(16), 4159–4164 (2000)
14. de Oliveira, M., BottaF, W.J., Yavari, A.R.: Mater. Trans. **41**(11), 1501–1504 (2000)
15. Castellano, R.N., Feinstein, L.G.: J. Appl. Phys. **50**, 4406–4411 (1979)
16. Krupanidhi, S.B., Maffeo, N., Sayer, M., El-Asselet, K.: J. Appl. Phys. **54**, 6601–6609 (1983)
17. Sreenivas, K., Sayer, M., Garrett, P.: Thin Solid Films **172**(2), 251–267 (1989)
18. Otsubo, S., Maeda, T., Minamikawa, T., Yonezawa, Y., Morimoto, A., Shimizu, T.: Jpn. J. Appl. Phys. **29**(Part 2 (1)), L133–L136 (1990)
19. Lasagni, A., Holzapfel, C., Weirich, T., Mücklich, F.: Appl. Surf. Sci. **253**(19), 8070–8074 (2007)
20. Kwak, B.S., Boyd, E.P., Erbil, A.: Appl. Phys. Lett. **53**(18), 1702–1704 (1988)
21. Budd, K.D., Dey, S.K., Payne, D.A.: Br. Ceram. Proc. **36**, 107–121 (1985)
22. Saotome, Y., Itoh, K., Zhang, T., Inoue, A.: Scr. Mater. **44**, 1541–1545 (2001)
23. Saotome Y., Inoue A.: Proceeding of IEEE 13th International Conference on Micro Electro Mechanical Systems, 288–293, (2000)
24. Saotome, Y., Iijima, T., Zhang, T., Inoue, A.: J. Metast. Nanocryst. Mater. **15–16**, 655–658 (2003)
25. Saotome, Y., Okaniwa, S., Kimura, H., Inoue, A.: J. Mater. Sci. Forum **539–543**, 2088–2093 (2007)
26. Fukuda, Y., Saotome, Y., Kimura, H., Inoue, A.: Materials Transactions **52**(2), 239–242 (2011)
27. Fukuda, Y., Saotome, Y., Nishiyama, N., Saidoh, N., Makabe, E., Inoue A. Japanese,: J. Appl. Phys. **51**, 086702 (2012)
28. Takenaka, K., Togashi, N., Nishiyama, N., Inoue, A.: J. Non-Cryst. Solids **356**, 1542–1545 (2010)
29. Saotome, Y., Miwa, S., Zhang, T., Inoue, A.: J. Mater. Process. Technol. **113**, 64–69 (2001)
30. Saotome, Y., Noguchi, T., Zhang, T., Inoue, A.: J. Mat. Sci. Eng. A **375–377**, 389–393 (2004)
31. Kato, H., Wada, T., Hasegawa, M., Saida, J., Inoue, A., Chen, H.: Scr. Mater. **54**, 2023–2027 (2006)
32. Saotome, Y., Okaniwa, S., Kimura, H., Inoue, A.: Mater. Sci. Forum **539–543**(PART 2), 2088–2093 (2007)
33. Fukuda, Y., Saotome, Y., Kimura, H and Inoue, A., Materials Transactions, 52, 239–242 (2011)
34. Chen, N., Yang, H.A., Caron, A., Chen, P.C., Lin, Y.C., Louzguine-Luzgin, D.V., Yao, K.F., Esashi, M., Inoue, A.: Glass-forming ability and thermoplastic formability of a $Pd_{40}Ni_{40}Si_4P_{16}$ glassy alloy. J. Mater. Sci. **46**(7), 2091–2096 (2011)
35. Zhang, T., Zhang, X., Zhang, W., Jia, F., Inoue, A., Hao, H., Ma, Y.: Mater. Lett. **65**, 2257–2260 (2011)
36. Hofmann, D.C.: J. Mat, Article ID 517904 (2013)
37. Greer, A.L., Rutherford, K.L., Hutchings, I.M.: Int. Mater. Rev. **47**, 87–112(26) (2002)
38. Togashi, N., Ishida, M., Nishiyama, N., Inoue, A.: Rev. Adv. Mater. Sci. **18**, 93–97 (2008)
39. Inoue, A., Shen, B., Takeuchi, A.: Mat. Sci. Eng. A **441**, 18–25 (2006)
40. Ishida, M., Uehara, T., Arai, T., Takeda, H., Yamaguchi, T., Taniguchi, T., Katsumi, T., Kobayashi, M., Ofune, H.: Intermetallics **10**, 1259–1263 (2002)
41. Mamoru, I., Hideki, T., Daichi, W., Kenji, A., Nobuyuki, N., Kazuhiko, K., Yasunori, S., Akihisa, I.: Mater. Trans. **45**, 1239–1244 (2004)
42. Saotome, Y., Ishizaki, H.: J. Mater. Proc. Tech. **119**, 307–311 (2001)
43. Sbia, R., Piramanayagam, S.N.: Recent Pat. Nanotechnol. **1**, 29–40 (2007)
44. Chou, S.Y., Wei, M.S., Krauss, P.R., Fischer, P.B.: J. Appl. Phys. **76**, 6673–6675 (1994)
45. Lodder, J.C.: J. Magn. Magn. Mater. **272–276**, 1692–1697 (2004)
46. Naito, K., Hieda, H., Sakurai, M., Kamata, Y., Asakawa, K.: IEEE Trans. Magn. **38**, 1949–1951 (2002)
47. Yang, X., Xiao, S., Wu, W., Xu, Y., Mountfield, K., Rottmayer, R., Lee, K., Kuo, D., Weller, D.: J. Vac. Sci. Technol. B **25**, 2202–2209 (2007)

48. Yang, J.K.W., Chen, Y., Huang, T., Duan, H., Thiyagarajah, N., Hui, H.K., Leong, S.H., Ng, V.: Nanotechnology **22**, 385301 (2011)
49. Solak, H.H., Ekinci, Y.: J. Vac. Sci. Technol. B **25**, 2123–2127 (2007)
50. Wu, W., Cui, B., Sun, X., Zhang, W., Zhuang, L., Kong, L., Chou, S.Y.: J. Vac. Sci. Technol. B **16**, 3825–3829 (1998)
51. Chou, S.Y., Krauss, P.R., Renstrom, P.J.: Appl. Phys. Lett. **67**, 3114–3116 (1995)
52. Yamamoto, R., Yuzawa, A., Shimada, T., Ootera, Y., Kamata, Y., Kihara, N., Kikitsu, A.: Jpn. J. Appl. Phys. **51**(Part 1), 046503 (2012)
53. Hosaka, S., Tanaka, Y., Shirai, M., Mohamad, Z., Yin, Y.: Jpn. J. Appl. Phys. **49**(Part 1), 046503 (2010)
54. Hosaka, S., Mohamad, Z., Shirai, M., Sano, H., Yin, Y., Miyachi, A., Sone, H.: Microelectron. Eng. **85**, 774–777 (2008)
55. Fukuda, Y., Saotome, Y., Nishiyama, N., Saidoh, N., Makabe, E., Inoue, A.: Jpn. J. Appl. Phys. **51**(Part 1), 086702 (2012)
56. Matsui, S., Kaito, T., Fujita, J., Komuro, M., Kanda, K., Haruyama, Y.: J. Vac. Sci. Technol. B **18**, 3181–3184 (2000)
57. Fukuda, Y., Saotome, Y., Nishiyama, N., Takenaka, K., Saidoh, N., Makabe, E., Inoue, A.: J Vac Sci Technol B: Microelectron Nanometer Struct **30**, 061602 (2012)
58. Inoue, A., Nishiyama, N., Kimura, H.M.: Mater. Trans. JIM **38**, 179–183 (1997)
59. Nakayama, K.S., Yokoyama, Y., Ono, T., Chen, M.W., Akiyama, K., Sakurai, T., Inoue, A.: Adv. Mater. **22**, 863–866 (2010)
60. Carmo, M., Sekol, R.C., Ding, S., Kumar, G., Schroers, J., Taylor, A.D.: ACS Nano **5**, 2979–2983 (2011)
61. Keating, C.D., Natan, M.J.: Adv. Mater. **15**, 451–454 (2003)
62. Walter, E.C., Penner, R.M., Liu, H., Ng, K.H., Zach, M.P., Favier, F.: Surf. Interface Anal. **34**, 409–412 (2002)
63. Tanase, M., Felton, E.J., Gray, D.S., Hultgren, A., Chen, C.S., Reich, D.H.: Lab Chip **5**, 598–605 (2005)
64. Lee, J.-W., Lin, Y.C., Chen, N., Louzguine, D.V., Esashi, M., Gessner, T.: Jpn. J. Appl. Phys. **50**, 087301 (2011)
65. Chen, N., Frank, R., Asao, N., Louzguine-Luzgin, D.V., Sharma, P., Wang, J.Q., Xie, G.Q., Ishikawa, Y., Hatakeyama, N., Lin, Y.C., Esashi, M., Yamamoto, Y., Inoue, A.: Acta. Mater. **59**, 6433–6440 (2011)

# Chapter 13
# Nanostructured Materials Driven by Dielectrophoresis on Nanoelectrodes Patterned by Focused Ion Beam

**Vera La Ferrara**

**Abstract** In this chapter the combined use of focused ion beam and dielectrophoresis is reported to show the powerful tool for realizing nanodevices based on nanostructured materials.

This combination allows to positioning by dielectrophoresis between nanoelectrodes deposited by focused ion beam single palladium nanowire, ZnO nanostructures, and graphene multilayer flakes dispersed in appropriate solutions. Devices based on single Pd nanowire operate as hydrogen sensor at room temperature. Nanostructures of ZnO are assembled and characterized as photodetector.

Graphene multilayer flakes are assembled by DEP towards electrode gap milled by FIB, and its electrical resistance is measured.

## 13.1 Introduction

The fabrication of nanoelectronic circuits needs to downscale the dimensions of electrodes. Usually device electrodes are built using electron beam or optical lithography, metallization, and liftoff. The focused ion beam (FIB) is often used as alternative method for realizing the nanoelectrodes on electronic devices. The use of high energy FIB is an already well-established technique in microelectronics and microfabrication for mask-less patterning [1–3]. This apparatus is able to micromachine the electrode directly onto substrate becoming a "direct-write" tool. The electrodes can be patterned in few minutes and be a good alternative to electrodes prefabricated by multistep and longtime procedure. FIB operating mechanism relies on a highly focused and energetic gallium ion beam that interacts with a platinum

---

V. La Ferrara (✉)
Portici Research Center, Piazzale Enrico Fermi, Portici, Naples 80055, Italy
e-mail: vera.laferrara@enea.it

Z.M. Wang (ed.), *FIB Nanostructures*, Lecture Notes in Nanoscale Science and Technology 20, DOI 10.1007/978-3-319-02874-3_13,

organometallic precursor to deposit nanocontacts [4, 5]. It can be used in different ways. It is possible to make nanoelectrode patterning defining the region where nanostructures and single nanowires are subsequently located: growing, positioning, and aligning them between the contacting electrodes. This method is commonly named *grow in place* [3, 6–9]. Another approach *pick and place* consists of capturing the single nanowire or nanostructure in a random dispersion of solution, positioning it onto the substrate, and FIB is employed to make electrical contacts at the nanowire end [10–13].

The fabrication and the alignment nanowires represent a challenge for the current technology. Recently the feasibility of *grow in place* process has been demonstrated by realizing nanostructure-based devices. First Pt electrodes are deposited by FIB then nanoparticles solution is dropped onto electrodes and dielectrophoresis (DEP) is applied [3, 14]. It uses a nonuniform electric field to polarize small particles suspended in solution forcing them to move in a well-controllable way. This process has emerged both as an effective process to handle and position the nanostructures/nanowires in precise site onto devices and as a deposition step to assemble the nanowires starting from nanoparticles suspended in solution [15–18].

The peculiarity of FIB and DEP processes is shown in this chapter, showing different devices based on single Pd nanowire, ZnO nanostructures, and single flake of graphene multilayers realized in the previous work [3, 20, 22, 28, 29]. For device fabrication, as a first step, FIB is used for nanoelectrode patterning onto a silicon/silicon nitride or glass substrates. Electrodes are designed in such a way that, by means of a DEP process, under suitable defined operating conditions, Zno or Pd nanowires are grown or a single flake of graphene multilayer is driven between the nanocontacts. Different devices have been realized: single Pd nanowire has been tested at room temperature as hydrogen sensor, ZnO nanostructures have been characterized as photodetector and graphene flakes have been characterized measuring the electrical resistance.

## 13.2 Experimental

### *13.2.1 Electrode Pattern Configurations*

FIB column, integrated in a FEI Quanta 200 3D dual-beam system, has been used for Pt nanoelectrodes patterning. During deposition the precursor, methylcyclopent adienyl[trimethyl] platinum ($C_9H_{16}Pt$), is fragmented with the high focused gallium ion beam resulting in the deposition of platinum nanoelectrodes composed of the elemental constituents of the precursor and Ga+ ions from the ion beam [19]. In Fig. 13.1, two device architectures are schematically shown [3, 20]. As a starting substrate, silicon or glass substrates are used. For single Pd nanowire and ZnO nanostructures assembly, the flowchart of device fabrication is reported in Fig. 13.1a.

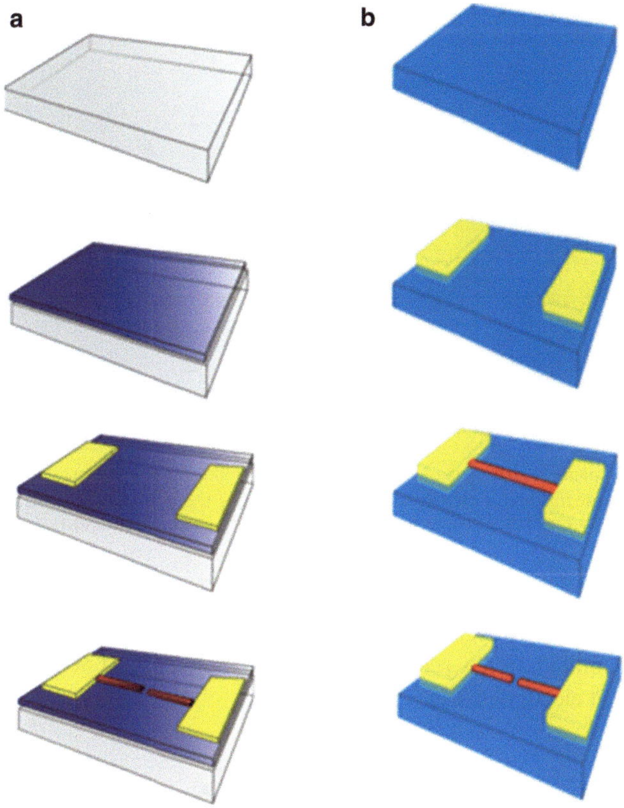


**Fig. 13.1** Flowchart of the process for single Pd nanowire assembly and ZnO nanostructures (**a**) and graphene flakes (**b**). (**a**) Silicon substrate as starting material, deposition of $SiO_2$ and Cr/Au pads deposited by e-beam evaporation, then Pt microelectrodes deposited by FIB [3]. (**b**) Glass substrate as starting material, Cr/Au pads deposited by e-beam evaporation, Pt microelectrode deposition by FIB, and gap milling by FIB to create two microelectrodes [20]

After a standard cleaning, a 100 nm thick $SiO_2$ layer and two Cr/Au pads with the area of 0.25 $cm^2$ and thickness of 20 nm/180 nm are deposited by e-beam-assisted evaporation onto silicon. The metal pads have 1 mm gap size. To reduce the gap size and allow the nanowire assembly, the substrate is inserted into the FIB vacuum chamber for the subsequent deposition of platinum microelectrodes between the Cr/Au pads, obtaining a new gap size of 8 μm. Using 30 kV as accelerating voltage and 10pA as ion beam emission current, Pt nanoelectrodes are deposited onto the substrate. Each platinum electrode has a width of 300 nm and a height of 500 nm.

For graphene flake assembly, the flowchart for device fabrication is shown in Fig. 13.1b. Glass is the starting substrate and Cr/Au pads are deposited with the same previous conditions. The metal pads have 100 μm gap size.

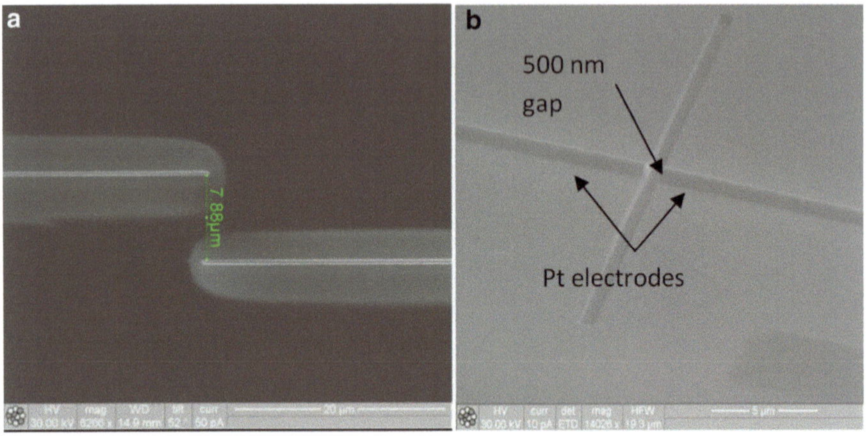


**Fig. 13.2** (**a**) Pt nanoelectrodes, made by FIB. The thin Pt halo, due to platinum deposition spreading out around the electrodes, is clearly visible. Halo reduces the gap size, set initially at 8 μm (**b**) 500 nm gap created by FIB milling of platinum electrode deposited previously by FIB [20]

A single Pt nanoelectrode is deposited, with the same previous conditions, realizing it on the same line. To create a gap useful to drive the graphene flake, the Pt electrode is then milled, in few seconds, by FIB creating a gap of 500 nm and forming two aligned Pt electrodes (Fig. 13.2b).

### *13.2.2 Different Feed Solutions for the Nanostructures Assembly*

For DEP processing, the nanostructures (palladium, ZnO, and graphene flakes) have been dispersed in a solution. Few microliters of solution have been deposited by casting onto the substrate, previously washed in isopropyl alcohol and deionized water and dried in a nitrogen flow. Following assembly, the chips are rinsed in successive baths of acetone, isopropanol, and deionized water to remove all remaining traces of the suspension [3]. A scheme of the apparatus used for assembly is shown in Fig. 13.3. A function generator (Agilent 33220A) and an oscilloscope (Tektronix TDS 3032) have been used as the alternating current (AC) source and for monitoring the applied electrical signal, respectively. Moreover, the assembly is monitored through an optical microscope that captures the movement of nanoparticles in real time by means of a video camera.

#### 13.2.2.1 Palladium Nanoparticles

The feed solution for DEP processing is prepared dissolving crystalline Pd(acetate)$_2$ ($(C_2H_3O_2)_2Pd$, Sigma Aldrich) in 10 mM HEPES ($C_8H_{18}N_2O_4S$, Sigma Aldrich) buffer solution in order to obtain a saturated solution. To ensure complete and

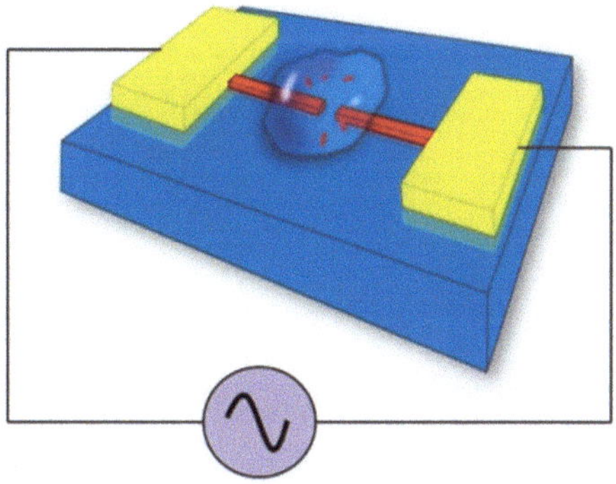

**Fig. 13.3** Scheme of circuit used for nanoparticles assembly [3]

homogeneous dissolution, the solution is maintained hermetically closed and in the dark on a magnetic stirrer for 3 h and then centrifuged at 13,000 rpm to separate the precipitate. The sinusoidal electrical signal 10 $V_{\text{peak-peak}}$ ($V_{\text{pp}}$) is monitored over time applying frequencies of 300 kHz and 300 Hz [3].

#### 13.2.2.2 ZnO Nanostructures

The solution has been prepared dissolving 0.01 g of ZnO nanopowders in deionized water to have a 0.3 mg/ml concentration. To ensure complete and homogeneous dissolution, the solution is maintained hermetically closed and in the dark on a magnetic stirrer for 10 min and then centrifuged at 13,000 rpm to separate the precipitate. This resulting solution has been inserted in an ultrasonic bath for 30 min to avoiding the agglomeration. The applied electrical signal is fixed to 10 $V_{\text{pp}}$ with frequencies of 500 KHz and 1 MHz [22].

#### 13.2.2.3 Graphene Multilayer Flakes

Liquid phase production of chemically unmodified graphene is obtained by natural graphite dispersion in ultrasonic bath. The solvent employed to facilitate the layer exfoliation and stabilize the obtained solution is the *N*-Methyl pyrrolidone (NMP) [23, 24]. The graphite (Sigma Aldrich P.N. 332461) is dispersed in NMP at a concentration of 1 mg/ml by sonicating in a low power bath (40 W, 40 Hz) for 30 min to cause exfoliation. The solution is finally centrifuged at 500 rpm for 90 min. 1 μl solution, get from the half-top of the dispersion, is dropped onto the nanogap. The applied electrical signal (6 $V_{\text{pp}}$ at 1 MHz) is monitored over time [20].

### *13.2.3 Electric Field Modeling of Electrode Pattern*

Platinum electrode geometry affects the electric field gradient present when DEP working conditions are applied. To better understand the starting point of nanostructures growing under DEP conditions, numerical simulations have been performed

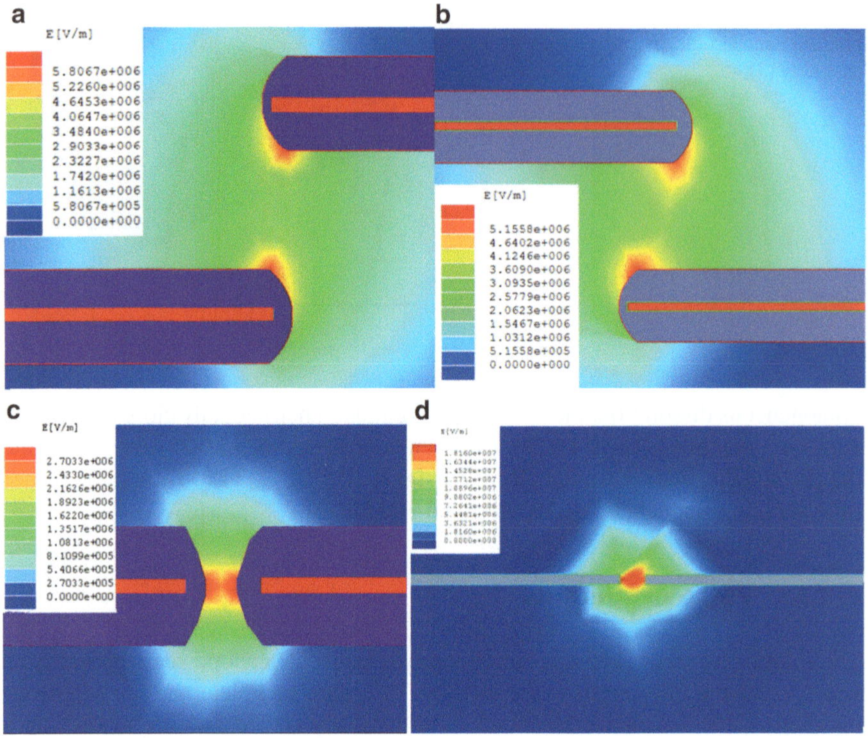


**Fig. 13.4** Simulation of the electric field for the different geometries at 10 $V_{pp}$ and 300 kHz: (**a**) parallel electrodes with 6 μm gap, (**b**) parallel electrodes with overlapping of 1 μm and 6 μm gap, (**c**) aligned electrodes with a gap of 2 μm, and (**d**) aligned electrodes with a gap of 500 nm

through the program Maxwell 2D by AnSoft. Taking into account both electrode and its Pt halo, when electrode pattern and working conditions are defined, the electric field streamlines are obtained. Figure 13.4 shows the electric field existing in the region between typical microelectrodes realized with different geometries.

In the figure the higher field gradient, which reaches values of $10^6$ V/m, when a voltage of 10 $V_{pp}$ and 300 kHz are applied, is represented with a red color. The streamlines are calculated under the real operating conditions.

### 13.2.4 Dielectrophoresis

The assembly process is based on the mobility and interaction of nanostructures suspended in solution, caused by an alternating current and nonuniform electric field commonly referred to dielectrophoresis. When an electric field is applied, the nanostructures dispersed in a fluid polarize, producing charge at all interfaces, particularly the interface between the particle and the fluid. The net charge has opposite sign on either side of the particle, leading first and foremost to a net-induced effective dipole moment.

This dipole moment depends on the frequency of the applied electric field and the dielectric properties (permittivity and conductivity) of the particle and the fluid [25, 26].

The general expression for a force due to DEP is expressed as

$$F_{DEP} = \frac{3}{2}\varepsilon_m V \operatorname{Re}\left[K(\omega)\right]\nabla E^2 \tag{13.1}$$

where $\varepsilon_m$ is the absolute permittivity of the surrounding medium, $V$ is the volume of particle, $E$ is the electric field, and $\omega$ is the angular frequency of the applied electrical field $E$. $\operatorname{Re}[K(\omega)]$ is the real part of the polarization factor, $K(\omega)$, also called the Clausius–Mossotti (CM) factor, which is a complex function of the complex permittivity of both medium and particle.

where $\operatorname{Re}|K(\omega)|$

$$\operatorname{Re}\left|K(\omega)\right| = \frac{\varepsilon_p - \varepsilon_m}{\varepsilon_p + 2\varepsilon_m} + \frac{3\left(\varepsilon_m\sigma_p - \varepsilon_p\sigma_m\right)}{\tau_{MW}\left(\sigma_p + 2\sigma_m\right)^2\left(1+\omega^2\tau_{MW}^2\right)} \tag{13.2}$$

is the real part of CM factor and indicates the effective polarizability of the particle, related to the particle dielectric constant $\varepsilon_p$ and medium dielectric constant $\varepsilon_m$. In the above formulae $\tau_{MW} = \frac{\varepsilon_p + \varepsilon_m}{\sigma_p + 2\sigma_m}$ is the charge relaxation time of Maxwell–Wagner.

The $K(\omega)$ factor is very important in the DEP process. The real part of CM factor in fact depends on the dielectric permittivity and conductibility of medium and particle dispersed in solution, and it is inversely dependent on the frequency. Generally particles in a solution move towards regions of high electric field strength if the polarizability of the particles is greater than the suspending medium polarizability (real part of CM factor is positive); otherwise they are repelled (real part of the CM factor is negative) [27]. When it is negative, the force changes from attractive to repulsive, at a crossover frequency $\omega_c = \tau_{MW}^{-1}$, and the particles move towards smaller electric field regions, and this is termed negative dielectrophoresis (Fig. 13.5).

Figure 13.5 shows the behavior of a particle subjected to dielectrophoresis, when the real part of the CM factor is positive or negative. The force changes sign, causing the motion with respect to areas of highest field gradient. In particular, it is important to obtain positive values of the CM factor to assemble the material in suspension in precise sites.

When external field is applied to the electrodes, a dipole is formed on nanoparticles dissolved in a polar liquid. DEP force is more intense around the local artifacts present onto the electrodes, due to fabrication defects. Dipoles are forced to move towards these sites where oxidation/reduction reactions can occur, starting the nucleation process. As a result the most intense electric field gradient settles at a nucleating tip, and a wire can grow until electrodes short-circuit. The aggregations of nanoparticles are irreversible, and the assembled structures remain stable after the voltage has been turned off.

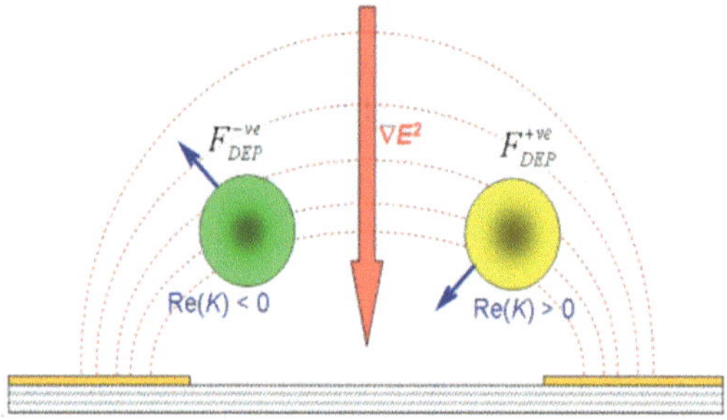


Fig. 13.5 Schematic representation of a particle subjected to DEP force [26]

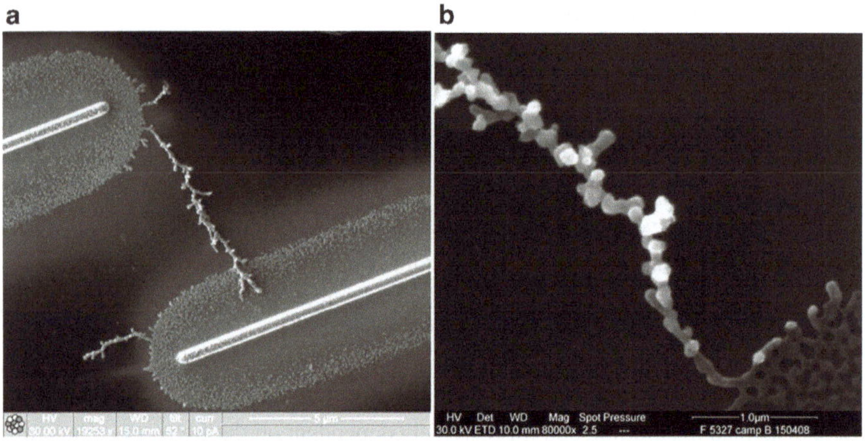


**Fig. 13.6** (**a**) Ionic image of device based on single Pd nanowire. Device is formed by a main nanowire that connects the electrodes. Two secondary nanowires are also formed. Nanowires appear branched [28]. (**b**) A SEM image at high resolution

## 13.3 Results and Discussions

### *13.3.1 Single Palladium Nanowire-Based Devices*

#### 13.3.1.1 Effects of Electrode Geometry

The palladium nanowires have been grown by applying a sinusoidal signal voltage of 10 $V_{pp}$ at a fixed frequency, 300 kHz, starting from different electrode patterns. In Fig. 13.6 platinum electrodes are 1 μm overlapped, showing the grown single nanowire and its magnification with a branched shape and 80 nm thickness. The electric field generated by this electrode pattern favors the growth of aggregated grains. As the nanowire grows close to the other electrode, more intense electric fields are produced at the edge of the palladium growth, and the trunk-structure growth appears dominant [29].

A single main nanowire is formed, although the growth of a secondary shorter nanowire with a different nucleation site is observed when electrode partly

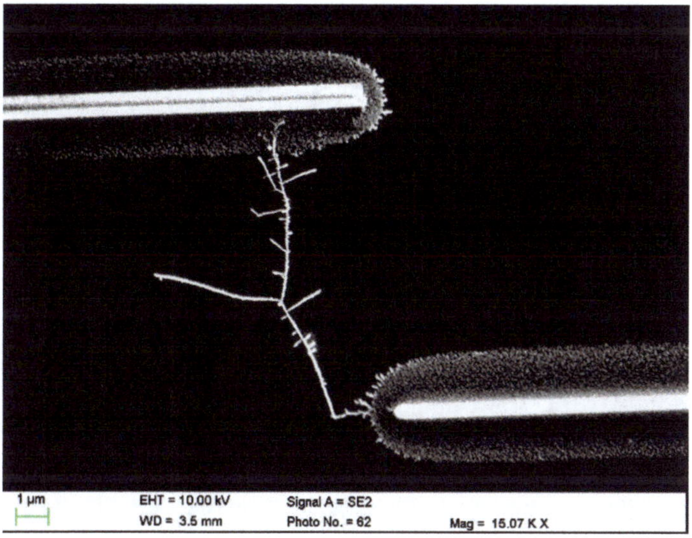


**Fig. 13.7** SEM image of single Pd nanowire grown using an electrode pattern with no overlap [29]

overlapped are used. The secondary nanowire cannot complete its growth by connecting the opposite electrode, because the growth of the main nanowire is faster and the process stops when it reaches the opposite electrode. In Fig. 13.7 the electrodes are not overlapped. The resulting nanowire is thinner (60 nm thickness) but always branched, while the growth of a secondary nanowire is not evident.

In Fig. 13.8, the electrodes are 2 μm overlapped and the main nanowire presents an unusual morphology composed of two shapes: the topside consists of aggregated grains ranging from 100 to 200 nm, while the underside shows a central trunk with many branches, which is thin with 60 nm width.

Changing the electrode pattern depositing Pt in alignment way, with gap size of 2 μm, Pd nanowire has been grown in 5 s (Fig. 13.9).

From electrode A, there is only one point of nucleation for nanowire growth. Instead, from the electrode B, different sites of growth are presented, probably due to a greater number of electrode imperfections. The nanowire magnification of device is shown in the inset of the figure, evaluating the nanowire diameter equal to 120 nm.

#### 13.3.1.2 Effects of Frequency

Fixing the geometry of the electrodes, realized by FIB and changing the frequency at 300 Hz, the nanowire morphology changes: no ramification respect to nanowire realized at 300 kHz (see Fig. 13.7) with a thickness varying from 180 nm to 600 nm. In Fig. 13.10 the single nanowire and its magnification are shown.

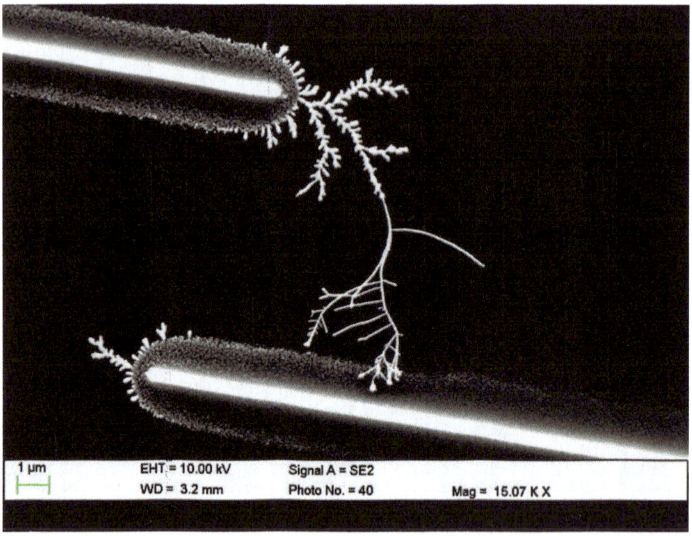


**Fig. 13.8** A main nanowire and a secondary nanowire are formed when electrode patterns are partly overlapped [29]

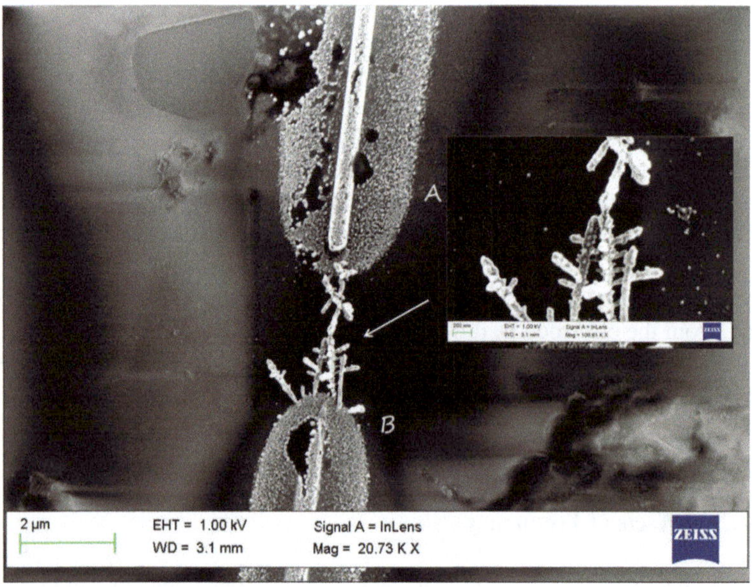


**Fig. 13.9** SEM image of palladium nanowire grown between aligned platinum electrodes (A and B) and its magnification (*inset*) [22]

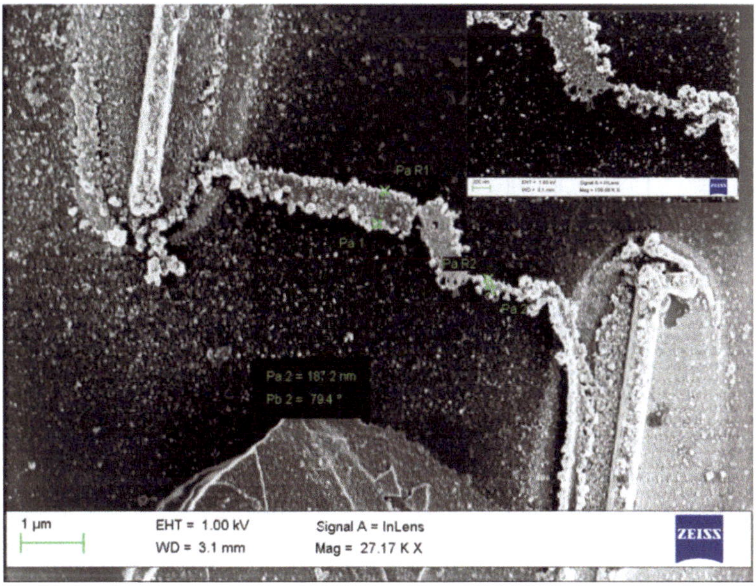


**Fig. 13.10** SEM image nanowire realized at 300 Hz and its magnification (*inset*) [3]

Wire assembly is always observed when field strength and particle concentration exceed experimental threshold values, suggesting that a minimum DEP force is required for its formation. Several parameters determine the threshold values for the assembly: frequency, concentration of particles in suspension, viscosity, and dielectric constant of the media. Once the applied field and the electrolyte concentration have been tuned so as to exceed the threshold values, the AC frequency becomes an important parameter conditioning the wire morphology.

At low frequencies (300 Hz) thick palladium nanowires, with no branches, are assembled, while at high frequencies (300 kHz), thinner and branched nanowire are produced when high frequency fields are applied. Branched structures may originate because Pd ions can diffuse towards secondary sites different from the main nucleating tip, bringing to the formation of new nucleating points [30].

#### 13.3.1.3 Electrical Characterization

Devices have been characterized electrically in air at room temperature environment, showing the same resistive type behavior in current–voltage with different values of electrical resistance in the range of 20–150 kΩ. The difference in resistance between the devices is mainly due to their morphology. Devices with ramified and granular morphology are more resistive than the device without ramification, more thickness with a planar conformation.

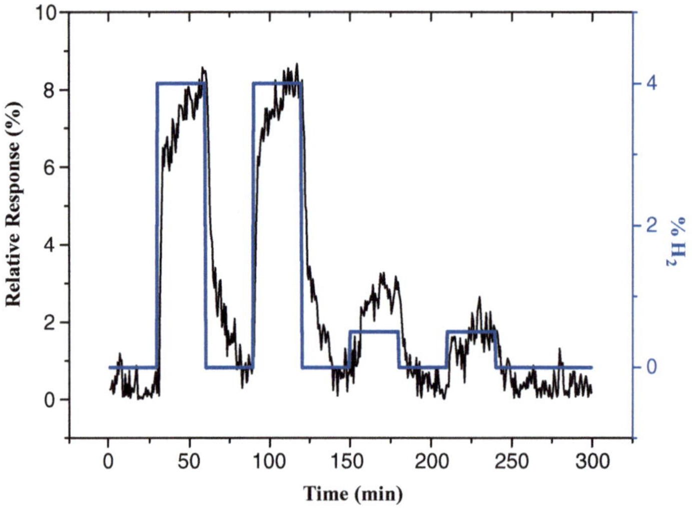


**Fig. 13.11** Branched single palladium nanowire: sensor output current (*left axis*, *squares+line*) versus time, recorded at room temperature and at 4 % and 0.5 % $H_2$ (*solid line*) concentrations in dry air (*right axis*) [28]

Volt-amperometric technique, at constant bias, has been then employed for sensor DC electrical characterization in a controlled gas flow environment, premixed with dry carrier in the desired percentage by mass flow meters, and continuously controlled by means of an online Fourier transform infrared spectrometer [21]. All the tested devices have been biased at 5 mV. Total gas flow has been set at 500 sccm.

Devices based on single palladium nanowire have been characterized as hydrogen sensors. They are kept for 30 min under nitrogen flow, to monitor baseline current. A concentration of 4 % hydrogen is introduced more for 30 min followed by an identical cycle of nitrogen/hydrogen. Two additional cycles are made to 0.5 % hydrogen. Devices show a current decreasing in presence of hydrogen. Upon exposure to hydrogen, palladium reacts to form the more resistive palladium hydride ($PdH_x$ where x is the atomic ratio H/Pd); as a consequence the electric current decreases.

To describe the sensor response, for the devices, the time responses have been obtained. $I_0 = \frac{I_0 - I_S}{I_0} \times \%$ is the device relative response. $I_0$ is the maximum current under nitrogen gas, while $I_s$ is the maximum current under hydrogen gas. The response time is the time required for the electrical current to increase from 10 to 90 % of $I_s$. In Figs. 13.11 and 13.12 the relative responses versus time in presence of different concentrations of hydrogen are reported for branched single palladium nanowire and thick palladium nanowire, respectively.

The branched single palladium-based device relative response is higher than thick palladium device because of higher surface/volume ratio due to the numerous ramifications. Moreover, branched single Pd follows better the concentration

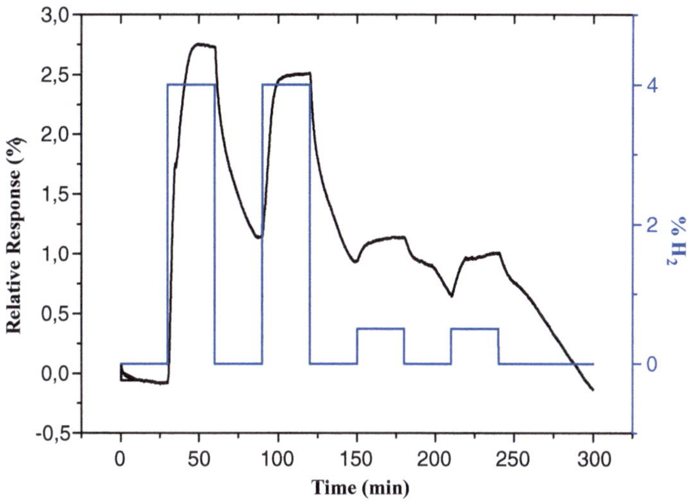


**Fig. 13.12** Thick single palladium nanowire: sensor output current (*left axis*, *squares+line*) versus time, recorded at room temperature and at 4 % and 0.5 % $H_2$ (*solid line*) concentrations in dry air (*right axis*) [28]

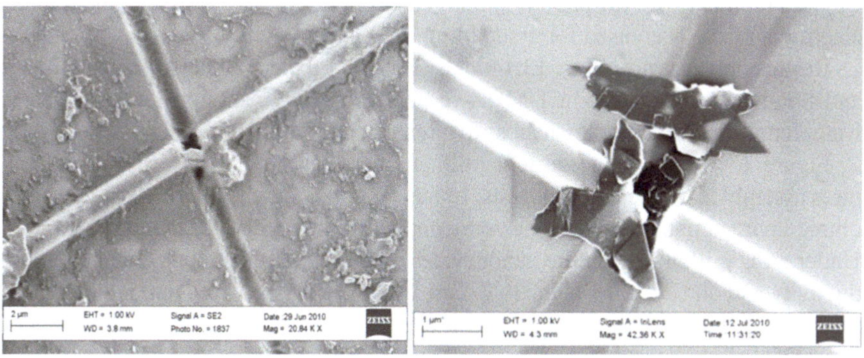

**Fig. 13.13** SEM images of two devices where flakes are assembled between nanogap [20]

change, and when hydrogen is removed from the test chamber, it has a lower drift current and greater stability than the device based on thick nanowire.

## *13.3.2 Graphene Multilayer Flake Assembly*

Preliminary devices are shown in Fig. 13.13. Comparing the images with the simulation of Fig. 13.4 is clear that the flakes are deposited mainly in the areas of highest field gradient, thus confirming the goodness of the mathematical modeling.

The alternate electrical signal used in the experiment is 6 $V_{peak-peak}$ and 1 MHz applied for 1 min, because these parameters permit to have a positive DEP force. The electrical resistance of devices is also measured finding 700 kΩ. As it is evident from SEM images, DEP force drives thicker flakes between the nanogap. Raman analysis confirms that only thicker flakes are assembled. Exfoliation of graphite in organic solvents, although it produces high-quality single-layer graphene, it suffers a high polydispersity and low single-layer yield (~1 %) [24]. When dielectrophoretic force is applied to such a nonhomogeneous solution, only flakes with larger thickness are carried out on the electrodes although single-layer graphene may be present in the solution. The single-layer graphene are not able to compete against larger flakes subjected to high dielectrophoretic force. It, in fact, increases proportionally to the particles volume, which is directly related to the flake thickness. These results indicate that to achieve single-layer graphene dielectrophoretic deposition, higher solution qualities must be available, consisting largely of single-layer graphene sheets, as already reported in the literature [23].

### *13.3.3 ZnO Nanostructure-Based Devices*

Devices realized with different platinum electrodes geometry (parallel and aligned electrodes) have been used for assembling ZnO nanostructures [22].

Results are shown in Figs. 13.14 and 13.15 where the applied signal was 10 $V_{pp}$ and frequency of 500 kHz e 1 MHz, respectively. ZnO nanopowder dispersed in water tends to form complex nanostructures when DEP is applied.

Figure 13.16 shows the current–voltage characteristics of the ZnO nanostructures assembled by dielectrophoresis. It may be noted that the curves exhibited non-ohmic characteristic showing the semiconducting behavior of ZnO nanoparticles. Under an UV illumination the nanodevices exhibited a tremendous change in conductivity and on UV removal the I–V retraces the curves observed under dark within a few seconds of measurement. This behavior shows that the fabricated device can be used as photodetector.

## 13.4 Conclusions

The feasibility of combined use of dielectrophoresis and focused ion beam for manufacture nanostructured devices is presented. Results show that FIB and DEP allow positioning between nanoelectrodes single palladium nanowire, ZnO nanostructures, and graphene multilayer flakes dispersed in appropriate solutions. Nanodevice can be fabricated in well-controllable conditions. Devices based on single Pd nanowire operate as hydrogen sensor at room temperature. Nanostructures of ZnO are assembled and characterized as photodetector.

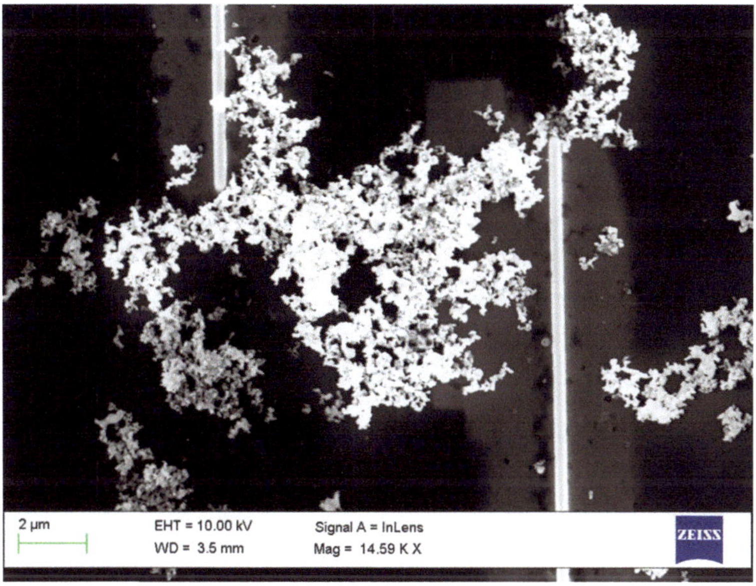


**Fig. 13.14** SEM image ZnO assembly at 500 kHz

**Fig. 13.15** SEM image of ZnO assembly at 1 MHz

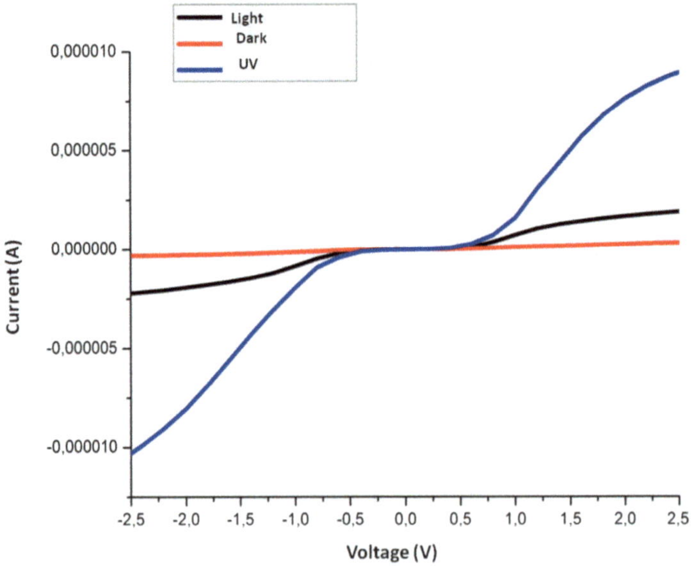


**Fig. 13.16** I–V characteristics of ZnO nanostructures under different illumination

Graphene multilayer flakes are assembled by DEP towards electrode gap milled by FIB, and electrical resistance is measured. These results show how electrodes can be patterned in few minutes by direct-write FIB technique and how DEP can be used as a scalable method for the growth of nanowires directly onto devices.

# References


1. Cronin, S.B., Lin, Y.-M., Rabin, O., Black, M.R., Ying, J.Y., Dresselhaus, M.S., Gai, P.L., Minet, J.P., Issi, J.P.: Making electrical contacts to nanowires with a thick oxide coating. Nanotechnology **13**, 653–658 (2002)
2. Gopal, V., Radmilovic, V.R., Daraio, C., Jin, S., Yang, P., Stach, E.A.: Rapid Prototyping of Site-Specific Nanocontacts by Electron and Ion Beam Assisted Direct-Write Nanolithography. Nano Lett. **4**, 2059–2063 (2004)
3. La Ferrara, V., Alfano, B., Fiorentino, G., Polichetti, T., Massera, E., Di Francia, G.: Nanopatterned platinum electrodes by focused ion beam in single palladium nanowire based devices. Microelectron. Eng. **88**, 3261–3266 (2011)
4. Ke, J.J., Tsai, K.T., Dai, Y.A., He, J.H.: Contact transport of focused ion beam-deposited Pt to Si nanowires: from measurement to understanding. Appl. Phys. Lett. **100**, 053503–053505 (2012)
5. Ho, C.Y., Chiu, S.H., Ke, J.J., Tsai, K.T., Dai, Y.A., Hsu, J.H., Chang, M.L., He, J.H.: Contact behavior of focused ion beam deposited Pt on p-type Si nanowires. Nanotechnology **21**, 134008–134012 (2010)
6. Rotkina, L., Lin, J.-F., Bird, J.P.: Nonlinear current–voltage characteristics of Pt nanowires and nanowire transistors fabricated by electron-beam deposition. Appl. Phys. Lett. **83**, 4426–4428 (2003)
7. Choi, T.-Y., Kang, B., Poulikakos, D.: Focused Ion Beam in thermal science and engineering. Microsc. Microanal. **13**, 1498–1499 (2007)

8. Yun, M., Myung, N.V., Vasquez, R.P., Lee, C., Menke, E., Penner, R.M.: Electrochemically Grown Wires for Individually Addressable Sensor Arrays. Nano Lett. **4**, 419–422 (2004)
9. Mitsuishi, K., Noda, T., Mano, T., Tanaka, M., Furuya, K., Koguchi, N.: Structure of Nanowires Fabricated by Electron Beam Induced Deposition to Connect Self-Assembled Quantum Structures. Japanese J. Appl. Phys. **46**, 6277–6281 (2007)
10. Wu, S.-E., Liu, C.-P.: Direct writing of Si island arrays by focused ion beam milling. Nanotechnology **16**, 2507–2511 (2005)
11. Blom, T., Welch, K., Stromme, M., Coronel, E., Leifer, K.: Fabrication and characterization of highly reproducible, high resistance nanogaps made by focused ion beam milling. Nanotechnology **18**, 285301–285306 (2007)
12. Romano-Rodriguez, A., Hernandez-Ramirez, F.: Dual-beam focused ion beam (FIB): A prototyping tool for micro and nanofabrication. Microel. Engin. **84**, 789–792 (2007)
13. La Ferrara, V., Nasti, I., Alfano, B., Massera, E., Di Francia, G.: Focused Ion Beam nanopatterning for carbon nanotube ropes based sensor. Sensors Transducers Journal **85**, 1708–1713 (2007)
14. Cheng, C., Gonela, R.K., Gu, Q., Haynie, D.T.: Self-Assembly of Metallic Nanowires from Aqueous Solution. Nano Lett. **5**, 175–178 (2005)
15. Hermanson, K.D., Lumsdon, S.O., Williams, J.P., Kaler, E.W., Velev, O.D.: Dielectrophoretic Assembly of Electrically Functional Microwires from Nanoparticle Suspensions. Science **294**, 1082–1086 (2001)
16. Bhatt, K.H., Velev, O.D.: Control and Modeling of the Dielectrophoretic Assembly of On-Chip Nanoparticle. Wires Langmuir **20**, 467–476 (2004)
17. Lumsdon, S.O., Scott, D.M.: Assembly of Colloidal Particles into Microwires Using an Alternating Electric Field. Langmuir **21**, 4874–4880 (2005)
18. Walter, E.C., Favier, F., Penner, R.M.: Palladium Mesowire Arrays for Fast Hydrogen Sensors and Hydrogen-Actuated Switches. Anal. Chem. **74**, 1546–1553 (2002)
19. Telari, K.A., Rogers, B.R., Fang, H., Shen, L., Weller, R.A., Braski, D.N.: Characterization of platinum films deposited by focused ion beam-assisted chemical vapor deposition. J. Vac. Sci. Technol. B **20**, 590–595 (2002)
20. La Ferrara, V., Fiorentino, G., Miglietta, M., Nasti, I., Polichetti, T., Massera E., Di Francia, G.: Focused Ion Beam and dielectrophoresis for graphene assembly into devices, in Proc. of I workshop ENEA on Graphene, Portici Research Centre, 15 July 2010 ISSN/0393-3016 (Italy) 56-61 (2010)
21. Quercia, L., Cerullo, F., La Ferrara, V., Baratto, C., Faglia, G.: Fabrication and Characterization of a Sensing Device Based on Porous Silicon. Phys. Status Solidi **182**, 473–477 (2000)
22. Fiorentino G.: Fabrication and Characterization of nanostructured materials by means of FIB and DEP, Thesis discussed at University "Federico II" of Napoli (Italy) (supervisor: Vera La Ferrara) (2009)
23. Burg, B.R., Schneider, J., Maurer, S., Schirmer, N., Poulikakos, D.: Dielectrophoretic integration of single- and few-layer graphenes. J. Appl. Phys. **107**, 034302–034307 (2010)
24. Hernandez, Y., Nicolosi, V., Lotya, M., Blighe, F.M., Sun, Z., De, S., et al.: High-yield production of graphene by liquid-phase exfoliation of graphite. Nat. Nanotechnol. **9**, 563–568 (2008)
25. Pohl, H.A.: Dielectrophoresis. Cambridge University Press, Cambridge UK (1978)
26. Lin, Y., PhD Thesis: Modeling of dielectrophoresis in micro and nano systems (2008)
27. Morgan, H., Green, N.G.: AC electrokinetics: colloids and nanoparticles, Research Studies Press (2003)
28. La Ferrara, V., Alfano, B., Massera, E., Di Francia, G.: Single palladium nanowire growth in place assisted by dielectrophoresis and focused ion beam. J Nanosci. Nanotechnol **9**, 2931–2936 (2009)
29. La Ferrara, V., Alfano, B., Massera, E., Di Francia, G.: Focused Ion Beam patterning to dielectrophoretically assemble single nanowire based devices. J Phys **209**, 012052–012055 (2010). conference series
30. Ranjan, N., Mertig, M., Cuniberti, G., Pompe, W.: Dielectrophoretic Growth of Metallic Nanowires and Microwires: Theory and Experiments. Langmuir **26**, 552–559 (2010)

# Chapter 14
# Focused Ion Beam-Assisted Nanoscale Processing and Thermoelectrical Characterization

**Kyung-Min Lee and Tae-Youl Choi**

**Abstract** In this report, we introduce a nanomanipulation and fabrication technology, which is useful to characterize thermoelectrical properties of individual one-dimensional nanosystems such as metallic or semiconducting carbon nanotubes (CNTs) and nanowires (NWs). For such characterization, a one-stop measurement platform was constructed by focused ion beam (FIB) nanolithography after which a freestanding NW was picked up from a bundle of NWs and placed on the platform using a nanomanipulator. As a unique and unparalleled control for nanoscale one-dimensional systems, FIB-assisted nanomanipulator could make a direct access to nanoscale materials and structures. Subsequently, the four-point 3-ω method combined with a nanoheater was used to obtain electrical conductivity, thermal conductivity, and Seebeck coefficient, with which one can estimate the figure of merit of β-silicon carbide (SiC) NWs. We found that the thermal conductivity of a single β-SiC NW was 82 ± 6 W/mK. The Seebeck coefficient was also successfully measured to be −1.21 mV/K. With a measured electrical conductivity of the NW, the dimensionless thermoelectrical figure of merit (ZT) was estimated to be 0.12.

## 14.1 Introduction

As a unique and unparalleled manipulation, fabrication, and characterization tool, focused ion beam (FIB) has played an important role in nanotechnology because of its site-specific working range up to a few nanometers and direct and straightforward access to a nanoscale structure. The first commercial instrument was

K.-M. Lee • T.-Y. Choi (✉)
Department of Mechanical and Energy Engineering, University of North Texas,
Denton, TX 76203-1098, USA
e-mail: KyungminLee@my.unt.edu; choi@egw.unt.edu

Z.M. Wang (ed.), *FIB Nanostructures*, Lecture Notes in Nanoscale Science and Technology 20, DOI 10.1007/978-3-319-02874-3_14,

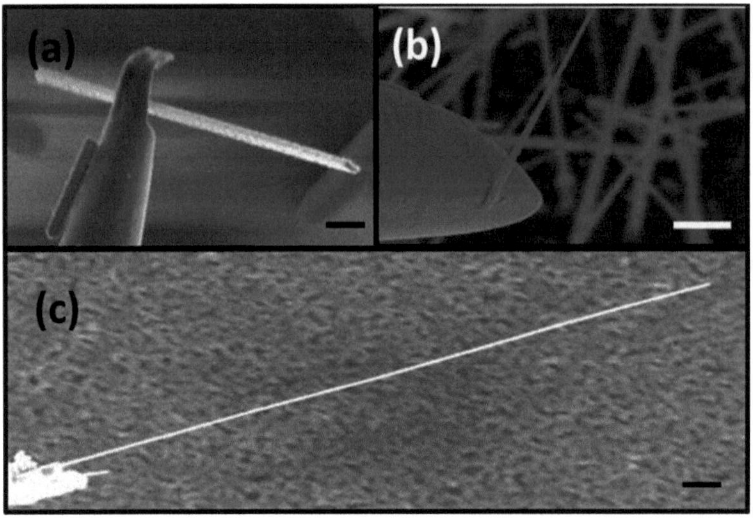


**Fig. 14.1** Direct manipulation of various NWs with a nanomanipulator annexed to FIB. (**a**) Ag NW attached to a nanomanipulator; (**b**) a single Si NW extracted from a bundle of Si NW by a nanomanipulator; (**c**) a β-SiC NW attached to a nanomanipulator. Each scale bar in the images indicates 1 μm

introduced in the market more than decades ago [1] since the prototype dual-beam technique was developed during the late 1970s and the early 1980s.

As liquid-metal ion sources (LMISs) are most commonly used, a gallium ion source was adopted in our study. A heated, thus liquid molten metal in contact with a tungsten needle wets the tungsten, and an electric field greater than $10^8$ V/cm initiates ionization and field emission of the gallium ions. Then the emitted ions are accelerated to energy of 5–50 keV and focused onto the sample through electrostatic lenses. Working at such small scale, FIB is able to modify tiny circuits and small surface areas and to deposit and etch materials with chemical reactions of $Ga^+$ ions (which are ejected from an FIB column) and reactants. FIB is now broadly used in various technological fields such as semiconductor and nanotechnology. FIB employs a nanomanipulator (attached to FIB), which has played and plays an important role to manipulate nanostructures such as picking (lifting) up and placing FIB-fabricated nano- and microstructures for transmission electron microscope (TEM) sample preparation. The nanomanipulator whose tip end is at a submicron size is small enough to control nanostructures such as NWs. Figure 14.1 shows various previous results of manipulated NWs (β-SiC, silicon (Si), and silver (Ag) NWs). This way, direct access to such a tiny structure with a finely focused ion beam and manipulator has made FIB to be a powerful and necessitated tool to work in the area of nanotechnology. Also, the relevant technologies such as fine arts of machining, drawing, and structuring have been much emphasized as more demands on well-defined nanostructures arise. This motivated us in the research to employ the FIB technology to characterize the thermoelectrical properties of NWs [2–4].

We used an Omniprobe (AutoProbe™ 200) nanomanipulator attached to a dual-beam FIB machine that enables fine control of nanostructures, e.g., creating mechanical repositioning of NWs. The nanomanipulator used in our study has 100 nm positioning resolution. It has been served as a fully automated and multipurpose nano-positioning tool that is able to make in situ lift-out, electrical measurements, nano-mechanical testing, and charge neutralization in FIB and SEM [5]. FIB used in our study was FEI Nova 200 NanoLab, which has been operated since January 2005 in the Center for Advanced Research and Technology (CART) at UNT. The secondary electron image resolution is 1.5 nm at 15 kV. The focused ion beam has better than 7 nm resolution at 30 kV.

In this article, we report thermoelectrical characterization of a single β-SiC NW. Being a material possessing superior physical properties such as high thermal conductivity, high mechanical strength, and wide bandgap structure, SiC has been used as a substantial material for applications in microelectronic, thermoelectric, and optoelectronic devices [6–9]. Mostly, the material has been used in a harsh environment such as high power, high frequency, and high temperature because of its physical and chemical durability and stability. Thus, it has a potential to replace Si in the near future. Owing to its excellent physical properties, this one-dimensional NW structure has been greatly interesting for scientific and technological research. Following this trend, SiC NWs have attracted much attention because of the excellent low-dimensional feature as well as their aforementioned intrinsic properties. For this reason, SiC NWs have been considered as important electric devices such as field-effect transistors (FETs) [10–12]. SiC crystallizes in more than 200 polytypes but the variants that have received the most research attention are the β-SiC (3C-SiC) zincblende form (studied in this article) and the 6H-, 4H-, hexagonal polymorphs.

## 14.2 Nanofabrication of One-Stop Measurement Platform

### *14.2.1 Focused Ion Beam Milling Process*

FIB can locally etch and mill a sample surface with nanoscale precision. The sputtering rate of a sample in FIB milling is affected by various parameters explained below and material properties as in other sputtering systems. The beam current, sample density, sample atomic mass, and incoming ion mass are factors to affect the FIB milling. The removal of a sample material is achieved by an energetic Ga+ beam that is used in our FIB. The result is physical sputtering of the sample material, as illustrated schematically in Fig. 14.2. By scanning the focused Ga+ beam over the substrate, a predefined shape can be obtained through milling and etching the surface of the substrate.

For the thermal conductivity measurement based on the 3-ω method, a pre-patterned metal electrode structure was made by standard photolithography

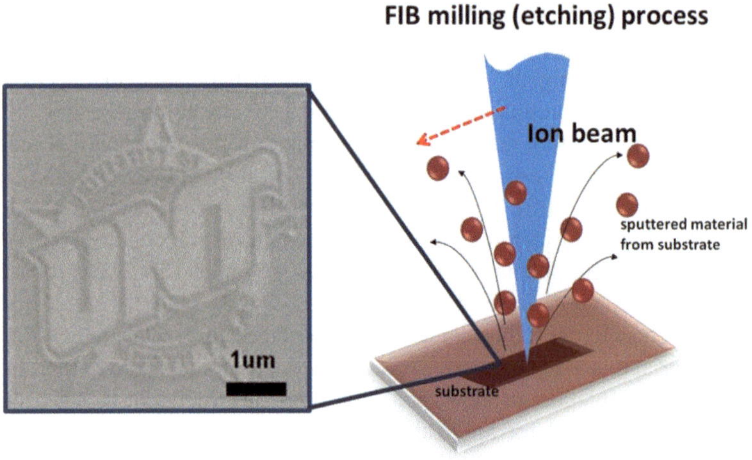


**Fig. 14.2** Schematic FIB milling process and UNT logo created by FIB milling (CART at UNT acknowledged for its permission to reproduce the UNT logo)

processes. Then, FIB was utilized to create a fine nanoscale structure consisting of four-point contact probes by milling. Figure 14.3 shows the process of fabricating the four-point probe structure for the 3-ω thermal conductivity method.

A silicon nitride ($Si_3N_4$) membrane with 500 nm thickness that had been grown on a Si wafer was used as a substrate in this study. Potassium hydroxide (KOH) was used for selectively etching Si at the center of the substrate to create a thin membrane structure with dimension of $300 \times 300\ \mu m^2$. A microelectrode pattern was created by sputtering deposition of Au/Cr on top of the membrane. The standard photolithography technique was employed to pattern electrodes for an electrical interface. The reason for creating the membrane structure is to make a through hole that serves as thermal insulation in the thermal conductivity and Seebeck coefficient measurement. This hole can prevent heat generated at the NW due to Joule heating from leaking through the substrate. The trench and four-point probe structures were created with an aid of FIB milling.

On the pre-patterned electrode, a focused ion beam of 30 kV (at 0.1 nA current and at 5,000 magnification) was applied for shaping a four-point probe structure. In situ observation with SEM was made to determine when to stop the milling process. This helped us save the processing time and avoid undesired damages on electrodes. The trench was created sequentially with the same ion beam intensity. At this time, we carefully observed the beam current change on a controlling panel in an FIB operation software instead of direct monitoring of the milling process with in situ SEM observation. This allowed us to create a well-defined and finely tailored trench with minimal damage of adjacent electrodes. We believe that in situ SEM observation can induce ion beam distortion and drift as a result of ion and electron beam interaction; thus, it is suitable for relatively large-scale milling.

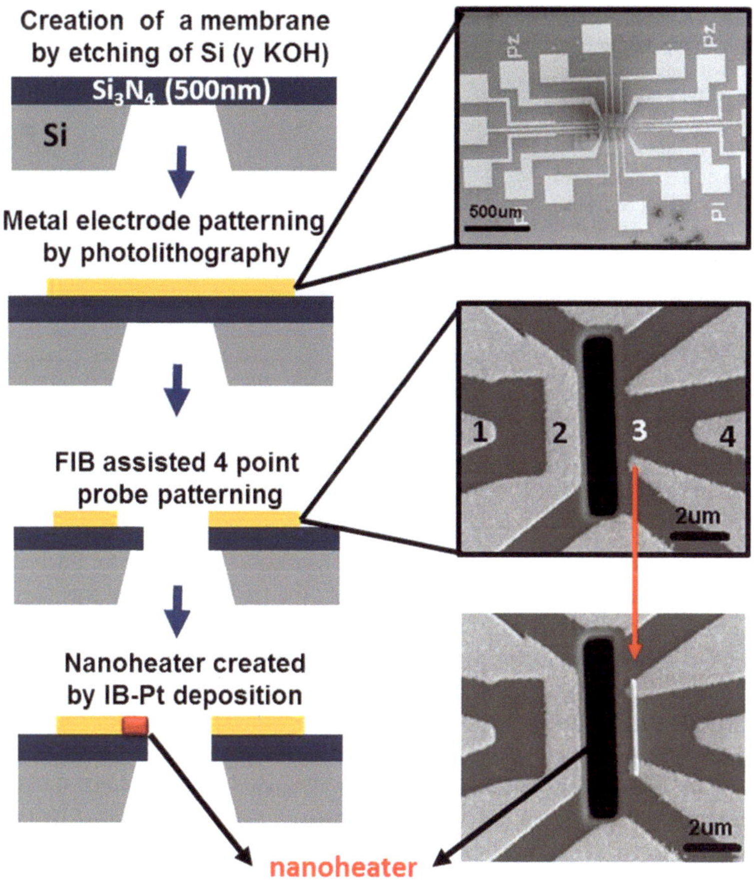


**Fig. 14.3** Fabrication of four-point probe platform for the 3-ω thermal conductivity measurement

## *14.2.2 Focused Ion Beam Deposition Process*

Technological advancement in making nanoscale devices is nowadays more demanded as novel and superior physical phenomena in the nanostructure have been discovered. In particular, excellent thermal transport characteristics in nanoscale devices have been revealed and thus attracted a particular interest because of its potential in thermal management applications. Herein, we introduce a new way of inducing nanoscale heating to create a highly localized heat source. This allowed us to successfully measure Seebeck coefficient (hence thermoelectric figure of merit) through accurate temperature calibration of the heater and measurement of temperature across the suspended SiC NW [13].

In Fig. 14.3, the number 3 probe was coated with an ion beam-induced platinum (IB-Pt) thin stripe for measuring the Seebeck coefficient. The IB-Pt stripe was

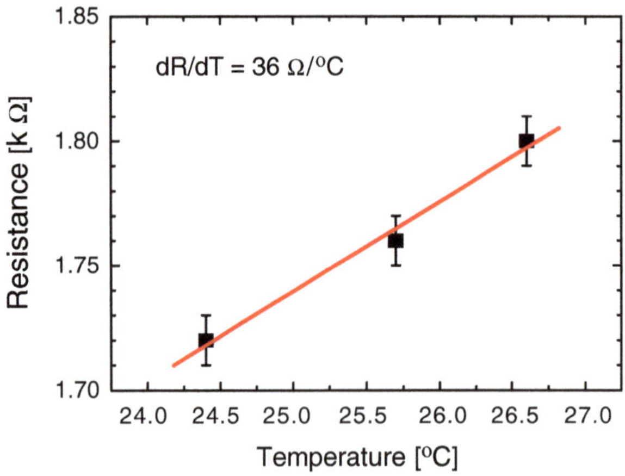


**Fig. 14.4** The resistance change by temperature variation of the nanoheater created by IB-Pt deposition. It shows that the resistance change by temperature variation of the nanoheater is 36 Ω/°C (reproduced from [15], ©Nano Science and Technology Institute)

manufactured by FIB deposition process at 30 kV/10 pA intensity for a finely defined structure. Because there should be a localized heat source to a NW for Seebeck coefficient measurement (for temperature gradient), the IB-Pt thin stripe was found to be a better heating source than the thin stripe of Au because higher resistance (~1.7 kΩ) could be achieved across the IB-Pt heater stripe. This high resistance is needed to accurately calibrate the nanoheater (resistance vs. temperature); the change of lead resistance of Au electrodes during calibration will be ignored. This type of nanoheater can also increase the sensitivity of temperature measurement due to the high temperature coefficient. The high resistance is attributed to an intrinsic electrical property of IB-Pt as it contains a higher amount of carbon that makes IB-Pt highly resistive to electrical conduction than a pure metal; the literature value of resistivity of IB-Pt is 10–20 μΩm depending on deposition condition [14].

The coated IB-Pt thin line (i.e., nanoheater) had a dimension of $200 \times 100 \times 2{,}300\ \text{nm}^3$. One of the strengths of this structure is that its resistance can be easily controlled by its geometrical configuration. We have tested the resistance change of the IB-Pt thin line due to the change of its geometry and obtained 17 kΩ for a nanoheater with a size of 200 nm × 100 nm × 25 μm. If it was scaled down to 2.3 μm which is the same length of the nanoheater shown in Fig. 14.3, the resistance could be 1.56 kΩ. This value was close to the measured value of the nanoheater that was used for Seebeck coefficient characterization. Therefore, it was concluded that the IB-Pt deposition process could control the resistance of the nanoheater by defining its geometry. The resistance of one particular nanoheater fabricated by FIB deposition process was 1.72 kΩ at room temperature which was a good localized heating source for accurate measurement of Seebeck coefficient. Figure 14.4 shows a calibration curve which is the result of resistance change by temperature variation of the nanoheater. We fabricated various nanoheaters with specific resistance values. One of them was used to measure the Seebeck coefficient of a single β-SiC NW. Figure 14.5 shows three nanoheaters with different thickness and width created by IB-Pt deposition.

**Fig. 14.5** IB-Pt thin lines (nanoheaters) whose lengths were ~2 μm and diameters were a few hundred nanometers. Each bar in the images is 2 μm long

### *14.2.3 Control of Nanomaterials Using a Nanomanipulator*

Controlling NWs is one of the most challenging tasks, and still novel methods are being developed including the methods assisted by molecular, electrostatic, and shear forces [16–18]. Another well-known method is to use nonuniform electric field and polarizability of NW to place a single NW on a desired location. The so-called dielectrophoresis (DEP) uses a phenomenon that a polarizable material tends to move in a converging electrical field [19–22]. Methods employing DEP are required to have specifically designed electrodes (to create nonuniform electric field) and optimization of applied field strength and frequency. Due to the geometrical constraints in the configuration of the four-point probe electrodes, it was revealed that DEP was not appropriate for our work, resulting in a low yield (less than 10 % success). Another concern is on the cleaning after using the DEP method for NW control, where NW-dissolved solution should be dropped on the measurement platform before applying the electric field. The force created by the applied field attracted not only NWs but also other particles and dust inside of a droplet of NW solution. Careful cleaning and rinsing residues and contaminants in the measurement platform should be unavoidable.

In order to overcome drawbacks in the DEP method, we used a direct approach to place a single NW on the measurement platform using a nanomanipulator. We used the van der Waals interaction between a nanomanipulator and an NW when a submicron-sized tip of a nanomanipulator approaches a single NW within a range of molecular force being effective. Before a nanomanipulator touches the NW, for clear viewing and easy controlling, the tip end of the nanomanipulator was sharpened by a high-intensity FIB at 30 kV/1 nA. Once the nanomanipulator contacted a NW, an electron beam-induced Pt (EB-Pt) deposition was made between the nanomanipulator and the NW for secure adhesion by using focused electron beam (5 kV/98 pA e-beam intensity). Then, we could obtain a single NW by withdrawing the nanomanipulator with high acceleration. Omniprobe (AutoProbe™ 200) nanomanipulator (Omniprobe, Inc.) attached to the FIB machine (FEI Nova 200 NanoLab) was employed for this technique, and a motion control system of the nanomanipulator was extremely accurate and repeatable within 100 nm of stage

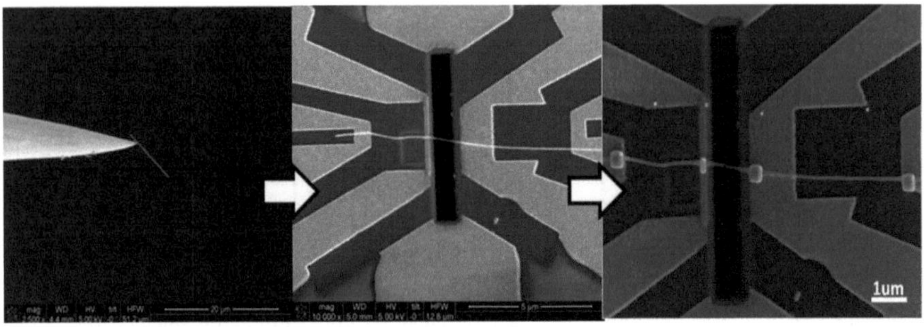


**Fig. 14.6** A single β-SiC NW was obtained from a host of bundles or substrates using the nanomanipulator and placed on the four-point probe measurement platform. At the last step, the contact points with the probe were covered with EB-Pt deposition

displacement over the entire range of motion. Figure 14.6 shows SEM images of a single NW obtained and placed on the four-point probe measurement platform by the nanomanipulation technique in series of steps. This direct manipulation technique was also applied to obtain specific NWs of the same material to compare their geometrical influences to thermal conductivities [23].

The four contact points between the NW and the probe electrodes were fully covered by EB-Pt deposition. This allowed for the reduction of the contact resistance between the NW and the electrode at each contact point. After the EB-Pt deposition on the contact points, the NW sample was annealed at 400 °C under high vacuum ($10^{-6}$ Torr) to further minimize the electrical contact resistance. It has been reported that the annealing step could remove the voids in the contact points between the NW and probe electrodes [24].

## 14.3 Electrical and Thermal Conductivity

The one-stop measurement setup is depicted in Fig. 14.7. Electrical and thermal conductivities and Seebeck coefficient could be measured in this one-stop measurement platform in situ. The measurement platform consists of the four-point probe and nanoheater. A lock-in amplifier (Stanford Research System SR850) was connected to the measurement system to read 3-ω signals for thermal conductivity measurement. An alternating current (AC) source (Keithley 6221) was used for a stable current supply. In addition to providing a stable current, the compliance setting of the current source could prevent any undesirable overshooting as any unwanted high current density might damage the placed NW.

In order to measure the thermal conductivities of individual NWs, microfabrication-assisted device structures have been suggested [25–29]. We used the 3-ω method in this study. The 3-ω method is based on a four-point probe system, which enables measurements of magnitude and phase of third-harmonic signal for one-dimensional sample subject to AC modulated at the fundamental frequency of ω. As a

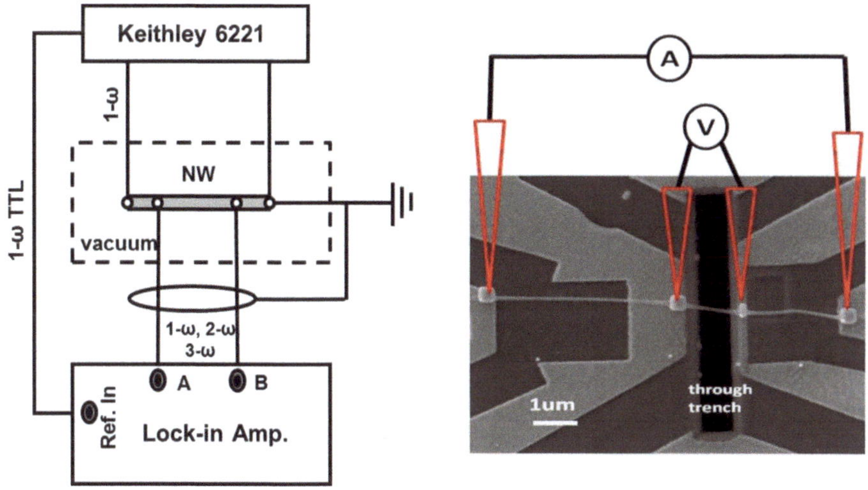


**Fig. 14.7** Four-point probe measurement setup for electrical and thermal conductivities and Seebeck coefficient (reproduced from [13], ©Springer)

time-independent frequency-domain technique, the 3-ω method can effectively eliminate spurious signals because a narrow-band detection technique is employed. This in turn enhances signal-to-noise ratio which allows for more accurate signal reading. Thus, the 3-ω method has been employed in various fundamental thermal characterizations including measurement of thermal conductivity for CNTs and bulk structure [30–32]. The third-harmonic amplitude created as a response to an applied AC current with a fundamental frequency has information about the 1D sample's thermal conductivity as shown in the following equation [31, 33]:

$$V_{3\omega,rms,vac} \cong \frac{\sqrt{2}\, I_0^3 R R' L}{\pi^4 k A} \tag{14.1}$$

where $L$, $R[=R_0+R'(T-T_0)]$, and $A$ are length, electrical resistance, and cross-sectional area of NW, respectively. $R'$ is the temperature coefficient of resistance at room temperature defined as $(\mathrm{d}R/\mathrm{d}T)_{\text{room temp.}}$ and $k$ is the thermal conductivity of NW. The advantage of this method can minimize the uncertainty due to the contact resistance which is present in the conventional two-point probing method.

The electrical resistivity of the β-SiC NW was measured to be ~$4.45 \times 10^{-3}$ Ωcm from the first-harmonic AC signal (1 kHz) and geometry of the NW (diameter of 60 nm and length of 1.5 μm). From the measured third-harmonic signals, we have estimated the thermal conductivity to be 82 ± 6 W/mK. The third-harmonic signals as a function of AC current amplitude at 1 kHz detected by the lock-in amplifier are shown in Fig. 14.8a. The data were best-fitted according to (14.1) and the result of the matching exponent of 2.83 was obtained. This value was very close to the theoretical value of 3.

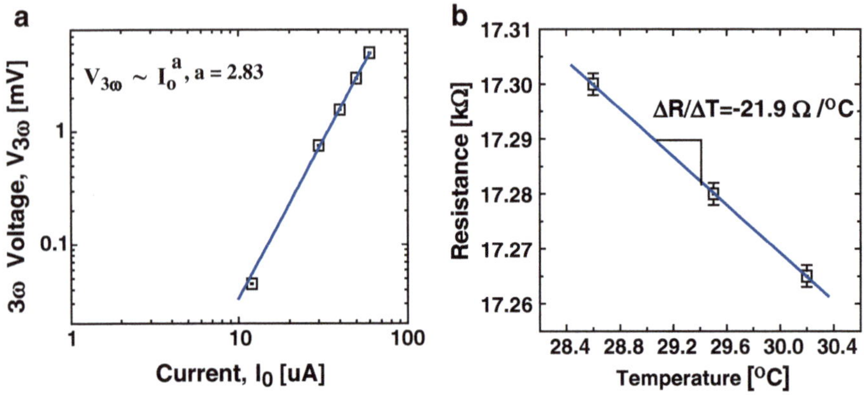


**Fig. 14.8** (**a**) 3-ω voltage signals as a function of current, $I_0$, obtained at room temperature; (**b**) the resistance change by temperature gradient of a single β-SiC NW (reproduced from [13], ©Springer)

In order to obtain the thermal conductivity of NW by the 3-ω method, the temperature coefficient (resistance change by temperature variation) should be known according to (14.1). In the temperature range of 28.6–30.2 °C, the resistance change by temperature variation was measured at −21.9 Ω/°C (Fig. 14.8b). The negative sign is verified by the fact that SiC nanowire is semiconducting.

## 14.4 Seebeck Coefficient and Figure of Merit

Seebeck coefficient, *S*, was measured immediately after characterizing the thermal conductivity. For a semiconductor, Seebeck coefficient can be written as [13]

$$S = -\frac{k_B}{e}\left(\frac{E_F - E_c}{k_B T} + const.\right) \cong \frac{(E_c - E_F)}{eT} \tag{14.2}$$

where $E_c$ is the conduction band energy, $E_F$ is the Fermi energy, *e* is the electron charge, $k_B$ is the Boltzmann constant, and *T* is the temperature. As seen in (14.2), Seebeck coefficient can be maximized when $E_F$ is located inside the bandgap. However, lowering $E_F$ to increase the gap induces reducing electrical conductivity. This will have the adverse effect on the thermoelectric figure of merit, which can be written as $Z\ (=S^2\sigma/\kappa)$ where $\sigma$ is the electrical conductivity and $\kappa$ is the thermal conductivity. Therefore, in an n-type semiconductor such as β-SiC NW, the optimal condition for a large value of Seebeck coefficient is that Fermi level should be close to the band edge without lowering an electrical conductivity. In this study, Seebeck coefficient was obtained from a voltage change by a thermal gradient across the NW such as $S=-(\mathrm{d}V/\mathrm{d}T)=-(\Delta V/\Delta T)$. The thermal gradient was established by a localized heating source created by IB-Pt. The localized heating source (nanoheater) was made of a thin line through IB-Pt deposition. The calibration for the resistance change by temperature change of the nanoheater was conducted to find an actual temperature of the heater when an electrical current is applied. Keithley 6221 AC/DC current source and Keithley 2182A nanovoltmeter were used to find the resistance change of the heater as the temperature increases. An electrical power from a power supply was provided to an aluminum (Al) block through the cartridge heater. A temperature sensor was used to read the temperature of the sample and Al block when an electric power was given to the cartridge heater. A high-speed A/D (analog-digital) converter (AD7760) was used to transfer and control analog signals such as power and temperature via LabVIEW® program in a computer. The data flows between a computer and A/D converter were made through a serial (e.g., USB) interface.

The measured value of resistance change by temperature gradient of the nanoheater was revealed to be 36 Ω/°C. From the calibrated value of resistance change by temperature gradient of the nanoheater, we could determine the temperature vs. applied current. Once the temperature of the nanoheater was known, the voltage drop was measured between the two inner probes out of the four-point probes. This revealed the voltage change by temperature gradient between the two inner probes (i.e., the nanoheater and the probe across the trench). Voltage and temperature difference, $\Delta V$ and $\Delta T$, were measured to be $2.015 \times 10^{-3}$ V and 1.67 °C. Thus, the Seebeck coefficient of the placed NW was −1.21 mV/K from the definition, $S = -(\mathrm{d}V/\mathrm{d}T) = -(\Delta V/\Delta T)$. Finally, the dimensionless thermoelectric figure of merit, $ZT$ ($= S^2 \sigma T/\kappa$), was estimated to be 0.12 from all the measured values of $\sigma$, $\kappa$, and $S$ at room temperature.

## 14.5 Conclusions

For measurement of thermal properties of a single NW, we developed a one-stop measurement platform consisted of four-point probes and nanoheater. Combining FIB and nanomanipulator, we placed a freestanding single NW (i.e., β-SiC NW) on the specific location of the measurement platform. Subsequently, its thermoelectrical properties such as electrical and thermal conductivities and Seebeck coefficient were measured. From the obtained data, the dimensionless thermoelectric figure of merit was also estimated. This FIB-assisted nanoscale processing would allow us to make more accurate thermoelectrical measurements of nanostructures such as NW and nanotube and to have better fundamental understanding of their thermal behaviors.

## References

1. Melngailis, J.: Critical review: focused ion beam technology and applications. J. Vac. Sci. Technol. B **5**, 469 (1987)
2. Reyntjens, S., Puers, R.: A review of focused ion beam applications in microsystem technology. J. Micromech. Microeng. **11**, 287 (2001)

3. Reyntjens, S., De Bruyker, D., Puers, R.: Focused ion beam as an inspection tool for microsystem technology. Proceedings of the 1998 Microsystem Symposium, vol. 125, Delft, the Netherlands (1998)
4. Ward, B.W., Economou, N.P., Shaver, D.C., Ivory, J.E., Ward, M.L., Stern, L.A.: Microcircuit modification using focused ion beams. Proc. SPIE **923**, 92 (1988)
5. Orloff, J., Utlaut, M., Swanson, L.: High resolution focused ion beams: FIB and its applications. Springer Press, New York (2003)
6. Casady, J.B., Johnson, R.W.: Status of silicon carbide (SiC) as a wide-bandgap semiconductor for high-temperature applications: a review. Solid-State Electron. **39**, 1409 (1996)
7. Wong, E.W., Sheehan, P.E., Lieber, C.M.: Nanobeam mechanics: elasticity, strength, and toughness of nanorods and nanotubes. Science **277**, 1971 (1997)
8. Liang, C.H., Meng, G.W., Zhang, L.D., Wu, Y.C., Cui, Z.: Large-scale synthesis of b-SiC nanowires by using mesoporous silica embedded with Fe nanoparticles. Chem. Phys. Lett. **329**, 323 (2000)
9. Shenai, K., Scott, R.S., Baliga, B.J.: Optimum semiconductors for high-power electronics. IEEE Trans. Elec. Dev. **36**, 1811–1823 (1989)
10. Dai, H.J., Wong, E.W., Lu, Y.Z., Fan, S.S., Lieber, C.M.: Synthesis and characterization of carbide nanorods. Nature (London) **375**, 769 (1995)
11. Li, Z.J., Li, H.J., Chen, X.L., Meng, A.L., Li, K.Z., Xu, Y.P., Dai, L.: Large-scale synthesis of crystalline β-SiC nanowires. Appl. Phys. A Mater. Sci. Process **76**, 637 (2003)
12. Rogdakis, K., Lee, S.Y., Bescond, M., Lee, S.K., Bano, E., Zekentes, K.: Theoretical comparison of 3C-SiC and Si nanowire FETs in ballistic and diffusive regimes. IEEE Trans. Elec. Dev. **55**, 1970–1976 (2008)
13. Lee, K.M., Lee, S.K., Choi, T.Y.: Highly enhanced thermoelectric figure of merit of a $\beta$-SiC nanowire with a nanoelectromechanical measurement approach. Appl. Phys. A **106**, 955–960 (2012)
14. Platinum deposition Technical note (FEI Company, Hillsboro, OR) PN 4035 272 21851-D (2003)
15. Lee, K.M., Choi, T.Y., Lee, S.K.: Measurement of figure of merit for a single β-silicon carbide nanowire by the four-point three-ω method. TechConnect World 2010 Proceedings (Nanotechnology 2010) vol. 2, pp. 202–205 (2010)
16. Lee, J., Wang, A.A., Rheem, Y., Yoo, B., Mulchandani, A., Chen, W., Myung, N.V.: DNA assisted assembly of multisegmented nanowires. Electroanalysis **19**(22), 2287 (2007)
17. Heo, K., Cho, E., Yang, J.E., Kim, M.H., Lee, M., Lee, B.Y., Kwon, S.G., Lee, M.S., Jo, M.H., Choi, H.J., Hyeon, T., Hong, S.: Large-scale assembly of silicon nanowire network-based devices using conventional microfabrication facilities. Nano Lett. **8**(12), 4523 (2008)
18. Deegan, R.D., Bakajin, O., Dupont, T.F., Huber, G., Nagel, S.R., Witten, T.A.: Contact line deposits in an evaporating drop. Phys. Rev. E **62**(1), 756 (2000)
19. Liu, Y., Chung, J., Liu, W.K., Ruoff, R.S.: Dielectrophoretic assembly of nanowire. J. Phys. Chem. B **110**, 14098–14106 (2006)
20. Chung, J., Lee, K., Lee, J., Ruoff, R.S.: Toward large-scale integration of carbon. Nanotubes (Langmuir) **20**(8), 3011–3017 (2004)
21. Boote, J.J., Evans, S.D.: Dielectrophoretic manipulation and electrical characterization of gold nanowires. Nanotechnology **16**, 1500–1505 (2005)
22. Schwamb, T., Choi, T.Y., Schirmer, N., Bieri, N.R., Burg, B., Tharian, J., Sennhauser, U., Poulikakos, D.: Dielectrophoretic method for high yield deposition of suspended, individual carbon nanotubes with 4-point electrode contact. Nano Lett. **7**, 3633 (2007)
23. Lee, K.M., Choi, T.Y., Lee, S.K., Poulikakos, D.: Focused ion beam-assisted manipulation of single and double β-SiC nanowires and their thermal conductivity measurements by the four-point-probe 3-ω method. Nanotechnology **21**, 125301 (2010)
24. Gopal, V., Radmilovic, V.R., Daraio, C., Jin, S., Yang, P., Stach, E.A.: Rapid prototyping of site-specific nanocontacts by electron and ion beam assisted direct-write nanolithography. Nano Lett. **4**, 2059 (2004)

25. Shi, L., Hao, Q., Yu, C., Mingo, N., Kong, X., Wang, Z.L.: Thermal conductivities of individual tin dioxide nanobelts. Appl. Phys. Lett. **84**, 2638 (2004)
26. Li, D., Wu, Y., Kim, P., Shi, L., Yang, P., Majumdar, A.: Thermal conductivity of individual silicon nanowires. Appl. Phys. Lett. **83**, 2934 (2003)
27. Wang, J., Wang, J.-S.: Carbon nanotube thermal transport: ballistic to diffusive. Appl. Phys. Lett. **88**, 111909 (2006)
28. Bryning, M.B., Milkie, D.E., Islam, M.F., Kikkawa, J.M., Yodh, A.G.: Thermal conductivity and interfacial resistance in single-wall carbon nanotube epoxy composites. Appl. Phys. Lett. **87**, 161909 (2005)
29. Vavro, J., Llaguno, M.C., Satishkumar, B.C., Luzzi, D.E., Fischer, J.E.: Electrical and thermal properties of $C_{60}$-filled single-wall carbon nanotubes. Appl. Phys. Lett. **80**, 1450 (2002)
30. Choi, T.Y., Poulikakos, D., Tharian, J., Sennhauser, U.: Measurement of thermal conductivity of individual multiwalled carbon nanotubes by the 3-$\omega$ method. Appl. Phys. Lett. **87**, 013108 (2005)
31. Choi, T.Y., Poulikakos, D., Tharian, J., Sennhauser, U.: Measurement of the thermal conductivity of individual carbon nanotubes by the four-point three-ω method. Nano Lett. **6**, 1589 (2006)
32. Lu, L., Yi, W., Zhang, D.L.: 3$\omega$ method for specific heat and thermal conductivity measurements. Rev. Sci. Instrum. **72**, 2996 (2001)
33. Yi, W., Lu, L., Dian-lin, Z., Pan, Z.W., Xie, S.X.: Linear specific heat of carbon nanotubes. Phys. Rev. B **59**, R9015 (1999)

# Chapter 15
# FIB Design for Nanofluidic Applications

**R. Fulcrand, N.P. Blanchard, A.-L. Biance, A. Siria, P. Poncharal, and L. Bocquet**

**Abstract** In this chapter we briefly review the techniques available to researchers in the nanofluidic domain to fabricate nanopores and nanochannels. In this context the focused ion beam (FIB) technique will be introduced as a useful and versatile tool for nanofluidics. We illustrate it with two specific examples involving nanopores as building blocks for nanofluidics.

## 15.1 Introduction

Nanofluidics is an emerging topic that has come into its own quite recently from the more established field of microfluidics. Nanofluidics is a branch of nanoscience and has great potential for numerous applications such as biotechnological devices [1], fluidic operations, and even energy conversion [2]. From the point of view of "lab on a chip" systems, decreasing the length scale considerably increases the sensitivity of analytic techniques, with the ultimate goal being isolating and studying individual macromolecules (e.g., DNA elongation [3], DNA capture [4]). Moreover, nanometric length scales allow new fluidic functionalities to be developed, which benefit from the predominance of surface sites: water desalination [5], pre-concentration phenomena [6], nanofluidic transistors [7], as well as capillary filling.

R. Fulcrand (✉) • N.P. Blanchard • A.-L. Biance • A. Siria • P. Poncharal • L. Bocquet
Institut Lumière Matière, UMR5306 Université Lyon 1-CNRS, Université de Lyon, 69622 Villeurbanne cedex, France
e-mail: remy.fulcrand@univ-lyon1.fr; Anne-Laure.Biance@univ-lyon1.fr; alessandro.siria@univ-lyon1.fr; philippe.poncharal@univ-lyon1.fr

L. Bocquet
Department of Civil and Environmental Engineering,
Massachusetts Institute of Technology, Cambridge, Massachusetts 02139, USA
e-mail: lbocquet@MIT.EDU

Z.M. Wang (ed.), *FIB Nanostructures*, Lecture Notes in Nanoscale Science and Technology 20, DOI 10.1007/978-3-319-02874-3_15,

From a different perspective, nanofluidics also carries the hope that new properties will emerge by benefiting from the specific phenomena occurring at small length scales: new solutions may be obtained where the behavior of matter departs from common expectations.

Finally, it is also important to note that nanofluidics has appeared recently as a scientific field, thanks to the substantial development of nano-fabrication technologies, now allowing specifically designed devices to be fabricated, for example, electron beam lithography (EBL), focused ion beam (FIB), nano-imprint lithography (NIL), or UV and X-ray lithography. There was also a considerable progress made over the last two decades in developing new instruments and tools that give the possibility of investigating fluid behavior at the nanometer scale. It is now possible to control/design what is occurring at micronic and submicronic length scales and observe/measure their effects.

## 15.2 Tools for Nanosystems Prototyping

Conventional photolithography is undoubtedly the most widely used manufacturing method for the production of micro- and nanosystems. This parallel process provides very high efficiency for high-volume production but can be very expensive for the realization of micro- and nanochannels into glass or silicon substrate. The use of this method requires the fabrication of a photomask (PM) and the deposition of a photoresist (PR) (Fig. 15.1a), which involve several technological steps: PM conception/realization and PR deposition, baking, insulating, and developing [8]. Typically, nanofluidic channels are made by using a conventional lithography step followed by dry or wet etching to define the depth of the channel [9]. Due to the

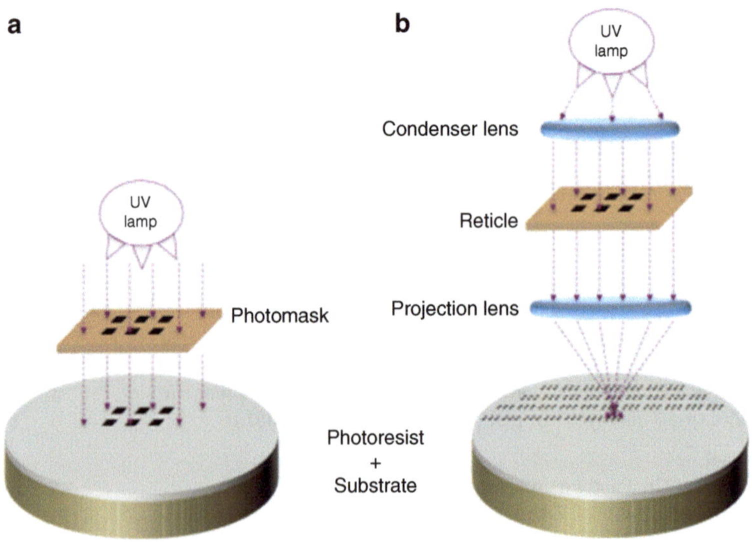


**Fig. 15.1** Brief schematic description of (**a**) conventional lithography and (**b**) projection lithography

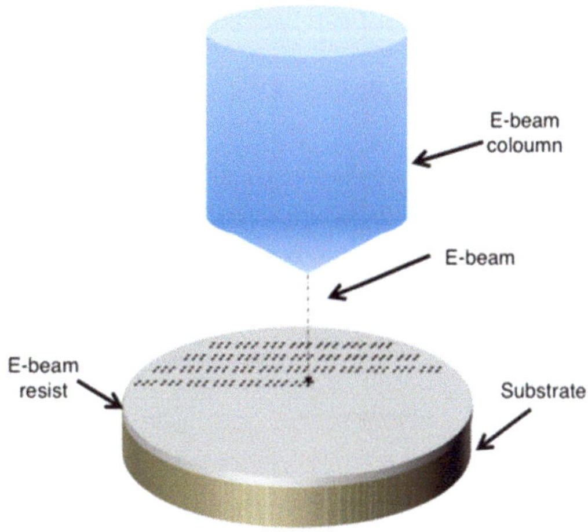


**Fig. 15.2** Brief schematic description of electron beam lithography

limitations inherent to conventional lithographic processes, almost all nanofluidic devices have simple geometries with only one nanometric dimension, the channel depth. Indeed, the minimum lateral dimension that we can reach with this method is a function of the wavelength of the UV lamp (in general between 365 and 405 nm), $R=3(\lambda \cdot S)^{1/2}/2$, with $R$ as the resolution, $\lambda$ the wavelength, and $S$ the distance between PM and PR, corresponding to a resolution between 300 and 500 nm. Significant advances have been made to improve the resolution of the conventional photolithographic process by using phase-shifting masks in place of the conventional binary intensity masks. Currently, the projection optical photolithography (POL or stepper photolithography) is the most common industrial technique for making integrated circuits [10]. Briefly, the stepper consists of several parts, as we can see in Fig. 15.1b:

- A light source (usually a mercury-vapor lamp)
- A condenser lens to image the light source on the mask (also called reticle)
- The mask, which contains the pattern
- The projection lens, which reduces (generally 5×) the image of the mask onto the substrate
- The substrate

Improvement of the resolution will only be possible with the development of new patterning tools not limited by the UV wavelength. For example, electron beam lithography (EBL) has been developed to write and design reticles directly onto a PR. EBL uses a focused electron beam to expose an electron-sensitive resist and has been applied to fabricating nanochannels with submicron sizes in the "x" and "z" directions (Fig. 15.2) [11]. This direct-write lithography approach confers great versatility and power: no need for a physical mask, rapid prototyping of small chips, and the ability to produce nano-devices smaller than the state-of-the-art conventional photolithography. EBL could be a tool that combines a reasonable writing speed with a high level of integration. Recently, a sub-10 nm half-pitch resolution using a negative PR has been successfully demonstrated with this method [12]. Moreover, a 30 nm pitch array of $SiO_2$ pillars by combining EBL, lift-off process,

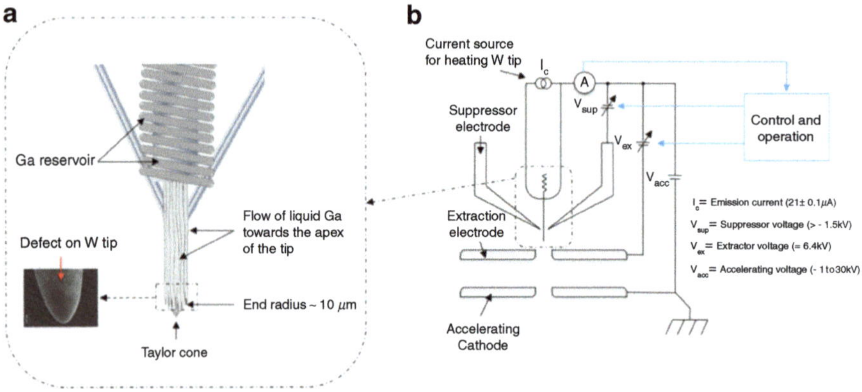


**Fig. 15.3** Schematic descriptions of (**a**) the gallium source and (**b**) the gun

and reactive ion etching process has been demonstrated [13]. Like conventional lithography, special resists are required and a wet or dry etching step is used to transfer a pattern onto a substrate. Generally, as it is a sequential process, the fabrication protocol itself limits the structural control and the resolution to the final structures. In fact, the control of the homogeneity and stability of the PR film, the development step, and the etching process are the most critical steps to improve the resolution of designs. Another effect that can limit the resolution of this technique is the interaction between electron scattering with each other and with atoms in the resist.

An alternative means to improve resolution is to directly machine devices onto the surface by using ion milling. The development of the liquid metal ion source (LMIS) in the late 1960s [14] and middle 1970s [15] and its application for focused ion beam (FIB) in the late 1970s [16] has resulted in a powerful tool for research and for industry. Several liquid metal ion sources have been made and most FIB LMIS use gallium. In practice, the usual method is to place a supply reservoir onto a tungsten needle that has been electrochemically etched to a suitable shape (Fig. 15.3a). Small natural defects on the surface of the needle carry the liquid metal from the reservoir to the end of the needle. The emission of ions is then initiated by placing a voltage between the needle and an extraction electrode, which results in the formation of a Taylor cone (Fig. 15.3b).

The FIB tool exhibits two main characteristics: material removal from a sample's surface by sputtering and material deposition onto a sample's surface by ion-induced reactions at submicron dimensions. Not surprisingly, FIB has become the tool of choice in failure analysis, defect characterization, design modification, and process control in a variety of industries. The three major historical categories of use are (1) integrated circuit review and modification, (2) transmission electron microscopy (TEM) sample preparation, and (3) thin film head (TFH) manufacturing. There are of course other uses of FIB and in particular for ion beam lithography. FIB lithography has significant advantages over the electron beam counterpart in terms of resist sensitivity, backscattering, and proximity effects. As a consequence, a higher resolution can be obtained with FIB as the pixel size is quasi equal to the beam spot size and no exposure occurs between pixels, hence allowing a short dwell time on each pixel. Another specificity of the FIB tool is the possibility for maskless and resistless structuring. It allows the three-dimensional patterning of various shapes directly onto a substrate due to a very fine control of the local ion dose during the surface scan. Practically, the exposure time at each point, or dwell time, determines the sputtering depth so that 3D structures can be made.

## 15.3 FIB for Manufacturing Nanochannels and Nanopores

As we have shown in the previous section, patterning with FIB is a direct process to create nano-devices. Although this technique is slower than the conventional lithography, it should be expected to provide a methodology for research laboratories for the production of nanofluidic devices with a great reproducibility, versatility, and a very fast conception/fabrication. The use of FIB for the direct patterning of nanofluidic systems has been investigated. Nanofluidic channels with lateral dimensions as low as 20 nm have been fabricated by milling various substrates (silicon or glass) with a focused $Ga^+$ ion beam [17, 18]. Several strategies have also been used to generate nanochannels for nanofluidic studies in the 5–300 nm range [19–21].

Equally, nanopores have also received a lot of attention in recent years for their potential applications as low-cost, high-throughput biosensors and membranes for filtering, desalination, and energy generation [5, 22–24]. Solid-state nanopores open the way to the study of the confinement, dynamics, and transport properties of single macromolecules at nanometric scales with a very high spatial resolution and high sensitivity. It has been demonstrated that by monitoring the ionic current across a nanopore, the translocation of nano-particles or biomolecules such as DNA, proteins, and viruses can be detected. In fact, a partial blockage of a pore by a target object leads to a transient downward pulse of ionic current, the magnitude and duration of which is directly related to the physical properties of the target (e.g., diameter and length) or its interaction with the pore surface. Today, the most promising technique for the realization of such nanopores is the local/direct milling obtained by FIB.

## 15.4 FIB Processing for Nanopore Fabrication

FIB instruments can be used for milling, deposition, implantation, and imaging depending on the peripheral equipment in addition to the ion source and column. However, in this chapter only the milling aspect of FIB will be developed and discussed. For material removal, "milling," when the ion beam is incident on a solid surface, a transfer of kinetic energy from the incident ions to the surface and near surface atoms will take place in the irradiated zone. If this energy is greater than the surface binding energy, the surface atoms will be sputtered away from their initial location. This knock-on sputtering is the primary mechanism for material removal and its efficiency is normally represented by the sputter yield, defined as the number of atoms ejected from the target surface per incident ion. The sputter yield is

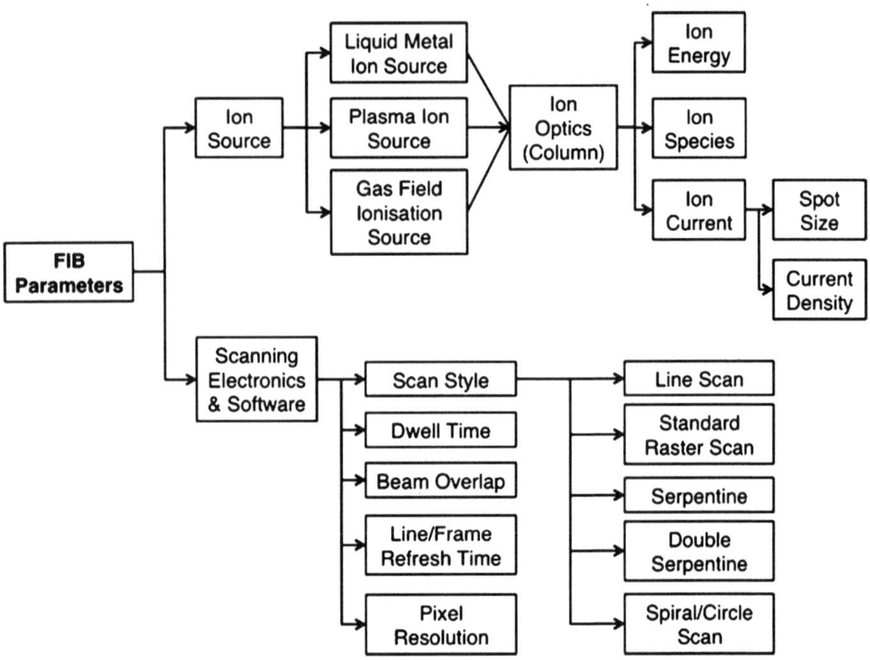


**Fig. 15.4** Key parameters for direct ion beam milling

material dependent and is also a function of the ion species, ion energy, ion beam/surface angle of incidence, and beam/crystallite orientation [25].

In this chapter only one material system will be considered, specifically thin silicon nitride (SiN) membranes. The examples presented below have been performed on a FIB/SEM system equipped with a Ga liquid metal ion source producing $Ga^+$ ions with an emission current of 2.1 μA. The ion column is a SIINT Zeta column operated at 30 kV and able to generate beam currents from 0.1 pA to 45 nA. The electron column is a Zeiss Gemini column and the instrument belongs to the Zeiss NVision 40 product line.

Figure 15.4 shows the different parameters that govern the milling efficiency and precision for a given material system. The ion species, energy, and probe current are determined by the choice of ion source and ion optics with the user of a given instrument being able to choose the energy and probe current of the ion beam. As a general rule high energies and low beam currents give the highest milling precision (spot size); however, low ion currents mean longer milling times. The scanning software and electronics allow a further control over the milling precision by controlling the number of increments the beam passes through within the scanned area (pixel resolution), how long the beam rests at each step or pixel (dwell time), the beam diameter with respect to the pixel size (beam overlap), the number of times the beam scans over a given line and/or frame, and finally the path that the beam follows

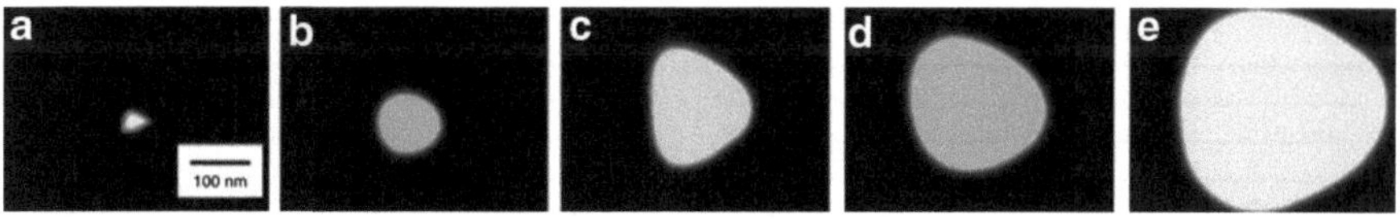


**Fig. 15.5** Pattern milling in SiN thin membrane (50 nm) by using various spot sizes: (**a**) 0–0,3 pA, (**b**) 1 pA, (**c**) 10 pA, (**d**) 40 pA, and (**e**) 80 pA

(scan style). In the following example the effect of these parameters on the form of nanopores milled into a thin SiN membrane, 50 nm, and a thick SiN membrane, 1 μm, will be demonstrated.

### *15.4.1 Thin Silicon Nitride Membrane*

The simplest strategy to fabricate nanopores using an ion beam is the "one-spot milling" where the ion beam is positioned at the location where the pore is to be pierced and remains there until the pore traverses the membrane. The final diameter of the pore is controlled by the ion beam diameter (beam energy/current couple), the dwell time per pore, and the thickness and composition of the membrane. For a 50 nm thin SiN membrane, the pores of increasing diameter have been pierced by increasing the beam current from 0.1 pA up to 80 pA. As can be seen from the TEM micrographs in Fig. 15.5, the pores are not perfectly circular. This is in part due to a non-optimized focus and astigmatism of the ion beam. The non-circularity is particularly noticeable for the lowest ion beam current (0.1 pA) and higher ion beam currents (>1 pA) due to the poor image signal at low currents and the rapid milling of the thin membrane during ion beam imaging for the higher currents. From a practical operator's point of view, the 1 pA beam current offers enough secondary electron signal to optimize the focus and astigmatism without over-milling the sample.

To achieve the ultimate milling resolution for nanopores in membrane materials (pore diameter <10 nm) with a Ga LMIS ion column, then the one-spot method offers the best results. A single-column FIB system with an optimized Ga LMIS source and ion optics (short ion beam focal distance, non-tilted sample, and high beam energy, 35 keV) has been developed [26]. By combining the accurate sample stage control developed for electron lithography instruments with an optimized ion beam, arrays of nanopores with a diameter of 3 nm can be fabricated [27].

However, for more fundamental studies of fluidic phenomena at the nanoscale, commercial FIB/SEM systems offer a highly flexible tool to create and measure nanopores as demonstrated for the case of solid-state nanopores detailed in the following section. Rather than keeping the beam at an established position and varying the probe current to define the diameter of the pore, the probe current is fixed at 1 pA (for the reasons described above) and then scanned over the membrane in the form of a circle with a diameter defined by the scan software. Using this method, one is able to create circular pores with a good control of the diameter in the 100–500 nm range.

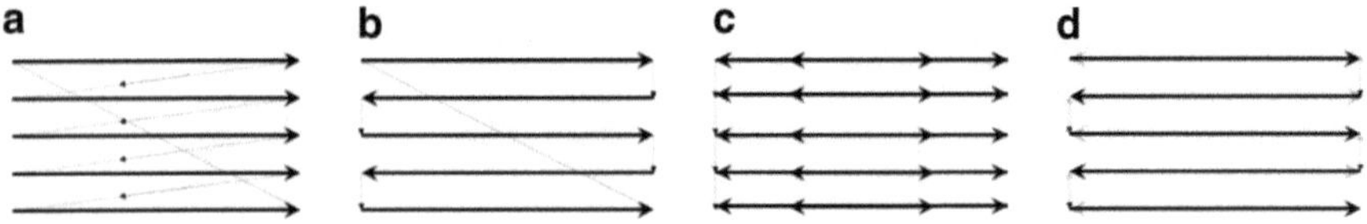


**Fig. 15.6** Various scan methods: (**a**) standard raster scan (**b**) serpentine scan, (**c**) line scan, and (**d**) double serpentine scan

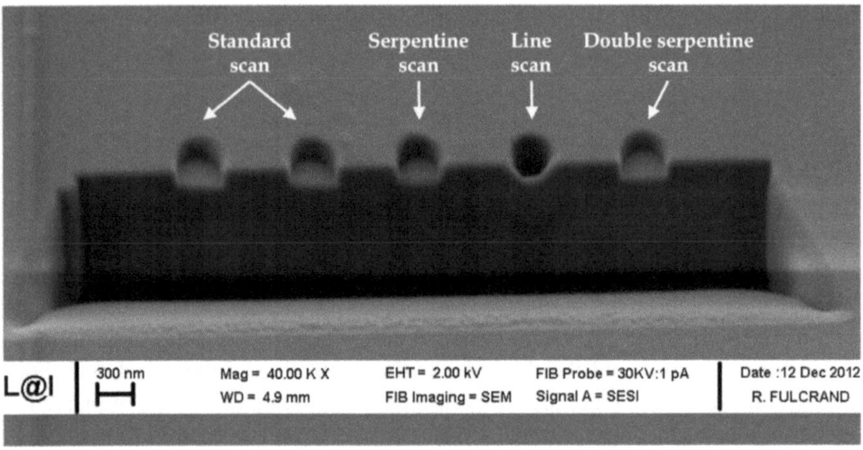


**Fig. 15.7** Pattern milling in SiN thick membrane (1 μm) at 1 pA by using the same ion dose (1 nC/μm$^2$). Comparison of cross sections between standard, serpentine, double serpentine, and line scan

## 15.4.2 *Thick Silicon Nitride Membrane*

Before developing further the study of a nanopore in a thin SiN membrane, the case of thick membrane (1 μm) will be very briefly reviewed. For thicker membranes the importance of the ion–solid interaction is closer to that of bulk samples where one might "mill" nano- or micro-channels. For this type of sample, the parameters illustrated in Fig. 15.4 play a more critical role on the final "quality" of the milled form. For the case of a homogeneous and amorphous material, the spread of the beam in the sample, the sputtering yield, and related redeposition will have an effect on the final aspect ratio of the pore/channel. In this chapter only the case of the scan style will be shown; however, the reader is encouraged to refer to numerous review papers in the literature for more detailed discussions of the influence of the various parameters on ion beam milled structures [28].

Figure 15.6 shows various scan protocols for charged particle beams, with the standard raster scan being the most familiar to those with experience in scanning electron microscopy. The cross-sectional SEM image in Fig. 15.7 showing five milled cylinders using the four available scan methods of the ion beam on the FIB/

SEM used for this work demonstrates how the scan method for the same dwell time, probe current, and ion dose alters the shape of the cylinder. The standard scan results in a nonplanar floor as the beam always scans in the same direction (left to right in Fig. 15.6a). The serpentine scan gives a level floor but results in an increased redeposition on the sidewalls, resulting in a more conical shape. This effect is further enhanced for the line scan method, whereas the double serpentine scan gives the best results for this particular material system.

## 15.5 Nanopores as Nanofluidic Building Blocks: Two Examples

We now illustrate the versatility of FIB-designed nanopores as building blocks for nanofluidic applications. By essence, nanofluidics involves the transport of fluid inside nanoscale pipes. FIB-designed nanopores thus constitute a model system with nanometric dimension, in which water and ion transport may be studied. We will exemplify it by exploring ion conduction in solid-state nanopores and demonstrate the existence of anomalous entrance effects in pores of nanometric size. In a second example, we will demonstrate that drilled nanopores may be used as an elementary item to build hierarchical transmembrane nanofluidic devices.

### *15.5.1 Large Apparent Electric Size of Solid-State Nanopores*

In this section, we present experimental results that show unexpectedly large ionic conduction in solid-state nanopores, originating from anomalous entrance effects [29].

Recently, ion transport in nanochannels with a channel length much larger than the channel height or diameter has been exhaustively investigated as a function of size and ionic strength (see, for example, [30, 31]), and a good understanding has now emerged. Altogether ion conduction was shown to be a combination of bulk and surface contributions; while bulk conduction is attributed to ions inside the bulk of the pore, surface conduction arises due to excess counter ions close to the surface and screening the surface charge, the so-called electric double layer (EDL) [31, 32]. Accordingly, at high salt concentration the ion conduction is mainly determined by the bulk conductivity, $\kappa_b$, but at low salt concentration, the surface conduction, determined by the surface conductivity $\kappa_s$, becomes dominant over the bulk conduction, leading to a saturation of the measured ion conductance. This crossover occurs for nanochannel cross sizes lower than the so-called Dukhin length, defined as the surface to bulk conductivity ratio, $l_{Du}=\kappa_s/\kappa_b$ [32, 33].

In contrast, ion transport in nanopores pierced in thin membranes is paradoxically far more complex than in nanochannels. Indeed, due to a much smaller aspect ratio, ion transport cannot be reduced to a *one-dimensional* view as for a nanochannel, and the *three-dimensional* nature of a nanopore makes it difficult to

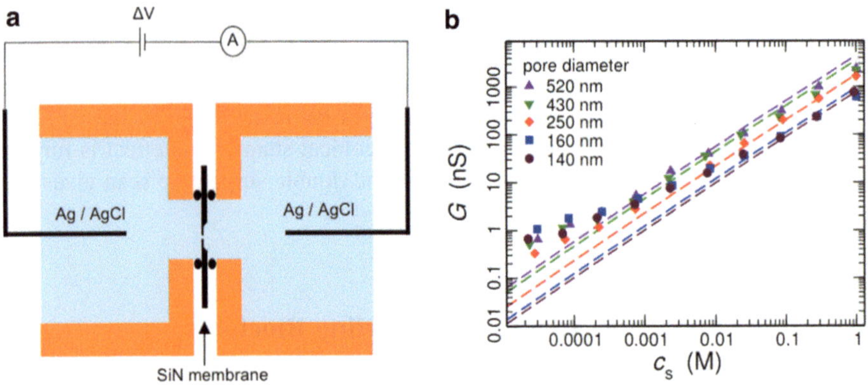


**Fig. 15.8** (**a**) Schematic of a flow cell and (**b**) measured conductance over varying salt concentrations in nanopores with different diameters. As a comparison, the predicted conductance in (15.1) is plotted together as *dashed lines*

comprehend ion transport in a nanopore on the basis of simple electric analogies. Several approximated forms were proposed in an attempt to account for the complex three-dimensional (3D) entrance geometry of the pore. For example, Hall [34] suggested that the access resistance of a pore could be accounted for by considering a semispherical cupola as an effective electrode [34, 35] leading to a nanopore conductance behaving as

$$G_0 = \kappa_b \left[ \frac{4L}{\pi D^2} + \frac{1}{D} \right]^{-1} \tag{15.1}$$

The second term in the right-hand side of this equation accounts for the entrance effect; $\kappa_b$ is the bulk conductivity and $L$ and $D$ are the length and diameter of the nanopore, respectively. This result assumes that the resistance of the nanopore is the sum of the resistance *inside the pore*, plus the entrance resistance that behaves typically like $(D\kappa_b)^{-1}$. Therefore, this entrance effect has merely a geometrical origin.

The access resistance has received more attention in recent years, as nanopores pierced in ultrathin membrane, like graphene, emerged as a new class of nanopore with enhanced performance [36, 37]. On the other hand, there have been few studies as to how surface effects contribute to the conductance in a nanopore [38, 39]. We show here that the 3D nature of the nanopore opens a new conductance path that plays a key role for the conductance at low ionic concentration.

We fabricated nanopores by drilling SiN membranes, 100 nm in thickness, with pore diameters D of 140, 160, 250, 430, and 520 nm. The ion conductance through the nanopore was measured in a home-built flow cell with the pierced membrane separating two fluidic reservoirs, Fig. 15.8a. Potassium chloride solutions with concentrations varying between $5\times10^{-5}$ and 1 M (mol/L) were considered (with fixed pH 6), and the voltage was imposed using Ag/AgCl electrodes. For all experimental data points, the salt concentration Cs was obtained from independent bulk

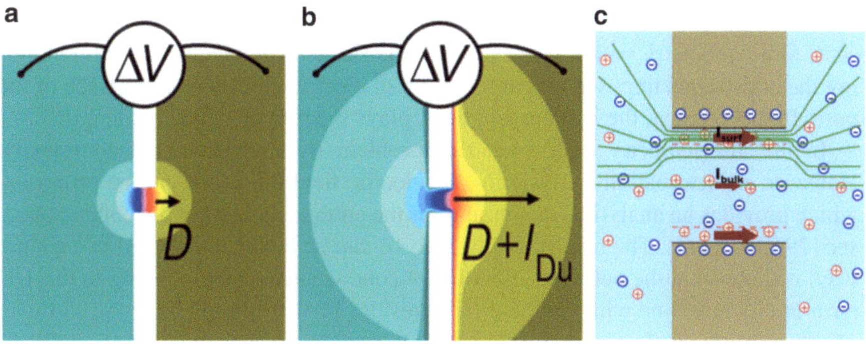


**Fig. 15.9** (**a**) Contour lines of the electric potential across a typical pore ($L=D=100$ nm) (**a**) without surface charge and (**b**) with surface charge (Dukhin length $l_{Du}$ ~3D); (**c**) sketch of the electric field lines inside and outside of the pore

conductivity measurements. The experimental data for the measured conductance are shown in Fig. 15.8b for a wide range of salt concentrations and various nanopore diameters. An interesting feature of these results is that the conductance depends markedly on the pore diameter for large salt concentrations but appears to converge to a similar value for low salt concentrations. The predicted conductance accounting for the geometric access effect in (15.1), using the bulk conductivity $\kappa_b$ measured independently, is plotted in Fig. 15.8b as a comparison with experimental data.

At high salt concentrations, the experimental results are in good agreement with the prediction, involving the geometrical entrance effect discussed in (15.1). However, at salt concentrations lower than ~$10^{-2}$ to $10^{-3}$ M, a deviation from the predicted conductance becomes apparent and (15.1) underestimates strongly the measured conductance by up to one order of magnitude. Therefore, this first approach fails to reproduce the experimental results.

The observation of a large conduction in a nanopore at small concentration points to a more complex scenario than that of nanochannels. The understanding of this effect requires fully revisiting the 3D conduction paths inside and outside the nanopore and points to an unforeseen conduction mechanism.

As we showed in [29], because of the 3D nature of the nanopore, surface conduction effects do extend spatially *outside* of the pore. The key point is that field lines outside the pore are perturbed due to surface conduction effects inside the pore. To highlight this effect, Fig. 15.9 compares typical electrical map without (a) and with (b) a surface charge inside the pore, as obtained with finite-element calculation [29]. This shows that the surface charge perturbs the electric field and the corresponding electric current streamlines far from the pore, well into the bulk of the reservoirs. The typical extension of the perturbation is typically given by the Dukhin length $l_{Du}$ introduced above.

The origin of this phenomenon is the following. Surface transport of ions requires that current and ions should be supplied within the diffuse interface to obey charge conservation, thus bending the electric streamlines toward the pore surface, as sketched in Fig. 15.9c. Khair and Squires pointed out this mechanism in a simple semi-finite geometry with a step charge in the surface conductance [33] highlighting the constraints imposed by conservation of electric current at the surface of the pore. In this context, the Dukhin length $l_{\mathrm{Du}}$ plays the role of a "healing length."

In the 3D geometry of a pore, this problem is of considerable complexity. However, a perturbation analysis for the electric field equations can be proposed, which provides an analytical expression for the entrance effect for the conductance; see [29] for details. The results of this analysis can be *in fine* interpreted in a simple way: it shows that the entrance effect in the conductance, corresponding to the 1/*D* term in (15.1) in line with Hall and Hille result [34, 40], should be replaced by an *effective entrance effect*, with the diameter replaced by an apparent one: $D_{\mathrm{app}}=D+\beta^{\square}\, l_{\mathrm{Du}}$ (with $\beta^{\square}\sim 1$ some numerical prefactor). This term results from the 3D perturbation of the electric field lines outside of the pore. More precisely the expression for the global conductance of the nanopore is found as

$$G=\kappa_{\mathrm{b}}\left[\frac{4L}{\pi D^2}\times\frac{1}{1+4\frac{l_{Du}}{D}}+\frac{2}{\alpha D+\beta l_{Du}}\right] \tag{15.2}$$

with $\alpha$ and $\beta$ as two numerical constants (~2). This expression was checked exhaustively against numerical finite-element calculations [29]. Some limiting behaviors of the above results are interesting to discuss. In the limit of large Dukhin lengths $l_{\mathrm{Du}}\gg D$, which occurs for low salt concentration, $c_{\mathrm{s}}\rightarrow 0$, and/or large surface charge $|\Sigma|$, then the conductance behaves as

$$G\left(l_{Du}\rightarrow\infty\right)=\kappa_{\mathrm{s}}\left(\frac{4L}{\pi D}+\frac{2}{\beta}\right)^{-1} \tag{15.3}$$

Therefore, the present mechanism indeed predicts a saturation of the conductance for small concentration, which stems from the fact that $D_{\mathrm{app}}\gg D$ in this limit. Furthermore, the conductance in this limit is shown to depend only on the aspect ratio *L*/*D* of the pore but not explicitly on the diameter *D*. In a counterintuitive way, small pores can conduct as efficiently as large pores.

We have compared the above predictions with the results of the experimental work. In Fig. 15.10, the measured conductance is compared with (15.2). We used $\alpha=\beta=2$ and the bulk conductivity $\kappa_{\mathrm{b}}$ measured independently [29]. The only remaining free parameter, $\kappa_{\mathrm{s}}$, was tuned to $\kappa_{\mathrm{s}}=0.8$ nS to obtain best agreement. As seen in Fig. 15.10, several key features observed in the experimental results such as a transition from bulk conductance dominated regime to surface conductance dominated regime and a saturation of the conductance at low salt concentration are well predicted by (15.2). Overall, the agreement between the predicted value and experimental results is good. A slight deviation at intermediate salt concentrations is observed for the 140 nm diameter pore, which might be attributed to its observed conical shape.

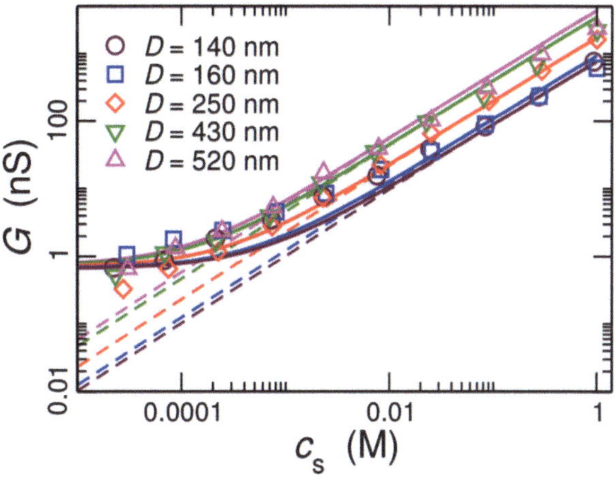


**Fig. 15.10** Comparison between the measured conductance (*solid line*) and the predictions (15.1) (*dashed line*), in nanopores with different diameters of 140, 160, 250, 430, and 520 nm. The only free parameter to fit all experimental data is the surface conductance, which was fixed to $\kappa_s = 0.8$ nS obtain the best agreement

Altogether our results show that the transport of ions in solid-state nanopores exhibits complex features, which strongly depart from what is observed in nano-channels. The three-dimensional nature of the system couples to the surface conduction inside the nanopore to induce unexpected electrical access effects that cannot be neglected. This is shown to lead to an anomalously large ion conductance at small ion concentration and/or large surface charge, whenever the Dukhin length is larger than the diameter D.

Such a situation will occur for highly charged and very small nanopores. This is the case of most biological transmembrane systems, even for relatively large salt concentration. Furthermore, electric entrance effects should be also taken into account when interpreting quantitatively translocation experiments of nano-particles and molecules through solid-state or biological nanopores. Indeed our results show that the global conductivity of the pore involves information over a region which extends on a size $D_{app}$, which can be much larger than the bare geometric size of the pore. Consequently, the sensitivity to translocation events, which is usually measured via ion current variations, should therefore integrate the corresponding information outside the pore. In particular, one expects that charged molecules or nano-particles will follow the perturbed electric field lines at the entrance of the pore. Their influence is accordingly expected to be monitored far beyond the pore entrance, thus limiting the spatial resolution of such a technique. This is particularly relevant in view of the recent efforts to use ultimately thin nanopore devices such as pierced graphene sheets [37, 41, 42].

### *15.5.2 Composite Nanosystems: Transmembrane Nanotubes*

We now turn to a second example of the versatility of FIB-drilled nanopores, showing how these may be used as a building block for new nanofluidic devices on the

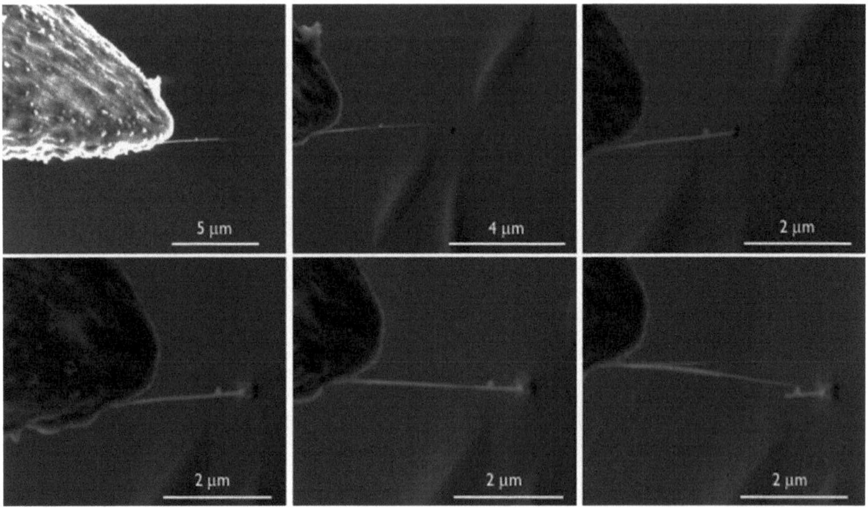


**Fig. 15.11** Insertion steps of a boron nitride nanotube inside a drilled nanopore to obtain a transmembrane nanotube for nanofluidic measurements

basis of nanostructure assembly. Among various nanosystems for nanofluidic applications, nanotubes represent ideal structures which combine the nanoscale dimension with a perfect crystalline structure. In particular hydrodynamic flow inside carbon nanotubes was shown to be nearly frictionless [43], a phenomenon which could be interpreted theoretically in terms of the superlubricity of the water–carbon interface [44]. In order to make progress in this field, exploring fluid transport inside a single nanotube is therefore highly desirable. However, a key difficulty is to integrate these nanotubes inside a fluidic system, amenable for fluid transport measurements.

Here, we show as an example a new route leading to hierarchical transmembrane nanofluidic devices. The idea is to use a nanopore pierced in a silicon nitride (SiN) membrane as a matrix to host a nanotube, like a mortise and tenon joint.

The first step of the assembly is to drill a hole with a diameter between 100 and 200 nm through the SiN membrane by using FIB. Then we use an electrochemically etched tungsten tip as a support, at the extremity of which we glue an individual multiwalled nanotube. This system is manipulated using a homemade nanomanipulator consisting of five axe piezo-inertial motors (Attocube ANPx-z51 and Attocube ANR51), which allow to realize the insertion of the nanotube into the membrane with nanometer resolution. These insertion and manipulation steps are carried out in situ in a scanning electron microscope (SEM). Figure 15.11 presents an example of an insertion of a nanotube (here, a boron nitride nanotube, BNNT) inside a nanopore by nano-manipulation. Once the BNNT is introduced, the tube is sealed to the hole using cracked naphthalene ($C_8H_8$) inserted using a gas injection system (GIS) and decomposed under e-beam irradiation. We checked independently that this sealing is water impermeable and electrically insulating. Retracting the tungsten tip from this position forces a telescopic sliding of the internal walls of the nanotube (see Fig. 15.11).

At the end of this nano-manipulation process, we obtain a transmembrane nanotube, with both extremities opened. The quality and inner diameter of the nanotubes are checked and measured a posteriori using transmission electron microscopy (TEM). The transmembrane nanotube is then squeezed between two reservoirs, allowing for nanofluidic measurements.

Using this hierarchical system, the ion transport was probed inside single BNNT under voltage drop, pressure drop, and salt concentration difference and combination of these. Measurements of ion transport under differences of salinity in the two reservoirs demonstrated the generation of giant electric current induced by osmotic salt gradients [45]. This suggests that osmotic energy conversion through BNNT membranes constitutes a new asset in the search of renewable source of energy.

Beyond the questions of energy, the route proposed here could be extended to develop new hierarchical nanofluidic systems with increased complexity.

## 15.6 Conclusion

The focused ion beam technique has clearly found its place among the more commonly used photo and electron beam lithography as a tool for researchers in the field of nanofluidics. FIB combined with an SEM in commercial two column FIB/SEM systems offers a highly flexible method to fabricate nanometric structures in a one-step process as well as to characterize their spatial dimensions via the high-resolution SEM. The versatility of this method gives the opportunity to the researchers to create an infinite variety of shapes in the nanoscale range, thus opening up the possibility to tailoring molecule-specific porous membranes. Whether focused ion beams will have a role to play in industrial scale manufacturing is yet to be seen, but the development of new ion sources or projection methods means that the question remains a legitimate one. We hope that this chapter convinces the reader that FIB is and will continue to be among the fabrication techniques in a nanofluidics researcher's "toolbox."

**Acknowledgements** We thank CLYM for providing access to FIB.

## References

1. Xia, D., Yan, J.C., Hou, S.F.: Fabrication of nanofluidic biochips with nanochannels for applications in DNA analysis. Small **8**, 2787–2801 (2012)
2. Daiguji, H., Yang, P., Szeri, A.J., Majumdar, A.: Electrochemomechanical energy conversion in nanofluidic channels. Nano Lett. **4**, 2315–2321 (2004)
3. Guo, L.J., Cheng, X., Chou, C.F.: Fabrication of size-controllable nanofluidic channels by nanoimprinting and its application for DNA stretching. Nano Lett. **4**, 69–73 (2004)
4. Mihovilovic, M., Hagerty, N., Stein, D.: Statistics of DNA capture by a solid-state nanopore. Phys. Rev. Lett. **110**, 028102 (2013)

5. Cohen-Tanugi, D., Grossman, J.C.: Water desalination across nanoporous graphene. Nano Lett. **12**, 3602–3608 (2012)
6. Ko, S.H., Song, Y.A., Kim, S.J., Kim, M., Han, J., Kang, K.H.: Nanofluidic preconcentration device in a straight microchannel using ion concentration polarization. Lab Chip **12**, 4472–4482 (2012)
7. Karnik, R., Fan, R., Yue, M., Li, D., Yang, P., Majumdar, A.: Electrostatic control of ions and molecules in nanofluidic transistors. Nano Lett. **5**, 943–948 (2005)
8. Howard, R.E., Liao, P.F., Skocpol, W.J., Jackel, L.D., Craighead, H.G.: Microfabrication as a scientific tool. Science **221**, 117–121 (1983)
9. Chauvet, F., Geoffroy, S., Hamoumi, A., Prat, M., Joseph, P.: Roles of gas in capillary filling of nanoslits. Soft Matter **8**, 10738–10749 (2012)
10. Rothschild, M.: Projection optical lithography. Mater. Today **8**, 18–24 (2005)
11. Reccius, C.H., Mannion, J.T., Cross, J.D., Craighead, H.G.: Projection optical lithography, compression and free expansion of single DNA molecules in nanochannels. Phys. Rev. Lett. **95**, 268101 (2005)
12. Hu, W., Sarveswaran, K., Lieberman, M., Bernstein, G.H.: Sub-10 nm electron beam lithography using cold development of poly (methylmethacrylate). J. Vac. Sci. Technol. **22**, 1711–1717 (2004)
13. Vieu, C., Carcenac, F., Pépin, A., Chen, Y., Mejias, M., Lebib, A., Manin-Ferlazzo, L., Couraud, L., Launois, H.: Electron beam lithography: resolution limits and applications. Appl. Surf. Sci. **164**, 111–117 (2000)
14. Mahoney, J.F., Yahiku, A.Y., Daley, H.L., David Moore, R., Perel, J.: Electrohydrodynamic ion source. J. Appl. Phys. **40**, 5101–5106 (1969)
15. Krohn, V.E., Ringo, G.R.: Ion source of high brightness using liquid metal. Appl. Phys. Lett. **27**, 479 (1975)
16. Seliger, R.L., Kubena, R.L., Olney, R.D., Ward, J.W., Wang, V.: High-resolution, ion-beam processes for microstructure fabrication. J. Vac. Sci. Technol. **16**, 1610–1613 (1979)
17. Campbell, L.C., Wilkinson, M.J., Manz, A., Camilleri, P., Humphreys, C.J.: Electrophoretic manipulation of single DNA molecules in nanofabricated capillaries. Lab Chip **4**, 225–229 (2004)
18. Arscott, S., Troadec, D.: A nanofluidic emitter tip obtained by focused ion beam nanofabrication. Nanotechnology **16**, 2295 (2005)
19. Menard, L.D., Ramsey, J.M.: Fabrication of sub-5 nm nanochannels in insulating substrates using focused ion beam milling. Nano Lett. **11**, 512–517 (2010)
20. Maleki, T., Mohammadi, S., Ziaie, B.: A nanofluidic channel with embedded transverse nanoelectrodes. Nanotechnology **20**, 105302 (2009)
21. Fanzio, P., Mussi, V., Manneschi, C., Angeli, E., Firpo, G., Repetto, L., Valbusa, U.: DNA detection with a polymeric nanochannel device. Lab Chip **17**, 2961–2966 (2011)
22. Humplik, T., Lee, J., O'Hern, S.C., Fellman, B.A., Baig, M.A., Hassan, S.F., et al.: Nanostructured materials for water desalination. Nanotechnology **22**, 292001 (2011)
23. Guo, W., Cao, L., Xia, J., Nie, F.Q., Ma, W., Xue, J., Song, Y., Zhu, D., Wang, Y., Jiang, L.: Energy harvesting with single-ion-selective nanopores: a concentration-gradient-driven nanofluidic power source. Adv. Funct. Mater. **20**, 1339–1344 (2010)
24. Kalman, E.B., Sudre, O., Vlassiouk, I., Siwy, Z.S.: Control of ionic transport through gated single conical nanopores. Anal. Bioanal. Chem. **394**, 413–419 (2009)
25. Giannuzzi, L.A., Stevie, F.A.: Introduction to Focused Ion Beams: Instrumentation, Theory, Techniques and Practice. Springer, New York (2005)
26. Gierak, J., Madouri, A., Biance, A.L., Bourhis, E., Patriarche, G., Ulysse, C., et al.: Sub-5 nm FIB direct patterning of nanodevices. Microelectron. Eng. **84**, 779–783 (2007)
27. Gierak, J., Bourhis, E., Faini, G., Patriarche, G., Madouri, A., Jede, R., et al.: Exploration of the ultimate patterning potential achievable with focused ion beams. Ultramicroscopy **109**, 457–462 (2009)
28. Kim, C.S., Ahn, S.H., Jang, D.Y.: Review: developments in micro/nanoscale fabrication by focused ion beams. Vacuum **86**, 1014–1035 (2012)

29. Lee, C., Joly, L., Siria, A., Biance, A.L., Fulcrand, R., Bocquet, L.: Large apparent electric size of solid-state nanopores due to spatially extended surface conduction. Nano Lett. **12**, 4037–4044 (2012)
30. Stein, D., Kruithof, M., Dekker, C.: Surface-charge-governed ion transport in nanofluidic channels. Phys. Rev. Lett. **93**, 035901 (2004)
31. Schoch, R.B., Lintel, H.V., Renaud, P.: Effect of the surface charge on ion transport through nanoslits. Phys. Fluids **17**, 100604 (2005)
32. Bocquet, L., Charlaix, E.: Nanofluidics, from bulk to interfaces. Chem. Soc. Rev. **39**, 1073–1095 (2010)
33. Khair, A.S., Squires, T.M.: Surprising consequences of ion conservation in electro-osmosis over a surface charge discontinuity. J. Fluid Mech. **615**, 323–334 (2008)
34. Hall, J.E.: Access resistance of a small circular pore. J. Gen. Physiol. **66**, 531–532 (1975)
35. Kowalczyk, S.W., Grosberg, A.Y., Rabin, Y., Dekker, C.: Modeling the conductance and DNA blockade of solid-state nanopores. Nanotechnology **22**, 315101 (2011)
36. Wanunu, M., Dadosh, T., Ray, V., Jin, J., McReynolds, L., Drndic, M.: Rapid electronic detection of probe-specific MicroRNAs using thin nanopore sensors. Nat. Nanotechnol. **5**, 807–814 (2010)
37. Schneider, G.F., Kowalczyk, S.W., Calado, V.E., Pandraud, G., Zandbergen, H.W., Vandersypen, L.M.K., Dekker, C.: DNA translocation through graphene nanopores. Nano Lett. **8**, 3163–3197 (2010)
38. Ho, C., Qiao, R., Heng, J.B., Chatterjee, A., Timp, R.J., Aluru, N.R., Timp, G.: Electrolytic transport through a synthetic nanometer-diameter pore. Proc. Natl. Acad. Sci. U. S. A. **102**, 10445–10450 (2005)
39. Smeets, R.M., Keyser, U.F., Krapf, D., Wu, M.Y., Dekker, N.H., Dekker, C.: Salt dependence of ion transport and DNA translocation through solid-state nanopores. Nano Lett. **6**, 89–95 (2006)
40. Hille, B.J.: Pharmacological modifications of the sodium channels of frog nerve. J. Gen. Physiol. **51**, 199–219 (1968)
41. Merchant, C.A., Healy, K., Wanunu, M., Ray, V., Peterman, N., Bartel, J., et al.: DNA translocation through graphene nanopores. Nano Lett. **10**, 2915–2921 (2010)
42. Howorka, S., Siwy, Z.: Nanopore analytics: sensing of single molecules. Chem. Soc. Rev. **38**, 2360–2384 (2009)
43. Holt, J., et al.: Fast mass transport through sub-2-nanometer carbon nanotubes. Science **312**, 1034–1037 (2006)
44. Falk, K., Sedlmeier, F., Joly, L., Netz, R.R., Bocquet, L.: Molecular origin of fast water transport in carbon nanotube membranes: superlubricity versus curvature dependent friction. Nano Lett. **10**, 4067–4073 (2010)
45. Siria, A., Poncharal, P., Biance, A.-L., Fulcrand, R., Blase, X., Purcell, S., Bocquet, L.: Giant osmotic energy conversion measured in a single transmembrane boron-nitride nanotube. Nature **494**, 455–458 (2013)

# Chapter 16
# FIB Patterning of Stainless Steel for the Development of Nano-structured Stent Surfaces for Cardiovascular Applications

**Michael Schmidt, Feroze Nazneen, Paul Galvin, Nikolay Petkov, and Justin D. Holmes**

**Abstract** Stent implantation is a percutaneous interventional procedure that mitigates vessel stenosis, providing mechanical support within the artery and as such a very valuable tool in the fight against coronary artery disease. However, stenting causes physical damage to the arterial wall. It is well accepted that a valuable route to reduce in-stent re-stenosis can be based on promoting cell response to nano-structured stainless steel (SS) surfaces such as by patterning nano-pits in SS. In this regard patterning by focused ion beam (FIB) milling offers several advantages for flexible prototyping. On the other hand FIB patterning of polycrystalline metals is greatly influenced by channelling effects and redeposition. Correlative microscopy methods present an opportunity to study such effects comprehensively and derive structure–property understanding that is important for developing improved patterning. In this chapter we present a FIB patterning protocol for nano-structuring features (concaves) ordered in rectangular arrays on pre-polished 316L stainless steel surfaces. An investigation based on correlative microscopy approach of the size, shape and depth of the developed arrays in relation to the crystal orientation of the underlying SS domains is presented. The correlative microscopy protocol is based on cross-correlation of top-view scanning electron microscopy, electron

M. Schmidt (✉) • N. Petkov • J.D. Holmes
Materials Chemistry and Analysis Group (MCAG), Electron Microscopy and Analysis Facility (EMAF), Tyndall National Institute, University College Cork, Cork, Ireland
e-mail: michael.schmidt@tyndall.ie; nikolay.petkov@tyndall.ie; j.holmes@ucc.ie

F. Nazneen • P. Galvin
Life Sciences Interface (LSI) group, Tyndall National Institute, University College Cork, Cork, Ireland
e-mail: feroze_nazneen@yahoo.co.in; paul.galvin@tyndall.ie

Z.M. Wang (ed.), *FIB Nanostructures*, Lecture Notes in Nanoscale Science and Technology 20, DOI 10.1007/978-3-319-02874-3_16,

backscattering diffraction, atomic force microscopy and cross-sectional (serial) sectioning. Various FIB tests were performed, aiming at improved productivity by preserving nano-size accuracy of the patterned process. The optimal FIB patterning conditions for achieving reasonably high throughput (patterned rate of about 0.03 $mm^2/h$) and nano-size accuracy in dimensions and shapes of the features are discussed as well.

## 16.1 Stents Overview: Design, Desired Properties, Materials

A stent is generally a metal mesh tube inserted into a natural passage/conduit in the body to prevent or counteract a disease-induced localised flow constriction or to temporarily hold such a natural conduit open to allow access for surgery [1]. The mesh structure allows the stent to bend and follow the blood or hollow vessels to the point it is required. The surface of stents is geometrically complex because of its mesh structure (Fig. 16.1). In a first approximation one can assume the surface that is in strongest contact with the body material to be flat—like an unrolled straw. To simplify the conditions we are using flat substrates in this study.

The key characteristics of stents are strength to withstand the peristaltic movements of blood vessels, flexibility and minimal size to reach the tiniest body tubes. Austenitic type 316L stainless steel can supply these desired properties and is commonly used for manufacturing medical implants [2], and so it is also commonly used as stent material, though in recent years there also have been studies on other alloys, containing Co, Cr and Pt [3–6] for higher strength and Mg as base metal for biodegradable stents [7].

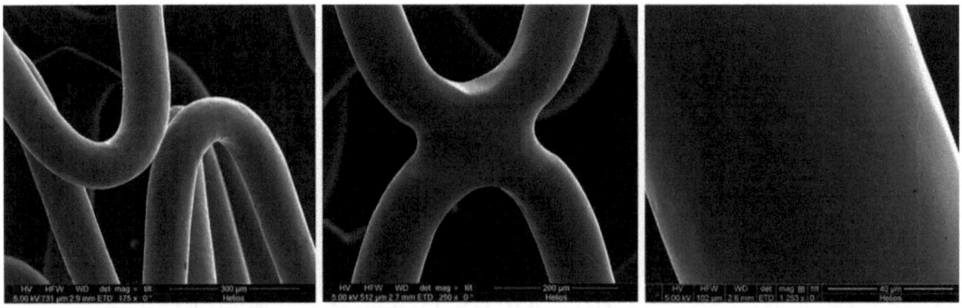

**Fig. 16.1** Various views on complex stent surfaces

## 16.2 Surface Treatments to Gain Desired Properties: Coating, Polishing, Patterning

There are various ways to improve the stent design in order to make it more effective. One way is coating the surface with organic and inorganic materials. This is not trivial with the complex geometry of the stainless steel mesh. Nevertheless this route is extensively studied given its benefit of incorporating drugs into the polymer-based surface coating which can be eluted over time into the surrounding vessel material. However, there are several reasons, including concerns about increased risks of late stent thrombosis when using drug-eluting stents [8, 9], why the development focus is likely to return to bare metal or polymer-free stent technologies [3].

While from the invention of stents it was aimed for a surface as smooth and polished as possible in order to minimise abrasion and inflammation of the body tubes, it becomes ever more apparent that a somewhat roughened/textured surface might be a better fit for the task.

## 16.3 Biomaterial–Cell Interaction: Advantages of Rough/Patterned Surfaces

The influence of textured material surfaces on the behaviour of cells has been studied for many years by now [10–16]. On one hand, theoretical studies show that cells prefer to grow on rough surfaces in general as it imitates best naturally occurring surfaces [17]. On the other hand, in order to have a better control over and to minimise the complexity of the experimental conditions, the natural urge to study regular patterns led scientists from rough to micro- to nano-patterned surfaces. Especially in tissue engineering where the tissues involved require certain mechanical and structural properties for proper functioning, the trend from micro- to nano-structured surfaces serving as artificially created support systems has become evident within the last decade [18–24]. Also for drug delivery the control over biointerfacial interactions is often the key to biomedical applications [14].

In particular endothelial, smooth muscle and fibroblast cells play an important role in the healing process and maintenance of cardiovascular systems and thus are likely to be in contact with biomedical implants such as stents and grafts. During a surgical procedure involving the introduction of a stent, vascular tissues in the arteries may be damaged. Healing of vascular tissues is promoted by the formation of an endothelial cells lining on the stent substrate [25], while the presence of smooth muscle cells and fibroblasts may cause re-stenosis. Micro- and nano-textures on substrates may provide control of cell functions. Such structures could promote better vascular cell adhesion, decrease the need for systemic administration of drugs and reduce the requirement for secondary surgery after stent implantation.

## 16.4 Patterning: Lithography, Electrodeposition, Dip Pen, Pulsed Laser, FIB

Recent advances in micro- and nanotechnology have allowed the patterning of surfaces with the desired textures for cell scaffolding [26–29]. Nano-texturing involves the creation of patterns or features with nanometre precision. The choice of the texturing method depends largely on the nature of the substrate that needs to be modified and on the dimensions of the features expected. Indeed, photolithography was particularly successful for the patterning of features of microscopic dimensions on elastomers such as polydimethylsiloxane (PDMS) [30–33] and on polymers such as polystyrene (PS) [34–37]. E-beam lithography was used for the patterning of submicron features on silicon substrates [38, 39] and on poly(methyl methacrylate) (PMMA) [40]. Features of 350 nm were patterned on PDMS and PMMA substrates by nanoimprint lithography [41]. Metal substrates such as Ti were also textured using micro-machining for feature dimensions in the microscopic range [10, 42]. Cell adhesion, migration, elongation, proliferation and gene expression on textured substrate can be greatly altered depending on the shape and the dimension of the features [40].

The different techniques are compared in Table 16.1. In indirect photolithography methods, patterns are formed over a large area using a mask [43]. Such lithography processes are time consuming with many steps and inherently inappropriate for prototype designs and processes. Electrodeposition is a simple, fast and cost-effective method of reproducing nano-structures on many materials using templates made of polymers and metals. However, this method is applicable only for electrically conductive substrates. Imprint lithography is a high-resolution direct technique for nano-patterning of large surfaces, but it requires moulds and is restricted to polymeric materials [44], but this could then be used as etch masks or filled with metal electrodeposition. E-beam lithography and lithography based on scanning tunnel microscopy (STM), atomic force microscopy (AFM) or dip pen are high-resolution mask-less procedures, but with a very low throughput and unsuitable for wide surface nano-patterning [45]. Interference lithography can be utilised to create or transfer array patterns on various metallic and polymeric surfaces, but only patterned features can be reproduced. Microtexturing of surfaces has also been reported by pulsed laser patterning [46, 47]. The feature sizes are however limited to the micron range.

Patterning by FIB milling is direct and offers several advantages for flexible prototyping: (1) practically any substrate material that is able to withstand high vacuum conditions of the microscope chamber can be used, (2) there is high flexibility in the obtainable shapes and geometries by modulating the ion beam current and the patterning conditions, (3) reduced complexity of the patterning process, e.g. it is a single-step process with a possibility of real-time monitoring of the milling progression. Thus for any particulate type of substrate, various depths as well as lateral dimensions including the optimal feature size can be obtained at minimum number of processing steps.

**Table 16.1** Advantages and disadvantages of indirect and direct nano-structuring techniques

| Mask/ moulds required | Nano-fabrication techniques | Advantages | Drawbacks |
|---|---|---|---|
| Yes | Photolithography | Well-controlled features | Requires photoresist, spin coaters and organic solvents Low aspect ratio |
| | | High throughput | Limited to set of materials |
| | Electrodeposition | Precise geometries and patterns | Require templates for creating of nano-structures |
| | | Large surface area | Limited to electrically conducting substrates |
| | Imprint lithography | High resolution | Requires moulds |
| | | High aspect ratio | Applied to polymers only |
| | | Large surface | |
| No | E-beam | High resolution | No direct writing on substrate |
| | | Precise geometry and patterns | Multistep process |
| | | | Low throughput |
| | | | Requires vacuum |
| | | | Time consuming |
| | | | Small surface coverage |
| | | | Expensive |
| | Interference lithography | No complex optical systems | Limited to patterned array features only |
| | | | Multistep process |
| | STM/AFM/ dip pen | Very high resolution | Low aspect ratio |
| | | | Very low throughput |
| | | | Very small surface area |
| | Nanoindentation | High aspect ratio | Wide and shallow features |
| | | Control over features depth | Slow process |
| | | Less expensive than FIB or e-beam writer | |
| | Laser patterning | Any material | Wide and shallow features |
| | | High throughput with high power laser | Micron resolution |
| | FIB milling | High resolution | Time consuming |
| | | High aspect ratio | Process requires vacuum |
| | | High etch rate | Very expensive |
| | | Any material | Low throughput |

## 16.5 FIB Advantages: Fast for Prototyping

Except FIB, none of the texturing techniques mentioned above were suitable to achieve features in the nanoscopic range on a hard substrate as 316L stainless steel. Nanoimprint lithography (NIL) and e-beam lithography (EBL) are able to pattern submicron features, which with NIL were achieved on soft substrates such as polymers and elastomers only and with EBL is a very time-consuming and expensive multistep process. In this research work, FIB milling was used to create nano-structures onto stainless steel because it is a direct writing process with simple steps, high resolution and aspect ratios. Nano-structured features such as pits were created on 316L steel surfaces. The optimal FIB patterning conditions for achieving reasonably high throughput (patterned rate of about 0.03 $mm^2/h$) and nano-size accuracy in dimensions and shapes of the features are discussed. Additionally, a characterisation protocol for analysis of such structures by combination of electron backscattering diffraction (EBSD), FIB, scanning electron microscopy (SEM), atomic force microscopy (AFM) and serial FIB–SEM sectioning is detailed. Furthermore, this chapter reports the comparison of in vitro EC adhesion and growth on FIB nano-structured, unpolished and electropolished 316L steel surfaces.

## 16.6 FIB Overview: $Ga^+$ Beam, Maskless, Pattern Design

FIB systems usually employ a finely focused beam of gallium ions ($Ga^+$) that can be operated at high beam currents for milling or low beam currents for imaging. It can be utilised to remove material locally in a highly controlled manner to the nanometre scale. When the high-energy Ga ions impinge the sample, atoms from the sample surface are sputtered. In addition, Ga atoms from the ion beam are also implanted into the top few nanometres of the surface, and the surface is made amorphous. As can be seen in Fig. 16.2, the $Ga^+$ primary ion beam hits the sample surface and sputters a small amount of material. The primary beam also generates secondary electrons. When the primary beam strikes the sample surface, the signal from secondary electrons or sputtered ions is collected to form an image of the surface [48].

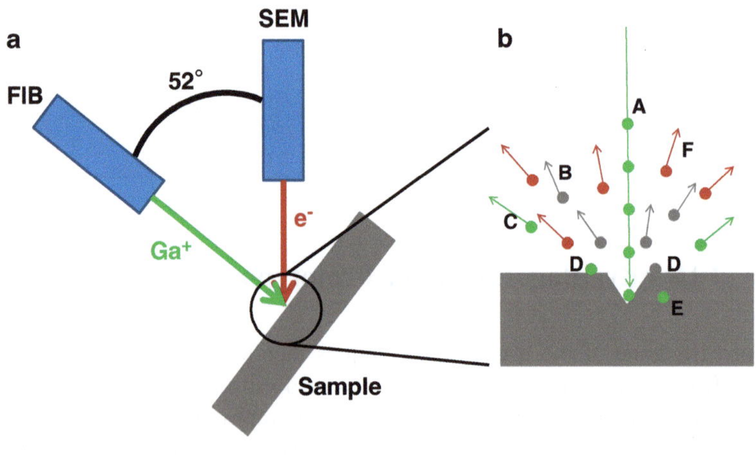


**Fig. 16.2** (**a**) Dual-beam FIB schematics. (**b**) Beam sample interactions: *A*—incident $Ga^+$ ions, *B*—sputtered substrate atoms, *C*—scattered $Ga^+$ ions, *D*—re-deposited $Ga^+$ ions and substrate atoms, *E*—in substrate trapped $Ga^+$ ions and *F*—secondary electrons

At a low primary ion beam current, a minuscule amount of material is sputtered, and with existing FIB methods, 5 nm resolution can be attained using Ga ions. However, at higher primary ion beam current, a considerable quantity of material can be taken out, which allows sub-micrometre- to nanometre-scale precision milling of sample surface.

FIB originated in the semiconductor industry and has become an important tool for a wide array of applications, ranging from circuit editing, reverse engineering, sample preparation for transmission electron microscopy (TEM), microstructural analysis and prototype nanomachining to name just a few [49]. Many modern FIB instruments supplement the FIB column with an additional SEM column so that it becomes a versatile dual-beam platform as depicted in Fig. 16.2. In nano-patterning, FIB has been used to create nano-structures on Si [50], silicon nitride [48, 51] and glass substrates [52] and to fabricate platinum nano-structures on peptide-based soft surfaces [53]. Only one study reported the protein adsorption on FIB patterned glass surfaces [52]. To date, no cellular studies have been reported on FIB-structured surfaces. Moreover, this and other aforementioned techniques have not been employed for patterning the key vascular stent material 316L stainless steel for vascular cell functions. Studies do not exist that determine the EC response on 316L steel with nano-pit features. Endothelial cell studies on unpatterned 316L stainless steel substrates have shown that the grain size and grain boundaries have an impact on their adhesion and morphology [54]. Chemically etched substrates with 16 μm grain size etched have demonstrated cell densities significantly higher than with grain sizes of 31, 47 and 66 μm. The authors attribute this increased cell density to greater boundary area and associated higher surface free energy [54]. Cell proliferation was also subject to another study discussing different materials. There the grain sizes varied from 320 nm to 22 μm. Again cell proliferation was reciprocal dependent on the grain size [55].

## 16.7 Crystal Structure of Stainless Steel 316L

Austenitic type 316L stainless steel is commonly used for manufacturing medical implants [56] and was hence selected as the substrate of choice for this study. Austenitic stainless steels have face-centred cubic (fcc) crystal structure, in which the unit cell is a cube with atoms located at the corners and middle of each side (Fig. 16.3a). The presence of higher concentration of Ni in austenitic stainless steels stabilises the fcc crystal structure, because Ni is a fcc crystal itself. This enhances the ductility, i.e. it can sustain large plastic deformation without fracture compared to other stainless steels (maternistic and ferritic phases).

Most metallic materials are composed of many small single-crystalline planes called grains. These materials are referred to as polycrystalline materials (e.g. steel), in which individual grains have identical arrangement of atoms but the orientation of the atom arrangement or crystal structure is different from each adjoining grain (see Fig. 16.3b for visualisation). The interfaces between these grains are grain boundaries, the surface that separates the individual grains [57].

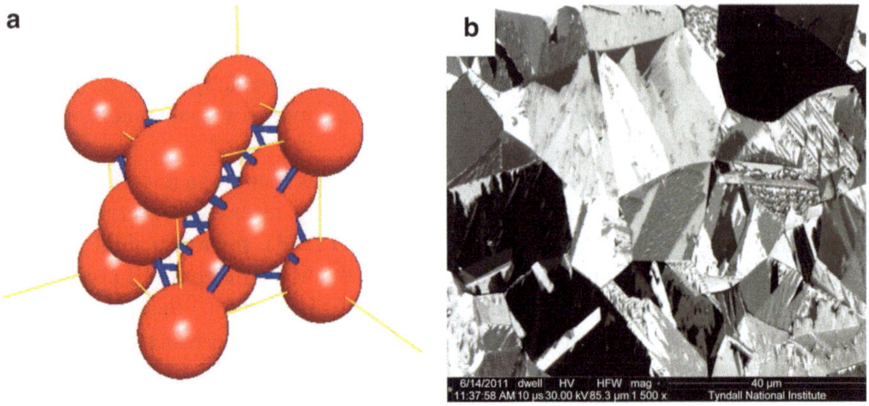


**Fig. 16.3** (**a**) Schematic face-centred cubic crystal structure and (**b**) FIB image of polycrystalline 316L used in this study

The austenitic stainless steel function can be affected by two microstructure features: grain (or crystal) size and shape. The general grain size suggested for 316L is 100 μm or less [56]. This is because smaller grains have more grain boundaries, which provide resistance to plastic deformation as they are responsible for slip deformation by dislocations.

Depending on the process conditions such as annealing and cold-working, the shape of austenitic stainless steel grains varies. Annealing is a heat treatment process where a material is modified, resulting in changes in its properties, for example, strength and hardness. It is a method that generates conditions via heating to above the recrystallisation temperature, maintaining an appropriate temperature, and subsequently cooling. This method is applied to reduce internal tensions, soften material, enhance ductility, improve the structure by creating it uniform and enrich cold-working properties. Austenite grains of the stainless steels under an annealed condition exhibit an equiaxial granular shape (i.e. the grains having axes of equal length).

Cold-working produces plastic deformation in the steels and generates a strain hardening effect, which improves both yield strength and tensile strength of steel considerably. However, in cold-worked steel, depending on the amount of cold work, the grains are elongated (i.e. longer in the rolling direction). During large plastic deformation, textured grain structures are produced and preferentially align the grains in specific crystallographic orientations. Hence, cold-worked steel with textured structures demonstrates anisotropic mechanical properties. When employing a cold-worked steel for implant fabrication, microstructure analysis is suggested as implants can be better prepared if the loading direction is concurrent to the high strength direction in the steel [56]. Hence, it is clear that the microscopic and crystalline structure can play a strong role on the nano-structuring of the stainless steel surface.

## 16.8 Substrate Preparation: Electropolishing, Cleaning

Electropolished 316L steel substrates were used for this study. Figure 16.4 illustrates the steps involved in the whole sample process flow. Electropolishing of steel was performed with a view to the preparation of these surfaces for nano-texturing. For such applications, a smooth surface is crucial. The composition of this electrolyte solution was 11 M $H_3PO_4$ + 4.5 M $H_2SO_4$ in water. The electropolishing procedure was conducted in two steps at 80 °C and 5 mV $s^{-1}$ [58]. The first step involved scanning of the potential from the open-circuit potential up to the point where the diffusion-limited current region was reached. The linear sweep voltammetry was then stopped, and the selected potential was maintained 10 min using chronoamperometry. This resulted in a smooth and relatively defect-free surface. XPS analysis of the electropolished surface has shown that the stainless steel was enriched with Cr, P, S, O, Mo and Ni elements. Prior to nano-structuring, the polished specimens were cleaned in acetone, in ethanol and finally in ultrapure water via an ultrasonic treatment for 10 min.

## 16.9 FIB Tests: Challenges with Anisotropic Milling

The FIB system used in the current study is the FEI Helios NanoLab 600i, which is a dual-beam FIB for localised milling and deposition, transmitting a 30 keV beam of $Ga^+$ ions combined with a high-resolution SEM. In this work, the working current was tuned between 0.28 nA (for 120 nm pits) and 0.92 nA (180 nm) depending on size of the nano-texture. The available pit sizes and the large range of obtainable working currents could make the FIB technique an ideal device for nanomachining in the range from 10 nm to a few micrometres. Nano-structured features (pits/holes) ordered in rectangular arrays were patterned on 316L steel surfaces using FEI Helios NanoLab 600i FIB system. This system was used because of high beam quality and stage stability.

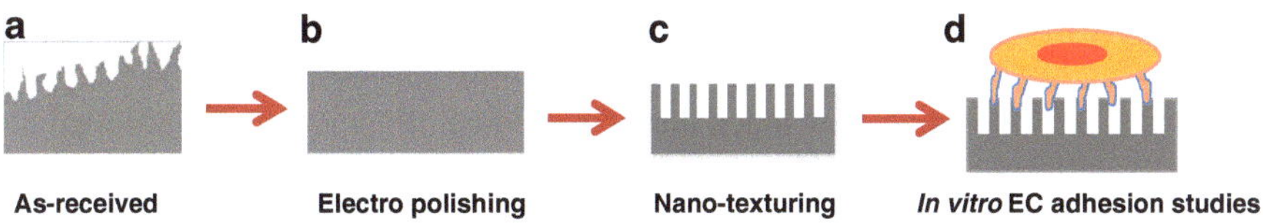


**Fig. 16.4** Sample process flow in total. (**a**) Samples as received with rough surface. (**b**) Electropolishing to obtain a smooth surface suitable for nano-structuring. (**c**) Nano-texturing of 316L stainless steel via a focused ion beam (FIB) milling. (**d**) In vitro endothelial cell adhesion studies

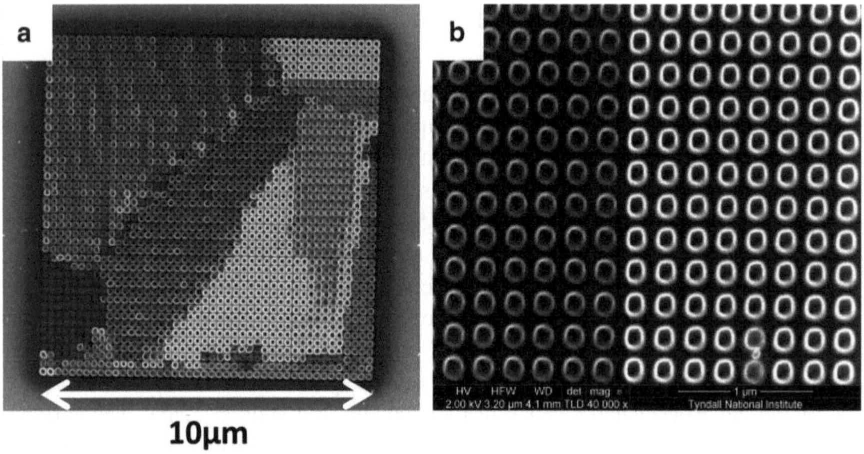


**Fig. 16.5** SEM images of preliminary FIB tests to determine the feasibility of the prototyping approach. (**a**) 10 × 10 µm area pattern by FIB in preliminary tests with nominal 120 nm wide pits at 240 nm pitch. (**b**) Detail of another area exhibiting the same pattern. Clearly visible are the differences in appearance of the patterned surface depending on the polycrystallinity of stainless steel

From the literature survey, promising cell responses to nano-structured features were identified including nano-pit features [40, 59]. However, to date no EC studies have been reported on nano-pit structures. Based on this, two pit designs were patterned on three electropolished 316L stainless steel samples on areas of 400 µm × 400 µm using FIB: design A, pits of 120 nm diameter with a pitch of 240 nm and intended depth of 50–100 nm, and design B, pits of 180 nm diameters with pitch of 360 nm and intended depths of 50–100 nm [60, 61].

Before attempting prototyping on large 400 × 400 µm areas used for the biological tests, we have performed optimisation tests on relatively small test patterns; one such area is shown in Fig. 16.5a. From the known polycrystalline nature of the 316L stent material, one can assume that when subjected to ion milling or imaging, it will show pronounced channelling contrast. It is well known for about a century for W, Ag or Cu which are all fcc metals that they etch and sputter faster in preferred directions [62–65], as well as Si [66, 67]. Similarly polycrystalline fcc austenitic stainless steel will show milling rates that are varying by the different orientation of grains towards the incoming beam.

Figure 16.5 illustrates how much this anisotropic milling affects the desired outcome of uniform concaves. Shown in Fig. 16.5 are examples from the pretests on 10 µm × 10 µm areas with 120 nm diameter holes at 240 nm pitch. The structures that appear with the brightest contrast showed deeper and sharper edges than the structures that appear darker in contrast. This will be discussed in more detail in the following section describing the correlative microscopy approach. A later section will focus on the study on the patterned substrates used for actual cell adhesion tests and the preliminary FIB tests.

## 16.10 Correlative Microscopy: EBSD, FIB, SEM, AFM, and Serial Sectioning FIB–SEM Towards Better Understanding of Beam–Substrate Interaction

In order to gain a better understanding of the beam–substrate interaction during patterning, a correlative microscopy approach was used to illuminate the patterning process from many possible angles [68, 69]. This was also done on a batch of samples that could not be used for the cell adhesion studies basically because of the destructive nature of the last step, the serial FIB–SEM sectioning.

The different techniques used for the characterisation of nano-textured, unpolished and electropolished stainless steel surfaces were electron backscattering diffraction (EBSD), FIB, SEM, atomic force microscopy (AFM) and serial FIB–SEM sectioning. Figure 16.6 shows the detailed process flow of the correlative approach.

EBSD technique was used to analyse the crystallographic structured surfaces of the polished stainless steel. EBSD imaging was performed in a Hitachi SEM SU-70 equipped with an Oxford Instruments EBSD attachment AztecHKL at 10 kV under 70° tilt angle and step size 2 μm.

FIB technique was also used to visualise the crystallographic structured surfaces of the polished stainless steel. FIB imaging was performed as a part of monitoring of the milling at 30 kV accelerating voltage.

SEM was used to analyse the topography of nano-textured surfaces of the polished stainless steel. The SEM images presented were obtained using a SEM at the FEI Helios NanoLab 600i at an electron beam current of 5 kV and 86 pA beam current.

A commercial atomic force microscope (MFP 3DTM, Asylum Research) in AC mode was used for topography mapping of the films. Olympus AC160TS silicon cantilevers (Al reflex coated, ~300 kHz resonant frequency) were used for imaging. Optimal results were achieved with a medium scan rate of 1 Hz and 256×256 pixels image resolution.

Samples for serial sectioning were prepared using protective carbon and Pt layers [70]. The electron beam-induced (EBID) carbon deposition supplied necessary contrast difference between the protective Pt and the stainless steel surface, hence enabling accurate determination of the concave's shape and depth.

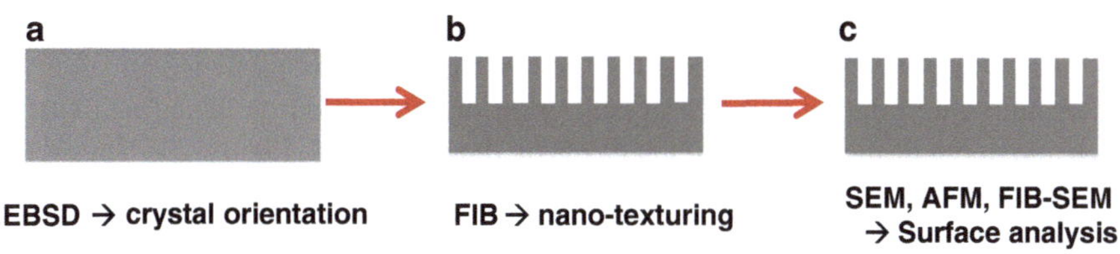


**Fig. 16.6** Process flow in the correlative microscopy approach. (**a**) The crystallographic grain orientations are measured by EBSD before patterning. (**b**) Surface patterning using FIB, also gaining FIB SE images as part of the monitoring process. (**c**) Extensive analysis of the textured steel surface with SEM, AFM and finally destructive serial FIB–SEM sectioning

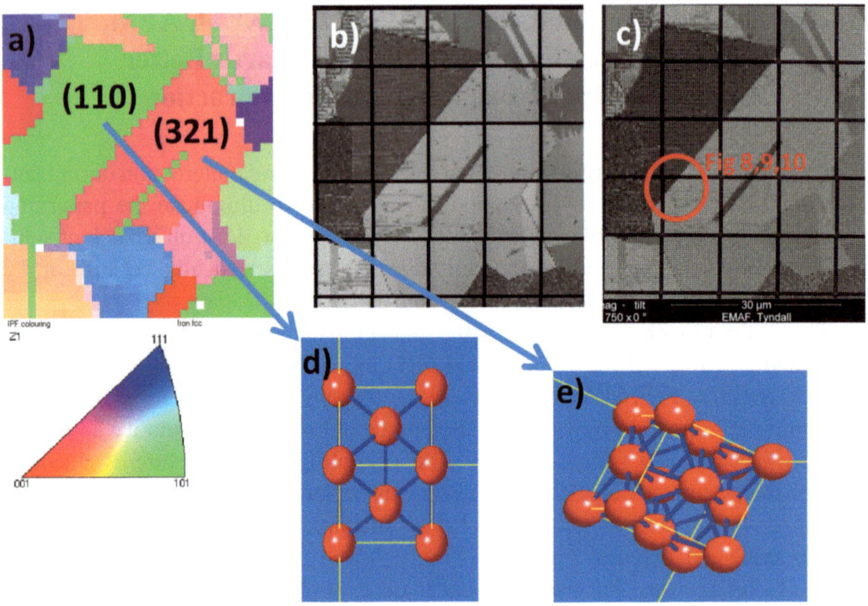


**Fig. 16.7** Correlative microscopy on the exact same sample location. (**a**) Inverse pole figure (IPF) as measured by EBSD before surface texturing with (110) and (321) crystal orientation labelled and legend underneath. (**b**) FIB SE image obtained during patterning as part of monitoring. (**c**) SE image taken after the patterning—the *red circle* indicates the region used for AFM and serial sectioning (see Figs. 16.8, 16.9 and 16.10 for details). (**d**) Graphical visualisation of (110) orientation. (**e**) Graphical visualisation of (321) orientation

EBSD is an SEM-based microstructural–crystallographic technique to measure individual grain orientations, local texture, point-to-point orientation correlations and phase identification and distributions on the surfaces of bulk polycrystals. It is also known as backscatter Kikuchi diffraction (BKD), and both acronyms are being used interchangeably in the literature. EBSD patterns are generated on a phosphor screen by backscatter diffraction of a stationary beam of high-energy electrons from the typically 70° tilted sample surface. Because of the high tilt angle—or the shallow 20° angle of the incident beam towards the surface—this technique is very surface sensitive and gives information of the top 20 nm down into the substrate [71].

EBSD mapping was accomplished before the area was patterned in order to determine a correlation between crystal grain orientation on one hand and shape, size and depth of the FIB-milled concaves on the other hand. Figure 16.7 shows the random size and orientation of the crystal grains and illustrates the correlative microscopy approach of three techniques combined in the exact same sample location, EBSD, FIB and SEM. In general the intensity of the emitted SE depends on the different inclination of the sample surface towards the incoming beam and crystal orientation [72]. Thus, the grey levels in the SEM image are directly linked to the surface topography, e.g. the shape of the pits and sidewall profile and the crystal orientation of the surface material. In this way we can correlate the EBSD data to the grey levels in the SEM images. The FIB reveals not surprisingly the same contrast in the SE image as the one taken afterwards in the SEM. The SE yield is independent of the type of the beam; hence the same contrast is achieved. The additional information of the SEM SE image lies in the much higher resolution. The FIB which was run as a monitoring tool only during the patterning process produced one pixel every 360 nm in *X* and *Y* direction. The SEM on the other hand was used afterwards as an analysis tool with an image resolution of 4,096 × 3,775 pixels which calculates at roughly 3 nm image resolution. Using a low current of 86 pA ensured that the real resolution is not far from this theoretical limit.

Two different grains were chosen for additional correlation with AFM and the serial FIB–SEM sectioning based on the crystallographic orientation, a low index grain with (110) orientation and a high index grain with (321) orientation. From as early as the 1920s, it is known that the sputter yield is dependent on the crystal orientation [62]. It is also known that the SE yield is dependent on the crystal orientation [72]. Based on this fact the chosen grains should display a very different behaviour when exposed to the ion beam during sputtering and also to the electron beam in the SEM study.

Studying the marked region from Fig. 16.7 across the grain boundary using the AFM (see Fig. 16.8), it appears that there is a difference in hole depth and diameter depending on crystallographic orientation of the patterned surface. Even the shape of the rim is evidently not circular but rather rhombohedral. Because of the high aspect ratio of the pits, the tip could not reach down and probe the full depth of the pits; therefore, the pit depth must be confirmed by the serial sectioning.

The first that comes to mind when seeing the rhombohedral shape of the pits is the directional dependence in which atoms are ejected when sputtering at threshold energy [63]. In our study however the energy used to create the patterned surfaces in the FIB was way beyond the threshold energy, which for Cr, Ni and Fe as main elements in 316L lie in the range of 60–90 eV. The directional dependence decreases with higher sputter energies and has no influence on the direction of the sputtered atoms at the 30 keV used here. It is also obvious when looking in detail at the (111) oriented grain at the bottom of the AFM overview scan that shows the same rhombohedral-shaped pits as the whole area around this region instead of the expected triangular shape. As can be seen below in the detailed AFM and SEM studies on the 400 μm × 400 μm patterns used for the cell adhesion studies, the shape of the holes is solely determined by the ion beam quality (focus, stigmation) at the place of impact on the sample surface.

In the SEM surface study as depicted in Fig. 16.9a, it appears as if the holes in both grains seem equal in diameter with only the higher SE yield obvious for the higher index grain. In order to clarify this impression, the region was imaged again after depositing a carbon layer as shown in Fig. 16.9b. Because secondary electrons are emitted from an area very close to the surface of the sample, this amorphous carbon layer masks the crystal orientation of the sample surface, and the image is

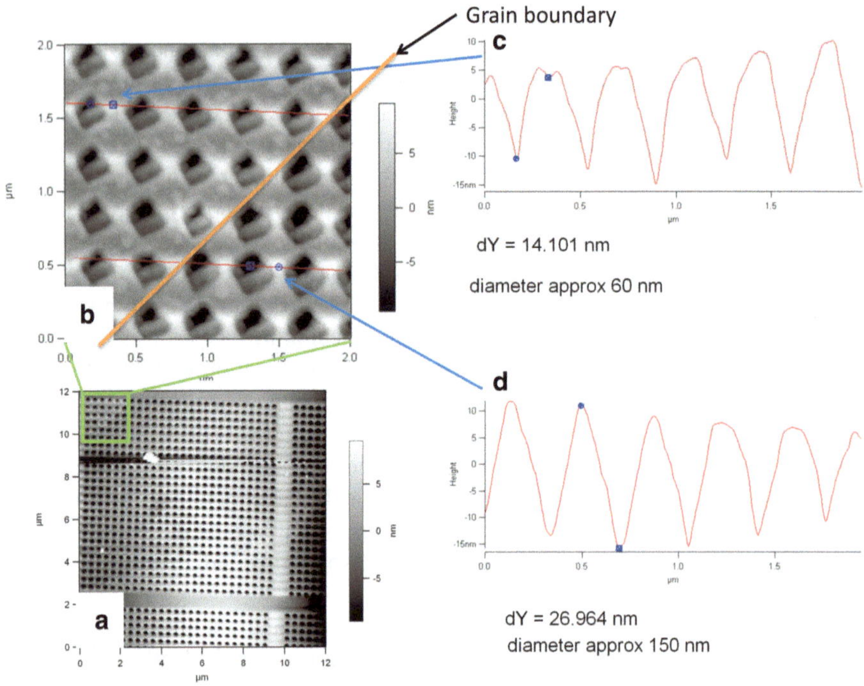


**Fig. 16.8** Representative AFM scans of the marked region from Fig. 16.7. (**a**) Overview scan over the whole 12 μm × 12 μm region. (**b**) Detailed 2 μm × 2 μm scan of the area around the crystal grain boundary. (**c**) Line profile along one row of five holes determining depth and diameter of the holes in the (110) oriented grain. (**d**) Same line profile determining the depth and diameter of the holes in the (321) oriented grain

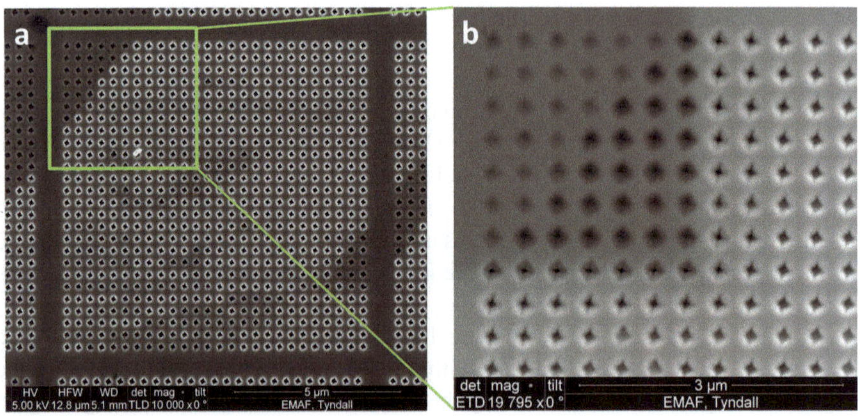


**Fig. 16.9** Detailed SEM images of patterned area marked by the *red circle* in previous picture. (**a**) Freshly patterned surface displays high SE yield for the high index grain and low yield for the low index grain, though the holes seem equal in diameter. (**b**) Detail of (**a**) after carbon deposition, the "true" diameter shines through as crystal orientations are hidden behind the amorphous carbon layer

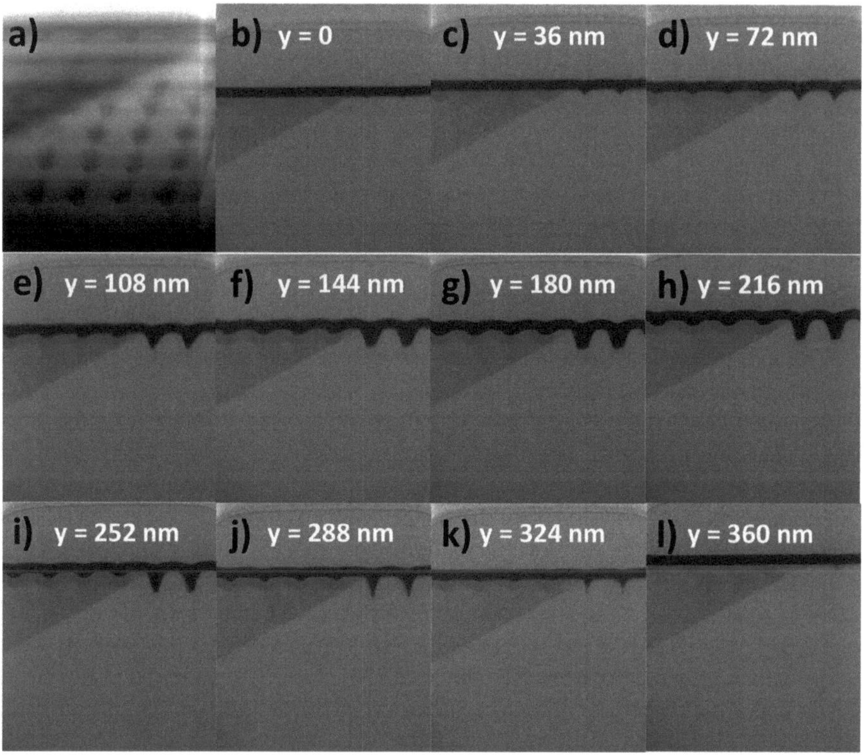


**Fig. 16.10** Serial sectioning as final destructive step of the correlative microscopy. (**a**) 3D reconstruction of all the 250 slices. (**b–l**) Image series illustrating one row of six holes taken every 36 nm (every third image). The *darker left* region is the low index grain (110) oriented, the *brighter right hand side* is the higher index (321) oriented grain. Obvious is the difference in depth by more than a factor of 3 between the two patterned grains

more related to topographical features. Indeed, the difference in diameter of the holes becomes pronounced, and it is visible that the higher index grain has apparently much larger holes than the lower index grain.

Serial sectioning of again the same region was performed composing 250 images of 8 rows by 5 holes, with one image taken every 12 nm. This detailed analysis verifies the difference not only in the diameter but even more pronounced in the depth. Though Fig. 16.10a presents a 3D reconstruction of all the 250 slices, more details can be observed when looking at the individual slices of Fig. 16.10b–l. The diameter of the pits in the low index (110) oriented grain are 150 ± 10 nm, while the diameter in the (321) oriented grains is only slightly bigger with 170 ± 10 nm. The depth however is much more influenced by the differences of sputter yield depending on the crystal orientation, and hence the (110) grain shows only 55 ± 5 nm depth in contrast to the 200 ± 20 nm depth of the (321) grain, nearly a factor of 4 between them.

## 16.11 Prototyping: Samples Prepared for Cell Adhesion Tests and Statistical Pattern Analysis

After the initial tests for the feasibility of the patterning was finished successfully on the batch of samples with small-scale pattern, the “real” prototyping on the 400×400 μm patterned areas had been started. As consensus between statistical needs for ideally a high number of samples on one hand and the slow process of patterning with the FIB on the other hand, five samples with each one area of 120 nm pits/240 nm pitch and one area of 180 nm pits/360 nm pitch were manufactured.

The two pit designs, A and B, were created on electropolished stainless steel samples by FIB and are presented in Fig. 16.11. Three things can be observed. First, the square areas that have been milled by the FIB are very different from the electropolished areas. Second, the triangular areas in the centre are much better defined than the areas at the edges. Finally, within and outside these triangular areas, the different colour tones observed are due to the polycrystallinity of the stainless steel as described in detail in the previous section.

Alternatively, AFM was used to evaluate statistically in more depth the information on topography, pit feature dimensions and as a comparison of the results obtained via serial FIB–SEM sectioning. AFM examinations for this analysis were performed in tapping mode using a Dimension 3100 with a Nanoscope IIIa controller equipped with a phase imaging extender (Digital Instruments, Santa-Barbara, CA, USA). The silicon cantilevers (purchased from Windsor Scientific, UK) have a tip radius of less than 10 nm and a 40 N $m^{-1}$ spring constant. AFM images were

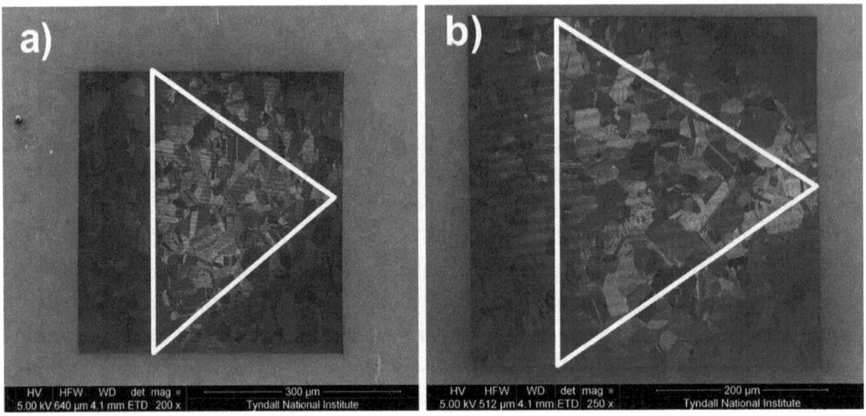


**Fig. 16.11** Overview of SEM images of (**a**) design A and (**b**) design B showing (1) differences in milled (*inside* the *square* area) and electropolished surfaces (*outside* the *square* area), (2) *triangular* areas covering more than half of the 400 μm×400 μm *square* are much better defined than the areas at the corners in the *squares* and (3) different colour tones illustrate the polycrystallinity of the stainless steel

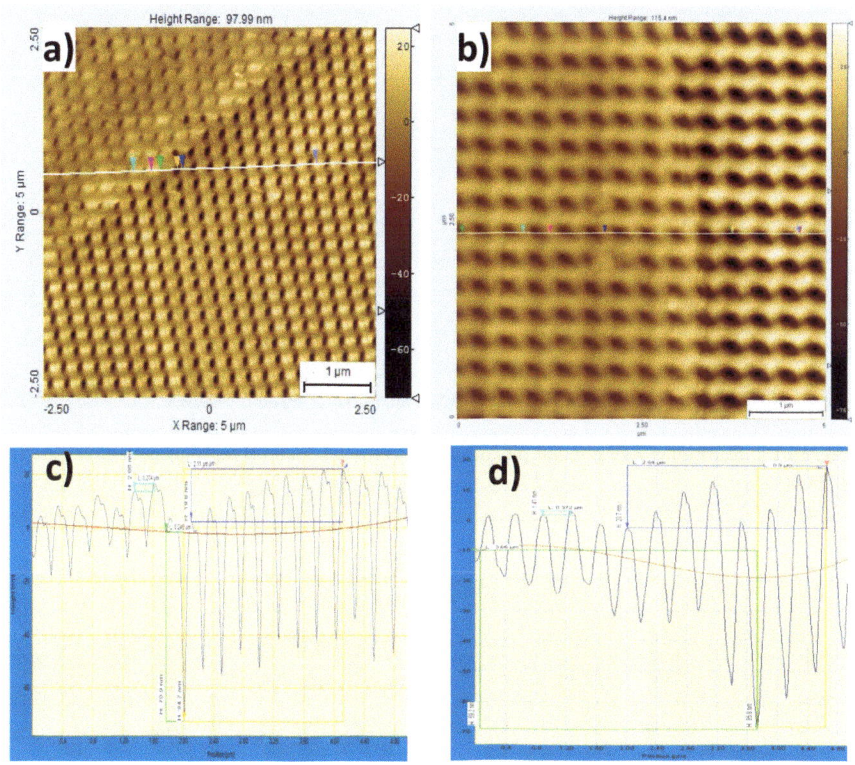


Fig. 16.12 AFM top view and cross-sectional images of pits (**a**, **c**) design A and (**b**, **d**) design B. Variation in depth dimensions from AFM profiles in region I and II is clearly noticed

recorded at a scan rate of 0.5 Hz and a resonance frequency of 300 kHz. Each sample was evaluated over scan fields of $5 \times 5$ μm², $1 \times 1$ μm² and $500 \times 500$ nm². The pit dimensions of the resulting images were evaluated using scanning probe imaging processor software (version 5.1.5).

Four nano-structured samples were used for the measurements. Four topographic measurements were performed in different random fields per substrate. Figure 16.12 shows the 2D AFM images and profiles of designs A and B on $5 \times 5$ μm² areas. A similar trend was noticed with AFM data. Region I showed shallower structures than region II, Fig. 16.12a, b. From the AFM profiles across a 5 μm width as depicted in Fig. 16.12c, d, the average depth recorded in region I for design A and design B was $24 \pm 18$ nm and $26 \pm 20$ nm ($N = 115$), whereas in region II the average depth was $68 \pm 48$ nm and $64 \pm 8.5$ nm ($N = 115$). This variation of more than four times indicates that an exact match of the desired depth of 50–100 nm can only be approximated.

The images presented in Fig. 16.13a, b are of a closer SEM inspection of designs A and B on triangular areas of $5 \times 5$ μm². The pit structures within the triangular zones are circular as wanted for pit designs A and B (Fig. 16.13a, b).

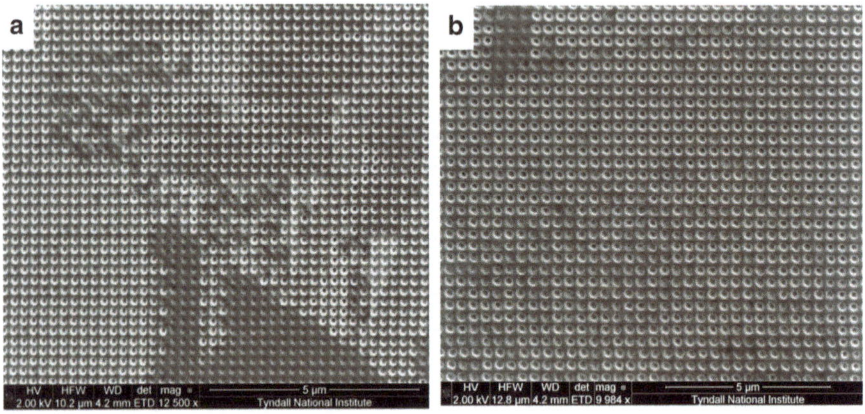


**Fig. 16.13** SEM images of circular pits of (**a** and **b**) design A and B reproduced within the triangular areas

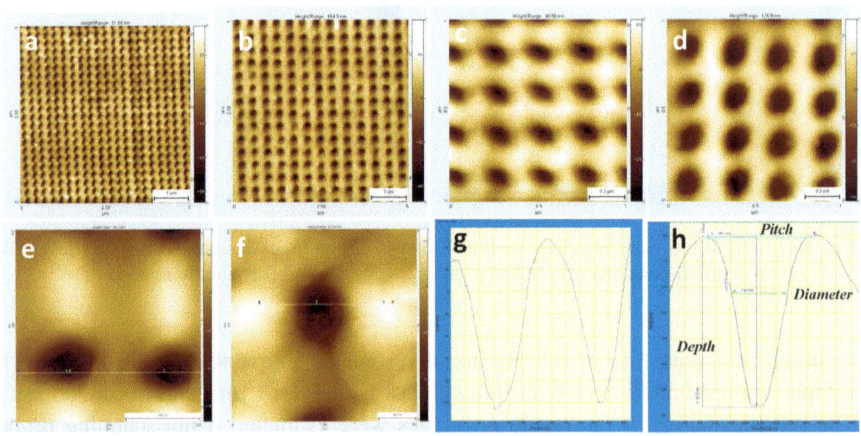


**Fig. 16.14** AFM topographies of uniformly fabricated structures of (**a, c, e**) pits design A and (**b, d, f**) design B pits reproduced within the triangular areas (scale: 5×5 µm, 1×1 µm and 500×500 nm). Panels (**g**) and (**h**) represent the AFM profiles of (**e**) and (**f**) across the length of 500 nm

Smaller areas of about 150×150 µm$^2$ showed circular shapes when imaged in top-down direction in the whole patterned area in the preliminary tests.

However, from the AFM images the shapes of the pits in design A in triangular areas are more elliptical (Fig. 16.14a, c, e), whereas those of design B are closer to perfect holes (Fig. 16.14b, d, f). The pit structures for designs A and B outside the triangular areas exhibited deformed circular shapes as demonstrated in Fig. 16.15a, b. From the stigmated nature one can conclude that there are problems with the focusing and astigmatism caused by the FIB optics. This is not surprising as the 200 times magnification is close to the lower limit of the FIB Helios column.

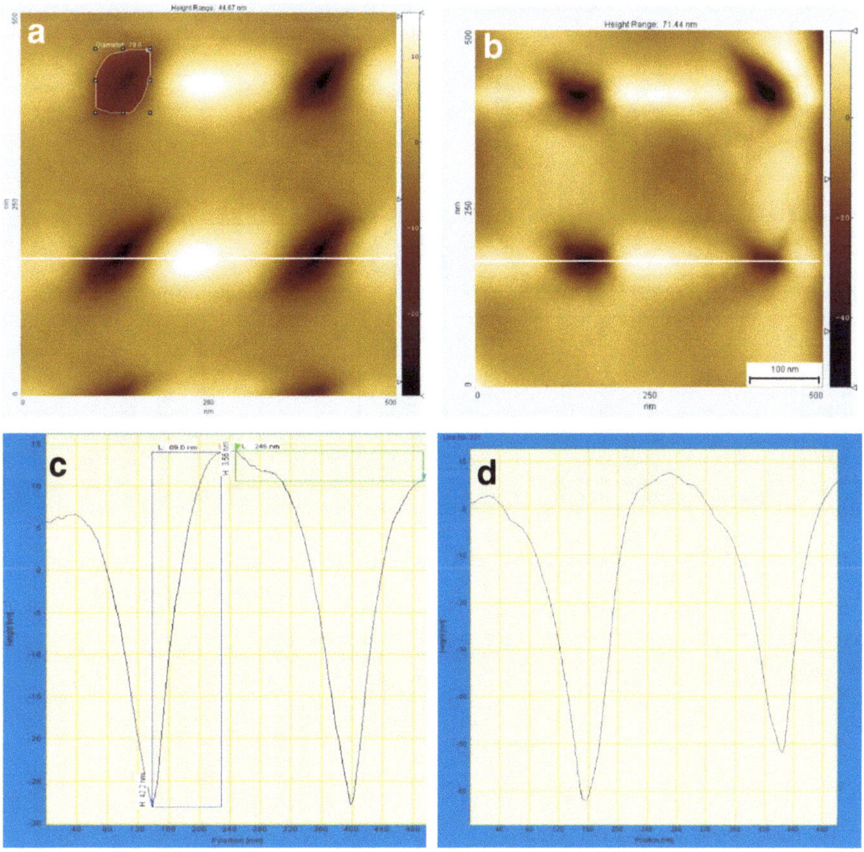


**Fig. 16.15** AFM pictures of unevenly patterned pit structures of (**a** and **b**) designs A and design B reproduced outside the triangular areas (scale: 500×500 nm) and (**c** and **d**) profiles of (**a**) and (**b**) across the length of 500 nm

From the AFM profiles, the uniformity and distribution of the pits A and B parameters (or dimensions) such as diameter, depth and pitch produced by FIB technique were evaluated. The A and B pit dimensions were measured using a Nanoscope imaging probe software as described above.

The distribution curves of A and B pit dimensions were plotted and presented in Fig. 16.16. Figure 16.16a shows that when seeking to produce design A diameter pits ($N=162$), the result was far from uniform. Although the highest number of pits did fall within the 116–120 nm range, the graph shows there were considerable variations in resulting pit sizes.

In seeking to produce design B diameter pits ($N=90$), Fig. 16.16b shows a similar pattern to the result shown in the previous graph. The highest number of pits was between 181 and 185 nm range with the lowest amount of pits registering in the ranges above 180 nm.

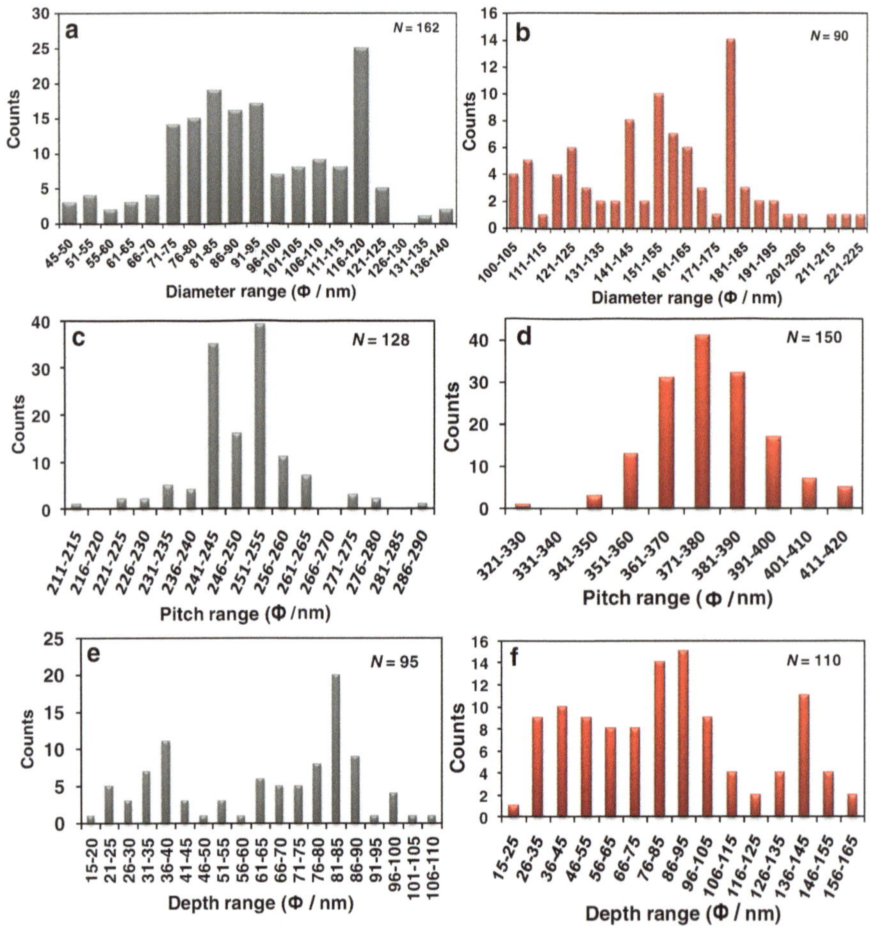


**Fig. 16.16** Distribution curves of pit diameter of (**a**) and (**b**) 120 and 180 nm with a pitch of (**c**) and (**d**) 240 and 360 nm and depth of (**e**) and (**f**) 50–100 nm

The required A pitch range here (Fig. 16.16c) was 240 nm ($N=128$). Most pits registered between 251–255 and 241–245 nm then at 246–250 nm range. However, only four pits were reported at 240 nm.

In looking for a design B pitch level of 360 nm ($N=150$ nm) presented in Fig. 16.16d, only 13 pits fell within this range. The majority of pits registered at pitches higher than the required level of 360 nm, falling between 371 and 380 nm, and one pit range as high as 411–420 nm.

In searching for a design A depth of 50 or 100 nm ($N=95$), Fig. 16.16e shows that one pit reported at 46–50 nm range and four pits reported at 96–105 nm range. However, considerable variation is still noted in the dimensions of all pits.

Again looking for a design B depth of 50 or 100 nm ($N=110$), nine pits were recorded within the 46–55 nm range (Fig. 16.16f). Nine were found to have a depth

**Table 16.2** Dimensions of 120 nm pits obtained with FIB patterning

| Dimensions | Sites number | Mean ± SD (nm) | Lowest dimension value (nm) | Highest dimension value (nm) |
|---|---|---|---|---|
| Diameter | $N=162$ | 97 ± 8.5 | 46 | 138 |
| Pitch | $N=128$ | 250 ± 11 | 213 | 290 |
| Depth | $N=95$ | 65 ± 24 | 19 | 116 |

**Table 16.3** Dimensions of 180 nm pits obtained with FIB patterning

| Dimensions | Sites number | Mean ± SD (nm) | Lowest dimension value (nm) | Highest dimension value (nm) |
|---|---|---|---|---|
| Diameter | $N=90$ | 155 ± 11 | 10 | 221 |
| Pitch | $N=150$ | 375 ± 14 | 323 | 411 |
| Depth | $N=110$ | 84 ± 36 | 15 | 165 |

in the 96–105 nm range. However, some pits were also recorded above the desired 100 nm range which should have minor influence as long as they are deep enough to influence the cell adhesion.

The variations in pit dimensions demonstrated that FIB milling rate is greatly influenced by the polycrystalline structure of stainless steel and beam quality.

Tables 16.2 and 16.3 summarise the obtained pit dimensions of design A and B diameter ($N=162$ and 90; scan area, $1 \times 1$ μm$^2$) and pitch and depth ($N=128$ and 150 and 95 and 90; scan area, $5 \times 5$ μm$^2$) created by FIB milling. For the required 120 nm pit, the average pit had a diameter of 97 ± 8.5 nm ($N=162$), pitch of 250 ± 11 nm ($N=128$) and depth of 65 ± 24 nm ($N=95$), whereas for the 180 nm pit, the average pit had a diameter of 155 ± 11 nm ($N=90$), pitch of 375 ± 14 nm ($N=150$) and depth of 84 ± 36 nm ($N=110$).

Nevertheless, correlation of FIB–SEM and AFM cross-sectional imaging with the top-down appearance of known patterned area (grain) was essential in establishing accurate size and shape distribution of the formed pit structures.

All these variations in pit dimensions reported in stainless steel samples can be due to the atomic arrangement or random orientation of crystallographic structures and nonuniformity in grain size, though the shape of the incident beam at the point of impact on the sample surface will have the greatest effect of all the factors, followed by the crystallographic orientation of the grain at the surface.

## 16.12 In Vitro Cell Studies: Endothelial Cell Adhesion Tests

Finally, in vitro human EC culture and EC adhesion and densities studies were performed on unpolished, electropolished and nano-textured stainless steel surfaces.

Figure 16.17 shows the fluorescence images of the EC adhesion and densities after 1–5 days on unpolished, electropolished and design A and B surfaces. Very little significant difference in EC adhesion between these surfaces is revealed.

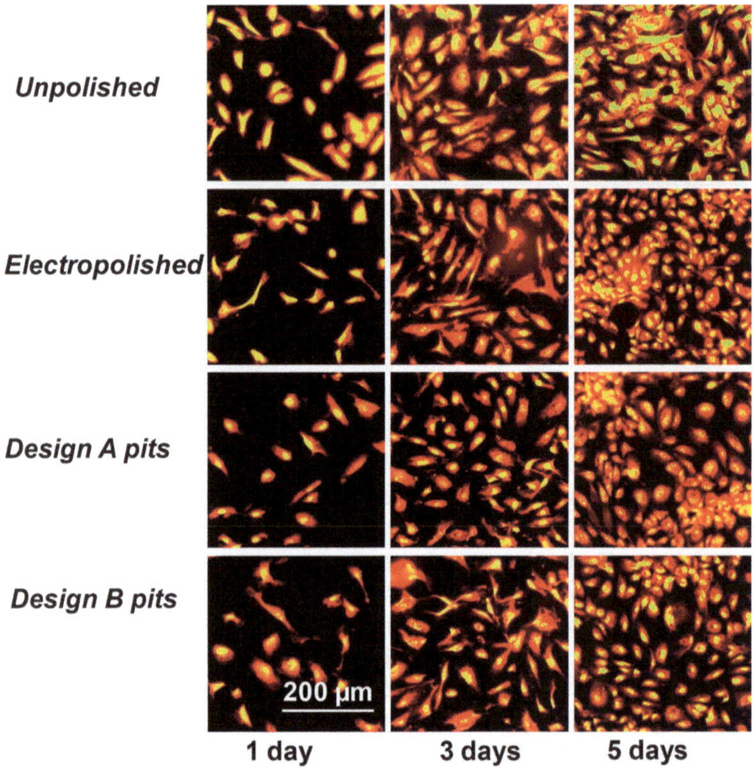


**Fig. 16.17** Fluorescent images of EC cultured on (**a**) unpolished (**b**) electropolished, (**c**) design A pits and (**d**) design B pits on stainless steel surfaces after 1, 3 and 5 days

However, the morphology of ECs appears to be greatly defined on pits design A and B relative to electropolished and unpolished samples. After 1 day, EC densities were significantly lower on nano-structured 316L steel substrates when compared to unpolished and electropolished control samples. EC adhesion density was significantly greater after 5 days compared to 1 day for all substrates tested.

To confirm the significant difference in EC adhesion and densities after 1–5 days involving these surfaces as demonstrated by fluorescence data, EC counts were performed on each three substrate of interest ($N = 12$).

## 16.13 Conclusion

FIB has compelling advantages for flexible prototyping compared to other traditional techniques; however, the milling rates and the corresponding shape and size of the formed structures are largely affected by the grain size of the polycrystalline

316L stainless steel and stability of the ion beam quality over large areas. Moreover this method is practically limited to 120 nm resolution for the desired pit depth and uniform scan size of 200 μm × 200 μm. Nevertheless formed structures show large variation of pit depths and shapes and as such surfaces might serve as a resourceful platform for screening large variations of cell/pattern stainless steel interactions. However, the FIB nano-pits design A and B created on polycrystalline stainless steel surfaces demonstrated low EC adhesion and proliferation relative to unpolished and electropolished specimens. There was no significant difference in EC adhesion and proliferation between unpolished–electropolished samples and design A and B pits. Further morphological examination of EC response on nano-structured steel surfaces would verify the mechanism for low EC adhesion and proliferation on these surfaces. Nano-patterning the stainless steel surfaces by FIB is time consuming and expensive, especially when patterning large areas. The precision and reproducibility of this technique is greatly affected by the polycrystallinity of stainless steel and a stable beam quality over large sample areas.

**Acknowledgements** This work was supported through a Starting Investigator Research Grant (09/SIRG/I1621) of the Science Foundation Ireland (SFI), the National Biophotonics and Imaging Platform, Ireland (NBIPI) and the Integrated NanoScience Platform for Ireland (INSPIRE) initiatives funded by the Irish Government's Programme for Research in Third Level Institutions, Cycle 4, National Development Plan 2007–2013. The authors are grateful to Dr Shanthi Muttukrishna (Department of Obstetrics and Gynaecology, University College Cork) for the gift of the human umbilical vein endothelial vein. Dr Lynette Keeney is gratefully acknowledged for performance of the AFM scans and line profiles for the correlative microscopy part of this chapter. Dr Calum Dickinson is gratefully acknowledged for contributing the EBSD measurements.

## References

1. Dugdale, D.C.: Stent, in Medicine Plus, 2012. http://www.nlm.nih.gov/medlineplus/ency/article/002303.htm
2. Balamurugan, A., Rajeswari, S., Balossier, G., Rebelo, A.H.S., Ferreira, J.M.F.: Corrosion aspects of metallic implants—an overview. Mater. Corrosion **59**(11), 855–869 (2008)
3. O'Brien, B., Carroll, W.: The evolution of cardiovascular stent materials and surfaces in response to clinical drivers: a review. Acta Biomater. **5**(4), 945–958 (2009)
4. Clerc, C.O., Jedwab, M.R., Mayer, D.W., Thompson, P.J., Stinson, J.S.: Assessment of wrought ASTM F1058 cobalt alloy properties for permanent surgical implants. J. Biomed. Mater. Res. **38**(3), 229–234 (1997)
5. Craig, C., Friend, C., Edwards, M., Cornish, L., Gokcen, N.: Mechanical properties and microstructure of platinum enhanced radiopaque stainless steel (PERSS) alloys. J. Alloys Compd. **361**(1), 187–199 (2003)
6. Craig, C., Friend, C., Edwards, M., Gokcen, N.: Tailoring radiopacity of austenitic stainless steel for coronary stents. In: Medical Device Materials: Proceedings from the Materials & Processes for Medical Devices Conference 2003, 8–10 September 2003, Anaheim, California. 2004. American Society for Metals
7. Heublein, B., Rohde, R., Kaese, V., Niemeyer, M., Hartung, W., Haverich, A.: Biocorrosion of magnesium alloys: a new principle in cardiovascular implant technology? Heart **89**(6), 651–656 (2003)

8. Lüscher, T.F., Steffel, J., Eberli, F.R., Joner, M., Nakazawa, G., Tanner, F.C., Virmani, R.: Drug-eluting stent and coronary thrombosis biological mechanisms and clinical implications. Circulation **115**(8), 1051–1058 (2007)
9. Daemen, J., Wenaweser, P., Tsuchida, K., Abrecht, L., Vaina, S., Morger, C., Kukreja, N., Jüni, P., Sianos, G., Hellige, G.: Early and late coronary stent thrombosis of sirolimus-eluting and paclitaxel-eluting stents in routine clinical practice: data from a large two-institutional cohort study. Lancet **369**(9562), 667–678 (2007)
10. Chou, L., Firth, J.D., Uitto, V.-J., Brunette, D.M.: Substratum surface topography alters cell shape and regulates fibronectin mRNA level, mRNA stability, secretion and assembly in human fibroblasts. J. Cell Sci. **108**(4), 1563–1573 (1995)
11. Boyan, B.D., Hummert, T.W., Dean, D.D., Schwartz, Z.: Role of material surfaces in regulating bone and cartilage cell response. Biomaterials **17**(2), 137–146 (1996)
12. Craighead, H.G., James, C.D., Turner, A.M.P.: Chemical and topographical patterning for directed cell attachment. Curr. Opin. Solid State Mater. Sci. **5**(2–3), 177–184 (2001)
13. Curtis, A., Wilkinson, C.: Nantotechniques and approaches in biotechnology. Trends Biotechnol. **19**(3), 97–101 (2001)
14. Koegler, P., Clayton, A., Thissen, H., Santos, G.N.C., Kingshott, P.: The influence of nanostructured materials on biointerfacial interactions. Adv. Drug Deliv. Rev. **64**(15), 1820–1839 (2012)
15. Nikkhah, M., Edalat, F., Manoucheri, S., Khademhosseini, A.: Engineering microscale topographies to control the cell–substrate interface. Biomaterials **33**(21), 5230–5246 (2012)
16. Chen, L., Han, D., Jiang, L.: On improving blood compatibility: from bioinspired to synthetic design and fabrication of biointerfacial topography at micro/nano scales. Colloids Surf. B Biointerfaces **85**(1), 2–7 (2011)
17. Gentile, F., Tirinato, L., Battista, E., Causa, F., Liberale, C., di Fabrizio, E.M., Decuzzi, P.: Cells preferentially grow on rough substrates. Biomaterials **31**(28), 7205–7212 (2010)
18. Kasemo, B.: Biological surface science. Surf. Sci. **500**(1–3), 656–677 (2002)
19. Duncan, A.C., Weisbuch, F., Rouais, F., Lazare, S., Baquey, C.: Laser microfabricated model surfaces for controlled cell growth. Biosens. Bioelectron. **17**(5), 413–426 (2002)
20. Berry, C.C., Campbell, G., Spadiccino, A., Robertson, M., Curtis, A.S.G.: The influence of microscale topography on fibroblast attachment and motility. Biomaterials **25**(26), 5781–5788 (2004)
21. Yim, E.K.F., Leong, K.W.: Significance of synthetic nanostructures in dictating cellular response. Nanomedicine **1**(1), 10–21 (2005)
22. Falconnet, D., Csucs, G., Grandin, H.M., Textor, T.: Surface engineering approaches to micropattern surfaces for cell-based assays. Biomaterials **27**(16), 3044–3063 (2006)
23. Yu, L.M.Y., Leipzig, N.D., Shoichet, M.S.: Promoting neuron adhesion and growth. Mater. Today **11**(5), 36–43 (2008)
24. Roach, P., Parker, T., Gadegaard, N., Alexander, M.R.: Surface strategies for control of neuronal cell adhesion: a review. Surf. Sci. Rep. **65**(6), 145–173 (2010)
25. Nazneen, F., Herzog, G., Arrigan, D.W., Caplice, N., Benvenuto, P., Galvin, P., Thompson, M.: Surface chemical and physical modification in stent technology for the treatment of coronary artery disease. J. Biomed. Mater. Res. B Appl. Biomater. **100**, 1989–2014 (2012)
26. Norman, J.J., Desai, T.A.: Methods for fabrication of nanoscale topography for tissue engineering scaffolds. Ann. Biomed. Eng. **34**(1), 89–101 (2006)
27. Buzea, C., Beydaghyan, G., Elliott, C., Robbie, K.: Control of power law scaling in the growth of silicon nanocolumn pseudo-regular arrays deposited by glancing angle deposition. Nanotechnology **16**(10), 1986 (2005)
28. Dolatshahi-Pirouz, A., Hovgaard, M.B., Rechendorff, K., Chevallier, J., Foss, M., Besenbacher, F.: Scaling behavior of the surface roughness of platinum films grown by oblique angle deposition. Phys. Rev. B **77**(11), 115427 (2008)
29. Tseng, A.A., Notargiacomo, A.: Nanoscale fabrication by nonconventional approaches. J. Nanosci. Nanotechnol. **5**(5), 683–702 (2005)

30. Sarkar, S., Dadhania, M., Rourke, P., Desai, T.A., Wong, J.Y.: Vascular tissue engineering: microtextured scaffold templates to control organization of vascular smooth muscle cells and extracellular matrix. Acta Biomater. **1**(1), 93–100 (2005)
31. Houtchens, G.R., Foster, M.D., Desai, T.A., Morgan, E.F., Wong, J.Y.: Combined effects of microtopography and cyclic strain on vascular smooth muscle cell orientation. J. Biomech. **41**(4), 762–769 (2008)
32. Biela, S.A., Su, Y., Spatz, J.P., Kemkemer, R.: Different sensitivity of human endothelial cells, smooth muscle cells and fibroblasts to topography in the nano–micro range. Acta Biomater. **5**(7), 2460–2466 (2009)
33. Green, A.M., Jansen, J.A., Van der Waerden, J.P.C.M., Von Recum, A.F.: Fibroblast response to microtextured silicone surfaces: texture orientation into or out of the surface. J. Biomed. Mater. Res. **28**(5), 647–653 (1994)
34. Walboomers, X., Croes, H., Ginsel, L., Jansen, J.: Contact guidance of rat fibroblasts on various implant materials. J. Biomed. Mater. Res. **47**(2), 204–212 (1999)
35. Walboomers, X., Croes, H., Ginsel, L., Jansen, J.: Growth behavior of fibroblasts on microgrooved polystyrene. Biomaterials **19**(20), 1861–1868 (1998)
36. Walboomers, X., Ginsel, L., Jansen, J.: Early spreading events of fibroblasts on microgrooved substrates. J. Biomed. Mater. Res. **51**(3), 529–534 (2000)
37. Walboomers, X., Monaghan, W., Curtis, A., Jansen, J.: Attachment of fibroblasts on smooth and microgrooved polystyrene. J. Biomed. Mater. Res. **46**(2), 212–220 (1999)
38. Loesberg, W.A., te Riet, J., van Delft, F.C., Schön, P., Figdor, C.G., Speller, S., van Loon, J.J., Walboomers, X.F., Jansen, J.A.: The threshold at which substrate nanogroove dimensions may influence fibroblast alignment and adhesion. Biomaterials **28**(27), 3944–3951 (2007)
39. Dalby, M.J., Gadegaard, N., Riehle, M.O., Wilkinson, C.D.W., Curtis, A.S.G.: Investigating filopodia sensing using arrays of defined nano-pits down to 35 nm diameter in size. Int. J. Biochem. Cell Biol. **36**(10), 2005–2015 (2004)
40. Dalby, M.J., Gadegaard, N., Wilkinson, C.D.: The response of fibroblasts to hexagonal nanotopography fabricated by electron beam lithography. J. Biomed. Mater. Res. A **84**(4), 973–979 (2008)
41. Yim, E.K.F., Reano, R.M., Pang, S.W., Yee, A.F., Chen, C.S., Leong, K.W.: Nanopattern-induced changes in morphology and motility of smooth muscle cells. Biomaterials **26**(26), 5405–5413 (2005)
42. Lee, S.W., Kim, S.Y., Rhyu, I.C., Chung, W.Y., Leesungbok, R., Lee, K.W.: Influence of microgroove dimension on cell behavior of human gingival fibroblasts cultured on titanium substrata. Clin. Oral Implants Res. **20**(1), 56–66 (2009)
43. Ito, T., Okazaki, S.: Pushing the limits of lithography. Nature **406**(6799), 1027–1031 (2000)
44. Chou, S.Y., Krauss, P.R., Renstrom, P.J.: Imprint of sub-25 nm vias and trenches in polymers. Appl. Phys. Lett. **67**(21), 3114–3116 (1995)
45. Raffa, V., Castrataro, P., Menciassi, A., Dario, P.: Focused Ion Beam as a Scanning Probe: Methods and Applications. Applied Scanning Probe Methods II, pp. 361–412. Springer, New York, NY (2006)
46. Iordanova, I., Antonov, V., Gurkovsky, S.: Changes of microstructure and mechanical properties of cold-rolled low carbon steel due to its surface treatment by Nd:glass pulsed laser. Surf. Coat. Technol. **153**(2–3), 267–275 (2002)
47. Duncan, A.C., Rouais, F., Lazare, S., Bordenave, L., Baquey, C.: Effect of laser modified surface microtopochemistry on endothelial cell growth. Colloids Surf. B Biointerfaces **54**(2), 150–159 (2007)
48. Kalantar-Zadeh, K., Fry, B.: Nanotechnology Enabled Sensors. Springer, New York, NY (2007)
49. Volkert, C., Minor, A.: Focused ion beam microscopy and micromachining. MRS Bull. **32**(5), 389–395 (2007)
50. Lanyon, Y.H., De Marzi, G., Watson, Y.E., Quinn, A.J., Gleeson, J.P., Redmond, G., Arrigan, D.W.: Fabrication of nanopore array electrodes by focused ion beam milling. Anal. Chem. **79**(8), 3048–3055 (2007)

51. Choi, C.H., Heydarkhan-Hagvall, S., Wu, B.M., Dunn, J.C., Beygui, R.E., Kim, C.J.: Cell growth as a sheet on three-dimensional sharp-tip nanostructures. J. Biomed. Mater. Res. A **89**(3), 804–817 (2009)
52. Raffa, V., Vittorio, O., Pensabene, V., Menciassi, A., Dario, P.: FIB-nanostructured surfaces and investigation of bio/nonbio interactions at the nanoscale. IEEE Trans. NanoBiosci. **7**(1), 1–10 (2008)
53. Joshi, K., Singh, P., Verma, S.: Fabrication of platinum nanopillars on peptide-based soft structures using a focused ion beam. Biofabrication **1**(2), 025002 (2009)
54. Choubey, A., Marton, D., Sprague, E.A.: Human aortic endothelial cell response to 316L stainless steel material microstructure. J. Mater. Sci. Mater. Med. **20**(10), 2105–2116 (2009)
55. Misra, R.D., Nune, C., Pesacreta, T.C., Somani, M.C., Karjalainen, L.P.: Understanding the impact of grain structure in austenitic stainless steel from a nanograined regime to a coarse-grained regime on osteoblast functions using a novel metal deformation–annealing sequence. Acta Biomater. **9**, 6245–6258 (2013)
56. Russell, A., Lee, K.L.: Structure-Property Relations in Nonferrous Metals. Wiley-Interscience, New York, NY (2005)
57. Shi, D.: Introduction to Biomaterials. World Scientific, London (2006)
58. Nazneen, F., Galvin, P., Arrigan, D.W., Thompson, M., Benvenuto, P., Herzog, G.: Electropolishing of medical-grade stainless steel in preparation for surface nano-texturing. J. Solid State Electrochem. **16**(4), 1389–1397 (2012)
59. Dalby, M.J., Berry, C.C., Riehle, M.O., Sutherland, D.S., Agheli, H., Curtis, A.S.G.: Attempted endocytosis of nano-environment produced by colloidal lithography by human fibroblasts. Exp. Cell Res. **295**(2), 387–394 (2004)
60. Schmidt, M., Nazneen, F., Georgiev, Y., Herzog, G., Galvin, P., Petkov, N.: FIB patterning of stainless steel for the development of nano-structured stent surfaces for cardiovascular applications. J. Phys. Conf. **371**(1), 012065 (2012)
61. Nazneen, F., Schmidt, M., McLoughlin, E., Petkov, N., Herzog, G., Arrigan, D., Galvin, P.: Nano-texturing of medical-grade 316L stainless steel by focused ion beam for endothelial cell studies. J. Nanosci. Nanotechnol. **13**, 5283–5290 (2013)
62. Feitknecht, W.: Über den angriff von krystallen durch kanalstrahlen. Helv. Chim. Acta **7**(1), 825–842 (1924)
63. Wehner, G.K.: Controlled sputtering of metals by low-energy Hg ions. Phys. Rev. **102**(3), 690–704 (1956)
64. Wehner, G.K., Rosenberg, D.: Angular distribution of sputtered material. J. Appl. Phys. **31**(1), 177–179 (1960)
65. Sigmund, P.: Theory of sputtering. I. Sputtering yield of amorphous and polycrystalline targets. Phys. Rev. **184**(2), 383–416 (1969)
66. Bean, K.E.: Anisotropic etching of silicon. IEEE Trans. Electron Dev. **25**(10), 1185–1193 (1978)
67. Seidel, H., Csepregi, L., Heuberger, A., Baumgärtel, H.: Anisotropic etching of crystalline silicon in alkaline solutions I. Orientation dependence and behavior of passivation layers. J. Electrochem. Soc. **137**(11), 3612–3626 (1990)
68. Schmidt, M., Nazneen, F., Herzog, G., Arrigan, D., Galvin, P., Keeney, L., Petkov, N., Holmes, J.D.: to be submitted
69. Schmidt, M., Nazneen, F., Herzog, G., Arrigan, D., Galvin, P., Dickinson, C., de Silva, J.P., Scanlan, D., O'Hara, N., Cross, G.L.W.: Correlative microscopy study of FIB patterned stainless steel surfaces as novel nano-structured stents for cardiovascular applications. MRS Proc 2012. 1466(1)
70. Giannuzzi, L., Stevie, F.: A review of focused ion beam milling techniques for TEM specimen preparation. Micron **30**(3), 197–204 (1999)
71. Schwartz, A.J., Kumar, M., Adams, B.L., Field, D.P.: Electron Backscatter Diffraction in Materials Science. Springer, New York, NY (2009)
72. Stark, Y., Fromter, R., Stickler, D., Oepen, H.P.: Sputter yields of single-and polycrystalline metals for application in focused ion beam technology. J. Appl. Phys. **105**(1), 013542-1–013542-5 (2009)

# Chapter 17
# Evaluation of Damages Induced by $Ga^+$-Focused Ion Beam in Piezoelectric Nanostructures

**A. Ferri, D. Rémiens, R. Desfeux, A. Da Costa, D. Deresmes, and D. Troadec**

**Abstract** The impact of $Ga^+$-focused ion beam (FIB) about functional properties of continuous and nanostructured piezoelectric thin films of lead zirconate titanate ($Pb(Zr_xTi_{1-x})O_3$) was investigated. A suitable way to fabricate piezoelectric

---

A. Ferri (✉)
Univ Lille Nord de France, F-59000 Lille, France

CNRS UMR 8181, Unité de Catalyse et de Chimie du Solide – UCCS, Lens, France

UArtois, UCCS, F-62307, Lens, France
e-mail: anthony.ferri@univ-artois.fr

D. Rémiens
UVHC, IEMN, F-59313 Valenciennes, France

Univ Lille Nord de France, F-59000, Lille, France

Institut d'Électronique, de Microélectronique et de Nanotechnologie – IEMN, CNRS UMR 8520, Villeneuve d'Ascq, France
e-mail: denis.remiens@univ-valenciennes.fr

R. Desfeux • A. Da Costa
CNRS UMR 8181, Unité de Catalyse et de Chimie du Solide – UCCS, Lens, France

UArtois, UCCS, F-62307, Lens, France

Univ Lille Nord de France, F-59000, Lille, France
e-mail: rachel.desfeux@univ-artois.fr; antonio.dacosta@univ-artois.fr

D. Deresmes • D. Troadec
Univ Lille Nord de France, F-59000 Lille, France

Institut d'Électronique, de Microélectronique et de Nanotechnologie – IEMN, CNRS UMR 8520, Villeneuve d'Ascq, France

Institut d'Électronique, de Microélectronique et de Nanotechnologie – IEMN, F-59652 Villeneuve d'Ascq, France
e-mail: dominique.deresmes@isen.iemn.univ-lille1.fr; david.troadec@iemn.univ-lille1.fr

Z.M. Wang (ed.), *FIB Nanostructures*, Lecture Notes in Nanoscale Science and Technology 20, DOI 10.1007/978-3-319-02874-3_17,

nanocapacitors was studied, based on the amorphous or crystallized state of the film before etching. Strong modification of structural and electrical behavior for area exposed to ion irradiation is observed when the film is etched in the crystallized state. Both the implantation of $Ga^+$ ions and the film amorphization highlighted by Raman spectroscopy and Kelvin force microscopy analyses can explain this result. The piezoactivity detected by piezoresponse force microscopy is fully destroyed even after a post-annealing treatment. In the case of amorphous etched film, no significant degradation is observed. The latter process is used to successfully fabricate $Pb(Zr_xTi_{1-x})O_3$-based nanocapacitors by means of FIB method. In 50-nm-size capacitors, the local electromechanical behavior is measured at similar level that the one obtained for the un-etched film, evidencing no manifest sidewall effect or FIB-induced damages. This further evidences that amorphous FIB lithography process can reduce the etching damages, demonstrating this is an effective alternative method and very beneficial to pattern such low-dimensional structures, which is a significant result in view of the development of functional nanostructures in the field of nanoelectromechanical systems applications.

## 17.1 Introduction

Piezoelectrics have been extensively investigated due to their attractive electromechanical properties. Piezoelectric-based systems have great potential for integrations into high-performance devices with various applications such as actuators, high-sensitivity sensors, micro-/nanoelectromechanical systems (M/NEMS), ultrasonic transducers, energy harvesting, or high-frequency applications [1–5].

In the pursuit of the trend to the miniaturization of existing devices in electronic industry, the scaling of the dimension of the piezo-/ferroelectric materials down to the nanometer scale has become a fundamental issue [6–8]. In particular, the nanopatterning of such materials, when grown in thin film form, allows reaching highly miniaturized systems. In this frame, both control and investigation of physical properties of the matter on the nanoscale are essential. Indeed, nanostructured thin films display unique phenomena and enhanced functional properties, enabling the improvement of traditional applications or the development of new generation of electronic nanodevices [9–12]. In piezoelectric nanostructures, properties are strongly influenced by the lateral size effect, and they significantly change from those in thin films [13, 14].

From an application point of view, one of the main process objectives in the integration of piezoelectric capacitors into M/NEM devices is developing the ability to pattern these films and associated electrode materials by etching. In particular, for the fabrication of metal-insulator-metal (MIM) structures, the obtaining of a process which is common for active layer and top/bottom electrodes and also compatible with silicon technology (CMOS process) is very essential. Several techniques have been developed to pattern the films, such as wet chemical etching [15], ion beam etching (IBE) [16, 17], reactive ion etching (RIE) [18, 19], electron cyclotron resonance (ECR) etching [20], or inductively coupled plasma etching [21]. Among these various methods, focused ion beam (FIB) technology is applied for advanced lithography and one of the most common methods for nanofabrication of piezo-based structures [22–26]. Major advantages are the high resolution, maskless lithography, uniform size for fabricated structures, and great flexibility and simplicity. It is possible to pattern thin films leading to nanostructures with very small lateral size (sub-100 nm). However, if the FIB process appears as a very suitable method in order to get discrete piezoelectric nanoislands from thin films, and although such lateral confinement allows enhancing of piezoelectric activity in ferroelectric thin films [27, 28], two major drawbacks exist which are the ion implantation (e.g., gallium) into the etched material and the material amorphization induced by the FIB milling [29]. These effects induce a degradation of electrical behavior of devices, and they are particularly pronounced when the FIB systems use gallium ions. Such kind of element is usually chosen owing to its low melting point, low volatility, and high mass. The damaged layer that appears on the film surface was thoroughly characterized in terms of the composition and charge implantation, but a more detailed characterization seems to be necessary. Besides, a specific post-annealing treatment (oxygen, vacuum) was commonly proposed for at least partially recovering functional properties [29–33].

In the present study we evaluate the damages induced by gallium-FIB etching on $Pb(Zr_xTi_{1-x})O_3$-based ferroelectric thin films. Lead zirconate titanate (PZT) compound is recognized as leading material for piezoelectric micro-/nanodevices, due to its excellent piezo-/ferroelectric properties. Both the implantation of $Ga^+$ ions and the film amorphization will be particularly investigated [25, 26, 34]. We will focus this work on the development of a suitable process of nanostructures manufacture without introducing piezoelectricity deterioration, by establishing a strategy to overcome the ion implantation and amorphization effects. This more effective alternative method to solve the problem of FIB-induced damages will be presented.

## 17.2 Material and Experimental Setup

### *17.2.1 Thin Film Structure*

Two types of heterostructures were studied in this work, i.e., Pt/PZT/Pt(111)/$TiO_x/SiO_2$/Si in which the PZT layer is in either amorphous or crystallized state before etching. Such substrate was preferred to enable integration of these

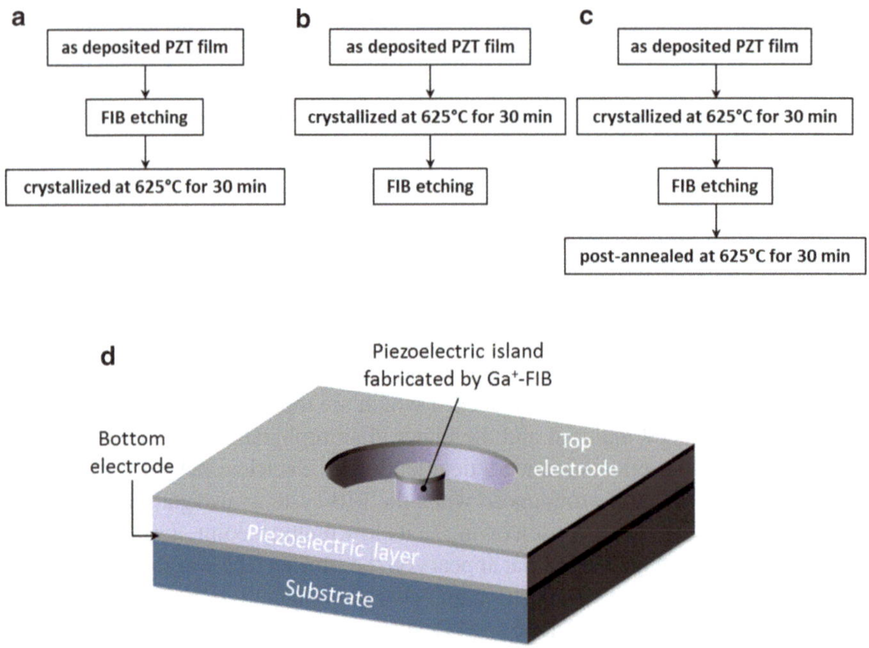


**Fig. 17.1** Flow chart of (**a**) amorphous FIB etching process, (**b**) conventional FIB etching process, (**c**) conventional FIB etching process then post-annealing at 625 °C. (**d**) Schematic view of the piezoelectric capacitor fabricated by $Ga^+$ FIB process from crystallized or amorphous PZT thin layer

materials into silicon devices and circuit technology, as well as the specific MIM structure. The thicknesses of the titanium oxide and platinum bottom electrode layers are 15 and 120 nm, respectively. Pt electrode was used for allowing macroscopic electrical measurements. Similar MIM capacitors with $LaNiO_3$ bottom electrode have been fabricated by FIB, and the local electrical properties have also been probed [25]. Such oxide conductive electrodes are known to improve the electrical fatigue resistance of ferroelectrics [22, 35]. Regardless of the state, the ferroelectric films are deposited by radio frequency magnetron sputtering technique without substrate heating, leading to amorphous layer [36]. The film composition is fixed to Zr:Ti = 54:46 near the so-called morphotropic phase boundary where the piezoelectric activity is larger. To crystallize the PZT film in perovskite phase, a conventional post-annealing treatment at 625 °C in air during 30 min was required [37]. The film thickness was ranging from 100 to 300 nm depending on the island diameter. Indeed, for a small island diameter, the film will be thinner in order to reduce the etching time. Figure 17.1a–c shows various processes for etching the samples, and Fig. 17.1d displays a schematic view of the nanocapacitor fabricated by $Ga^+$ FIB process from PZT thin layer.

### 17.2.2 FIB System

A focused beam of $Ga^+$ ions (STRATA DB 235—FEI Company) was used to pattern the PZT films. The system was equipped with both the electron and ion beams taking advantage of being able to monitor the etching process in real time. In this work, the acceleration voltage was 30 kV, and the ion current was included between 300 and 10 pA (corresponding to 25 nm spot size) depending on the island diameter. The ion beam was positioned at an angle of 90° to the etched surface.

### 17.2.3 Nanoscale Electrical Characterization

Electrical characterizations of the structures on the nanoscale were performed by means of complementary tools of atomic force microscopy (AFM). In particular, specific electrical modes of AFM, namely, piezoresponse force microscopy (PFM), Kelvin force microscopy (KFM), and conductive atomic force microscopy (C-AFM), were successfully employed to address electrical properties of piezoelectric nanostructures. For such investigations, commercial AFM microscopes (Multimode for PFM and Dimension 3100 for KFM/C-AFM—Veeco Digital Instruments) operating under environmental conditions were used.

PFM was used in spectroscopic mode, i.e., recording of hysteresis piezoloops, in view of investigating the local electromechanical behavior of the PZT islands. Briefly, PFM technique is based on the detection of surface deformation due to the converse piezoelectric effect induced by an alternating voltage applied to the tip. This technique is a powerful tool and the most widely used for locally studying piezo-/ferroelectric materials due to its versatility and its high resolution (measurement of deformation at the sub-picometer scale and imaging ferroelectric domains down to several nanometers) [38–41]. Local piezoresponse loops were obtained by plotting the amplitude of piezoelectric vibration as a function of the DC voltage in the so-called infield mode in order to provide information about electromechanical activity. The frequency and the amplitude of the alternating voltage were set to values of $f=2$ kHz and $V_{ac}=1.5$ V, respectively. The applied DC voltage was gradually swept between −10 V and +10 V within the period of 200 s [42]. For such electrical experiments, $Si_3N_4$ tips with conductive PtIr coating were chosen, and a stiff cantilever ($k\sim 40$ N/m) was used to minimize the electrostatic contribution to the PFM signal [43].

KFM is a noncontact technique based on the electrostatic interaction between the tip and the surface. The resulting electrostatic signal provides information related to surface potential. The surface potential maps are easily detected due to work function variation as a result of surface charges, electric dipoles, or absorption layer. KFM experiments were carried out by using a Pt-coated tip with a resonant frequency of 70–80 kHz and a spring constant of 1–5 N/m [25].

Like PFM, C-AFM is a contact mode technique where a metallic cantilever is connected to a current amplifier. This tool allows to probe the local electrical conductivity of the material by placing a conducting cantilever on the film surface and applying a DC bias between the cantilever and the sample [26].

## 17.3 Results and Discussion

In order to evaluate the impact of irradiation on the ferroelectric layer, the degradation induced by the $Ga^+$ ion beam on the crystallized PZT films deposited on the entire substrate was studied prior the fabrication of Pt/PZT/Pt islands. In this experiment, large areas of $10 \times 10\ \mu m^2$ were exposed to the ion beam using different ion doses, as seen on the scanning electron microscopy (SEM) image in Fig. 17.2. The $Ga^+$ dose is controlled by the etching time and is directly related to the path number of the beam. More precisely, the number of ion beam paths was varied between 1 and 20, corresponding to $4 \times 10^{13}$ and $3 \times 10^{15}$ ions/cm$^2$, respectively [44].

The first characterization was realized at the macroscale by analyzing the eventual structural defects induced in the PZT film by Raman spectroscopy. The Raman spectra were carried out in the 200–1000 cm$^{-1}$ range with a Dilor XY-800 confocal micro-Raman spectrometer, and they were recorded using the 514.5 nm line of an Ar/Kr ion laser. The incident laser beam was focused on the FIB-exposed area with 1 μm spot size and 10 mW power. The Raman measurements were directly conducted on the etched areas described above. However, different PZT films were investigated: (i) un-etched crystallized PZT film, (ii) hardest etched amorphous PZT film (then post-annealed at 625 °C), (iii) lightest etched crystallized PZT film, and

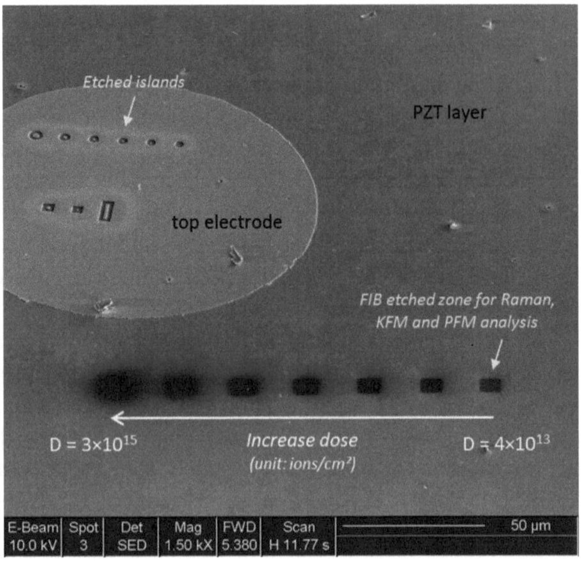


**Fig. 17.2** SEM image of FIB-exposed PZT layer with different $Ga^+$ ion doses (*square areas*) and FIB-etched nanoislands onto Pt top electrode

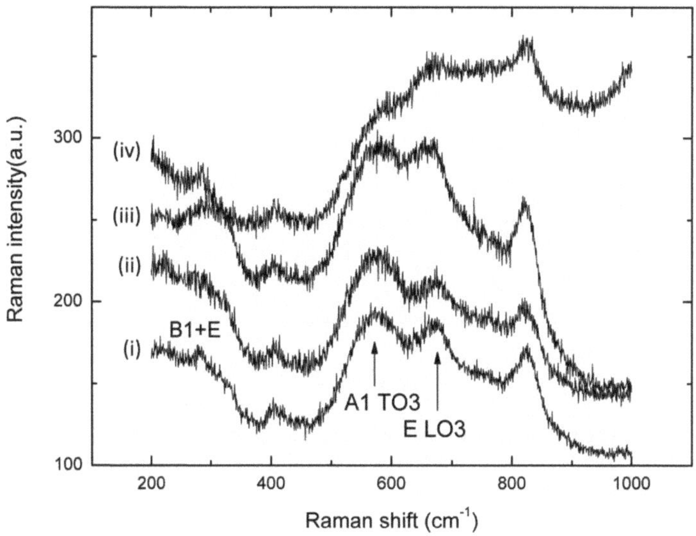


**Fig. 17.3** Raman spectra of (*i*) un-etched crystallized PZT film, (*ii*) amorphous PZT film after hard FIB etching, (*iii*) lightest etched crystallized PZT film, and (*iv*) hardest etched crystallized PZT film. Figure reprinted with permission from [25]. Copyright 2009, American Institute of Physics

(iv) hardest etched crystallized PZT film. The results are depicted in Fig. 17.3. From the spectra, the broadband centered at 280, 572, and 676 $cm^{-1}$ can be assigned as $B_1+E$, $A_1$ ($TO_3$), and $E$($LO_3$) according to the modes in the single PZT crystal [45], while the band at 820 $cm^{-1}$ is tentatively assigned to Ti–O bond [46]. For the region exposed to the lightest "conventional" (i.e., crystallized material) FIB etching (case (iii)), the spectrum is rather similar to that of the un-etched zone (case (i)), except the higher intensity for the Raman peak at 676 $cm^{-1}$. However, for the hardest conventional etching case ((iv)), Raman spectra recorded in the region exposed to the FIB show obvious change. The sharp increase in the intensity of the Raman band at 676 $cm^{-1}$ [$E$($LO_3$) mode] may be correlated to the distortion of the tetragonal symmetry of the PZT crystalline lattice due to the ion implantation and the modification of the stress state [45, 47, 48]. The broadening of Raman scattering in 550–900 $cm^{-1}$ region may be attributed to the existence of amorphized material, which is usually expected during the conventional FIB etching process. On the other hand, the Raman signal recorded on the amorphous PZT film after similar hard FIB etching does not show apparent change compared to spectrum of the un-etched PZT film surface. Thus the amorphous state seems to be benefit for the preservation of crystalline structure of the film without structural defects after FIB etching. Also, the peak position of Raman shift is similar regardless of the FIB etching process, which means the crystalline structure still exhibits perovskite phase.

Then, piezoactivity in etched zones of crystallized films was probed by recording amplitude piezoresponse hysteresis loops via the spectroscopic tool of PFM. As seen in Fig. 17.4, the absence of typical butterfly shape for hysteresis loop points

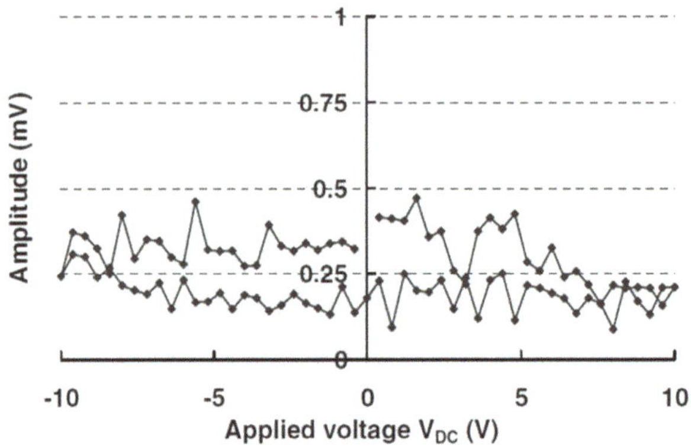


**Fig. 17.4** Amplitude piezoresponse loop recorded on etched zone of crystallized PZT film after one path of the Ga⁺ ion beam. Figure reprinted with permission from [26]. Copyright 2010, American Institute of Physics

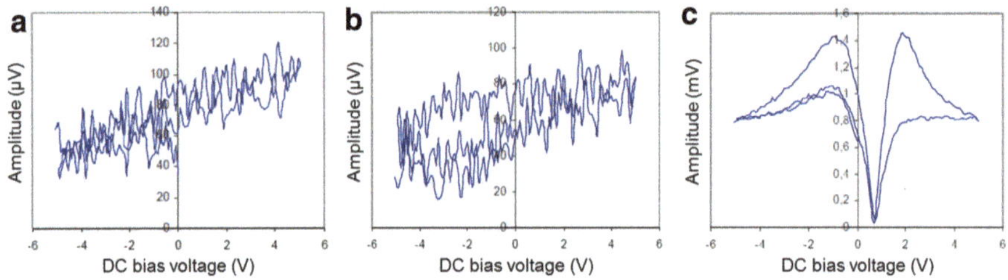


**Fig. 17.5** Local probing of piezoelectric behavior in crystallized PZT films at various distances from the $Ga^+$ ion beam-exposed area. Piezoloops recorded (**a**) in the etched zone and (**b**) at 20 µm and (**c**) at 100 µm beyond the etched region

out the strong degradation of electromechanical behavior whatever the ion dose applied during etching. Indeed, one path is sufficient to destroy piezoactivity, and no ferroelectric domains can be imaged by PFM. Therefore, the FIB etching process is detrimental to functional properties of PZT films. Local probing of piezoelectric behavior outside the directly exposed area also evidences the existence of some defects, signifying the degradations induced by the $Ga^+$ FIB are not limited to the etched region. This phenomenon is highlighted on the Fig. 17.5 where piezoloops recorded at various distances from etched regions with conventional FIB process (crystallized PZT) are presented. Degradation of piezoactivity is observed as far away as 20 µm from the exposed zone (ion dose of $3\times10^{15}$ ions/cm$^2$), while at 100 µm the PFM signal is fully restored. The explanation can be related to the spatial extension of the ion beam, depending on experimental parameters such as ion beam current, overlap, and dwell time. It is well known that the ion beam presents a Gaussian form and thus a zone outside the directly etched region is also impacted by the $Ga^+$ ion beam.

For the crystallized samples, a post-annealed treatment was achieved in order to try to partially recover the piezoelectric properties, as usually performed on such etched ferroelectric films [22, 23, 29, 30, 49], but irrespective of the experimental conditions (air or $O_2$ atmosphere, different temperatures, and times), we were unable to restore the initial activity of the films. These results suggest that a damaged layer exists on the film surface which strongly affects the piezoelectric performance of PZT films. These damages could be an amorphization of the film surface and/or a modification of the film composition. Also, this upper layer of damaged material could present some structural defects or a lattice disorder which is usually expected during the FIB etching experiments.

Instead, in the case of PZT etched before crystallization, the electromechanical activity (measured after crystallization) is similar as the one obtained for an un-etched crystallized PZT sample, regardless of the ion dose applied. Indeed, even if the degradations are induced in the amorphous film, the annealing treatment (crystallization of the film) permits to eliminate the etching damages.

This behavior confirms that the strong degradation occurs on the initially crystallized samples and no degradation happens if the FIB process is conducted on amorphous films.

In order to get further understanding on the impact of $Ga^+$ FIB about the PZT films, surfaces exposed to the beam were characterized by means of KFM technique. Both KFM and C-AFM (see later) measurements were made at the same time that means with C-AFM configuration the tip is virtually connected to the ground and the bias is applied to the sample. The evolution of the tip grounded CPD (contact potential difference) response as a function of the path number of the $Ga^+$ ion beam on the PZT film etched after crystallization is displayed on the Fig. 17.6. The change of contrast is directly related to the variation in the surface properties such as the composition and/or amorphization of the material and induced structural defects, due to the strong impact of gallium ions which locally leads to structural damaging at the sample surface. Moreover, the ion beam implantation into the film or ion beam-induced transport of chemical species such as oxygen in the ferroelectric layer [22], as well as the creation/injection of charges during FIB lithography, cannot be excluded. From the KFM image, a clear observation is that after 20 paths the etched region is strongly modified, but no more definitive conclusion can be made at this step. The CPD profile (here the sample is grounded) associated to the three etched areas does not exhibit significant modification of the surface potential for one path, while an apparent decrease down to –350 and –500 mV for 10 and 20 paths, respectively, is observed. The observed modifications can be explained by several factors such as a lead loss or a gallium implantation. Some authors have observed the substitution of Ga ions for Pb ions, i.e., PZT before FIB etching becomes GZT (G stands for Ga) after etching [29]. Such substitution occurs because lead compound is preferentially etched (high sputtering yield) and gallium is implanted.

To compare impact of FIB lithography on PZT films etched in amorphous and crystallized states, we performed additional experiment under the same protocol (tip, instrumentation configuration, and sample grounding). The KFM measurements were made on the two films with the same tip during the same day to exclude

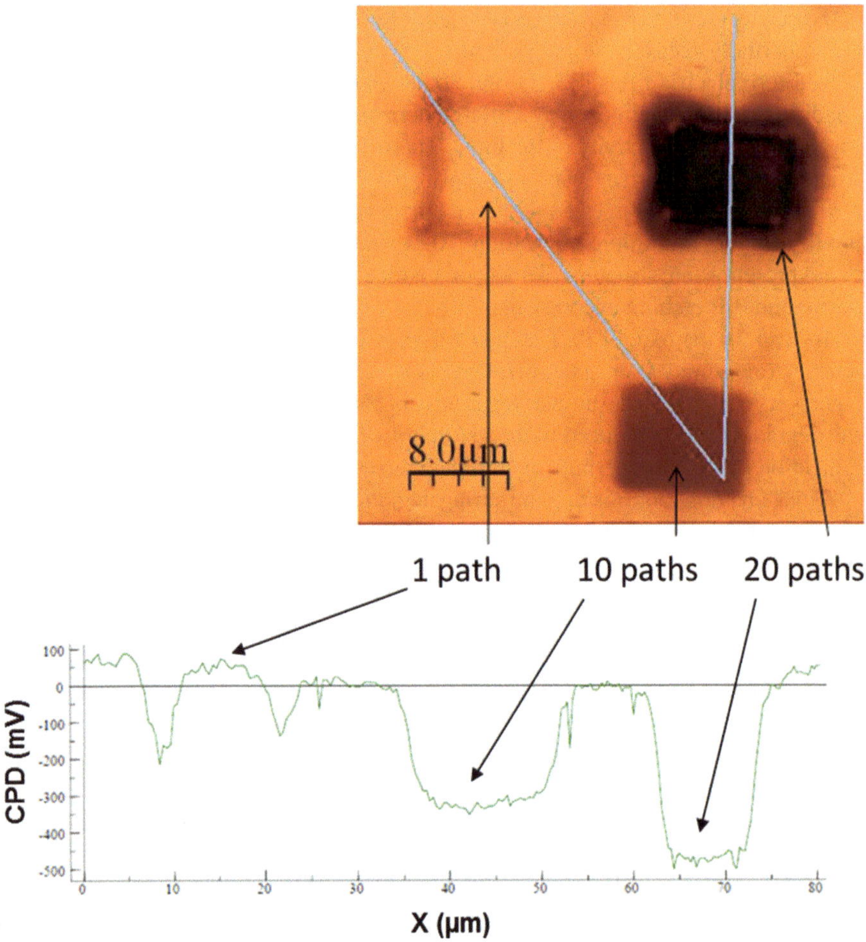


**Fig. 17.6** KFM signal measured on the PZT film etched after crystallization presenting regions exposed at various path number of the $Ga^+$ ion beam. Corresponding profiles for each etched areas are presented. Figure reprinted with permission from [26]. Copyright 2010, American Institute of Physics

measurements perturbations by extrinsic factor. The results are presented in Fig. 17.7. The high-contrast difference obtained for the etched film in crystallized state (Fig. 17.7b) between the exposed and unexposed zone confirms the strong damage induced by the ion beam and the consequent modification of the surface potential. On the contrary, only weak damage (low-contrast variation) is observed for the etched film before crystallization (etched in the amorphous state), as seen in Fig. 17.7a. The profiles of the contact potential acquired across the exposed regions for the PZT films etched in both states are depicted in Fig. 17.7c. It confirms that the increase of surface potential between the un-etched surface and the surface of PZT etched after crystallization is very large (~350 mV). The implanted charges are

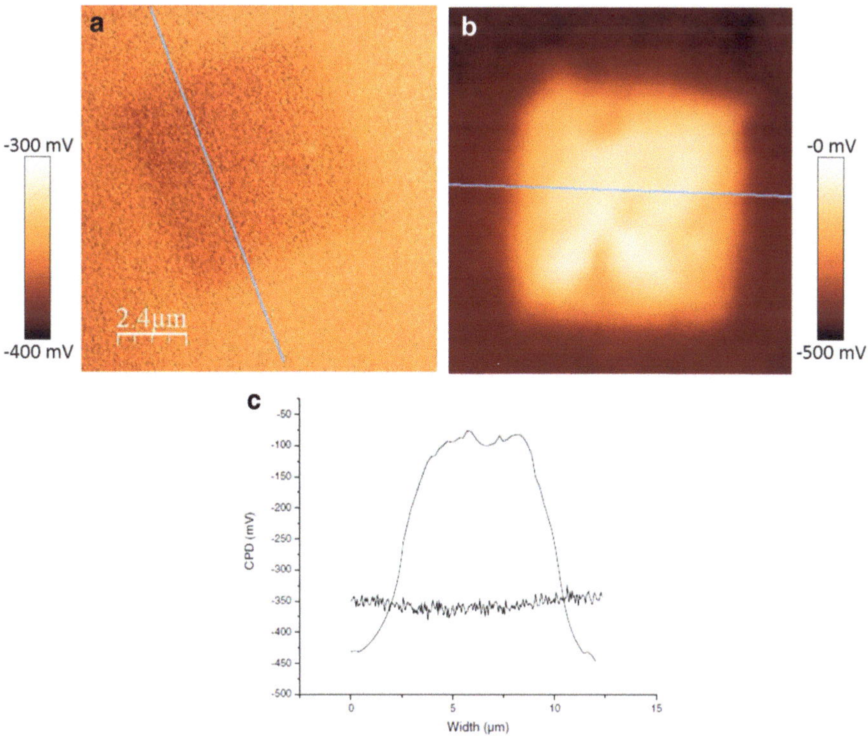


**Fig. 17.7** KFM images recorded for PZT film FIB etched in the (**a**) amorphous and (**b**) crystallized states. (**c**) Profile of the contact potential corresponding to the PZT films etched prior and after crystallization. The position of the scans is specified in both KFM images. Figures reprinted with permission from [26]. Copyright 2010, American Institute of Physics

similar in both samples, but for amorphous film we can presume that the annealing treatment leads to the neutralization of these charges, inducing a slight decrease of the CPD (~10 mV). However, depending on the etching process, the properties of the etched regions have to certainly differ. In terms of material, the difference must not to be only due to the presence of charges on the sample surface, but some crystal defects (dislocations), substitution between Pb and Ga elements [29], and implantation of metallic $Ga^+$ were probably maintained on crystallized etched film but not in the amorphous etched sample, owing the annealing treatment. To highlight this result, the absolute surface potential value across the etching areas is depicted in Fig. 17.8. We remark a very small potential difference (~15 mV) between the un-etched surface and the amorphous etched region, while a large difference (about 300 mV) is measured between the un-etched and conventional FIB-etched surfaces. Consequently, a clear modification of the surface potential is observed after etching, especially for the region exposed to conventional FIB, i.e., on crystallized PZT film. These findings point out the surface after conventional FIB etching is dramatically changed and damaged, while the surface exposed to amorphous FIB process is

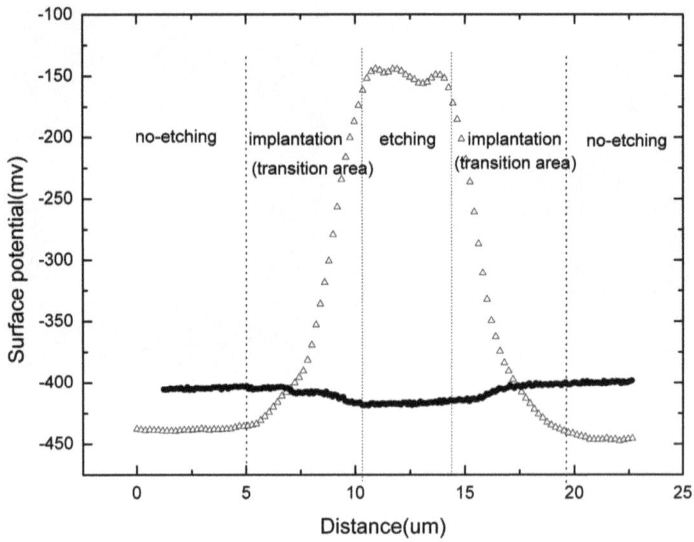


**Fig. 17.8** Absolute surface potential measured by KFM across the regions after hardest amorphous FIB etching (*filled square*) and hardest conventional etching process (*open triangle*). Figure reprinted with permission from [25]. Copyright 2009, American Institute of Physics

almost unchanged, which is corresponding to the above Raman results. As a result, it is clearly observed that such amorphous etching can reduce the etching damage on the surface. We can relate these results to the three following reasons. First, this method can reduce the surface roughness induced by etching. In previous work, the grain boundaries were found to be favorably etched because of the lead excess at these boundaries and the etching rate of lead oxide is higher than titanium or zirconium oxides [50]. Moreover, lateral etching which leads to the grain boundary widening can add to the etching process. The excess of physical atom bombardment can occur at the grain boundary, which results in the quick etching rate at this local area, and then increases the roughness after etching. On the other hand, the etching rate about the crystalline structure is anisotropic. Indeed, the effect of Ga ion bombardment depends on the atom density. Different orientations for crystalline planes induce several atom densities; therefore the etching rate is different and the roughness increases after etching. In the case of amorphous etching, there is neither grain boundary nor crystalline structure, which allows to consider the etching rate as isotropic and thus the roughness will be reduced. In our experiments, the roughness of the ion-exposed regions is about 1.2 and 5 nm after amorphous and conventional FIB etching, respectively. This roughness reduction can improve the electrical properties of PZT film. Second, for amorphous etching process, the damaged layer on amorphous PZT film surface induced by Ga ion bombardment can be almost crystallized into the ferroelectric perovskite phase by the post-annealing treatment. This leads to a minor degradation at the surface, which is shown in the above Raman and KFM results. Finally, there is no ferroelectric domain in the amorphous PZT film, and it is well known that the charged defects at the film surface induced by the ion

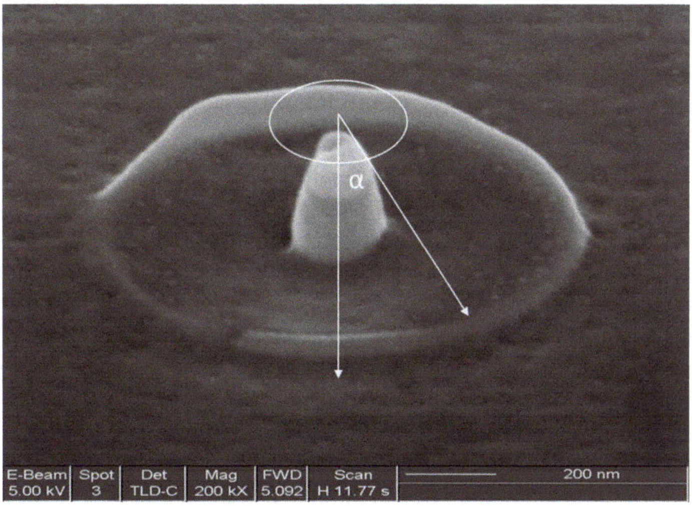


**Fig. 17.9** SEM image of a PZT nanoisland of 50 nm in diameter fabricated by $Ga^+$ FIB etching. The angle α is related to the etching selectivity between Pt and PZT compounds. Figure reprinted with permission from [26]. Copyright 2010, American Institute of Physics

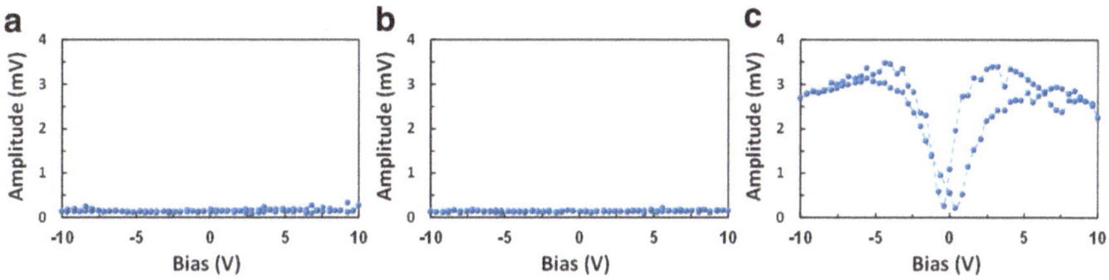


**Fig. 17.10** Local amplitude piezoloops measured on PZT islands of 50 nm in diameter after (**a**) conventional etching process, (**b**) post-annealing at 625 °C under $O_2$, and (**c**) amorphous FIB etching process (then crystallized)

bombardment can limit the domain mobility and decrease the amplitude vibration (which will be proved by the results in Fig. 17.10). This effect will not exist in amorphous FIB etching case.

In this second part, we focused on the results and discussions about the fabrication of PZT nanostructures (named nanoislands). The typical MIM capacitor imaged by SEM is shown in Fig. 17.9, where the diameter is 50 nm. The profile of the islands is almost the same for the different etching process; thus only one nanostructure is presented here. The etching anisotropy observed (α angle ~45° in Fig. 17.9) is probably due to the selectivity between Pt and PZT compounds. It should be noted this effect could be a limitation of the FIB process and it seems difficult to scale down the island diameter with this technique. By considering the results discussed above and prior to pattern the film, a Pt top electrode used as a mask was sputter-deposited to protect the top of the island. In this case, only the few defects introduced into the sidewall layer of islands are taken into account. All the structures etched onto the Pt top electrode can be observed on the Fig. 17.2. Local piezoactivity was directly measured onto the etched PZT nanoislands by PFM. The comparative results of the PFM measurements on the islands of 50 nm of size are depicted in Fig. 17.10. We can observe that no electromechanical response is obtained when the islands are fabricated from crystallized PZT sample, even after a recovery treatment in $O_2$ atmosphere (we also used a Pt mask in this process), as seen in Fig. 17.10a, b, respectively. Instead, a strong piezoelectric response is still detected in the case of nanoislands patterned from amorphous film then post-annealed at the crystallization temperature (see Fig. 17.10c). Interesting is that the local piezoactivity (~3 mV) obtained for island as small as 50 nm is similar to values measured on 100 nm islands, but higher than values determined on larger nanocapacitors (~2 mV) whose diameter was ranging from 100 nm to 1 μm [17]. The notable enhancement of piezoactivity when the size of the PZT island decreases, and especially to 100 nm, could be due to a size effect owing the reduction of the substrate clamping effect. Another explanation would be the existence of parasitic effect due to "in air" electrostatic interaction between the AFM tip and the film surface. This effect is more pronounced when the PZT island diameter decreases [30]. From the achieved piezoloops, it can be seen that if FIB etching is performed before PZT crystallization, the nanoscale electrical properties of islands can be preserved compared to conventional etching process. We attribute this phenomenon to the possibility that the sidewall damages can be diluted if the PZT island is fabricated by amorphous FIB etching process. Indeed, charged defects on the amorphous PZT surface after etching may also be reduced with the thermal process, which would avoid the possible electrical breakdown effects on the sidewall. During the island etching, the implantation of $Ga^+$ ions is preferentially on the sidewalls of the nanostructure, but after annealing, gallium can migrate into the crystalline lattice of PZT island and fewer charged defects left on the sidewall. Hence, from these results and those obtained on etched continuous film, we demonstrate that the roughness reduction, the small degradation of the surface, and no ferroelectric domain etching can effectively reduce the sidewall effect and make the amorphous FIB etching process a very suitable and promising etching way for the fabrication of MIM nanostructures with no degradation in electrical properties. We can presume the post-annealing treatment (at temperature corresponding to the perovskite phase formation) following the etching process could suppressed the eventual small degradations induced is the amorphous one. Besides, as observed above, the signal measured on crystallized etched islands cannot be restored after usual post-annealing. These results contradict to those published by Stanishevsky et al. [29] where the acceleration voltage of $Ga^+$ ions was larger (50 kV) and thus it should induce stronger damages. To explain the fully destruction of the piezosignal, we performed C-AFM experiments by directly recording local current–voltage I(V) curves onto the similar 250 nm PZT islands fabricated by the conventional FIB process (etching on crystallized film). For these measurements, a DC bias was applied to the substrate holder and the Pt top surface was grounded. Typical I(V) curve measured is depicted in Fig. 17.11. The current rapidly increases with the applied voltage indicating a strong local conductivity related to the existence of an

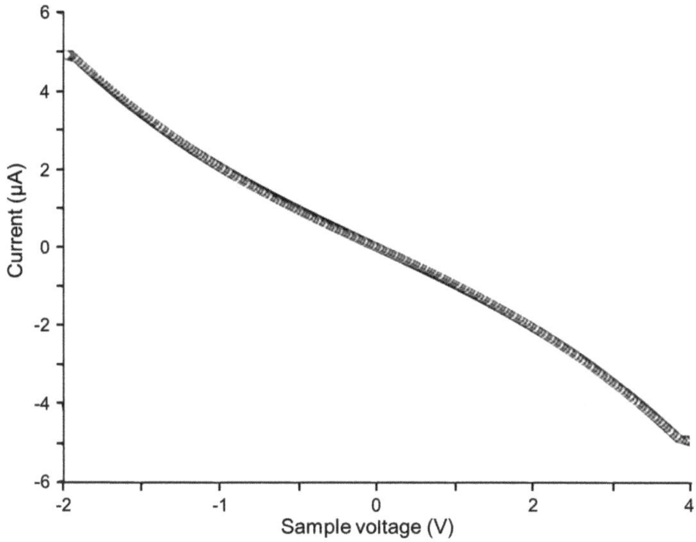


**Fig. 17.11** Local I(V) curve measured by C-AFM technique on crystallized etched PZT island with 250 nm in diameter. Figure reprinted with permission from [26]. Copyright 2010, American Institute of Physics

electrical conducting channel. This evidences the very strong damages due to $Ga^+$ ion implantation in the crystallized PZT nanoislands, in particular on the island walls, and can point out the associated loss of electromechanical behavior. In addition, due to this sidewall effect, we know the leakage currents are more important when the island diameter decreases to nanometer dimensions; thus higher leakage currents should exist for 50 nm islands, which could be considered for explaining the complete absence of piezoactivity for such small MIM nanocapacitors obtained by conventional $Ga^+$ FIB process. As a remark, we note the current density is in the order of maximum density of metallic material without degradation but depends on the joule heating. We believe it does not damage the sample because the characteristics were reproducible. Besides, if it is destructive with I(V) curves, then it should be also destructive under PFM loop experiments (samples are biased in the same manner without protection; in the case of C-AFM, the current is limited by the amplifier gain to 500 nA). This observed conductivity improvement is thus responsible of the absence of electromechanical activity previously observed on the etched zone of the crystallized film (see Fig. 17.4).

## 17.4 Conclusion

The damages induced by $Ga^+$-FIB technique on two kinds of nanostructured PZT films, i.e., in amorphous or crystallized state before etching, were evaluated in view of obtaining nanocapacitors. These damages are found to be very strong in the case of crystallized PZT films, and the piezoactivity is fully destroyed even after a post-annealing treatment. Both the implantation of $Ga^+$ ions and the film amorphization highlighted by Raman spectroscopy and KFM analyses can explain this result. However, when the etching is performed in the amorphous state, no significant degradation is observed. The latter process is thus used to successfully fabricate PZT-based nanocapacitors by means of FIB method. In 50-nm-size capacitors, the local electromechanical behavior is measured by PFM at similar level that the one obtained for the un-etched film evidencing no manifest sidewall effect. Consequently, this further confirms that amorphous FIB lithography process can reduce the etching damages, demonstrating this is a suitable alternative method and very beneficial to the development of the piezoelectric-based devices in the field of NEMS applications.

**Acknowledgments** This work was partially supported by the postdoctoral project provided by the French Minister for Research and New Technologies. The authors want to thank the Université de Valenciennes et du Hainaut Cambrésis and the Université d'Artois. The authors gratefully acknowledge Dr. J. F. Blach, Dr. R. H. Liang, Dr. C. Soyer, and Dr. S. Quignon for the collaboration in this work. The authors also wish to thank L. Maës for technical support.

## References

1. He, R.R., Feng, X.L., Roukes, M.L., et al.: Self-transducing silicon nanowire electromechanical systems at room temperature. Nano Lett. **8**, 1756–1761 (2008)
2. Bhaskaran, M., Sriram, S., Ruffell, S., et al.: Nanoscale characterization of energy generation from piezoelectric thin films. Adv. Funct. Mater. **21**, 2251–2257 (2011)
3. Wang, Z.L., Song, J.: Piezoelectric nanogenerators based on zinc oxide nanowire arrays. Science **312**, 242–246 (2006)
4. Trolier-McKinstry, S., Muralt, P.: Thin film piezoelectrics for MEMS. J. Electroceram. **12**, 7–17 (2004)
5. Eom, C.B., Trolier-McKinstry, S.: Thin-film piezoelectric MEMS. MRS Bull. **37**, 1007–1021 (2012)
6. Evans, P.R., Zhu, X., Baxter, P., et al.: Toward self-assembled ferroelectric random access memories: hard-wired switching capacitor arrays with almost Tb/in.$^2$ densities. Nano Lett. **7**, 1134–1137 (2007)
7. Xu, S., Hansen, B.J., Wang, Z.L.: Piezoelectric-nanowire-enabled power source for driving wireless microelectronics. Nat. Commun. **1**, 93 (2010)
8. Lee, W., Han, H., Lotnyk, A., et al.: Individually addressable epitaxial ferroelectric nanocapacitor arrays with near Tb inch$^{-2}$ density. Nat. Nanotechnol. **3**, 402–407 (2008)
9. Ahn, C.H., Rabe, K.M., Triscone, J.M.: Ferroelectricity at the nanoscale: local polarization in oxide thin films and heterostructures. Science **303**, 488–491 (2004)
10. Gruverman, A., Kholkin, A.: Nanoscale ferroelectrics: processing, characterization and future trends. Rep. Prog. Phys. **69**, 2443–2474 (2006)
11. Han, H., Kim, Y., Alexe, M., et al.: Nanostructured ferroelectrics: fabrication and structure-property relations. Adv. Mater. **23**, 4599–4613 (2011)
12. Nagarajan, V., Roytburd, A., Stanishevsky, A., et al.: Dynamics of ferroelastic domains in ferroelectric thin films. Nat. Mater. **2**, 43–47 (2003)
13. Scott, J.F.: Dimensional effects on ferroelectrics: ultra-thin single crystals, nanotubes, nanorods, and nano-ribbons. Ferroelectr **316**, 13–21 (2005)
14. Morozovska, A.N., Eliseev, E.A., Glinchuk, M.D.: Ferroelectricity enhancement in confined nanorods: direct variational method. Phys. Rev. B **73**, 214106 (2006)

15. Mancha, S.: Chemical etching of thin-film PLZT. Ferroelectr **135**, 131–137 (1992)
16. Kawagughi, T., Adachi, H., Setsune, K., et al.: PLZT thin-film waveguides. Appl. Opt. **23**, 2187–2191 (1984)
17. Legrand, C., Da Costa, A., Desfeux, R., et al.: Piezoelectric evaluation of ion beam etched Pb(Zr, Ti)$O_3$ thin films by piezoresponse force microscopy. Appl. Surf. Sci. **253**, 4942–4946 (2007)
18. Saito, K., Choi, J.H., Fukuda, T., et al.: Reactive ion etching of sputtered $PbZr_{1-x}Ti_xO_3$ thin-films. Jpn. J. Appl. Phys. Pt. 2 **31**, L1260–L1262 (1992)
19. Blach-Legrand, C., Saitzek, S., Da Costa, A., et al.: Comparative analysis for the local piezoelectric properties of ion beam and reactive ion beam etched Pb(Zr, Ti)$O_3$ thin films. Integr. Ferroelectr. **98**, 230–240 (2008)
20. Yokoyama, S., Ito, Y., Ishihara, K., et al.: High-temperature etching of PZT/PT/TIN structure by high-density ECR plasma. Jpn. J. Appl. Phys. Pt. 1 **34**, 767–770 (1995)
21. Chung, C.W., Kim, C.J.: Etching effects on ferroelectric capacitors with multilayered electrodes. Jpn. J. Appl. Phys. Pt. 1 **36**, 2747–2753 (1997)
22. Ganpule, C.S., Stanishevsky, A., Su, Q., et al.: Scaling of ferroelectric properties in thin films. Appl. Phys. Lett. **75**, 409–411 (1999)
23. Ganpule, C.S., Stanishevsky, A., Aggarwal, S., et al.: Scaling of ferroelectric and piezoelectric properties in Pt/$SrBi_2Ta_2O_9$/Pt thin films. Appl. Phys. Lett. **75**, 3874–3876 (1999)
24. Schilling, A., Byrne, D., Catalan, G., et al.: Domains in ferroelectric nanodots. Nano Lett. **9**, 3359–3364 (2009)
25. Liang, R.H., Rémiens, D., Deresmes, D., et al.: Enhancement in nanoscale electrical properties of lead zirconic titanate island fabricated by focused ion beam. J. Appl. Phys. **105**, 044101–044107 (2009)
26. Rémiens, D., Liang, R.H., Soyer, C., et al.: Analysis of the degradation induced by focused ion $Ga^{3+}$ beam for the realization of piezoelectric nanostructures. J. Appl. Phys. **108**:042008–6 (2010).
27. Roytburd, A.L., Alpay, S.P., Nagarajan, V., et al.: Measurement of internal stresses via the polarization in epitaxial ferroelectric films. Phys. Rev. Lett. **85**, 190–193 (2000)
28. Li, J.H., Chen, L., Nagarajan, V., et al.: Finite element modeling of piezoresponse in nanostructured ferroelectric films. Appl. Phys. Lett. **84**, 2626–2628 (2004)
29. Stanishevsky, A., Nagaraj, B., Melngailis, J., et al.: Radiation damage and its recovery in focused ion beam fabricated ferroelectric capacitors. J. Appl. Phys. **92**, 3275–3278 (2002)
30. Tiedke, S., Schmitz, T., Prume, K., et al.: Direct hysteresis measurements of single nanosized ferroelectric capacitors contacted with an atomic force microscope. Appl. Phys. Lett. **79**, 3678–3680 (2001)
31. Schilling, A., Adams, T., Bowman, R.M., et al.: Strategies for gallium removal after focused ion beam patterning of ferroelectric oxide nanostructures. Nanotechnology **18**, 035301 (2007)
32. Hambe, M., Wicks, S., Gregg, J.M., et al.: Creation of damage-free ferroelectric nanostructures via focused ion beam milling. Nanotechnology **19**, 175302 (2008)
33. Morelli, A., Johann, F., Schammelt, N., et al.: Ferroelectric nanostructures fabricated by focused-ion-beam milling in epitaxial $BiFeO_3$ thin films. Nanotechnology **22**, 265303 (2011)
34. Rémiens, D., Soyer, C., Troadec, D., et al.: Integration and optimisation of PZT piezoelectric thin films in micro and nano dimensional structures. Micro. Nanosyst. **1**, 214–225 (2009)
35. Gatoux, A., Ferri, A., Detalle, M., et al.: Characterizing nanoscale electromechanical fatigue in $Pb(Mg_{1/3}Nb_{2/3})O_3$-$PbTiO_3$ thin films by piezoresponse force microscopy. Thin Solid Films **520**, 591–594 (2011)
36. Detalle, M., Wang, G., Rémiens, D., et al.: Comparison of structural and electrical properties of PMN-PT films deposited on Si with different bottom electrodes. J. Cryst. Growth **305**, 137–143 (2007)
37. Herdier, R., Detalle, M., Jenkins, D., et al.: Piezoelectric thin films for MEMS applications-A comparative study of PZT, 0.7PMN-0.3PT and 0.9PMN-0.1PT thin films grown on Si by r.f. magnetron sputtering. Sens. Actuators. A. Phys. **148**, 122–128 (2008)
38. Gruverman, A., Auciello, O., Tokumoto, H.: Imaging and control of domain structures in ferroelectric thin films via scanning force microscopy. Annu. Rev. Mater. Sci. **28**, 101–123 (1998)

39. Alexe, M., Gruverman, A. (eds.): Nanoscale Characterisation of Ferroelectric Materials – Scanning Probe Microscopy Approach. Springer, Berlin (2004)
40. Balke, N., Bidkin, N., Kalinin, S.V., et al.: Electromechanical imaging and spectroscopy of ferroelectric and piezoelectric materials: state of the art and prospects for the future. J. Am. Ceram. Soc. **92**, 1629–1647 (2009)
41. Soergel, E.: Piezoresponse force microscopy (PFM). J. Phys. D: App. Phys. **44**, 46400–46417 (2011)
42. Ferri, A., Da Costa, A., Desfeux, R., et al.: Nanoscale investigations of electrical properties in relaxor $Pb(Mg_{1/3}Nb_{2/3})O_3$-$PbTiO_3$ thin films deposited on platinum and $LaNiO_3$ electrodes by means of local piezoelectric response. Integr. Ferroelectr. **91**, 80–96 (2007)
43. Desfeux, R., Ferri, A., Legrand, C., et al.: Nanoscale investigations of switching properties and piezoelectric activity in ferroelectric thin films using piezoresponse force microscopy. Int. J. Nanotechnol. **5**, 827–837 (2008)
44. Ziegler, J.F. (ed.): The Stopping and Range of Ions in Solids. Pergamon, New York (1985)
45. Meng, J.F., Katiyar, R.S., Zou, G.T., et al.: Raman phonon modes and ferroelectric phase transitions in nanocrystalline lead zirconate titanate. Phys. Status Solidi A **164**, 851–862 (1997)
46. Lou, X., Hu, X., Zhang, M., et al.: Phase separation in lead zirconate titanate and bismuth titanate during electrical shorting and fatigue. J. Appl. Phys. **99**, 044101–044107 (2006)
47. Frantti, J., Lantto, V., Nishio, S., et al.: Effect of A- and B-cation substitutions on the phase stability of $PbTiO_3$ ceramics. Phys. Rev. B **59**, 12–15 (1999)
48. Nomura, K., Takeda, Y., Maeda, M., et al.: In situ observation of the crystallization process of ferroelectric thin films by Raman microspectroscopy. Jpn. J. Appl. Phys. Pt. 1 **39**, 5247–5251 (2000)
49. Kang, M.G., Kim, K.T., Kim, C.I.: Plasma-induced damage in PZT thin films etched by inductively coupled plasma. Thin Solid Films **435**, 222–226 (2003)
50. Soyer, C., Cattan, E., Rémiens, D.: Ion beam etching of lead–zirconate–titanate thin films: correlation between etching parameters and electrical properties evolution. J. Appl. Phys. **92**, 1048–1055 (2002)

# Chapter 18
# Instabilities in Focused Ion-Beam-Patterned Nanostructures

**A.K. Raychaudhuri**

**Abstract** Fabrication and rapid prototyping of nanolines for different nanoelectronics or NEMS applications using FIB is an important science and technology issue. The nanolines due to large surface energy are susceptible to liquid-like instabilities. The process of nanofabrication by FIB leads to modifications of these instabilities. These instabilities arise due to fundamental reasons and can break up nanowires/nanolines when the lateral dimension (diameter/width) typically falls below 40–50 nm. In this chapter the physics concepts that can lead to the instabilities have been discussed. The experimental observations of the instabilities in metal nanowires/nanolines produced using FIB etch on a variety of substrates have been discussed to bring out such effects as the effect of ion-beam-induced roughening, de-wetting, charging and adhesion of the films to the substrates. We show that such instabilities also occur in FIB patterned nanolines of complex oxides. The presentation done in this chapter is a review of the state of art of the activity although not all inclusive or extensive.

## 18.1 Introduction

One of the central themes in modern-day fabrication is miniaturization down to submicron length scales. Focussed ion beam (FIB) provides a versatile enabling tool for top–down fabrications down to submicron levels and often down to even few nanometre scales [1]. Use of FIB as a tool for nanofabrication has a number of important physical attributes; foremost, however, is the process of sputtering by the incident ion beam that leads to size reduction. Though the process of sputtering is the

A.K. Raychaudhuri (✉)
Theme Unit of Excellence in Nanodevice Technology, S.N. Bose National Centre for Basic Sciences, Salt Lake, Sector-III, Kolkata 700098, India
e-mail: arup@bose.res.in

Z.M. Wang (ed.), *FIB Nanostructures*, Lecture Notes in Nanoscale Science and Technology 20, DOI 10.1007/978-3-319-02874-3_18,

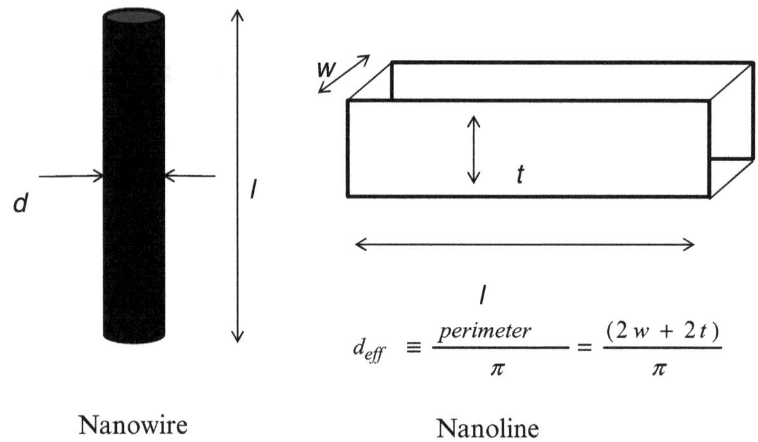


**Fig. 18.1** One-dimensional structures that have been investigated for instability formation

principal process, there are a number of other physical effects like roughening- and smoothening-induced processes, surface energy and capillary effects, de-wetting and local heating which start playing important role as the size is taken below 100 nm and particularly below 50 nm. Different aspects of the fabrication of nanostructure using FIB will be discussed in different chapters of the book.

In this chapter our focus is on nanostructures like nanowires and nanolines and instability that sets in these nanostructures on size reduction using ion beams. We will use the name nanowires and nanoline interchangeably to mean long objects with narrow widths and will refer to them generally as NW. There is, however, a clear distinction that one would like to note (see Fig. 18.1). A nanowire is a long cylindrical object with well-defined diameter ($d$) and length ($l$) and $l >> d$ and the $d \leq 100$ nm. Nanowires are fabricated by bottom–up methods using vapour phase deposition or chemical methods including such process as electrochemical growth in nanoporous templates [2]. A nanowire can be self-supported. A nanoline is generally made by a top–down method using one or more methods of nanolithography and is supported on a substrate (although one can have short segments of self-supported nanolines). For instance, electron beam lithography with vapour phase deposition and lift-off is a popular method for fabricating a nanoline in sub-100 nm regime. The method of fabrication leads to a rectangular cross-section defined by a width ($w$) and thickness ($t$) where $w \sim t$ and length $l >> w, t$. To define a single scale for lateral dimension as an effective diameter ($d_{\text{eff}}$), one can use the definition $d_{eff} \equiv \frac{perimeter}{\pi} = \frac{(2w + 2t)}{\pi}$ (see Fig. 18.1). In the rest of the chapter, we will treat $d$ and $d_{\text{eff}}$ at par for an NW and use $d$ to mean both diameter of a nanowire and effective diameter of a nanoline.

In this chapter our focus is on these instabilities that become prominent in NWs with small diameters ($d < 50$ nm) and lengths ($l$) that are much larger than the diameter ($l >> d$). The surface energy-driven instability (referred to as Rayleigh–Plateau instability [3, 4]) can occur spontaneously even in the absence of ion beam, when the diameter of the NW is taken below a critical size $d_c$. However, in the presence of ion irradiation (particularly low-energy ion irradiation) where the reduction in size is due to the sputtering process, other competing interactions come into play and contribute to the instability process. This is most prevalent in metal NWs and has been extensively studied in them as discussed below. However, they also can occur in insulating materials like oxides, which make the occurrences of the instability a general phenomenon. For applications, the NW needs to be supported on a substrate. This brings into considerations factors related to the surfaces that are technologically important like Si (with native oxide or treated with buffered HF), $SiO_2$/Si ($SiO_2$ grown on Si with typical thickness of few hundreds of nm) and $Si_3N_4$.

The study of the instability that occurs in ion-beam-processed NW is important not only from the view point of fundamental knowledge, but it is an important phenomenon that can limit applications of particularly metallic NWs of small diameter in nanoelectronics. In order to use metallic NW in these applications, for example, as interconnects, they must have well-defined stable morphology and well-defined geometries. Si microfabrication technology is used for making NEMS/MEMS devices. Study of stability of NW on substrates of Si or related materials is thus an important knowledge area to develop methodology for producing circuits for nanoelectronic applications or nanodevice fabrications.

The chapter has six sections including the current section. In Sect. 2 we discuss some of the fundamental physics issues that give rise to/control the process of formation of instabilities in NWs. This will be a background information that is needed for the rest of the chapter. In Sects. 3 and 4, we discuss some of the reported observations in this area that elucidate the phenomena. While in Sect. 3 we discuss observation of liquid-like instability in the absence of ion beams, in Sect. 4 we discuss the observation of the instability in the presence of ion beams. This is followed by a discussion of effects of substrate in such instabilities to highlight such issues as adhesion or charging. Section 6 gives a brief observation of the effects as seen in complex oxides.

## 18.2 Background Information

In this section we discuss some of the basic physics that we need to understand the instabilities in FIB fabricated nanostructures. To be specific, we will focus on nanoline or nanowire structures. Some of the ideas below have been developed for 2-dimensional structures like films. It is possible to extrapolate the ideas developed to nanolines. Working on nanowires-/nanoline-type structures can make the mathematics simpler in selected cases and one can achieve quantitative understanding.

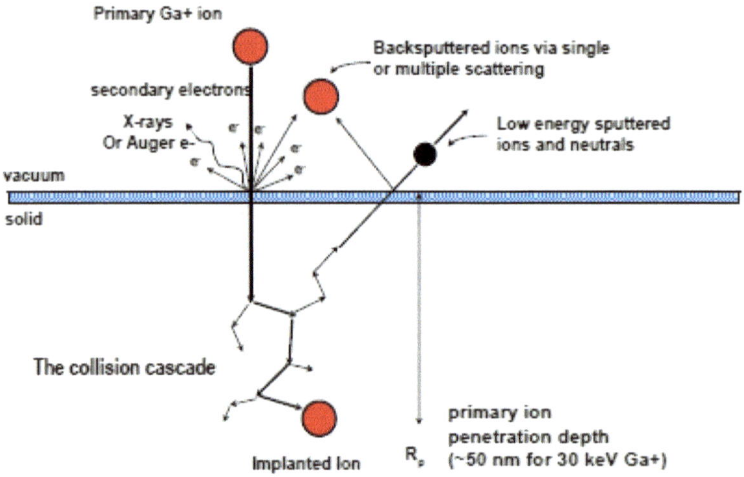


**Fig. 18.2** A cartoon of various processes that occur due to ion–material interaction

The instabilities that are observed in the NWs have some important ingredients. One of the basic causes for instability in NWs arises from the predominant contribution of surface energy or the surface tension that gives rise to liquid-like instabilities often referred to as Rayleigh–Plateau instabilities [3, 4]. This instability is sustained by surface diffusion and thus has a temperature dependence. In the presence of ion bombardment, the primary effect is sputtering which adds momentum to the atoms of the nanostructures and leads to roughening [5]. (In Fig. 18.2 we show a simple schematic that highlights different processes that take place when high-energy heavy ions impinge on a solid.)

The initiation of roughening by radiation introduces additional length scales for instabilities. When they are present together, they can act in tandem, and in most practical situations, both the liquid-like Rayleigh–Plateau instability and the radiation-induced effects are present although one of them may dominate. At elevated temperatures or due to localized heating produced by ions, de-wetting of the line structure can occur which can modify the instabilities [6]. (For de-wetting see the review in Ref. [7].) In the following subsections, we provide simple descriptions of the above effects. For details the original papers can be referred. The scope of our discussion is not to be exhaustive but to provide adequate background that would help us in various observations.

The process of creating a nanoline or similar nanostructures by downsizing larger structures (top–down) using FIB (which we term as “nanopatterning”) is an involved process. The subsections below bring out these complexities. The main purpose of the chapter is to make the reader aware of the complexities and the various factors that determine the processes and how effectively they can be controlled.

### 18.2.1 Rayleigh–Plateau Liquid-Like Instabilities

When the size is reduced to below 100 nm in number of nanomaterials, due to the predominance of surface energy, instabilities of different types can set in. For instance, it may lead to an instability in crystal structure in closed packed metallic system between structures that are energetically similar (e.g., transition between hcp and fcc structures [8]). In some systems there can be spontaneous shape transition or faceting [9]. These instabilities are "solid like" that they occur only in solid systems. However, when the force due to surface energy becomes comparable or more than the Yield strength, the solid can break up leading to surface energy-dominated "liquid-like" instabilities.

A cylindrical object is intrinsically unstable towards liquid-like instability when the surface energy dominates over the Yield strength and the system breaks down to nearly spherical droplets. This has been initially proposed by Plateau and elaborated later on by Rayleigh in case of a liquid cylinder [3, 4]. Rayleigh considered instabilities due to surface tension and explained the modulation of its free surface when the modulation preserves the volume.

The modulation in the form of oscillations in the diameter of the liquid cylinder ($d$) (radius $r$) occurs when the modulation wavelength ($\lambda$) is larger than the circumference of the cylinder ($\lambda > \pi d$). The driving force for this surface modulation is derived from the decrease of the surface area which makes the modulation energetically favourable. The modulation (or the instability) can lead to break-up of the liquid cylinder when the diameter is below some critical value ($d_c$).

The above ideas can be extended even for a solid cylinder [10]. We can make the idea quantitative with a simple picture that will explain the existence of a critical diameter ($d_c$) [11]. For a wire of length $l$ and diameter $d$ and made from a material with Yield strength $\sigma_Y$, the maximum tension it can sustain before the tension breaks the cylinder is $F_Y = \frac{\pi}{4} d^2 \sigma_Y$. This balances the force due to surface tension $F_s = -\frac{\pi}{2} d \sigma_s$. When the surface tension wins over ($|F_s| > F_Y$), the material undergoes plastic flow (like a liquid), and if the length $l \geq \lambda > \pi d$, the Rayleigh instability will set in leading to breakdown of the solid wire like a liquid cylinder. From the relation above, a critical diameter $d_c^R$ can be defined so that when the diameter falls below this value, the Rayleigh instability is expected to set in. The critical diameter is given as

$$d_c^R \equiv \frac{2\sigma_s}{\sigma_Y} \tag{18.1}$$

Generally in a solid $\sigma_Y$ is larger than that of a liquid by few orders (nearly 6–7 orders) of magnitude. As a result in a solid the instability occurs for $d_c^R$ typically in the range of few tens to few nanometre range. In FCC materials like Cu, Ag, or Au (which are the materials from which NW are often made), $\sigma_Y \approx 1 - 1.5$ N/m and their Yield strength $\sigma_Y \approx 100 - 200$ MPa. The critical $d_c^R$ below which $|F_s| > F_Y$ | will be $\approx$ 1–30 nm. Thus a metallic NW with its diameter approaching this range will

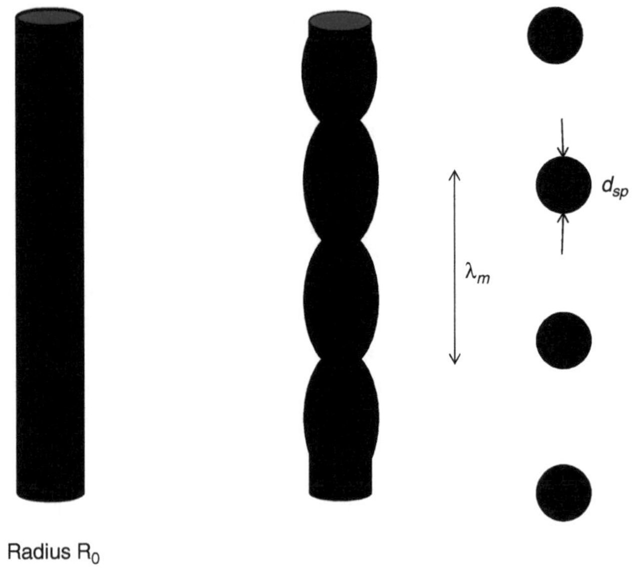


**Fig. 18.3** A cartoon of Rayleigh–Plateau instability leading to breakdown of a cylinder into spheres

show Rayleigh–Plateau instability. This analysis implies that for metal nanowires with lateral dimension of a few tens of nm, liquid instabilities should occur. In Fig. 18.3 we show a simple schematic of a cylinder that develops a modulation of a given wavelength leading to breakdown into spherical droplets as a result of Rayleigh–Plateau instability.

Nichols and Mullins [10] have extended Rayleigh's perturbation approach to solids. Consider a cylinder of length and uniform initial radius $R_0 \equiv \frac{d}{2}$ subjected to a varicose (volume preserving) perturbation of wave vector $k \equiv \frac{2\pi}{\lambda}$, ($\lambda$ is the wavelength). The radius $R(y,t)$ of the wire as a function of $y$ along its axis can be expressed as

$$R(y,t) = R_0\left[1 + b\cos(ky)\right]\exp(\omega t) \tag{18.2}$$

where $b \equiv \frac{\delta R}{R_0} << 1$ is the fractional change in the radius due to the perturbation $\delta R$ and the last term is the time-dependent term that describes the temporal growth of different modes of perturbation (see Fig. 18.4 for a schematic). For cylindrical nanowires of length $l$ undergoing volume conserving modulation of the radius, the fractional change in radius $b$ and the wave number of modulation $k$ are not independent quantities and are related by the relation:

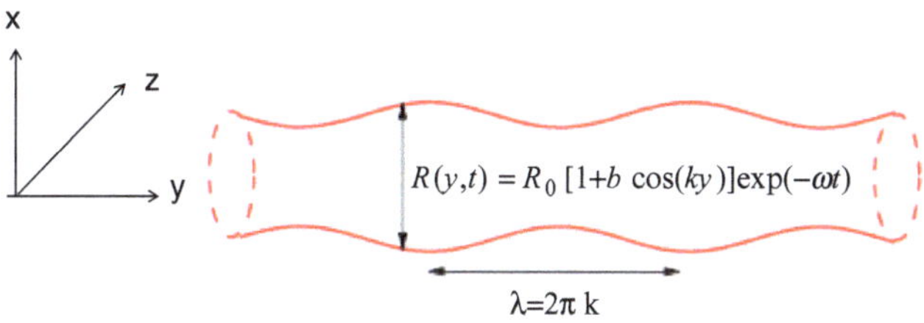


**Fig. 18.4** A cartoon of the perturbation of the radius at the onset of liquid-like instability

$$b = -\frac{8\sin(kl)}{2kl + \sin(2kl)} \tag{18.3}$$

Perturbations as stated earlier will have a kinetics of growth (or decay), and for wavelengths that lead to instability, they should grow in time. Perturbations with wavelength $\lambda > \pi d$ are expected to increase spontaneously in amplitude becoming more pronounced with time [11]. In particular, those perturbations with $\lambda_m \approx 9R_0$ would grow fastest. These perturbations would eventually cause the solid cylinder to break up into a row of spheres with an average spacing $\lambda_m$ and diameter $d_{sp} \approx 3.78R_0$.

The dispersion relation between $k$ and $\omega$ is given as [12]

$$\omega^2 = \frac{\sigma_s(kR_0)}{\rho R_0^3} \frac{I_1(kR_0)}{I_0(kR_0)} \left[1 - k^2 R_0^2\right] \tag{18.4}$$

The dispersion relation as given by (18.4) is plotted in Fig. 18.5. $\omega > 0$ for $kR_0 < 1$. When $\omega > 0$ and large for that mode, the radius quickly decays to zero, a node develops, and this acts as the point where the cylinder breaks down. From Fig. 18.5 we can also see that the fastest growing mode that occurs at the wave vector is given by $k_m = \frac{0.693}{R_0}$ which corresponds to $\lambda_m \approx 9R_0$, the wavelength which defines the spacing of the spheres when the cylinder breaks down into spheres (see Fig. 18.3).

While classical considerations apply for the onset of Rayleigh–Plateau instability, there are important contributions that can occur due to quantum effects from the conduction electrons in metallic nanowires. It has been observed that metallic nanowires of smaller dimensions with diameter $< d_c^R$ can be made stable [13]. In addition to the surface tension, the electron confinement in nanowires of very small diameter gives rise to quantum-size effects which become more important as wire diameter is reduced closer to the atomic scales [11]. In fact the liquid-like instability can be completely suppressed at some of the magic radii of the nanowire. This arises due to “shell effects” which is an electronic effect in cylindrical nanowires similar to electronic shells in atoms. Competition between surface tension and electron-shell effects produces a complex pattern of stability, with regions of stability for certain “magic” diameter values. These magic sizes are defined by the scale

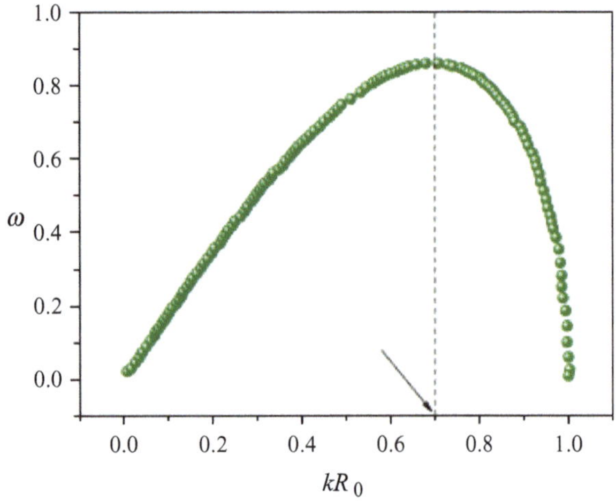


**Fig. 18.5** Dependence of the rate of growth of the perturbation on the wave vector for a varicose perturbation. Courtesy Dr. Avik Bid

of the Fermi wave vector $k_F$ and it occurs for certain values of the product $k_F R_0$. Since for metallic systems typical $k_F^{-1} \sim 0.1$ nm and the shell effects are predicted for $k_F R_0 \leq 25$, these effects will arise when the value of $R_0$ reaches few nm.

The above scenario is valid for a self-standing nanowire in the absence of ion beams. For a supported nanowire where the effect of the substrate can alter the surface tension, additional considerations apply. The Rayleigh criterion should be valid for all wire sizes, but due to kinetic limitations, its effects have not yet been observed for macroscopic objects.

### *18.2.2 Effects of Ion Beam*

The presence of ion beam, particularly those with energy range that can cause sputtering, can lead to significant modification to the instabilities and in some parameter space can even dominate. The interaction of the ion beam with a material depends on its energy. In the energy regime the FIB instruments work (beam energy < 100 keV), the primary effect of ion beams is to sputter away the material and thus induce roughening that can grow with time. In this range, the heavy ions that are used in FIB machines have a stopping distance $R_P < 100$ nm. In Fig. 18.6 we show the stopping range of $Ga^+$ ions of energy 30 keV, as it impinges on Au film on Si, obtained by using SRIM code. This is a structure (Au on Si) that is relevant for this chapter that we will discuss later on. It can be seen that at this energy, the ions stop within the film. This happens when the range $R_P <$ film thickness. For higher energy and/or thinner films when $R_P >$ film thickness, the ions will cross the film and will stop inside the substrate below.

The sputtering of the material also breaks down an important condition for the Rayleigh instability that the volume is conserved. There are other effects like

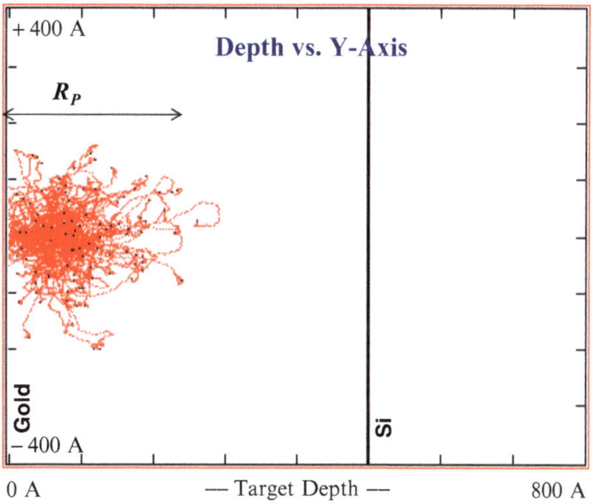


**Fig. 18.6** SRIM code-generated simulation of Ga ion (30 keV) range in Au film on Si substrate. In the example shown, the range ~20 nm is less than the film thickness of 50 nm (Data courtesy Dr. K. Das)

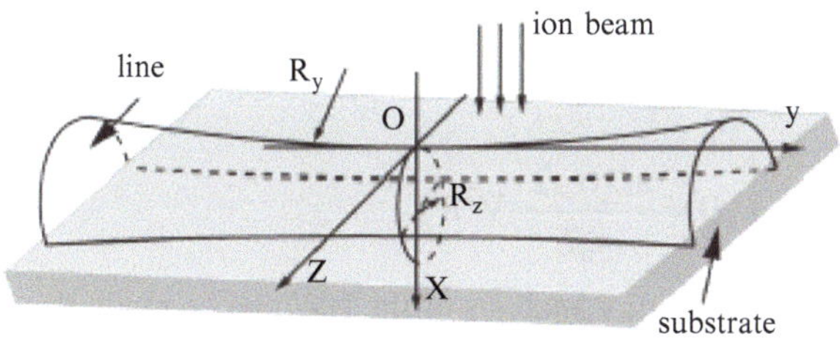


**Fig. 18.7** Schematic of the model used. Taken from Ref. [6] (Reproduced with permission from AIP)

de-wetting which may arise due to localized heating from ion beams. We discuss these specific effects in the next subsection.

An effective model for the simultaneous effect if ion beams and Rayleigh instability has been worked out recently [6]. We describe the model briefly below. In this model, sputtering has been treated in the framework of Sigmund theory [5]. When the ions in the ion beam penetrate the material, it spreads its kinetic energy to the neighbouring sites. This leads to ejection of material whose rate (yield) of ejection from a point is proportional to the energy accumulated there by all ions. In Sigmund theory of sputtering, the distribution of kinetic energy is assumed to be a Gaussian. A schematic of the geometry used for the model is shown in Fig. 18.7.

In Sigmund theory the sputtering rate depends on the local curvature. The erosion rate of the radius is given by the equation [6]

$$\frac{\partial R_z}{\partial t} = -v_0 + \frac{\eta}{R} \tag{18.5}$$

where $1/\langle R\rangle = (1/R_z - 1/R_y)$ is the mean surface curvature and the two terms are given as

$$v_0 = \frac{\Omega\Lambda_{target}\varepsilon_{ion}F}{\sqrt{2\pi}\alpha_t}\exp\left(-\frac{a^2}{2\alpha_t^2}\right) \quad (18.6)$$

$$\eta = \frac{\Omega\Lambda_{target}\varepsilon_{ion}F\alpha_l^2}{\sqrt{2\pi}\alpha_t^3}\exp\left(-\frac{a^2}{2\alpha_t^2}\right) \quad (18.7)$$

$\Omega$ = atomic volume, $\Lambda_{target}$ = constant for target, $\varepsilon_{ion}$ = ion energy, $F$ = ion flux, $a$ = ion penetration depth and $\alpha_t$ and $\alpha_l$ are transverse and longitudinal cascade sizes, respectively.

The above equation shows the uniform rate of sputtering given by the first term that is modified by the curvature-dependent term. The above equation, in the presence of capillary effects (the surface energy contribution), is modified by an additional term that depends on double derivative of the curvature $\langle R \rangle^{-1}$ and the surface tension $\sigma_s$, surface diffusion coefficient $D_s$ and the number of diffusing atoms per unit surface area $n$.

The differential equation in the presence of both effects can be written as [6]

$$\frac{\partial R_z}{\partial t} = -v_0 + \frac{\eta}{R} + D\nabla^2 R^{-1} \quad (18.8)$$

where $D = \frac{nD_s\sigma_s\Omega^2}{k_B T}\nabla^2 R^{-1}$

The stability analysis of the NW can be made from the above differential equation. The perturbation in radius $\delta R$ will have a kinetics of growth given approximately by the first-order differential equation [6]:

$$\frac{1}{b}\frac{\partial b}{\partial t} \approx \left[R_0^{-2} - \left(m^2 R_0^{-2} + k^2\right)\right]\left[D\left(m^2 R_0^{-2} + k^2\right) - \eta\right] \quad (18.9)$$

The modes of the modulation are given by m which is a positive integer. From the above equation, it can be seen that the perturbation will grow and will contribute to the instability if the right-hand side of (18.9) is positive. This can happen if both the terms are positive or both negative. When both the terms are positive, $R_0^{-2} > \left(m^2 R_0^{-2} + k^2\right) > \frac{\eta}{D}$ so that $R_o < \sqrt{\frac{D}{\eta}}$; when both terms are negative, one obtains the opposite situation $R_o > \sqrt{\frac{D}{\eta}}$. Importantly for both the limits, the perturbation can grow leading to instability. Since $D \propto T^{-1}$ (see 18.8), even one may start from the situation $R_o > \sqrt{\frac{D}{\eta}}$, and by increasing the temperature crossover to a regime with $R_o < \sqrt{\frac{D}{\eta}}$. $\eta$, the radiation-related term is mostly independent of $T$. Similarly, for a fixed $T$, one can increase the ion energy ($\varepsilon_{ion}$) and/or the ion flux ($F$)

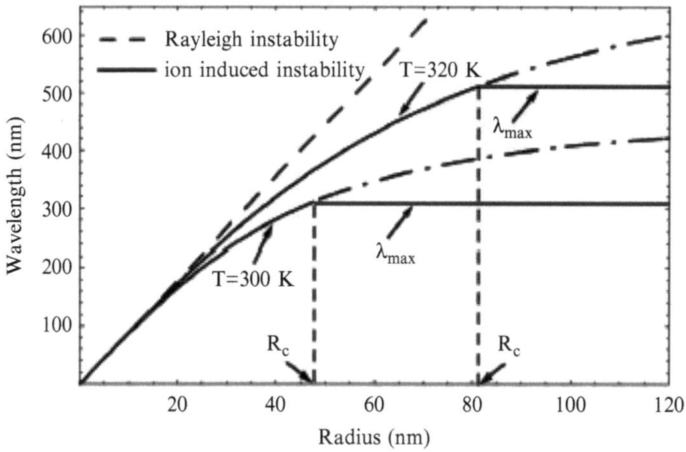


**Fig. 18.8** Dependence of the wavelength of the modulation in radius as a function of the radius of the wire (see 18.9). Taken from Ref. [6] (Reproduced with permission from AIP)

and start from a regime $R_o < \sqrt{\frac{D}{\eta}}$ and cross over to the regime $R_o > \sqrt{\frac{D}{\eta}}$ as the parameter $\eta$ increases on enhancing the ion-beam-related parameters. When the relation $R_o < \sqrt{\frac{D}{\eta}}$ is satisfied, the maximum wave length is given by

$$\lambda_m \approx \frac{8.89R_0}{\sqrt{1+\frac{\eta}{D}R_0^2}} \tag{18.10}$$

When the radiation is absent ($\eta \approx 0$) or its effect is small ($\eta R_0^2 << D$), $\lambda_m \rightarrow 8.89R_0$, which is the value obtained for capillary effect (Rayleigh instability) alone. When the nanowire breaks under capillary effect, $\lambda_m$ gives the spacing of the spheres that results from the instability of the wire and its eventual breakdown. For smaller $R_0$, $\lambda_m \propto R_0$ as is expected from Rayleigh instability. However, the ion bombardment inhibits the instability and the growth of the $\lambda_m$ and makes the dependence on $R_0$ sublinear (see Fig. 18.8).

### *18.2.3 Effects of Ion-Beam-Induced De-wetting*

The presence of an underlying substrate can make qualitative difference between a nanowire that is free standing (e.g., as those grown from templates by electrochemical growth) and a nanowire that is supported by a substrate. The differences will primarily come from the nature of the substrate whether it promotes wetting (by adhesion) or it leads to de-wetting. The other two effects that are important will be

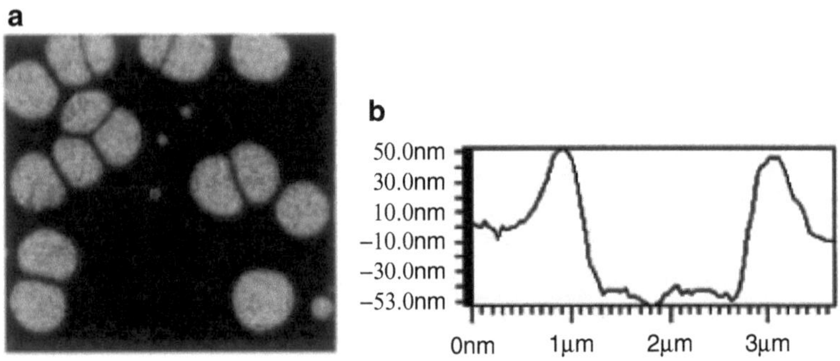


**Fig. 18.9** Panel (**a**) shows de-wetting of a film of Cu (thickness) on fused silica. Panel (**b**) shows a line scan taken by a scanning near-field optical microscope (Ref. [14]) (Reproduced with permission from APS)

heat conduction and also effect of charging that may come from charged ions as they deposit themselves within the body of the film. The effect of charging will be discussed in the next subsection. The local heating would lead to effects like local enhancement of the parameter $D$ leading to inhomogeneous parameter values and it can promote de-wetting, a factor that we would discuss in this subsection.

De-wetting is a surface-tension-driven phenomena [7], and for a metal film of surface tension $\sigma_s$ when it is heated, a de-wetting will occur if $\sigma_s > \frac{1}{2} w_{adh}$, where $w_{adh}$ is the work of adhesion. The de-wetting of metal films on substrates like glass/silica has been studied [14]. A typical example of de-wetting created by heating is shown in Fig. 18.9.

It has been found that this also can introduce its own length scale as a periodic pattern with wavelength $\lambda_{dewetting} \approx \frac{5.13t^2}{\zeta}$ [15], where $t$ is the thickness of the film and $\zeta$ is the characteristic of the material and depends on combination substrate and film. For a Au film on fused silica of thickness $t \approx 50$ nm, the measured $\lambda_{dewetting} \approx 2$ μm [15] and $\lambda_{dewetting}$ reduces to $\approx 0.4$ μm for $t \approx 20$ nm. It can be seen that for a real system, the length scale of de-wetting $\lambda_{dewetting}$ is generally comparable to but larger than $\lambda_m$. If the scale of instability in a metal film on a substrate leading to breakdown into droplets is determined by the smaller of the two scales, it is expected that $\lambda_m$ will determine the scale.

While the phenomena of de-wetting can be driven thermally at elevated temperatures, at lower temperatures ($T << T_m$), it can be induced by ion beam. The ion beam deposits energy within the solid which causes local heating (thermal spike model [16]) which then can lead to de-wetting. There is kinetics associated with this phenomenon. The time scales of thermal spike (heating and subsequent cooling) need to be faster than the time scales of de-wetting for the Rayleigh instability to grow.

Though not important at low keV or few tens of keV range, at higher energy ($\geq 1$ MeV), there are other factors, like strain in the film/strain relaxation due to ion beams, which become important [17]. These issues are not discussed here. However, the reader needs to be aware of these effects.

### 18.2.4 Effects of Charging

When film or a nanowire is exposed to ion beam, it can accumulate charges due to the charged state of the ions. In low-energy ion beams, the charged state is a singly charged state, but due to low energy, the ion stops within the bulk of the material. For instance, in Fig. 18.6, we show the simulation (using SRIM code) of $Ga^+$ ions of energy 30 keV in a film of Au on Si. It can be seen that the range of the ions is ~20 nm. If the films used have a thickness >20 nm, the ions will stop within the bulk of the film and will deposit its charge. The charge will relax to the substrate which in turn will relax it to the ground. If the film is in a substrate that is insulating, then there is a possibility that there will be accumulation of charge in the film. This can create a stress within film. The effect of this can be seen when the liquid-like instabilities set in.

When an electrically conducting liquid is charged electrically or is exposed to an electric field, the shape of the liquid deviates from its equilibrium shape determined by the surface energy. Briefly, when the liquid is static, there are two competing forces (surface tension and the electrostatic stress) that contribute to the stable shape of the liquid. The force on the droplet leads to formation of a cone called Taylor cone. If the two forces are balanced everywhere on the cone (the equilibrium shape), then one can write

$$\frac{1}{2}\varepsilon_0 E^2 = \frac{2\sigma_s}{R_{cone}} \tag{18.11}$$

where $E$ is the electric field on the surface due to the charges and $R_{\text{cone}}$ is the principal radius of the cone. The potential is constant on the surface, and since $E \propto R_{\text{cone}}^{-2}$, the surface potential has the form $V \propto R_{\text{cone}}^{-2} P_{1/2}\cos(\theta)$ where $P_{1/2}$ is the Legendre polynomial of order ½ and $\theta = \pi - \varphi/2$, $\varphi$ is the cone apex angle. The free surface becomes an equipotential when the Legendre polynomial of order ½ has its zero for $\theta = 130.7°$ corresponding to $\varphi/2 = 49.3°$ and with a very specific whole angle of the cone of 98.6°. In this situation the free surface of the liquid takes its equilibrium shape of a cone called the Taylor cone. Beyond a critical applied potential, this leads to liquid jet that carries material out from the liquid [18]. This is schematically shown in Fig. 18.10. Interestingly Taylor cone formation plays a deciding role in liquid metal ion sources like Ga sources in FIB machines. Recent experiments with FIB-produced metal lines clearly show that formation of a Taylor cone can indeed occur due to charging and the jet can carry material out of the film breaking the condition of volume conservation as required for Rayleigh instability.

Another way the charging effects can nucleate instability is the process of structural instability of localized kind initiated by the electrostatic force of the accumulated charge when it exceeds the Yield strength of the film. This is similar to the concept of "Coulomb explosion" that happens in the context of swift heavy ions in solids [19]. In this case local charging of a localized region creates enough potential energy gradient that can lead to the instability. We can make a simple model for this in the following way in an insulator:

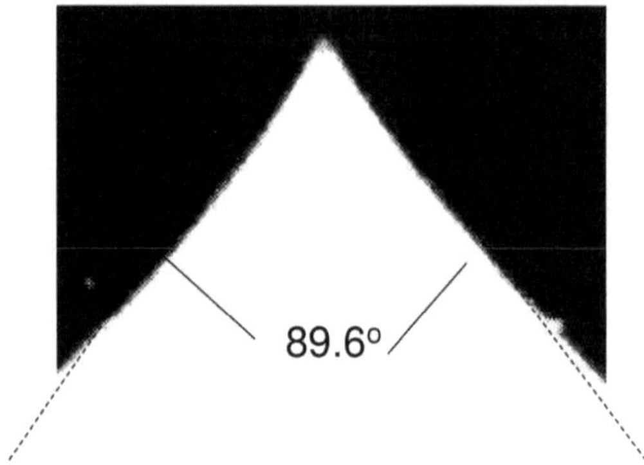


**Fig. 18.10** A schematic of formation of Taylor cone in a charged liquid. The cone apex angle is ~89.6°

Suppose there is accumulation of charges in a localized area in a material with average lattice spacing $a$. Suppose there are $N$ units of electronic charge ($e$) caused by ionization of atoms at the lattice site. This will give rise to an electrostatic stress $\sigma_{electro} \approx \dfrac{N^2 e^2}{4\pi\varepsilon_0 \varepsilon a^4}$, where $\varepsilon$ is the relative dielectric constant of the medium. Taking the approximate Yield strength $\sigma_Y \sim \dfrac{1}{10} Y$, we can obtain an estimate of $N$ for which such an instability can occur which is given as $N \approx \sqrt{\dfrac{4\pi\varepsilon_0 \varepsilon a^4 Y}{10 e^2}}$. This estimate is valid for an insulator, and in a metal nanoline where the charge can relax, this will not be an issue. However, in a case where the film thickness is smaller than the ion range, the ions will stop in the insulating substrate below. The charging-induced instability in the substrate then can be a spot where the instability can nucleate. This particular issue has not been addressed to in any form when nanolines are exposed to ion beams of relatively lower energy (few tens of keV).

## 18.3 Experimental Observation of Rayleigh Instability in Nanowires in the Absence of Ion Beams

Though the instability in the presence of ion beam is the main focus of the paper, it is worthwhile to consider the observation of the instability in the absence of ion beam. This is to mainly make us aware of the pure Rayleigh instability in metal nanowires (free standing) or metal nanolines on a substrate.

Liquid-like instabilities were studied in Ag nanobelts with thickness down to 10 nm [20]. Thermal stability of Cu nanowires (diameter ~30–50 nm) was studied in the temperature range of 400–600 °C [21]. It was found that the nanowires were unstable against Rayleigh instability even at a temperature as low as 400 °C which is much less than its melting point of Cu. The onset of Rayleigh instability at such

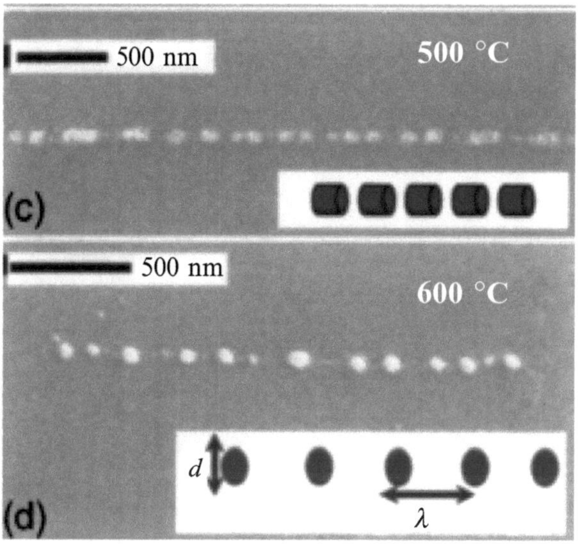


**Fig. 18.11** Breakdown of a nanoline on thermally induced instability. Taken from Ref. [21]. Reproduced with permission from AIP

a low temperature occurs due to onset of large-scale diffusion that is needed to sustain the Rayleigh instability that can occur at temperatures as low as $0.2-0.3T_{\mathrm{m}}$. The nanowires exhibit thermal stability above 50 nm and show diameter-dependent fragmentation below this value. An example of breakdown of an NW due to liquid-like instability is given in Fig. 18.11.

While nanowires of smaller diameter break down into spherical droplets due to the Rayleigh instability, wires of somewhat larger diameters often show formation of necks as can be seen in Fig. 18.12 for a 100 nm Ag nanowires grown in a template and exposed to electrons in a transmission electron microscope. Even nanowires constrained by templates show fluctuations in diameter leading to corresponding variations in local resistance. The resistance fluctuations occur due to fluctuations in radius in the area of cross-section of the wire due to the instability [22].

Morphological instabilities in Au nanowires (diameter ~25 nm) heated up to 500 °C were studied [23]. At this temperature range, the wires break up into spheres. A quantitative evaluation was made of the wavelength of the instability $\lambda_{\mathrm{m}}$ and the diameter spheres so formed $d_{\mathrm{sp}}$ and their dependence on the starting wire radius $R_0$. While both $\lambda_{\mathrm{m}}$ and $d_{\mathrm{sp}}$ have a linear dependence on $R_0$ as is expected, the observed values of both the parameters deviate from the theoretically predicted values (like $\lambda_{\mathrm{m}} \approx 9R_0$) and show larger values. While the presence of substrates and surface energy anisotropy due to crystallographic orientations has been the suggested reason, no clear cause for the deviation has been established.

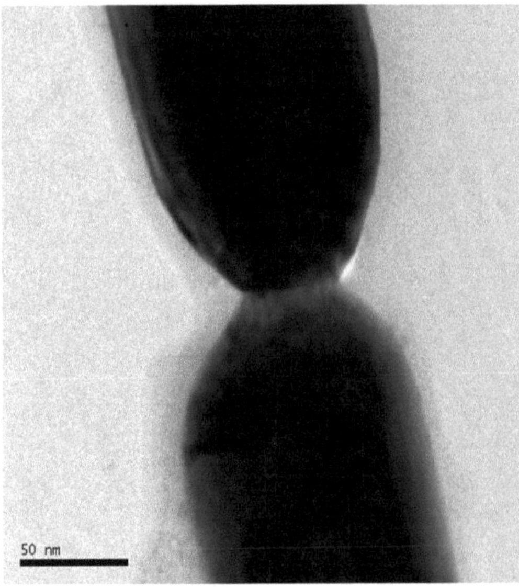


**Fig. 18.12** Formation of neck in a silver nanowire (diameter 100 nm) due to onset of instability (Data courtesy Dr. Avik Bid)

## 18.4 Instability in Metallic Nanowires Under Ion Beams

The investigations of instability in nanowires under ion-beam exposure have been carried out mostly using $Ga^+$ ions in a FIB machine. These ions have energy ≤30 keV. However, there are studies using $Ar^+$ or $Xe^+$ ions at lower energy ≤1 keV or $He^+$ ions with energy in the range of 1–2 MeV. The change of ion energy changes the depth of penetration as well as the sputter yield. The nature of the energy transfer from the ion to the nanowire/film/substrate changes as the ion energies increase. At lower energies the dominant mode of energy transfer is through nuclear collisions, and at higher energies (Energy >1 MeV) the transfer of energy occurs through electronic loss. In the following we present the reported studies on the ion-beam-induced instabilities. It can be seen that there is clear evidence of formation of Rayleigh-type liquid-like instabilities in ion-beam-irradiated metal lines. In addition, there are also other effects that are present as expected based on the background information in Sect. 2. The ion-beam-induced effects are often intimately connected to the phenomena of de-wetting which facilitates the formation of instability. The ion beams also bring forward the effect of charging which we discuss separately in the next section.

One of the first studies that used ion-beam-induced Rayleigh instability and de-wetting to pattern metallic nanostructures was done by Lian et al. [24]. They used vacuum-evaporated and FIB-written Co lines (initial width 200 nm) for ion-beam irradiation using $Ga^+$ ions at 30 keV and at current of 100 pA in a FIB machine. Si (100) was used as a substrate. Progressive ion-beam irradiation breaks the nanolines into regular spherical chains. In this case the ions were scanned over the whole surface of the film. Ion-beam sputtering over the film surface leads to thinning of the width, and when the width reaches ~50 nm, the instability occurs leading to breakdown of the lines. The wavelength of the modulation in the wire width (a quantitative measure of the instability) and the sphere diameter was found to be comparable to the prediction from Rayleigh instability.

When the ion-beam irradiation is performed on Co lines of widths that are larger (>400 nm), the ion-beam-induced sputtering leads to splitting of the lines at the middle (along the long axis) into two wires which on further progress of the sputtering breaks each of the parallel lines into spheres. The splitting of the wires with larger width into two parallel wires with smaller widths occurs due to ion-beam-induced de-wetting. The sputtering rate is curvature dependent. There is preferential sputtering at the centre of lines with larger width leading to initiation of de-wetting along the centre region of the line. The value of $d_{sp}$ in this experiment was found to be less than that what is expected from pure Rayleigh instability. The discrepancy was proposed to arise from sputtering.

Comprehensive investigations of morphological instability of Cu lines (with widths down to 100 nm and thickness in the vicinity of 100 nm)) exposed to Ga ions with energy ~10 keV were carried out [6] and compared with a theoretical model that includes both effect of ion beam and capillary effects. The films were made on Si (100) with native oxide. In this case also the film was exposed to uniform ion irradiation across the film surface. At the beginning of the ion treatment when the width is larger, the line width decreases linearly with sputtering time and there is no periodic pattern that forms. When the line width reaches a critical value, the modulation forms and eventually there is breakdown into spheres. The data are shown in Fig. 18.13.

The experiment clearly shows the existence of a critical diameter ($d_c$ < film thickness) so that when the line width falls below that diameter, the liquid-like instabilities set in and the line breaks into spherical particles. However, there is a clear distinction between the case of breakdown of lines into spheres in the case when the Rayleigh instability is initiated by thermal de-wetting alone and the case when the ion beam is used. In the first case the pattern formed is regular, while in the later case it is somewhat irregular with diameter of spheres thus formed showing large variability. This variability can be due to the randomness that is inherent in the ion-beam bombardment.

Ion-beam-induced de-wetting has been studied on nanolines of Au and Pt (width ~60–400 nm, thickness 8–25 nm) [25]. $Ga^+$ ions with 30 keV energy in a FIB machine have been used. They observe that the ion beam can induce directed de-wetting. The de-wetting process leads to formation of modulation of the width at the rim. For both Pt and Au, the instability sets in at around 35 nm for wider lines (~200 nm) and 22 nm for lines with less width (~85 nm). The formation of the de-wetting-initiated instability involves flow of material from rims to interior. They find that the break-up of lines (instability) has all the signature of Rayleigh instability where the most probable wavelength $\lambda_m \approx 9R_0$. They estimate from the energy balance of the ion-beam-deposited energy that the region where the ion beam falls is a molten region and the de-wetting starts at the molten regions. In this case also the formation of the spherical particles from the instability has random spacing as has been seen in other studies that used ion beam to initiate instability.

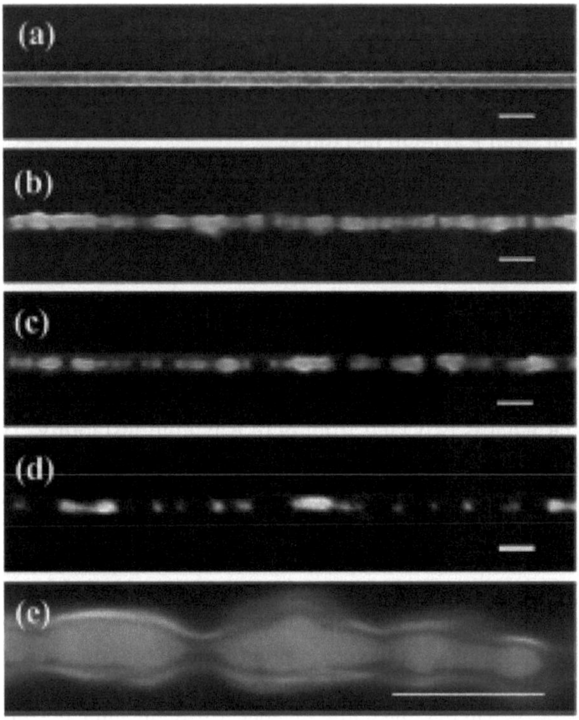


**Fig. 18.13** Radiation-induced initiation of instability in Cu nanolines. As the wire is progressively exposed to the ion beam, the initial stable line forms modulation of its width with eventual breakdown to spheres (from Ref. [6]). Reproduced with permission of AIP

Use of low-energy exposure (energy <1 keV) for producing nanolines is of importance as this may have direct application in sputtering of nanolines. Low-energy ion beams have been used to produce Au nanowires of effective diameter <10 nm by "downsizing" larger nanowires through ion-beam sputtering [26]. They found that Ar + irradiation at 200 eV beam energy produced stable nanowires by sputter "downsizing" of 60 nm wide, 50 μm long wires on a silicon wafer. Strongly contrasting results were obtained using a 500 eV beam of Ga + ions focused to a spot of diameter 11 μm. Using $Ar^+$ ions of low energy, one could obtain stable nanowires even down to an effective diameter less than 10 nm (width 16 nm, thickness 5 nm) as shown in Fig. 18.14.

In case of $Ga^+$ (energy ~500 eV), there is some degree of Ga implantation near the surface. The Ga content of the downsized structure was found to have a destructive effect, causing the ensuing nanowire to break up into a collection of islands. This was explained in terms of the interaction between the Au deposit and nanocondensates of Ga in the form of liquid droplets. That this occurs at room temperature was explained by the fact that Ga nanoparticles have lower melting point than bulk Ga.

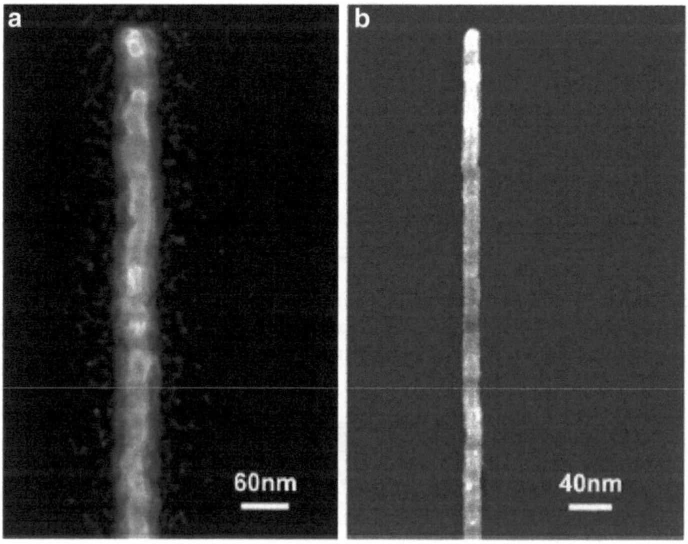


**Fig. 18.14** (**a**) Au line on Si initial effective diameter 60 nm. (**b**) Thinned down to effective diameter 9 nm, width 16 nm and thickness 5 nm with 200 eV $Ar^+$ ions. The low energy of the ions ensures a smaller effective diameter without being affected by Rayleigh instability. Data from Ref. [26]. Reproduced with permission of IOP

Ion beam that induced the instability in Au nanolines on $Si_3N_4$ membranes has been investigated [27]. These membranes are often used in making MEMS/NEMS devices. The typical substrate used was 200 μm × 200 μm window of $Si_3N_4$ membrane (200 nm thick). The membrane is supported by a silicon frame. This is also used as electron-transparent substrate for use in transmission electron microscopy. The Au film (50 nm thick) was grown on the $Si_3N_4$ membrane. Nanopatterning to a nanoline (nanowires) was done using 30 keV $Ga^+$ in a FIB machine.

The length of the nanolines is around 5 μm. The patterning was done by sputtering the Au film from two sides instead of uniform exposure to the ion beam as has been done in the studies mentioned above (see Fig. 18.15 for a cartoon).

The EDAX data shown in Fig. 18.16 show the distribution of Ga ions in the nanoline so formed. NW widths ranged from more than 200 nm down to 20 nm with rectangular cross-section.

The NW on a beam of the nitride membrane is stable for width in the vicinity of 100 nm or so (see Fig. 18.17a). The instability starts to appear once the width is taken below ~50 nm (Fig. 18.17b, c) and the breaking down to spherical droplets occurs for width ~20 nm. For a 20 nm width nanoline, the effective radius $R_0 \approx 22$ nm and the expected modulation wavelength for a Rayleigh instability without the presence of ion beam are $\lambda_m \approx 9R_0 \approx 198$ nm. The observed modulation wavelength is ~165 nm. This smaller value is due to the presence of the ion beam as expected from (18.10) for $\frac{\eta}{D} R_0^2 \approx 0.44$.

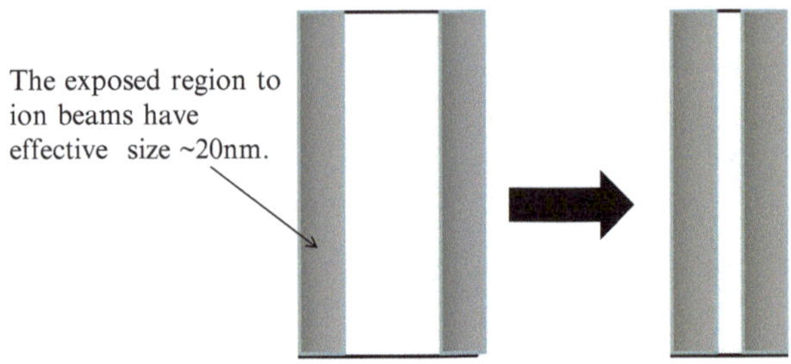


**Fig. 18.15** Schematic of nanopatterning nanolines used by Naik et al. [27] by milling the line from edges as opposed to uniform exposure to ion beam. When the width ~2× effective ion beam size (~20 nm), the exposure is over the whole nanoline

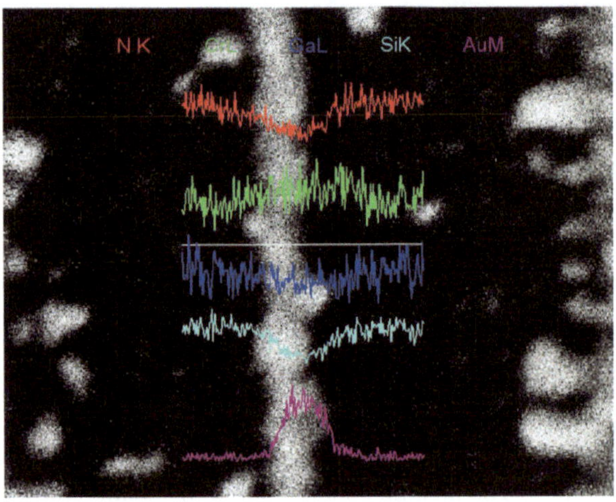


**Fig. 18.16** Analysis by EDAX shows concentration of Ga ions at the edges for nanolines with width >50 nm. For smaller width ~2× beam size, the Ga can be detected over the film width (Data of EDAX courtesy Dr. K. Das)

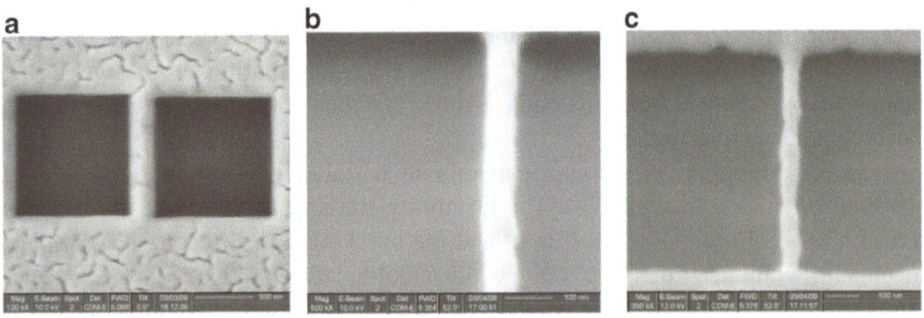


**Fig. 18.17** Progressive narrowing of a Au nanoline made on a $Si_3N_4$ membrane. (**a**) Stable nanoline on the bar with width ~100 nm. (**b**) Appearance of small modulations as the width reaches 50 nm. (**c**) The liquid-like instability sets in when the width is <40 nm (Data from Ref. [27]) (Reproduced with permission from Elsevier)

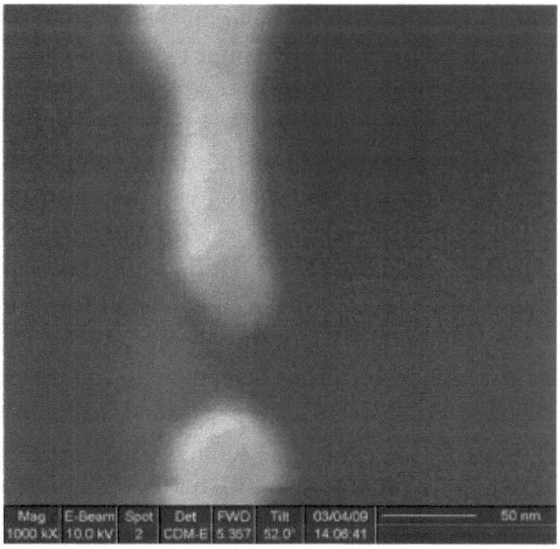


**Fig. 18.18** Fragmentation of 30 nm Au nanoline on the membrane leading to formation of a 20 nm nanogap (Data from Ref. [27]) (Reproduced with permission from Elsevier)

The presence of a 4 nm adhesion layer inhibits the instability which sets in at a much smaller width of 15 nm. In this case, by controlling the breakdown of the nanoline, one can make nanogaps that can be used as electrodes. As shown in Fig. 18.18, one can obtain a gap of width ~20–25 nm.

## 18.5 Effects of the Substrates on Ion-Beam-Induced Instabilities

The previous sections brought out the important role played by the ion beam in modifying the liquid-like instability that sets in metal nanowires. In actual applications the nanowires are supported by substrates of different types like Si with or without native oxides, Si with thick grown oxide (~300 nm) often used in making back-gate FET and $Si_3N_4$ membranes that are used to make NEMS/MEMS cantilevers. The substrate can play its role mainly through two effects. First, it can modify the surface tension $\sigma_S$ through the presence/absence of adhesion layer and second, it can enhance/inhibit charge build-up during ion bombardment through its electrical conductivity. In a similar fashion, the substrate thermal conductivity can influence formation of thermal spike induced de-wetting phenomena discussed in the section before. However, there is no clear experimental study or simulation that addresses this very important issue. In the following two subsections, we discuss the effects of surface passivation and the surface electrical conductivity. It is observed that both factors can influence the liquid-like instability and can lead to formation of NW of smaller diameters when the adhesion is better and the substrates have higher electrical conductivity ensuring better charge relaxation. Better surface adhesion through passivation and better electrical conductivity inhibit the instability-driven spherical droplet formation and lead to stable short NW of smaller effective diameter approaching sub-10 nm regime [28].

The experiments for these investigations were carried out with physical vapour-deposited Au films (50 nm), with an underlying Cr adhesion layer of thickness 4 nm. The presence of the Cr layer on substrates like $SiO_2$ promotes adhesion and inhibits formation of the instability. The nanowires in these experiments were taken down to the desired line width using FIB sputtering. The lines were obtained by sputtering materials from the edges rather than by uniform erosion by ion beam as has been done in some past work (see Fig. 18.15). This avoids damage of 30 keV $Ga^+$ ions (beam current ~30 pA) at an energy of 30 keV. The nanowires produced have rectangular cross-section with widths ranging from more than 200 nm down to 10 nm; their starting length in each case is 2 μm. For these experiments where the wires were patterned by sputtering from edges, the ion spot employed should have much smaller size (<30 nm). The spot size can be tested by checking the width of a milled section on a substrate like Si. Though this spot diameter is of the same order as the width of the nanowires, the dose of Ga ions received by the bulk of the nanowires can be tested by using EDAX. A line scan EDX measurement was carried out across the gold wire along the line shown in Fig. 18.16 to reveal the distributions of Au, Cr, Si, N and Ga. The N and Si peaks appear from the substrate while Au and Cr are due to the wire and its adhesion layer. As expected, the concentration of Au shows a steep rise in the wire region where there is also a small increase in the concentration of Cr. The Ga contamination is localized at the outer edges of the NW of larger width where the effects of the ion-beam milling are concentrated with some limited spread into the nanowire due to the non-Gaussian tails on the focused ion beam (FIB). However, for smaller widths of NW (width ~2× spot size), there is accumulation of Ga in the film. This method of sputtering/milling the wire from edges as opposed to uniform sputtering has an advantage that it helps to make nanowires of smaller dimensions.

### 18.5.1 *Effect of Adhesion*

The effect of adhesion can be seen by the presence or absence of the native oxide layer on Si substrates. The adhesion can also be improved for films like Au using Cr. It has been seen that the presence of Cr layer inhibits the onset of the instability. As a simple rule it can be stated that anything that promotes the wetting of the film on the substrate will reduce the instability and anything that facilitates the film to de-wet will promote the instability.

The effect of FIB reduction of wire width on two nanowires made on a Si wafer (100) with 4 nm native oxide can be seen in Fig. 18.19a, b. The starting wire had a

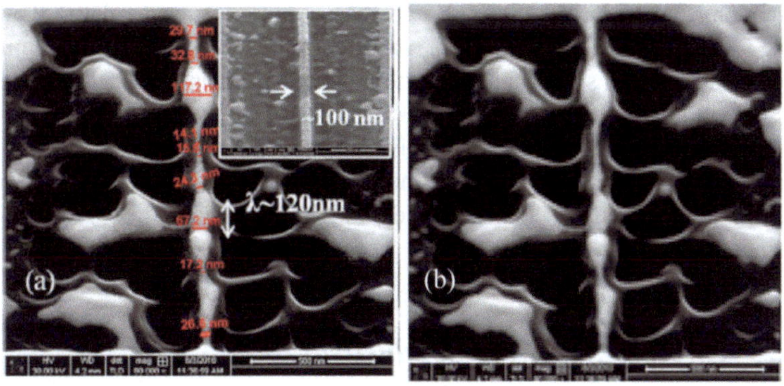


**Fig. 18.19** Formation of instability on Au nanoline on Si wafer with native oxide. (**a**) Shows formation of the instability with approximate wavelength of 120 nm when the width is taken down to below 35 nm (Figure from Ref. [28], reproduced with permission of AIP) and (**b**) on further exposure after the onset of instability short stretches of nanowires with widths ~15 nm or less form (Data courtesy Dr. K. Das)

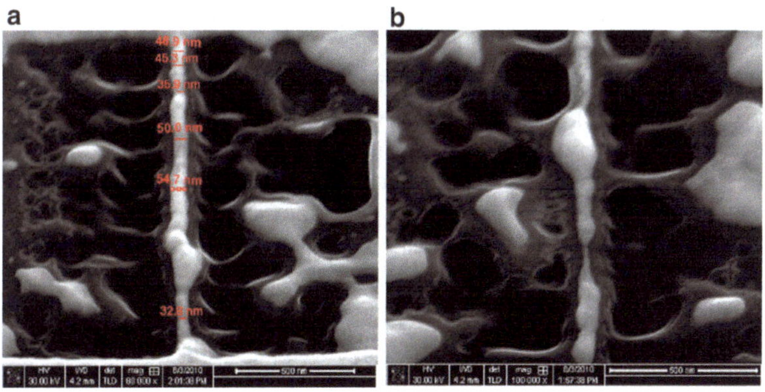


**Fig. 18.20** Formation of instability on Au nanoline on Si wafer passivated by buffered HF. (**a**) Instability sets in with width <30 nm. (**b**) On further ion beam exposure, short strands of nanowires with width ≤10 nm can be seen (Data courtesy Dr. K. Das, jointly taken with Dr. J. Naik)

width of 100 nm and shows no instability. Thinning the wire width to 35 nm initiates the onset of instability as shown in Fig. 18.19a. The instability leads to the formation of Au spheres in a number of spots on the nanowires which gives it a rather irregular appearance. On lowering the width further, a nanowire of width ~15 nm was achieved as shown in Fig. 18.19b with the formation of more spherical regions due to fluid flow.

However, when the native oxide is removed by buffered HF and the surface is passivated with F, the Au NW retains its morphology down to 40 nm with occurrence of limited instabilities (see Fig. 18.20a, b). For width below 40 nm, the radial perturbation of the wire becomes pronounced with appearance of spheres that appear along the length of the wire, which sometimes can have large diameters varying from 50 to 70 nm. However, in some occasions, the NW does not completely fragment into isolated spheres: instead the fragments remain connected by thin filaments of width down to $\leq 10$ nm (Fig. 18.20b). This is distinctly of smaller width than that seen on a film made on unpassivated surface.

The possibility of having stable nanowires of smaller width using passivated surface by inhibiting the instability is a distinct possibility and needs to be investigated more.

### *18.5.2 Effect of Substrate Conductivity*

The effect of charge transport from the film due to irradiation can nucleate/facilitate formation of instability as discussed in Sect. 2 above. The two physical effects that can cause this are (a) electric stress leading to formation of Taylor-like instability and (b) Coulomb explosion from strong Coulomb potential of accumulated charge. The charge transport from the film during irradiation is controlled by conductivity of the underlying substrate. Though this can be an important factor, barring a recent study, there are no comprehensive experimental investigations of the effect of substrate conductivity yet. The initial results show that the substrate conductivity can indeed be a factor that affects the instability.

The experiments reported here were done on Au films with underlying Cr layer on substrates of different conductivity like Si with different level of doping (as stated before), Si substrate with $SiO_2$ grown on it and on $Si_3N_4$. The experiments show that as the substrate conductivity goes up, facilitating transport of charge, there is lowering of the critical diameter when the instability sets in. This implies the nanowires are stable to less diameter when they are grown on substrate with more conductivity. However, the dependence on the substrate conductivity is rather weak and the dependence of the minimum width $w$ on the conductivity $\sigma$ is logarithmic in nature as shown in Fig. 18.21.

It is also observed that after the instability occurs, there is formation of short lengths (0.5–1 μm) of wires of small diameter (see Sect. 5.1). This diameter also reduces as the conductivity is enhanced. Some of the details of the results are shown below.

Figure 18.22a, b shows the results for Au NW fabricated on heavily doped Si wafers of conductivity $\sigma \approx 100$ S/cm in the presence of native oxide and following buffer HF treatment, respectively. Figure 18.22a shows much thinner NW formation (width below 40 nm) without break-up due to liquid instabilities for the wafer with higher conductivity. For the highly conducting wafer (see Fig. 18.22b), after the instability sets in finally, a wire of minimum width of 8 nm is obtained (approximate continuous length ~250 nm). The results demonstrate a noticeable change in the stability of the wire depending on the doping of the substrates and the conductivity.

The conductivity of the substrate can be further reduced using Si substrate with thermally grown $SiO_2$ (thickness 300 nm). Figure 18.23 shows the results for 300 nm $SiO_2$/Si substrate. Similar results are obtained for 200 nm $Si_3N_4$/Si substrate. The fluid flow instability effects are apparent even for the wider wires in this case.

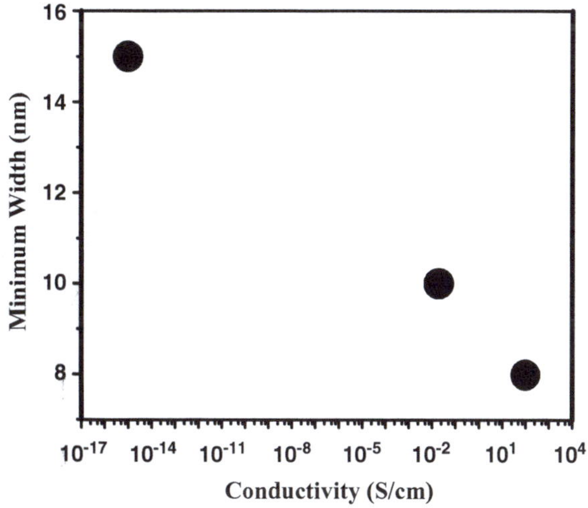


**Fig. 18.21** Dependence of the smallest width of nanowires on substrate conductivity

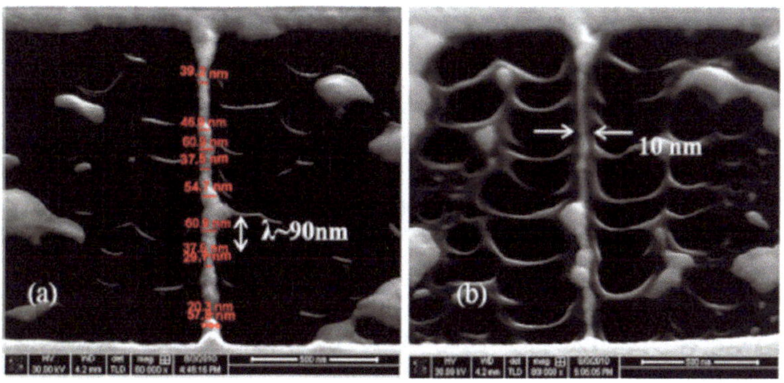


**Fig. 18.22** Formation of instability on Au nanoline on heavily doped Si wafers with conductivity ~100 S/cm. (**a**) Instability on a wafer with native oxide. The instability sets in when width is well below 50 nm. (**b**) Stability of wires with narrow width ~10 nm coexisting with spherical droplets of larger size formed on wafers with passivated surface. (Data from Ref. [28]) Reproduced with permission from AIP

For instance, the instability starts to occur when the width of the nanowire is taken down to even 80 nm. In addition, Taylor cones, associated with charging due to the insulating substrate, are clearly seen on the sides of the NW. The magnified part is shown inset of Fig. 18.23. The half angles of the cone are seen to be very close to the value of 49.3°. In this case, after the instability sets in, one may observe formation of stable NW of width ~15 nm which is much smaller than the width of ~8 nm observed for ion-beam-processed NW grown on more conducting wafers.

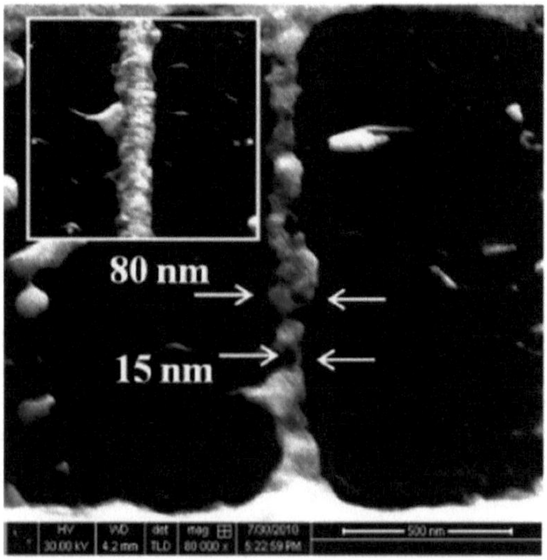


**Fig. 18.23** Formation of instability on Au nanoline on Si wafer (lightly doped) with 300 nm grown oxide. The low conductivity of the wafer leads to formation of instability even when the width is ~80 nm. The inset shows the formation of the Taylor cone liquid jet. Data from Ref. [28]. Reproduced with permission from AIP

## 18.6 Instability in Nanopatterning of Complex Materials

Nanopatterning of metal nanolines and the resulting instabilities, as discussed above, have been investigated by a number of groups. However, the issue of instability in complex materials like oxides has not been addressed to yet in a comprehensive way. In the following we present some preliminary results on observation of onset of instabilities in complex oxide films (e.g., $WO_3$). The sample used in this study was a thin film of $WO_3$ grown on oxidized Si ($SiO_2$/Si) and on $SrTiO_3$ (STO). The films were exposed to 30 keV $Ga^+$ ions in FIB machine with an ion current ~28 pA. These are similar parameters of ion beams which have been used in studies of metal nanolines. The films used had a thickness $t \approx 300$ nm. The grown films were initially made into nanolines of width $w \approx 150$ nm. The lines are then exposed to ion-beam milling at the edges. As shown in Fig. 18.24, the nanolines are stable against the milling at the edges at least till $w \approx 50$ nm on both the substrates. However, at smaller width, instabilities appear in the nanolines, which also leads to pronounced variation in the film thickness as can be seen in Fig. 18.24b, where images are shown with a tilt. The results qualitatively are similar for both the substrates although there are important quantitative differences in the value of $w$ at which the instability starts. It appears that in the $SiO_2$/Si substrate, the instability starts at somewhat smaller values of the width.

It has also been observed that it is possible to get smaller segments of even smaller width as in the metallic systems when the films are patterned on $SiO_2$/Si substrate. As shown in Fig. 18.24a, there is a nanoline with smallest diameter ~11 nm and length 2 μm produced by ion-milling the edges. The modulation along the edges can be seen but the line is stable. The data shown here are preliminary and the instability in nonmetallic systems during ion-beam patterning needs to be investigated with more rigorousness.

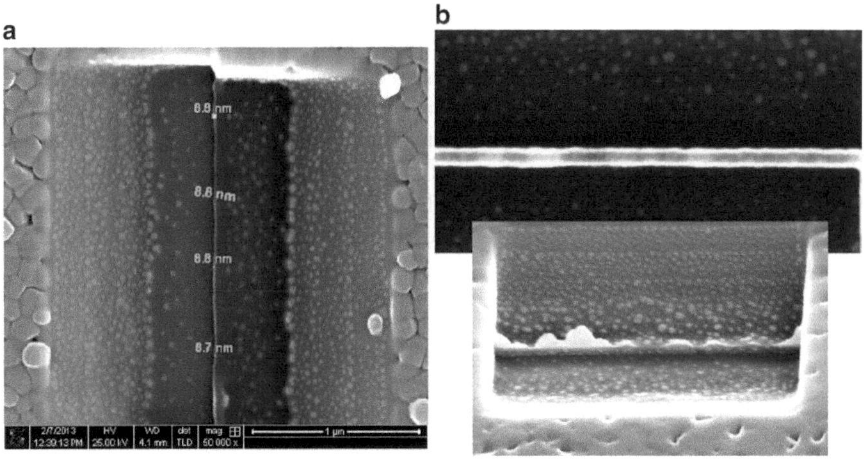


**Fig. 18.24** Formation of instability on complex oxide ($WO_3$) formed on two substrates (**a**) $SiO_2$/Si and (**b**) $SrTiO_3$ (STO) formed by pulsed laser deposition. The films of thickness ~300 nm were patterned by FIB (Ga, 30 keV) (Data courtesy Dr. B. Ghsoh and Dr. K. Das)

## 18.7 Conclusions

The morphological stability of nanowires and nanolines, fabricated by FIB milling, is an important issue that may limit the applicability of the method in nanofabrications. These instabilities have basic physical reason for their appearances. It is important to appreciate that these instabilities appear even in the absence of ion-beam irradiation mainly due to predominance of surface energy. The effect of the ion beam is to modify the instabilities. We presented a brief summary of the various physical effects that contribute to the instability. The theory of conventional surface energy-driven Rayleigh–Taylor instability was developed for nanowires where the instability leading to oscillations in diameter occurs under the constraint of conserved volume. The ion-beam-induced sputtering breaks that important constraint and adds additional important factors.

The experimental observations suggest that Rayleigh–Taylor instability is one of the principal causes for these instabilities. However, the observed instabilities show effects of ion-beam-induced roughening, de-wetting and charging of the substrate. We have also shown example of different substrate-induced modifications of the instability. For instance, the presence of an adhesion layer of Cr in case of Au films and passivation of the Si surface by buffered HF treatment are some of the factors that can change the extent of instability and the critical diameter where the instability appears. Recent experiments have also shown the importance of the substrate conductivity in determining the value of the critical diameter where the instability sets in. Nanostructures grown on substrate that reduce charging facilitate fabrications of smaller wires.

Finally, we have shown that the instabilities induced by FIB are not confined to metallic nanostructures only and can be observed even in nanolines of complex oxides. These observations would need further investigations.

The important issue of instabilities in nanolines formed by FIB needs further investigation that can address the parameters involved in a controlled way and also thus can lead to better understanding. In fact the possibility that one may get nanolines with width <10 nm is intriguing and production of such sub-10 nm nanolines in controlled way reproducibly will add tremendous value to the use of these nanolines.

**Acknowledgements** The author would like to thank Dr. Kasutuv Das for some of the experimental data that are given here prior to publications and helpful discussions. The author also thanks Prof. P.D. Prewett and Mr. J. Naik, University of Birmingham, for encouraging discussion. The work was initiated as part of UKIERI project. The data in some of the figures were taken at the S.N. Bose National Centre for Basic Sciences by Dr. K. Das and Mr. J. Naik. The work was also supported by DST and sponsored by the Centre of Nanotechnology and Theme Unit of Excellence in Nanodevice Technology.

## References

1. For a review see Tseng A.A.: J. Micromech Microengg, **14**, R15 (2004)
2. Raychaudhuri A.K.: In: Rao, C.N.R., Muller, A., Chetham, A.K. (eds.) The chemistry of nanomaterials, p 688, Wiley-VCH, Germany (2004)
3. Plateau, J.: Transl. Annual Reports of the Smithsonian Institution, p 1863 (1873)
4. Rayleigh, L.: Proc. London Math. Soc. **10**, 4 (1878)
5. Sigmund, P.: Phys. Rev **184**, 383 (1969)
6. Wei, Q., Li, W., Sun, K., Lian, J., Wang, L.: J. Appl. Phys. **103**, 074306 (2008)
7. De Gennes, P.G.: Rev. Mod. Phys. **57**, 827 (1985)
8. Jiu X, Luo J and Zhu J , Nanoletters 6, 408 (2006)
9. Ghosh, M., Raychaudhuri, A.K.: Nanotechnology **19**, 445704 (2008)
10. Nichols, F.A., Mullins, W.W.: Trans. Metall. Soc. AIME **1965**, 233 (1840)
11. Zhang, C.-H., Kassubeck, F., Stafford, C.A.: Phys. Rev. B **68**, 165414 (2003)
12. Chandrasekhar, S.: Hydrodynamic and hydromagnetic stability. Dover, New York (1981)
13. Rodrigues, V., Bettini, J., Rocha, A.R., Rego, L.G.C., Ugarte, G.: Phys. Rev. B **65**, 153402 (2002). and references there in
14. Bischof, J., Scherer, D., Herminghaus, S., Leiderer, P.: Phys. Rev. Lett. **77**, 1536 (1996)
15. Brochard Wyart, F., Daillant, J.: Can. J. Phys. **68**, 1084 (1990)
16. Toulemonde, M., Dufour, C., Paumier, E.: Phys. Rev. B **46**, 14362 (1992)
17. Mayr, S.G., Averback, R.S.: Phys. Rev. B **68**, 214105 (2003)
18. Taylor, G.I.: Proc. R. Soc. London AV. **280**, 383 (1964)
19. Fleischer, R.L., Price, P.B., Walker, R.M.: J. Appl. Phys. **36**, 3645 (1965)
20. Qin Y, Lee S.M, Pan A, Gosele U, Knez M Nanoletters **8**, 114 (2008)
21. Toimil-Molares, M.E., Balogh, A.G., Cornelius, T.W., Neumann, R., Trautmann, C.: Appl. Phys. Lett. **85**, 5337 (2004)
22. Bid, A., Bora, A., Raychaudhuri, A.K.: Phys. Rev. B **72**, 113415 (2005)
23. Karim, S., Toimil-Molares, M.E., Balogh, A.G., Ensinger, W., Cornelius, T.W., Khan, E.U., Neumann, R.: Nanotechnology **17**, 5954 (2006)
24. Lian, J., Wang, L., Sun, X., Yu, Q., Ewing, R.: Nanoletters **6**, 1047 (2006)

25. Zhao, K., Averback, R.S., Cahill, D.G.: Appl. Phys. Lett. **89**, 053103 (2006)
26. Tuboltsev, V., Raisanen, J.: Ion beam processing of gold nanowires. Nanotechnology **20**, 335302 (2009)
27. Naik, J.P., Prewett, P.D., Das, K., Raychaudhuri, A.K.: Microelectr. Eng. **88**, 2840 (2011)
28. Naik, J.P., Das, K., Prewett, P.D., Raychaudhuri, A.K.: Appl. Phys. Lett. **101**, 163108 (2012)

# Chapter 19
# Nanostructures by Mass-Separated FIB

**Lothar Bischoff, Roman Böttger, Peter Philipp, and Bernd Schmidt**

**Abstract** The introduction of mass-separated systems in the field of focused ion beams significantly increases the area of application in nanotechnology due to the availability of a broad spectrum of ions with the same advantages compared to classical Ga instruments. A short description of the configuration of a mass-separated FIB tool is given as well as the fundamentals of alloy liquid metal ion sources. Examples of application include patterned tailoring of functional surfaces and ion-induced phase transformation in thin layers, in particular the Si nanowire fabrication by FIB implantation and subsequent wet-chemical, anisotropic etching and the FIB lithography of thin ta-C films. Furthermore, the ion beam synthesis (IBS) of $CoSi_2$ nanostructures by Co-FIB writing and annealing, and the modification of surface morphology by various mono- and polyatomic projectiles in a broad energy and temperature range in different materials are described and discussed.

## 19.1 Introduction

Ion beam nanotechnology evidently provides an impressive variety of tools available for many applications, offering precision and predictability, together with the special virtue of nanometer dimensional control. Focused ion beams (FIB) in the ion energy range 1–50 keV are able to locally modify or pattern solid-state surfaces with nanometer resolution directly, without any lithographic processes. Fundamental processes of ion beam interaction with materials, like doping by ion implantation, material removal and redeposition by sputtering, secondary particle generation for

L. Bischoff (✉) • R. Böttger • P. Philipp • B. Schmidt
Helmholtz-Zentrum Dresden-Rossendorf, Institute of Ion-Beam Physics
and Materials Research, P.O.B. 510119, Bautzner Landstraße 400, 01328 Dresden, Germany
e-mail: l.bischoff@hzdr.de; r.boettger@hzdr.de; p.philipp@hzdr.de; bernd.schmidt@hzdr.de

Z.M. Wang (ed.), *FIB Nanostructures*, Lecture Notes in Nanoscale Science and Technology 20, DOI 10.1007/978-3-319-02874-3_19,

ion microscopy or analysis, beam writing for ion lithography or gas assisted ion beam etching and material deposition from precursor gases are used in modern FIB tools having a resolution down to about 5 nm for $Ga^+$ ions at beam currents in the range of 1 pA [1, 2].

Today, modern FIB machines are applied as powerful tools in cross section sample preparation for high-resolution transmission electron microscopy (X-TEM) investigations by local ion beam sputtering (milling). In contrast to conventional TEM-sample preparation techniques, FIB milling opens the possibility to prepare the TEM specimens at the place of interest from large-area samples with high spatial resolution. Furthermore, cross-sectional cuts prepared by FIB milling can be easily investigated by methods of analytical scanning electron microscopy (SEM) [3]. Nowadays, most of the FIB columns are combined with SEM columns in one system as so-called dual-beam or cross-beam machines, firstly proposed by Sudraud et al. in 1988 [4] which are at present one of the workhorses in nanotechnology.

Ion sources used in FIB systems are mostly liquid metal or liquid metal alloy ion sources (LMIS) providing a very high brightness of ~$10^6$ A $cm^{-2}$ $sr^{-1}$. The point like ion emission from an LMIS has the ability to provide an ion beam spot of ~5–10 nm in diameter (FWHM). In general, the LMIS has the advantage of high ion beam intensities and relatively long lifetimes. Furthermore, they have a simple design and are easy to operate. Micro- and nano-focused ion beams have a wide field of applications nowadays, for example, in microfabrication and nanofabrication, materials analysis, and space propulsion. Consequently, this type of ion source has been studied by many groups and there exist many review articles [5–7] and books [1–3, 8] giving an overview of the development and the technology as well as about the physics and application of LMIS in FIB tools. For special applications, like high rate milling, ECR sources [9] or ICP sources [10] for noble gases (Xe) delivering currents in the microampere range are used in FIB systems today. For highest resolution, ion microscopy field emission ion sources [11] for He or Ne are applied able to facilitate a spot size of 0.35 nm. But these types of sources are not the issue of this chapter.

This work is concentrated on the application of LMIS, especially alloy type sources, in modern FIB columns with mass separation (ExB filter), which are able to provide several kinds of ions, isotopes, and clusters in addition to the usual $Ga^+$ ions. These alloy LMIS are successfully used for materials research and development in microfabrication and nanofabrication [12]. Roughly, the emerging applications of mass-separated FIB in nanofabrication can be divided into following classes:

- Maskless ion implantation on sub-micrometer scale to fabricate prototype devices, e.g.:
  - Si-based nanostructures (NEMS) using different kinds of ions
  - $Si_{1-x}Ge_x$-based nanostructures using $Si^+$ and $Ge^+$ ions from one LMIS
  - Nano-optical devices with rare earth ions, e.g., $Nd^+$ and $Er^+$
  - Doping of magnetic nanodevices with $Co^+$ and $Ni^+$ ions

- High-rate contamination free FIB milling (not $Ga^+$ ions!) in semiconductor industry
- Low-damage FIB imaging using, e.g., $Li^+$ and $Si^+$ ions
- Three-dimensional microfabrication and nanofabrication

Furthermore, the unique properties of the mass-separated FIB together with alloy LMIS providing high ion fluxes for a lot of ion species, from light ions (Li, C, Si) up to heavy ions (Au, Bi) and even polyatomic ions (cluster), enables time-saving investigation of different material science issues. Examples among many, which have been researched in detail, include, e.g., fast measurements of ion sputtering yields (milling rates) for different ion-target combinations [13], ion flux-dependent defect evolution in crystalline materials (Si, Ge, SiC, etc.) [14], strain analysis of laterally ion beam patterned semiconductor surfaces [15], local defect engineering and local tailoring of functional properties of materials [16], writing ion beam-induced phase transformation and separation in different materials [17], surface patterning in the form of self-assembled periodic dots or ripples after heavy ion impacts [18].

In most of these examples different kinds of ions and high fluences are required to achieve the desired materials modifications, which can easily be realized by mass-separated FIB. Meanwhile, the use of mass-separated FIB with optimized feature definition and correlation distances to generate functional nanopatterns seems to be a likely strategy. Future development of direct patterning, in 3D or for direct surface functionalization, including self-assembly, in particular for prototyping may well be more amenable to future manufacturing requirements.

In this contribution the design principle and operation of the FIB tool with ion mass-separation inserted into the ion column will shortly be described, followed by an overview of available alloy LMIS with their main characteristics developed in the Institute of Ion Beam Physics and Materials Research of the Helmholtz-Zentrum Dresden-Rossendorf. The selected examples for applications of the mass-separated FIB in nanoscience and nanotechnology fall into the following categories:

- Local tailoring of functional surface properties and patterned ion-induced phase transformation and separation in thin layers by Ga FIB implantation.
- FIB assisted synthesis of solid state phases on the nanometer scale.
- Self-organization of regular surface pattern under FIB irradiation with heavy ions.

## 19.2 Configuration of Mass-Separated FIB Tools

The ion optics of modern FIB columns is quite similar. The main components are the ion source, mostly an LMIS, a first lens (condenser), a set of beam defining apertures, correction and deflection units (octupoles), and finally a focusing lens (objective). For imaging and analytical purposes different detectors for secondary electrons or ions can be applied. A comprehensive overview about FIB column design is given by Orloff [19]. Most of the present FIB columns work with a Ga LMIS due to its convenient handling (see next section). To enlarge the application

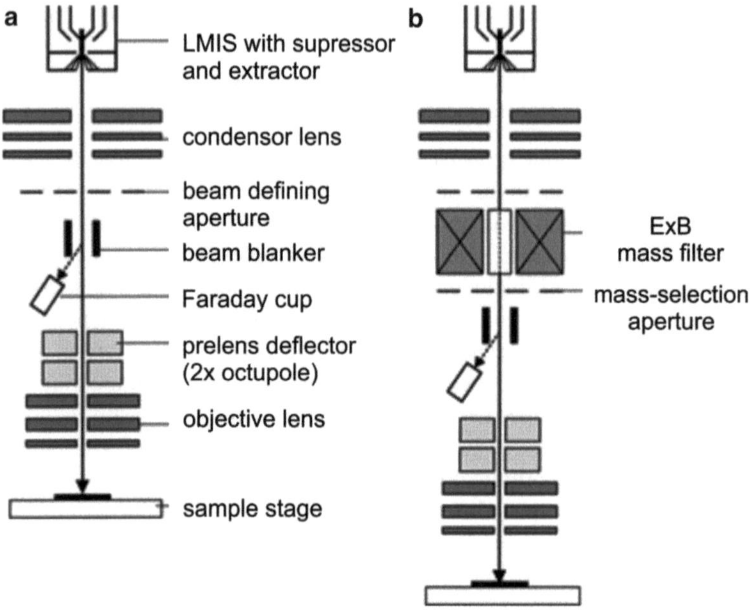


**Fig. 19.1** Typical FIB column geometry without (**a**) and with mass-separator (**b**)

field of FIB to other species than Gallium, alloy LMIS can be applied, but they require a mass separation system, i.e., an ExB mass filter (Wien-filter), with a selection aperture to choose a certain ion mass and the ion charge state. The two basic configurations of ion optics are shown schematically in Fig. 19.1 [20].

The FIB column can be operated in different modes. For high current applications a crossover mode is used as well as bigger apertures, resulting in a weaker resolution. For high-resolution operation at low currents, i.e., small aperture, a parallel beam characteristic is preferred, because beam broadening by Coulomb interaction can be avoided. The beam spot size can be estimated by

$$d = \sqrt{\left(Md_q\right)^2 + d_S^2 + d_C^2}, \tag{19.1}$$

where $d$ is the beam spot diameter at the sample measured as FWHM, $d_q$ is the virtual source size, which is a function of the ion mass and current according to $(m^{0.5}\ i)^{0.45}$ and is in the range of 20–50 nm [21]. $M$ denotes the magnification of the column. The quantities $d_S$ and $d_C$ represent the contribution due to the spherical and axial chromatic aberrations, respectively, which are defined as:

$$d_S = \frac{1}{2} C_S \alpha^3, \quad d_C = C_C \frac{\Delta E}{E} \alpha. \tag{19.2}$$

$C_S$ and $C_C$ are the spherical and chromatic aberration coefficients of the ion optical column, respectively, $\alpha$ is the acceptance half-angle on the target and can be determined by $\alpha = [i/\pi m^2(\mathrm{d}i/\mathrm{d}\Omega)]^{1/2}$ with $m$ the ion mass, $i$ the ion current, and $\mathrm{d}i/\mathrm{d}\Omega$

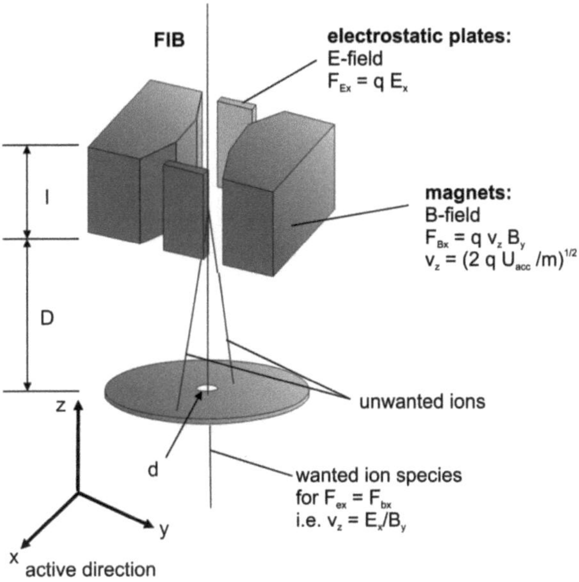


**Fig. 19.2** Operation principle of an ExB mass filter

the angular intensity. $\Delta E$ is the energy spread of the LMIS and $E$ is the final ion energy. For the above described column, operating with an ion current in the range lower than 1 nA, the chromatic aberration determines the spatial resolution with a $d \sim i^{1/2}$ dependency, whereas for the high current operation the spherical aberration becomes the dominant contribution according to $d \sim i^{3/2}$ [22]. This formalism is in principle also valid by switching to a mass-separating column (Fig. 19.1b). The main difference is the introduced ExB mass filter, which is explained schematically in Fig. 19.2. An electric field is applied to the beam and leads to a bending. On the same place a magnetic field is located, which brings a charged particle perpendicular to a circular orbit. The radius is dependent on the mass of the ion species. Now, the fields are adjusted in a manner that for a given mass the divergence of the trajectories from a straight line are compensated and the wanted particles pass the separation aperture. The mass resolution of an ExB filter can be written in the following form [5]. The geometric quantities may be derived from Fig. 19.2.

$$\frac{m}{\Delta m} = \frac{E_x}{2U_{acc}} \frac{l}{d} \left( \frac{1}{2} l + D \right) \quad (19.3)$$

$U_{acc}$ is the acceleration voltage of the ion beam and $E_x$ the electric field strength of the filter perpendicular to the ion optical axis. The ExB acts as a velocity filter and separates the beam according to the mass-to-charge ratio, which has to be taken into account for $U_{acc}$.

For a given geometry, the sensitivity of the filter increases with decreasing ion energy as shown in Fig. 19.3. $m/\Delta m$ is in the range from 30 to about 100 in commercially available FIB columns. So light elements like Li up to heavy polyatomic ions of more than 1,000 amu $q^{-1}$ (e.g., $Bi_5^+$) can be separated from the source emitting spectrum.

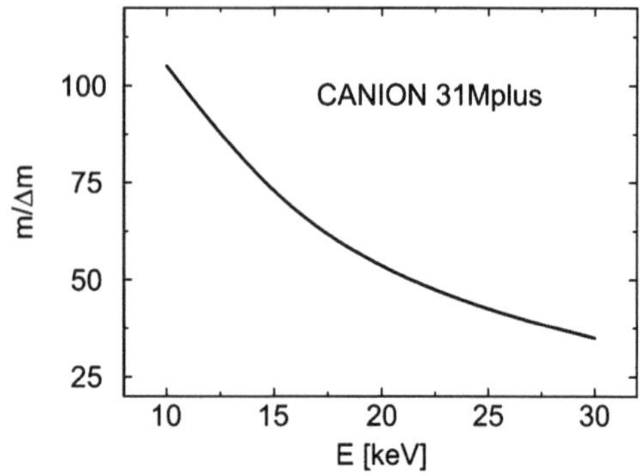


**Fig. 19.3** Dependence of the mass resolution on the ion energy for the ExB filter of a CANION 31MPlus FIB Column (Orsay Physics)

If the ExB filter can be switched off, a mass-separating column can also be used for single element LMIS, which focuses all species in one spot, but a marginal reduction in the spatial resolution is obtained by the transversal velocity components in the longer drift space. Nevertheless, the next section is focused on alloy ion sources in addition to single element LMIS.

## 19.3 Liquid Alloy Ion Sources: Fundamentals and Applications

Liquid metal (or metal alloy) ion sources (LMIS) have found their wide application in FIB tools due to their characteristics as a high brightness point source. In order to focus an ion beam into a spot size of a diameter much smaller than 1 μm, a source is needed, which emits the ions from a very small area into a limited solid angle. To meet this demand several types of LMIS such as needle-type LMIS and capillary-type LMIS have been proposed. The needle-type ionizer [23] is most commonly used in FIB tools. As an example an SEM image of an AuGeSi alloy wetted tungsten emitter is shown in Fig. 19.4 [12].

The application of LMIS was initiated about 50 years ago with the aim of developing ion-thrusters for satellite navigation, but finally, a very suited point-like source for FIB was created. A detailed picture of the history of LMIS is given elsewhere [1, 2].

Demands on metals on metal alloys used in LMIS are: (1) a relative low melting point to minimize any reaction (or interdiffusion) between the liquid and the W needle, (2) a low vapor pressure at the melting point conserving the supply of metal and yielding in a long source life, (3) a low surface free energy (high surface tension) promoting viscous wetting behavior on the W needle, (4) good mechanical, electrical, and vacuum properties, and (5) emission characteristics enabling high angular intensity with a small energy spread [12]. Therefore, Ga with a melting point of $T_m = 29.8$ °C is considered most suitable and is mostly used in FIB tools. Additionally, compared to other suitable low melting elements (e.g., Rb, Cs) Ga has the advantage

Fig. 19.4 SEM image of an AuGeSi alloy wetted tungsten emitter

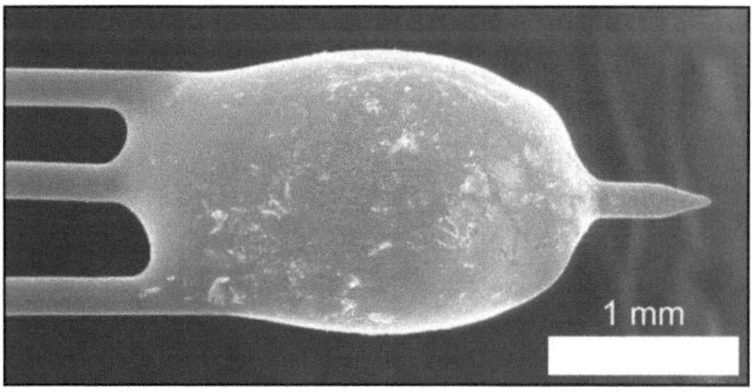


that the emitted ion mass spectrum contains about 99 % singly charged ions. Therefore, no mass filter is needed and the ion column design becomes easier due to the use of electrostatic ion optics, which is mass-independent. Thus, the majority of commercial FIB systems operating worldwide are suited only for Ga ions.

The energy spread $\Delta E$ of a LMIS is a crucial parameter in view of its influence on chromatic aberrations and thus on the final spatial resolution of a FIB system. $\Delta E$ of the near-Gaussian-shaped ion-energy distribution represents the work expended to create an ion from the surface of a solid or liquid in the presence of the electric field, $E$. In a surface field-ionization, e.g., field-evaporation, mechanism, it is given as [24]

$$\Delta E(n) = \Lambda + \sum_n I_n - n\varphi_c, \tag{19.4}$$

if the ion is $n$-fold ionized. $\Lambda$ is the binding energy (heat of evaporation) of the atom, subsequently ion. $I_n$ is the $n$th ionization potential of the atom; $\varphi_c$ is the work function of the retarding electrode. The evaporation field is the value of the electric field for which the field-reduced potential energy barrier, seen by an escaping ion, is equal to zero. It can be written as [24]

$$E(n) = \frac{4\pi\varepsilon_0}{n^3 e^3}\left\{\Lambda + \sum_n I_n - n\varphi\right\}^2. \tag{19.5}$$

Here $\varphi$ is the work function of the emitter and $e$ the fundamental electronic charge. According to Brandon [25], if $E(n=2) < E(n=1)$, the ion is likely to come out as doubly charged, or dominate in the beam, and vice versa. For singly charged ions the predicted dependence of the energy spread on the ion current, the mass and the temperature is proportional to $\Delta E \sim I^{2/3} m^{1/3} T^{1/2}$ [26, 27], which was reasonably confirmed for a broad spectrum of elementary as well as alloy LMIS. For a consistent emission current of 15 μA the energy spread for singly charged, doubly charged and several polyatomic ions obtained from retarding field measurements is plotted in Fig. 19.5 [28].

Note that $\Delta E$ is much smaller for the minimum stable current of the certain emitters, i.e., 5 eV for $Ga^+$ at 1 μA [8]. For sources, emitting a manifold of ion species of different charge states, mono- and polyatomic ions other dependencies may be

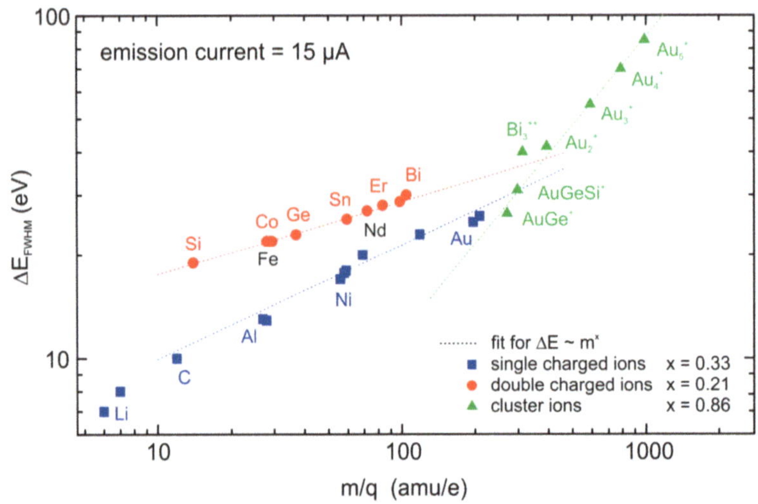


**Fig. 19.5** Energy spread of singly charged, doubly charged, and polyatomic ions (cluster) as a function of their mass-to-charge ratio obtained from alloy LMIS like AuGeSi, AuGe [30], AlCeC [31], ErFeNiCr [32], CoNd [33], GaBiLi [34] and from pure metal LMIS for Ga, Sn [35], and Bi [36] at a constant emission current of 15 μA

obtained [29]. For practical application the LMIS should operate at the lowest stable emission current in order to minimize the chromatic aberration [8].

Another significant parameter of the LMIS behavior is the *I–V* curve giving an advice about the impedance of the source. The expected current from an emitter can be calculated following the model of Mair [37].

$$i = 3\pi \sqrt{\frac{2e}{m}} \frac{r\gamma \cos\varphi}{\sqrt{V_0}} \left\{ \frac{V}{V_0} - 1 \right\} \tag{19.6}$$

$r$ is the Taylor cone radius, $\gamma$ the surface tension coefficient, $\varphi$ the Taylor cone half-angle of 49.3°; $m$ the atomic mass of the source material, $e$ the electronic charge, $V_0$ the onset voltage, and $V$ the extraction voltage.

The emission-temperature behavior is also important, in particular for alloy sources at elevated melting temperatures to find a stable operating point. Furthermore, the mass spectrum is giving information about which species are available at certain intensities [28]. A typical mass spectrum of a GaBi alloy LMIS as a function of the emission current is shown in Fig. 19.6 [38].

With the development of modern FIB columns with mass-separation (ExB filter), it became possible to provide several kinds of other ions from metal alloys in LMIS, which are successfully used for research and development in microfabrication and nanofabrication [1, 3, 7, 28]. The utilization of these unique point-like sources in FIBs

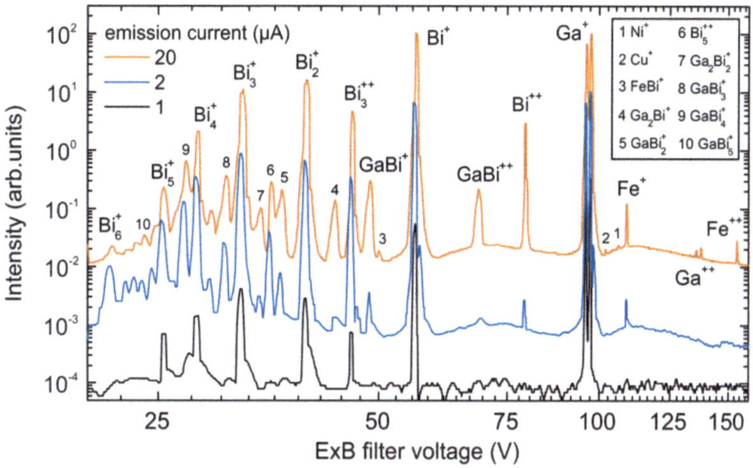


**Fig. 19.6** High-resolution mass spectrum of a GaBi alloy LMIS for different emission currents

**Table 19.1** Elemental and eutectic alloy materials for LMIS

| Selected ions | Liquid element or alloy | $T_m$ (°C) | Reference |
|---|---|---|---|
| *Without mass-separator* | | | |
| $Ga^+$ | Ga | 29.6 | [28, 39] |
| $In^+$ | In | 157 | [40, 41] |
| $Bi^+$ | Bi | 271 | [36, 42] |
| $Rb^+$ | Rb | 39 | [43] |
| $Cs^+$ | Cs | 26 | [44] |
| *With mass-separator* | | | |
| $In^+$, ($Ga^+$) | $In_{14}Ga_{86}$ | 14.2 | [28] |
| $Sn^+$, $Pb^+$ | $Sn_{74}Pb_{26}$ | 183 | [35] |
| $Bi^+$, ($Ga^+$) | $Ga_{38}Bi_{62}$ | 222 | [38] |
| $Bi^+$, $Li^+$, ($Ga^+$) | $Ga_{35}Bi_{60}Li_5$ | 250 | [34] |
| $Au^+$, $Ge^+$ | $Au_{73}Ge_{27}$ | 365 | [45] |
| $Au^+$, $Ge^+$, $Si^+$ | $Au_{77}Ge_{14}Si_9$ | 365 | [45] |
| $Au^+$, $Si^+$ | $Au_{82}Si_{18}$ | 365 | [46] |
| $Co^+$, $Nd^+$ | $Co_{36}Nd_{64}$ | 566 | [33] |
| $Mn^+$, $Ge^+$, $Si^+$ | $Mn_{45}Ge_{54}Si_1$ | 720 | [47] |
| $Pr^+$, $Si^+$ | $Pr_{89}Si_{11}$ | 732 | [48] |
| $Er^+$, $Ni^+$ | $Er_{69}Ni_{31}$ | 765 | [49] |
| $Co^+$, $Ge^+$ | $Co_{27}Ge_{73}$ | 817 | [50] |
| $Er^+$,$Fe^+$,$Ni^+$,$Cr^+$ | $Er_{70}Fe_{22}Ni_5Cr_3$ | 862 | [32] |
| $Al^+$, $Ce^+$, $C^+$ | $Al_{10}Ce_{70}C_{20}$ | 660 | [31] |

with a diameter of less than 10 nm and current densities up to more than 20 A $cm^{-2}$ opens a broad field of new applications. Possible other elemental metals and eutectic alloys, mostly investigated and tested in the own lab, are listed in Table 19.1.

## 19.4 Mass-Separated FIB Applications in Nanotechnology

Ion beams are principally well suited to nanofabrication, because ions exhibit a relatively small scattering in solids. They are used to modify the topography as well as the electronic, optical, magnetic and mechanical properties of materials. Using finely focused ion beams (FIB), and writing them across the substrate, the creation of feature sizes in the 10 nm range is possible. Nanofabrication with scanned FIBs is a sequential process and is therefore much slower than the parallel processing by optical lithography. It cannot be expected to provide FIB-based technology for mass-production of nanoelectronic structures or devices. Nonetheless, FIB technology has already found applications in IC manufacturing for local, micrometer and nanometer scale operations on individual wafer chips. Examples are failure analysis, mask repair or trimming of magnetic read-heads. One can expect that FIB will find further niches in prototyping, small-volume fabrication and customization of individual nanodevices in a research environment.

The main applications of FIB technology today are concerned with local ion sputtering for material removal. This requires local ion fluences of $\geq 10^{16}$ cm$^{-2}$, usually of Ga ions (with energies of tens of kiloelectronvolts). At present, Ga-FIB and ion mass-separated FIB systems no longer rely on sputtering effects, but rather on local surface modification and defect creation also at lower fluences. For example, many target materials (III–V opto-electronic compound semiconductors, magnetic thin films, or weakly bonded chemicals) exhibit very high ion sensitivity, allowing fluences in the $10^{12}$–$10^{16}$ cm$^{-2}$ range. Consequently, FIB patterning and tailoring of surfaces, using chemical phase or structural (crystal) modification by local ion bombardment, becomes an effective process, which offers new possibilities for selective functionalization of different materials.

In this relation some selected examples for FIB-assisted modification of chemical, structural and electrical properties as well as the synthesis of solid-state phases will be described in this chapter. Surface nanopattern formation under irradiation with heavy ions and ion clusters is a further topic of this chapter due to an increasing interest in self-organization of nanostructures by ion beam processing.

### *19.4.1 Si Nanowires by FIB Writing and Subsequent Wet-Chemical Etching*

The monolithic integration of micro- and nano-electro-mechanical system with microelectronic components figures prominently in the modern Si-based technology [51]. Micro- and nano-electro-mechanical system, like nanowires, bridges or cantilevers are important, constitutive parts for this purpose, as they show an enormous sensitivity towards temperature, applied forces, or added masses. Therefore, they are ideally suited for application as sensors.

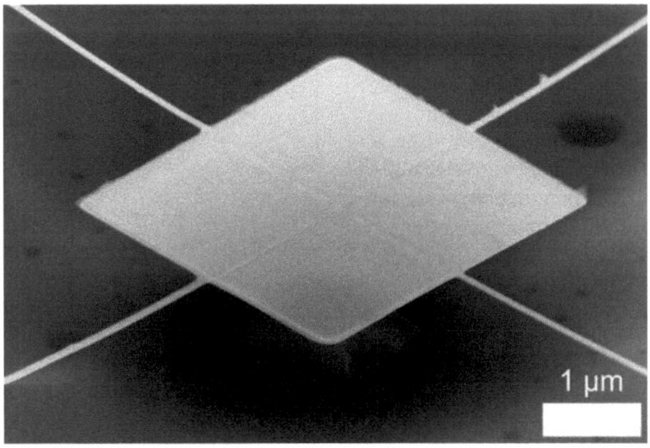


**Fig. 19.7** SEM image of a free-hanging pad (3 × 3 μm$^2$) suspended by four nanowires, which was fabricated by high-fluence Ga implantation of Si and subsequent anisotropic and selective, wet-chemical etching

Nanoscale patterning of materials can be achieved by different conventional top-down techniques, e.g., electron beam lithography [52], deep ultraviolet light lithography [53], nanoimprint lithography [54], stencil lithography [55], or by FIB [7]. Resistless methods like the FIB can simplify nanostructure fabrication especially for prototyping. Moreover, three-dimensional structures can be fabricated by combination of high-concentration p-type doping of Si by writing FIB implantation and subsequent anisotropic and selective, wet-chemical etching [56–58]. An example of a freely suspended pad fabricated using this technique is shown in Fig. 19.7.

This section gives an overview of the electrochemical fundamentals, the actual fabrication, size tuning as well as electrical and structural properties of Si nanowires, which can fabricated by high fluence implantation of Si and subsequent anisotropic and selective wet-chemical etching.

#### 19.4.1.1 Anisotropic and Selective Etching of Si

Anisotropic etchants for crystalline Si have been known for a long time [59, 60]. After recognizing its unique potential for micromachining and nanomachining of three-dimensional structures, increasing attention has been paid to this etching technology. An enormous variety of Si structures can be fabricated in a highly controllable and reproducible manner due to the strong dependence of the etch rate on crystal directions and dopant concentrations.

Dependence on the Si Crystal Planes

All anisotropic etchants of Si are aqueous, alkaline solutions, whereas the main component can be either organic, e.g., in TMAH, $(CH_3)_4NOH$ [61], $NH_2(CH_2)_2NH_2$ or $CH_6H_4(OH)_2$ [62], or inorganic, e.g., in KOH or NaOH [63–65]. In general,

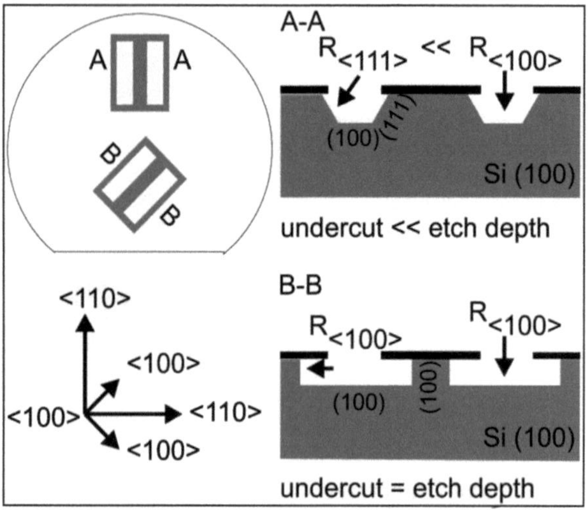


**Fig. 19.8** Schematic presentation of FIB patterning by $Ga^+$ implantation into ⟨100⟩-oriented Si wafers marked by *gray areas* and the corresponding cross section of the Si structures after selective, anisotropic, wet chemical etching

solutions containing hydroxides of alkali metals perform in a comparable way. Among the inorganic solutions KOH is most commonly used, mainly due to its cost efficiency and convenience of handling. An extensive overview of the anisotropic etching process of crystalline Si has been given by Seidel et al. [62]. It has been found that the etch selectivity of the Si planes {110}:{100}:{111} varies from 50:30:1 at room temperature to 160:100:1 at 100 °C. The effect of etch selectivity for differently aligned masks is schematically shown in Fig. 19.8.

The anisotropic etching behavior of Si is usually explained in the following electrochemical model [62]: Crystalline Si has a bandgap of 1.12 eV. Depending on the dopant concentration, the Fermi level is located above mid gap for n-type and below mid-gap for p-type Si. The electron affinity of Si (distance between vacuum level and the lower edge of the conduction band) is 4.05 eV. Due to different bonding situations of atoms near the crystal's surface, surface states arise within the forbidden band gap. The relative position of the surface states with respect to the band edges is independent of the dopant type within wide concentration levels.

After immersion of the Si into the electrolyte, a negative excess charge on the Si surface builds up, because the $H_2O$–$OH^-$ redox couple has a higher original Fermi level than the solid. This results in a downward bending of the energy bands on the solid surfaces for both, p- and n-type Si.

In an oxidation step, a Si atom is removed from the surface by reaction of four hydroxide ions forming $Si(OH)_4$, which is then converted to $SiO_2(OH)_2^{2-}$ due to the high pH environment. This step is accompanied by the injection of four electrons into the conduction band, which originates from the reacting hydroxide ions. For {100} surfaces, two of these electrons are transferred via dangling bond surface states, and the other two via back bond surface states. However, these injected electrons stay localized near the Si surface due to the potential well provided by the downward bended energy bands.

In a reduction step, the injected electrons leave the solid and react with water forming hydroxide ions and hydrogen. These hydroxide ions are considered to be the ones reacting in the oxidation step, as they are generated directly at the surface. Hydroxide ions of the electrolyte only play a minor role in the dissolution reaction due to the repellent force of the negatively charged crystal.

Finally, the breaking of the back bonds, which is induced by thermal excitation of electrons from the orbitals of the respective surface states into the conduction band is considered to be rate limiting. Thus, the anisotropy is due to the different energy levels of these back bond surface states for different crystal orientations, being lowest for {111} planes.

Dependence on the Dopant Concentration

Etching of crystalline Si by aqueous KOH solutions is known to depend on the dopant concentration [56, 66, 67]. After implantation of, e.g., B, P, or Ga, altered regions acts as an etch stop mask depending on the dopant concentration. This etch stop effect is usually quantified by the critical fluence, which is defined as the minimum fluence implanted into Si that will inhibit etching of the implanted regions for given etch conditions.

The mechanism by which implantation of Si provides an etch stop is not well understood. However, there is evidence that heavily doped regions grow a nonvolatile oxide layer during the etch process, which etches at much slower rates than Si [67]. This would be in agreement with the observation that the {111} planes oxidizes at a higher rate than other crystallographic orientations [68]. This passivation hypothesis implies that implanted Si oxidizes at a greater rate than non-implanted Si.

Another possible explanation is related to the large concentration gradient of the implanted species at the boundary of doped regions. The gradient would produce a depletion layer that can affect the electrochemical behavior of the etching process. The electrically active sheath of implanted Si may surround the beam damaged areas. This would explain the etch stop by an activated junction.

#### 19.4.1.2 Fabrication of Si Nanowires

One way to fabricate freely suspended Si nanowires, which are electrically isolated from the Si substrate, is to use 4″ silicon-on-insulator (SOI) wafers fabricated by smart-cut processing (e.g., ⟨100⟩-oriented, n-type, 1–10 Ω cm, top device layer thickness of 2 μm) [65]. A buried $SiO_2$ layer (200 nm thick) allows to control the etching process, because it is acting as an etch stop.

Initially, wafers are pre-patterned by optical lithography and conventional $Ga^+$ broad beam implantation ($E=30$ keV, $\Phi=3\times10^{16}$ cm$^{-2}$). In this way large area contact pads ($400\times400$ μm$^2$) with edges parallel and perpendicular to the ⟨110⟩-direction of the wafer are formed. Then, metallic contact pads are prepared by Cr/Au evaporation and standard lithographic lift-off process. Nanowires are defined by

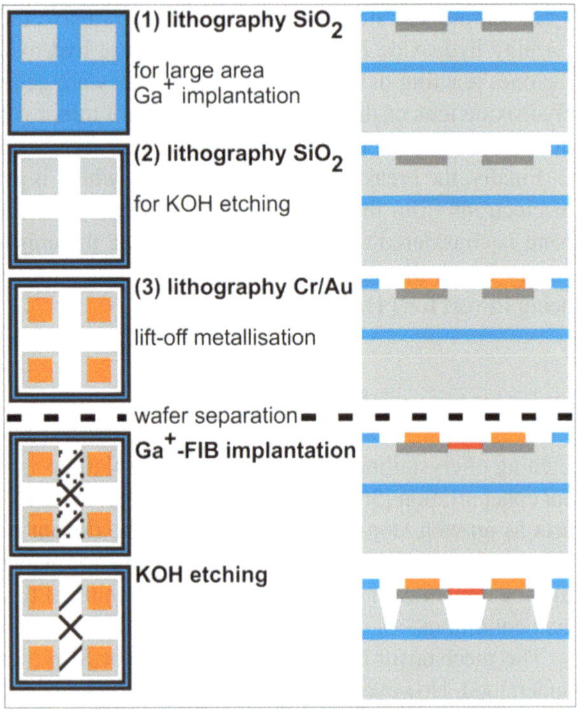


**Fig. 19.9** Illustration of the nanowire fabrication process: n-type SOI wafers with a top device layer of 2 µm Si are pre-patterned by optical lithography and conventional $Ga^+$ implantation to from large area contact pads. Metallic contacts are created by Cr/Au evaporation and standard lithographic lift-off process. Non-implanted Si stripes are treated by FIB writing to define nanowires that are free-standing after KOH etching

FIB writing between the contact pads areas over non-implanted Si stripes. Line scans of the FIB implantation should be aligned parallel to the ⟨100⟩-direction to achieve a fast under-cut during etching. Finally, the selective and anisotropic etching is done in a 30 % KOH/$H_2O$ solution at 80 °C for 3 min. The corresponding etch rate on non-implanted Si in ⟨100⟩-direction is 1.1 µm $\text{min}^{-1}$. The whole fabrication process is illustrated in Fig. 19.9.

#### 19.4.1.3 Size Tuning of Si Nanowires

In the following, experimental results are related to the implanted line fluence $\Phi_L$, i.e., the number implanted ions per centimeter line scan, and not to the typically used areal fluence, which is defined as

$$\Phi_A = \frac{\Phi_L}{w} = \frac{It}{elw}, \tag{19.7}$$

where $I$ is the FIB current, $t$ the implantation time, $e$ the elementary charge, $l$ the length, and $w$ the width of the irradiated area.

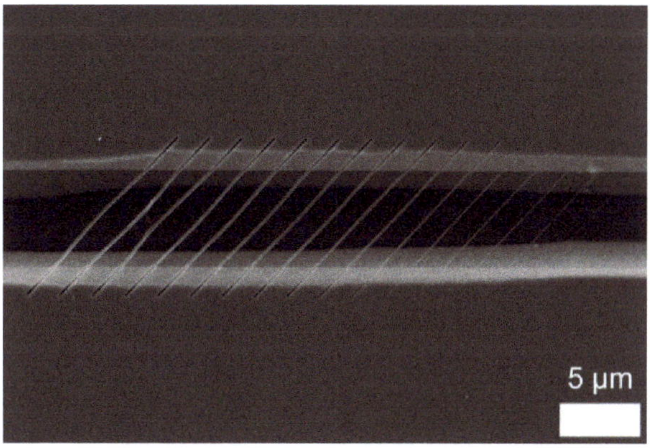


**Fig. 19.10** SEM image of nanowires implanted with an ion beam current of 10 pA. The line fluence is increasing from $1\times10^{11}$ to $4\times10^{12}$ cm$^{-1}$ from *right* to *left*, apparent by the increase of the nanowire width (from [65])

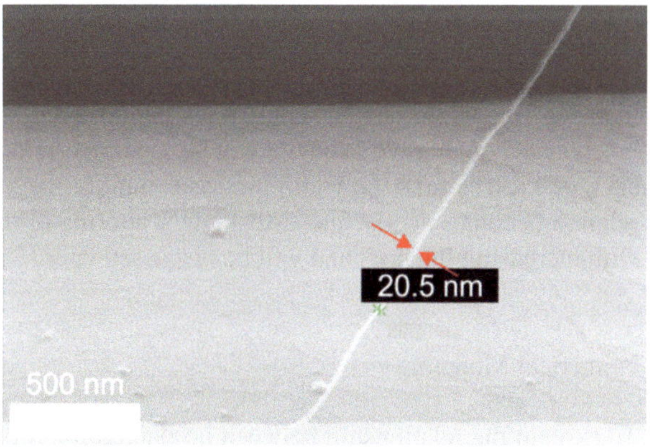


**Fig. 19.11** SEM image of one of the narrowest fabricated Si nanowires of 20.5 nm width. The line fluence for preparation was $6.6\times10^{10}$ cm$^{-1}$ (from [65])

### Nanowire Width

The line fluence was varied between $1\times10^{11}$ and $4\times10^{12}$ cm$^{-1}$ leading to different nanowire widths. Thus, wire dimensions as narrow as 20–200 nm in width and 7 and 15 µm in length were achieved (cf. Figs. 19.10 and 19.11). The nanowire thickness can be estimated to about 54 nm from the depth of the implantation profile and the from the critical Ga concentration for etch retardation.

Nanowires implanted with computational areal fluences below $2\times10^{15}$ cm$^{-2}$ did not withstand the etching process. This is in good agreement with results of Berry et al. [66], Chen et al. [69], and Steckl et al. [70], which reported the initiation of the etch stop at fluences exceeding $1\times10^{15}$ cm$^{-2}$, under similar etching and implantation conditions, but for larger implanted areas.

At fluences $\geq 9\times10^{11}$ cm$^{-1}$ sputtering effects appear during the line scanning, which leads to V-shaped cross sections of the nanowires. These sputtering grooves can be nicely seen at the ends of the implanted lines in Fig. 19.10, where the nanowires are suspended at bulk Si.

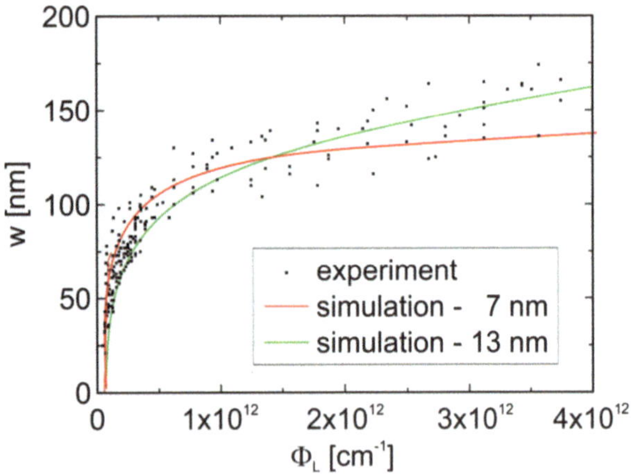


**Fig. 19.12** Width of Si nanowires as a function of the implanted line fluence. The experimental data were obtained using different FIB currents and spot sizes. *Solid curves* show the results obtained by the numerical model for FIB currents of 1 and 10 pA corresponding to spot sizes of 7 and 13 nm (from [65])

Figure 19.12 shows the nanowire width as a function of the implanted line fluence. The width of the nanowires is increasing with increasing line fluence, whereas the qualitative behavior can be separated into two regions: up to a line fluence of $1 \times 10^{12}$ cm$^{-1}$ the width increases rapidly; beyond this line fluence the correlation becomes nearly lines. Absolute scattering of the measured width increases with increasing fluences and will be discussed later.

### Numerical Modeling

To explain the relationship between implanted line fluence and resulting nanowire width the following numerical model was applied. The probability distribution for 30 keV $Ga^+$ ions to stop at a certain position in the Si sample, while line scanning the FIB over it, was simulated to determine the range of several ions. Afterwards, by examination of the Ga concentration per volume, the volume, which would withstand the etching process, i.e., having a concentration corresponding to a fluence higher than the critical threshold, was estimated. Effects of sputtering and backscattering were neglected.

The shape of the FIB intensity profile was approximated by superposition of two Gaussians [57]. Therefore, the probability for an impinging ion to hit the sample at a point in the *x*-*y*-plane can be was modeled by

$$P(x,y) = \frac{1}{\sqrt{2\pi}}\left[\frac{1}{\sigma_{peak}^2}\exp\left(-\frac{x^2+y^2}{2\sigma_{peak}^2}\right)+\frac{1}{\sigma_{tail}^2}\exp\left(-\frac{x^2+y^2}{2\sigma_{tail}^2}\right)\right]. \quad (19.8)$$

Variances of this distribution were calculated from measured FWHM of the FIB spot size for certain FIB currents, e.g., 1 pA yields $\sigma_{peak}{}^2 \approx 8.84$ nm$^2$ and

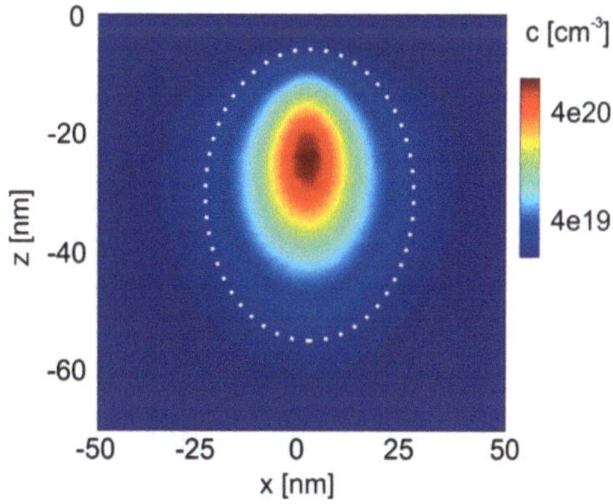


**Fig. 19.13** Cross section of the simulated implantation profile showing the Ga concentration. The *dashed line* shows the Ga concentration of about $4\times10^{19}$ cm$^{-3}$ for etch retardation (from [65])

$\sigma_{tail}^2 \approx 30.94$ nm$^2$. The case of defocused implantation could be obtained by scaling of the variances. The probability of an incident $Ga^+$ ion to reach a certain point in the Si bulk was then modeled by SRIM simulations [71] to obtain a concentration profile.

The cross section of the resulting implantation profiles was then used to estimate the nanowire dimensions by examination of the critical Ga concentration. Comparison between experimental and simulation (solid lines) is shown in Fig. 19.13. The onset of the curves and the exponential behavior for lower and linear relationship for higher fluences are reproduced by the simulation. For line fluences up to $1\times10^{12}$ cm$^{-1}$, the resulting nanowire width is slightly overestimate, while for higher line fluences the resulting wire width is underestimated.

## Conclusion of the Nanowire Size Evaluation

The increase of the nanowire width with increasing fluence can be attributed to the increasing amount of Ga implanted into the Si, because an increasing volume exceeds the critical Ga concentration to initiate the etch stop [67, 72]. The rapid increase of the nanowire width for line fluences up to $1\times10^{12}$ cm$^{-1}$ can be related to the shape of the FIB profile. Numerical simulation showed that in this regime the Gaussian peak fraction dominates the distribution of ions in the bulk. For higher line fluences, the lateral distribution of implanted ions is more homogeneous. Therefore, the slope of the measured curves decreases to an almost constant level. Furthermore, for higher fluences, significant sputtering has to be taken in the account. This leads to removal of sample material and thus to a deeper implantation of Ga ions and a further broadening of the lateral distribution of implanted ions. The stronger scattering of the nanowire widths for higher line fluences is due to deviations from the ideal ion beam focus during implantation. This effect accumulates over time leading to a significant variation of the nanowire width. Additionally, with increasing implantation time, the stability of the sample holder stage in the nanometer range becomes another issue.

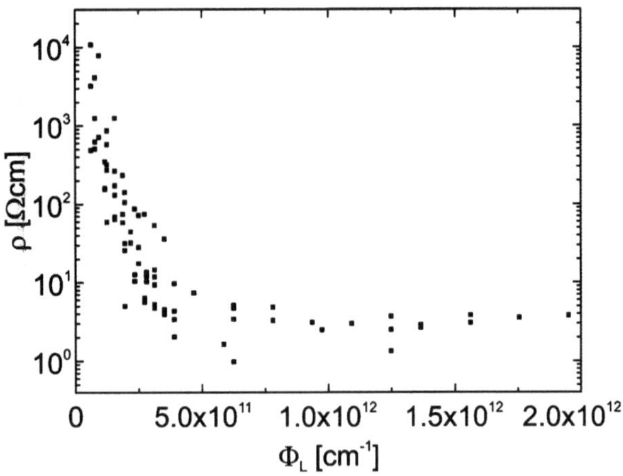


**Fig. 19.14** The resistivity of the as-fabricated nanowires is rapidly decreasing increasing line fluences and reaches saturation (from [65])

#### 19.4.1.4 Electrical Performance of Si Nanowires

As-Fabricated Nanowires

To assess the electrical performance of the fabricated nanowires, current–voltage (I–V) characteristics was measured using a semi-automatic prober (PA200 SÜSS MicroTech Test Systems) combined with a semiconductor measuring system (Keithley Instruments, 4200-SCS). Therefore, via two terminal measurements, voltage sweeps from −5 to +5 V were done while measuring the current. All nanowires revealed linear I–V characteristics, i.e., an ohmic behavior. The slope of the I–V characteristics and the nanowire dimensions, i.e., length $l$, width $w$, and thickness $b$, were used to estimate their resistivities $\rho$ using the following relationship.

$$R = \rho \frac{l}{wb} \tag{19.9}$$

The resistivities of the as-implanted nanowires ranged over several orders between 6 and 11,000 Ω cm. The estimated resistivities as a function of the implanted line fluence are given in Fig. 19.14. The resistivity is vastly increasing with decreasing line fluence, i.e., for thinner nanowires. For higher line fluences exceeding $\approx 5 \times 10^{11}$ cm$^{-1}$, i.e., for broader nanowires, the resistivity is in the range of the bulk Si wafer (1–10 Ω cm). It is likely that the rapid resistivity increase for low line fluences originates from a larger surface to volume ratio of the corresponding nanowires.

Additionally, the electrical performance was also tested at different temperatures in the range from −15 to 95 °C. This is exemplarily depicted in Fig. 19.15 for a nanowire with a width of 90 nm. The resistance of the all the fabricated nanowires is decreasing with increasing temperature, i.e., they are showing a semiconducting behavior in the as-fabricated stage (highly doped amorphous Si layer).

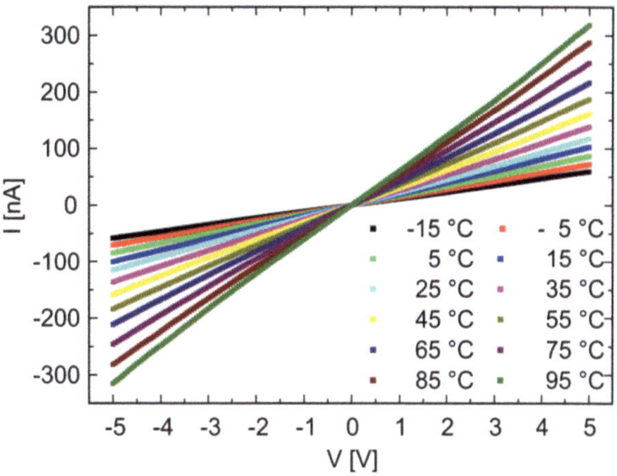


**Fig. 19.15** I–V characteristics of a single as-fabricated nanowire ($w = 90$ nm) at different temperatures. The behavior is ohmic in the chosen temperature range. Resistance is decreasing with increasing temperature (from [65])

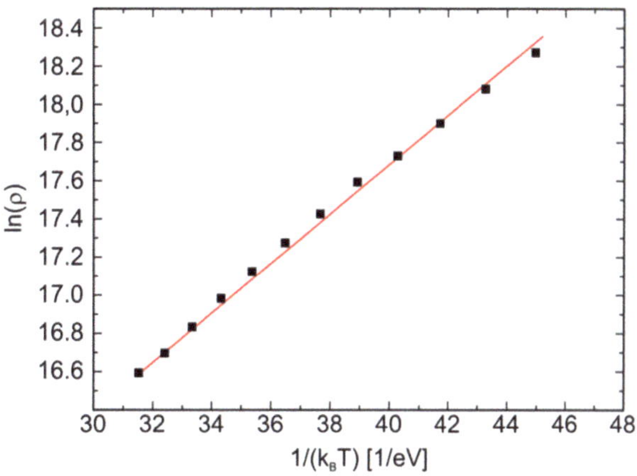


**Fig. 19.16** Arrhenius plot of the temperature dependence of the resistivity one nanowire ($w = 90$ nm) showing a semiconducting behavior. The slope of the linear fit represents the activation energy of the conduction mechanism (from [65])

The resistivity as a function of temperature can be expressed as follows, if an activation behavior is assumed [73]:

$$\rho(T) = \rho_0 \exp\left(\frac{E_A}{k_B T}\right), \tag{19.10}$$

where $\rho_0$ is a pre-exponential factor, $k_B$ is the Boltzmann constant, $T$ the absolute temperature, and $E_A$ the activation energy for the conduction mechanism. The slope of Arrhenius plots was used to estimate the activation energies of the as-implanted nanowires. For example, as depicted in Fig. 19.16., a nanowire having a width of 90 nm ($\Phi_L = 6 \times 10^{11}$ cm$^{-1}$) revealed an activation energy of 126 meV.

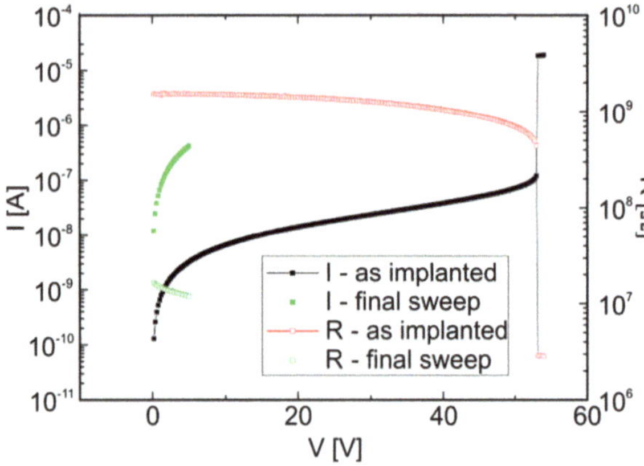


**Fig. 19.17** I–V characteristic and resistance of a single nanowire ($w = 74$ nm) during and after direct current annealing. According to the expected self-heating, there is a rise of current of some orders of magnitude (from [65])

Measured activation energies are increasing for increasing line fluences. This qualitatively reproduces the correlation between resistivity and line fluence. Activation energies range between 110 and 290 meV, which is in the expected range of amorphous Si [74]. However, no systematic correlation between the activation energies and the line fluence used for the fabrication could be observed.

Direct Current Annealing

The resistivity of Si nanowires can be significantly reduced by annealing processes. However, thermal movement of nanowires between the SOI islands and their consequent rupture at temperatures ≥150 °C excludes conventional furnace annealing (details of the thermal behavior were investigated by Sainiemi et al. [75]). Nevertheless, it was reported that current-driven self-heating processes can be applied to recrystallize Si nanowires [76]. Therefore, to lower the resistivity of the fabricated Nanowires a two terminal set-up providing voltage while measuring the current was used to investigate possible DC annealing. Voltage was raised from 0 V up to higher voltages of 30–50 V with sweep rates between 0.5 and 1.0 V $s^{-1}$. The hold times per voltage value was set between 1 and 500 μs.

I–V characteristics of a nanowire during and after DC annealing are shown in Fig. 19.17. The plot shows a temperature depending behavior, which corresponds to the increasing power deposition in the nanowire with increasing voltage. Exceeding a critical voltage, the current through the nanowire rises drastically. This indicates a modification of the nanowire, i.e., annealing of the nanowire. Remeasurement of room temperature I–V characteristics showed that the resistivity of annealed nanowires decreased by one to two orders of magnitude. Achieved resistivities ranged between 0.4 and 3.8 Ω cm. This is less than the resistivity of the initial Si wafer. Activation energies of the annealed nanowires ranged between 180 and 210 meV. While for narrower nanowires an increase of the activation energy could be found, for broader nanowires no significant change occurred, which might be related to the already mentioned surface-to-volume ratio.

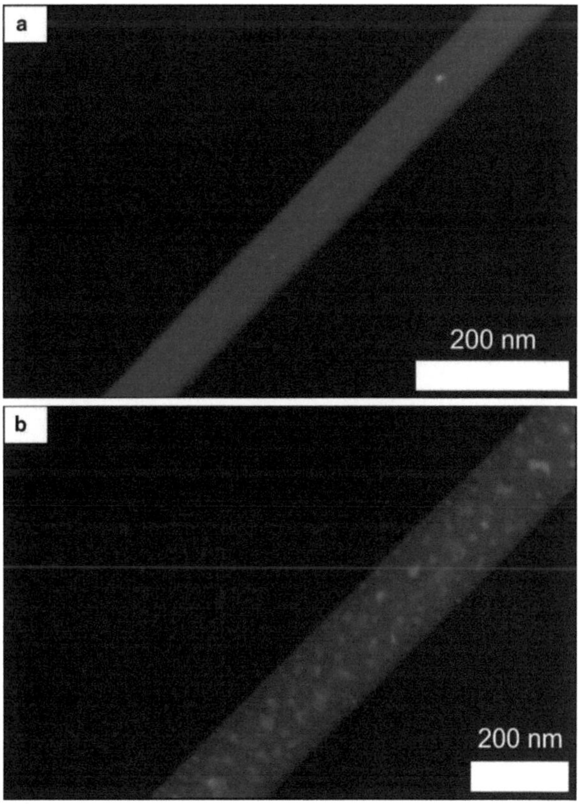


**Fig. 19.18** SEM images of (**a**) an as-fabricated, amorphous nanowire and (**b**) a nanowire after direct current annealing showing a granular structure (from [65])

A change of the annealing ambient from air to vacuum ($10^{-6}$ mbar) using the same annealing parameters did not result in any further improvement of the resistivity. Due to this, it is likely that the thermal conductivity of the ambient as well was chemical reaction, like oxidation, are negligible for this annealing process.

The change of the electrical properties of the Si nanowires can be summed up as follows: with increasing applied voltage the current through the nanowires increases. Current densities at the moment of the sudden current rise (see Fig. 19.17) range between 10 and 300 kA $cm^{-2}$. Thus it can be assumed that due to the self-heating process the amorphous nanowires underwent a phase transition from the amorphous to a nano-/polycrystalline state. This can also readily be concluded from Fig. 19.18, which shows an as-fabricated nanowire with a homogeneous surface (Fig. 19.18a) and an annealed nanowire with a granular structure (Fig. 19.18b).

A further increase of the annealing current leads to the formation of larger grains (Fig. 19.19) and with it to a fluctuation of the nanowire width. Furthermore, Fig. 19.19 also provides some evidence for an occurring Rayleigh instability, before finally local melting takes place at pinched-off regions leading to the rupture of nanowires. This can be seen in the inset of Fig. 19.19, where a Si slug remained after local melting and rupture of the nanowire.

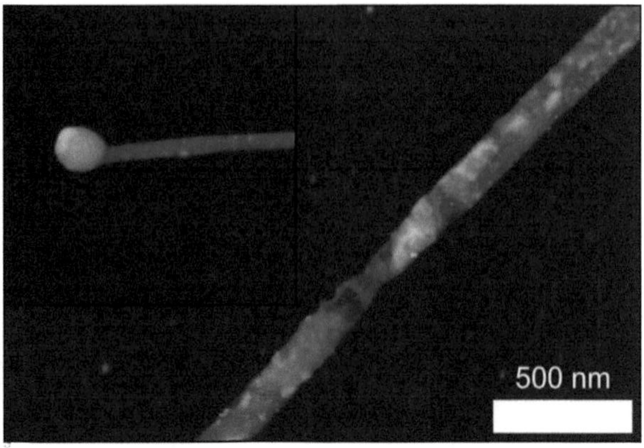


**Fig. 19.19** SEM images of the onset of the rupture due to thermal stress and electromigration during direct current annealing of a single nanowire. The *inset* shows a Si slug at one end of a nanowire after its rupture due to excessive annealing (from [65])

Nanowire resistance below the resistivity of the initial Si wafer, reveal a change in the conduction mechanism. Due to the recrystallization of the nanowires, it can readily be assumed that a certain fraction of the implanted Ga is activated during the annealing process. This results in a change from n- to p-type conduction. Estimation shows that 0.1–1 % of the implanted Ga is actually activated during the direct current annealing.

Summarizing the annealing behavior it can be estimated that the nanowire temperature during the direct current treatment was between 400 and 900 °C, due to the activation of the Ga and as there were no thermal radiation effects during the vacuum annealing.

#### 19.4.1.5 Structural Investigations by Raman Scattering

The transition of the as-fabricated amorphous nanowires to a nano-/polycrystalline state can be proven by Raman measurements. Therefore, a micro-Raman 180° backscattering configuration on a Labram HR spectrometer equipped with a holographic 18,000 lines $mm^{-1}$ grating and a liquid nitrogen cooled CCD detector (Horiba-Jobin-Yvon, France) was used [77, 78]. For excitation a frequency doubled Nd:YAG laser with a wavelength of 532 nm was used. The spectral resolution was 1.5 $cm^{-1}$. Stray light rejection was provided by holographic notch filters. A wave number accuracy of ±0.5 $cm^{-1}$ was assured. The laser beam was focused using a long working distance objective with a 100-fold magnification to a focus diameter of about 400 nm. Laser power and accumulation times were set to 100 μW and 2 min, respectively. Heating or degradation effects can be excluded at these conditions. For Raman mapping, a motorized microscope stage with 100 nm lateral resolution and 1 μm reproducibility was used.

Line scans with a step width of 100 nm were performed across a bunch of as-fabricated nanowires and one single annealed nanowire. The path of the line scans and the resulting Raman spectra are presented in Fig. 19.20 resulting from a bunch of nanowires with decreasing widths from 140 to 40 nm. The signal from the underlying bulk Si at a wave number of 521 $cm^{-1}$ is dominant in all spectra, since the $SiO_2$

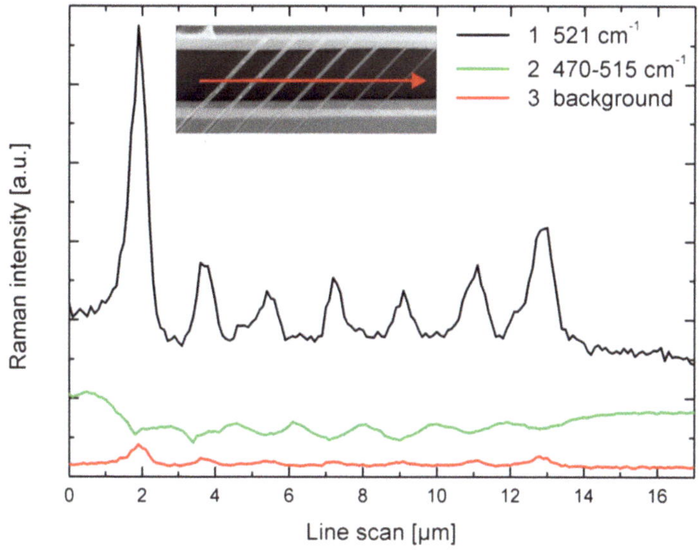


**Fig. 19.20** Comparison of the Raman intensity of the crystalline Si peak at 521 cm$^{-1}$ (*black line*), the summed range from 470 to 515 cm$^{-1}$ (*green line*) and the background intensity (*red line*). The *inset* shows the path of the lines scan across the nanowires with decreasing width (from [65])

**Fig. 19.21** Normalized Raman spectra of as-fabricated nanowire (*green*, *1*), annealed nanowire (*red*, *2*) and crystalline reference Si (*3*, *black*) (from [65])

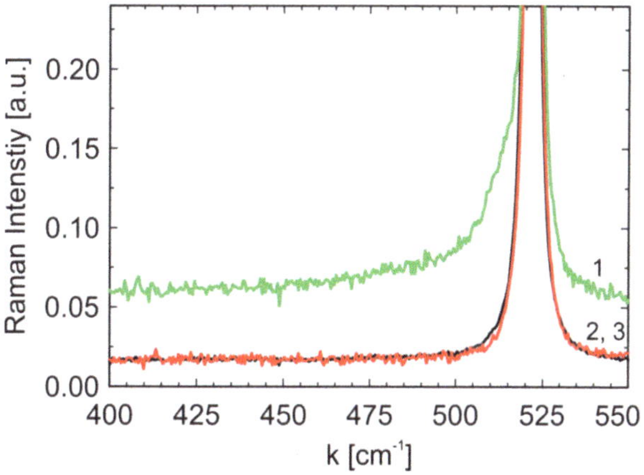


layer between the nanowires and the bulk is transparent for the used laser light. However, this intensity drops significantly whenever the laser spot crosses a nanowire. Simultaneously, the Raman intensity at wave numbers between 470 and 515 cm$^{-1}$ increases slightly (see Fig. 19.20). The period of the intensity fluctuation and the real distance of the nanowires coincide. The intensity drop of the Si crystal peak can thus be readily identified with the crossing of a nanowire.

Raman spectra obtained from an as-fabricated and an annealed nanowire at the central position, i.e., at minimum bulk Si intensity, were compared to the spectrum of reference Si. In Fig. 19.21 it is shown that the as-fabricated nanowire shows a significant Raman intensity between 470 and 515 $cm^{-1}$, besides the line of crystalline Si. Line shape analysis shows two lines at 492 and 513 $cm^{-1}$ as origin of this intensity. The lower energetic line can be attributed to amorphous Si, whereas the latter originates from nanocrystals or grain boundaries [79]. After annealing the nanowires do not show these additional Raman lines. Hence, the as-fabricated nanowires have an amorphous structure containing nano-/polycrystalline grains, whereas the annealed nanowires show no significant difference to the reference Si spectrum. This supports the assumed structural change of the nanowires due to direct current annealing.

#### 19.4.1.6 Summary

Freely suspended Si nanowires can be fabricated by FIB Ga implantation and subsequent wet-chemical etching. Adjustment of the line fluence allows control over the nanowire width and is mostly limited by the resolution of the FIB. Electrical properties of the nanowires depend strongly on their size, i.e., the line fluence during fabrication. Such fabricated nanowires show a semiconducting behavior with activation energies for the electrical conduction in the range of amorphous Si. Moreover, self-heating by application of direct current can be used to anneal and recrystallize the nanowires and, in this way, to improve their electrical properties.

### *19.4.2 Conducting Pathways and Ga Nanoparticles by FIB Lithography on ta-C*

Tetrahedral amorphous carbon (ta-C) is a metastable form of carbon consisting of a high fraction of $sp^3$ hybridized atoms (70–80 %) and a minority of $sp^2$ cluster sites. Such ta-C thin film coatings show a strong diamond-like character like high electrical resistivity, optical transparency in the visible range, high atomic density and a high hardness. All these properties arise from the $sp^3$ carbon network and the structure of the embedded $sp^2$ clusters. Consequently they can be tuned over a wide range by adjusting the $sp^3/sp^2$ ratio during film deposition [80–82]. The formation of films with low $sp^3/sp^2$ ratios (around 0.2) results in a more graphite-like material, which consists of a $sp^2$ matrix and some incorporated $sp^3$ cluster sites. In terms of electrical properties the unique ability of amorphous carbon to reflect either diamond-like (insulating) or graphite-like (conductive) behavior, makes it very promising for post deposition processing and modification. Therefore, ion irradiation is an appropriate method to destroy the matrix of $sp^3$ rich ta-C layers that enforces a reordering of carbon atoms and a phase transformation to $sp^2$ bonded carbon. Reaping the benefits of a focused ion beam (FIB), above all the very small beam spot size of several nm, ta-C layers can be transferred to $sp^2$ rich graphite-like amorphous carbon in small lateral dimensions. This allows a direct FIB lithography on ta-C that is simply writing of conductive nanostructures in the high resistive $sp^3$ matrix without the use of photo resist, subsequent chemical etching or film deposition.

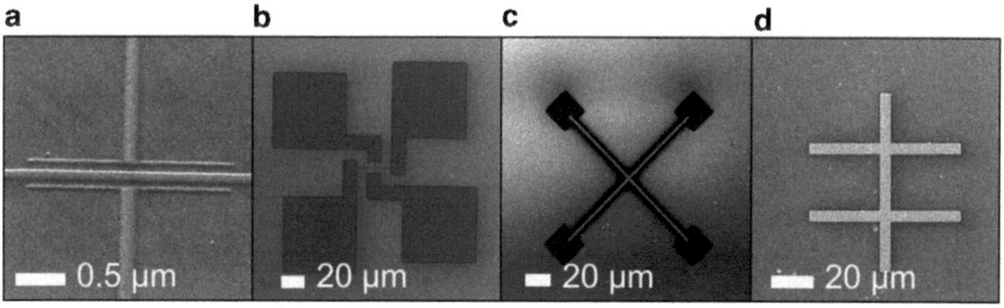


**Fig. 19.22** SEM images of graphitic structures made by FIB lithography on ta-C. (**a**) FET gate, (**b**) and (**c**) van der Pauw structures, (**d**) 6-terminal Hall bar

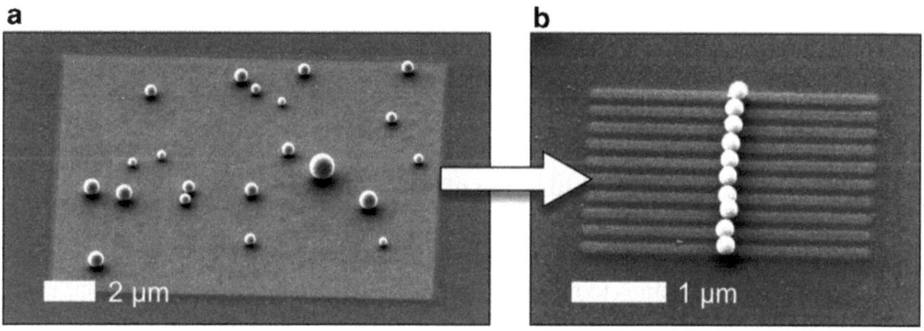


**Fig. 19.23** SEM images of Ga nanoparticles on ta-C made by Ga implantation and subsequent annealing. (**a**) Various arrangement and (**b**) tailored evolution to a nanoparticle chain

Figure 19.22 shows some examples for defined layouts which were transferred as electrical active structures into the ta-C layer by FIB lithography. Whereas the achievable feature size is mainly dependent on the ion beam spot diameter, the electrical resistivity can be tuned over a wide range by adjusting the irradiation conditions, like the primary ion energy and the ion fluence that is ions per $cm^{-2}$.

Besides the above mentioned insulating-to-conductive phase transformation of carbon, the chemical inertness of carbon can be used to enforce a phase segregation between carbon and the implanted species. The incorporation of Ga into ta-C layers followed by thermal treatment of the sample causes clustering and the diffusion of Ga to the surface. The high mobility and surface tension of Ga supports the formation of Ga nanospheres on irradiated sites. It will be shown that well-considered FIB implantation and annealing conditions are sufficient for the fabrication of Ga nanoparticles with a high reproducibility and a very small divergence in the particle size.

Utilizing this promising technique of $Ga^+$-FIB implantation in ta-C and subsequent annealing, metal nanoparticles with a defined size can be arranged in chains (see Fig. 19.23) and two-dimensional arrays that are prospective structures for plasmonic applications.

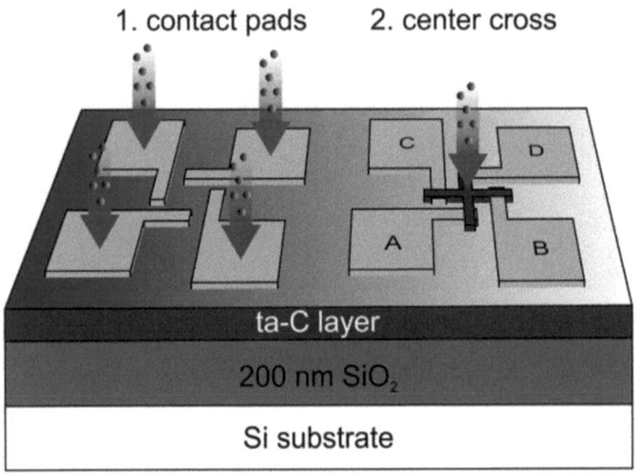


**Fig. 19.24** FIB lithography of a VDP measurement structure. (*1*) Implantation of contact pads and (*2*) Implantation of the center cross

#### 19.4.2.1 Ion Induced Graphitization of ta-C

The graphitization of pure diamond by ion bombardment was discovered by Vavilov in 1974 [83] and so it stands to reason that the conversion mechanism from $sp^3$ to $sp^2$ hybridized carbon occurs in diamond-like materials as well. Fundamental investigations concerning the nature of defects generated by ion irradiation of ta-C were done in the 1990s by McCulloch [84], Kalish [85], Prawer [86], and Stanishevsky [87]. The main findings are a decrease of the electrical resistivity of several orders of magnitude, a decrease of the film density, which is accompanied by surface swelling of the irradiated regions and a decrease of the optical transparency. In order to tailor the electrical properties of implanted ta-C sites, the dependence of the ion fluence on the resistivity was intensively studied for the case of irradiation with 30 keV $Ga^+$ ions. The ta-C films used in this work are hydrogen free diamond-like carbon layers deposited by filtered cathodic vacuum arc (FCVA) (provided by the Fraunhofer IWS in Dresden). Amorphous carbons show a good linear correlation between the film density and the $sp^3$ content thus the density, measured by X-ray reflectivity, of about 2.5 g $cm^{-3}$ corresponds to a $sp^3$ content of ~60 %. A 200 nm thick thermally grown $SiO_2$ layer on a silicon wafer was used as substrate. This oxide interlayer avoids any short circuits through the bulk Si during the electrical characterization.

For an effective conductivity measurement of graphitized ta-C formed by various irradiation conditions, the direct conversion of a Van der Pauw (VDP) [88] measurement structure by FIB lithography is the method of choice. The fabrication process is shown in Fig. 19.24. Firstly the structure periphery consisting of four contact pads with small connection extensions is implanted with 30 keV $Ga^+$ ions and a fluence of $5 \times 10^{16}$ $cm^{-2}$. The four pads (A, B, C, and D) are $30 \times 30$ $\mu m^2$ in size and can be electrically connected with the narrow needles of a micro probe station. Secondly the connection extensions are linked to each other by a cross irradiated in the center and so the VDP structure is realized.

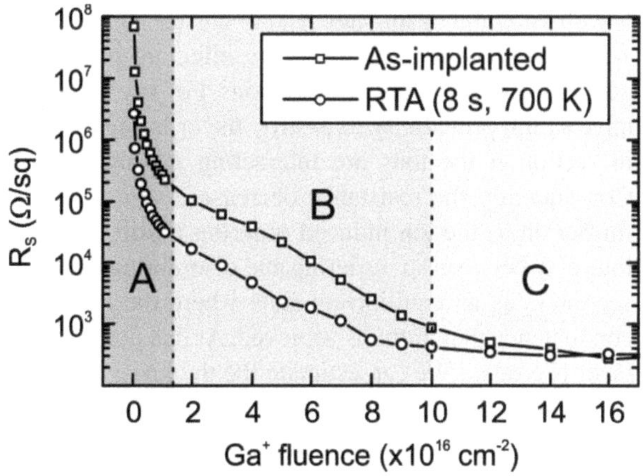


**Fig. 19.25** Sheet resistance of ta-C irradiated with 30 keV $Ga^+$ ions in dependence of the implantation fluence. The values are for as the as-implanted case and for structures treated by rapid thermal annealing (RTA)

The implantation parameters of the center cross can be varied for every single structure, whereas the processing of the contact pads relates always to the same fixed values. Performing a VDP measurement at this structure delivers the electrical sheet resistance $R_S$ of the region right in the middle of the center cross where the irradiated lines are overlapping. Such a measurement is simply done by applying a small current $I$, usually in the range of 0.1–10 μA, between two adjacent pads and measuring the potential difference $V$ between the opposite pads. The resulting resistance $R' = V/I$ is independent of all parasitic resistances from the contact pads and connections. Furthermore, the structure symmetry simplifies the VDP formula so that $R_S$ can be obtained in good approximation from

$$R_S = \frac{\pi}{\ln(2)} R'. \tag{19.11}$$

To further minimize errors caused by for example variations in the structure symmetry, the pads that correspond to current injection and voltage measurement are permuted. This yields in four values of $R'$ and their average is used for the calculation of $R_S$. In order to obtain the fluence dependence of the conductance of ion irradiated ta-C a series of embedded VDP structures was fabricated. The implantation of the center cross was done with 30 keV ion energy and fluences ranging from $5\times10^{14}$ cm$^{-2}$ to $1.6\times10^{17}$ cm$^{-2}$. The results are shown in Fig. 19.25 [89].

The general behavior of the as-implanted sheet resistance is a decrease by several orders of magnitude caused by the continuous formation of $sp^2$ sites during irradiation. To shed more light on this processes we can divide the graph into three sections. In the low fluence regime A, the resistance drops dramatically with increasing fluence conformal to a power law. In section B the correlation switches to an exponential decrease for the mid-to-high fluence irradiations. This can be attributed to the interplay between conversion of $sp^3$ to $sp^2$ sites and the destruction of $sp^2$ clusters by the continuous ion bombardment. Section A reflects the approximately complete transformation of the ta-C matrix to mainly $sp^2$ bonded amorphous carbon. It is well known that $sp^2$ sites tend to cluster and can achieve some near ordering like sixfold rings. Such structures are effective in increasing the conductivity by the generation of delocalized $\pi$ electrons. For any further irradiation (section B) the ions have a high probability to destroy the ordering of the just created $sp^2$ sites, whereas in section A the ions are interacting mainly with virgin ta-C and its $sp^3$ sites. Consequently the resistance decrease is alleviated from section A to B but goes further on as the ion induced ordering is still the dominant process. Based on this interplay between $sp^2$ ordering and disordering during ion irradiation we can explain section C as an equilibrium state where the resistance decrease saturates and the conductance maximum is achieved. At this point sputtering occurs and the graphitic layer is eroded. We can estimate the thickness of this modified ta-C layer by SRIM simulations [90] to approximately 40 nm. Whilst the ta-C film thickness is beyond this value continuous sputtering will drive the graphitic layer downwards until the $SiO_2$ interlayer is reached. At this prominent point the graphitic layer is vanishing and the resistance goes up which was observed for even higher fluences.

Once the $sp^2$ rich structures are created by ion implantation the graphitization can be enhanced by thermal treatment and thus annealing is effective to improve the electrical properties. This is shown in the circle data points in Fig. 19.25 which correspond to a rapid thermal annealing (RTA) of all structures at 700 K for 8 s. The conductance is pushed upwards and so all resistance values decrease round about one order of magnitude. It has to be noted that, for the case of the chosen 30 keV $Ga^+$ irradiation, the conductivity maximum is not affected by the annealing and consequently the ion induced processes ensure the optimal arrangement of the $sp^2$ sites at the point of saturation (see section C in Fig. 19.25). If we consider the effect of annealing on structured ta-C samples, the advantage of the ta-C's thermal stability can be used. ta-C with an $sp^3$ content above ~70 % is able to withstand heating up to 1,100 K [91]. This stability is caused by the $sp^3$ matrix and consequently any defects like $sp^2$ sites introduced by ion irradiation make the carbon contestable. Annealing of a ta-C layer consisting of graphitic structures below a critical temperature will result in an improvement of the structure's conductance without affecting the surrounding ta-C. A corruption of the electrical insulation is absent.

#### 19.4.2.2 Embedded Graphitic Nanowires

The discussions in the above section considered the basics of ion induced graphitization of ta-C and the electrical resistivity of $sp^2$ rich sites. Let us now proceed to conducting nanostructures fabricated by FIB lithography on ta-C layers. The small ion spot size enables a defined local graphitization of areas with a lateral dimension down to several tens of nanometers. If a focused beam of $Ga^+$ ions is scanned laterally in a single line over the ta-C surface the result is an $sp^2$ rich nanowire (NW) embedded in the virgin ta-C matrix that can provide a significant current transport. Firstly the influence of the ion fluence on the NW resistance will be discussed for 30 keV $Ga^+$ irradiations. Figure 19.26 shows the resistance of several NWs ($R_{NW}$) irradiated with line fluences ranging from $5 \times 10^7$ up to $1.2 \times 10^{11}$ $Ga^+$ ions per cm [89]. The line

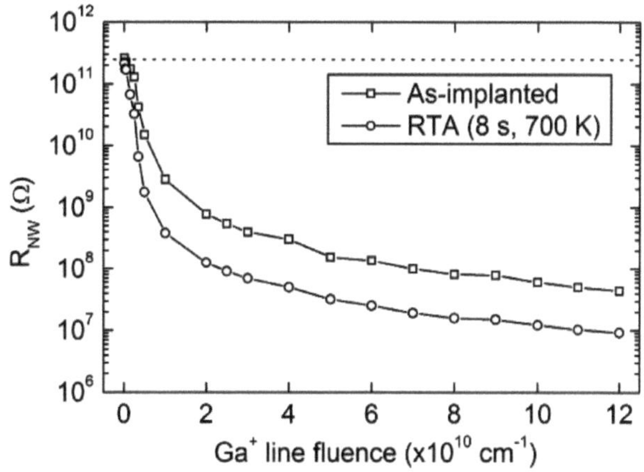


**Fig. 19.26** Resistance of nanowires ($R_{NW}$) fabricated by 30 keV $Ga^+$ irradiations in dependence of the line fluence. As-implanted wires and wires annealed by RTA. The *dashed line* indicates the resistance of the virgin ta-C

fluence is introduced in Sect. 19.4.1.3 and gives an adequate value for the amount of implanted ions in a one dimensional structure.

To perform the resistance measurements two contact pads with a lateral distance of 15 μm were created by FIB lithography. Therefore, the same irradiation conditions like for the periphery of the VDP structures (see Sect. 19.4.2.1) were used. The NW was then subsequently fabricated between the two pads and $R_{NW}$ was obtained by determining the slope of the current–voltage curve (I–V curve). It has to be noted that this resistance includes the contact resistance of the probe tips as well as the resistance of the contact pads. Due to the small NW cross section the wire resistance $R_{NW}$ is quite high in comparison with the pad and contact resistance and thus these two resistances can be neglected. The measurement yields in a value for $R_{NW}$ in good approximation. Figure 19.26 contains a dashed line that indicates the resistance of the virgin ta-C, which was gained by performing an I–V measurement between two contact pads without any NW in between. Low fluence irradiations show a very small increase of the conductivity and NWs fabricated with a line fluence below $10^9$ $cm^{-1}$ show no significant current transport and reflect the virgin ta-C resistance. Above this fluence the resistance drops dramatically but the rapid conductance enhancement gets damped at round about $2\times10^{10}$ $cm^{-1}$ to a slight resistance decrease for higher fluences according to the curve progression in Fig. 19.26. Saturation of the NW conductance is expected for even higher fluences, which is not shown here. The improvement of the current transport by annealing mentioned in the above section is effective to decrease the NW resistance of about one order of magnitude. The corresponding RTA process is again 700 K for 8 s. The graph demonstrates the adjustability of the NW conductivity over four orders of magnitude by varying the line fluence.

For the discussions of the electrical conductance it is important that we carefully distinguish between the sheet resistance $R_S$ obtained by VDP measurements and the nanowire resistance $R_{NW}$ gathered from two-terminal I–V curves. The former is independent of the structure design and their lateral dimensions and gives a value for the intrinsic materials conductivity, whereas the latter is strongly affected by the

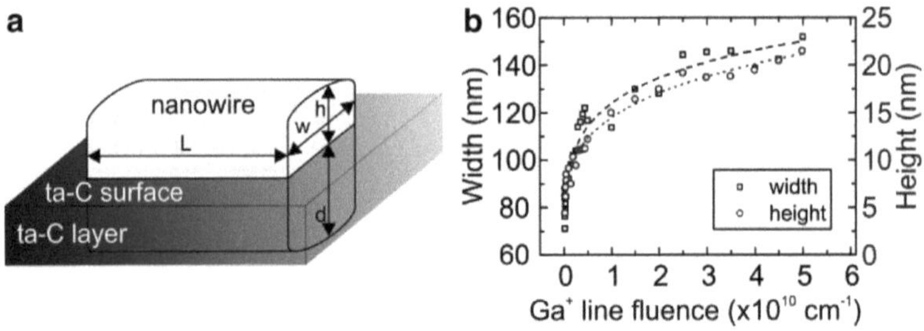


**Fig. 19.27** (**a**) Nanowire embedded in ta-C. Nanowire dimensions: *L*—length; *w*—width; *h*—height from the ta-C surface; *d*—depth in the ta-C layer. (**b**) Width and swelling height of nanowires fabricated by 30 keV $Ga^+$ irradiation on ta-C. The *lines* are to guide the eyes

NW dimensions that are the length *L* and the cross section (see Fig. 19.27a). *L* is predefined by the contact pad distance of 15 μm but the other parameters *w*, *h* and *d*, which defines the NW cross section, is predicted by the focused beam itself. The depth *d* of the graphitization in the ta-C layer is given by the ion energy and thus fixed for all NWs. The height *h* in Fig. 19.27a is the surface swelling height caused by the $sp^3$ to $sp^2$ conversion and the density decrease.

Whereas *d* is independent of the ion fluence, *h* rises during ongoing implantation. The wire width *w* shows a fluence dependence as well and is furthermore affected by the ion beam spot size and charging effects. Let us consider a FIB spot with a defined finite diameter of a few nanometers which is hitting the ta-C surface and propagates along it to form a NW. We will expect *w* to be larger than the spot size due to the lateral straggling of the implanted ions within the solid. The ion collision cascade will affect surrounding target atoms and the range of graphitization gets broader. Additionally it has to be taken into account that the virgin ta-C is high insulating and thus the incoming ions cause charging of the surface. This charging forces the FIB to spread and so the effective spot size on the surface increases. Nevertheless, a fine FIB focus is essential for structuring the ta-C in very small lateral dimensions. The fluence dependence of the NW width and the height are shown in Fig. 19.27b [89].

Like it was mentioned before the NW height, shown as rectangular data points in Fig. 19.27b, increases with increasing fluence due to the density decrease by the evolving $sp^2$ sites. The fluence dependence of *w* can be understood by considering the ion beam profile. The intensity distribution of a FIB can be described as the superposition of two Gaussians that reflects the high intense beam center and low intense wings (see Sect. "Numerical modeling"). For low fluences the amount of Ga incorporated by the wings can be neglected but rises consequently during irradiation and for higher fluences the regions exposed to the wings of the beam profile become significantly contaminated with gallium. We can summarize the two fluence-dependent effects, NW broadening by the beam profile wings and NW height increase by surface swelling, as an increase of the NW cross section with increasing ion fluence. If we compare the behavior shown in Fig. 19.27b with the resistance chart in Fig. 19.26, the dramatic drop of $R_{NW}$ for low fluences can be understood as the sum of conductivity increase by graphitization and current transport increase by a growth of the NW cross section.

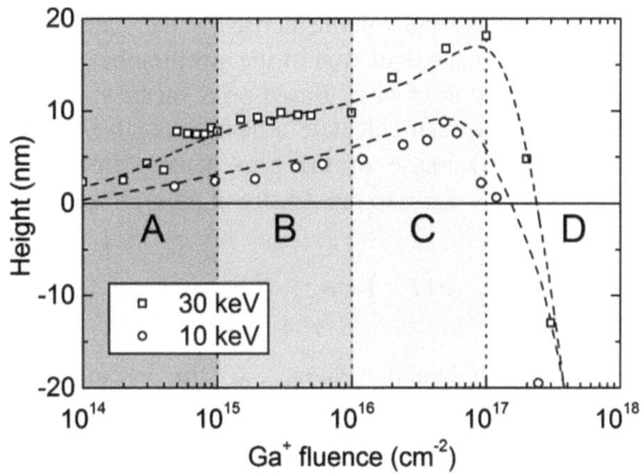


**Fig. 19.28** Surface swelling of ta-C by irradiation with $Ga^+$ ions in dependence on the ion fluence. The data points are AFM measurements and the *dashed lines* are analytical calculations

### 19.4.2.3 Surface Swelling

For a deeper investigation of the surface swelling, which occurs on irradiated ta-C areas, several $4 \times 4\ \mu m^2$ squares were implanted with 10 and 30 keV $Ga^+$ ions by FIB and the height was determined by atomic force microscopy (AFM). The results for the fluence dependence of the swelling height are shown in Fig. 19.28 [92].

The two swelling curves can be divided into four sections: Section A reflects the density decrease due to $sp^3$ to $sp^2$ conversion up to a fluence of $1 \times 10^{15}$ cm$^{-2}$. For the case of 30 keV ion energy the irradiated sites are outgrowing of about 10 nm with respect to the virgin ta-C surface. In the mid fluence regime in section B the density decrease saturates and the height remains more or less constant. Above $1 \times 10^{16}$ cm$^{-2}$ the outgrowth continues again up to a maximum swelling height that is dependent on the ion energy. This mechanism is addressed to gallium swelling that is the volume increase due to a significant amount of the incorporated material. For the corresponding fluences in section C the Ga content can excess 20 at% and so the volume occupied by the implanted ions cannot be neglected anymore. The maximum of the swelling curve indicates the complete saturation of the ta-C matrix with Ga ions and the resulting material can be described as a mixture of $sp^2$ rich amorphous carbon with some short range order and incorporated gallium. The fluence corresponding to the swelling maximum of the 30 keV energy curve agrees well with the fluence of the maximum conductivity shown in Fig. 19.25. Any further ion bombardment yields in sputtering of the ta-C layer that is shown in section D. The difference in the absolute swelling height between the two ion energies originates from the energy dependent penetration depth of the $Ga^+$ ions. Higher energies result in a larger projected ion range and so the target volume affected by graphitization increases with increasing energy.

Recent work on the surface swelling induced on diamond by the irradiation with light ions at high energies ($H^+$ and $He^+$ at several MeV) was theoretical described by Bosia and coworkers [93]. Vacancy creation during ion bombardment was introduced as the driving force for $sp^3$ to $sp^2$ conversion, which is accompanied by the density decrease and volume increase of the material. We will now use this approach for an analytical calculation of the swelling height in section A and B for the case of Ga irradiation of ta-C. Based on a vacancy depth profile obtained from SRIM simulations we can calculate a density depth profile. This is characterized by an exponential decrease of the film density from the value of virgin ta-C $\rho_{\text{ta-C}}$ (2.710 g cm$^{-3}$) down to the density of amorphous carbon $\rho_{\text{a-C}}$ (1.557 g cm$^{-3}$) [94]:

$$\rho(F,z) = \rho_{ta\ \mathrm{C}} - (\rho_{ta\ \mathrm{C}} - \rho_{\mathrm{a\ C}})\left[1 - \exp\left(-\frac{F\lambda(z)}{\alpha}\right)\right]. \tag{19.12}$$

$F$ is the implantation fluence, $\lambda(z)$ the vacancy depth profile and $\alpha$ is a critical vacancy density for graphitization. In our case $\alpha$ was used as fit parameter and amounts to $5\times10^{22}$ cm$^{-3}$. By simply integrating $\rho(F, z)$ over the depth $z$ we can calculate the volume increase and from that the swelling height. This analytical solution fits quite well with the experimental data for section A and B, but the swelling height comes to saturation at a certain maximum at which the density depth profile reaches the minimum density $\rho_{\text{a-C}}$. The additional Ga swelling in section C and the sputtering in D is not addressed to graphitization and consequently the swelling curve cannot be described analytically with (19.10) in the high fluence regime. Therefore, TRIDYN v. 4.0 [95] simulations have been performed, which deliver data for both, surface swelling by material incorporation and erosion by sputtering in dependence on the ion fluence. If we include the simulation data as additional term to the swelling height, calculated using (19.10), we end up with an extended model that is able to fully describe the swelling and sputtering behavior of ta-C implanted with Ga ions at different energies and covers a fluence range between $1\times10^{14}$ cm$^{-2}$ and beyond $1\times10^{17}$ cm$^{-2}$. The results are shown as the two dashed curves in Fig. 19.28, which fit very well to the experimental data.

#### 19.4.2.4 Ga Segregation by Thermal Treatment

In the above sections we treated the phenomena of ion induced graphitization of ta-C layers induced by the bombardment of $Ga^+$ ions that originate from a FIB. Electrical active nanostructures have been shown, which enable current transport through the insulating ta-C matrix. Besides the effect of ta-C modification the incorporation of Ga itself can be used as a powerful tool for the fabrication of nanoparticles on ta-C layers. Under room temperature conditions the implanted Ga remains within the ta-C but does neither serve as dopant nor bond to the carbon atoms of the matrix. Taking the low melting temperature of Ga of 29.5 °C into account together with the high mobility of incorporated Ga atoms, which is known from the Ga diffusion in Si [96], it stands for reason that segregation between Ga and C occur during thermal treatment. The consequence is out diffusion of an amount of implanted Ga to the surface of the ta-C layer, which can be observed after irradiation under elevated temperatures like it is shown in Fig. 19.29 [92].

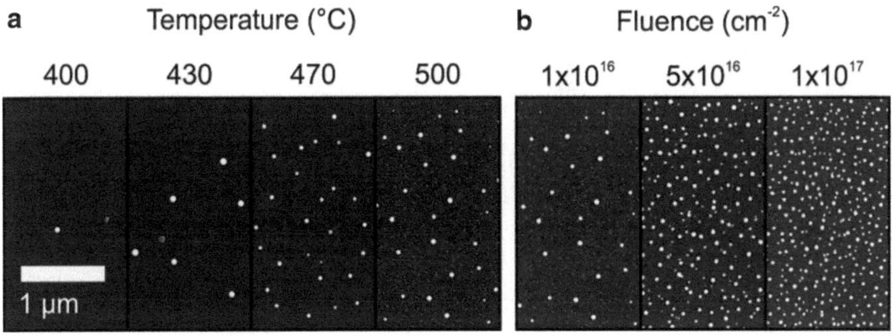


**Fig. 19.29** Ga segregation on ta-C by 10 keV $Ga^+$ irradiation under elevated substrate temperatures. (**a**) Temperature dependence for a fluence of $1 \times 10^{16}$ $cm^{-2}$ and (**b**) fluence dependence for a substrate temperature of 500 °C of the areal density of Ga droplets

The result of Ga segregation is the formation of small droplets on the surface, which shapes are ranging from disc like up to spheres dependent on the irradiation temperature. Furthermore, the SEM images in Fig. 19.29 indicate clearly the possibility of tuning this process in terms of the areal density and size of the droplets. If a higher thermal budged is provided, the formation of nearly perfect spherically droplets can be achieved by post implantation thermal treatment of Ga irradiated ta-C, like it was shown at the beginning of this section in Fig. 19.23a. The spherical shape caused by the high surface tension of gallium opens the way to a defined production of gallium nanoparticles (Ga-NPs) by FIB implantation and subsequent annealing, once the tailoring of the particle's arrangement and their size can be mastered [97].

Let us consider a $Ga^+$-FIB implantation on ta-C within an area $A_{imp}$ with a finite size. Vacuum annealing of the sample at 500 °C for 1 h will result in an arrangement of Ga-NPs with random positioning and various diameters (see Fig. 19.23a). If we now shrink $A_{imp}$ the area will undergo a critical value at which only one single droplet evolves caused by the finite Ga diffusion length and ripening processes at the surface. For fixed implantation and annealing conditions a further decrease of $A_{imp}$ can now be used to define the Ga-NP diameter by adjusting the amount of Ga provided for the NP evolution. This precise correlation between the implanted area and the size of the single Ga-NP is shown in Fig. 19.30 [97].

The linear relationship gives access to a fabrication of single spherical Ga-NPs with a size precisely defined by the implanted area (in the following mentioned as the material reservoir). Utilizing the high FIB resolution a ta-C layer can be patterned with a dens matrix of small material reservoirs. After annealing, the resulting structure is a two dimensional array of well-ordered Ga-NPs that are all equal in size, which is shown in Fig. 19.31a [97].

For the structuring of the 2D array, $50 \times 50$ $nm^2$ reservoirs were chosen that get broadened during implantation due to the effects discussed in Sect. 19.4.2.2. The fluence depending increase of the structure width is the main limitation for a successful downscaling of such Ga-NP structures. Once the single material reservoirs

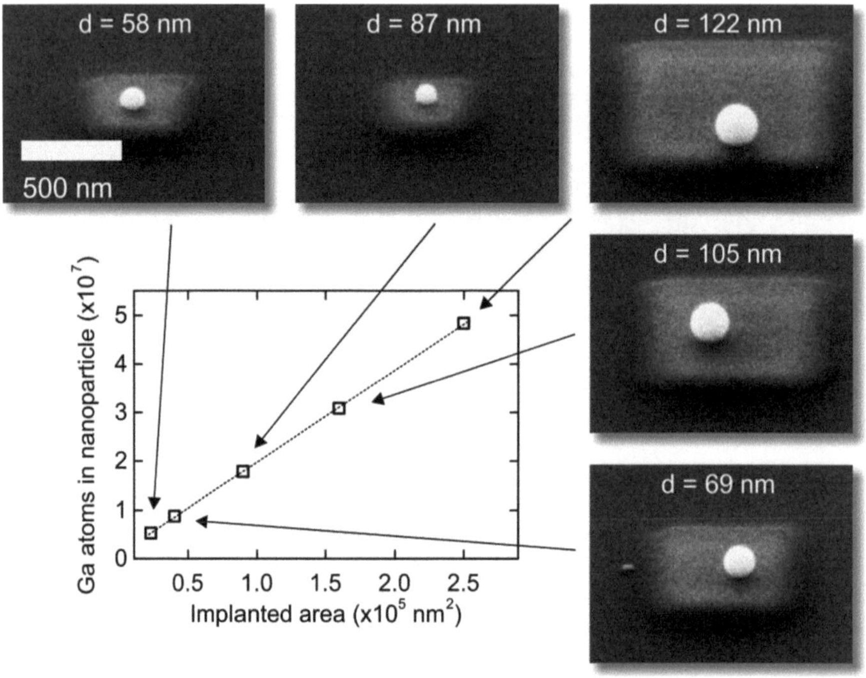


**Fig. 19.30** The graph shows the atoms within one Ga-NP in dependence of the implanted area. Implantation: 30 keV $Ga^+$, $F = 1.4 \times 10^{17}$ $cm^{-2}$; Annealing: RTA at 430 °C for 10 s. The surrounding pictures are SEM images (recorded under 63° with respect to the sample normal) of the corresponding Ga-NPs

are overlapping the segregation process is not restricted to each square anymore and only a few NPs with various sizes will evolve in this case.

#### 19.4.2.5 Ga Nanoparticles by FIB Templating and Annealing

It is shown in the above section that single Ga-NPs can be fabricated with a size controlled by the amount of implanted Ga and the arrangement of 40 nm particles in a 2D matrix was demonstrated. However, beside these benefits it is clear from the SEM images in Figs. 19.30 and 19.31a that the NPs position is still undefined and the Ga droplet evolves randomly within the material reservoir. To overcome this drawback a second FIB implantation of the so-called "template seed" was developed to gain control of the NP position. At first the material reservoir is changed from squares to thin lines with a width in the order of 70–100 nm similar to the nanowires discussed above. Annealing of such a structure delivers a single Ga-NP which size is now defined by the line length and its position is confined in one dimension along the line. Let us now consider the segregation of gallium as a process driven by

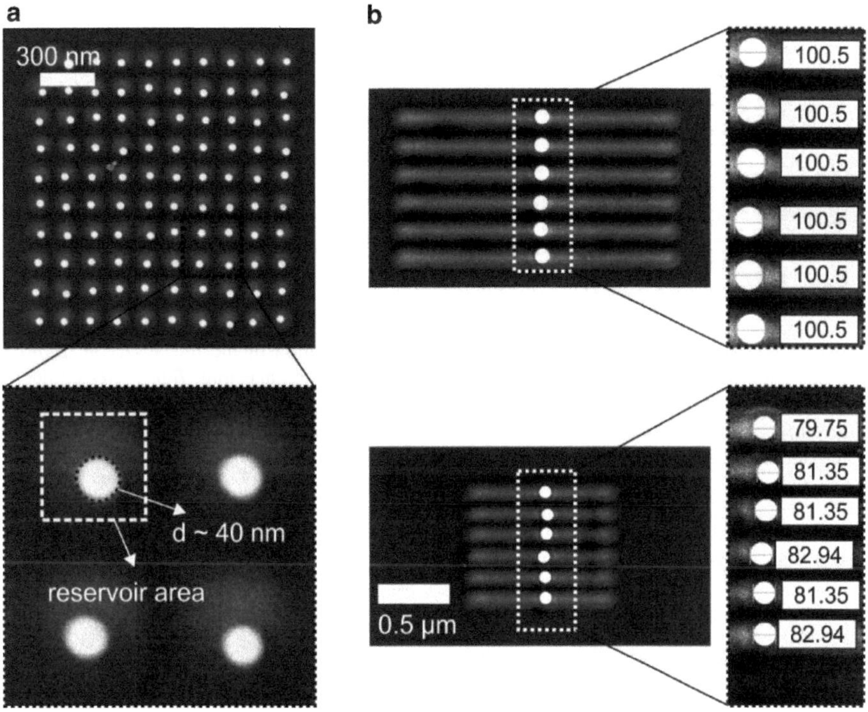


**Fig. 19.31** SEM images of Ga-NPs fabricated by FIB implantation (30 keV $Ga^+$) and annealing by RTA (430 °C, 10 s). (**a**) 2D array of 40 nm particles with 150 nm spacing and (**b**) Ga-NP chains with a distance of 200 nm (*top*) and 150 nm (*bottom*). The *numbers* at the NP chains are the single NP diameters in nanometer

clustering in the bulk, surface diffusion and ripening of atoms, small clusters, and nano droplets that result in a single large Ga-NP. During clustering and ripening the gallium will always tend to accumulate in the largest cluster or the largest droplet respectively. Utilizing this behavior the preferential evolution of clusters and a large primal droplet can be forced by implanting an additional amount of Ga within the reservoir at the desired place for the Ga-NP appearance. We concluded the process of this position controlled droplet formation schematically in Fig. 19.32.

The main findings and advantages of the demonstrated technique can be summarized as follows: The line reservoir length gives precise control of the NP diameter. The template seed defines the NP position during the segregation process. The reservoir line width can be chosen as small as possible and consequently an arrangement of Ga-NPs in chains with a narrow spacing can be realized. In Fig. 19.32b two different chains of well aligned Ga-NPs are demonstrated. The structures are featuring a very high reproducibility in the NP diameter but a slight misalignment in the positioning. The latter one requires improvement of the FIB irradiation settings for the template seed.

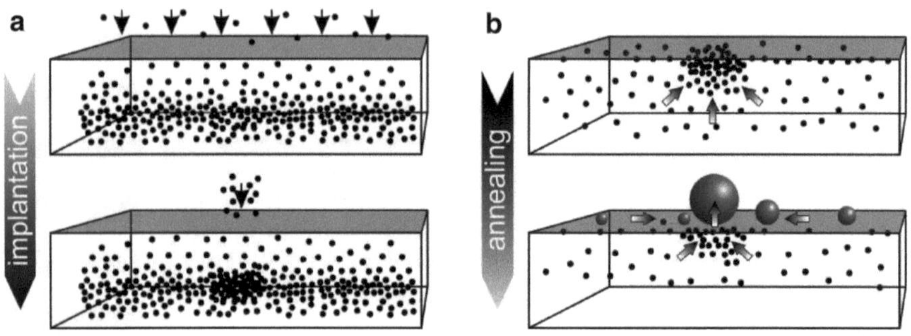


**Fig. 19.32** (**a**) Ga$^+$-FIB lithography on ta-C. *Top*: Implantation of the line reservoir. *Bottom*: Implantation of the template seed. (**b**) Ga-NP evolution during annealing. *Top*: Preferred nucleation of Ga atoms at the seed and diffusion to the surface. *Bottom*: Droplet formation on the surface and surface diffusion of all atoms, clusters, and nano droplets to the large primal droplet at the pre-patterned seed position

### 19.4.2.6 Conclusions

In Sect. 19.4.2 we have demonstrated applications of Ga$^+$-FIB processing on tetrahedral amorphous carbon thin films. The ion impact causes a destruction of the sp$^3$ carbon matrix and a preferential rearrangement in sp$^2$ carbon sites. This is the basis for a direct FIB lithography on ta-C and conducting structures can be written embedded in an electrical insulating sp$^3$ rich matrix. The fundamental dependence of this graphitization on the ion fluence was investigated in terms of the conductivity and the surface swelling of the evolving sp$^2$ layer. Besides fundamental investigations the small spot size of a FIB can be used to produce graphitic nanowires that are conductors embedded in the ta-C layer with a length of several micrometers and a width down to 70 nm. The dependence of the ion fluence on the nanowire broadening and their swelling height was studied and is caused by the double Gaussian FIB profile.

As a second application the segregation of gallium implanted in ta-C, driven by thermal treatment was used to fabricate spherical gallium nanoparticles. The presented technique relies on the correlation between the size of a Ga droplet that evolves on an implanted area after annealing and the amount of Ga provided for the segregation. The material reservoir is defined by the fluence and the size of the implanted area. A subsequent annealing procedure yields in single Ga nanoparticles with a precisely defined diameter. Furthermore, control of the nanoparticle positioning was gained by implanting an additional amount of Ga within the reservoir area, which acts as a template seed for preferential nucleation. FIB implantation and subsequent annealing was demonstrated as a powerful technique for the production of well-defined arrangements of spherical Ga nanoparticles with an accurate size control on the nanometer scale. The procedure can be usefully employed in the field of plasmonics for the fabrication of waveguides, plasmonic crystals or substrates for surface enhanced spectroscopy. Even the conversion of the technique to other ion target-systems is prospective.

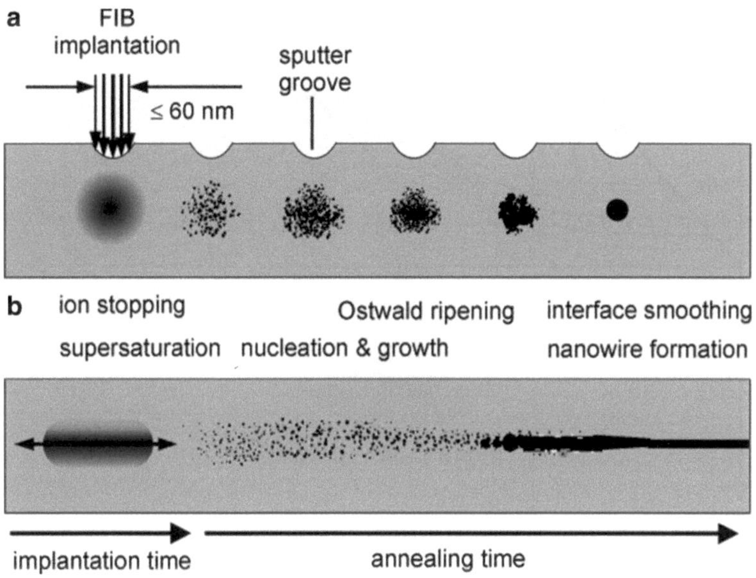


**Fig. 19.33** Schematic representation of the processes contributing to the evolution of FIB-based synthesis of NW structures: (**a**) dots and (**b**) wires

## *19.4.3 FIB Assisted Synthesis of New Solid-State Phases*

In the last two decades ion beam techniques have attracted much interest due to their possibilities for ion beam synthesis (IBS) of nanostructures (nanodots, nanowires, and nanolayers) with unique physical properties. In the field of ion beam processing of nanostructures focused ion beams (FIB) are of special interest because nanodots and nanowires can be fabricated directly by local ion implantation on the nanometer scale. Conventional ion implantation (broad beam, $x$, $y$-scanned) have been successfully applied to large area IBS of nanocrystal (NC) ensembles as well as to tailoring of NC size distribution and depth location, respectively [98, 99].

According to conventional IBS the modified scheme of synthesis of nanowire (NW) structures using FIB implantation (cf. Fig. 19.33) originates at first in the advent of alloy LMIS, making a variety of elements other than Ga available for FIB implantation. Secondly, FIB line scans can result in implantation profiles of particular shape from which elongated nanoscale structures like wires or rods can emerge in a self-organizing manner during phase separation. Therefore, this nonconventional method of controlled nanostructure synthesis represents a combination of a top-down with a bottom-up approach: on the one hand the implantation of energetic ions with an ion beam that is laterally focused down to a few tens of nanometers, on the other hand the self-organized phase separation of a supersaturated solid solution in the course of which elementary atomistic processes cause the formation of mesoscopic structures. Combining self-assembly of nanostructures during IBS

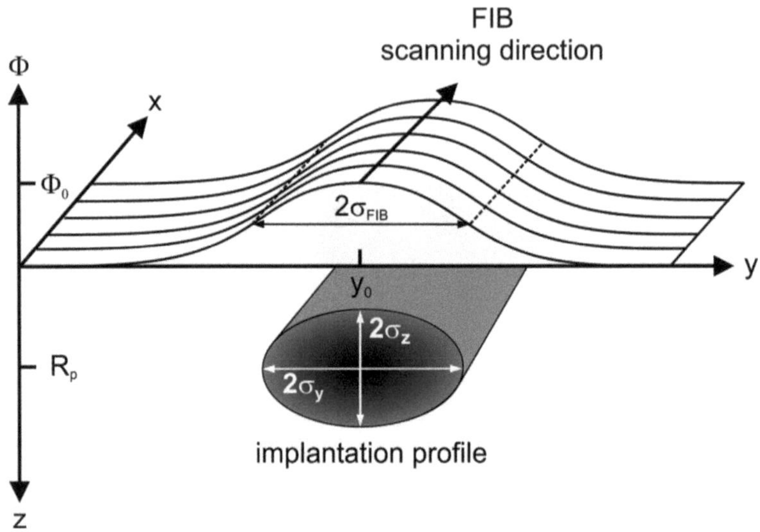


**Fig. 19.34** FIB line scan implantation along the $z$-direction with a Gaussian FIB current density profile centered at $y_0$. The resulting implantation profile has a cylindrical Gaussian shape with an elliptical cross section

(bottom-up approach) with fine-focused ion beam (FIB) implantation (top-down approach) the fabrication of ordered chains or arrays of monodispersive NCs as well as of NWs seems to be practicable.

As shown in Fig. 19.33, writing high fluence FIB implantation ($>10^{16}$ cm$^{-2}$) leads to a laterally confined (on nanometer scale) surface-near concentration profile of the implanted species in the host matrix with a maximum concentration well above the solid solubility limit (supersaturated solid solution). For the synthesis of NW structures, the 3D distribution of implanted atoms in the host matrix represents approximately a cylindrical symmetry where the longitudinal axis is parallel to the scanning direction of the FIB.

As schematically shown in Fig. 19.34, FIB line scan implantation along the $z$-direction with a Gaussian FIB current density profile is centered at $y_0$. The resulting implantation profile has a cylindrical Gaussian shape with an elliptical cross section.

If the ion beam focused equally in both lateral directions, is linearly and homogeneously scanned over the target (in $z$-direction) at normal incidence with respect to the initial target surface the total ion fluence becomes a $y$-dependent function and is given by:

$$\Phi = \Phi(x,y) = \Phi_0 \exp\left[-\frac{(y-y_0)^2}{2\sigma_{FIB}^2}\right] \tag{19.13}$$

where $\Phi_0$ is the maximum fluence at the center of the FIB profile which scales linearly with ion current density at constant scanning speed (cf. Fig. 19.34) and $\sigma_{FIB}$ denotes the width of the FIB current density profile. The integration of $\Phi(y, z)$ from $y=-\infty$ to $y=+\infty$ gives the line density of implanted ions along $z$ if one neglects sputtering of implanted ions. If the ion beam is focused to such a degree that $2\sigma_{FIB}$ is in the range of the projected range $R_p$ an implantation profile is obtained with a particular shape resembling a cylindrical Gaussian distribution whose longitudinal axis coincides with the scanning direction of the FIB (cf. Fig. 19.34). The location of highest concentration of implanted ions in the matrix coincides laterally with the maximum of the FIB current density profile. Vertically, this location (projected range $R_p$) depends on further implantation parameters, e.g., ion energy, the mass ratio between the ions and the target atoms, and the sputtering coefficient. In a first approximation (neglecting sputtering) it can be stated that the corresponding 3D implantation profile can be perceived as a spatial convolution of the lateral FIB current density profile with the vertical depth profile of a broad area ion implantation, i.e., it resembles a cylindrical Gaussian distribution with an elliptical cross section. For the description of realistic FIB line scan profiles implanted at high fluences (e.g., $\geq 10^{16}$ cm$^{-2}$) sputtering effects has to be taken into account and can be modeled using the TRIDYN code based on the binary collision approximation [100]. Due to the Gaussian-like FIB current density profile the sputtering yield varies locally to a high degree which results in the formation of a sputtering groove (cf. Fig. 19.33) whose trough is located at $y_0$ (cf. Fig. 19.34). For example, in the high-fluence regime (stationary implantation profiles) the sputtering yield for 60 keV $Co^{2+}$ ions in silicon obtained by TRIDYN reads $Y_s=3.9$ (assuming the surface-near formation of $CoSi_2$ in the stationary state). Applying the TRYDYN simulation code to high fluence 60 keV $Co^{2+}$ ion implantation into Si at normal incidence the change of the depth profile due to surface layer removal by sputtering can by very well predicted, as shown in Fig. 19.35. This plot demonstrates that at the surface the maximal atomic fraction of Co in Si introduced by ion implantation is limited to 33 %. The black bars at the beginning of each profile indicate the location of the actual surface for each implanted fluence, or, with other words, the thickness of the sputter-removed Si layer in the bottom of the sputter groove (cf. Fig. 19.33) which amounts, for example, 90 nm at an implanted fluence of $2\times10^{17}$ cm$^{-2}$.

Target swelling effects in the case of high fluence Co implantation into c-Si possible can be neglected. Providing that Co is implanted at slightly elevated temperature which prevents amorphization of the crystalline Si target (8 atoms per unit cell) the formation of the Si-rich cobalt silicide phase $CoSi_2$ (12 atoms per unit cell: 8 Si and 4 Co) can be expected to happen directly during implantation [5]. Since $CoSi_2$ and c-Si have both a cubic crystal structure ($CaF_2$ and diamond structure, respectively) and a lattice mismatch of only −1.2 % ($g_{CoSi2}=0.5364$ nm; $g_{Si}=0.543$ nm) it can be assumed that the implanted Co is fully incorporated into the matrix as $CoSi_2$ precipitates within c-Si without changing the volume of the target significantly [102].

After FIB implantation, during subsequent thermal treatment the ion implanted system not only undergoes phase separation but also the accumulated radiation damage caused by energetic particle collisions during implantation anneals. After

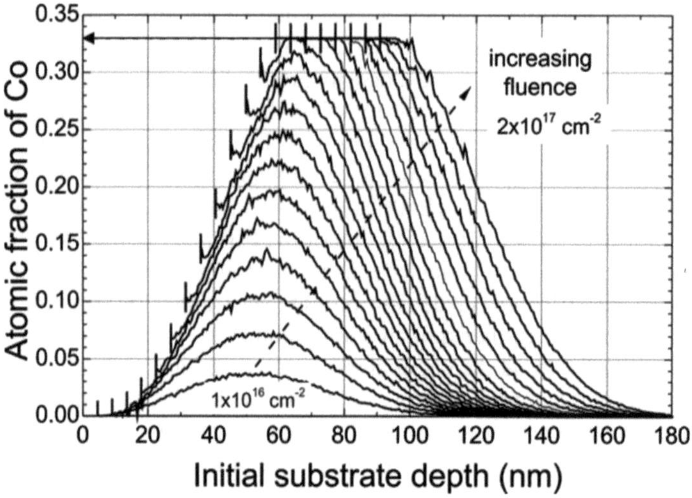


**Fig. 19.35** Change and shift of the Co concentration depth profile in dependence on implanted fluence in the range $(1–20) \times 10^{16}$ cm$^{-2}$ for 60 keV $Co^+$ implantation into silicon. The fluence increment between the profiles is $1 \times 10^{16}$ cm$^{-2}$ [101]

phase separation the complex ensemble of precipitates undergoes processes of coarsening and interface smoothing, which can finally result in the formation of elongated structures like rods or wires in the center of the implantation profile that are oriented parallel to the scanning direction of the FIB during implantation. Corresponding predictions of nanowire IBS were obtained from atomistic kinetic MC simulations [103] and are shown in Fig. 19.36.

However, the state at which a NW structure has formed does not correspond to the system's equilibrium state because NWs are inherently unstable from the viewpoint of thermodynamics. Consequently, it can be stated that the formation of NWs by FIB-based IBS is of transient nature, i.e., the state where a NW is formed is an intermediate state of the system on its pathway towards equilibrium. However, the NW structure can be "stabilized" for technological applications by cooling down the system so that its further evolution at much lower temperatures is hampered kinetically. Furthermore, MC simulations show that for long-lasting thermal treatment the continuous nanowire decays into nearly equidistant and monodispersive NCs (Fig. 19.37). Its decay is driven by the minimization of the interfacial energy and has been described in [103] as a result of pearling or Rayleigh instability.

As an example for FIB assisted synthesis of new solid-state phases the fabrication of cobalt disilicide NW structures will be described in the following. First attempts to apply FIB to nanowire synthesis are described elsewhere [104–106]. In contrast to usually applied Ga LMIS, alloy based LMIS [107] provide different ion species, i.e., $Au^+$, $Ge^+$, $Si^+$, for metallic and semiconducting nanowire synthesis in insulting materials, for example, in $SiO_2$, and, i.e., $Co^+$, $Ni^+$, $Pr^+$, for synthesis of silicide nanowires in Si. With alloy LMIS for ions other than Ga in combination with a ion mass-separated focused ion beam a beam spot diameter of about 30–60 nm (FWHM of the Gaussian beam profile) has been achieved [12]. Recently, the IBS of

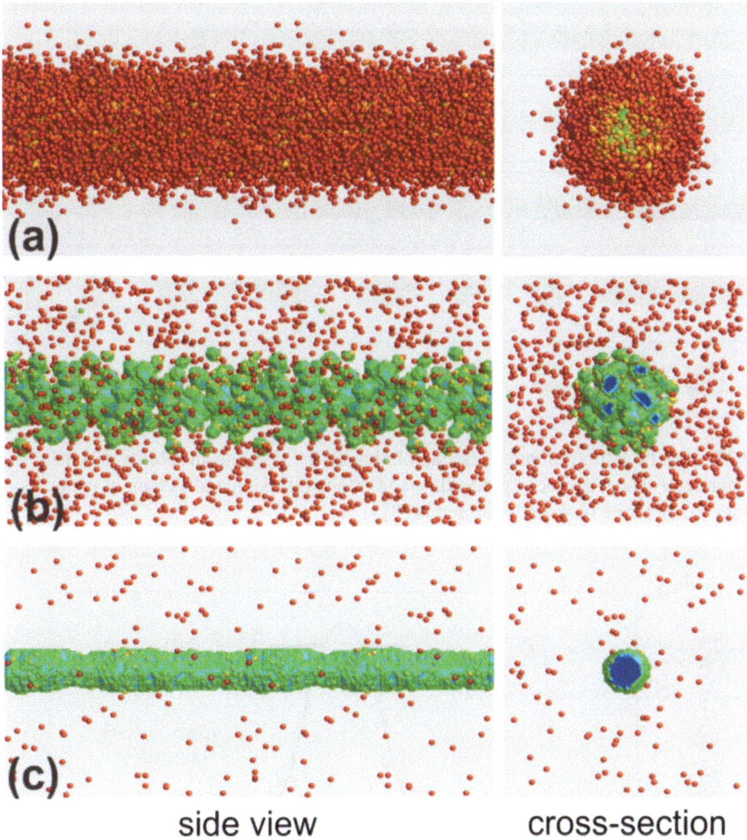


**Fig. 19.36** Kinetic lattice Monte Carlo simulation of the nanowire formation from an initially cylindrical Gaussian distribution of impurity atoms with a peak concentration of 31 at% and a width of 16 nm (95 % of atoms lay within this range). Shown are the side views and cross sections of (**a**) the initial, (**b**) an intermediate (10 kMCS), and (**c**) the final (490 kMCS) state. The atoms are colored according their coordination. The final wire diameter is around 5 nm [103]

crystalline $CoSi_2$ nanowires with diameters of 10–80 nm and lengths of some micrometer embedded in a crystalline silicon matrix has been demonstrated [17, 107]. $CoSi_2$ is of special interest due to its low resistivity, its metallic behavior, its low lattice mismatch to Si of only −1.2 %, the plasmon resonance wave length of approx. 1.2 μm, and its compatibility with modern CMOS technology.

The IBS of $CoSi_2$ nanowires (NW), by writing stoichiometric FIB implantation and subsequent annealing, has been studied using a scanned $Co^{2+}$ ion beam at a constant ion current of 42 pA. Figure 19.38 shows a typical profile of a 60 keV $Co^{2+}$ ion beam for the IMSA-Orsay Physics FIB at HZDR, showing that there is an exponential halo in superposition to the main beam. Doubly charged ions were chosen to increase the ion energy which has virtually no influence on the ion-solid interactions and to achieve a profile buried below the surface. Taking into account the measured current density profile of the $Co^{2+}$ FIB shown in Fig. 19.38 the ion fluence

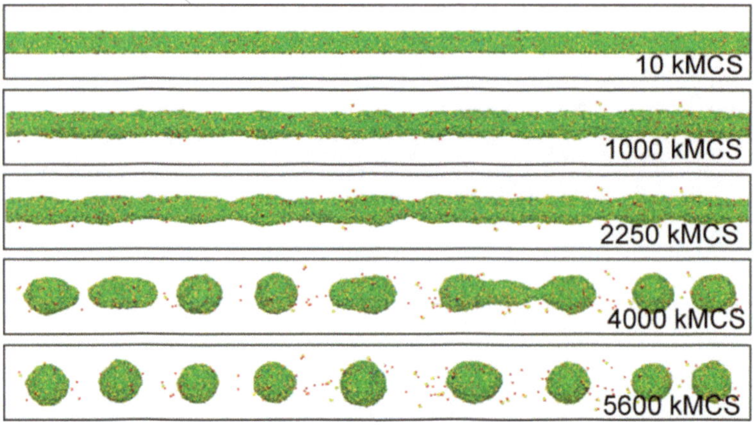


**Fig. 19.37** Kinetic lattice Monte Carlo simulation of the pearling instability of a nanowire with 8.6-nm diameter. The wire decay proceeds via interface diffusion and a chain of nanoclusters with an average diameter of 16.4 nm is formed [103]

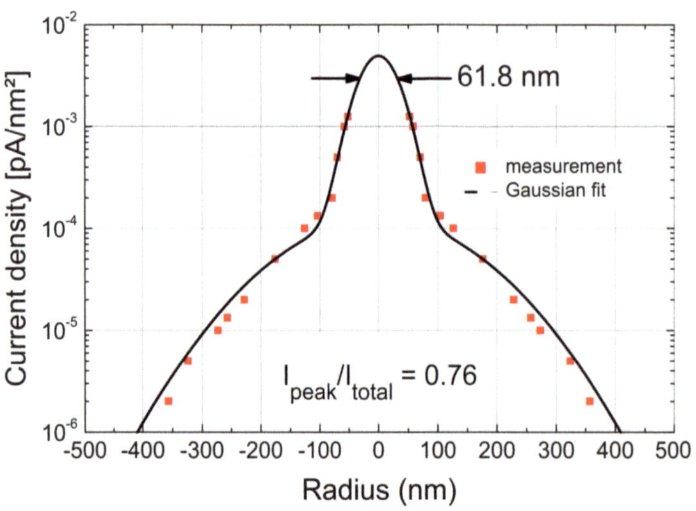


**Fig. 19.38** Profile of a 60 keV focused $Co^{2+}$ ion beam. Here, the total current was 42 pA. According to Gaussian fittings, the FWHM of the peak and the halo are 61.8 nm and 264.2 nm, respectively. The integrated peak current density is 76 % of the total one (cf. Fig. 1 of Ref. [12])

distribution for a digitally scanned beam can be expressed by the superposition of two Gaussian functions:

$$\Phi(x,,y,,t) = \frac{t_{\mathrm{d}}}{q} \cdot \sum_{k} \left[ \frac{I_{peak}}{2\sigma_{peak}^{2}} \cdot \mathrm{e}^{\frac{\left((x-k\cdot\Delta x)^{2}+y^{2}\right)}{2\sigma_{peak}^{2}}} + \frac{I_{halo}}{2\sigma_{halo}^{2}} \cdot \mathrm{e}^{-\frac{\left((x-k\cdot\Delta x)^{2}+y^{2}\right)}{2\sigma_{halo}^{2}}} \right] \quad (19.14)$$

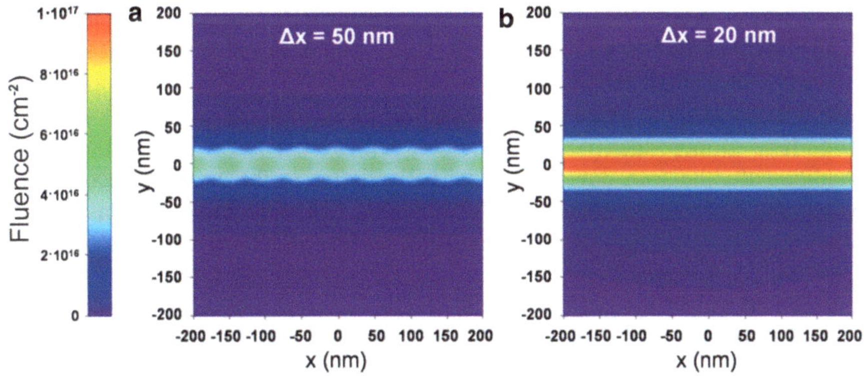


**Fig. 19.39** Calculated fluence distributions of 60 keV $Co^{2+}$ ions in the FIB trace for two different distances between two pixels, (**a**) $\Delta x = 50$ nm and (**b**) $\Delta x = 20$ nm. The widths $\sigma_{peak}$ and $\sigma_{halo}$ are 62 nm and 264 nm, respectively

where $q$ is the elementary charge and $\sigma_{peak}$ and $\sigma_{halo}$ denote the widths of the FIB current density profiles in the peak and halo, respectively. The increment $\Delta x$ represents the distance between two beam spot pixels, $k$ the number of pixels in the FIB line scan and $t_d$ the irradiation time in each pixel (dwell time). The calculated implantation fluences in the $Co^{2+}$ FIB trace for two different distances between two pixels are shown in Fig. 19.39.

The fluence distributions in Fig. 19.39 were calculated for an FIB ion current density in the peak of 1.4 A cm$^{-2}$ and a pixel dwell time of 1 µs. As can be seen for a FIB with $\sigma_{peak} = 62$ nm and a pixel distance of $\Delta x = 20$ nm one get a homogeneous ion distribution in the irradiated trace. Therefore, the beam spot increment $\Delta x$ has to be elected smaller than the beam width $\sigma_{peak}$. The digital scan parameters of pixel dwell time $t_d$ and pixel relaxation time $t_r$ (time between two cycles of FIB irradiation) determine mainly the ion induced defect accumulation in the FIB trace. These parameters are usually in the order of microsecond and millisecond for the pixel dwell time and relaxation time, respectively.

Regarding the atomistic modeling of the ion implantation profile and the subsequent phase separation, the exponential halo is not expected to change the simulation results due to its large FWHM and lower intensity by two orders of magnitude compared to the primary peak of the main beam. For the FIB profile a simple Gaussian shape of some tens of nanometers in width can therefore be assumed for IBS of the nanowires.

The cobalt ions were extracted from a $Co_{36}Nd_{64}$ alloy LMIS with a melting point of 566 °C [34], which was operated at a temperature of about 700 °C. The separation of the $Co^{2+}$ ions from the other ion species was done by means of an ExB-type mass filter in the FIB column (CANION 31Mplus, Orsay Physics) which allows the focusing of the $Co^{2+}$ ion beam into a spot diameter of 20–60 nm. The ion energy was chosen to be 60 keV, corresponding to a mean ion penetration depth of $R_p = 53$ nm. For the study of the conventional IBS of $CoSi_2$ NWs (phase separation through precipitation and wire ripening during thermal treatment) and their decay into chains of nanoparticles due to NW instabilities Si(111) and Si(100) samples were implanted with high fluences ($1\times10^{16}$–$2\times10^{17}$ $cm^{-2}$). To avoid defect accumulation and amorphization, the sample temperature was kept at about 420–450 °C. The implantation was done in a digital mode with a short pixel dwell time (<2 μs) and a long relaxation time of more than 100 μs (time between two exposures of the same pixel). Additionally, the scan direction of the FIB relative to the crystal directions was varied for study of the stability of NW growth. Normally, the FIB scan-direction was parallel to the [108] direction on the Si surface which correspond to the energetically favored growth direction of the $CoSi_2$ nanocrystals [108]. After FIB irradiation and wet chemical cleaning, the samples were treated by a two-step thermal annealing (600 °C, 30 min and 1,000 °C, 1 h in dry $N_2$). In order to improve the image contrast during SEM (Hitachi 4800) investigations or to apply AFM imaging of the nanowires, a short reactive ion etching in $CF_4$ was carried out for 20–40 s to remove the 15–20 nm thick silicon top layer covering the buried $CoSi_2$ nanostructures. Stable $CoSi_2$ nanowires with diameters of 100–150 nm and up to 10–20 μm long (Fig. 19.40a) were achieved when the FIB trace was precisely aligned along the [108] direction. As shown in Fig. 19.40b in the case of a small misaligned FIB trace, the nanowires are not stable and decay into shorter ones chains of $CoSi_2$ nanoparticles [109, 110]. Larger angles of misalignment result in the formation of chains of more or less prolonged $CoSi_2$ nanoparticles (Fig. 19.40c). This behavior is consistent with the 3D KLMC modeling of phase separation, coalescence and decay of nanowires which is demonstrated in Fig. 19.40d.

For the formation of single-crystalline cobalt disilicide, it is necessary to keep the silicon layer on top of the implanted region crystalline. This can be reached by varying the following parameters like pixel dwell time (the irradiation time in one pixel), relaxation time (time between following irradiation cycles of the same pixel), the temperature of the sample as well as the FIB current density and energy. In order to prevent damage accumulation, leading to an amorphous top layer, the pixel dwell time should be as short as possible with a long relaxation time. The fluence, necessary for IBS, in the range of $2\times10^{16}$–$2\times10^{17}$ $cm^{-2}$ requires a lot of repetitions for each pixel irradiation. Furthermore, the surface erosion by FIB sputtering must be taken into account in the high-fluence region (>$5\times10^{16}$ $cm^{-2}$). Especially, FIB-induced defects, local strain and surface erosion by sputtering presently have been identified as essential problems for reproducible IBS of continuous NWs with diameters well below 100 nm by FIB implantation. A further reduction of the NW diameter can be expected by decreasing the concentration of cobalt atoms implanted in silicon at the same FIB spot. NWs synthesized at lower fluences of $1\times10^{16}$–$3\times10^{16}$ $cm^{-2}$ are not stable and decay into shorter NW fragments with a diameter of about 50 nm (Fig. 19.41a). Moreover, because of the existence of the other ⟨110⟩ crystalline directions crossing the FIB trace, there is a certain probability of spontaneous and self-aligned NW growth in these directions (Fig. 19.41b).

In order to investigate the influence of locally FIB induced radiation defects on the $CoSi_2$ NW growth the samples were implanted at lower ion fluences ($10^{14}$–$10^{16}$ $cm^{-2}$) at room temperature by FIB line implantation of $Co^+$ and $Co^{2+}$ as well as of other ions ($Nd^{2+}$, $Ga^+$, $Au^+$, and $Si^{2+}$). Before the FIB implantation the rear side of

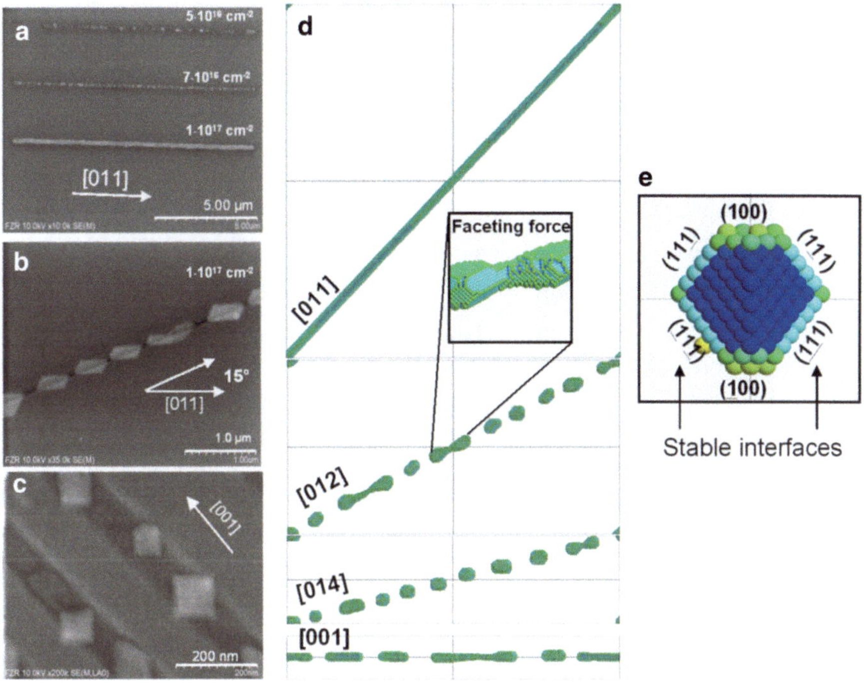


**Fig. 19.40** $CoSi_2$ nanowire growth and decay at FIB synthesis in (111)Si: (**a**) nanowire growth in dependence on the implanted cobalt fluence (FIB trace along ⟨110⟩ direction), (**b**) small misalignment of the FIB trace of about 1° leads to the decay of the long nanowire into short ones, and (**c**) transformation into a chain of crystalline $CoSi_2$ nanoparticles at a misalignment of the FIB trace relative to the ⟨110⟩ crystal direction of about 15°. The sample temperature during implantation was kept to 450 °C and the two-step annealing was done at 600 °C for 1 h and 1,000 °C for 30 min. (**d**) 3D KLMC simulations of NW growth and decay for different Si crystal orientations [111]. The *insert* in (**d**) shows the crystalline nanostructure surface for demonstration of the Co surface diffusion as a driving force for surface faceting of the $CoSi_2$ nanostructures towards the formation of stable (111) interfaces shown in (**e**)

the samples was covered with a 10 nm Co-layer by thermal evaporation to achieve a high concentration of Co in the Si bulk and a supersaturation of Co during cooling down of the samples in the thermal treatment. It was assumed, that the amount of Co atoms above the solubility limit will precipitate preferentially on $CoSi_2$ nanoparticles or on defects formed in the FIB trace. After FIB irradiation the samples were annealed at 1,000 °C for different times in dry $N_2$ atmosphere. At this temperature the solubility and the diffusivity of Co in Si are $5\times10^{14}$ cm$^{-3}$ and $1\times10^{-4}$ cm$^2$ s$^{-1}$, respectively. Already after 10 min of annealing short (200–500 nm) and 10–30 nm thin NWs self-aligned along the ⟨110⟩ directions were observed independent on the kind of implanted ions (Fig. 19.42a). This effect was also clearly seen in the case of a dot-like FIB implantation (without pixel overlapping). It is essential that at lower fluences ($10^{14}$–$10^{15}$ cm$^{-2}$) the NW growth occurs more effectively than at higher ones, presumably depending on the defect structure complexity. After 30 min

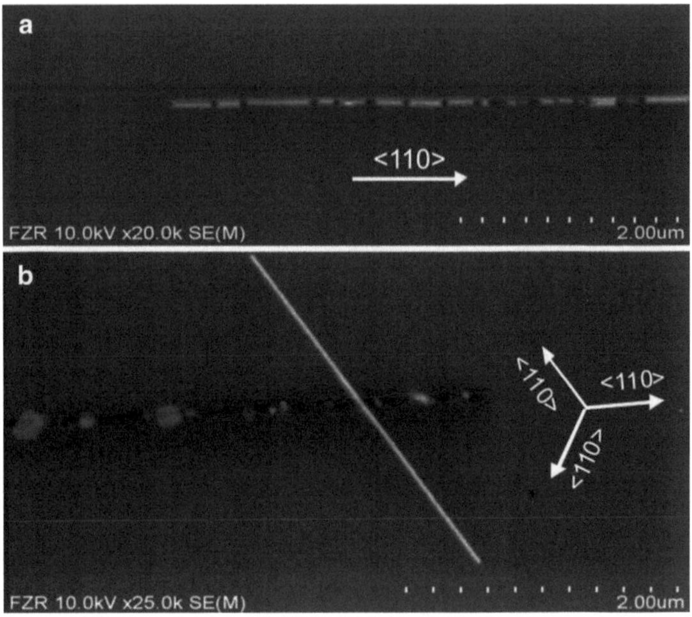


**Fig. 19.41** $CoSi_2$ NW in (100)-Si implanted with a fluence of $7\times10^{16}$ cm$^{-2}$ (**a**) and in (111)-Si, implanted with a fluence of $3\times10^{16}$ cm$^{-2}$ (**b**) [109]

annealing the wires reach several micrometers in length and 20–50 nm in diameter (Fig. 19.42b). Similar results were obtained using, for example, $Ga^+$ FIB irradiation (Fig. 19.42c) and irradiation with the other ions (not shown here). It should be mentioned here that these NWs are very stable (e.g., for some hours at 1,000 °C), but it is difficult to provide a better controlled growth of such nanowires because of the spontaneous nature of the defect generation initiating a spontaneous $CoSi_2$ NW growth along other in-plane ⟨110⟩ Si crystal directions. In contrast to conventional IBS of $CoSi_2$ NWs by high fluence FIB implantation along a narrow trace the defect-induced and self-aligned NWs seem to be more stable.

One of the possible reasons for spontaneous and self-aligned NW growth in ⟨110⟩ directions in (111)-oriented Si wafers should be related to the well-known {311} defects in silicon induced during FIB implantation [112]. It can be assumed that the generation of {311} defects in the silicon crystal plays a crucial role in the CoSi2 nanowire formation process. {311} defects consist of excess Si-interstitials coming from FIB induced point defects which rapidly agglomerate at the beginning of annealing (temperature ramp up) [113]. They are relatively stable at temperatures below 600 °C, at 800 °C they grow up to 150 nm in length via nonconservative Oswald ripening [114] and at 850–1,000 °C they anneal out in a few minutes forming stable dislocation loops and ejecting silicon interstitials. These defects are closed between two {311}-planes and are elongated in the ⟨110⟩ direction [113]. From transmission electron microscopy (TEM) investigations it is known that after

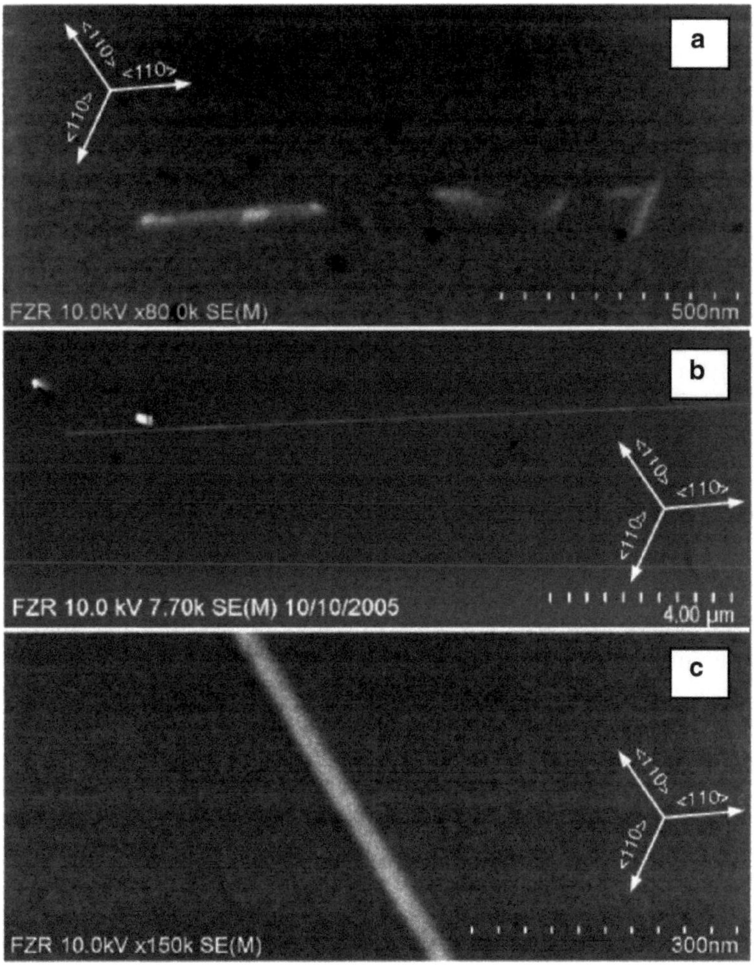


**Fig. 19.42** $CoSi_2$ NWs synthesized in (111)-Si after 60 keV $Co^{2+}$ FIB implantation and annealing at 1,000 °C for 10 min (**a**), for 30 min (**b**) and after 30 keV $Ga^+$ FIB implantation and annealing at 1,000 °C for 30 min (**c**)

annealing at 800 °C the {311} rod-like defects can also transform into similar ones embedded in {111} habit planes and elongated in the ⟨110⟩ direction. During annealing the implanted and solved Co atoms diffuse and can be gettered in the {311} defects. They react with silicon and form silicide precipitates along these defects, i.e., along the ⟨110⟩ direction which is the most preferable direction for the 1-D $CoSi_2$ crystalline structure growth. This process stabilizes the defect structure and hinders dissolution of the rod-like defects. Subsequent Oswald ripening of the silicide precipitates along this ⟨110⟩ direction leads to the formation and alignment of oblong initial $CoSi_2$ nanoparticles at the defect position. During further heating the solved cobalt atoms are still available around the nanoparticles and diffuse to them promoting the growth of crystalline $CoSi_2$ NWs. This finally results the formation of some micrometers long NWs embedded epitaxially within the silicon host.

The stages of defect engineering by FIB and the self-assembly of $CoSi_2$ NWs are specific to this particular system. However, they serve to illustrate how self-assembly and ripening processes can be leveraged to create ordered structures much smaller than those directly generated by the FIB. This example of FIB assisted nanoscale synthesis of new solid-state phases serves to highlight the potential for introducing a robust, well understood step for tailoring self-assembled, cleanly defined nanostructures, using FIB tools that are already familiar, and procedures that promise to be self-limiting, stable, and reliable. Applications of FIB-controlled nanoscale self-assembly in the future evidently merit further examination to overcome problems during fabrication of a nanowire-based device.

### *19.4.4 Surface Pattern Formation by Heavy Mon- and Polyatomic Ions*

Ion irradiation of solid materials' surfaces is a well-established and widely spread technique in material synthesis and modification. Application fields range from small-scale, laboratory studies and synthesis to large-scale production of materials and technological devices. Especially, low-energy ion irradiation (typically <100 keV) of solids can lead to the development of complex and diverse surface morphologies. Particularly, the fabrication of ordered surface patterns on the nanoscale is highly desired for a wide range of technological applications, e.g., storage of information bits by patterned media. Self-organized surface patterns by ion beam irradiation offers a route for the fabrication of ordered patterns over large areas generally in short processing times on the nanometer scale, beyond the limits of lithography [115, 116]. Especially, the formation of ordered quantum dots on semiconductors, e.g.,Si, Ge, or GaSb, can be achieved under certain irradiation conditions [18, 115, 117–121].

It is widely accepted that the morphological instability of a sputtered surface by a kinetic balance between roughening and smoothing mechanism is the basis for a self-organized pattern formation [122–125]. For a long time it has been expected that no formation of ordered quantum dots on surfaces of elemental semiconductors by normal-incidence ion beam irradiation can occur, if no silicide-forming metal is incorporated in the pattern formation process [126–128]. However, it was found that employing irradiation of elemental semiconductors with mono- and polyatomic Bi ion irradiation cuts out a new path for the fabrication of uniform, nanoscale dot patterns with aspect ratios close to one. This is exemplarily shown in Fig. 19.43.

In this section, it will be shown that normal-incidence irradiation of Ge with polyatomic $Bi_3^{2+}$, $Bi_3^+$, $Bi_2^+$ ions lead to the formation of hexagonally ordered dot patterns at room temperature (RT). For comparison, irradiation with monoatomic $Bi^+$ ions results in the formation of the well-known sponge-like patterns at RT, only

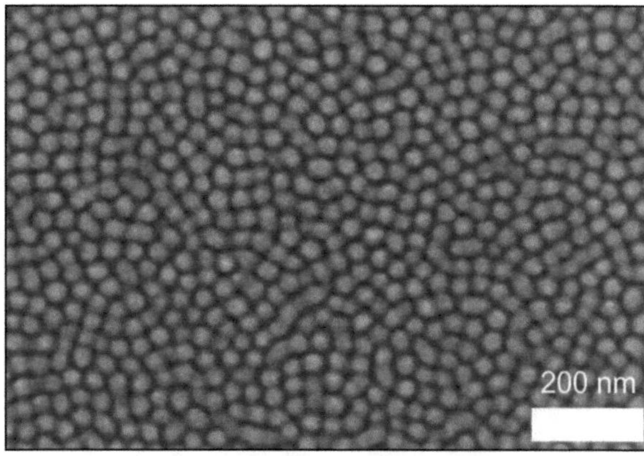


**Fig. 19.43** HR-SEM image of a Ge surface after the irradiation with 30 keV $Bi_3^+$ ions at RT. The fluence was $1 \times 10^{17}$ cm$^{-2}$

irradiation at temperatures well above RT leads to similar dot patterns. Whereas the formation of the sponge-like morphologies can be attributed to vacancy dynamics, the formation of the dot-like pattern is related to a liquid–solid phase transition.

#### 19.4.4.1 Formation of Sponge-Like Ge Morphologies

Depending on the ion energy, the irradiation of Ge at RT with heavy $Bi^+$ ions leads to the formation of porous, sponge-like surface morphologies. The sponge formation starts after irradiation with ~1–5 × $10^{15}$ ions cm$^{-2}$, i.e., well above the amorphization threshold for the irradiated surface layer. Beginning with some surface roughening, the surface pattern evolves with increasing fluence and reaches a "steady state" of the surface morphology after applying a fluence of ~1–5 × $10^{16}$ cm$^{-2}$, while the pattern is still being sputter into the Ge substrate by ion erosion. This pattern evolution is shown for 60 keV $Bi^+$ and increasing fluence in Fig. 19.44. The formation of such porous, sponge-like networks by ion irradiation of Ge is well known and exceptionally pronounced for heavy ion species like Ge [120], Ga [119], Au [129], Pb [130], or Bi [120].

This sponge-formation can be understood in the following model: The ion irradiation of Ge leads to defect creation in a surface layer. By the ongoing creation of interstitials and vacancies due to ion irradiation, i.e., increasing fluence, an amorphous Ge layer is formed.

The depth distribution of the vacancy generation is dependent on the ion energy. The mean of the distribution as well as the total amount of created vacancies per single ion impact increases with increasing ion energy. The formation of a porous surface layer is related to a strong vacancy creation and defect diffusion. Due to the high mobility of vacancies in Ge they can cluster and grow to form voids [131]. To understand the morphology evolution, a model of a vacancy kinetics based pattern formation, which was roughly sketched for ion implantation of GaSb and InSb can be elaborated [132, 133].

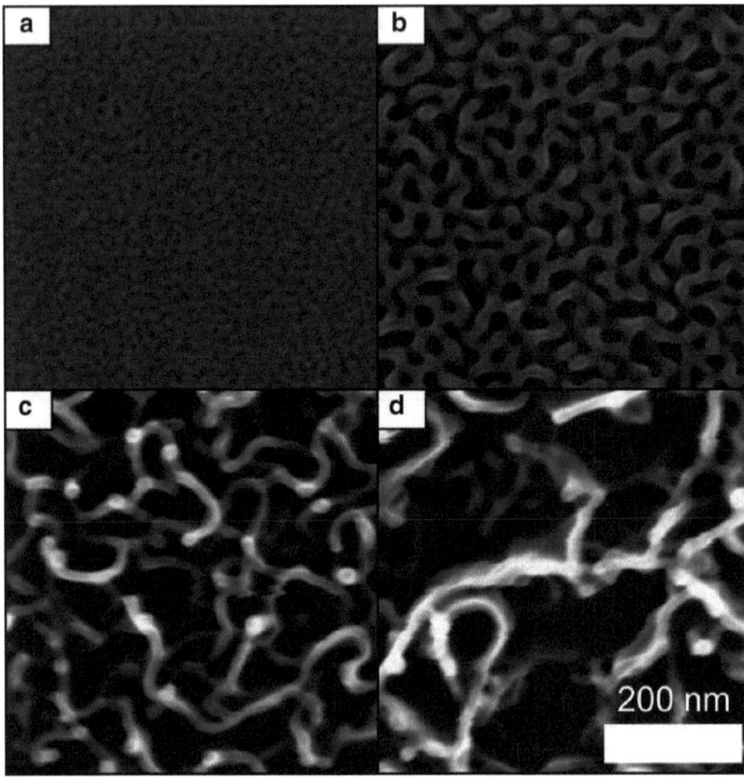


**Fig. 19.44** HR-SEM images of Ge surface morphology after irradiation with 60 keV $Bi^{+}$ at RT. The fluence is increasing from (**a**) to (**d**): $1\times10^{14}$, $1\times10^{15}$, $1\times10^{16}$, $1\times10^{17}$

Even in the amorphized Ge, where the atomic structure only has a very local short-range order, further ion impacts in the amorphized Ge create further vacancies and interstitials, which correspond to Ge atoms having dangling or floating bonds, respectively. Vacancies are stabilized in the amorphous Ge network by dangling bond switching processes [132] and thus have a certain, finite lifetime. Due to their high mobility, the vacancies' free energy can be reduced [131], i.e., within their finite lifetime diffusion allows the vacancies to travel a certain distance. This leads to a certain probability that vacancies are trapped in a vacancy cluster or reach the surface, instead of being annihilated. Once vacancy cluster are created, those can grow in size by attachment of further vacancies. In this way, voids can be formed, which are also, like the sample surface, acting then as a sink for the diffusing vacancies.

Depending on the generation depth of the vacancies, i.e., depending on the ion energy, different morphologies can be obtained. This is schematically depicted in Fig. 19.45, where ions (black arrows from the top) are incident on an amorphous Ge layer (solid bottom part). Vacancies (empty circles) are generated by the ions.

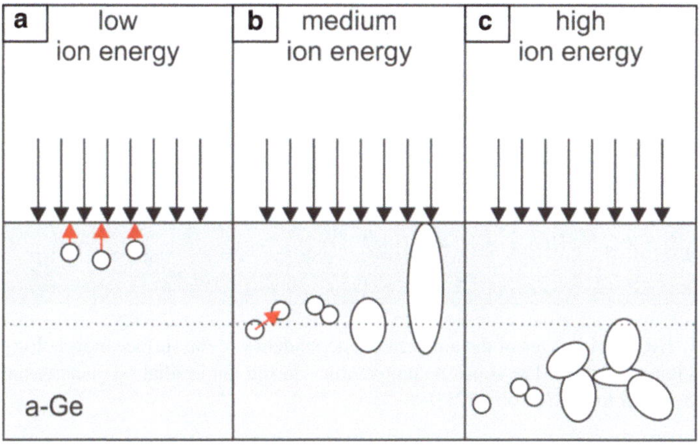


**Fig. 19.45** Schematic of the surface evolution by vacancy kinetic for (**a**) smooth, (**b**) hole-like and (**c**) sponge-like surface morphologies. Vacancies (*empty circles*) are generated by ion impacts (*black arrows* from the *top*) in the Ge bulk. (**a**) For low ion energies, vacancy generations happens in a very surface near volume, diffusion to the surface is most probable. (**b**) For medium ion energies, vacancies are generated more deeply in the bulk, and thus vacancy clustering and void formation is favorable. (**c**) High ion energies lead to void formation in the bulk, where void coagulation then leads to the formation of a porous network. Ongoing ion erosion sputters the original sample surface down to the created structures (*dotted line*)

For low ion energies, vacancies are created close to the surface. Therefore, out-diffusion of the vacancies to the sample surface is most likely and due to ongoing sample erosion (dotted line Fig. 19.45), no surface patterning will be visible (Fig. 19.45a). For medium ion energies, vacancies will be generated deeper in the Ge bulk. Therefore, it is more probable that vacancy clusters are formed compared to the out-diffusion to the surface. Once vacancy clusters are formed, they will grow in size by attachment of further vacancies and finally form voids. By ongoing ion erosion, the sample surface will reach those voids and intersect them. In this way hole-like patterns can be formed on Ge (Fig. 19.45b), which was also observed experimentally for Ga implantation of Ge [135, 136]. For even higher ion energies void formation is happening deeper in the bulk. In this case, voids coagulate, which is then visible as a porous sponge-like network, when erosion moves down the original sample surface (Fig. 19.45c). Moreover, the whole process is enhanced by the increased vacancy creation with increasing ion energy. A stabilization of the pattern formation is likely to be obtained by interstitial induced stress in the amorphous layer. Because, as more and more interstitials accumulate, stress in the surface layer increases. This in turn leads to a reduction of the vacancies' lifetime [134].

A shift of the energy regime for different surface morphologies between different ion species is related to different vacancy creation depths and displacements per impinging ion.

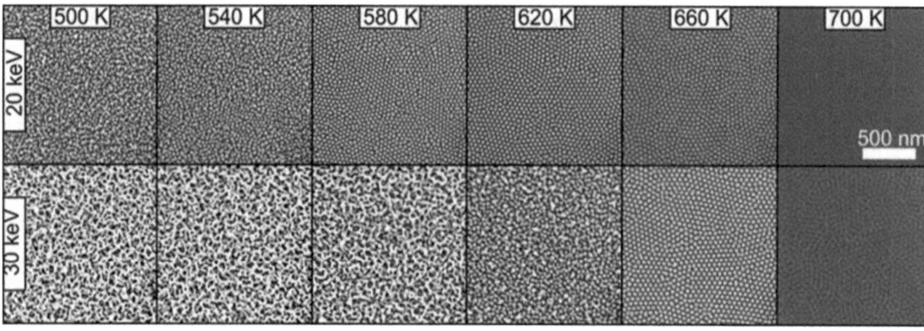


**Fig. 19.46** HR-SEM images of the temperature dependency of the surface morphology for 20 and 30 keV Bi ion irradiation. The substrate temperature during ion irradiation is increasing from 500 to 700 K from *left* to *right* (from [121])

#### 19.4.4.2 Formation of Dot-Like Ge Patterns by Liquid–Solid Phase Transition

Elevation of the Ge substrate temperature during ion irradiation with heavy ions like Bi changes the resulting surface morphology drastically. The effect of an increased substrate temperature between 460 and 700 K during ion irradiation is depicted for ion irradiation with Bi ions having kinetic energies of 20 and 30 keV in Fig. 19.46.

At low temperature the resulting surface morphology equals that at room temperature, i.e., a sponge-like structure. However, an increase of the substrate temperature changes the surface morphology to a dot-like pattern. There is an energy-dependent temperature window, where hexagonally ordered dot-like patterns are formed. The short-range order increases with increasing substrate temperature up to an optimum in the temperature window and decreases at a further temperature decrease. Moreover, a further temperature increase leads to a flattening of the dot patterns. Above a critical temperature, smooth surface are obtained.

The substrate temperature to obtain smooth surfaces depends only slightly on the ion energy, whereas the temperature for the beginning of the dot formation increases significantly with increasing ion energy. Thus, the temperature window for dot formation gets smaller for higher ion energies.

Having a look at the structure of some of the dot-like patterns by HR-XTEM (Fig. 19.47), reveals that irradiated surface layer have sinusoidal shape. Moreover, even at irradiation temperatures above the recrystallization temperature of Ge, the irradiated surface layer stays amorphous. Because, even if the material is molten by the high energy deposition of the heavy ion impacts (see below), the molten volume inside the ion impact cascade cools down below the melting temperature in some tens of picoseconds. This exceeds the maximum epitaxial solidification speed according to the Frenkel–Wilson law, i.e., the material is quenched in its amorphous state.

To identify the underlying mechanisms for the change of the surface morphology with increasing substrate temperature, experimental data for $Bi^+$ irradiation of Ge at ion energies of 10–30 keV and substrate temperature between RT and 780 K can be composed

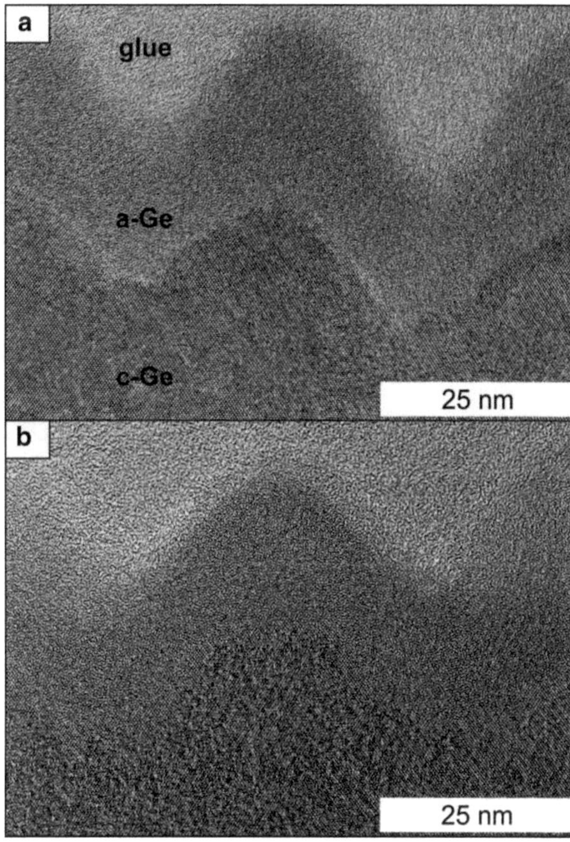


**Fig. 19.47** HR-XTEM images of (**a**) 20 keV $Bi^+$ irradiated Ge at 600 K and (**b**) 30 keV $Bi^+$ irradiated Ge at 700 K substrate temperature. The dot-like surface layer stays amorphous during ion irradiation even at elevated substrate temperatures (according to [121])

into a phase diagram, which is shown Fig. 19.47. The color bar indicated the hexagonal short-range order according to [120]. Blue areas indicate ion energy-temperature combinations, where smooth surfaces or sponge-like morphologies are obtained.

The formation of the dot pattern can be identified with ion impact-induced local melting: Local melting occurs if the energy, which is deposited by an ion into its collision cascade volume, exceeds the latent heat of fusion plus the energy needed to heat up the collision cascade volume from the substrate temperature to the corresponding melting temperature. Using the thermodynamic parameters of Ge [121] and approximating the collision cascade volume from ion implantation parameters from TRIM simulations [137], the fulfillment of a melting condition in the ion collision cascade volume can be estimated.

This estimated melting threshold is represented by the white curve in Fig. 19.48, separating the sponge-like from the dot-like surface morphologies. The onset of dot formation coincides approximately with the melting condition. Moreover, agreement increases, if it taken into account that the melting condition overestimated the needed ion energy for local melting, because heat conduction is neglected in this simple estimation. For all ion energies, dot pattern disappear at temperatures above ~720 K.

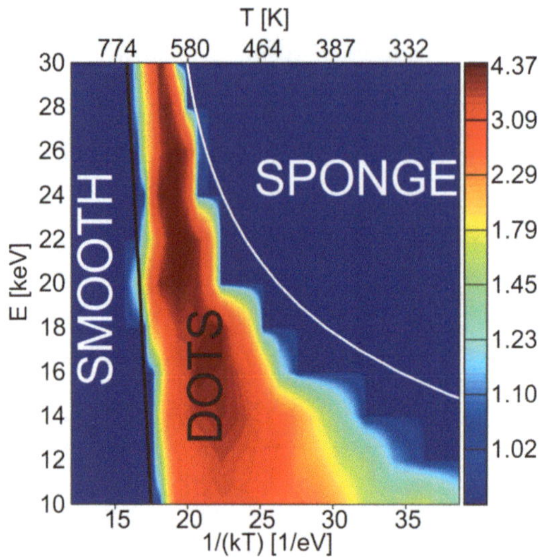


**Fig. 19.48** Phase diagram of the surface morphologies in the plane of $Bi^+$ ion energy vs. substrate temperature. *Blue* areas have no dot-like morphology, i.e., at this energy-temperature combinations smooth or sponge-like morphologies are obtained. The *color bar* indicated the short-range order of dot pattern according to [120], a value of 6 would correspond to an ideal ordering. High substrate temperatures result in smooth surfaces. Below the *white curve* local transient melting in the collision cascade can be predicted, i.e., dot pattern formation is related to ion-induced nanomelt kinetics (from [121])

## 19.5 Conclusions

In this chapter the mass-separated focused ion beam system was introduced and adapted to several fields of applications in nanotechnology. Providing a wide spectrum of different ion species for FIB processing, such systems open up new horizons in terms of material modifications at the nanometer scale. The configuration of a mass-separated FIB apparatus containing an ExB mass filter was briefly described as well as the fundamentals of liquid metal alloy ion sources. The alloy LMIS and their progressing development are the basis for operating FIB systems with multiple ion species while the benefits of conventional Ga based machines are maintained. To demonstrate the wide range of applications, four different techniques of FIB based material processing were presented.

A direct lithography process on $sp^3$ rich tetrahedral amorphous carbon layers was introduced. Ion irradiation of ta-C causes the conversion of $sp^3$ to $sp^2$ bonded carbon atoms, which is an insulating-to-conductive transition of the material. The FIB can therefore be used to fabricate locally conductive structures embedded in the highly resistive ta-C matrix. This technique does not rely on additional process steps like resist deposition, resist development, chemical etching and film deposition. Thus, any structure design can be easily transferred by FIB implantation in ta-C films. The fundamental dependence of the irradiation conditions, especially the ion fluence, on the electrical properties of ion implanted ta-C was intensively studied. Furthermore, the small FIB spot diameter makes it possible to fabricate conducting nanowires that provide a significant current transport through the high resistive diamond-like carbon. Besides the effect of $sp^3$ to $sp^2$ conversion, the segregation of implanted Ga induced by thermal treatment after FIB irradiation can be used effectively to produce spherical Ga nanoparticles. It was shown that the annealing of well-chosen microstructures, made by Ga-FIB lithography on ta-C, results in the growth of Ga droplets with a defined size and accurate positioning. In this manner, well-ordered arrangements of Ga nanoparticles with a size of several tens of nanometer were produced on the ta-C surface, which is prospective for plasmonic applications.

The implantation of Ga in single crystalline Si strongly affects the selectivity during wet-chemical etching with KOH. By applying the FIB, well defined etch-resistive sites in the Si were produced. Furthermore, the use of silicon-on-insulator substrates enables the self-aligned fabrication of free-standing Si structures by Ga-FIB implantation and subsequent KOH etching. The dependence on the photoresist processing or multistage etching processes, which are usually essential for the structuring of free-standing elements, is absent. It was shown that the electrical properties and the width of Si nanowires can be tailored by the FIB implantation parameters. Additionally, the fabrication of silicon based micro mechanical resonators is prospective.

One advantage of a FIB is the possibility to locally incorporate a defined amount of material into a solid. The formation of new solid phases can be enforced by thermal treatment after implantation of well-chosen ion species into certain target materials. This effect was demonstrated for the case of Co implantation in Si, which results in the synthesis $CoSi_2$ nanowires. The influence of the irradiation and annealing conditions on the nanowire growth and their properties was intensively discussed. Experimental observations were supported by kinetic Monte Carlo simulations. Here, the variety of different elements provided by the mass-separated FIB is essential for the fabrication of insulating, semiconducting or metallic phases, which can be formed in various substrates.

Self-organized surface patterning by ion irradiation is of great interest from the fundamental and technological point of view. The bombardment of Ge substrates by heavy Bi cluster ions was shown as an effective method for creating well-defined, hexagonally ordered dot arrays. The evolution of nanodots during irradiation was related to the very high energy deposition by implantation of such massive particles. A local temperature increase during the impact event that creates a melt pool consisting of liquid Ge was proposed, which induces densification and material transport. Consequently, the substrate temperature during ion irradiation and the ion energy are the crucial factors that determine the pattern formation. This was summarized in a phase diagram, which reflects the optimal conditions for the fabrication of well-ordered dot patterns. Analytical calculations were presented that support the theoretical assumptions.

After consideration of these four peculiarly diverse examples of FIB application, it stands to reason that the mass-separated FIB is a powerful tool in the entire field of material science, ranging from fundamental physics to pinpointed nanoprocessing.

**Acknowledgments** The authors would like to thank the staff of the IBC and the clean room team of the HZDR for technical support, as well as K.-H. Heinig, S. Akhmadaliev, W. Pilz, S. Facsko, R. Hübner, and M. Krause for assistance in the experiments and helpful discussions.

# References

1. Orloff, J., Utland, M., Swanson, L. (eds.): High Resolution Focused Ion Beams. Kluwer Academic, New York (2003)
2. Utke, I., Moshkalev, S., Russell, P. (eds.): Nanofabrication Using Focused Ion and Electron Beams. Oxford University Press, New York (2012)
3. Giannuzzi, L.A., Stevie, F.A. (eds.): Introduction to Focused Ion Beams: Instrumentation, Theory, Techniques and Practice. Springer, New York (2005)
4. Sudraud, P., Ben Assayag, G., Bon, M.: Focused-ion-beam milling, scanning-electron microscopy, and focused-droplet deposition in a single microcircuit surgery tool. J. Vac. Sci. Technol. B **6**, 234–238 (1988)
5. Melngailis, J.: Focused ion beam technology and applications. J. Vac. Sci. Technol. B **5**, 469–495 (1987)
6. McKenzie, R.A.D., Smith, G.D.W.: Focused ion beam technology: a bibliography. Nanotechnology **1**, 163–201 (1990)
7. Reyntjens, S., Puers, R.J.: A review of focused ion beam applications in microsystem technology. J Micromech. Microeng. **11**, 287–300 (2001)
8. Prewett, P.D., Mair, G.L.R.: Focused Ion Beams from Liquid Metal Ion Sources. Wiley, Chichester (1991)
9. Otis, C.E., Graupera, A., Laur, D., et al.: Mass filtered plasma focused ion beam system. J. Vac. Sci. Technol. B **30**, 06F604-1–06F604-1-4 (2012)
10. Delobbe, A., Salord, O., Sudraud, P.: i-FIB—The ECR—FIB. EFUG Meeting, Bordeaux, 3 Oct 2011
11. Winston, D., Manfrinato, V.R., Nicaise, S.M., et al.: Neon ion beam lithography. Nano Lett. **11**, 4343–4347 (2011)
12. Bischoff, L.: Alloy liquid metal ion sources and their application in mass separated focused ion beams. Ultramicroscopy **103**, 59–66 (2005)
13. Bischoff, L., Teichert, J., Heera, V.: Focused ion beam sputtering investigations on SiC. Appl. Surf. Sci. **184**, 372–376 (2001)
14. Posselt, M., Bischoff, L., Teichert, J., Ster, A.: Dose rate and temperature dependence of ion-beam-induced defect evolution in Si and SiC: influence on the shape of channeling implantation profiles. J. Appl. Phys. **93**, 1004–1013 (2003)
15. Grenzer, J., Bischoff, L., Pietsch, U.: Grazing-incidence diffraction strain analysis of a laterally patterned Si wafer treated by focused Ge and Au ion beam implantation. Phys. Status Solidi A **202**, 1009–1016 (2005)
16. Martin, J., Bischoff, L., Wannemacher, R.: Microscopy of ion-beam generated fluorescent color center patterns in LiF. Opt. Commun. **188**, 119–128 (2001)
17. Akhmadaliev, C., Bischoff, L., Schmidt, B.: $CoSi_2$ nanostructures by writing FIB ion beam synthesis. Mater. Sci. Eng. C **26**, 818–821 (2006)
18. Bischoff, L., Heinig, K.-H., Schmidt, B., et al.: Self-organization of Ge nanopattern under erosion with heavy Bi monomer and cluster ions. Nucl. Instrum. Methods. B **272**, 198–201 (2012)
19. Orloff, J. (ed.): Handbook of Charged Particle Optics, 2nd edn. CRC Press, Boca Raton (2009)
20. Schmidt, B., Wetzig, K.: Ion Beams in Material Processing and Analysis. Springer, Wien (2013)
21. Schwind, G.A., Swanson, L.W.: Emission characteristics of $Au_{60}Be_{40}$ and $Au_{62}Si_{23}Be_{15}$ liquid metal ion sources. J. Vac. Sci. Technol. B **25**, 2586–2592 (2007)

22. Prewett, P.D., Mair, G.L.R.: Focused Ion Beams from Liquid Metal Ion Sources. Research Study Press, Taunton (1991)
23. Swanson, L.W.: Use of the liquid metal ion source for focused beam applications. Appl. Surf. Sci. **87/88**, 1–11 (1994)
24. Forbes, R.G.: Field-evaporation theory: a review of basic ideas. Surf. Sci. **61**, 221–240 (1995)
25. Brandon, D.G.: The structure of field-evaporated surfaces. Surf. Sci. **3**, 1–18 (1964)
26. Knauer, W.: Energy broadening in field emitted electron and ion beams. Optik **59**, 335–354 (1981)
27. Kim, Y.G., Kim, Y.S., Choi, E.H., et al.: Temperature effects on the energy spreads in liquid metal ion sources. J. Phys. D Appl. Phys. **31**, 3463–3468 (1998)
28. Bischoff, L.: Application of mass-separated focused ion beams in nano-technology. Nucl. Instrum. Methods B **266**, 1846–1851 (2008)
29. Mair, G.L.R.: Electrohydrodynamic instabilities and the energy spread of ions drawn from liquid metals. J. Phys. D Appl. Phys. **29**, 2186–2192 (1996)
30. Bischoff, L., Teichert, J., Hausmann, S., et al.: Investigation and optimization of the emission parameters of alloy liquid metal ion sources. Nucl. Instrum. Methods B **161–163**, 1128–1131 (2000)
31. Mazarov, P., Wieck, A., Bischoff, L., et al.: Alloy liquid metal ion source for carbon focused ion beams. J. Vac. Sci. Technol. B **27**, L47–L49 (2009)
32. Bischoff, L., Teichert, J.: Liquid metal ion source working with an $Er_{70}Fe_{22}Cr_5Ni_3$ alloy. J. Phys. D Appl. Phys. **33**, L69–L72 (2000)
33. Hesse, E., Bischoff, L., Teichert, J.: Development of a cobalt liquid alloy ion source. J. Phys. D Appl. Phys. **27**, 427–428 (1994)
34. Bischoff, L., Akhmadaliev, C.: An alloy liquid metal ion source for lithium. J. Phys. D Appl. Phys. **41**, 052001-1–052001-3 (2008)
35. Bischoff, L., Akhmadaliev, C., Mair, A.W.R., et al.: Investigation of a tin liquid metal ion source. Appl. Phys. A **79**, 89–92 (2004)
36. Bischoff, L., Pilz, W., Mazarov, P., et al.: Comparison of bismuth emitting liquid metal ion sources. Appl. Phys. A **99**, 145–150 (2010)
37. Mair, G.L.R.: Theoretical determination of current–voltage curves for liquid metal ion sources. J. Phys. D Appl. Phys. **17**, 2323–2330 (1984)
38. Bischoff, L., Pilz, W., Ganetsos, T., et al.: GaBi alloy liquid metal ion source for microelectronic research. Ultramicroscopy **107**, 865–868 (2007)
39. Krohn, V.E., Ringo, G.R.: Ion source of high brightness using liquid metal. Appl. Phys. Lett. **27**, 479–481 (1975)
40. Swanson, L.W.: Liquid metal ion source: mechanism and applications. Nucl. Instrum. Methods. **218**, 347–353 (1983)
41. Tajmar, M., Vasiljevich, I., Grienauer, W.: High current liquid metal ion source using porous tungsten multiemitters. Ultramicroscopy **111**, 1–4 (2010)
42. Swanson, L.W., Schwind, G.A., Bell, A.E., et al.: Emission characteristics of gallium and bismuth liquid metal field ion sources. J. Vac. Sci. Technol. **16**, 1864–1867 (1979)
43. Bhaskar, N.D., Klimcak, C.M., Fraukolz, R.P.: Liquid metal ion source for cluster ions of metals and alloys. Rev. Sci. Instrum. **61**, 366–368 (1990)
44. Umemura, K.: Measurement of energy spread and angular intensity of a cesium liquid metal ion source. Appl. Phys. A **54**, 115–119 (1992)
45. Bischoff, L., Pilz, W., Ganetsos, T., et al.: On the temperature dependence of the mass spectra of AuGe and AuGeSi liquid metal alloy ion sources. J. Phys. Conf. Ser. **10**, 214–217 (2005)
46. Mair, G.L.R., Bischoff, L., Londos, C.A., et al.: An in-depth investigation into the temperature dependence of the mass spectra of an $Au_{82}Si_{18}$ liquid metal field emitter. Appl. Phys. A **81**, 385–388 (2005)
47. Akhmadaliev, C., Pilz, W., Bischoff, L.: European Focused Ion Beam Users Group EFUG Annual Meeting, Wuppertal, Germany, 02 Oct 2006
48. Pilz, W., Bischoff, L.: Emitter for an ion source and method of producing the same, European patent application 04,017,894.9 (2011)

49. Bischoff, L., Mair, G.L.R., Aidinis, C.J., et al.: Fundamental properties of erbium-ions-producing liquid metal alloy ion sources. Nucl. Instrum. Methods B **197**, 282–287 (2002)
50. Driesel, W., Dietzsch, C., Hesse, E., et al.: In situ observation of the tip shape of Co-Ge liquid alloy ion sources in a high voltage TEM. J. Vac. Sci. Technol. B **14**, 1621–1629 (1996)
51. Kelsall, R.W., Hamley, I.W., Geoghegan, M. (eds.): Nanoscale Science and Technology. Wiley, Chichester (2005)
52. Yamazaki, K., Yamaguchi, H.: Flexible nanofabrication in three-dimensional electron beam lithography enhanced by suppression of proximity effect. APEX **1**, 097001-1–097001-3 (2008)
53. Silverman, P.J.: The Intel lithography roadmap. Intel Technol. J. **6**, 55–61 (2002)
54. Chou, S.Y., Krauss, P.R., Renstrom, P.J.: Imprint lithography with 25-nanometer resolution. Science **272**, 85–87 (1996)
55. Brugger, J., Berenshot, J.W., Kuiper, S., et al.: Resistless patterning of sub-micron structures by evaporation through nanostencils. Microelectron. Eng. **53**, 403–405 (2000)
56. LaMarche, P.H., Levi-Setti, R., Wang, Y.L.: Focused ion-beam microlithography using an etch-stop process in gallium-doped silicon. J. Vac. Sci. Technol. B **1**, 1056–1058 (1983)
57. Schmidt, B., Bischoff, L., Teichert, J.: Writing FIB implantation and subsequent anisotropic wet chemical etching for fabrication of 3D structures in silicon. Sens. Actuator A Phys. **61**, 369–373 (1997)
58. Brugger, J., Belkjakovic, G., Despont, M., et al.: Silicon micro/nanomechanical device fabrication based on focused ion beam surface modification and KOH etching. Microelectron. Eng. **35**, 401–404 (1997)
59. Shimkunas, A.R.: Advances in X-ray mask technology. Solid State Technol. **27**, 192–199 (1984)
60. Bartelt, J.L.: Masked ion-beam lithography—an emerging technology. Solid State Technol. **29**, 215–220 (1986)
61. Sievilä, P., Chekurov, N., Tittonen, I.: The fabrication of silicon nanostructures by focused ion beam implantation and TMAH wet etching. Nanotechnology **21**, 145301-1–145301-6 (2010)
62. Seidel, H., Csepregi, L., Heuberger, A., et al.: Anisotropic etching of crystalline silicon in alkaline-solutions. 1. Orientation dependence and behavior of passivation layers. J. Electrochem. Soc. **137**, 3612–3626 (1990)
63. Heuberger, A. (ed.): Mikromechanik. Springer, Berlin (1989)
64. Bischoff, L., Schmidt, B., Lange, H., et al.: Nano-structures for sensors on SOI by writing FIB implantation and subsequent anisotropic wet chemical etching. Nucl. Instrum. Methods B **267**, 1372–1375 (2009)
65. Böttger, R., Bischoff, L., Schmidt, B., et al.: Characterization of Si nanowires fabricated by Ga+FIB implantation and subsequent selective wet etching. J. Micromech. Microeng. **21**, 145301-1–145301-8 (2011)
66. Berry, I.L., Caviglia, A.L.: High-resolution patterning of silicon by selective gallium doping. J. Vac. Sci. Technol. B **1**, 1059–1061 (1983)
67. Seidel, H., Csepregi, L., Heuberger, A., et al.: Anisotropic etching of crystalline silicon in alkaline-solutions. 2. Influence of dopants. J. Electrochem. Soc. **137**, 3626–3632 (1990)
68. Kendall, D.L.: Vertical etching of silicon at very high aspect ratios. Ann. Rev. Mater. Sci. **9**, 373–403 (1979)
69. Chen, W., Chen, P., Madhukar, P., et al.: Creation of 3D patterns in Si by focused Ga-ion beam and anisotropic wet chemical etching. Mater. Res. Soc. Symp. Proc. **279**, 599 (1993)
70. Steckl, A.J., Mogul, H.C., Mogren, S.: Localized fabrication of Si nanostructures by focused ion beam implantation. Appl. Phys. Lett. **60**, 1833–1835 (1992)
71. Ziegler, J.: SRIM—the stopping and range of ions in matter. http://www.srim.org (2008)
72. Schmidt, B., Oswald, S., Bischoff, L.: Etch rate retardation of $Ga^+$-ion beam-irradiated silicon. J. Eletrochem. Soc. **152**, G857–G879 (2005)
73. Jiles, D.: Introduction to the Electronic Properties of Materials. Chapman and Hall, London (1994)

74. Ohno, K., Murayama, K., Matsuda, A.: Hopping transport of electrons and holes at localized band tails states in amorphous hydrogenated silicon and amorphous heavy-hydrogenated silicon. Jpn. J. Appl. Phys. **44**, 4764–4769 (2005)
75. Sainiemi, L., Grigoras, K., Kassamakov, I., et al.: Fabrication of thermal microbridge actuators and characterization of their electrical and mechanical responses. Sens. Actuator A Phys. **149**, 305–314 (2009)
76. Bakan, G., Cywar, A., Silva, H., et al.: Melting and crystallization of nanocrystalline silicon microwires trough rapid self-heating. Appl. Phys. Lett. **94**, 251910-1–251910-1-3 (2009)
77. Krause, M., Virsek, M., Remskar, M.P., et al.: Raman characterization of $MoS_2$ microtube. Phys. Status Solidi B **246**, 2786–2789 (2009)
78. Krause, M., Virsek, M., Remskar, M.P., et al.: Diameter and morphology dependent Raman signatures of $WS_2$ nanostructures. ChemPhysChem **10**, 2221–2225 (2009)
79. Lu, Z.Q., Quinn, T., Reehal, H.S.: Polarization-dependent Raman spectra of thin crystalline silicon films. J. Appl. Phys. **97**, 033512-1–033512-6 (2005)
80. Aisenberg, S., Chabot, R.: Ion-beam deposition of thin films of diamondlike carbon. J. Appl. Phys. **42**, 2953–2958 (1971)
81. Miyazawa, T., Misawa, S., Yoshida, S., et al.: Preparation and structure of carbon film deposited by a mass-separated $C^+$ ion beam. J. Appl. Phys. **55**, 188–193 (1984)
82. Witke, T., Siemroth, P.: Deposition of droplet-free films by vacuum arc evaporation—results and applications. IEEE Trans. Plasma Sci. **27**, 1039–1044 (1999)
83. Vavilov, V.S., Krasnopevtsev, V.V., Yu, V., Miljutin, Y.V., et al.: On structural transitions in ion-implanted diamond. Radiat. Eff. Defect Solid **22**, 141–143 (1974)
84. McCulloch, D.G., Gerstner, E.G., McKenzie, D.R., et al.: Ion implantation in tetrahedral amorphous carbon. Phys. Rev. B **52**, 850–857 (1995)
85. Kalish, R., Prawer, S.: Graphitization of diamond by ion impact: fundamentals and applications. Nucl. Instrum. Methods B **106**, 492–499 (1995)
86. Prawer, S., Kalish, R., Adel, M., et al.: Effect of heavy ion irradiation on amorphous hydrogenated (diamondlike) carbon films. J. Appl. Phys. **61**, 4492–4500 (1986)
87. Stanishevsky, A.: Modification of hydrogen-free amorphous carbon films by focused-ion-beam milling. J. Appl. Phys. **86**, 7052–7058 (1999)
88. Van der Pauw, L.J.: A method of measuring the resistivity and Hall coefficient on lamellae of arbitrary shape. Philips Res. Rep. **13**, 1–9 (1958)
89. Philipp, P., Bischoff, L.: Investigation of nano structures on ta-C films made by gallium FIB lithography. Diam. Relat. Mater. **23**, 140–143 (2012)
90. Ziegler, J.F., Biersack, J.P.: The Stopping and Range of Ions in Matter. Pergamon Press, New York (1985)
91. Kalish, R., Lifshitz, Y., Nugent, K., et al.: Thermal stability and relaxation in diamond-like-carbon. A Raman study of films with different $sp^3$ fractions (ta-C to a-C). Appl. Phys. Lett. **74**, 2936–2938 (1999)
92. Philipp, P., Bischoff, L.: Investigation of conducting nano-structures on ta-C films made by FIB lithography. Nucl. Instrum. Methods B **282**, 121–124 (2012)
93. Bosia, F., Olivero, P., Picollo, F.: Finite element analysis of ion-implanted diamond surface swelling. Nucl. Instrum. Methods B **268**, 2991–2995 (2010)
94. Wei, Y.X., Wang, R.J., Wang, W.H.: Soft phonons and phase transition in amorphous carbon. Phys. Rev. B **72**, 0122031 (2005)
95. Möller, W., Eckstein, W.: TRIDYN—A TRIM simulation code including dynamic composition changes. Nucl. Instrum. Methods B **2**, 814–818 (1984)
96. Falta, J., Copel, M., LeGoues, F.K., et al.: Surfactant coverage and epitaxy of Ge on Ga terminated Si(111). Appl. Phys. Lett. **62**, 2962–2964 (1993)
97. Philipp, P., Bischoff, L., Schmidt, B.: Taming of Ga droplets on DLC layers—size tuning and local arrangement with nanometer accuracy. Nanotechnology **23**, 475304 (2012)
98. Schmidt, B.: Nanostructures by ion beams. Radiat. Eff. Defect Solid **162**, 171–184 (2007)
99. Heinig, K.H., Müller, T., Schmidt, B., et al.: Interfaces under ion irradiation: growth and taming of nanostructures. Appl. Phys. A **77**, 17–25 (2003)

100. Möller, W., Eckstein, W., Biersack, J.P.: TRIDYN-binary collision simulation of atomic collisions and dynamic composition changes in solids. Comput. Phys. Commun. **51**, 355–368 (1988)
101. Röntsch, L.: Private communication
102. Mantl, S.: Ion beam synthesis of epitaxial silicides: fabrication, characterization and applications. Mater. Sci. Rep. **8**, 1–95 (1992)
103. Müller, T., Heinig, K.H., Schmidt, B.: Template-directed self-assembly of buried nanowires and the pearling instability. Mater. Sci. Eng. C **19**, 209–213 (2002)
104. Bischoff, L., Teichert, J., Hesse, E., et al.: $CoSi_2$ microstructures by means of a high current focused ion beam. J. Vac. Sci. Technol. B **12**, 3523–3527 (1994)
105. Bischoff, L., Heinig, K.H., Teichert, J., et al.: Submicron $CoSi_2$, structures fabricated by focused ion beam implantation and local flash lamp melting. Nucl. Instrum. Methods B **112**, 201–205 (1996)
106. Hausmann, S., Bischoff, L., Teichert, et al.: Single-crystalline $CoSi_2$ layer formation by focused ion beam synthesis. Jpn. J. Appl. Phys. **38**, 7148–7150 (1999)
107. Bischoff, L., Schmidt, B., Akhmadaliev, C., et al.: Investigation of FIB assisted $CoSi_2$ nanowire growth. Microelectron. Eng. **83**, 800–803 (2006)
108. Adams, D.P., Yalisove, S.M., Eaglesham, D.J.: Interfacial and surface energetics of $CoSi_2$. Appl. Phys. **76**, 5190–5194 (1994)
109. Akhmadaliev, C., Schmidt, B., Bischoff, L.: Defect induced formation of $CoSi_2$ nanowires by focused ion beam synthesis. Appl. Phys. Lett. **89**, 23129-1–23129-3 (2006)
110. Bischoff, L., Akhmadaliev, C., Schmidt, B.: Defect induced nanowire growth by FIB implantation. Microelectron. Eng. **84**, 1459–1462 (2007)
111. Röntsch, L., Heinig, K.H.: Private communication
112. Eaglesham, D.J., Stolk, P.A., Grossmann, H.J., et al.: Implantation and transient B diffusion in Si: the source of the interstitials. Appl. Phys. Lett. **65**, 2305–2307 (1994)
113. Claverie, A., Colombeau, B., de Mauduit, B., et al.: Extended defects in shallow implants. Appl. Phys. A **76**, 1025–1033 (2003)
114. Calvo, P., Claverie, A., Cherkashin, N., et al.: Thermal evolution of 113 defects in silicon: transformation against dissolution. Nucl. Instrum. Methods B **216**, 173–177 (2004)
115. Facsko, S., Dekorsy, T., Koerdt, C., et al.: Formation of ordered nanoscale semiconductor dots by ion sputtering. Science **285**, 1551–1553 (1999)
116. Gago, R., Vázquez, L., Plantevin, O., et al.: Order enhancement and coarsening of self-organized silicon nanodot patterns induced by ion-beam sputtering. Appl. Phys. Lett. **89**, 233101-1–233101-3 (2006)
117. Gago, R., Vazques, L., Cuerno, R., et al.: Production of ordered silicon nanocrystals by low-energy ion sputtering. Appl. Phys. Lett. **78**, 3316–3318 (2001)
118. Ziberi, B., Frost, F., Rauschenbach, B.: Pattern transitions on Ge surfaces during low-energy ion beam erosion. Appl. Phys. Lett. **88**, 173115-1–173115-4 (2006)
119. Bischoff, L., Pilz, W., Schmidt, B.: Amorphous solid foam structures on germanium by heavy ion irradiation. Appl. Phys. A **104**, 1153–1158 (2011)
120. Böttger, R., Bischoff, L., Facsko, S., et al.: Quantitative analysis of the order of Bi ion induced dot patterns on Ge. EPL **98**, 16009-p1–16009-p5 (2012)
121. Böttger, R., Bischoff, L., Heinig, K.H., et al.: From sponge to dot arrays on (100) Ge by increasing the energy of ion impacts. J. Vac. Sci. Technol. B **30**, 06FF12-1–06FF12-5 (2012)
122. Bradley, R.M., Harper, J.M.E.: Theory of ripple topography induced by ion bombardment. J. Vac. Sci. Technol. A **6**, 2390–2395 (1988)
123. Cuerno, R., Barabàsi, A.L.: Dynamic scaling of ion-sputtered surfaces. Phys. Rev. Lett. **74**, 4746–4749 (1995)
124. Bradley, R.B., Shipman, P.D.: A surface layer of altered composition can play a key role in nanoscale pattern formation induced by ion bombardment. Appl. Surf. Sci. **258**, 4161–4170 (2012)
125. Carter, G., Vishnyakov, V.: Roughening and ripple instabilities on ion-bombarded Si. Phys. Rev. B **54**, 17647–17653 (1996)

126. Ziberi, B., Frost, F., Rauschenbach, B.: Self-organized dot patterns on Si surfaces during noble gas ion beam erosion. Surf. Sci. **600**, 3757–3761 (2006)
127. Ziberi, B., Cornejo, M., Frost, F., et al.: Highly ordered nanopatterns on Ge and Si surfaces by ion beam sputtering. J. Phys. Condens. Matter **21**, 224003-1–224003-17 (2009)
128. Macko, S.: Mechanisms and Manipulation of Ion Beam Pattern Formation on Si(001). Dr. Hut, Cologne (2011)
129. Yanagisawa, J., Takarabe, K., Ogushi, K., et al.: Nanoporous structure formations on germanium surfaces by focused ion beam irradiations. J. Phys. Condens. Matter **19**, 445002-1–445002-10 (2007)
130. Campisano, S., Baeri, P., Ciavola, G., et al.: Lattice reordering in Pb implanted Ge crystals. Appl. Phys. A **13**, 101–103 (1977)
131. Wilson, I.H.: The effects of self-ion bombardment (30–500 keV) on the surface topography of single-crystal germanium. J. Appl. Phys. **53**, 1698–1705 (1982)
132. Nitta, N., Taniwaki, M., Hayashi, Y., et al.: Formation of cellular defect structure on GaSb ion-implanted at low temperature. J. Appl. Phys. **92**, 1799–1802 (2002)
133. Nitta, N., Taniwaki, M.: Novel nano-fabrication technique utilizing ion beam. Nucl. Instrum. Methods B **206**, 482–485 (2003)
134. Edler, T., Mayr, S.G.: Mechanisms of stress generation during bombardment of Ge with keV ions: experiments and molecular dynamics simulations. New J. Phys. **9**, 325-1–325-11 (2007)
135. Wei, Q., Zhou, X., Joshi, B., et al.: Self-assembly of ordered semiconductor nanoholes by ion beam sputtering. Adv. Mater. **21**, 2865–2869 (2009)
136. Fritzsche, M., Mücklich, A., Facsko, S.: Nanohole pattern formation on germanium induced by focused ion beam and broad beam $Ga^+$ irradiation. Appl. Phys. Lett. **100**, 223108-1–223108-4 (2012)
137. Biersack, J.P., Haggmark, L.G.: A Monte Carlo computer program for the transport of energetic ions in amorphous targets. Nucl. Instrum. Methods **174**, 257–269 (1980)

# Index



Z.M. Wang (ed.), *FIB Nanostructures*, Lecture Notes in Nanoscale Science and Technology 20, DOI 10.1007/978-3-319-02874-3,

## E



## F



## G



## H



## I



## K



## L



## M

MIX
Papier aus verantwortungsvollen Quellen
Paper from responsible sources
FSC® C105338

If you have any concerns about our products,
you can contact us on
**ProductSafety@springernature.com**

In case Publisher is established outside the EU,
the EU authorized representative is:
**Springer Nature Customer Service Center GmbH**
**Europaplatz 3, 69115 Heidelberg, Germany**

Printed by Libri Plureos GmbH
in Hamburg, Germany